International Colloquium on Computational & Experimental Mechanics (ICCEM 2020)

IOP Conference Series: Materials Science and Engineering
Volume 1062

Online
25 - 26 June 2020

ISBN: 978-1-7138-4574-4
ISSN: 1757-8981

Printed from e-media with permission by:

Curran Associates, Inc.
57 Morehouse Lane
Red Hook, NY 12571

Some format issues inherent in the e-media version may also appear in this print version.

Institute of Physics
Dirac House, Temple Back
Bristol BS1 6BE UK

Phone: 44 1 17 929 7481
Fax: 44 1 17 920 0979

techtracking@iop.org

Additional copies of this publication are available from:

Curran Associates, Inc.
57 Morehouse Lane
Red Hook, NY 12571 USA
Phone: 845-758-0400
Fax: 845-758-2633
Email: curran@proceedings.com
Web: www.proceedings.com

TABLE OF CONTENTS

Preface ICCEM 2020

Peer Review Declaration

The Development of Cutlery Tool to Reduce Parkinson Movement........ 1
Reem Musab Abdul Satar, Azizan As'arryi, Mohd Faiz Bin Abdullah, Mohd Zarhamdy Bin MD Zain

Bearing Fault Diagnosis Using Deep Sparse Autoencoder 8
S R Saufi, Z A B Ahmad, M S Leong, L M Hee

Frequency Domain Decomposition Method: A Comparative Study on Signal Processing for Unbiased Damping Ratio Estimates 18
M Danial A Hasan, M Salman Leong, L M Hee, Z. A. B. Ahmad

Modal Analysis of Keropok Keping Drying Machine........ 28
M S Z Mohamad Suffian, S Mohamaddan, A M A Abang Mohamad Mohtar, M N Leman

Simulation of Active Force Control Using MR Damper in Semi Active Seat Suspension System........ 34
R Rosli, Z Mohamed, G Priyandoko

Study of nozzle distance and oil flow rate effects on the droplets size of minimum quantity lubrication........ 41
K L J Kiat, N I K Ismail, N Rosli, E A Alias

Vertical Vibrations Effect on Forced convection heat transfer from a Longitudinal Finned Tube 53
Saleem Khalefa Kadhim, Sinan Abdulghafar Ali, N F D Razak, M S M Sani

Study of modal properties and crashworthiness performance of crash box numerical model with different joining modelling strategies........ 65
N.A.Z. Abdullah, M.S.M. Sani, M.S. Salwani

A Study of Soil Characteristics of Road damages in Jeram, Kuala Selangor........ 76
A S Intan Najiha, S I Doh

Spatial Patterns of Peak Ground Acceleration (PGA) Earthquake on Final Disposal Sites Plan in Samosir Regency 83
Della Ayu Lestari, Dewi Susiloningtyas, Muchamad Hartanto, Taufik Ejaz Ahmad, Dhea Rahma Azhari, Amien Rais

Monitoring and Optimizing Solar Power Generation of Flat-Fixed and Auto-Tracking Solar Panels with IoT System 93
Ku Azmie Ku Husin, Norfatihah Mohd Adenam, Muhamad Yuzaini Azrai Mat Yunin, Khairul Nizar Syazwan Wan Salihin Wong, Siti Zaiton Mohd Hashim, Hasyiya Karimah Adli

Effects of Various Alcohol Blends on Performance and Emissions of Small Compression Ignition Engine Fuelled with Diesel-20% Palm Oil Methyl Ester 100
Helmisyah Ahmad Jalaludin, Nik Rosli Abdullah, Muhammad Syafiq Roslee

Effect of Fuel Additives (Ethanol) on the Engine Performance Emissions of Single Cylinder Diesel Engine Fueled with Palm Oil Methyl Ester (PME) (B20) 111
Asiah Ab. Rahim, Nik Rosli Abdullah, Mohamad Hafsham Mazlan, Hazim Sharudin, Helmisyah Ahmad Jalaludin

Enhancement of the heat transfer in radiator with louvered fin by using Graphene-based hybrid nanofluids 119
M Sandhya, D Ramasamy, K Sudhakar, K Kadirgama, W S W Harun

Experimental investigation on noise and vibrations of biodiesel-butanol blends for diesel engine 132
Erdiwansyah, M.S.M. Sani, Mahidin, M. Zaki, Asep Kadarohman, Muhibbuddin, R.E Sardjono, Asri Gani

Engine Emissions Analysis of Emulsified Fuel of Different Blend Ratios 139
Hazman Abu Hassan, Ftwi Y Hagos, Asep Kadarohman, R E Sardjono, Fitri Khoerunnisa, Bo Zhang, X X Jiang

Secondary Metabolites from Colletotrichum phyllanthi, an Endophytic Fungi from Stem of Physalis peruviana and Evaluation of Media Culture 148
Amelinda Pratiwi, Elvira Hermawati, Yana Maolana Syah

Susceptibility of Sesbagrandiflorain B against Chlorination: A DFT Study 157
Mita Nurhayati, Hafiz Aji Aziz, Muhamad Abdulkadir Martoprawiro

Experimental Determination of Water, Water/Ethylene Glycol and TiO2-SiO2 Nanofluids mixture with Water/Ethylene Glycol to Three Square Multilayer Absorber Collector on Solar Water Heating System: A Comparative Investigation 165
M A Fikri, W M Faizal, H K Adli, Z Bo, X X Jiang, A I Ramadhan

Investigation on stability of TiO2-SiO2 nanofluids with ratio (70:30) in W/EG mixture (60:40) 174
M A Fikri, W M Faizal, H K Adli, Z Bo, X X Jiang, A I Ramadhan

Effect of n-Butanol Blends on Engine Performance and Exhaust Emission of Compression Ignition Engine Fuelled with Diesel-Palm Oil Methyl Ester (B20) 181
Mohd Azrul Ahmad, Nik Rosli Abdullah, Zaky Zuhairi Zawawi, Asiah Ab Rahim, Hasannuddin Abdul Kadir

Experimental Investigation on the Optical and Stability of Aqueous Ethylene Glycol/Mxene as a Promising Nanofluid for Solar Energy Harvesting 194
K Kadirgama, L Samylingam, Navid Aslfattahi, M Samykano, D Ramasamy, R Saidur

Optimization of Performance and Exhaust Emissions of a Spark-Ignition (SI) Engine Fueled with Bioethanol-Gasoline Blends using TOPSIS Methodology 203
I Baratian, Muhammad Yasar, G Najafi, B Ghobadian, Bo Zhang, X X Jiang

Investigation on the Fouling and Heat Transfer Characteristics of Novel EGR Cooler (Semi-spiral) for Diesel Engine Fueled with Biodiesel 214
F. Jamali, Mustaqimah, B. Ghobadian, G. Najafi, Bo Zhang, X.X Jiang

Implementation of the UTAUT Model to Understand the Technology Adoption of MOOC at Public Universities 223
Hafiza Haron, Supyan Hussin, Ahmad Rizal Mohd Yusof, Hafiza Samad, Hafidzan Yusof

Implementation of Anti-Profanity Words in Mobile Application Platform 231
F Razali, Z Jusoh, A Salleh@Omar, N Azizan

Hole Introduction Effect on The E-Glass WR 185 Fibre Reinforcement Polyester 238
Ilhamdi, D Hidayat, D E Sepriyani

Physical and Mechanicals Characteristic of the Nano Particulate TiO2-Epoxy Composite 245
Ilhamdi, A Nugroho, D Hidayat, R A Nur

Slowloris DoS Attack Based Simulation 252
Shima Sabri, Noraini Ismail, Amir Hazzim

Secure Database on Attendance Verification System by using Data Masking Technique 259
Siti Faizah Miserom, Shima Sabri, Fatin Farina Aida, Noraini Ismail

Adhesion Strength of Hydroxyapatite Coating on Titanium Alloy (Ti-6Al-4V ELI) for Biomedical Application 268
Gunawarman, Jon Affi, Agus Sutanto, Dian Mustika Putri, Dian Juliadmi, Nuzul Ficky Nuswantoro, Hidayatul Fajri, Djong Hon Tjong, Menkher Manjas

The Effect of Wing Spar Cross-Sectional Profile Variation on the Unmanned Aerial Vehicle (UAV) Natural Frequency 277
Lovely Son, Arif Rahman, Mulyadi Bur

Numerical Study on Various Position of Multiple Water Tanks toward Earthquake on High-rise Building 286
Izzul Syazwan Ishak, Norhayati Abdul Hamid, Norliyati Mohd Amin

Simulation Modeling The Performance of Ocean Thermal Energy Conversion Power Cycle 294
N Samsuri, N Sazali, A S Jamaludin, M N M Razali

Axial Crush Response of Aluminium Square Tube with Origami Patterns 305
Amir Radzi Ab Ghani, M Amirul Syafiq Zulfikry, Ramzyzan Ramly, Mohd Zaid Othman

Vulnerability Assessment of Tsunami-Affected Inundated Area Using Geospatial Analysis Based Tsunami Run-Up Simulation 315
Rina Febrina, Lusmelia Afriani, Rr M I Retno Susilorini, Tri Prastio

Implementing Student Attendance System Using Fingerprint Biometrics for Kolej Universiti Poly-Tech Mara 335
N Zakiah Lamin, W N Asnida Wan Jusoh, Juanita Zainudin, Hafiza Samad

Modelling and Validation of Seismic Performance for Corner and Interior Beam-column Joint Using HYSTERES Program 345
N H Hamid, N D Hadi

Parcel Tracking System Using Barcode Scanner with Verified Notification 356
Juanita Zainudin, Hafiza Samad, Faizah Miserom, Shima Sabri

Performance of Ocean Thermal Energy Conversion Closed Rankine Cycle Using Different Working Fluids 366
N Samsuri, N Sazali, A S Jamaludin, M N M Razali

The Development of a Web-based Credit Transfer Application (CTA) for Higher Academic Institution: From Feasibility Study to Testing Phase 377
N Azizan, R Isa, F Farzana, A Aziz, M Amiruddin

Techno-economic efficiencies and environmental criteria of Ocean Thermal Energy Conversion closed Rankine cycle using different working fluids 384
N Samsuri, N Sazali, A S Jamaludin, M N M Razali

The Implementation of Electronic Voting System for Student Representation Council using reCAPTCHA 396
Amirul Asyraf Khairul Salleh, Harlinawati Abdul Kadir, Aslimariah Ahmad, Mohd Azahari Mohd Yusof

Cutting Forces and Flank Wears Analysis for End Mill Processes Using HSS Tools Cryogenic Treatment When Cutting PFRP Material 405
F Ridwan, D Noerhamzah

Design of indenter rig for measuring seat cushion dynamic stiffness 414
AM Hassan, KAM Rezali, NA Abdul Jalil, A As'arry, MA Azizan

Stress Analysis on Cracking Plate Structure For Reducing Stress Concentration 422
F Huda, R A Safrudin, S Utomo, M Rusli

Effect of Forging Temperature on Biodegradable Mg-0.7%Ca Alloy Properties for Implant Application 430
Is Prima Nanda, Zahran Ali, Nur Farah Bazilah Wakhi Anuar, Mohd Hasbullah Idris, Andril Arafat

Validation of Modal Analysis of Two-Storey Reinforced Concrete Frame 440
I F Azmi, N H Hamid, W I I Wan Iskandar Mirza, M N Abdul Rani, M A Yunus

Buckling Analysis of Curve Stiffened Fuselage Panel of Very Light Jet Aircraft 449
Devi Chandra, Eka Satria, Nukman Yusoff

Filter Component Impact on EGS002-based Inverter Circuit Performances 460
Hardisal, R A Candra, D N Ilham, D P Saragi, R Harahap, S Suherman, Z. Yahya

Nanovoids in fracture surface of unsaturated polyester/vinyl ester blends resulting from disruption of the cross-linking of the polymer chain networks 466
Hairul Abral, Rahmat Fajrul, Melbi Mahardika, Dian Handayani, Eni Sugiarti, Ahmad Novi Muslimin, Santi Dewi Rosanti

Microstructure and Corrosion Rate of the Hot Rolled Mg- 1.6Gd Alloys as Prospective Degradable Implant Materials 471
O Susanti, D Dwiputra

Thermogravimetric Analysis of Densified Coco Peat Briquettes 476
A Pratoto, A J Abhen

Mechanical Properties and Machinability of Magnesium Alloy AZ31 and AZ91 – A Comparative Review 484
Louis Denis Kevin Catherine, Darulihsan Abdul Hamid

Pattern Recognition of Facial Electromyography (FEMG) Signal for Aceh Language Speech using Naïve Bayes and Learning Vector Quantization (LVQ) 490
Darma Setiawan Putra, Yuril Umbu Woza Weru, Fardiansyah, Fitriady, Fakhruddin, Z. Yahya

Candlenut Shell Charcoal Yield in Every Chamber Produced by Candlenut Shell Carbonization Tool Using a Vertical Multi Chamber 498
Asbahrul Amri, Oktalia Triananda Lovita, Devi Satria Saputra, Ihsan, Fitriani, Mahdi Arifan, Miki Saputra, Z. Yahya

Distillation Equipment Design of Nutmeg Oil using Hybrid System 507
Nuzuli Fitriadi, Balkhaya, Edi Saputra, S. Othman

Movement Parameters for Evaluating the Two-legged Humanoid Robot 517
Dirja Nur Ilham, Rudi Arif Candra, Hardisal, Erwinsyah Sipahutar, Abdullah Abdullah, Suherman Suherman, S Othman

Finite Element Modelling and Updating of a Thin Plate Structure using Normal Mode Analysis 522
A R Bahari, M A Yunus, M N Abdul Rani, W I I Wan Iskandar Mirza, M A S Aziz Shah, Z Yahya

Author Index

IOP Conf. Series: Materials Science and Engineering 1062 (2021) 011001 doi:10.1088/1757-899X/1062/1/011001

Preface ICCEM 2020

The International Colloquium on Computational & Experimental Mechanics (ICCEM 2020) is the first international colloquium organized by Universiti Teknologi MARA and Universiti Malaysia Pahang. The goal of the colloquium is to facilitate researchers around the world to publish and to share their current research.

As you may be aware, the World Health Organization officially declared the novel coronavirus COVID-19 a pandemic. This global health crisis is a unique challenge that has impacted many authors. We would like to express our concern and support for all the authors and families affected by this outbreak. Considering this situation, ICCEM 2020 will be held virtually as a virtual conference.

These proceeding contains the selected scientific manuscripts submitted to the colloquium. The call for papers attracted submissions of 86 abstracts from both local and international universities. Thanks to University Andalas Padang Indonesia and Kolej Universiti Poly-Tech MARA Kuala Lumpur Malaysia to become as Co-Organizer for this colloquium.

The scientific papers published in these proceeding have been revised and approved by the technical committee of the ICCEM 2020. All of the papers exhibit clear, concise, and precise expositions that appeal to a broad international readership interested in science, technology and engineering. The topics of the colloquium were: (1) Mechanical and Industrial Engineering; (2) Civil Engineering; (3) Computer and Informatics Engineering; (4) Development and Environmental Engineering; (5) Electrical Engineering; and (6) Agricultural Engineering.

The reports present original ideas or results of general significant supported by clear reasoning and compelling evidence, and employ methods, theories and practices relevant to the research. The authors state clearly the questions and the significance of their research to theory and practice, describe how the research contributes to new knowledge, and provide tables and figures that meaningfully add to the narrative. In this edition of ICCEM 2020 representatives coming from academia, industry, governmental and private sectors. The plenary and invited speakers present, discuss, promote, and disseminate

ICCEM 2020 would like to thank the Organizing Committee, Co-Organizer and Sponsors supporting the colloquium and everyone who contributed along the colloquium and publication of these proceeding.

Assoc. Prof. Ir. Dr. Mohd Azmi Yunus, Chairman of ICCEM 2020
Dr. Muhamad Norhisham Abdul Rani, Co-Chairman of ICCEM 2020
Assoc. Prof. Dr. M.S. M. Sani, Chief Editor of ICCEM 2020

IOP Conf. Series: Materials Science and Engineering 1062 (2021) 011001 doi:10.1088/1757-899X/1062/1/011001

Organizing Committee of The International Colloquium on Computational & Experimental Mechanics (ICCEM 2020)

Patron 1
Prof. Emeritus Datuk Ir. Dr. Mohd Azraai Kassim (Universiti Teknologi Mara)

Patron 2
Prof. Dr. Mohd Nazip Suratman (Universiti Teknologi Mara)

Advisor 1
Prof. Ir. Dr. Hj. Muhammad Azmi Ayub (Universiti Teknologi Mara)

Advisor 2
Prof. Dr. Rizalman Mamat (Universiti Malaysia Pahang)

Chairman
Assoc. Prof. Ir. Dr. Mohd Azmi Yunus (Universiti Teknologi Mara)

Vice Chairman 1
Dr. Muhamad Norhisham Abdul Rani (Universiti Teknologi Mara)

Vice Chairman 2
Assoc. Prof. Dr. Eng. Meifal Rusli (Universitas Andalas)

Vice Chairman 3
Assoc. Prof. Ts. Dr. Zahrah Yahya (KUPTM)

Editor in Chief
Assoc. Prof. Dr. Mohd Shahrir Mohd Sani (Universiti Malaysia Pahang)

Treasurer
Wan Imaan Izhan Wan Iskandar Mirza (Universiti Teknologi Mara)

Publication Unit
Noor Amzura Abdullah (Universiti Malaysia Pahang)
Muhammad Zikri Japri (Universiti Malaysia Pahang)
Muhammad Nur Arif Asri (Universiti Malaysia Pahang)

Secretariat, Technical and Logistics Unit
Wan Imaan Izhan Wan Iskandar Mirza (Universiti Teknologi Mara)
Crystal Peter (Universiti Teknologi Mara)
Mohamad Azam Shah Bin Aziz Shah (Universiti Teknologi Mara)
Abdul Rahim Bahari (Universiti Teknologi Mara)

IOP Conf. Series: Materials Science and Engineering 1062 (2021) 011001 doi:10.1088/1757-899X/1062/1/011001

Reviewers

Prof. Dr. Rizalman Mamat
College of Engineering
Universiti Malaysia Pahang

Assoc. Prof. Dr. Eng Meifal Rusli
Faculty of Engineering
Universitas Andalas

Assoc. Prof. Dr. Mohd Shahrir Mohd Sani
Faculty of Mechanical and Automotive Engineering Technology
Universiti Malaysia Pahang

Dr. Eng. Rina Febrina
Faculty of Engineering
Universitas Malahayati

Dr Muhamad Norhisham Abdul Rani
Faculty of Mechanical Engineering
Universiti Teknologi MARA

Assoc. Prof. Ir. Dr. Amir Radzi Ab. Ghani
Faculty of Mechanical Engineering
Universiti Teknologi MARA

Dr. Mohd Yusof Taib
College of Engineering
Universiti Malaysia Pahang

Crystal Peter
Faculty of Mechanical Engineering
Universiti Teknologi MARA

Mr. Abdul Rahim Bahari
Faculty of Mechanical Engineering
Universiti Teknologi MARA

Mr Muhammad Nur Arif Asri
Universiti Malaysia Pahang

Mrs. Rosmaliza Mat Yaacob
DRB Hicom University

Prof. Dr. Norhayati Abd. Hamid
Faculty of Civil Engineering
Universtiti Teknologi Mara

Dr. Ahmad Razin Zainal Abidin
Faculty of Civil Engineering
Universiti Teknologi Malaysia

Assoc. Prof. Ts. Dr. Mohd Ruzaimi Rejab
Faculty of Mechanical and Automotive Engineering Technology
Universiti Malaysia Pahang

Assoc. Prof. Ir. Dr. Mohd Azmi Yunus
Faculty of Mechanical Engineering
Universiti Teknologi MARA

Dr Alias Mohd Saman
Faculty of Mechanical Engineering
Universiti Teknologi MARA

Dr. Norazlianie Sazali
College of Engineering
Universiti Malaysia Pahang

Dr. Januar Parluangan
College of Engineering
Universiti Malaysia Pahang

Mr. Mohamad Azam Shah Bin Aziz Shah
Faculty of Mechanical Engineering
Universiti Teknologi MARA

Mr. Wan Imaan Izhan Bin W. I. Mirza
Faculty of Mechanical Engineering
Universiti Teknologi MARA

Mr Muhammad Zikri Japri
Universiti Malaysia Pahang

Miss Noor Amzura Abdullah
Universiti Malaysia Pahang

Peer Review Declaration

International Colloquium on Computational & Experimental Mechanics (ICCEM2020)

All papers published in this volume of *IOP Conference Series: Materials Science and Engineering* have been peer reviewed through processes administered by the Editors. Reviews were conducted by expert referees to the professional and scientific standards expected of a proceedings journal published by IOP Publishing.

• **Type of peer review:** Double-blind

• **Conference submission management system:** conference email and online.

• **Number of submissions received:** 86

• **Number of submissions sent for review:** 70

• **Number of submissions accepted:** 64

• **Acceptance Rate (Number of Submissions Accepted / Number of Submissions Received X 100):** 74%

• **Average number of reviews per paper:** 2

• **Total number of reviewers involved:** 22

• **Any additional info on review process:**

Verification and approval of the papers is carried out in three stages (Stage 1 – Title, abstract and full text evaluation by Chief Editor; Stage 2 – Review process; Stage 3 – Correction Approval). Each paper goes plagiarism check by Turnitin Software and a double-blind peer review process. At least two reviewer per paper is being assigned to check the quality of paper. Also check on novelty and continuity of flow along with the clarity of thought and correctness concerning the data. Upon satisfying the above-mentioned criteria, the papers were submitted for final publication with IOP. Upon completion of the review process, the papers and their reviews are submitted to the Editorial board for approval.

IOP Conf. Series: Materials Science and Engineering 1062 (2021) 011002 doi:10.1088/1757-899X/1062/1/011002

• **Contact person for queries:** Assoc. Prof. M.S.M. Sani, Universiti Malaysia Pahang, Malaysia, E-mail: mshahrir@ump.edu.my

The Development of Cutlery Tool to Reduce Parkinson Movement

Reem Musab Abdul Satar[1*], Azizan As'arryi[1], Mohd Faiz Bin Abdullah[2] and Mohd Zarhamdy Bin MD Zain[2]

[1] Faculty of Mechanical and Manufacturing Engineering, Universiti Putra Malaysia (UPM), 43400 UPM Serdang, Selangor, Malaysia
[2] Faculty of Mechanical Engineering, Universiti Teknologi Malaysia, 81200 Skudai, Johor Bahru

*Corresponding e-mail: reemusab244@gmail.com

Abstract. The typical feeding skills are considered to be eating from a spoonful, and imbibition from a cup and simple activities are difficult for Parkinson's patients. The present study investigates the development of this acquirement in Parkinson's patients of feeding spoon thought out improves on the concept design of feeding spoon, where it can be helpful for patients. In return this can help to improve the future work in designing and fabrication for any instrument to control tremor trembling allowing the patient to practice their delay rotten. Feeding spoon device can provide accurate and diagnostically relevant information about postural tremor. Its portability and ease of use could help translate such techniques into routine clinic use or to the community. With this study it is shown that 70% of improvement in skill assistance spoon feeding compared with ordinary spoon. It is apparent that the developed model can improve the feeding process and make it easier on the patient.

Keyword: Parkinson, vibration, tremor spoon

1. Introduction

Parkinson's disease is a progressive neuro-degeneration of multiple systems damaging motor and non-motor functions, affects individual and societal dimensions negatively [1]. Parkinson disease can disturb a patient's life. Patients will face problem in pick up and retention target because of tremor or vibration of hands which are the most obvious and important symptoms of tremor diseases [2]. Introduction Professional person involved in assessment and treatment of infant with feeding problems must have knowledge of normal development of oral motor and eating accomplishment [3]. There is uncontrollable hand tremor in some neurological diseases sufferers like Parkinson's patients. [4-5]. There are various character of tremor, which differ from each other according to the circumstance (posture, action or resting), the amplitude and absolute frequency, ranging from 3 Hz to 12 Hz [6]. Tremor is prominent in the distal part of an extremity. The rest tremor at the hand is usually produce a high fast vibration at the whole arm [7]. Thus, it makes the affected person enjoy difficulties which intrude with acting every day sports together with writing, lifting objects, placing the object at the target factor and so forth [8]. In most system that related to mechanical vibration and mechatronics system, there are sources of detrimental vibrations that may significantly influence the mission performance, effectiveness and accuracy of operation. There are many techniques that have been develop to reduce this vibration [9]. A vibrational-control system, either as an isolator or an absorber, is to be active, passive, or semi active and passive depending on the amount of the vibration that need to be counter and the load or forces exerted on the device.

Previous study have been done to help Parkinson's disease sufferers and their difficulties with feeding. Studies with several projects were made regarding these type of spoon. First is self-levelling spoon. The invention is a spoon including a hand-grip able handle, a rotary coupler mounted to the handle, and a bowl mounted to the rotary coupler [10]. The bowl is relatively rotatable mounted to the handle through the rotary coupler, permitting the bowl to remain level even if the handle is rotated. In a preferred embodiment, the centre of gravity of the bowl is offset from the axis of rotation of the rotary coupler, and is most preferably near the bottom of the bowl. The centre of gravity of the bowl is low enough to for the bowl to remains parallel with the ground regardless of the rotation of the handle. The rotation of the head or bowl will be dependent on upon the force of the gravity. The position of the rotation does not affect the levelling of the bowl. The creation of this spoon is to prevent spilling the content of the lifting food during eating. With the design of hand gripped handle, free rotation coupler and bowl the reduction in vibration from patient can be reduce in use of spoon. Referring to the patent of self-levelling spoon, the main purpose of the spoon using is to develop a controllable position of the spoon since the attached mass at the end of the spoon will provided the balance of the spoon. The mass and rotary handle will take full advantages in design this spoon stability. When the handle is held in the hand in the act of manipulating the spoon, the bowl maintains a level position by reason of its weighted-bottom portion, the many advantages of which are sufficiently obvious without further description.

Another invention of spoon that can counter vibration is gyro spoon. Gyro Spoon is a spoon that can rotate 360 degrees so that the food inside will not be spilled. This helps people with Parkinson's disease and people with other disorders that cause them to have hand tremor, which leads to a difficulty in feeding themselves. The Gyro spoon helps alleviate that difficulty by using gyroscopic technology to keep the spoon part facing upward at all times. The technology in this spoon design is using concept of gyroscope. This spoon concept need some spacing and support. This is solid design of spoon with freely rotate bowl spoon head. This spoon go through analytical process and give the result that the reduction of vibration can be counter [7].

Next, the invention of the spoon that can reduce the hand trembling motion also been develop by Liftware Company using active control system namely as Liftware spoon. Lift Labs is the company which has designed the spoon for those suffering from Parkinson's disease and essential tremors. The company was acquired by Google in September 2014, then Google launched its version of the spoon in November 2014 [11]. The second part is the actuator section containing two motors in the handle to move the spoon in the opposite direction of any detected tremor. They are steered by the on board computer in order to counteract the tremors and to stabilize the spoon. This paper intention the development of this acquirement in Parkinson's patients of feeding spoon thought out improves on the concept design of feeding spoon, where it can be helpful for patients.

2. Methodology

The spoon conceptual design presents base on the design analysis, the ergonomics study and experimental observation gain from literature review. However, Solidwork software played a raw in designing the spoon after selecting the best conceptual designed spoon among all the models, which is using spring stiffness to control the spoon vibration. In the other hand, the spring is supported the connection where it can extend with the string on the both sides of the spoon as it shown in Figure 1. Taking into consideration reducing the spoon moment by making it as short as possible to eliminate the intensity of vibration concentrated at the tip of the spoon. The spoon vibration control is mass balancing and spring system. This is the minimal type of vibration control since the application is only spoon. The conceptual design is evaluated and the best design in term of mechanism and spoon ergonomics is proceed to the fabrication process. The fabrication process is done by assembling the spoon by using 3D printed model. The spoon model was printed in various size of handle from the conceptual design because to select the best size handle that can be used without constraint by the patient.

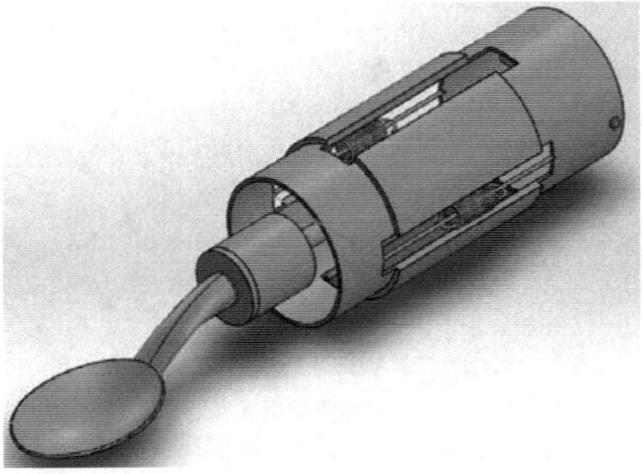

Figure 1. The conceptual design of spoon.

Using three axis accelerometer ADXL335 for data. Which is required 3.5 V or 5 V voltage supply, since it has dual connection to voltage. An Arduino board is used as the power supply and the connection is done by using analogue input from the Arduino. The data is collected by using two accelerometer, the accelerometer were attached at the hand of the patient and at the spoon head, where the reduction of vibration should be observed. The connection of the accelerometer was connected to 6 analogue signal input, the purpose here just to collect the raw data of the vibrational behaviour. Since the Arduino IDE cannot store the data automatically, the data is stored at using another program that is PLX-DAQ system. The data collected will automatically transfer into excel file. The experimental setup to collect data is shown in Figure 2.

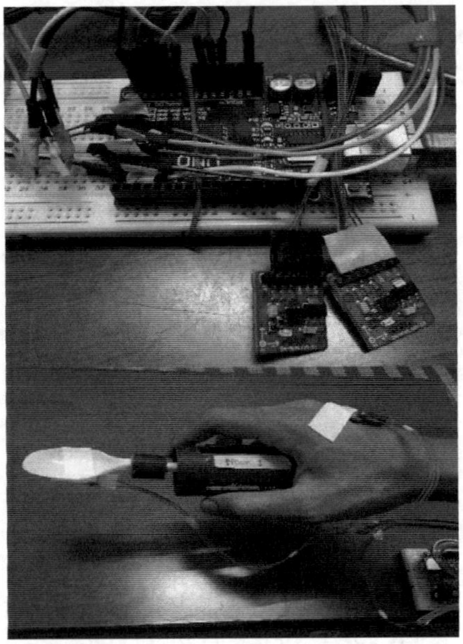

Figure 2. Configuration of accelerometer for data collection.

The data is collected using two condition that is when the patient hold the spoon without load and with load. This is done in order to find the accuracy of the mass and spring balance system. The spoon behaviour during first condition is different with the second condition because of the load applied to the spoon head.

IOP Conf. Series: Materials Science and Engineering 1062 (2021) 012001 doi:10.1088/1757-899X/1062/1/012001

The hand behaviour also show some different since when the patient lift the arm, in order to stay elevated the hand required more power to counter the gravitational force. The right hand was the body part that involve in this testing.

3. Result and discussion

All accelerations values produce by the vibration of the hand is converted into time and frequency domain. After the data collection, the pattern of both accelerometer sensor reading is interpreted by using Microsoft excel and MATLAB software. During the data analysis, all the recorded data were being compared between accelerometer attached at hand and accelerometer attach and the tremor spoon that have been assemble with the passive counter vibrational system. The purpose of comparing between these two values is to gain the result of vibration reduction. No load condition means that the patient is holding the spoon connected with accelerometer without lifting any load from the bowl. This testing is conducted to validate the spoon in this condition. Only critical axis is consider in this experiment that is X-axis.

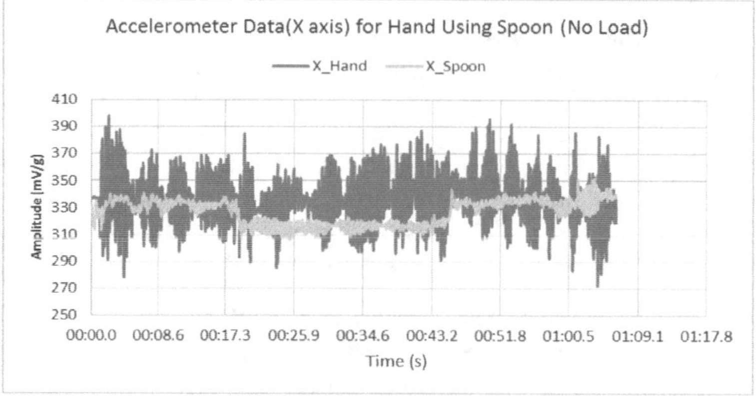

(a)

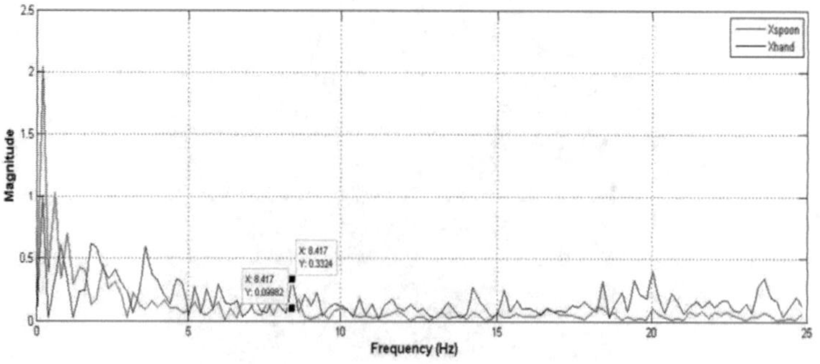

(b)

Figure 3. X-axis no load condition data plot (a) Plot of data sample in time domain (b) frequency domain.

With load condition is the testing of the spoon where the patient try to lift load using the spoon from the bowl. This testing was done to test the spoon behaviour of vibration reduction after load is being applied at the spoon head.

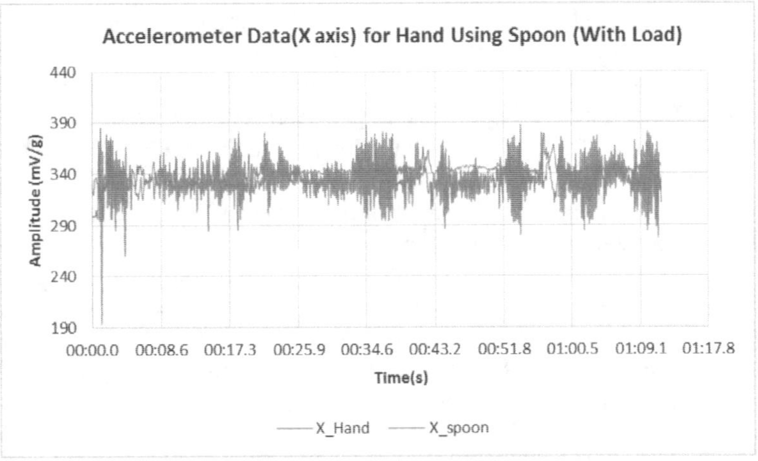

(a)

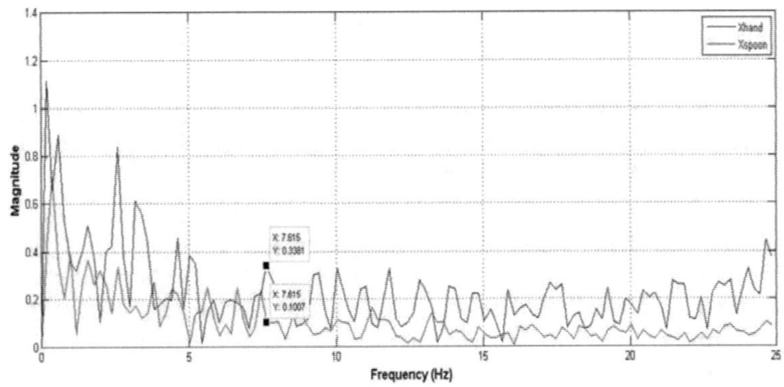

(b)

Figure 4. X-axis with load condition data plot (a) Plot of data sample in time domain (b) frequency domain.

Table 1. The interpreted data of the spoon accelerometer reading.

Axis	Amplitude				Frequency (Hz)	Magnitude different	Reduction
	Hand		Spoon				
	Max	Min	Max	Min			
X (No load)	398	272	352	306	8.417	0.23258	Yes
X (With load)	388	194	368	313	7.615	0.094	Yes

From time domain graph plot in Figure 3(a) raw data observation accepted from accelerometer attach at hand is more vibrant with amplitude rate that higher in the range of 398 mV/g to 272 mV/g. This show in the posture condition the rate of vibration transfer from source to the spoon in much lower because of the counter vibration of the mass. Rate of vibration at spoon is in the range of 306mV/g to 352 mV/g, this value is the magnitude of the spoon time domain graph. Difference between both amplitude of the two accelerometer in hand and spoon is 26 mV/g and 46 mV/g. The analyse frequency is at the peak of 8.417 Hz because of the frequency range for hand tremor is 6-12 Hz. The peak at this frequency is the highest magnitude gain from the frequency domain, for the hand frequency the peak magnitude value is 0.3324 and for the spoon the peak magnitude value is 0.09982, the different is 0.23258 and from the frequency domain the reduction of vibration is observe.

From the Figure 4, the time domain plot from the raw data collected using the accelerometer sensor show that, the vibration from hand is much higher than vibration at the spoon, the amplitude magnitude recorded at the hand accelerometer is in the range of 388mV/g and 194 mV/g. By comparing with the amplitude magnitude at the spoon result from the vibration transfer from hand to the spoon the value is in the range of 368 mV/g and 313 mV/g. Different between both amplitude of the two accelerometer is 194 mV/g and 55 mV/g respectively. The peak of this frequency is the highest in magnitude for the range of hand tremor frequency. For the analyse frequency, the peak value for both hand and spoon is 0.3381and 0.1007 respectively. The peak of this frequency is the highest value gain from the frequency domain, the different is 0.094 and from the frequency magnitude different, the reduction of vibration is observed.

4. Conclusions

The spoon successfully reduced the vibration and reduced the spillage of load during the patient uses the spoon. But there were several limitations for the patient used the spoon. The limitation was the spoon cannot be operated by someone with a very serious hand tremor, because the passive counter vibration system does not accurate enough to counter very fast vibration. Other than that, the development of the spoon model has been able to introduce the concept of passive reduction system even the best result obtains only at one axis. Based on previous research, usually the vibration generated by the hand tremor patient is between 6-12 Hz. The design is still not reliable to reduce the vibration or trembling motion occur. Based on the analysis, comparison and conclusion that have been made for this study, here are recommendation to be considered. The passive counter vibrational system may not so accurate to counter the trembling motion. For great result to reduce the motion is by implementing sensor to detect the hand vibration. To obtain best result, for future work a proper rig test system that behave same as the hand of the tremor patient can be develop in order to reduce unintended movement, rotating axis and the proper value of data.

Acknowledgments

The authors would like to thank Universiti Teknologi Malaysia, Universiti Putra Malaysia (UPM) and the Malaysian Ministry of Higher Education (MOHE) for their continuous support in the research work. This work was financial supported by the Geran Putra IPS fund under Grant Vot. Number 9540600.

References

[1] Cheng F K 2017 The use of acupuncture in patients with Parkinson's disease *Geriatr Nurs* **38(4)** 302-314

[2] McNamara P, Durso R and Harris E 2006 Life goals of patients with Parkinson's disease: a pilot study on correlation mood and cognitive functions *Clinical Rehabilitation* **20** 818-826

[3] Delaney A L and Arvedson J C 2008 Development of swallowing and feeding: prenatal through first year of life *Dev Disabil Res Rev* **14(2)** 105-17

[4] Hussein M, As'arry A, Zain M Z M, Mailah M and Abdullah M Y 2009 Experimental study of human hand-arm model response *J. Basic Appl. Sc.* **1(4)**

[5] Kazi S, As'arry A, Zain M Z M, Mailah M M and Hussein M 2010 Experimental implementation of smart glove incorporating piezoelectric actuator for hand tremor control *WSEAS Transactions on Sys. & Con* **5(6)** 443-453

[6] Manto M, Topping M, Soede M, Sanchez-Lacuesta J, Harwin W, Pons J, Williams J, Skaarup S and Normie L 2003 Dynamically responsive intervention for tremor suppression *Ieee Eng. in Medicine & Biology Magazine* **22120 -132**

[7] Theis J, Radisch C and Werner H 2014 Self-scheduled control of a gyroscope *The Int. Federation of Auto. Con.* 24-29
[8] Szabo Z, Storkova B, Hozman J and Zanchi V 2007 Objective assessment of patients with Parkinson's disease *WSEAS Int. Conf. on Remote Sensing* 39-42
[9] Jankovic J Parkinson's disease: Clinical features and diagnosis *J. of Neuro* **79(4)** 368-376
[10] George F, Carrier R, Courant, Rosenbloom P and Yang C N 1962 Applied mathematics: What is needed in research and education *US Patent* **4(4)**
[11] Noe R 2014 A vibrating fork, and an anti-vibration spoon. Which would you rather have design? *Core77*
[12] Belton S E, Sipahi R, Gouldstone A, Jaeger B K and HinckelCavalcante L 2015 Assistive writing device for tremor patients *NIH-IEEE Strategic*

Bearing Fault Diagnosis Using Deep Sparse Autoencoder

S R Saufi[1]*, Z A B Ahmad[1], M S Leong[1]and L M Hee[1]

[1] Institute of Noise and Vibration, Faculty of Mechanical Engineering, Universiti Teknologi Malaysia (UTM), Jalan Sultan Yahya Petra 54100 Kuala Lumpur, Malaysia

*Corresponding email: msramadhan93@yahoo.com

Abstract. Rolling element bearing is an important component in various machinery. Faulty on bearing cause severe equipment damage that lead to high maintenance cost. The development of deep learning has been paid a considerable amount of attention to fault diagnosis on rolling element bearing. Traditional machine learning such as Artificial Neural Network and Support Vector Machine have problems of lacking expression capacity, existing the curse of dimensionality, require manual feature extraction and require an additional feature selection. Deep learning model has the ability to effectively mine the high dimensional features and accurately recognize the health condition. In consequence, deep learning model has turned into an innovative and promising research in bearing fault diagnosis field. Thus, this paper tends to proposed Deep Sparse Autoencoder (DSAE) with Teager Kaiser Energy Operator (TKEO) to diagnose the bearing condition. DSAE is one of deep learning model which uses the architecture of neural network. During the analysis, the hyperparameter of DSAE model was optimized by Ant Lion Optimization. The analysis results show that the proposed TKEO-DSAE achieved 99.5% accuracy of the fault diagnosis. The comparative study between proposed model and ANN proved that deep learning model outperform traditional machine learning model on bearing fault diagnosis.

Keywords: bearing; fault diagnosis; teager kaiser energy operator; deep sparse autoencoder; ant lion optimization

1. Introduction

Rolling element bearing (REB) is broadly used in domestic and industrial applications. The bearing provides a physical support to a component and allows that components to rotate with less friction. Working condition of this component depends on the smooth and quiet running of the bearings. Bearing is extensively used in industry field such as induction machines [1] wind turbines [2], helicopters [3], and automotive [4]. Bearing is considered as a critical component in a majority of machines due to exposure on extreme conditions (dirty conditions, high temperature, overstress) continuously. Extreme conditions cause the generation of fault on bearing components such as outer race, inner race and rolling element. Therefore, Fault diagnosis is introduced in providing an effective maintenance strategy to diagnose the fault growth in bearing component. Various monitoring method have been introduced for bearing fault diagnosis such as vibration analysis, acoustic analysis, thermal analysis and motor current analysis. However, vibration analysis has been proven to be an effective monitoring method for bearing fault diagnosis [5- 8].

Presently, the integration of machine learning on bearing fault diagnosis has been paid a lot of attention by researchers as machine learning could produce more accurate outcomes with a consistent result. Artificial neural network (ANN) and Support Vector Machine (SVM) are among the popular machine learning model which widely used on bearing fault diagnosis. However, this machine learning needs a manual feature extraction from the bearing vibration signal. The extracted features require a selection process in order to choose the best feature that reflect the bearing conditions. The process of feature extraction and feature selection is time-intensive. In 2006, Hinton and Salakhutdinov develop a modification version of the neural network by increasing the number of hidden layer [9] which later called as deep learning. Deep learning model capable of extracting high dimensional feature without the needs of a human assistant. Convolutional Neural Network (CNN), Deep Belief Network (DBN), and Deep Sparse Autoencoder (DSAE) are the examples of deep learning model which is widely used in many applications [10- 12]. Among the deep learning model, DSAE model has the capability to be trained based on unsupervised manner [13]. The DSAE model is the integration of multiple neural network architecture. The DSAE aims to produce the output that contains similar characteristics to its input. DSAE is trained with the same mathematical model as feedforward neural network that utilized gradient descent algorithms for the back-propagation process.

Due to the complexity of the bearing signal, a lot of consideration is required to acquire the important feature from the bearing signal. This is due to the prediction performance of deep learning model is influenced by the quality of the signal. It is known that during data acquisition from the sensor, the captures signal contains the signal of interest and noise. Noise is a type of disturbance that bury the useful important feature in the signal. Therefore, a suitable signal processing method is required to filter the noise so that useful feature can be extracted from the signal. There is a lot of signal processing method utilized in bearing fault diagnosis field such as wavelet transform, empirical mode decomposition, variational mode decomposition, etc. In this study, the Teager Kaiser Energy Operator (TKEO) is used instead of another signal processing method due to its ability to pre-process the signal by filtering the low-frequency background signals [14]. TKEO has been paid less attention toward bearing fault diagnosis.

This paper aims to solve bearing fault diagnosis using Teager Kaiser Energy Operator (TKEO) and deep sparse autoencoder (DSAE). The DSAE model is optimized via Ant Lion Optimization. The performance of (DSAE) model is compared with Artificial Neural Network (ANN).

2. Teager Kaiser Energy Operator Theory

The Teager–Kaiser energy operator was introduced by Teager and Kaiser to measure the energy of a time domain signal that was generated by the mechanical process [15]. This method is proposed for AM-FM demodulation and to estimate the envelope of the amplitude of AM signals [16]. The TKEO process on continuous time signals x(t) is defined in the following Equation (1);

$$x(t) = (\frac{dx}{dt})^2 - x(t)(\frac{d^2x}{dt^2}) \tag{1}$$

Meanwhile, the Equation (2) defined the discrete model of TKEO;

$$x(n) = x^2n - x(n-1)x(n+1) \tag{2}$$

Several studies used TKEO for signal processing. For example, Rodriguez extracted important feature from the Amplitude Modulated (AM) signal using TKEO. The result shows that features extracted after the signal has been processed by TKEO outperform the diagnosis results of the extracted features from the raw time vibration signal [17]. Meanwhile, Kwak et al. used TKEO in their research to enhance the signal peak induced by the defect located on bearing component [14]. Tran et al. used to TKEO to reveal the fault pattern from the reciprocating compressor valves signals [18].

3. Deep Sparse Autoencoder

The autoencoders aim to produce the output that contains similar characteristics to its input. Sparse autoencoder used the architecture of neural network which is suitable for the dimensional reduction of the features. The hidden layer representation of sparse autoencoder network is illustrated in the following Equation (3);

$$h(x) = f(w_1 x_1 + b_1) \tag{3}$$

where f(z) is the nonlinear activation function to maps the input. Generally, the logistic sigmoid function $f(z) = 1/1 + e^{-z}$ is used as activation function which maps the input from zero to one. The network output maps the hidden representation h back to a reconstruction $\tilde{x} \in R^n$ is illustrated in Equation (4);

$$\tilde{x} = f(w_2 h(x) + b_2) \tag{4}$$

In general, autoencoder is used for feature dimensional reduction and do not contain any sparsity term on its model. The hidden node number of autoencoder should be less than the input size. By adding the sparsity term on the autoencoder cost function, the network of autoencoder can be more versatile since the hidden node number can be any number either lower, equal or higher than input node. Therefore, the cost function of sparse autoencoder is developed based on three terms which are defined in Equation (5);

$$E = \frac{1}{N} \sum_{n=1}^{N} \sum_{k=1}^{K} (x_{kn} - \hat{x}_{kn})^2 + \lambda * \Omega_{weights} + \beta + \Omega_{sparsity} \tag{5}$$

The integration of multiple sparse autoencoders is called deep sparse autoencoder (DSAE).

4. Ant Lion Optimization

The Ant Lion Optimization (ALO) is the optimization algorithms in finding the unknown parameters based on the objective function and fitness values. The ALO algorithms are proposed by mimicking the nature of Antlion hunting mechanism. This algorithm was proposed by Mirjalili in 2015 [19]. The antlion hunting mechanism is based on four following process; digging the trap, hide underneath the bottom cone and attack the prey once the prey falls into the trap. Finally, the antlion pulled the prey under the soil and consumed. The leftovers are thrown outside the trap and the antlions prepare the trap for the next hunt. The details regarding the algorithms may refer to Mirjalili works [19].

The ALO algorithms were used to optimized six hyperparameters of DSAE such as weight regularization, sparsity regularization and sparsity proportion that are associated with each sparse autoencoder. In this study, two components of sparse autoencoder are used to construct the DSAE network. Manual selection of six DSAE hyperparameter is time-intensive. The overall step of bearing fault diagnosis is shown in the following Figure 1.

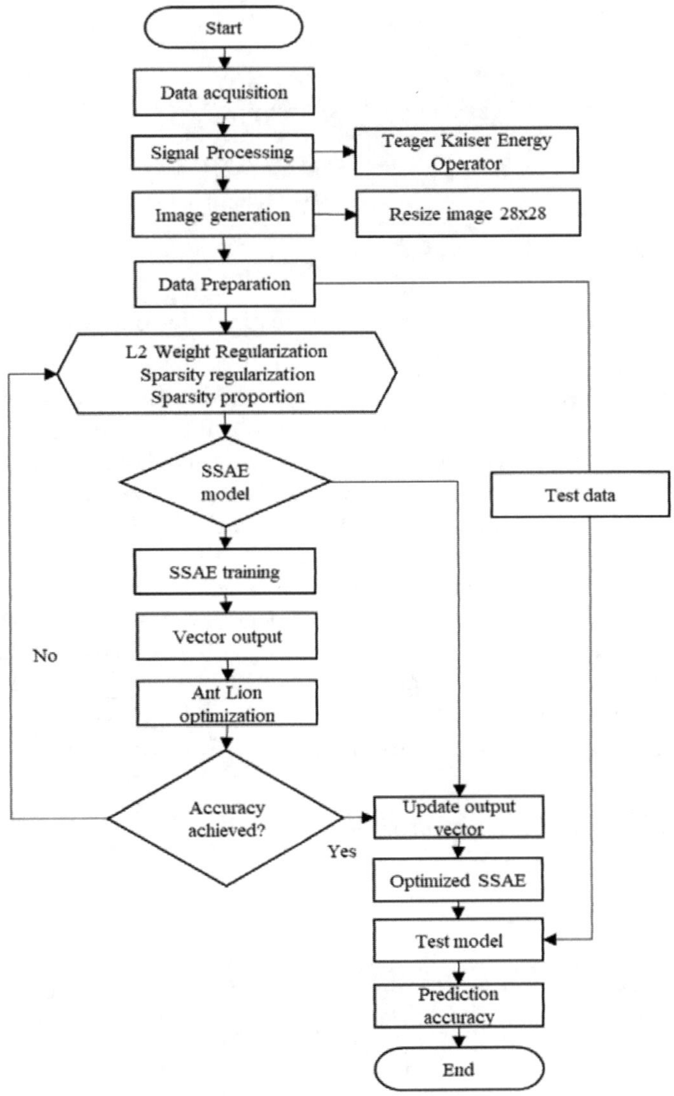

Figure 1. Bearing fault diagnosis of Proposed Method

5. Experimental setup

In this section, the details about experimentation from Case Western Reserve University is discussed. The data were collected with 48kHz from the experimental set up as shown in Figure 2. Every single fault is artificially made on each bearing components such as an outer raceway, inner raceway and rolling element. In this study, the fault diagnosis focus on the 1800rpm speed conditions and 0.007-inch fault size. Based on the data selection, we believed that if the proposed model could diagnose the 0.007-inch fault effectively, then there is no a problem for the proposed model to diagnose 0.014-inch and 0.021-inch fault size. The selected vibration signal is then processed by Teager Keiser Energy Operator (TKEO). The 2D-images of TKEO signal is resized to 28x28 image patches. In addition, the statistical feature from the filtered signal by TKEO is extracted for further analysis.

IOP Conf. Series: Materials Science and Engineering 1062 (2021) 012002 doi:10.1088/1757-899X/1062/1/012002

Figure 2. Experimental Rig for bearing fault simulation

There are four types of signal that represent bearing conditions as illustrated in table 1 which are normal, outer race fault, rolling element fault and inner race fault. The vibration signals are segmented to one rotation based on equation 6 and the segmented process is illustrated in Figure 3. The segmented signals are sampled as shown in Table 1 which means the signal is cut into 100 samples on each bearing conditions. The 100 samples are divided into a training dataset and testing dataset. Each segmented vibration signal is processed via TKEO as shown in Figure 4-7. These images from Figure 4-7 are used as input to our proposed method. The proposed model is based on image classification. These image patches are used as an input to DSAE model and ANN model for bearing fault classification analysis. Instead of TKEO images, the statistical features from the filtered signal has been extracted for a comparative study between the proposed method and ANN model with statistical features. There are nine statistical features that were extracted from the filtered signal such as amplitude, mean, variances, standard deviation, RMS, kurtosis, skewness, crest factor and clearance factor.

$$Segmented\ Signal\ for\ one\ rotation = \frac{number\ of\ sample\ data\ in\ one\ second}{speed\ of\ motor\ (RPM)/60} \quad (6)$$

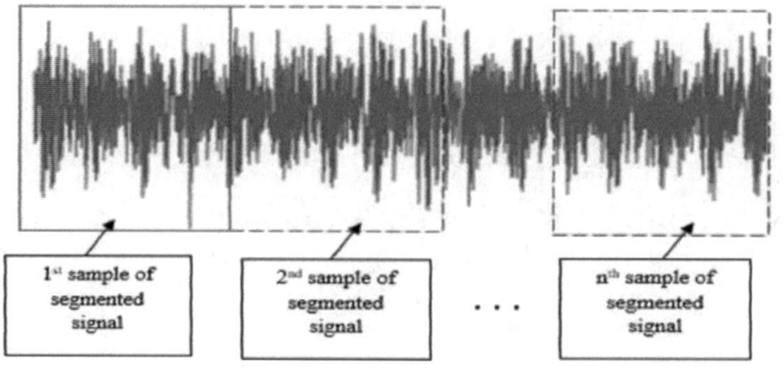

Figure 3. The segmented signal of one rotation

Table 1. Data distribution

Bearing Conditions	Training Data	Testing Data
Normal	50	50
Outer race fault	50	50
Rolling element fault	50	50
Inner race fault	50	50

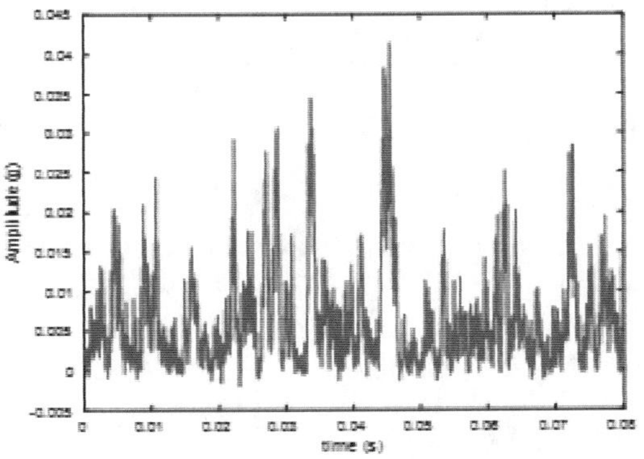

Figure 4. Rolling Element Fault

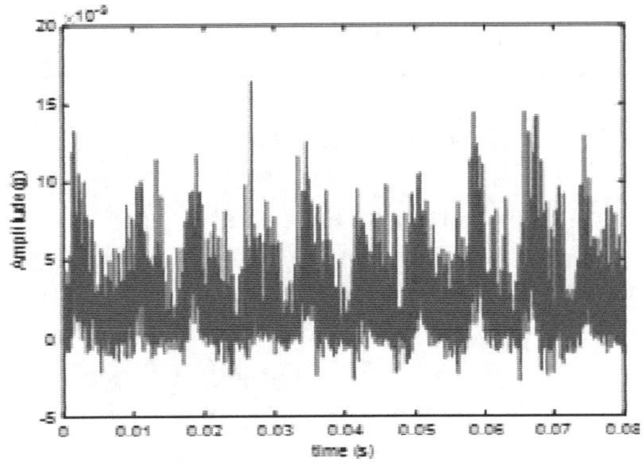

Figure 5. Normal state

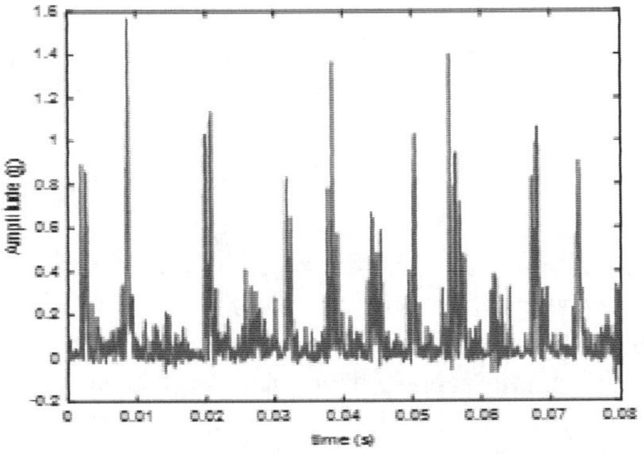

Figure 6. Inner Race Fault

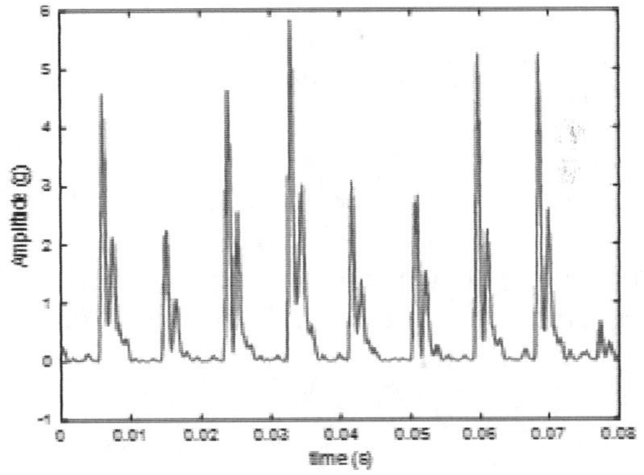

Figure 7. Outer Race Fault

The performance accuracy of DSAE training and supervised fine-tuning stages is affected by many hyperparameters. The hyperparameters of DSAE classifier are the number of the hidden nodes, number of layers, maximum iteration (epoch), sparsity proportion etc. In general, there is no standard method to determine the optimal hyperparameters of DSAE model. As mentioned earlier, three important hyperparameters of DSAE is optimized by ALO algorithms. The range of three hyperparameters that were optimized using ALO is tabulated in Table 2. In addition, the rest hyperparameter of DSAE model is tabulated in Table 2.

Table 2. Setting of parameters for DSAE

Sparse Autoencoder 1	Hidden node number: 300 Epoch: 300 L2 Weight Regularization: (0-0.004) Sparsity Regularization: (0-4) Sparsity Proportion: (0-1) Encoder and Decoder Transfer function: Logistic Sigmoid
Sparse Autoencoder 2	Hidden node number: 150 Epoch: 300 L2 Weight Regularization: (0-0.004) Sparsity Regularization: (0-4) Sparsity Proportion: (0-1) Encoder and Decoder Transfer function: Logistic Sigmoid

Softmax Layer	Loss function: Cross-entropy Epoch: 300

The comparative study between DSAE and ANN model is conducted. ANN is a traditional machine learning model which has been broadly used in bearing fault diagnosis. The ANN model is trained using two different datasets such as statistical features and image features. The prediction of ANN for both features are presented in the next section. The parameter setting regarding ANN model is illustrated in Table 3.

Table 3. The setting of the ANN hyperparameter

Feed Forward Neural Network	
Activation function hidden layer	Logistic Sigmoid
Number of Hidden Neuron	10-50
Loss function	Cross-entropy

6. Results and Discussion

In this section, the experimental results of TKEO-DSAE, TKEO-ANN, ANN with statistical features are presented. Matlab R2017a installed on a conventional computer with a 2.4GHZ CPU and 8GB memory is used to conduct the experiments. In order to compare the machine learning model, Receiver Operating Characteristics (ROC) curve is used to make the comparison more meaningful. ROC based on the measurement of specificity and sensitivity of all cut off point. The y-axis of ROC curves represents a true positive rate (TPR) while x-axis of ROC curves represents a false positive rate (FPR). TPR denote as the sample ratio that is classified correctly while FPR denotes the sample ratio that is incorrectly classified. The performance of the model is higher if the located on the top and left edges of the plot.

From figure 8, it is shown that DSAE model curve is located closer to the top and edge of the plot. Meanwhile, ANN with statistical features has higher performance compared to ANN model with image features since ANN with statistical features has steeper curve compare to ANN with image features. In order to present more accurate result regarding classification accuracy, the confusion matrix is tabulated in table 4-6. From Table 4 -6, it is clear that SSAE model outperforms ANN methods in term of overall classification accuracy using image patches of TKEO. TKEO-DSAE produced 99.5%. DSAE misclassify one rolling element fault image to normal bearing conditions. However, most of the image sample is correctly classified. Meanwhile, ANN produced 77.2% for image classification. The data used to train and test the ANN model is similar to DSAE dataset. From the confusion matrix in Table 5, ANN model misclassified 41 images which rolling element fault contribute to the highest misclassification rate. Images contain a large size of features which in this study, the size of features is 784. We believe that ANN unable to process the large size of features that make the ANN model unable to produce a satisfactory performance.

Therefore, further study is conducted by extracting nine statistical features from the filtered signal. The analysis shows that ANN model produced 89.0% prediction accuracy of bearing fault classification. The details of the classification are tabulated in Table 6. The ANN model suffers from the same problems which rolling element fault contribute to the highest misclassification among the four classes of bearing conditions. Based on the result, DSAE model outperforms ANN model for bearing fault classification.

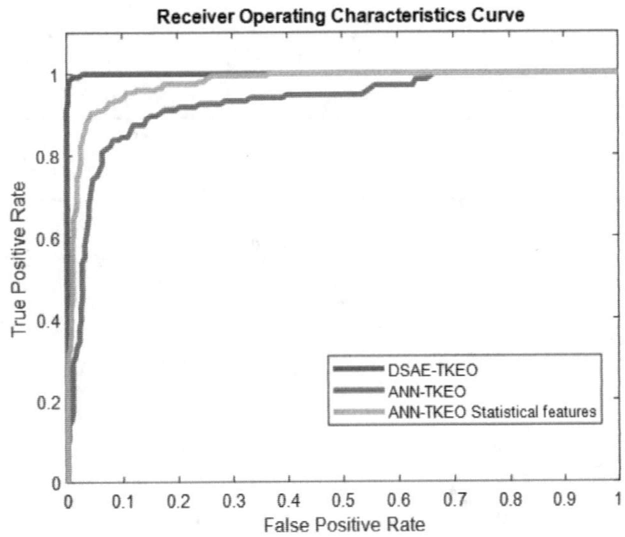

Figure 8. Comparison of ROC curve between three models

Table 4. Confusion matrix of DSAE prediction

Class		Actual			
		Normal	Outer Race Fault	Rolling element Fault	Inner Race Fault
Predicted	Normal	**50**	0	1	0
	Outer Race Fault	0	**50**	0	0
	Rolling element Fault	0	0	**49**	0
	Inner Race Fault	0	0	0	**50**
Sensitivity (%)		100	100	98	100
Accuracy (%)		**99.5**			

Table 5. Confusion matrix of ANN prediction

Class		Actual			
		Normal	Outer Race Fault	Rolling element Fault	Inner Race Fault
Predicted	Normal	**47**	0	10	0
	Outer Race Fault	0	**50**	0	5
	Rolling element Fault	3	0	**25**	13

IOP Conf. Series: Materials Science and Engineering 1062 (2021) 012002 doi:10.1088/1757-899X/1062/1/012002

	Inner Race Fault	0	0	15	**32**
Sensitivity (%)		94	100	50	64
Accuracy (%)			**77.2**		

Table 6. Confusion matrix of ANN prediction using statistical features

Class		Actual			
		Normal	Outer Race Fault	Rolling element Fault	Inner Race Fault
Predicted	Normal	**40**	0	8	0
	Outer Race Fault	0	**50**	0	0
	Rolling element Fault	10	0	**39**	1
	Inner Race Fault	0	0	3	**49**
Sensitivity (%)		80	100	78	98
Accuracy (%)			**89.0**		

The performance of ALO is shown in the following Figure 9. The ALO model is used to optimize the 6 hyperparameters of the DSAE model. The result indicated ALO is able to search the optimal value of hyperparameters as shown in Fig. 9. The ALO require 20 iterations to reach the lowest percentage error of DSAE which is 99.5%. Based on the analysis, we found that ALO requires less parameter setting in order to perform the optimization process of DSAE model.

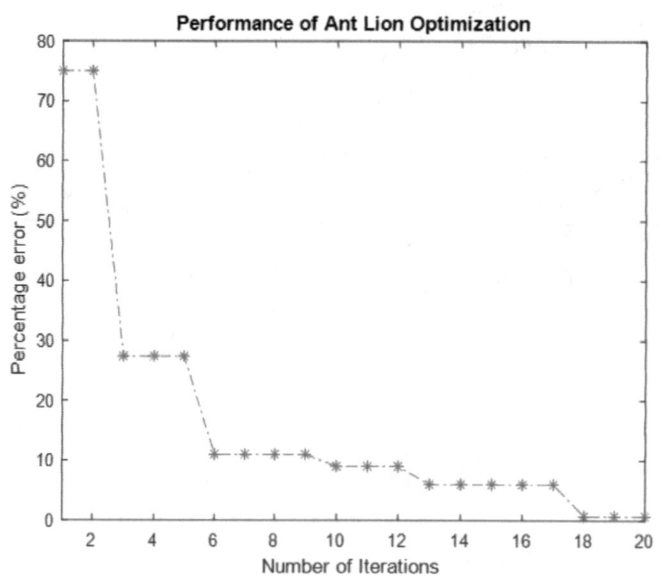

Figure 9. Performance of ALO on DSAE

7. Conclusion

In this paper, the method of applying TKEO and stacked autoencoder DSAE to bearing fault diagnosis problem shows the excellent classification performance compared to TKEO and ANN. During the analysis, the DSAE is optimized using ALO for automated DSAE hyperparameter selection. The proposed model capable of achieving 99.5% prediction accuracy on bearing fault classification. In addition, this DSAE model has good capability to extract a feature from the image compare to ANN model.

Acknowledgements

The authors would like to extend their greatest gratitude to the Institute of Noise and Vibration UTM for funding the study under the Higher Institution Centre of Excellence (HICoE) Grant Scheme R.K130000.7843.4J227. Additional funding for this research also comes from the UTM Research University Grant (Q.K130000.2543.11H36), and Fundamental Research Grant Scheme (R.K130000.7840.4F653) by The Ministry of Higher Education Malaysia.

References

[1] Bouzida A, Touhami O, Ibtiouen R, Belouchrani A, Fadel M and Rezzoug 2011 *IEEE Trans. Ind. Electron.*, vol. **58**, no. 9, pp. 4385–4395
[2] Liu W Y, Tang B P, Han J G, Lu X N, Hu N N, and He Z Z 2015 *Renew. Sustain. Energy Rev.*, vol. **44**, pp. 466–472
[3] Qi X, Theilliol D, Qi J, Zhang Y and Han J 2013 *2013 Int. Conf. Unmanned Aircr. Syst. ICUAS 2013 - Conf. Proc.*, pp. 1114–1118
[4] Zhao M, Lin J, Miao Y and Xu X 2016 *Meas. J. Int. Meas. Confed.*, vol. **91**, pp. 421–439
[5] Zhang W, Peng G, Li C, Chen Y and Zhang Z 2017 no. January, pp. 1–21
[6] Li W, Qiu M, Zhu Z, Wu B and Zhou G 2016 *Meas. Sci. Technol.*, vol. **27**, no. 3
[7] Dolenc B, Boškoski P and Juričić Đ 2016 *Mech. Syst. Signal Process.*, vol. **66**, pp. 521–532
[8] Grasso M, Chatterton S, Pennacchi P and Colosimo B M 2016 *Mech. Syst. Signal Process.*, vol. **81**, pp. 126–147, 2016.
[9] Salakhutdinov R and Hinton G 2009 *Aistats*, vol. **1**, no. 3, pp. 448–455.
[10] Shao H, Jiang H, Zhao H and Wang F 2017 *Mech. Syst. Signal Process.*, vol. **95**, pp. 187–204
[11] Shao H, Jiang H, Zhang X and Niu M 2015 *Meas. Sci. Technol.*, vol. 26, no. 11, p. 115002
[12] Zhang W, Li C, Peng G, Chen Y and Zhang Z 2018 *Mech. Syst. Signal Process.*, vol. **100**, pp. 439–453
[13] Qu Y, He M, Deutsch J and He D 2017
[14] Kwak D H, Lee D H, Ahn J H and Koh B H 2013 *Sensors (Basel).*, vol. **14**, no. 1, pp. 283–98
[15] Kaiser J F 1993 *IEEE Int. Conf. Acoust. Speech Signal Process.*, pp. 149–152 vol.**3**, 1993.
[16] Kumar J and Jena P 2017 *Arab. J. Sci. Eng.*, vol. **42**, no. 12, pp. 5003–5013, 2017.
[17] Henríquez Rodríguez P, Alonso J B, Ferrer M A and Travieso C M 2013 *ISA Trans.*, vol. **52**, no. 2, pp. 278–284
[18] Tran V T, Althobiani F and Ball A 2014 *Expert Syst. Appl.*, vol. **41**, no. 9, pp. 4113–4122, 2014.
[19] Mirjalili S 2015 *Adv. Eng. Softw.*, vol. **83**, pp. 80–98

Frequency Domain Decomposition Method: A Comparative Study on Signal Processing for Unbiased Damping Ratio Estimates

M Danial A Hasan[1*], M Salman Leong[1], L M Hee[1] and Z. A. B. Ahmad[2]
[1]Institute of Noise and Vibration, Universiti Teknologi Malaysia (UTM), 54100 Kuala Lumpur, Malaysia
[2]School of Mechanical Engineering, Universiti Teknologi Malaysia (UTM), 81310 Skudai, Johor Bahru, Malaysia

*Corresponding e-mail: muhd_danial200@yahoo.com

Abstract. Frequency domain decomposition (FDD) is one of OMA methods in the frequency domain and this method has become well-known among engineering community engaged in the system modal identification due to its capability as a user-friendly and fast processing algorithm. Though, this method has problems in offering an accurate estimation of modal damping ratios, even though natural frequencies and mode shapes can be accurately estimated. The accurate estimation of modal damping is still an open problem and often leads to biased estimates since the errors are stemming from each step in FDD procedures and primarily caused by signal processing. Therefore, the identification of modal damping ratio turns out to be immensely essential in structural dynamics since damping is one of the crucial parameters of resonance. This study is to determine the appropriate signal processing for FDD because signal processing such as the time window, correlation function (CF) and the spectral density (SD) are the main contributors to the bias estimate. The goal of this paper is to provide necessary information on modal damping for reliable estimation.

Keywords: Operational modal analysis; frequency domain decomposition; damping estimation.

1. Introduction

In the early twentieth century, Operational Modal Analysis (OMA) has become well-known among engineering community engaged in the system modal identification by replacing Experimental Modal Analysis (EMA) [1, 2]. The great interest of OMA is powered by its capability to carry out simple, quick and cost-effective tests that depend solely on the responses of the structure without compromising its normal operation [3].

The user-friendly and fast processing algorithm of frequency domain decomposition (FDD) proves that it can yield reliable results but with a good selection of parameters for spectra estimation and well-separated modes [4]. Though, this method has problems in offering an accurate estimation of modal damping ratios, even though natural frequencies and mode shapes can be accurately estimated [3]. The accurate estimation of modal damping is still an open problem and remain unsolved, and often leads to biased estimates since the errors are coming from each step in FDD procedures and primarily caused by signal processing [5]. Even though, this method is capable of dealing with closely spaced modes [6].

For damage detection, modal damping ratio is a very sensitive parameter to damage compared to natural frequency and mode shape. Therefore, the identification of modal damping ratio turns out to be immensely essential in structural dynamics since damping is one of the crucial parameters of resonance, particularly for the low-damped structure. This is because the dynamic responses of the system are primarily influenced by the structural behavior and input forces [7].

The goal of this paper is to provide necessary information on modal damping for a reliable estimation and to determine the appropriate signal processing for FDD. Signal processing approach that consists of the time window (windowing), correlation function (CF) and the spectral density (SD) are the main contributors to the bias estimate.

2. Numerical Simulation

2.1. Design of simulation

The algorithms are explored using simulated data, implemented autonomously within MATLAB from one simple model of two degrees of freedom (DOF) as shown in Figure 1.

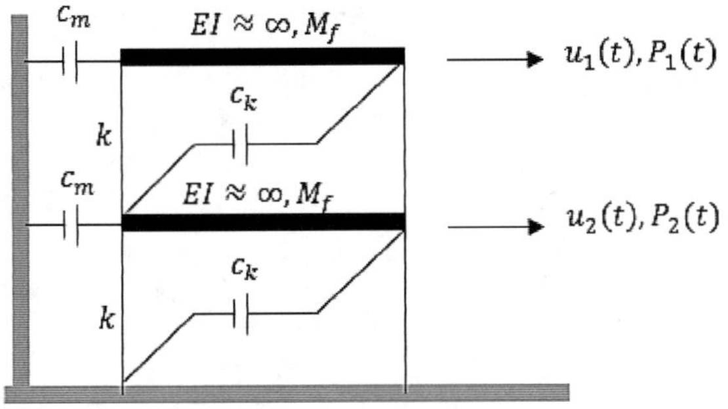

Figure 1. Simulated models of 2 DOF shear frame

The Characteristics of the simulated models of 2 DOF shear frame are characterized by Equation (1) and Equation (2) by the stiffness (*K*), and mass (*M*).

$$K = 10^8 \begin{bmatrix} 2.54 & -1.24 \\ -1.24 & 1.24 \end{bmatrix} (N/m) \tag{1}$$

$$M = 70{,}000 \begin{bmatrix} 1 & 0 \\ 0 & 1 \end{bmatrix} (kg) \tag{2}$$

Meanwhile, damping (C) matrices of the structure are described by mass and stiffness proportional Rayleigh damping as shown in Equation (3):

$$[C] = [C_m] + [C_k] = a_0[M] + a_1[K] \tag{3}$$

Where $[C_m]$, $[C_k]$, $[M]$ and $[K]$ are the mass proportional damping, stiffness proportional damping, mass and stiffness matrix, respectively and are the coefficient describing the mass and stiffness proportional damping. A complex eigenvalue analysis is carried out on the 2 DOF shear frame. The prescribed modal damping ratios, to determine the coefficients a_0 and a_1, are set to $\zeta_k = \zeta_j = 0.01$ (1%).

The broadband ambient excitation with zero mean, Gaussian white noise is replicated by time series with normally distributed random numbers assuming independent inputs for all DOF of the models. The white noise has a constant PSD and ensures a broad banded input [5]. The white noise input also simulates the assumption of the excitation system to be linear and time-invariant which characterized by constant parameters if all its fundamental properties are invariant with respect to time. The response of the system is simulated using Newmark's method with constant average acceleration (i.e., $\gamma = \frac{1}{2}$ and $\beta = \frac{1}{4}$) [8]. The adopted parameters for the two-storey shear frame in the processing are shown in Table 1.

Table 1. The adopted parameters for the two-story shear frame

Parameters	Two-story frame
Length of time series, t(s)	400
Sampling frequency, f_s(Hz)	200
Adopted Frequency resolution, Δf(Hz)	0.0025

In this numerical simulation, the eigenvalue problem analysis is used to validate the effectiveness of the proposed approach. By taking values of the stiffness K, mass M, and damping C, the modal parameters (modal damping ratio, natural frequency, and mode shape) can be estimated.

2.2. Classical FDD algorithm procedure

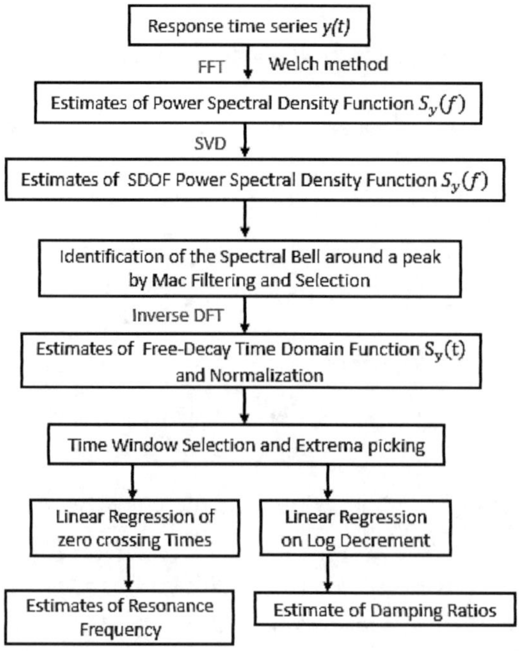

Figure 2. Schematic Illustration of classical FDD Procedure

Normally, signal processing serves as a tool to give a clear illustration of the physical problems we face. A good selection of signal processing techniques can provide all necessary information about the system that obtained from the response signals of the system itself. Signal processing consists of correlation functions (CF) and spectral densities (SD) function. Before estimating SD, the FFT algorithm requires time windows to reduce leakage by forcing the endpoints of each signal sample data to zero. Next, the

Singular Value Decomposition (SVD) is applied to decompose the output SD into auto SD that represents an SDOF system. The singular values that are determined around a resonance peak with an equivalent singular vector that possesses a sufficiently high MAC index value (for example 0.95 [9]) were transformed to SDOF of the time domain (TD) signal by the inverse FFT [10, 11]. All extrema of the free decay that within an appropriate time window (for example 90% to 20% of the maximum amplitude) were used to implement the subsequent linear regression operations for assessing the logarithmic decrement (LogDec), δ:

$$\delta = \frac{2}{k}\ln\left(\frac{r_0}{|r_k|}\right) \tag{4}$$

$$2\ln(|r_k|) = 2\ln(r_0) - \delta k \tag{5}$$

where k is an integer counter of the kth extreme of the auto-correlation function, $k = 1,2,3, \ldots$, while r_0 and r_k are the initial and the kth extreme value of the auto-correlation function, respectively. Then, modathe l damping ratio can be obtained.

$$\zeta_q = \frac{\delta_q}{\sqrt{4\pi^2 + \delta_q{}^2}} \tag{6}$$

Meanwhile, the damped natural frequency, ω_d was estimated by linear regression on the zero crossing times of the equivalent SDOF correlation function. Then, the undamped natural frequency, ω_n was computed by using the following equation:

$$\omega_n = \frac{\omega_d}{\sqrt{1-\varsigma^2}} \tag{7}$$

Further information on all alternative FDD method can be found in [9, 12, 13].

2.3. Signal processing

The step of denoted signal processing contains the tapering (windowing), correlation functions (CF) and the spectral densities (SD).

2.3.1. Tapering (windowing)

The various type of tapering on correlation function for 2 DOF models shear frame were used in this study with 10 datasets. In literature, there is a lot of tapering (windowing) functions introduced by researchers, but in this paper only concern the normally used tapering (windowing) function in OMA such as Hanning window with 50 % overlap, flat-triangular window with a value of $\alpha = 0.5$ and classical exponential window. The correlation function matrix was computed by using the direct estimation method via Equation (8) with a maximum time lag equivalent to 1024 times the sampling time step.

$$\hat{R}(k) = \frac{1}{N-k} Y_{(1:N-k)} Y^T{}_{(k+1:N)} \tag{8}$$

where the measured responses that arrange as a column in data series, N is the total number of data points in the time series, k corresponds to the time lag $\tau = k\Delta t$ and $(N-k)\Delta t$ corresponds to $T-\tau$, T is the total length of the time series.

2.3.2. Correlation function and spectral density estimate

There are various approaches to estimate correlation function (CF) and spectral density (SD). One of them is by direct estimates of the correlation function matrix via Equation (8) and then convert to spectral density (SD) by the FFT. The flat-triangular window with α = 0.5 was applied in correlation function before estimating SD in order to suppress sidelobe noise. Besides that, the conventional Welch approach with a Hanning window and 50% overlap, involves signal sectioning, windowing, and overlapping. Moreover, the random decrement (RD) was computed by using band triggering with an asymmetric band around zero and a width of $\pm 2\sigma$ where σ is the standard deviation of the triggering signal. This approach neglects the need of windowing due to the natural decay of the RD functions. Finally, the half-spectrum was obtained based on the estimated zero-padded direct correlation function with a flat-triangular window and then converted by FFT.

2.4. Statistical analysis

The 10-data set from the 2 DOF models shear frame are analyzed to estimate spectral densities. All spectral plots were in dB relative to the measurement unit in order to provide a clearer picture of the spectral density. Thus, all necessary information of singular values of spectral density can easily be captured in one single plot. In addition, all the analysis was done by applying a maximum time lag of the correlation functions equivalent to 1024 times the sampling time step.

The damping estimation is evaluated by examining the percentage error of the difference between the estimated and the assigned damping as shown in Equation (9),

$$\Delta\zeta_{k,j} = \frac{|\hat{\zeta}_i - \zeta_{k,j}|}{\zeta_{k,j}} \times 100(\%) \tag{9}$$

The difference $\Delta\zeta_{k,j}$ is referred to the error, where $\hat{\zeta}_i$ and $\zeta_{k,j}$ are the estimated damping mode i and the assigned damping respectively.

3. Results and Discussion

3.1. Tapering (windowing)

The results of the tapered correlation function and spectral density with the various application of tapering (windowing) are shown in Figure 3.

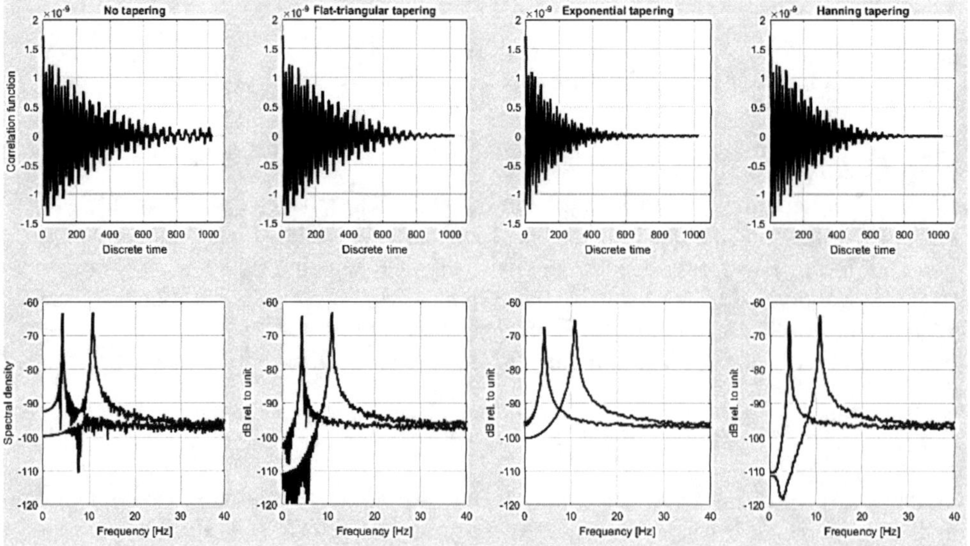

Figure 3. Result of various tapering (windowing) application. Top plots indicate the autocorrelation function with different cases of tapering. From the left: The first plot is without any tapering, the second plot is tapered by the flat-triangular window, following

IOP Conf. Series: Materials Science and Engineering 1062 (2021) 012003 doi:10.1088/1757-899X/1062/1/012003

with an exponential window, and lastly by Hanning window. Bottom plots illustrate the corresponding singular values of spectral density plots

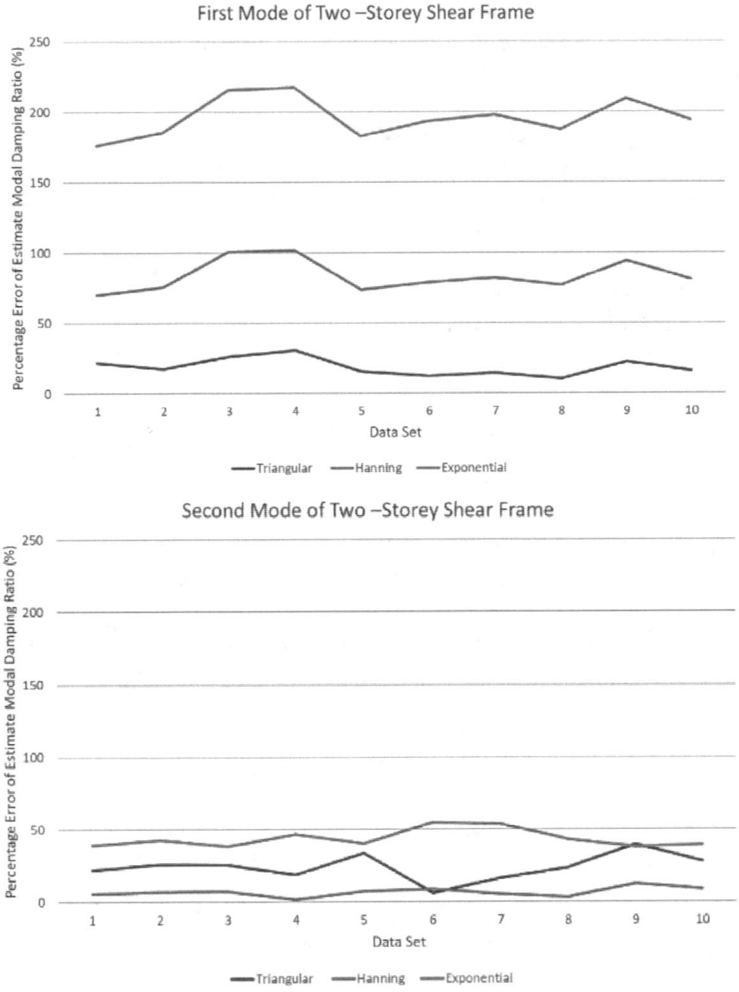

Figure 4. The percentage error (%) of modal damping ratio for the first and second mode of the two-story shear frame with the various application of tapering (windowing).

In the case of without any tapering (left side of Figure 3), there was significant side lobe noise on PSD which caused by the corrupted CF with noise tail.

Meanwhile, for application of flat-triangular window with α = 0.5 as shown in the second bottom plot in Figure 3, it can slightly minimize the side lobe noise of the SD. Instead, this windowing function able to consistently reduce the percentage error of modal damping ratio for both modes of the two-story shear frame with the mean percentage errors is 18.6% mode and 23.8% for the first mode and second mode respectively compared to other application of windowing.

Additionally, the adoption of the classic exponential window can minimize the side noise significantly by 5% of the initial value at the boundaries, but it will lead to a clear appearance of bias at the spectrum peaks. This can clearly see on the first and the second mode with the mean percentage error is 195.8 % and 43.3 % respectively. It was neither accurate nor precise.

Even though Hanning window with 50% overlap is a well-known among analysts especially when doing signal processing which can minimize leakage effects in spectrum estimation, but it will lead to bias error with respect to the actual damping value. This can be seen in the last plot in Figure 3. Besides that, the results of modal damping ratio were not consistent or precise, although the result for the second mode was quite accurate with 6.6% of the mean percentage error and 2.96% for standard deviation error, however for the first mode produced higher error with the mean percentage error up to 83.2 % and 11.4% for standard deviation error.

4. Correlation Function and Spectral Density Estimate

The computation of modal damping ratios relies heavily on the singular value plot, obtained from singular value decomposition (SVD) of the estimated SD. Figure 5 below depicted the various approaches for of singular values estimation.

The top plot of Figure 5 showed the SD which estimated using a direct estimation of the CF matrix with the application of the flat-triangular window ($\alpha = 0.5$) provided the significant reduction of mean percentage error for the first and second mode which were 18.6% and 23.8% respectively. The results of this approach showed a quite consistent and close to the true value of assigned modal damping ratio compared to other approaches.

The second plot from the top of Figure 5 is the traditional Welch approach with a Hanning window and 50% overlap, that involved signal sectioning, windowing, and overlapping. The results showed inconsistent for both modes. The first mode yielded higher error which is 63.1%, meanwhile, for the second mode, the result was opposite with the mean percentage error only 10.7% and with 4.9 % of standard deviation error. It proved that the application of signal sectioning, windowing, and overlapping occasionally may produce noisier SVs. Thus, it needs some additional process in order to enhance the quality of the results. Even though, it has a low computational effort.

Besides that, the third plot from the top of Figure 5 is the RD estimate. This approach has shown it neglects the need for windowing due to the natural decay of the RD functions. Thus, there was significant side lobe noise on PSD which was due to the corrupted CF with noise tail and yields 72.4% and 62.7% of mean percentage errors for the first and second mode respectively. It should be noted that the triggering condition that used in RD function had influenced the initial value and affected the scale of the RD spectral estimate.

Finally, the half-spectrum was obtained based on the estimated zero-padded direct correlation function with a flat-triangular window and then converted by FFT. Meanwhile, for results of half spectrum as shown at the bottom plot of Figure 5 which based on the estimated zero-padded direct correlation function with a flat-triangular window ($\alpha = 0.5$) and then converted by FFT to estimate SD. It yielded a smoother and clearer result of the desired physical information compared to other approaches. However, the results also produced a large error relative to actual values of assigned modal damping with the mean percentage error up to 176.7% and 21.5% of standard deviation error for the first mode, following with 72.3% and 15% of mean percentage error and standard deviation error respectively for the second mode.

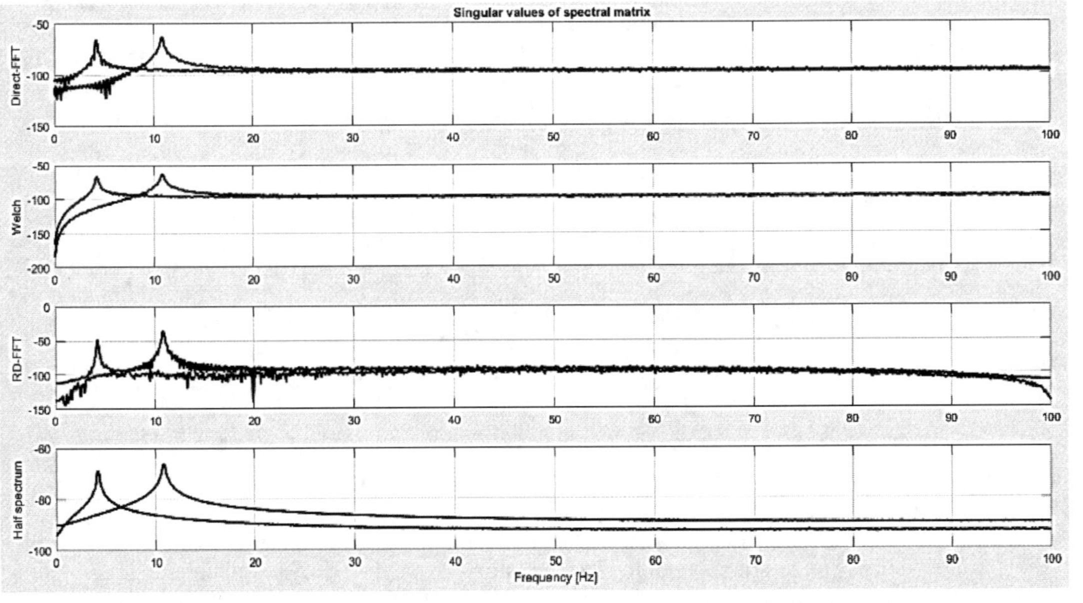

Figure 5. Singular values of the spectral density matrix with various signal processing approaches. From top plot is the Fourier transformed direct estimate of the correlation functions; the conventional Welch approach with a Hanning window and 50% overlap; following with the RD estimate; and finally, the half spectrum based on the zero-padded direct correlation function matrix estimate.

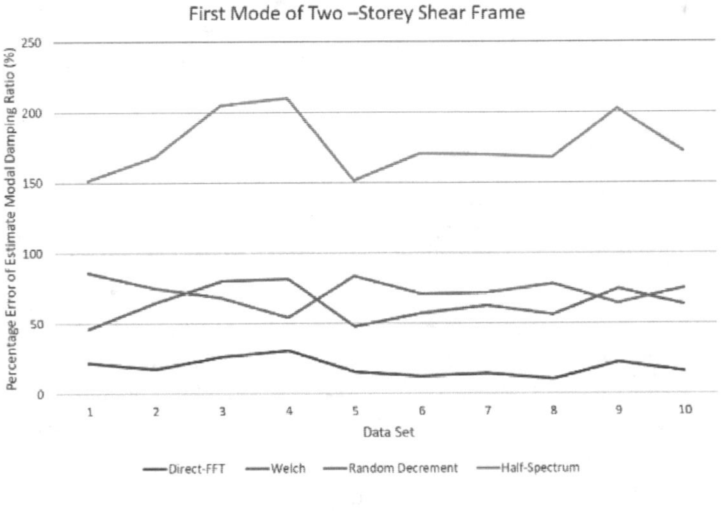

Figure 6. The percentage error (%) of modal damping ratio for the first and second mode of the two-story shear frame with a different application of signal processing.

5. Conclusion

The accurate estimation of modal damping ratio using frequency domain decomposition (FDD) is still an open problem, and often leading to bias estimates since the errors are primarily caused by signal processing [5]. The rapid development of FDD method proves the need for comparative studies, in order to find the best criteria for appropriate signal processing approaches such as the time window function, correlation function (CF) and the spectral density (SD) which are the main contributor to bias estimate. From this study, all necessary information about modal damping for a reliable estimation, reduce uncertainties and define error bounds also can be determined. Based on the results of the different application of tapering (windowing), the time window can illustrate the spectral peaks more clearly, instead, produces a clear bias of the spectral peaks. However, among this different application of time window, the flat-triangular window ($\alpha = 0.5$) depicted the significant reduction of mean percentage error for the first and second mode which are 18.6% and 23.8% respectively and showed a quite consistent and close to the true value of assigned modal damping ratio compare to other approaches. Meanwhile for different ways of estimating the SD, all approaches provided almost the similar information on spectral representations of the modes, however, the high suppression of the side lobe noise as in the half spectrum, will lead to the appearance of a clear bias at the spectral peaks since each modal peak corresponds to the damping. Thus, the application of direct FFT which involved direct estimation of the CF matrix with a flat-triangular window ($\alpha = 0.5$) turn out to be the best approach for signal processing with significantly reduce the mean percentage error for the first and second mode into 18.6% and 23.8% respectively and depicted quite consistent results. Further studies are required to enhance the exact computation of modal damping in FDD using necessary information provided by the current study.

Acknowledgments

The authors would like to extend their greatest gratitude to the Institute of Noise and Vibration UTM for funding the current study under the Higher Institution Centre of Excellence (HICoE) Grant Scheme (R.K130000.7843.4J227). Additional funding for this research came from the UTM Research University Grant (Q.K130000.2543.11H36) and the Fundamental Research Grant Scheme (R.K130000.7840.4F653) from The Ministry of Higher Education, Malaysia.

References

[1] Zhang L, Brincker R and Andersen P 2005 *Proc. 1st Int. Operational Modal Analysis Conf.* (Copenhagen: Denmark) pp 179–190

[2] Mironov A, Doronkin P, Priklonsky A and Kabashkin I 2015 *Transport and Telecommunication Journal* **16(4)** 305–319

[3] Rainieri C and Fabbrocino G 2015 *Proc. 6th Int. Operational Modal Analysis Conf.*(Gijon: Spain)

[4] Magalhães F, Cunha Á, Caetano E and Brincker R 2010 *Mech. Syst. Signal. Process.* **24(5)** 1274–1290

[5] Rainieri C and Fabbrocino G 2014 *Operational Modal Analysis of Civil Engineering Structures* (New York: Springer)

[6] Rainieri C, Fabbrocino G and Cosenza E 2010 *Proc. 10th Int. Conf. Computational Structures Technology* (Valencia: Spain) pp 1–18

[7] Chen B, Zhang Z, Hua X, Basu B and Nielsen S R K 2017 *Mech. Syst. Signal. Process.* **91** 198–214

[8] Chopra A K 2001 *Dynamics of Structures: Theory and Applications to Earthquake Engineering* (Prentice Hall)

[9] Pioldi F, Ferrari R and Rizzi E 2016 *Mech. Syst. Signal. Process.* **68–69** 265–291

[10] Gade S, Møller N B, Herlufsen H and Konstantin-Hansen H 2005 *Proc. 1st Int. Operational Modal Analysis Conf.* (Kissimmee: Florida) pp 261–271

[11] Zhang L and Tamura Y 2003 *21st International Modal Analysis Conference*

[12] Jacobsen N, Andersen P and Bricker R 2008 *Proc. 26th Int. Modal Analysis Conf.* (Orlando: Florida)

[13] Rodrigues J, Brincker R and Andersen P 2004 *Proc. 23rd Int. Modal Analysis Conf.* (Dearbon: Michigan)

Modal Analysis of Keropok Keping Drying Machine

M S Z Mohamad Suffian[1*], S Mohamaddan[1], A M A Abang Mohamad Mohtar[1], and M N Leman[2]
[1] Faculty of Engineering, Universiti Malaysia Sarawak (UNIMAS), 94300 Kota Samarahan, Sarawak, Malaysia
[2] Institut Latihan Perindustrian (ILP), 94300 Kota Samarahan, Sarawak, Malaysia

*Corresponding e-mail: msmsyazwan@unimas.my

Abstract. This project is about analyzing the effect of vibration towards the keropok keping drying machine. The structure frame of the drying machine has been analyzed by conducting modal analysis using ANSYS software. The results of this project consist of the mode shape (pattern of vibration) with the respective natural frequency and the total deformation that occur at the component. The structure of the component is being analyzed whether they are going to resonate or not by comparing the natural frequency obtained from the analysis results and the operating speed of the drying machine. This study will benefit the small and medium-sized enterprise (SMEs), whom the machine was originally developed for, by giving a better performance and low maintenance cost thus minimize the production cost of keropok keping.

Keywords: Keropok keeping drying machine, modal analysis, natural frequency, mode shape

1. Introduction

Cracker is a famous food product in Southeast Asia which known as *keropok* in Malaysia. *Keropok* usually made of seafood such as shrimp, fish or crab which then mixed with flour and flavoured with salt, sugar or monosodium glutamate (MSG). Due to high resource of the seafood, the coastal areas in Malaysia such as Kelantan, Johor, Kedah, Terengganu and Pahang actively operated small and medium enterprises (SMEs) based on keropok product.

Small and medium enterprises (SMEs) play a significant role in the economic development of a country. Thus, it is very important for the SMEs to hire human resources to ease the *keropok keping* production process since the production needs to undergo a lot of process. The most crucial process is the drying process. This process confronted a lot of issues where the temperature of the environment and the hygiene level need to be deliberated. Therefore, a drying machine has been developed for the SMEs for easing the process of drying. Unlike the old-style method, *keropok* commonly dried on a board which is called pemidai. This method is inefficient to be used since the bad weather can affect the drying process and it is also exposed to a very low level of hygiene. Way of solving the tedious problem of relating the process parameters to the performance measure.

The *keropok keping* drying machine consists of six main components namely drying chamber, moving tray, transmission system, heating element, cooling fan, and control system. The machine will improve the hygiene of *keropok keping* drying process in the SMEs. The drying machine that has been fabricated is shown in Figure 1.

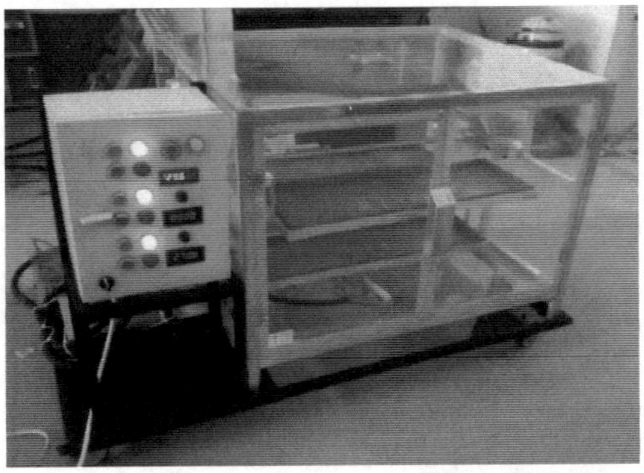

Figure 1. *Keropok Keping* drying machine.

2. Design and Fabrication Process

2.1. Conceptual Design

The *keropok keping* drying machine was designed and fabricated. The critical solution designs for this machine is divided into five elements, which are the drying chamber, type of tray design, transmission system, air source, and heat source design. The solution for each problem is identified into sub problem (Table 1). The number of possible combination ideas concepts can be calculated as 2x2x1x1x1=4 (based on the number of table row and column), which means four ideas of concepts can be generated through this table.

Table 1. Morphological chart for the drying machine.

Subproblems solution concept				
Drying chamber	Tray	Transmission system	Air source	Heat source
Drum type	Stationary tray	Roller chain type with sprocket	Fan	Heating element
Cabinet type	Moving tray			

The computer aided design software is applied in order to develop a 3D model for the selected idea [1-3]. The machine is made up of six main parts namely drying chamber, moving tray, transmission system, heating element, cooling fan and control system. The CAD model and design architecture for the machine can be illustrated as in Figure 2 and Figure 3.

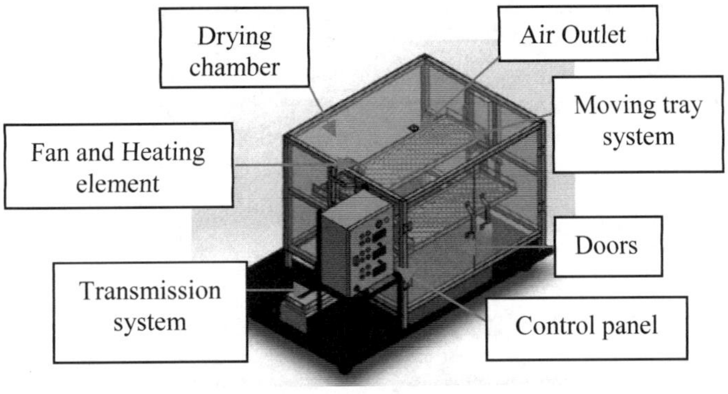

Figure 2. Conceptual design of *Keropok Keping* drying machine.

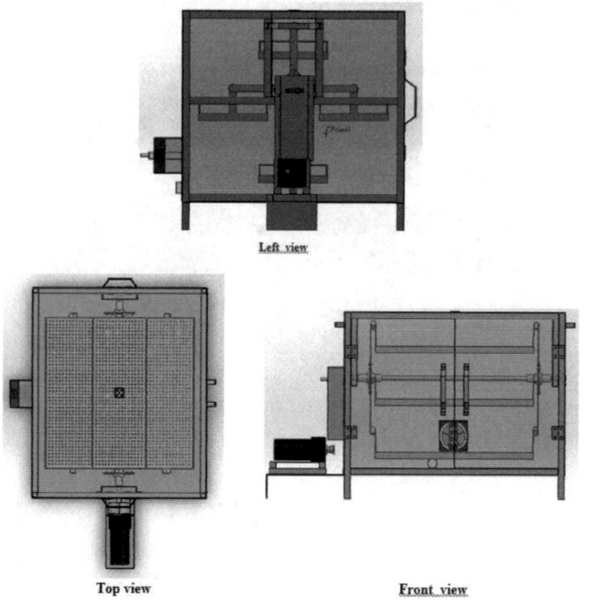

Figure 3. Different views of *Keropok Keping* drying machine.

2.2. Structure Frame Properties

The material properties and technical condition applied in the structure frame can be shown in Table 2. Tetrahedron mesh with refinement and structured hex-dominant mesh are applied to the geometry of the frame and generated by finite element analysis software [5-8]. The mesh generated can be illustrated as in Figure 4.

Table 2. Material properties of the structure frame.

Material properties	Technical Condition
Material	Alluminium alloy
Mass density	2770 kg/m^3
Elastic modulus	7.1e+10 Pa
Poisson's ratio	0.33
Yield limit	2.8e+0.8 Pa

Figure 4. Mesh generation of the structure frame.

3. Results and Discussion

3.1. Modal Analysis

The result and the value of the natural frequency and the structure mode shapes are shown as in Figure 5 until Figure 7. In the modal analysis, six mode shape of the theoretical natural frequency of the structural frame was investigated.

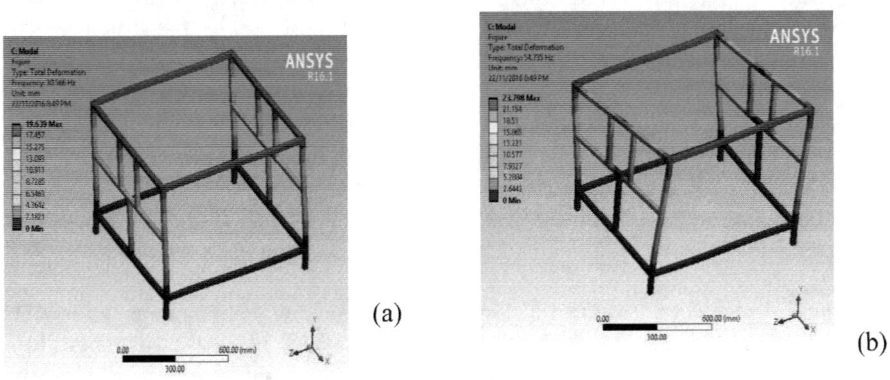

Figure 5. (a) 1st Vibration mode (30.56 Hz) (b) 2nd vibration mode (54.73Hz).

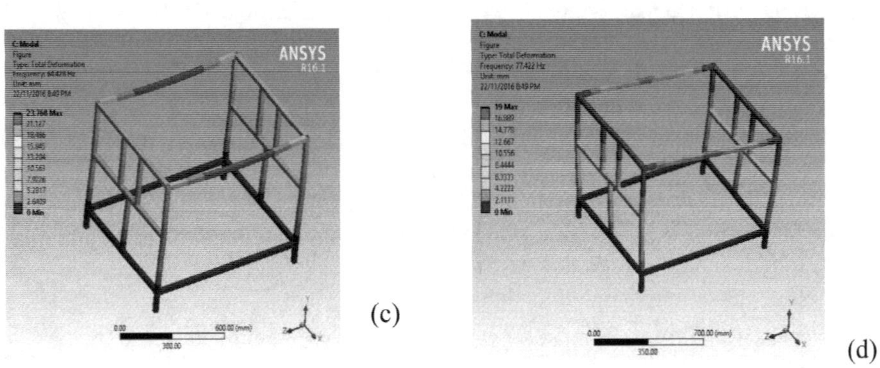

Figure 6. (c) 3rd Vibration mode (64.63Hz) (d) 4th Vibration mode (77.42Hz).

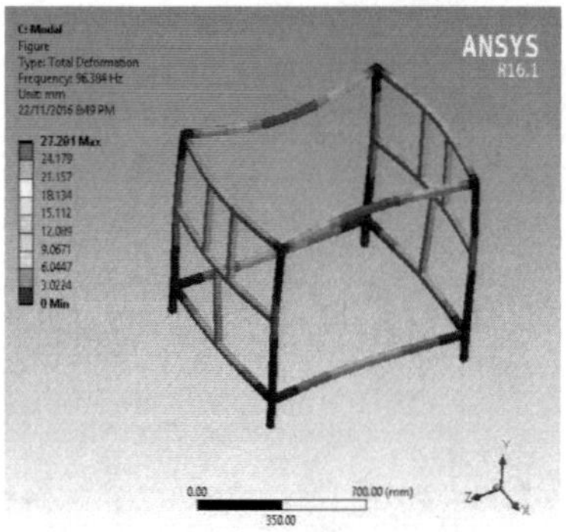

(e)

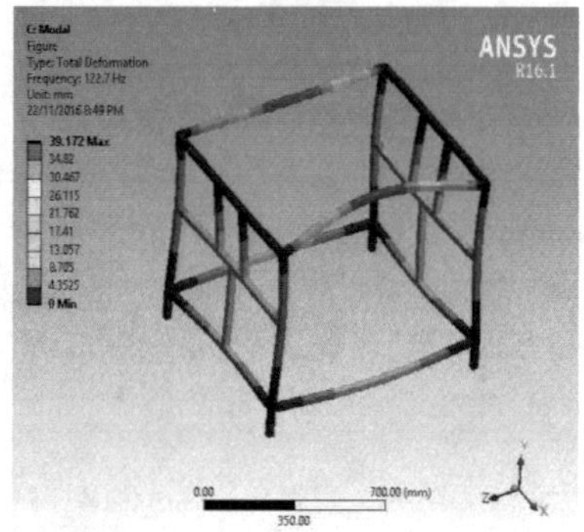

(f)

Figure 7. (e) 5thVibration mode (96.38 Hz) (f) 6th Vibration mode (122.7 Hz).

Based on the results, it was found that, the minimum natural frequency of the main frame was 30.56 Hz, which is much larger than the operating frequency of the motor to rotate the rotary tray during rotating. The motor was assumed to run at a constant speed in a range between 6 to 8 rpm equal to 0.1 to 0.13 Hz respectively. As a result, the frame will not produce resonance during operation and the dryer frame is considered safe from any vibration. The results are supported by Huang [4, 9-13], stated that the frame structure does not produce resonance if the natural frequency of the structure is much larger then operating frequency during machine operation.

4. Conclusions

Based on the modal analysis simulation, it can be concluded that resonance will not occur on the structure frame, which is the biggest component of the drying machine, as its first modal frequency is bigger than the operating frequency of the machine's motor. The structure can be considered safe for assembly and operation.

Acknowledgments

This research is funded by Research Acculturation Grant Schemes [Grant no: RAGS/TK01(4)/1318/2015(12)]. The authors would like to thank Universiti Malaysia Sarawak (UNIMAS) for providing facilities for this research.

References

[1] Suffian M S Z M, Leman, M N, Mohamaddan, S and Mohtar A M A A M 2017 In *Asian Simulation Conference* 12-25 (Singapore: *Springer*)

[2] Leman M N, Mohamaddan S, Suffian M S Z M, Mohtar A M A A M, Mohtadzar N A A, Junaidi N and Julaihi S A 2017 In *MATEC Web of Conferences* EDP Sciences **87** 02016

[3] Mohamaddan S, Mohtar A M, Junaidi N, Mohtadzar N A A and Suffian M M 2016 In *IOP Conference Series: Materials Science and Engineering IOP Publishing* **114** (1) 012037

[4] Huang Z 2012 *Research Journal of Applied Sciences, Engineering and Technology*, **4**(18), 3415-3424.

[5] Marjan J, Johari E, Abbas K A and Farshad A 2010. *American Journal of Applied Sciences*, **7**(2) 160-166.

[6] Yadollahinia A R, Omid M and Rafiee S 2008 *Int. J. Agri. Bio* **10** 61-65

[7] Ehiem J C, Irtwange SV and Obetta S E 2009 *Research Journal of Applied Sciences, Engineering and Technology* **1**(2) 44-53

[8] Golisz E, Jaros M and Kalicka M 2013 *Technical Sciences/University of Warmia and Mazury in Olsztyn*
[9] Ibrahim R & Dewi E N 2003 **6**(3) 145–151
[10] Sitesh K P, Subodh D and Richa S 2015 *International Journal of Current Engineering and Scientific Research* **2**(11) 96-107
[11] Izham M H M, Abdullah N A Z, Zahari S N and Sani M S M 2017 *MATEC Web of Conferences* 90 (01043)
[12] Abdullah N A Z, Sani M S M, Husain N A, Rahman M M, and Zaman I 2017 *International Journal of Automotive and Mechanical Engineering (IJAME)* **14** 3887-3897
[13] Fouzi M S M, Jelani K M, Nazri N A and Sani M S M 2018 *Int. J. Automot. Mech. Eng.* **15** 5874–89

Simulation of Active Force Control Using MR Damper in Semi Active Seat Suspension System

R Rosli[*1]**, Z Mohamed**[1]**, and G Priyandoko**[2]
[1] Advanced Structural Integrity and Vibration Research Group, Faculty of Mechanical Engineering, Faculty of Mechanical Engineering, Universiti Malaysia Pahang (UMP), 26600 UMP Pekan, Pahang, Malaysia
[2] Teknik Elektro, Universitas Widyagama, Malang, Indonesia

*Corresponding e-mail: rosmazi@ump.edu.my

Abstract. With the rapid development of electronic sensor and actuator technology, semi-active seat suspension system has become even more practical driven by lower power consumption. Magneto-rheological (MR) dampers are among the best and the most reliable semi active control devices that can produce controllable damping force in seat suspension system to further improve the ride comfort. This paper focus on a new controller scheme named Active Force Control (AFC) to control the damping force of the MR damper to achieve better ride comfort. The phenomenological Bouc-Wen model for MR damper has been simulated in Matlab Simulink to study the effectiveness of the new AFC controller. A sinusoidal signal simulated as vibration source is applied to the seat suspension system to investigate the improvement of ride comfort as well as to ascertain the new AFC controller robustness. Comparison of body acceleration signals from the passive suspension with AFC controller semi active seat suspension system shows improvement to the occupant ride comfort under different vibration intensities.

Keywords: Active Force Control; Magneto Rheological; Semi Active seat suspension

1. Introduction

The vehicle suspension system is responsible in providing the ride comfort to the passenger by optimally isolating the passenger from any vibration sources. The spring in a suspension system stores the energy from the vibration excitations as potential energy which in time is released back as kinetic energy as the spring bounce back and forth. Another element in the suspension system is the damper which can be installed in parallel with the spring to absorb the kinetic energy produced by the bouncing excitations of the spring [1].

Modern suspension system consists of three different categories which are passive, active, and semi-active suspension. Active and semi-active suspension systems are created to provide a better ride comfort for a wider range of ride conditions, with better adaptability to road profiles which passive suspension cannot do. With wide varieties of controller developed recently, it is also possible for the active and semi-active suspension systems to improve both handling and ride comfort without compromising either one [2].

A number of researchers have turn their attention to further improve ride comfort in this multiple degree of freedom system by introducing another level of vibrations control on top of standard suspension system which then contribute to the development of seat suspension system such as in [3] and [4]. With similar approach to the research done on suspension system, seat suspension systems are divided to passive, active and semi-active system with some researchers have already come up with robust controller technique.

The quarter car seat suspension model used in this study is shown in Figure 1 where the system is assumed to be a single degree of freedom suspended spring mass damper system. Total mass of the passenger and seat is given by m_b, k_s representing the stiffness of the seat, while c_s is the damping rate of the seat but in semi active seat system it is replaced by the f-v characteristic of MR damper.

The ride comfort which is the main objective of this research is quantified by the peak to peak vertical body acceleration experienced by the occupant of the seat. The degree of comfort is analysed by referring to the ISO standard as in Table 1 [5].

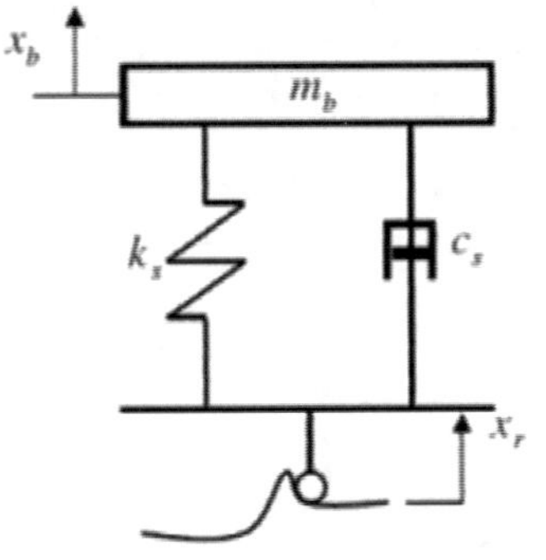

Figure 1. Quarter car seat suspension model.

Table 1. Comfort evaluation [ISO 2631-4].

Vertical Body Accelerations (ms-2)	Comfort Evaluation
< 0.315	Not uncomfortable
0.315 – 0.63	A little uncomfortable
0.5-1	Fairly uncomfortable
0.8-1.6	Uncomfortable

Active and semi-active suspensions have control systems which force the system to achieve optimized conditions. Semi-active suspension uses semi-active dampers whose force is commanded indirectly through a controlled change in the dampers' properties. This change is affected by a damper controller that receives information from the system controller. A semi-active suspension combines the advantages of both active and passive suspensions. It can be nearly as efficient as a fully active suspension in improving ride comfort and stability and is much more economical [6], and [7]. It is also safer to use since if the control system fails, the semi-active suspension can still work as a passive suspension system.

2. MR Damper

MR damper has been shown to be effective as semi-active damper that can satisfy the requirements of ride comfort and vehicle stability [8]. Moreover, an MR damper is an effectively fail-safe device from an electronic perspective. If any fault happened in the system, the MR damper still works as a passive damper within definite performance characteristics, depending on the off-state case of the MR damper [9].

Corresponding mathematical model for MR damper is still intensely researched as it is relatively complex because of its rheological and thermodynamics effects [10]. Complex model of MR damper had led the researcher to turn their heads onto numerous parametric phenomenological models, like the

IOP Conf. Series: Materials Science and Engineering 1062 (2021) 012005 doi:10.1088/1757-899X/1062/1/012005

Bingham plastic model [11,12], Bouc–Wen hysteretic model [13–16], Duffing's equation [17] sigmoid-based model [18], hyperbolic tangent function [19] or even intelligent method approach like neural networks in [20].

3. MR Damper Modelling

A parametric approach had been used in this research to model the MR damper using existing model from previous researchers. A comprehensive review of parametric models can be found in [21]. This paper will utilize the phenomenological Bouc-Wen model modified by spencer as the mathematical modelling for MR damper.

This model was proposed firstly in [16], as shown in Figure 2. This modifies the Bouc-Wen representation through the introduction of an extra internal degree of freedom which differ from the original Bouc-Wen model. This model was proven to better in adaptively tracks the force-velocity hysteresis loop from the experimental measurements.

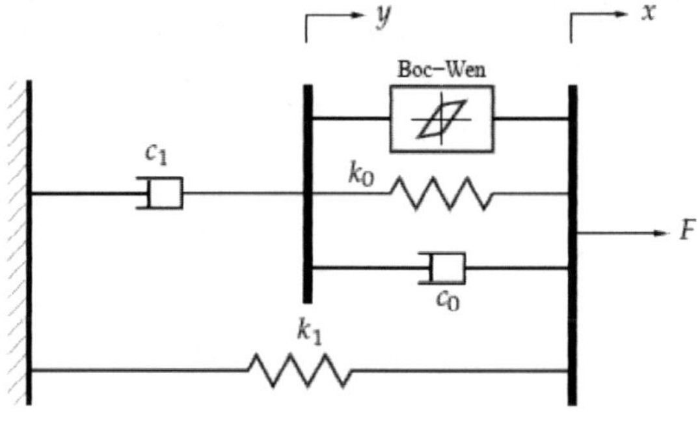

Figure 2. Modified Bouc-Wen model.

The force equation from Figure 2 is given by:

$$f = c_1\dot{y} + k_1(x - x_0) \quad (1)$$

where the evolutionary equation to produce the hysteretic characteristics of MR damper is given by:

$$\dot{z} = -\gamma|\dot{x} - \dot{y}|z\,|z|^{n-1} - \beta(\dot{x} - \dot{y})\,|z|^n + A(\dot{x} - \dot{y}) \quad (2)$$

$$\dot{y} = \frac{1}{c_0 + c_1}[\alpha z + c_0\dot{x} + k_0(x - y)] \quad (3)$$

Conventional Bouc-Wen model basically introduce only equation (2) while the modified Bouc-Wen model introduce equation (3) as another additional evolutionary equation which surprisingly improve the tracking ability of the overall model. The three equations are simulated to compose the MR damper model in Simulink. This bouc-wen model will be then replaced the damping parameter of the seat, c_s in Figure 1.

The optimized model of modified Bouc-Wen simulated in Simulink performance is shown in Figure 3. The F-v curve shows the hysteresis characteristics in MR damper for difference input current values. The peak to peak force values are matching with the datasheet provided by the manufacturer and the model is considered to track the real MR damper response.

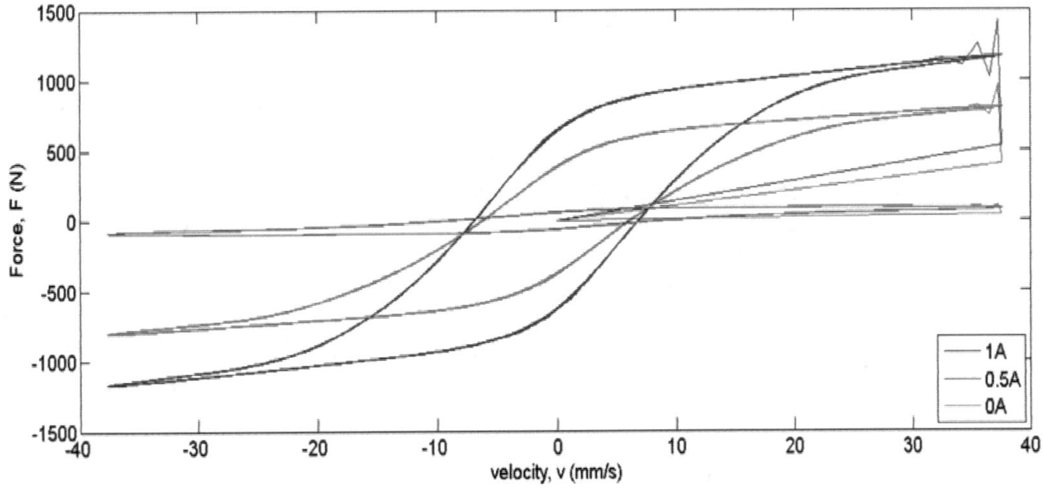

Figure 3. Force velocity curve generated from modified Bouc-Wen model.

The F-v curve shown in Figure 3 represents the significance of input current changes make to the force output in the MR damper at certain velocity value. The maximum and minimum values differences are about 1kN for the changes of only 1A in input current. This shows that small changes in input current has affected the changes of damping rate of the damper significantly which is much desired in controlling the damping force of the semi active seat suspension system.

4. Active Force Control

Active Force Control, AFC is known as a robust disturbance compensator and is superior compared to other conventional methods in dealing with various types of disturbances [22]. It is a very simple scheme that is suitable to be applied to a plant that we wish to control in the presence of disturbance. Applying a simple Newton Law, AFC will react to estimate the disturbance occur at some instant, and give an appropriate signal to the actuator to counter the corresponding disturbance force. There are two main parameters in AFC which needs some intelligent method to estimate. One is the estimated mass, and second is the inverse dynamic of the actuator.

Figure 4 shows the AFC loop in the controller scheme where it takes place to estimate the disturbance forces acting to the sprung mass of the vehicle. Base on the Newton Second Law we have the sum of all resultant forces including disturbance force acting to the sprung mass as:

$$\sum F = m\ddot{z}_s \tag{4}$$

The sprung mass acceleration, now is actually produced by both actuated force and also disturbance forces from various sources. The sprung mass acceleration can be taken from the accelerometer which will be attached to the sprung mass. The controller now needs the value of estimated mass to multiply with the acceleration value from the accelerometer to produce a calculated value of the resultant force in computer.

To get the estimated disturbance force, the controller now need to subtract the actuated force which can be measured easily using a force (or pressure) sensor, from the resultant force calculated by the controller.

$$D_f' = F' - m_s'\ddot{z}_s \tag{5}$$

Where

D_f': Estimated disturbance force
F' : Actuator force taken by using force (or pressure) sensor
$\ddot{z}_s$: Body acceleration taken using accelerometer.

The calculated disturbance force, then will be converted into an appropriate electrical signal through the inverse dynamic of the actuator. The real pneumatic actuator then will convert this signal into the force and inject it into the suspension system to overcome the disturbance force.

There were numbers of research that had successfully applied AFC into dynamic mechanical system. Similarly, [23] implemented AFC into suspension system but combined with skyhook controller. In his work, the AFC- based controller scheme has successfully improved ride comfort. [24] implemented AFC into robotic arm to control its motion. In their work the proposed controller has improved the motion of the mobile robot under the effect of disturbances. Kwek et.al., 2003 implemented AFC into 5-link biped robot [22]. Their work was done via simulation study.

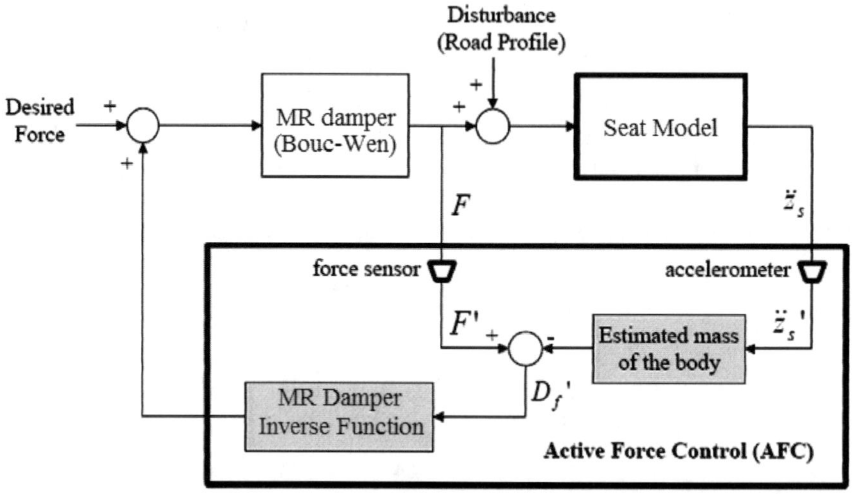

Figure 4. Active force control loop.

In other application such as in robotics, AFC has been researched widely such as in [25-27] to mention a few but there is very limited research done on AFC in translational application such as suspension system. While in seat suspension system, AFC approach has only been done by [28] in active seat suspension. The project focused on a helicopter seat suspension where AFC is coupled with Iterative Learning Algorithm control as the function of inverse actuator feed.

5. Results and Discussion

This section discusses the output of the newly design controller scheme, which will determine the improvement in ride comfort level achieved by this research. The main indicators will be the peak to peak body acceleration experienced by the seat occupant and the acceleration transmissibility of the passive seat suspension system. The indicators are compared to controller scheme over different frequencies of vibration.

In this paper, the inverse actuator model in the AFC loop is treated as a linear function which is the inverse of the linearized function for the modified Bouc-Wen model. The actual model of the inverse function need another comprehensive literature to cover or by using another parametric approach to get a better result. The estimated mass of the body for this simulation has been determined by using heuristic approach as it is only involving a simple single degree of freedom mass spring damper configuration.

Figures 5 show the body acceleration experienced by the seat occupant for both passive seat suspension and controller scheme seat suspension when subjected to 4mm of excitation at 1Hz and a sine chirp signal with varying frequencies of 0.1 to 1 Hz. It is clear that comparison in peak to peak value of body acceleration shows that the controller scheme managed to give useful input to MR damper hence providing the suspension system with better damping force compared to the uncontrolled damping force of a passive system.

(a)

(b)

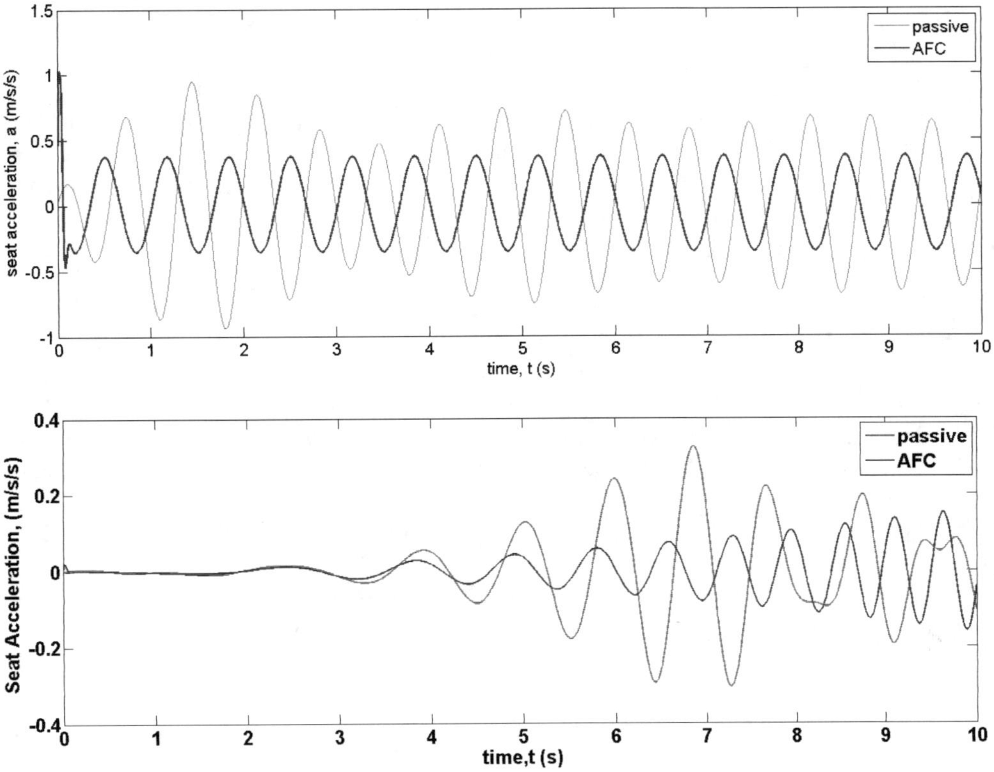

Figure 5. Seat acceleration against time (a) subjected to 4 mm 2 Hz sine wave, (b) sine chirp signal 0.1 Hz to 1 Hz Force velocity curve generated from modified Bouc-Wen model.

Based on comfort evaluation by ISO shown in Table 1, the degree of comfort has improved from fairly uncomfortable (nearly 1m/s2) to a little uncomfortable (0.3m/s2). Though the proposed semi active seat system shows better reduction in seat vertical acceleration in low frequency, but the opposite phenomenon happens at higher frequency where we can see the passive response has lower vertical seat acceleration. But since the lower frequency vibration that gives harm to human body, it is essential to better suppress vibration at low frequencies than at higher frequencies.

6. Conclusion

Active Force Control loop has been successfully adopted into a semi active seat suspension system in this paper. Using a modified Bouc-Wen hysteresis model, the MR damper has been shown to be able to act as actuator to the controller. Damping force has been successfully controlled in the simulation. Comfort evaluation of the semi active seat suspension system has been improved from fairly uncomfortable (nearly 1m/s2) to a little uncomfortable (0.3m/s2) when compared to passive system. Further improvement can be achieved if the inverse model of the MR damper can be formulated more accurately and further study on intelligent method to find estimated mass can be done in the future.

Acknowledgments
The authors would like thanks University Malaysia Pahang (www.ump.edu.my) and Advance Structural Integrity and Vibration Research Group (www.asivr.ump.edu.my) for the financial and research support.

References
[1] Sadok S, Khaled C, Lotfi M, Marc T and Asma K 2005 *Smart Materials and Structural* **14** pp 811–822
[2] Rajamani R 2006 *Vehicle Dynamics and Control* (New York, USA: Springer Science & Business Media)
[3] Ning D, Sun S, Li H, Du H and Li W 2016 *J. of Sound and Vibration* **384** pp 28–44
[4] Orecny M, Segl'a S, Hunady R and Ferkova Z 2016 *Procedia Engineering* **96** pp 338–344
[5] ISO 2631-4:2001 Mechanical vibration and shock – Evaluation of human exposure to whole body vibration – Part 4: Guidelines for the evaluation of the effects of vibration and rotational motion on passenger and crew comfort in fixed-guideway transport systems.
[6] Karnopp D, Crosby M J and Harwood R A 1999 *ASME Journal of Engineering for Industry* **96** pp 619–626
[7] Yi K and Song B S *Proceedings of the Institution of Mechanical Engineers Part D: J. of Automobile Engineering* **213 (3)** pp 293–303
[8] Lee H S and Choi B S 2000 *Intelligent Material Systems and Structures* **11** pp 80–87
[9] Yao G Z, Yap F F, Chen G, Li W H and Yeo S H 2002 *Mechatronics* **12 (7)** pp 963–973
[10] Graczykowski C and Pawłowski P 2017 *Applied Mathematical Modelling* pp 400–424
[11] Honga S R, Wereleya N M, Choia Y T and Choib S B 2008 *J. of Sound and Vibration* **312** pp 399–417
[12] Wereley N M, Pang L and Kamath G M 1998 *J. Intell. Mater. Syst. Struct.* **9** pp 642–649
[13] Bouc R 1971 *Acustica* **24** pp 16–25
[14] Wen Y K 1976 *J. Eng. Mech. Div. ASCE* **102** pp 249–263
[15] Ismail M, Ikhouane F and Rodellar J 1976 *Comput. Methods Eng.* **16** pp 161–188
[16] Spencer B F Jr, Dyke S J, Sain M K and Carlson J D 1997 *J. Eng. Mech. ASCE* **123**
[17] Jia-Ying T, Tsu-Yun C and Pei-Yang L 2014 *Proceedings of the Sixth World Conference on Structural Control and Monitoring* pp 15–17
[18] Ma X Q, Rakheja S and Su C Y 2007 *J. Intell. Mater. Syst. Struct.* **18** pp 487–502
[19] Kwok N M, Ha Q P, Nguyen T H, Li J and Samali B 2006 *Sens. Actuators A* **132** 441–451
[20] Wang D H and Liao W H 2005 *Smart Mater. Struct.* **14** 111–126
[21] Wang D H and Liao W H 2011 *Smart Mater. Struct* **20** 023001
[22] Kwek L C, Wong E K, Loo C K and Rao M V C 2003 *Journal of Intelligent and Robotic Systems* **37** pp 143–162
[23] Priyandoko G, Mailah M and Jamaluddin H 2009 *Mechanical Systems and Signal Processing* **23** pp 855–868
[24] Mailah M and Rahim N I A 2000 *IEEE International Conference on Intelligent Systems and Technologies* 291–296
[25] Jahanabadi H, Mailah M, Md Zain M Z and Hooi H M 2011 *J. of Bionic Engineering* **8** 474–484
[26] Noshadi A, Mailah M and Zolfagharian A 2012 *Applied Mathematical Modelling* **36** pp 2370–2383
[27] Ramli H, Meona M S, Mohameda T L T, Isaa A A M and Mohamed Z 2012 *Procedia Engineering* **41** pp 1389–1397
[28] Ahmadi S and Gohari M 2016 *6th International Conference on Computer and Knowledge Engineering*

Study of nozzle distance and oil flow rate effects on the droplets size of minimum quantity lubrication

K L J Kiat[1], N I K Ismail[1], N Rosli[1*] and E A Alias[2]
[1]Faculty of Manufacturing & Mechatronics Engineering Technology, Universiti Malaysia Pahang, 26600 Pekan, Pahang, Malaysia
[2]Faculty of Mechanical & Automotive Engineering Technology, Universiti Malaysia Pahang, 26600 Pekan, Pahang, Malaysia

*Email: nurrinarosli@ump.edu.my

Abstract. Minimum Quantity Lubrication (MQL) machining process has been widely applied to replace the traditional lubrication and cooling method that brings various drawbacks involving environment, human's health and manufacturing cost. Although the amount of oil being used in the MQL machining process is extremely little, oil mist formed in the cutting zone still can successfully prolong the cutting tool life and improve the surface quality of workpiece. However, the size of lubricant oil droplets must be fundamentally investigated to clarify the lubricating mechanism in the cutting zone. This paper is aimed to study the droplets size of lubricant oil delivered by a commercial MQL generator. Experiments were conducted to study the effects of vertical nozzle distance and lubricant oil flow rate to the droplets count and size. As a result, the smallest diameter of droplets were found to range in between 10 μm - 22 μm. With increasing vertical nozzle distance, the diameter became larger due to the increasing dominancy of air flow that influences the movement of droplets. With increasing oil flow rate, the droplets tend to merge together, forming multiple larger size of droplets. The size of lubricant oil droplets did not significantly change when varying the vertical nozzle distance starting from 35 mm and oil flow rate starting from 18 mL/h. Finally, the control dial to adjust oil flow rate when using an MQL generator with internal feeding system application must be handled properly to ensure precise amount of oil mist successfully being delivered to the cutting zone.

Keywords: nozzle, minimum quantity lubrication, machining

1. Introduction

Machining process is the most important operation in developing manufacturing industry across the globe. Monitoring the cutting process is a crucial one to ensure the manufactured products are made in desired size and shape. During the cutting process, the resistance to relative motion between cutting tool and workpiece which is commonly called as friction raise the temperature around the cutting zone due to the kinetic-to-heat energy conversion. Thus, the sharpness of cutting tool is declining. In turn, the usage of excessive cutting power increases while the surface of the workpiece becomes less smooth. Here, the usage of cutting fluids is vitally in demand to sustain the manufacturing performance. Since then, cutting fluids has been widely utilized in manufacturing operation to enhance the tribological properties of workpiece and cutting tool. The fluid is mainly functioned to lubricate and cool the cutting zone where temperature rise occurs as a result of the frictional effects arises in the workpiece and cutting tool interfaces. Cutting fluids such as petroleum-based oils are non-biodegradable [1] and hence inappropriate disposal of this oils has leads to environmental issues such as water and soil pollution [2]. Moreover, this oils are also being reported to be harmful and jeopardize the health of workers as they experienced skin and breathing problems due to continuous exposure in high volume [3]. The cancer breeding has also become the major concern if the oils being used were not well-refined beforehand. Alternatively, vegetable oils are increasingly being used and even reportedly can produce a better performance [4].

Apart from all the issues being raised, inevitable matters to date are the rising of manufacturing cost resulted from the increasing demand of the cutting fluids especially the neat oils as lubricant. It has been reported that 20 % out of total production cost is merely sourced from the usage of lubricant oils [2]. The cost increment can be attributed to the cost of storage system, labour cost for oils monitoring and maintenance procedure, purchase cost as biodegradable oils are much more expensive, water purifying cost [5] and drying cost of wet chips for recycling process [6].

Therefore, the effort to reduce the usage of lubricant oils in manufacturing operation has been frequently debated despite the requirement to enhance the quality of manufacturing products as well as cut down the cost of manufacturing operation is continuously in demand. In order to replace the traditional lubrication and cooling method due to the problem mentioned above, Near-Dry Machining (NDM) or better known as Minimum Quantity Lubrication (MQL) has been introduced. According to Boswell et al. [7], the word "Minimum Quantity Lubrication" has been initially used by Weck and Koch [8] decades ago through their study of bearings lubrication. Since then, greater attention has been addressed to explore the implementation of MQL in the machining operation.

MQL became popular by the usage of extremely little amount of lubricant oil, typically in 50 to 500 mL/h [9]. With this amount of oil being used, additional cost of disposal is needless [10] and thus the manufacturing cost can be cut down. Moreover, MQL is also considered as green manufacturing approach since it is human and environmental friendly [11]. The great performance of MQL in machining process has been strongly proven in various machining condition including drilling, milling, turning, grinding, tapping, reaming and grooving. Despite the limited amount of oil being used, oil mist formed in the cutting zone still can successfully prolong the cutting tool life and improve the surface quality of workpiece [12]. In drilling process, the implementation of MQL produced better quality of holes compared to flood cooling method [13]. Built-up edge formation at the cutting edge and adhesion of workpiece to the cutting tool have been also reduced in MQL drilling [14]. Crucial improvement has been observed to the surface roughness, cutting power and specific cutting force when MQL was utilized [15]. In milling process, tool wear and surface finish improvement has been achieved by the application of pulsed-jet MQL [16]. On the other hand, better cutting temperature, chip reduction coefficient, cutting forces, tool wear, surface finish, and dimensional deviation has been obtained in MQL turning process compared to the dry turning operation [17].

The lubricant oil from MQL generator basically delivered from the nozzle in tiny droplets of oil mist. This is where the volume of lubricant oil being consumed are always in little amount. In order to produce these tiny droplets, the atomization process between the oil and air supply of MQL generator plays an important role [18]. However, the atomization process taking place in the MQL generator presents in different mechanisms, depending on the type of feeding system.

Basically, there are two feeding systems i.e., external and internal. External feeding system employs lubricant storage tank where several tubes equipped with their own nozzles are attached to the tank. This external MQL system can be installed anywhere around the machine to facilitate the nozzle position for better penetration of lubricant oil to the cutting zone. This system also utilizes self-contained convertible air and lubricant flow to ensure steady oil supply throughout the operation. Moreover, the external feeding system is also more economical, convenient and practically suitable for all machining process such as sawing, milling, broaching, shaping, drilling and threading operations [19].

On the other hand, the oil mist in internal feeding system can be administered directly into the cutting zone. Spindle, tool revolver and tools internal cooling passage are used to deliver the lubricant oil.

Compared to external feeding system, dispersion for internal feeding system is unlikely to occur and feed nozzles are also needless for adjustment. In addition, proportion of oil and air can be calibrated by the control system of the MQL generator [20].

Although the MQL generator usually allows users to adjust the oil flow rate and air pressure for spray output manipulation, the configuration may be difficult for inexperience users. Thus, excessive lubricant will adversely affect the chip formation and be wasted, whereas inadequate lubricant oils reduce the tool life. Moreover, although the oil mist of MQL has been proven to enhance the machining quality, the tiny droplets of oil mist also easily evaporates thus can weaken the lubrication effects during the machining process. Former studies have found that smaller diameter size of oil droplets will evaporate faster than that of larger diameter size [16]. Recent report [17] has revealed that oil mist was found to hardly penetrate the cutting zone due to the chip accumulation in between the cutting zone and inappropriate droplet size of oil. An appropriate combination of nozzle distance, pressurized air and oil mist flow rate were also reported to help reducing the chip accumulation [18], [19].

Moreover, appropriate droplets size of oil was also proven to increase the penetration ability of the oil mist through the cutting zone. Since the surface roughness in between tool chip and tool work interfaces are relatively high, only oil with smaller droplets size can successfully passing through the cutting zone [11]. Hence, the size of lubricant oil droplets under MQL machining system must be fundamentally investigated to clarify the lubricating mechanism in the cutting zone. Nozzle angle, nozzle distance, oil flow rate and air pressure are examples of factors affecting the droplets size of lubricant oil [20]. A study of parametric optimization of MQL in turning of AISI 4340 using nano fluids [21] has been reported to observe the surface roughness by varying the pressure, flow rate and coolant type factors using the method of Taguchi and ANOVA analysis. From the ANOVA results, the flow rate of coolant was found to be the 2nd most contributing factor coming after the factor of cutting nano fluid coolant which gave the highest contribution to the results. Furthermore, another study using Taguchi optimization, S/N ratio and Anova analysis on machining under MQL with different variable [22] has found out that the flow rate operated in MQL generator appeared to be the 2nd most influencing parameter after the cutting speed.

Moreover, comparing different MQL conditions of the flow rate, speed and feed rate on the turning nickel based Nimonic 90 alloy, it was found that specific cutting energy decreased with pressure and flow rate increment [23]. Furthermore, the shear flow stress was found to reduce when a high pressure was applied during the MQL application. This is due to the number of droplets entering the contact zone at high pressure increases which in results in a better lubrication ability. Moreover, the frictional energy is observed to be halved when the flow rate increased from 50 ml/hour to 250 ml/hour as the wetting area increased when the volume of fluid increased in high flow rate. Furthermore, another study conducted for MQL drilling process [24] noticed that the attaching workpiece material on insert for MQL at the flow rate 15 ml/hour and 23 ml/hour can cause two little hills while there was only one little hill and whiter area which shows a higher wear on the substrate for air and MQL at the flow rate of 5 ml/hour. The whiter area was the results of higher temperature caused by the friction of the chip and inserts. This indicates that the higher flow rate conditions have a lower temperature as well as lower friction compared to the lower flow rate and air conditions.

Although there are quite a number of researcher exploring the investigation of MQL machining process, fundamental study merely focusing on the behaviour of lubricant oil droplets driven by the MQL generator are still need to be highlighted. This is of great importance area to discover the relationship between the machining performance and the behaviour of lubricant oil droplets behaviour. Moreover, since the operation mechanism of the MQL generator is comparatively complicated than that of conventional lubrication method, a proper handling of MQL generator also must be taken into account. Here, this paper is aimed to fundamentally investigate the effects of vertical nozzle distance and lubricant oil flow rate to the droplets counts and size of lubricant oil delivered by a MQL generator. The experiments were conducted to study the effects of vertical nozzle distance and lubricant oil flow rate to the droplets count and size. The images of lubricant oil droplets were captured by using a microscope and the droplets size were quantified by using ImageJ software.

2. Methodology

2.1 MQL Generator

CoolubricatorTM from Unist Coolube® was used as the MQL generator to provide the MQL oil mist. This generator utilizes the internal feeding system with separated air and fluid control. It is equipped with a magnetic mount for it to be flexibly mounted anywhere on the machine. There are three adjustable knobs, i.e. cycle rate, pump stroke and air flow knobs. From the combination of these knobs adjustment, the form of oil mist spray can be customized based on the suitable lubrication urgency. It uses a standard 1-drop pump that provides a pump output of 0.03 mL lubricant oil per stroke. Table 1 shows the MQL generator specification.

Table 1. MQL generator specification.

Specification	Value
Supply air pressure (bar)	4 – 7
Air flow rate (ft.3/minute)	0 – 4
Pulse generator frequency (pulses/minute)	5 – 50
Operating temperature range (˚C)	0 – 50
Storage temperature range (˚C)	-16 –70
Fluid reservoir capacity (mL)	473 – 1893

The desired pump stroke for the experiment was firstly set from the configuration system of the MQL generator. The pump piston will then deliver a narrow conical spray pattern of lubricant oil mist through the nozzle tip. Atomization process to generate the lubricant oil mist occurs when the air surrounds the lubrication oil to disperse it to the cutting zone. The air supply can be adjusted to control the size of lubricant oil droplets. In this study, the amount of air supply was roughly adjusted as long as it can satisfactorily produce oil mist form of lubricant spray. The lubricant oil being employed was from Unist Coolube® 2210 which is a vegetable-based oil with dynamic viscosity of 18.5 cSt.

2.2 Experimental Setup and Procedure

The experimental setup in this study is illustrated in Figure 1. As a preliminary experiment, it is important to obtain a filter plate with suitable size of hole that able to produce a single lubricant oil droplet. Here, a filter plate with diameter of hole, d at 0.7 mm, 0.8 mm, 0.9 mm and 1.0 mm, respectively were fabricated by using PCB print circuit board carbide micro drill bits. The filter plate was attached on a holding tray mounted to the cutting tool of a milling machine. Thus, the filter plate can be move along with the cutting tool in the feed direction during the lubricant oil is being sprayed on it.

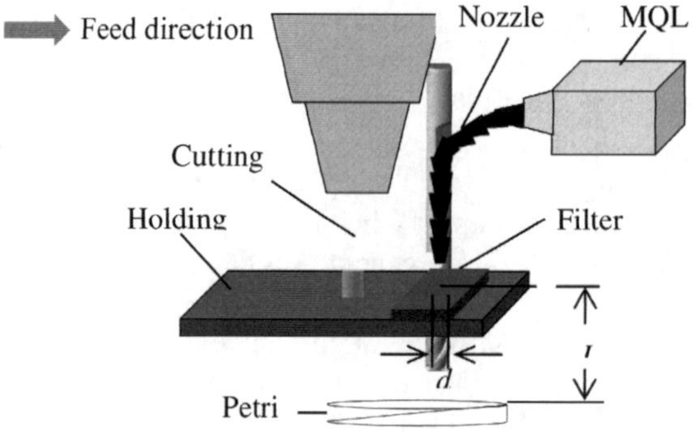

Figure 1. Illustration of experimental setup.

Upon experiment, the oil droplets from the nozzle passed through the filter plate and dropped onto a petri dish located on the machine worktable. Therefore, the overlap of multiple oil droplets at same location can be prevented. The images of lubricant oil droplets were then visualized by using a microscope connected to a personal computer. The images were then captured to quantify the droplets size by using ImageJ software. The droplets count categorized by its diameter was also obtained. A scale according to the microscope must be set beforehand to produce the scale bar for the image. Since there were no specification on the scale in the microscope, a ruler was used to determine the scale manually. The ruler was placed on the stage of the microscope under 4X magnification and the image of the ruler showing 1mm was captured simultaneously. The image was then measured on the Paint software to determine the length of the 1 mm in pixel unit.

The filter plate found to produce the smallest size of lubricant oil droplets was then selected for further analysis of oil droplets size affected by the parameters, i.e. vertical nozzle distance, L and lubricant oil flow rate, Q. Vertical nozzle distance, L is referred to the distance between nozzle tip to the petri dish. The smallest nozzle distance was manually set up in the beginning of the experiment by using a scale. The cutting tool mounted with the holding tray was then moved downward in z-axis to set the larger nozzle distance. Moreover, the distance between the nozzle tip and the filter plate was fixed at approximately 10 mm throughout the experiments.

On the other hand, the lubricant oil flow rate, Q can be controlled by adjusting the control dial of cycle rate. Specifically, air flow and pump stroke knobs were initially adjusted beforehand in order to obtain lubricant oil in mist form. After lubricant oil mist was successfully being produced, air flow and pump stroke knobs were fixed. Then, cycle rate knob was adjusted to the desired value of oil flow rate. Equation (1) shows the calculation of the lubricant oil flow rate based on the adjustment of cycle rate. Table 2 shows the experimental conditions set in this study. The experiments were conducted in 4 sets for each parameter.

$$\text{Oil flow rate}, Q = \frac{60 \times 60}{\frac{60}{\text{pumpstroke}}} \times \text{pump output} \tag{1}$$

Table 2. Experimental conditions.

Machining Condition	Value
Feed rate, V_f (mm/min)	520
Nozzle inner diameter (mm)	2
Diameter of filter hole, d (mm)	0.7, 0.8, 0.9, 1.0
Oil flow rate, Q (mL/h)	9, 18, 27, 54
Vertical nozzle distance, L (mm)	30, 35, 40, 55

3. Results and Discussion

3.1 Preliminary Results – Selection of Filter Plate

Figure 2 shows the sample images of lubricant oil droplets visualized by the microscope under various cases of vertical nozzle distance and oil flow rate. As can be seen from the figure, multiple size of droplets were observed. For analysis, only the droplets with proper round shape were selected for the diameter measurement. Figure 3 shows the results of smallest diameter of lubricant oil droplets passing through various types of filter holes under 30 mm, 35 mm, 40 mm and 55 mm of vertical nozzle distance. The oil flow rate was fixed to 9 mL/h. For 35 mm, 40 mm and 55 mm of vertical nozzle distance, it is evident from the figure that the smallest diameter of lubricant oil was found to increase between 0.7 mm to 0.8 mm of filter hole but significantly decrease at 0.9 mm of filter hole before bouncing back upon

1.0 mm of filter hole. Meanwhile, for 30 mm of vertical nozzle distance, a weak increase of the smallest diameter of lubricant oil droplets was found between 0.7 mm and 0.9 mm of filter hole. The diameter size was then discovered to increase similarly back to its previous value upon 1.0 mm of filter hole. Since all considered vertical nozzle distance exhibit smallest diameter of lubricant oil droplets at 0.9 mm of filter hole, this size of filter hole was selected to study the lubricant oil droplets count affected by the vertical nozzle distance.

	Oil flow rate, Q = 9 mL/h	Oil flow rate, Q = 18 mL/h	Oil flow rate, Q = 27 mL/h	Oil flow rate, Q = 54 mL/h
Vertical nozzle distance, L = 30 mm	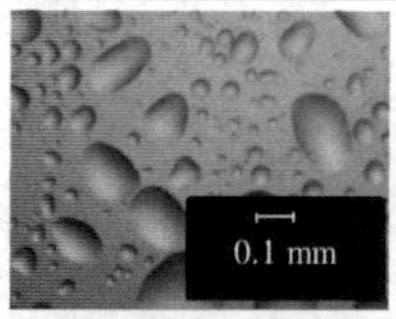	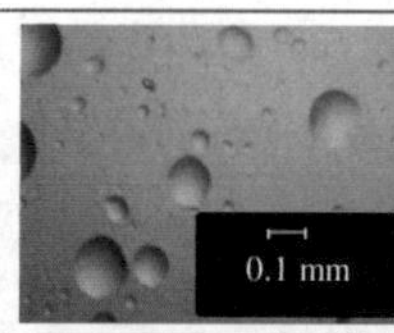	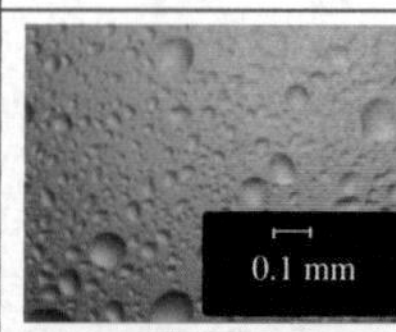	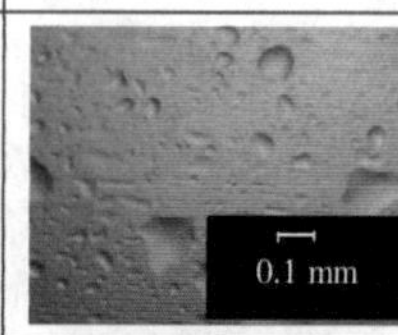
Vertical nozzle distance, L = 35 mm	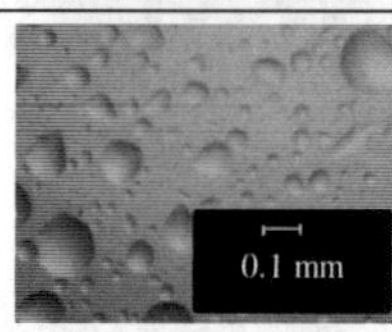	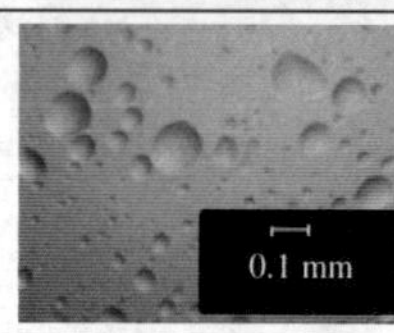	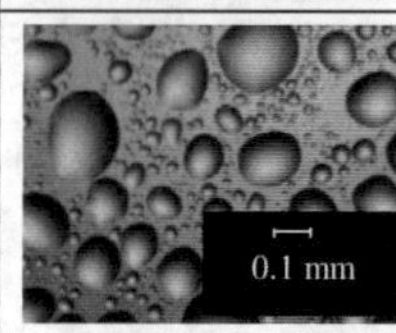	
Vertical nozzle distance, L = 40 mm	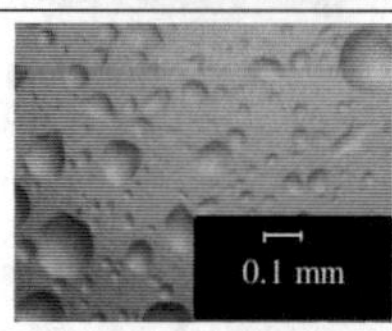	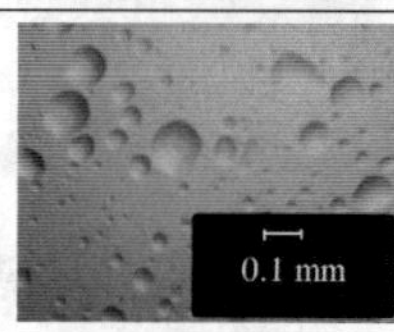	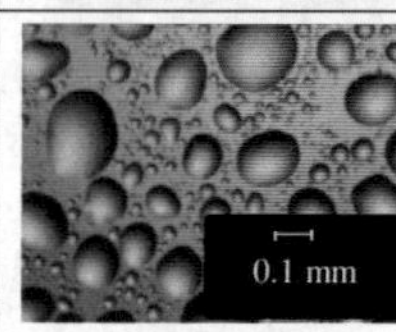	
Vertical nozzle distance, L = 55 mm	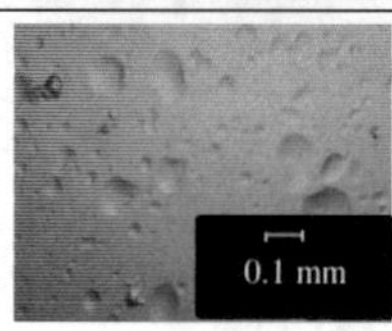	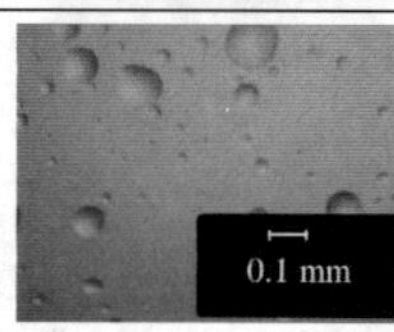	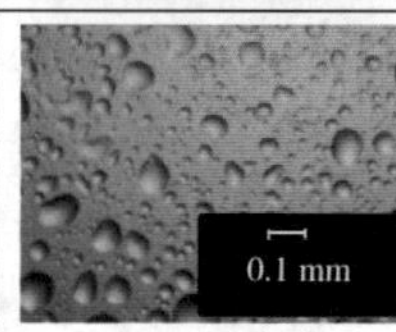	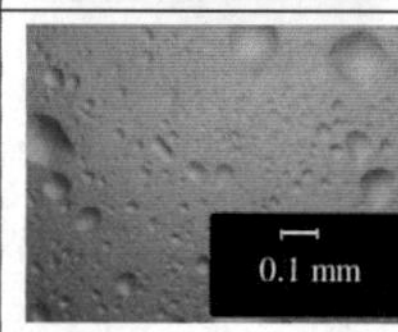

Figure 2. Images of lubricant oil droplets visualized by microscope.

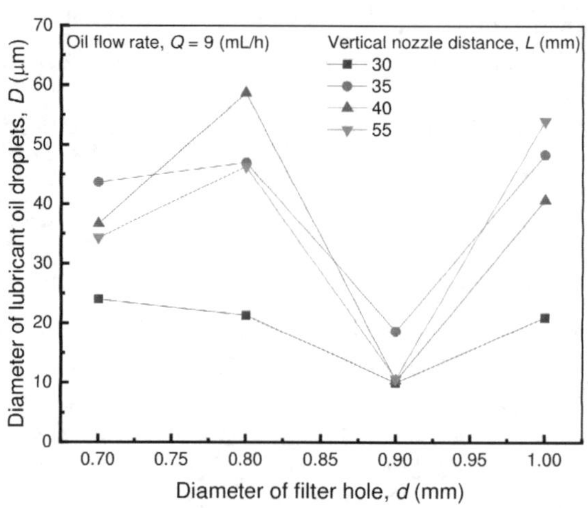

Figure 3. Smallest diameter of lubricant oil droplet passing through filter hole at different vertical nozzle distance.

Figure 4 shows the results of smallest diameter of lubricant oil droplets passing through various types of filter holes under 9 mL/h, 18 mL/h, 27 mL/h and 54 mL/h of oil flow rate. The diameter of lubricant oil droplets was found to significantly fall between 0.7 mm to 0.9 mm of filter hole. However, the smallest diameter of lubricant oil droplets between the same filter hole for lower oil flow rate, i.e. 9 mL/h and 18 mL/h was found to similarly present a weak decreasing trend. For filter with 1.0 mm of hole, the smallest diameter of lubricant oil droplets for all cases were found to bounce back nearly to its initial value when using the filter with 0.7 mm of hole. Since all considered flow rates show smallest diameter of lubricant oil droplets at 0.9 mm of filter hole, this size of filter hole was selected to study the lubricant oil droplets count affected by the oil flow rate.

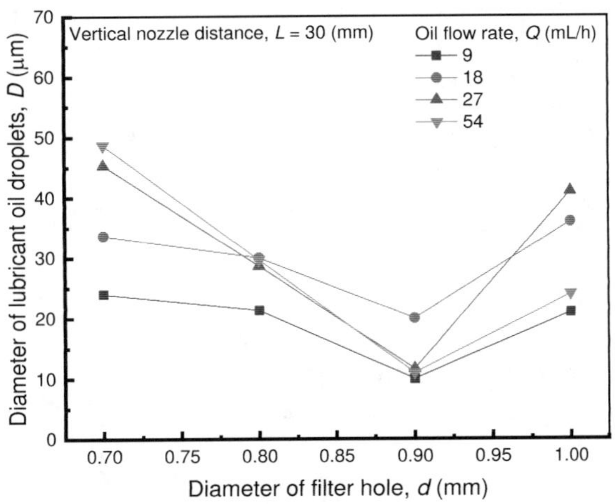

Figure 4. Smallest diameter of lubricant oil droplet passing through filter hole at different oil flow rate.

For the cases of different vertical nozzle distance and lubricant oil flow rate as shown in Figure 3 and 4, the filter plate whose diameter of hole larger than 0.9 mm was not suitable for analysis of droplets count since it produced larger size of smallest diameter of lubricant oil droplets. This may be resulted from multiple flying droplets that has combined together and successfully passed through the hole. At 0.9 mm of filter hole, the smallest size of droplets appeared. These kinds of droplets were believed to be the genuinely single droplets that was delivered from the nozzle of MQL generator and passed through the filter hole. However, when the filter plate whose diameter of hole smaller than 0.9 mm was being used, the smallest diameter of lubricant oil droplets were found in inconsistent value for cases of different vertical nozzle distance in Figure 3. This can be attributed to the collision of multiple droplets attempt to fit themselves inside the filter hole.

Upon entrainment through the filter hole, the pieces of each droplet have combined together and overlayed as a larger single droplet of lubricant oil on the petri dish. Hence, these kinds of droplets were not valid for further analysis. This phenomenon was strongly supported when looking through the results of shortest vertical nozzle distance, i.e. 30 mm where the diameter of lubricant oil droplets were found as the smallest of all other distances. This implies that the number of droplets collided before passing through the filter hole was much lesser than that of longer vertical nozzle distance. Similar occurrence can also be considered for filter plate whose diameter of hole is smaller than 0.9 mm for the cases of difference oil flow rate in Figure 4. As can be seen, the smallest diameter of lubricant oil droplets were found to increase with increasing oil flow rate. This suggests that with much higher flow rate, the number of droplets collided before passing through the filter hole became higher and hence larger diameter of lubricant oil droplets were formed.

3.2 Lubricant Oil Droplets Count and Size

Overall, the diameter size of lubricant oil droplets as shown in Figure 5 and 6 was found to range in between 10 μm - 100 μm. This range of diameter size was supported by the previous work of Park et al. [25], which agreed that the diameter of lubricant oil droplets in this range is not involved in major concerns of the human's health.

3.2.1 Effects of Vertical Nozzle Distance

Figure 5 shows the results of number of droplets categorized by the diameter size at 30 mm, 35 mm, 40 mm and 55 mm of vertical nozzle distance. The oil flow rate was fixed at 9 mL/h.

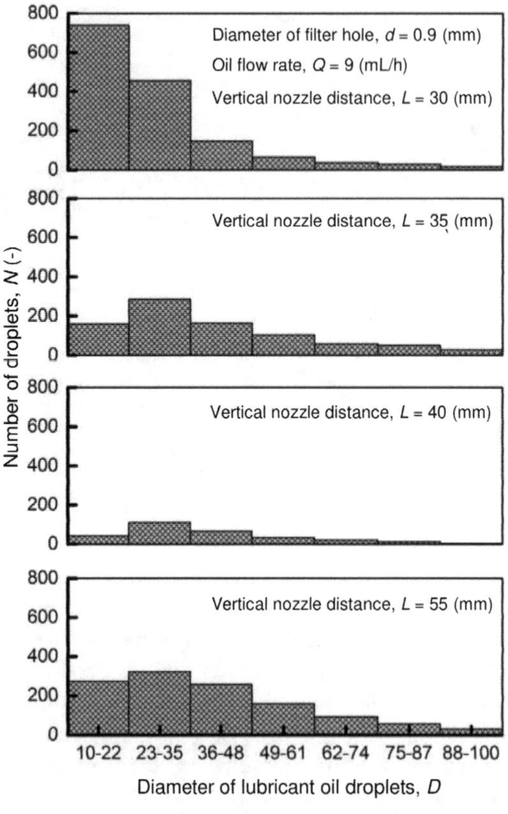

Figure 5. Number of droplets at different vertical nozzle distance.

For the case under 30 mm of vertical nozzle distance, the highest number of droplets were found at smallest range of diameter i.e., 10 μm - 22 μm. However, for the cases at 35 mm, 40 mm and 55 mm of longer vertical nozzle distance revealed the highest number of droplets in larger range of diameter, i.e 23 μm - 35 μm. The reason why larger diameter of droplets appeared the highest at this range can be attributed to the inertia of the droplets. As stated by Duchosal et al. [26], small diameter of droplets carrying less inertia at longer vertical nozzle distance could be easily driven by the air flow to fly off from the test surface. Therefore, only large diameter of droplets successfully reached the test surface without majorly conquered by the air flow.

Furthermore, the highest number of droplets at 35 mm, 40 mm and 55 mm of vertical nozzle distance were also found to decrease almost half than that of 30 mm of vertical nozzle distance. This can be attributed to the evaporation of some droplets when the vertical nozzle distance was set at this range, as reported by Rahim et al. [21]. Park et. al [27] also have stated that increasing the vertical nozzle distance increases the chances of droplets being exposed to the surrounding temperature and hence the oil droplets can be easily evaporated. Therefore, lengthening the vertical nozzle distance may result inadequate tiny oil droplets to penetrate the cutting zone. Najiha et al. [28] also agreed that adequate distance of oil channeled through the nozzle is crucial in order to generate ideal dispersion of oil mist to ensure effective lubrication.

Moreover, less than 200 number of droplets with diameter more than 36 μm were found for all considered vertical nozzle distance. A significant difference in number of droplets according to their diameter size can also be seen only for 30 mm of vertical nozzle distance. This suggests that changing the vertical nozzle distance between 35 mm, 40 mm and 55 mm does not significantly contribute to the change of droplets size for the case under 9 mL/h of oil flow rate.

3.2.2 Effects of Oil Flow rate

Figure 6 shows the results of number of droplets categorized by the diameter size at 9 mL/h, 18 mL/h, 27 mL/h and 54 mL/h of oil flow rate. The vertical nozzle distance was fixed at 30 mm.

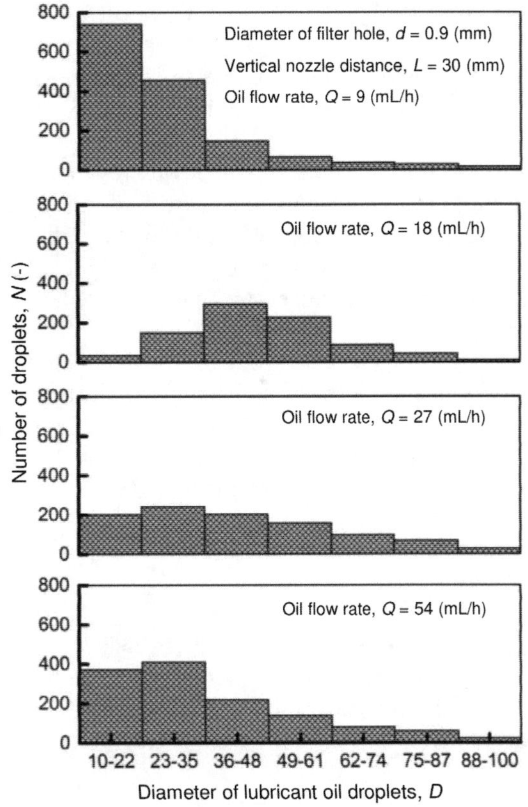

Figure 6. Number of droplets at different oil flow rate.

For the case under 9 mL/h oil flow rate, the highest number of droplets was found at smallest range of diameter i.e., 10 μm - 22 μm. However, the higher oil flow rate, i.e.18 mL/h, 27 mL/h and 54 mL/h gave the highest number of droplets in larger range of diameter. This outcome is corresponded to the past study conducted by Emami et al. [29]. They discovered that the large number of droplets gained from increasing oil flow rate tend to combine themselves to form another multiple huge droplets. Therefore, increasing oil flow rate probably less effective to enhance the lubricating effects since the oil droplets size become larger. Najiha et al. [30] had already stated that unsuitable oil flow rate will affect the infiltration of oil into the cutting zone.

Thoroughly looking into the trending exhibited for the results at 18 mL/h, 27 mL/h and 54 mL/h shows that the highest number of droplets for 18 mL/h of oil flow rate was found at third range of diameter, i.e. 36 μm - 48 μm. However, the highest number of droplets for 27 mL/h and 54 mL/h of oil flow rate were found to drop to smaller range of diameter, i.e. 23 μm - 35 μm. This inconsistency is likely to be affected by the human errors when controlling over the pump stroke in determining the amount of oil delivered from the nozzle. The scale in the control dial is too narrow that it probably caused the adjustment to become uncertainty for every experiment.

Moreover, less than 200 number of droplets with diameter more than 49 μm were also found for all considered oil flow rate. A significant difference in number of droplets according to their diameter size was found significantly for 9 mL/h of oil flow rate. This suggests that changing the oil flow rate between 18 mL/h, 27 mL/h and 54 mL/h is less important in changing the lubricant oil droplets size for the case under 30 mm of vertical nozzle distance.

4. Conclusion

The droplets size of lubricant oil delivered by a commercial MQL generator was fundamentally investigated under different vertical nozzle distance and lubricant oil flow rate. Several conclusion remarks have been made as follows:

i. Under the shortest vertical nozzle distance and lowest operated oil flow rate, the smallest diameter of droplets were found to range in between 10 μm - 22 μm.
ii. With increasing distance, the air flow increasingly influences the movement of the droplets. Hence, the diameter became larger by ranging in between 23 μm - 35 μm. With increasing oil flow rate, the droplets tend to merge together, forming multiple larger size of droplets that range in between 23 μm - 48 μm.
iii. The size of lubricant oil droplets did not significantly change when varying the vertical nozzle distance starting from 35 mm and more at fixed 9 mL/h of oil flow rate and varying the oil flow rate starting from 18 mL/h and more at fixed 30 mm of vertical nozzle distance.
iv. Lengthening the vertical nozzle distance and increasing oil flow rate probably less effective to enhance the lubricating effects since the oil droplets size become larger.
v. When using an MQL generator with internal feeding system application, the control dial to adjust oil flow rate must be handled properly to ensure precise amount of oil mist successfully being delivered to the cutting zone.

Acknowledgement
The authors gratefully acknowledge the financial support from Universiti Malaysia Pahang through internal grant of RDU1803129 and Ministry of Higher Education Malaysia through Fundamental Research Grant Scheme (FRGS) of RDU190124.

References

[1] Baradie M A E 1996 Cutting fluids: Part I. characterisation *Materials Processing Technology.* **56** 786-97
[2] Davim J P 2009 Sustainable manufacturing (ISTE Ltd, 27-37 St George's Road London SW19 4EU UK)
[3] Mijanovic K and Sokovic M 2001 Ecological aspects of the cutting fluids and its influence on quantifiable parameters of the cutting processes *Journal of Materials Processing Technology.* **109** 181-89
[4] Belluco W and Chiffre L D 2002 Surface integrity and part accuracy in reaming and tapping stainless steel with new vegetable based cutting oils *Tribology International.* **35** 865-70
[5] Boswell B, Islam M N, Davies I J, Ginting Y R and Ong A K 2017 A review identifying the

effectiveness of minimum quantity lubrication (MQL) during conventional machining *International Journal of Advanced Manufacturing Technology*. 321-22

[6] Boubekri N, Shaikh V and Foster P R 2010 A technology enabler for green machining : Minimum quantity lubrication (MQL) *Journal of Manufacturing Technology Management*. **21(5)** 556-66

[7] Freiler C, Hörner D, Buss W, Kubicki F, Losc A and Mang T. Lubricants 7: Metalworking fluids and forming lubricants. Ullmann's Encyclopedia of Industrial Chemistry

[8] Weck M and Koch A 1993 Spindle-bearing systems for high-speed applications in machine tools *Annals of the CIRP*. **42 (I)** 445-48

[9] Diniz A E, Ferreira J R and Filho F T 2003 Influence of refrigeration / lubrication condition on SAE 52100 hardened steel turning at several cutting speeds International *Journal of Machine Tools & Manufacture*. **43** 317-26

[10] Lubrication EF. Uni-Max Revolution Retrieved from *https://unist.com/mql-machining.html* 23 October, 2018

[11] Dhar N R, Islam M W, Islam S and Mithu M A H 2006 The influence of minimum quantity of lubrication (MQL) on cutting temperature, chip and dimensional accuracy in turning AISI-1040 steel *Journal of Materials Processing Technology*. **171(1)** 93-9

[12] Rahim E A, Ibrahim M R, Rahim A A, Aziz S and Mohid Z 2015 Experimental investigation of minimum quantity lubrication (MQL) as a sustainable cooling technique *Procedia CIRP*. **26** 351-54

[13] Braga D U, Diniz A E, Miranda G W A and Coppini N L 2002 Using a minimum quantity of lubricant (MQL) and a diamond coated tool in the drilling of aluminum-silicon alloys *Journal of Materials Processing Technology*. **122** 127-38

[14] Bhowmick S and Alpas A 2008 Minimum quantity lubrication drilling of aluminium - silicon alloys in water using diamond-like carbon coated drills *International Journal of Machine Tools & Manufacture*. **48** 1429-43

[15] Kumar A and Davim J P 2009 A study of drilling performances with minimum quantity of lubricant using fuzzy logic rules *Mechatronics*. **19(2)** 218-32

[16] Thepsonthi T, Hamdi M and Mitsui K 2009 Investigation into minimal-cutting-fluid application in high-speed milling of hardened steel using carbide mills International *Journal of Machine Tools & Manufacture*. **49(2)** 156-62

[17] Dhar N R, Ahmed M T and Islam S 2007 An experimental investigation on effect of minimum quantity lubrication in machining AISI 1040 steel *International Journal of Machine Tools & Manufacture*. **47(5)** 748-53

[18] Weinert K, Inasaki I, Sutherland J W and Wakabayashi T 2004 Dry machining and minimum quantity lubrication *Manufacturing Technology*. **53(2)** 511-37

[19] Khan W A, Hoang N M, Tai B and Hung W N P 2018 Through-tool minimum quantity lubrication and effect on machinability *Journal of Manufacturing Process*. **34** 750-57

[20] Madhukar S, Shravan A, Vidyanand P and Reddy G S 2016 A critical review on minimum quantity lubrication (MQL) coolant system for machining operations *International Journal of Current Engineering and Technology*. **6(5)** 1745-51

[21] Patole P B and Kulkarni V V 2018 Parametric optimization of minimum quantity lubrication in turning of AISI 4340 using nano fluids *Materials Today: Proceedings*. **5(5)** 12419-425

[22] Dureja J S, Singh R, Singh R T, Singh P, Dogra M and Bhatti M S 2015 Performance evaluation of coated carbide tool in machining of stainless steel (AISI 202) under minimum quantity lubrication (MQL) *International Journal of Precision Engineering and Manufacturing-Green Technology*. **2(2)** 123-29

[23] Ghosh S and Rao P V 2018 Specific cutting energy modeling for turning nickel-based Nimonic 90 alloy under MQL condition International *Journal of Mechanical Sciences*. **146-147** 25-38

[24] Tasdelen B, Wikblom T and Ekered S 2007 Studies on minimum quantity lubrication (MQL) *Journal of Materials Processing Technology*. **200** 339-46

[25] Park K, Olortegui-Yumel J A, Joshil S and Kwon P 2008 Measurement of droplet size and distribution for minimum quantity lubrication (MQL) *Int. Con. on Smart Manufacturing Application, Gyeonggi-do, Korea*. 447-54
[26] Duchosal A, Leroy R, Vecellio L, Louste C and Ranganathan N 2013 An experimental investigation on oil mist characterization used in MQL milling process *International Journal of Advanced Manufacturing Technology*. **66** 1003-14
[27] Park K H, Olortegui-Yume J, Yoon M C and Kwon P 2010 A study on droplets and their distribution for minimum quantity lubrication (MQL) *International Journal of Machine Tools & Manufacture*. **50(9)** 824-33
[28] Najiha M S, Rahman M M, Yusoff A R and Kardigama K 2012 Investigation of flow behavior in minimum quantity lubrication nozzle for end milling processes *International Journal of Automotive and Mechanical Engineering (IJAME)*. **6** 768-76
[29] Emami M, Sadeghi M H and Sarhan A A D 2013 Investigating the effects of liquid atomization and delivery parameters of minimum quantity lubrication on the grinding process of Al_2O_3 engineering ceramics *Journal of Manufacturing Process*. **15(3)** 374-88
[30] Najiha M S and Rahman M M 2014 A computational fluid dynamics analysis of single and three nozzles minimum quantity lubricant flow for milling *International Journal of Automotive and Mechanical Engineering (IJAME)*. **10** 1891-900

Vertical Vibrations Effect on Forced convection heat transfer from a Longitudinal Finned Tube

Saleem Khalefa Kadhim[1*], Sinan Abdulghafar Ali[1], N F D Razak[2], M S M Sani[2,3]

[1]Control and System Engineering Department, University of Technology-Iraq, Baghdad, Iraq

[2]Faculty of Mechanical and Automotive Engineering Technology, Universiti Malaysia Pahang Pekan Campus, 26600 Pekan, Pahang, Malaysia

[3]Automotive Engineering Centre, Universiti Malaysia Pahang Pekan Campus, 26600 Pekan, Pahang, Malaysia

*Corresponding author: 60014@uotechnology.edu.iq

Abstract. Thermal systems more efficient by emanating from industrial applications and space program galvanized interest in ways by increasing heat transfer is the system. The main aim of this paper is to investigate experimentally the effects of the vertical vibration and force convection on the average Nusselt number in a longitudinal finned tube. The finned tube was located inclined or horizontally in different angles of 0°, 30° and 45°. The effect of the excitation frequency covers below 16 Hz with various heat fluxes ranged from 500-1500 W/m^2. It was noticed that the good agreement between the experimental measurement and the previous experimental studies with deviation of 5%. The results showed that average Nusselt number values at angle 45° from longitudinal finned tube were higher by up to 14%, 16% compared with the angles 30°, and 0°, respectively. Furthermore, it can be summarized that the vertical vibration significantly affects to the average Nusselt number from longitudinal finned tube cylinder and the influence on the heat transfer coefficient for this system should be considered.

Keywords. Vertical Vibration; Forced Convection; Longitudinal finned tube

1. Introduction

Heat exchangers are utilized in generally in the oil, chemical, refrigeration, control and electronic ventures by heating or cooling. One technique for increment that has been explored has incorporated the utilization of vibration [1]. The effect of vibration on forced convective heat transfer has been explored for cylinders, fin arrays, and flat plate. This vibration for these surfaces are either vertical or horizontal vibration affect, and different range of amplitude and frequency, also, the different of heat flux. There has been a difference in these investigations in terms of the effects either to be large or small on the average Nusselt number [2].

Experimentally, Takahashi, and Endoh [3] studied the effect of vibration on forced convection heat transfer occurring on the sphere, a cylinder and a square-section tube to water. They found that the average Nusselt number with the vibration effect is good correlated in terms of the energy dissipation calculated from the fluid drag acting on the vibrating bodies. Experimental investigation was conducted by Eid and Gomaa [4] vibration of the thin planner fins to improve of heat transfer rates. they found that the excitation vibration lead to improvement of heat transfer may however equal or slightly more than that of the vibration due to pulsating flow.

Tian et al. [5] performed experimental study on heat transfer enhancement by crossflow-induced vibration. They conclude that the increase of the average Nusselt number is registered at vibrations of low and average excitation frequencies for a heat pipe with large diameter. Also, they also found out that the vibrations of higher excitation frequencies indicate an increase of the average Nusselt number for the heat pipes with small diameters.

Jin et al. [6] conducted experimental and studied the effects of the pulsating flow agitation on the heat transfer in a triangular grooved channel, had been considered for the ranges of $270 \leq Re \leq 910$. They summarized that the Nusselt number increase at Reynolds number decreases. Also, they found that the frequency of pulsation flow increase at Reynolds number decreases.

In the work of Hommema et al. [7], they performed experimentally investigation of the heat transfer in a condensing pulsating flow. Pulsate flow experiment at the Reynolds number range between 2600 to 4300, and frequency is 34 Hz. In their result they observed that the pulsate flow was shown to enhance Nusselt number to increase up to a factor of 1.8. Sun et al. [8] investigated the forced convection heat transfer from a circular cylinder with a flexible fin in laminar flow numerically using a Fluid Structure Interaction (FSI) solver. This process considering heat transfer, with Reynold number 200 and also investigated the vibration effect on their model occurred due to flow-induced vibration. The result stated that the flow-induced vibration considered, a maximum of 11.07% enhancement in Nusselt number is obtained by the flexible fin.

Experimentally study the effect of pulsatile flow on finned-tube heat exchangers was carried out by Michaud et al. [9]. They used Reynolds number range from 50 to 900. In their results stated that the heat transfer rate was inversely proportional with the pulsatile flow. Kore et al. [10] performed experimentally the heat transfer enhancement from dimpled pin fin. They used Reynolds number in the range from 150 to 208. They conclude that the maximum of 99.75% improvement in heat transfer coefficient is obtained by the dimpled pin fin.

Izadpanah et al. [11] carried out the effect of rotating and oscillating blade on heat transfer enhancement of Newtonian and non-Newtonian fluid flow in a channel. The study is performed analysis through numerical and used Reynolds number as 50, Prandtl number of 1. From the simulation result, the maximum heat transfer enhancement to straight channel is 58.7% for a stationary blade with $n= 0.4$. Also, the output shows that the presence of the blade for all power law indexes has a positive effect on the heat transfer enhancement. Kim et al. [12] studied experimentally the effects of mechanical vibration on critical heat flux in vertical annulus tube. It concludes that the critical heat flux increased by mechanical vibration up to 16.4%.

Testing related to investigate the effect of vertical vibrations upon heat transfer rate from a horizontal tube immersed in a tank of water was carried out by Martinelli and Boelter [13] through experimental work. The results revealed that the Nusselt number was not affected at low Reynolds numbers, but that Nusselt number increases by as much as 400% when comparison without vibrations. Li et al. [14] was carried out numerical work and simulated the novel flapping vortex generator mounted on a heatsink fin for airside heat transfer improvement. The findings stated that the flapping vortex generator can improve the average Nusselt number by 200%.

Mittal and Al-Mdallal [15] developed a two-dimensional mathematical model in order to investigate of forced convection from an isothermal cylinder performing rotational oscillations in a uniform stream. The finite difference method was applied which solved the Naiver–Stokes, momentum equations and factored in the relationship between Nusselt number and Reynolds number. The result show that the frequency and amplitude of oscillation have a noteworthy effect on the heat transfer mechanism. Sellapan and Pottebaum [16] experimental investigations the nusselt number from rotationally oscillating cylinders placed in a cross-flow at Renolds number 750. They found out that the forced oscillation frequency and amplitude are crucial parameter and play a significant role in the enhancement of heat transfer rates. Also, they found that significant enhancement in the heat transfer rates was reported in the lock-on regime.

Beskok et al. [17] numerical investigate to Nusselt number from the uniformly heated walls of a straight channel in presence of a rotationally oscillating cylinder at Reynolds number 100. They found that the probability of utilize rotationally oscillating cylindrical with air flow for electronic chip cooling applications. Pottebaum, and Gharib [18] investigated the effects of transverse oscillations on the heat transfer from a circular cylinder in cross-flow experimentally. Their results revealed that the heat transfer is significantly improved to be strongly dependent on synchronization with harmonics of the natural shedding frequency. Rahman et al. [19] experimentally study the investigated the effect of fin in the performance of closed loop pulsating heat pipe. Moreover, they compared the effects on the heat transfer performances of closed loop pulsating heat pipe with finned and un-finned condenser section with different angles from 0 (vertical), to 45). Their results indicate that the finned closed loop pulsating heat pipe at 45 degree exhibits the considerable enhancement of heat transfer compared with that of closed loop pulsating heat pipe without fin.

Past researchers performed their research based on the influence of vibration on the heat transfer coefficient of cylindrical surfaces and plate. To the best of our knowledge no research was performed to investigate vertical vibrations on the heat transfer rate in longitudinal finned tube. Consequence, this paper investigated the effect of vertical oscillation on the longitudinal finned tube, including considering different frequency (2 - 16 Hz), different heat flux and different inclination angle (0°, 30°, and 45°), and different Reynolds number (250 - 900) on the average Nusselt number.

2. Methodology

2.1. Experimental Setup

Experimental test rig is developed and setup in order to measure the temperatures at the base of the fins, amplitude and acceleration occurs at the longitudinal finned tube. This measurement is used to validate the previous experimental studies. The schematic of the test rig system and the experimental test rig components used as shown in Figure 1a. The test rig is divided into (test section, vibration circuit, and electrical circuit). The system schematic for the finned tube test rig shows the rig components, and instrumentations as illustrated in Figure 1a.

Table 1 shows the overall physical measurement of the test rig components. Eight triangular longitudinal fin (made of an aluminum) equally distributed by (45°) in model was used (Figure 1b) [20]. Varied of heat flux is applied and the heater has been placed inside the cylinder tube. In order to reduce the heating loss at its ends, the cylinder tube is fixed to connection arms using Teflon substances. The angle of the finned tube can be varied from 0° to 45°, using a lever mechanism which connects the holder and its base.

Table 1. Test Rig Measurements Rate

Components	Size (mm)
Length: Cylinder tube	320
Length: Finned tube	300
Outer Diameter: Finned tube	48
Height: Finned tube	13

A heating coils power of 1000W are applied in electric circuit. There are 12 thermocouples type k were installed at the fin surface and distributed uniformly. A thermocouple was used to measure ambient temperature which is located 15 cm away from the finned tube. The thermocouples were connected to a selector switch, then connected to digital thermometer. Test rig suspended horizontally inside air duct by oscillating arm which receives its oscillatory motion from exciter which had been fixed at inside the air duct working as guides to keep the movement of test model assembly horizontally, as shown in Figure (1a). However, the measured vibration amplitude for both the holder and the heated cylinder are not much different.

The air duct has been designed (1× 0.5 m) cross-section with 2 m length. The duct walls are built from wood of 2 cm thickness in order to improve the radiation heat transfer from the square cylinder to the walls. The air duct has been provided with centrifugal fan fixed horizontally to deliver proper parallel air flow, air duct at finned tube as shown in Figure 1a. A Pitot static tube is installed which is fixed at upstream of the finned tube. This is for measure the air flow velocity in the air duct. The air duct outlet is connected to the variable speed electric Centrifugal Blower inlet. The centrifugal blower [Brook Compton, Parkiuon Motor and Doncaster-England] is driven by A.C. motor of 550 W at 2800 r.p.m. The flow of air is controlled by using an adjustable throttling valve mounted just at the air outlet to control the velocity of air between (1-5 m/s).

The experimental work was performed by heating the finned tube and to make sure constant conditions was occurred after 1.5 to 2 hours [20]. Then, temperature measurement was recorded without vibration effect. After that, vibration effect was applied ranged between 2 to 16 Hz and temperature readings were obtained as well. Same procedure was implemented in order to achieve stable period, record the vibration amplitude and temperature. The vibration system consists of three main parts: The Random Function Generator, which generates and sends the signal to the digital Oscilloscope, and the Power Amplifier, which amplifies the signal and sends it to the Vibration Exciter, which vibrates according to the desired frequency. The test model and stand are mounted vertically by the vibration system inside the duct.

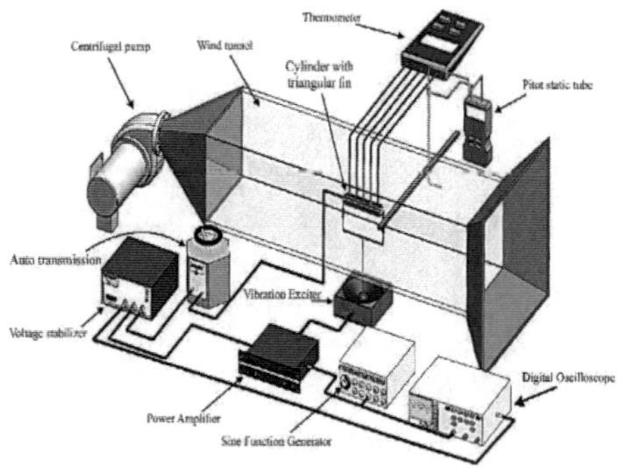

(a) Schematic diagram of the test rig.

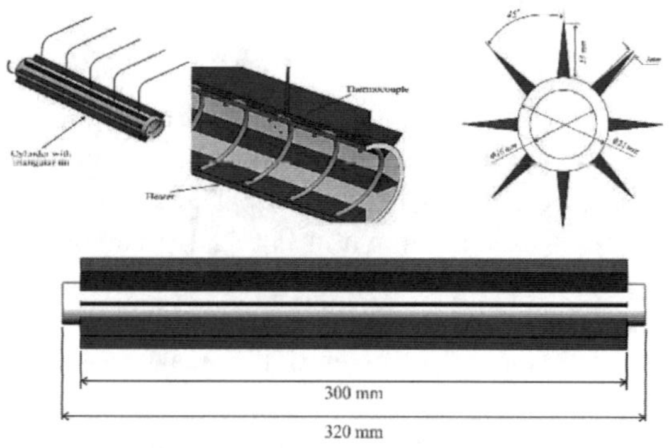

(b) Schematic diagram of the fined tube.

Figure 1. Schematic diagram of the test section, vibration circuit, and electrical circuit used in the test rig.

2.2. Processing and boundary conditions

The amount of heat generated due to electric heater through to test section (longitudinal finned tube) by conduction and then dissipate to the surroundings by force convection in addition to radiation.

When energy balance is implemented on the test section, the convection heat transfers from the longitudinal finned tube [20], which gives

$$Q_{gen} = Q_{con} + Q_{rad} \tag{1}$$

where Q_{gen} is represented heat generation in electrical heater (W), Q_{con} is the conduction heat transfer (W), Q_{rad} is represented radiation heat transfer (W). The heat generation (Qgen) due to electric heater related to the current (I) and voltage (V) in the electrical heater, as follows:

$$Q_{gen} = I \times V \tag{2}$$

where I is represented current (Amp), V is the voltage (Volt). The radiation heat transfers through a longitudinal finned tube is commonly expressed as:

$$Q_{gen} = \sigma \cdot A_f \cdot \epsilon \cdot (T_s^4 - T_a^4) \quad (3)$$

Where σ is the Steven poltzman constant, ε is represented emissivity, T_s is the surface temperature (°K), Ta is the air temperature (°K), A_f is represented fins area (m^2). However, the heat transfers coefficient through the finned tube may be determine using the following formula [20]:

$$h = \frac{Q_{con}}{A_f(T_s - T_a)} \quad (4)$$

where h is represented heat transfer coefficient ($W/m^2.°C$). The air properties have been calculated based on the on the average value of finned tube surface temperature and ambient temperature:

$$T_f = \frac{T_s + T_a}{2} \quad (5)$$

where T_f is the film temperature (℃). The vibration intensity (U_v) and amplitude of vibration is commonly expressed as:

$$U_V = a * f \quad (6)$$

$$a = \frac{a_{cc}}{\sqrt{2} * (2 * \pi * f)^2} \quad (7)$$

where a is the amplitude of vertical vibration (m), f represents the vibration frequency (Hz), a_{cc} the acceleration of vibration, (m/s^2).

An experimental work has been performed using air as a working fluid in order to create the effect of forced vertical vibration and force convection on the average Nusselt number of the longitudinally finned tube. Variation inclination angle was set and heat generation and vibration frequency were applied for each test. Furthermore, the heat transfer coefficient, vibration intensity, Reynolds number, and vibration Reynolds number, in addition to the amplitude of a vibrating system and their acceleration of vibration for different values of air flow velocity were determined for the analysis. The results of this longitudinally finned tube are obtained for forced convection and force vibration. As stated, the heat flux was set up with range between 500 to 1500 W/m^2.

3. Result and Discussion

3.1. *Validation of the current present work and literature*

As mentioned in section II, the experimental had validate against the result from experiment which is determined from the previous study. For presented work, the validation is performed by comparing the average Nusselt number from literature with measured average Nusselt number in the experiment.
The average Nusselt number is used to analyze the effect of vibrations on the heat transfer rate in longitudinal finned tube. The values of Reynolds number utilized in this study is from 250 to 900. In order to investigate the effect of vertical vibration on the heat transfer rate in longitudinal finned tube, firstly, the values of amplitude, used in this study is from 0.2 mm to 0.5 mm, and the range of heat generation from 500 to 1500 w/m2.

The current experimental was validated against previous experimental studies. Figure 2 shows the comparison of the current experimental result with the experimental result obtained by previous researcher [4-7]. Comparison was performed using average Nusselt number longitudinal finned tube. The average Nusselt number altitude for the finned tube from 74 to 93 (at Re = 250, Re = 900), respectively, obtained from the experiment. Tian et al. [5] found that the average Nusselt number altitude obtained from their experimental value was 53 to 75 at Re from 100 to 750. Also, Eid and Gomaa [4] found that the average Nusselt number altitude obtained from their experimental value was 67 to 70 at Re from 230 to 900. Figure 2 shows that the average Nusselt number obtained from the current experimental results was altitude 5% at Re from 250 to 900. Therefore, the current work possible give an improvement of about 84% in average Nusselt number over that from the steady state flow case (Eid and Gomaa [4]). Consequently, the present work could be used with confidence in this study to perform heat transfer analysis.

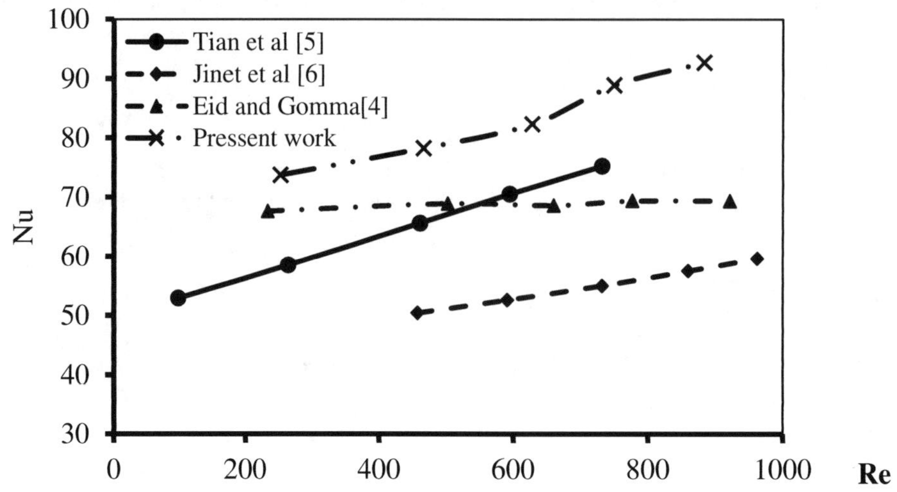

Figure 2. Comparison between current work and previous studies.

3.2. Effect of vibration on the nusselt number

Primary parameters that determine the performance of finned tubes are the amount of heat transfer from the surfaces, include heat transfer coefficient due to the external vibration of the finned tubes. A series of frequency at different angles and different heat flux are shown in Figures 3, 4, 5, 6 and 7 to present the performance of finned tubes under investigation. Figures 8 to 12 show the effect vibration intensity on the average Nusselt number at different angles.

The range of vibration frequency using as all model at a different angles (θ = 0°, 30°, 45°) from 2 Hz to 16 Hz, and the ranged of heat flux from 500-1500 W/m2. The values of Reynolds number, utilized in this study is from 250 to 900, as shown in figures 3 to 7. Figure 3 shows that the increase in the average Nusselt number continues to increase, Reynolds values increased from 250 to 900 at increase in the finned tube angles. At f =0 Hz, Re = 900, and Q= 1500 W/m2, it found that the average Nusselt number increase from 381, 425, and 481 occurring at θ = 0°, 30° and 45°, respectively.

Figure 4 show the average Nusselt number increase from 440, 461, and 482, at f = 4 Hz at different angles (0°, 30°, 45°). Also, figure 5, found that the average Nusselt number at angle 45° higher than from angle 0°, and 30°, where the values of average Nusselt numbers were 533, 471, and 451, respectively, at f = 8 Hz, and Re = 900. Figure 6 show the average Nusselt number increase from 480, 495, and 561, at f = 12 Hz at different angles (0°, 30°, 45°). At Re= 900, f =16 Hz, and different angles (θ = 0°, 30°, 45°) at heat flux 1500 W/m2, it found that, the maximum average Nusselt number values were 504, 520, and 606, respectively, as shown in Figure 7. Although, average Nusselt number occurred at angle 45°, Re = 900, and Q = 1500 W/m2. This is attributed to the large-amplitude motion forced air around the bottom and top of the finned tube causing the boundary layer to thin on the leading side of the cylinder which resulted in an increase in the heat transfer in this region [21]. Wherefore, increase in average Nusselt number occurred at θ = 45°.

Figures 3 to 7 illustrates the average Nusselt number that occurred at different frequency and air Reynold numbers, which corresponds to different angles. The average Nusselt number at the finned tubes at various Reynold numbers from 250 to 900 are presented in Figures 3 to 7. According to the figures, slight increment occurred in the Nusselt number values with increasing Reynolds number, when compared to the with and without vibration. Figures 3 to 7 shows average Nusselt number values increased when the Reynolds number increased from 250 to 900 and vibration frequency increased from 2 to 16 Hz, respectively. It can be concluded that the average Nusselts number valued are proportional to Reynolds number.

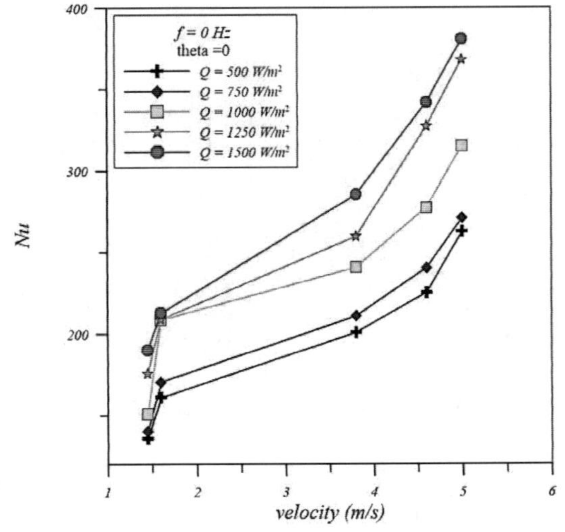

a) θ=0°

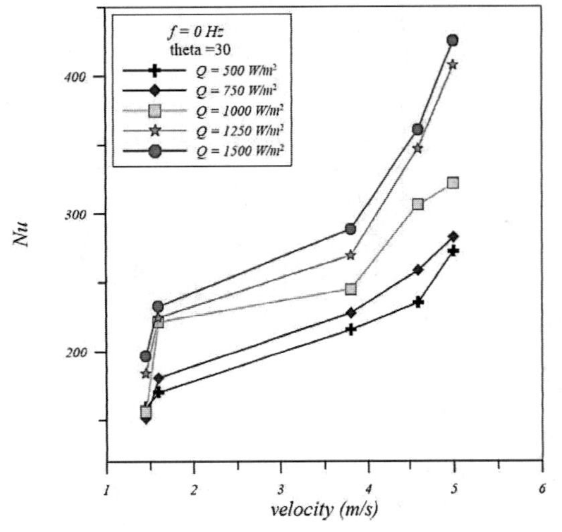

b) θ=30°

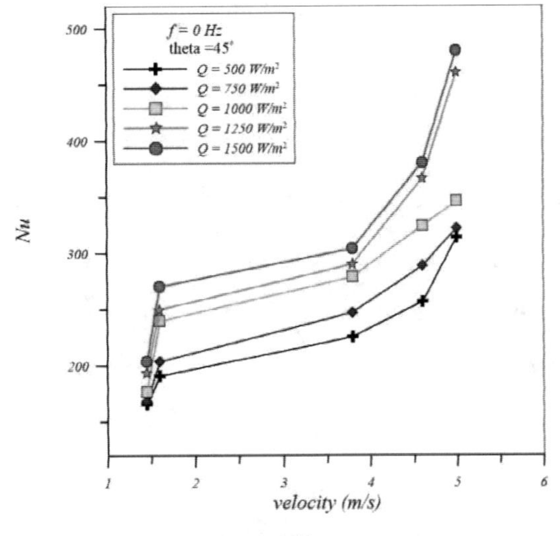

c) θ=45°

Figure 3. The effect of velocity on the Nu for the fin tube with $f = 0$ Hz at different heat rate and angle.

IOP Conf. Series: Materials Science and Engineering 1062 (2021) 012007 doi:10.1088/1757-899X/1062/1/012007

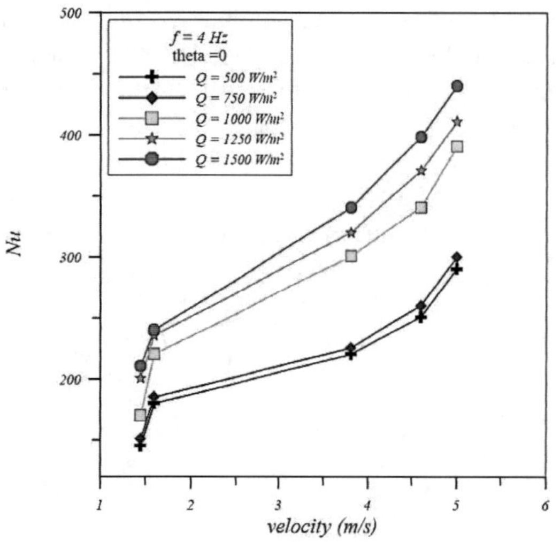

a) θ=0°

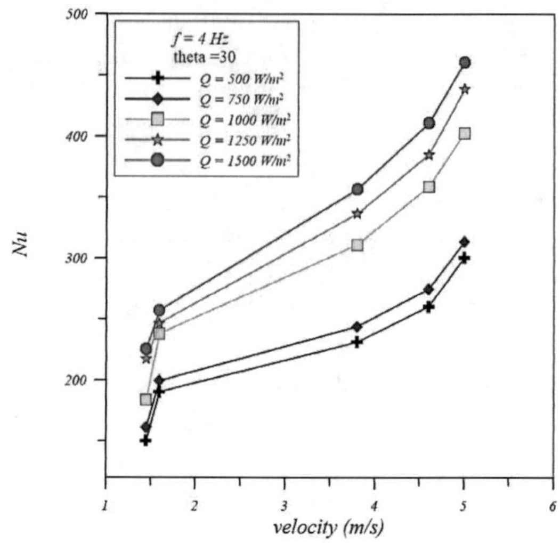

b) θ=30°

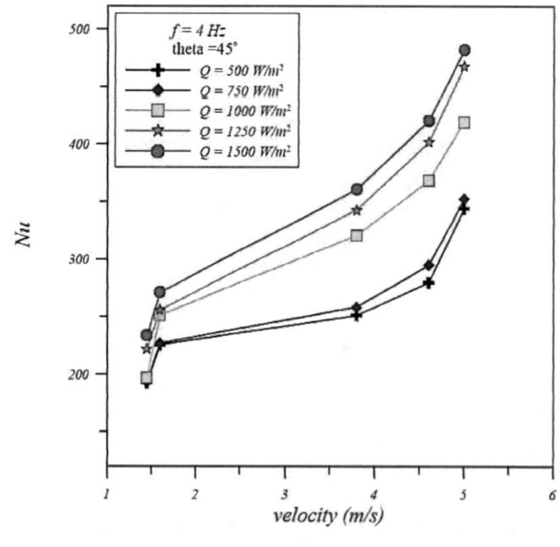

c) θ=45°

Figure 4. The effect Velocity on the Nu for the fin tube with *f* = 4 Hz at different heat rate and angle.

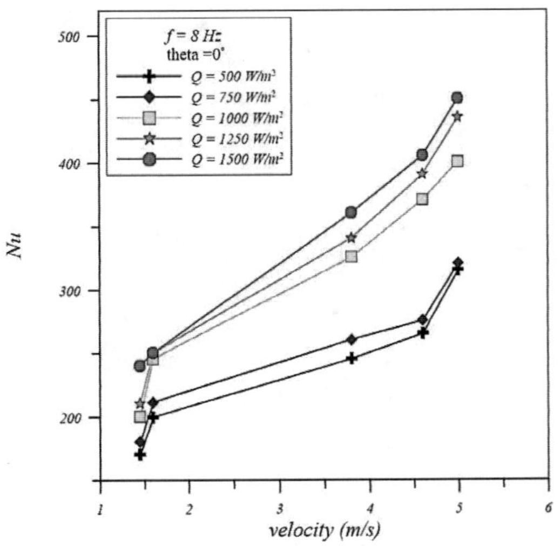

a) θ=0°

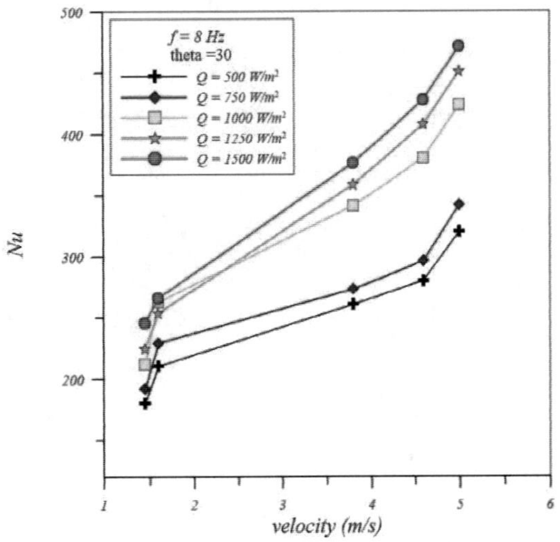

b) θ=30

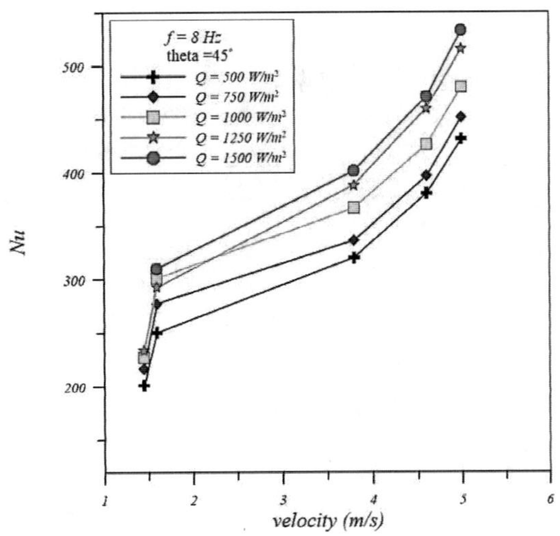

c) θ=45°

Figure 5 The effect Velocity on the Nu for the fin tube with $f = 8$ Hz at different heat rate and angle.

IOP Conf. Series: Materials Science and Engineering 1062 (2021) 012007 doi:10.1088/1757-899X/1062/1/012007

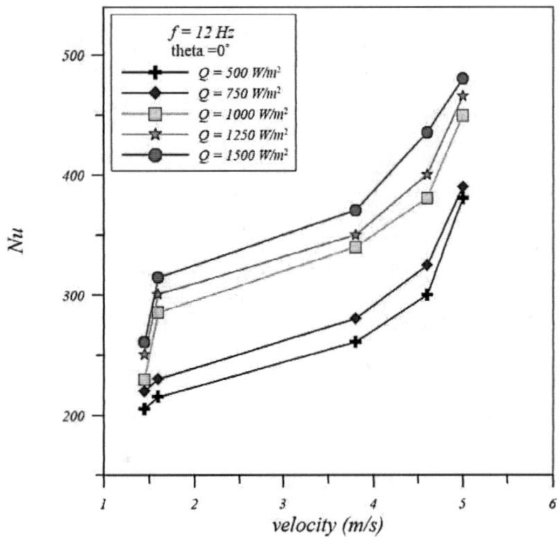

a) θ=0°

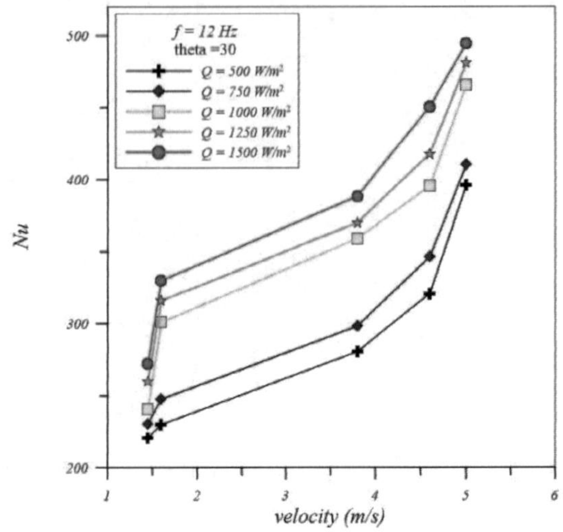

b) θ=30

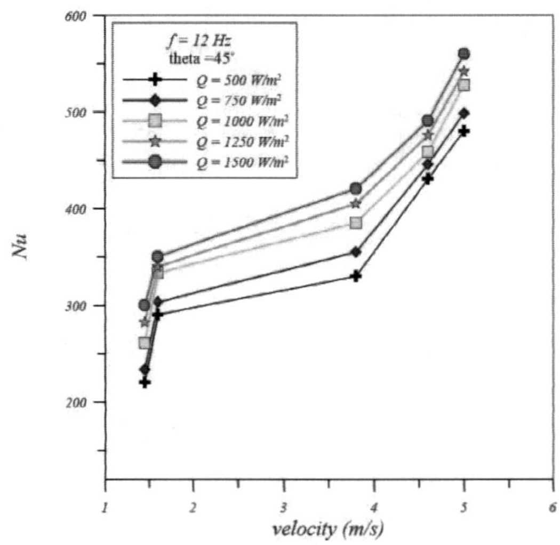

c) θ=45°

Figure 6. The effect Velocity on the Nu for the fin tube with f = 12 Hz at different heat rate and angle.

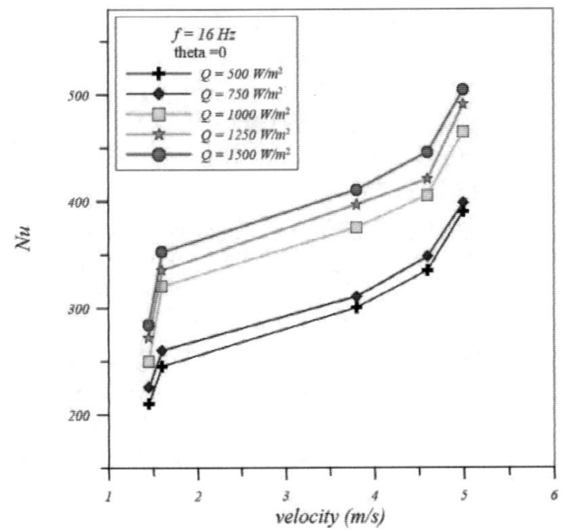

a) θ=0°

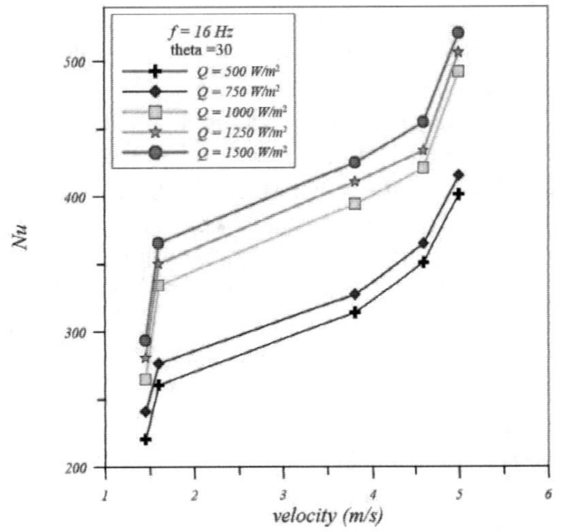

b) θ=30

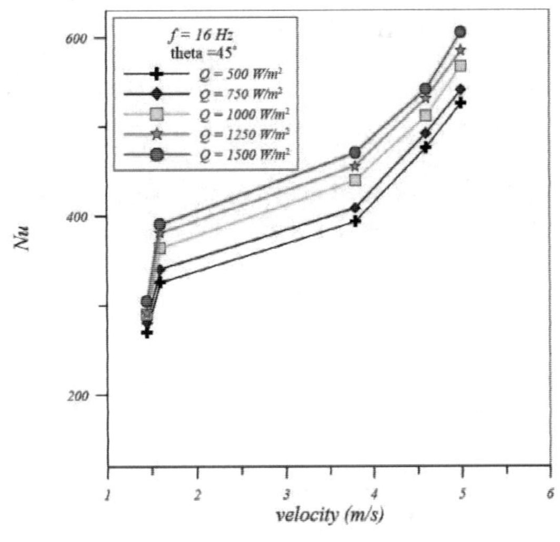

c) θ=45°

Figure 7. The effect Velocity on the Nu for the fin tube with f = 16 Hz at different heat rate and angle.

4. Conclusion

A longitudinal finned tube cylinder was developed to experimental investigated average Nusselt number at different angles and different excitation frequency. The experimental measurements are validated against the previous experimental studies. From the experimental measurements found that the quantitative comparison between the previous experimental studies of the vertical vibration effect on the average Nusselt number, showed reasonable agreement with deviation of 5%. The vibration intensity affected on the average Nusselt number of angle (45°) at the model were 14% greater than those for the (30°) and exceeds that of the (0°) from the zero angle by 16%, respectively. From all the current results in this work, the vibration can significantly affect the average Nusselt number from longitudinal finned tube cylinder and its effects on the heat transfer coefficient should be considered.

IOP Conf. Series: Materials Science and Engineering 1062 (2021) 012007 doi:10.1088/1757-899X/1062/1/012007

References

[1] Walunj A A, Daund V S and Palande D D 2014 *Int. Journal of Innovation and Applied Studies* **6(2)** 232-238
[2] Harahap F, Lesmana H and Sambegoro P L 2010 *Heat and mass transfer* **46** 929-942
[3] Takahashi K and Endoh K 1990 *Journal of chemical engineering of Japan* **23(1)** 45-50
[4] Eid E I and Gomaa M E 2009 *Heat and mass transfer* **45** 713–726
[5] Tian M, Cheng L, Lin Y and Zhang G 2004 *Heat Transfer - Asian Research* **3(44)** 211-218
[6] Jin D X, Lee Y P and Lee D Y 2007 *Heat Mass Transfer* **50** 3062–3071
[7] Hommema S E, Temple K A, Jones J D and Goldschmidt V K 2002 *Int. Journal Heat Mass Transfer* **45** 57-65
[8] Sun X, Ye Z, Li J, Wen K and Tian H 2019 *Heat and Mass Transfer* **128** 319-334
[9] Michaud J P, Speer C P, Miller D A and Nobes D S 2017 *Int. Journal of Mechanical and Mechatronics Engineering* **11(6)** 1230 - 1235
[10] Kore S S, Joshi S V and Sane N K 2017 *Int. Journal of Eng. Development and Research* **5(1)** 395 - 412
[11] Izadpanah E, Babaie Rabiee M, Sadeghi H and Talebi S 2016 *Applied Thermal Eng* **113** 1277-1282
[12] Kim D H, Lee Y H and Chang S H 2007 *Nuclear Engineering and Design* **237** 982–987
[13] Martinelli R C, Boelter L M K 1938 *Proceedings of 5th Int. Congress on Applied Mechanics* 578–586
[14] Li Z , Xu X, Li K, Chen Y, Huang G and Chen C 2018 *Heat and Mass Transfer* **118** 1340-1356
[15] Mittal H V R and Al-Mdallal Q M 2018 *Int. Journal of Heat and Mass Transfer* **127** 357–374
[16] Sellappan P and Pottebaum T 2014 *Journal of Fluid Mechanics* **748** 549-579
[17] Beskok A, Raisee M, Celik B, Yagiz B and Cheraghic M 2012 *Int. Journal of Thermal Sciences* 58 61-69
[18] Pottebaum T S and Gharib M 2006 *Int. Journal of Heat and Mass Transfer* **49** 3190–3210.
[19] Rahman M L, Sultan R A, Islam T, Hasan N M, Ali M 2015 *Procedia Engineering* **105** 137-144
[20] Kadhim S K and Nasif M S 2016 *MATEC Web of Conferences* **38** 01012
[21] Elshafei E A M 2007 *Applied Thermal Engineering* **27(13)** 2233-2242

Study of modal properties and crashworthiness performance of crash box numerical model with different joining modelling strategies

N.A.Z. Abdullah[1], M.S.M. Sani[1,2*], M.S. Salwani[2]
[1]Advanced Structural Integrity & Vibration Research (ASIVR), Faculty of Mechanical and Automotive Engineering Technology, Universiti Malaysia Pahang, 26600 Pekan, Pahang, Malaysia
[2]Automotive Engineering Centre, Universiti Malaysia Pahang, 26600 Pekan, Pahang, Malaysia

*Corresponding e-mail: mshahrir@ump.edu.my

Abstract. It is important to understand the natural frequencies and the mode shapes for automotive structure such as crash box as it will be subjected to the dynamic loading in its real application. In addition, it is also important to study other dynamic behaviour such as crash and buckling as crash box is one of energy absorbing member that is intended for car safety during collision. In this study, modal properties of three different crash box models joint by three different joint modelling strategies and also their crash and buckling behaviour was analysed and investigated. Computational frequency analysis, dynamic crash analysis and nonlinear buckling analysis was performed on all of crash box models. the crash box modelled with node equivalent method to represent the welded joint shows higher stiffness at the joining area and therefore shows higher value of natural frequencies that the model with CBEAM and CBAR elements as their joining elements. In term of crash behaviour, the model with CBEAM element shows higher reaction force compared to the CBEAM. This proves that different type of joint modelling strategies behaves differently.

Keywords: Top hat structure, crash analysis, nonlinear buckling analysis, joint modelling strategies, modal analysis

1. Introduction

A structural natural mode of vibration not only effect on the structure's NVH characteristics but also the knowledge on dynamic characteristics of any structure is very crucial during the design phase. It is more important to understand their dynamic characteristics especially for the structures that will undergo dynamic loading during the actual application. For instance, it is essential to know the natural frequencies and their respective mode shapes for many automotive structures as most of them will be subjected to dynamic loading in real application [1]. This knowledge is very important for making a number of improvements that can be implemented in a design in order for the structure to have better performance.

In making improvements towards various automotive parts and structures, researchers are setting many performances goals and objectives. Mostly, safety is the main criteria that are taking into consideration when making improvements. The study on the dynamic behaviour, the way of deformation of vehicle construction elements during impact and energy transferred during loading for frontal section of vehicle are among popular topic of research among scholars [2]. By gathering the information on the vehicle frontal parts, new implementation were developed and studied such as improving material mechanical properties and geometric characteristics of elements section [3–5].

Crash box, which is one of the frontal parts of a vehicle, is one of the popular topics of study. It is a component that is attached in between the car chassis and front bumper and supposed to deform and absorb energy during crash to provide comfort and safety to the car's passengers [6]. Various types of analysis and development was investigated and implemented on crash box structure in order to maximize its dynamic performance [7,8]. For example, many geometrical optimizations were tested to investigate the capability of many types of geometrical configuration with aim of finding the better performance structure. These studies of many configurations has cause the design of initially simple thin-wall column crash box to be evolved into more complex structure [9]. For example, classic crash box that was normally an extruded square column can now become adapting new profile such as top-hat or hexagonal. In addition, the application of triggers and joining has contributed to different dynamic performance for the crash box structure [10–13]. In fact, joining such as welding can influence the material properties of crash box. The location of joining can cause discontinuities in the stiffness of the box which leads to deformation into different direction. For instance, welding can helps to stiffen the material and makes the non-welded area to collapse easier [14]. In this case, the welding also plays the parts as trigger or collapse mechanism to the crash box and it is very essential to understand how the joining contribute to the dynamic response [14].

In studying the characteristics of structures, finite element method is one of the popular analysis tools to be used to gather as many knowledge and information on the structure. The behaviour of structure, for example a crash box, under dynamic loading can be observed and studied before validating the study with experimental work. For most structure with joining element, mostly, the modelling of joining element is neglected [15]. However, it has been stressed by many researchers that the modelling of joining element is important towards having more accurate analysis [13,16,17].

In this study, the application of several joining modelling strategies in top-hat crash box structure and how the models behave under modal and crushing analysis was investigated. The details on the modelling strategies and the analysis conducted on the crash box are explained in the following section of this paper.

2. Construction of crash box models

In order to perform a prediction analysis on the modal properties of the top-hat crash box structure, finite element analysis by using a software package was used. The finite element model was constructed based on actual top-hat structure as shown in **Figure 1**. The interest to study the top-hat shaped crash box with spot welded joint is due to findings that stated the flange that acts as stiffness which contribute towards the energy absorption capability. In addition, spot weld joint is widely used in automotive and therefore gaining a lot of research interest [18]. The finite element model was constructed and meshed to produce 979 quad-shell elements. The nominal values of material properties assigned on the crash box model are based on aluminium alloy 6061 properties which are as follows; Young's modulus ($\boldsymbol{E}$) is 69 GPa, density ($\boldsymbol{\rho}$) is 2800 kg/m^3 and Poisson's ratio is 0.33. The height and width of the crash box are 60 mm with the flanges are 10 mm on each side, the length is 200 mm and the wall thickness is 1.5 mm.

For replicating the stiff behaviour of the spot-welded joint that were available in the actual top-hat structure several modelling strategies were used. The stiffness of the joint was created by using connecting bar element such as CBAR and CBEAM, and also by constructing node equivalence at the welded location. The CBAR element is a general purpose beam that supported tension and compression, torsion, bending in two perpendicular planes, and shear in two perpendicular planes. The CBAR used two grid points and provided stiffness to all six DOFs of each grid point. With CBAR, its elastic axis, and shear centre all coincided. The displacement components of the grid points were three translations and three rotations. The CBEAM element provided all the capabilities of the CBAR element, plus the following additional capabilities; first, the neutral axil and shear centre did not need to coincide, which

IOP Conf. Series: Materials Science and Engineering 1062 (2021) 012008 doi:10.1088/1757-899X/1062/1/012008

were important for unsymmetrical sections, also, the effect of cross-sectional warping on torsional stiffness was included in CBEAM properties (PBEAM) only, and lastly, the effect of taper on transverse shear stiffness (shear relief) was included (PBEAM only) [19]. The CBAR and CBEAM element was using circular profile with diameter of 5mm in order replicate the spot weld nugget size. Very high stiffness value of was assigned in order to replicate the joint stiffness. On the other hand, node equivalencing is a method that combines two nodes from two neighbouring elements into one node. This is to stiffen the connectivity between those two elements which is sometimes used to replicate the stiffness of welded joint in actual structure. In the constructed model, the two nodes at the place of spot weld were equivalence into one single nodes.

The application of CBAR and CBEAM elements as connector elements was explained where different cross section for the element was used (see **Figure 2**) [20,21]. However, in this study, both CBAR and CBEAM elements were created by using same cross section. Location of CBAR and CBEAM elements are shown in **Figure 3**.

Figure 1. Top-hat crash box structure with spot-weld joint

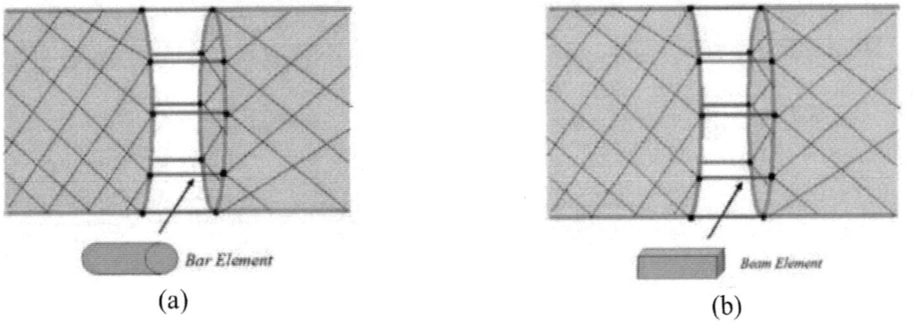

Figure 2. Cross section of (a) CBAR element and (b) CBEAM element [19]

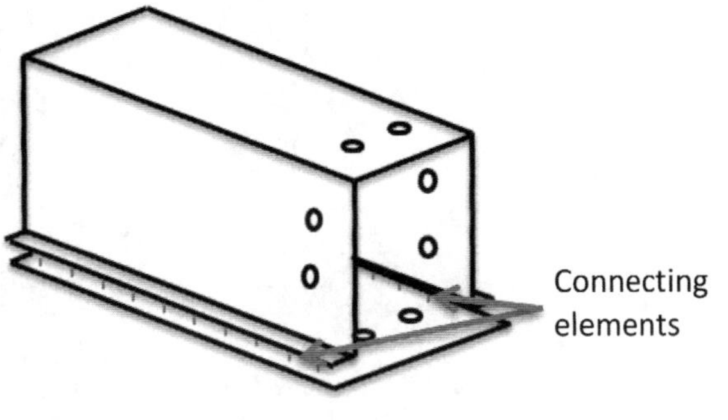

Figure 3. Location of CBAR and CBEAM elements on the crash box model

3. Computational modal analysis on crash box model

Computational modal analysis or normal mode analysis was performed in order to obtain the modal properties of each crash box model. No load or translational and rotational boundary conditions were applied to any node on the structure which left the model to be analysed under free-free boundary condition. The output of the analysis is as shown in the Table 1. Based on the table, the crash box model with node equivalent joint shows higher stiffness at the joining area and therefore, having higher value of natural frequency compared to crash box models with CBAR and CBEAM element. In addition, the modal properties of crash box models joined by CBAR and CBEAM showed similar characteristics of modal properties. This is due to the similar stiffness and mass properties assigned on both models with resulting on the similar modal properties to be generated.

Experimental modal analysis was conducted in order to validate and measure the discrepancies of respective models. Response from the impact hammer test was shown in Figure 4 where the obtained frequency response functions and their respective coherence signal was analysed. Mode indicator shows that the lower frequency peak is associated with rigid body mode. Correlation with all crash box models shows that crash box model with CBAR and CBEAM joining elements has lower discrepancies (see.Table 2).

Table 1. Modal properties of crash box models

Mode	Natural frequencies (Hz) for crash box model joint by			Mode shapes for crash box model joint by		
	Node equivalence	CBAR	CBEAM	Node equivalence	CBAR	CBEAM
1	472.64	395.18	395.18	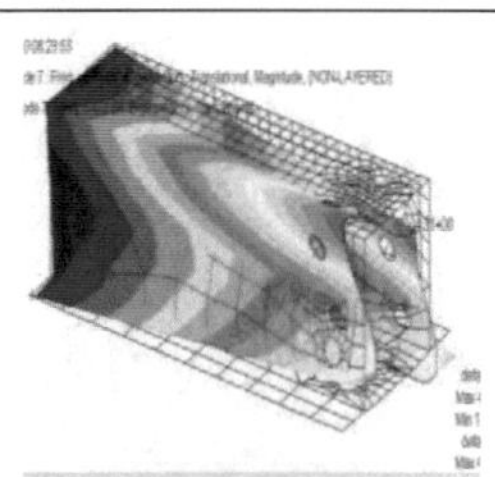	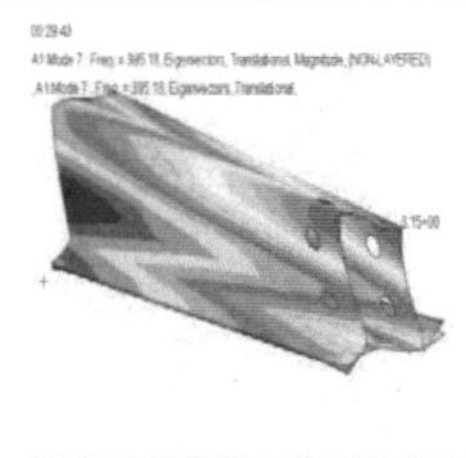	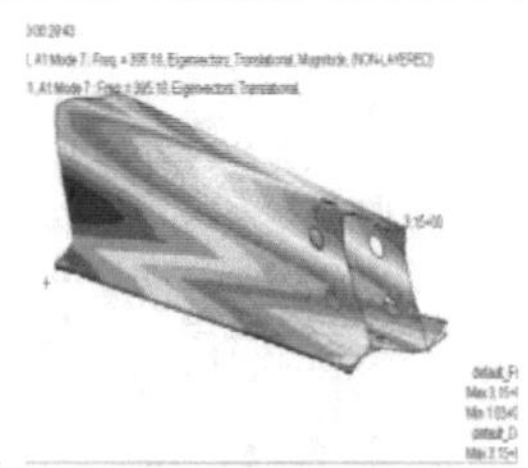
2	479.76	403.93	403.93	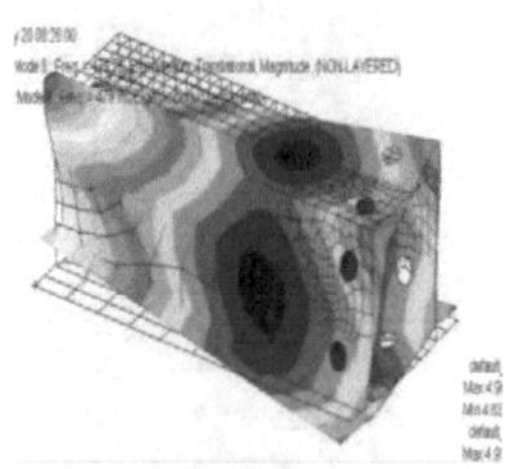	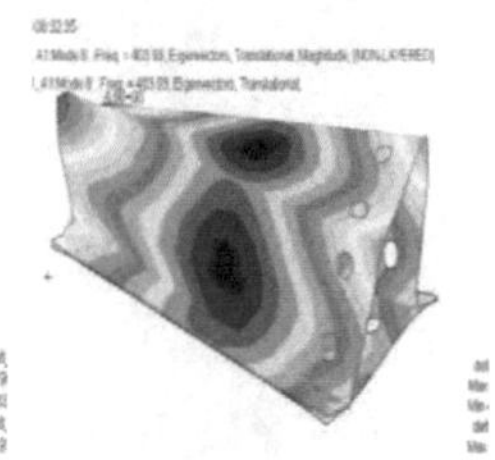	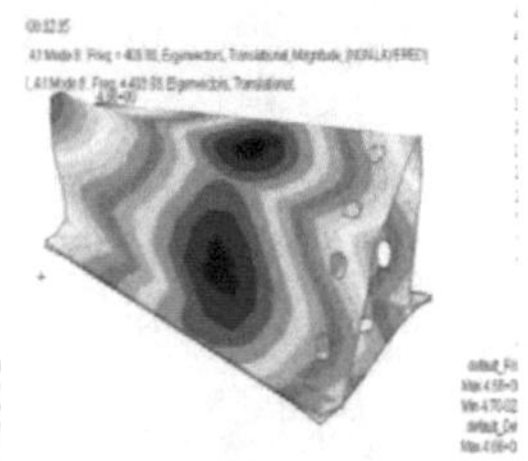
3	826.78	795.18	795.18	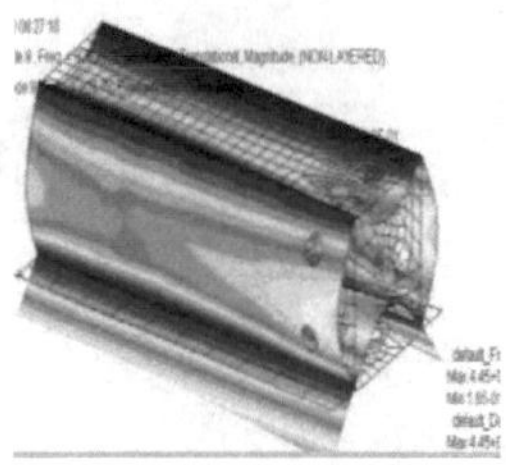	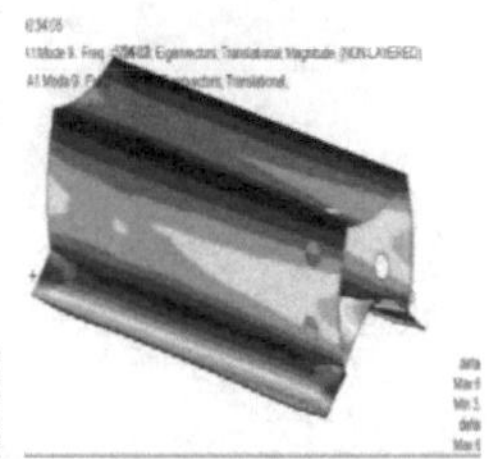	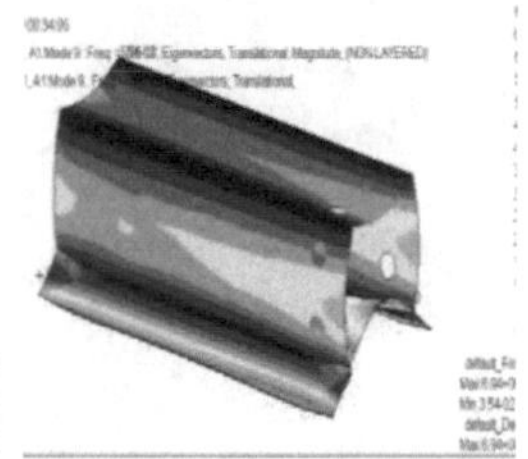

4	927.77	878.57	878.57	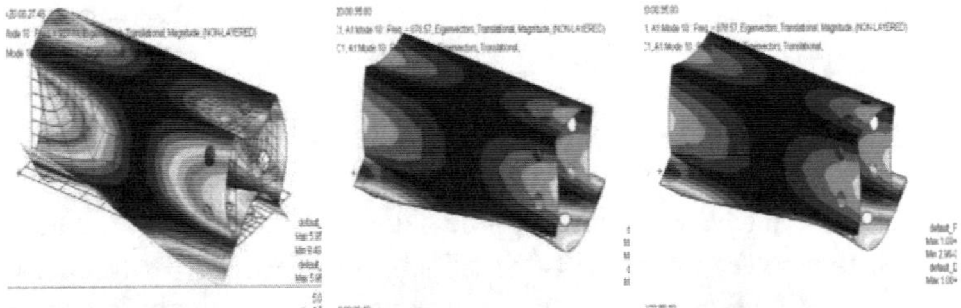
5	1217.3	1034.8	1034.8	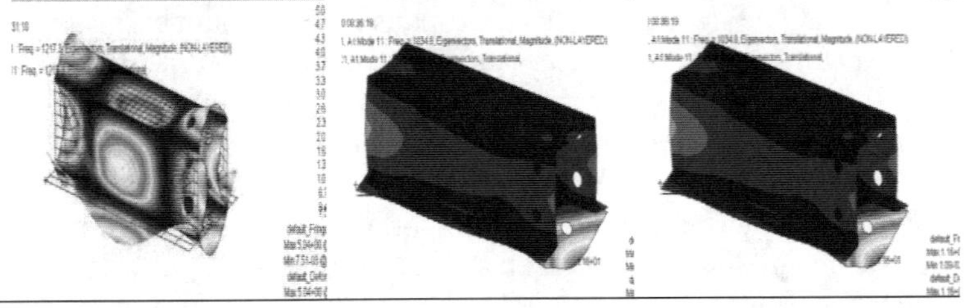

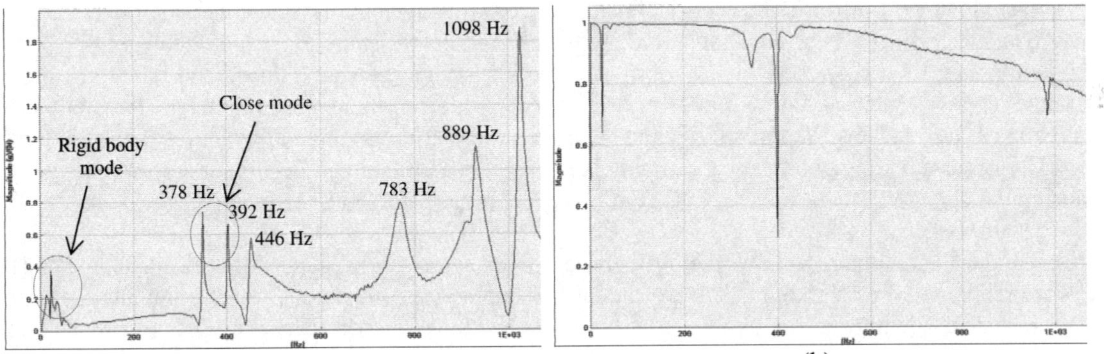

(a) (b)

Figure 4. Signal of (a) frequency response function and (b) coherence of the actual crash box structure

Table 2. Correlation of experimental natural frequencies and finite element models

Mode	Natural frequencies (Hz)					
	Experimental	Node equivalence	Error (%)	CBAR	Error (%)	CBEAM	Error (%)
1	392	472.64	20.57	395.18	0.81	395.18	0.81
2	446	479.76	7.57	403.93	9.43	403.93	9.43
3	783	826.78	5.59	795.18	1.56	795.18	1.56
4	889	927.77	4.36	878.57	1.17	878.57	1.17
5	1098	1217.3	10.87	1034.8	5.76	1034.8	5.76

4. Dynamic crash analysis on crash box model

The same finite element model the crash boxes that were used for modal analysis were used to perform dynamic crash analysis on the respective models. The properties for the models were remained to be same. However, plastic characteristics for the material was added by referring to the data provided by other research on the same material [22-23]. The crash box is considered to be fixed at the bottom and the impactor is considered to be rigid. The impactor was given a velocity of 25 m/s and movement constraint are given for the impactor to move along the z-axis only. Contact was defined between the impactor and the crash box as hard contact and coefficient of friction is defined as 0.3. The simulation set up of the crash analysis is as shown in Figure 5.

IOP Conf. Series: Materials Science and Engineering 1062 (2021) 012008 doi:10.1088/1757-899X/1062/1/012008

The equation of motion for nonlinear dynamic behaviour can be formulated as shown in Equation (1) below [22].

$$M(x)\ddot{Z}_N(t) + C(x)\dot{Z}_N(t) + K_N\big(x, Z_N(t)\big)Z_N(t) = f(t) \tag{1}$$

where M is the mass matrix, C is the damping matrix, K_N is the nonlinear stiffness matrix, $Z_N(t)$ is the displacement vector, $\dot{Z}_N(t)$ is the velocity vector, $\ddot{Z}_N(t)$ is the acceleration vector, and t is time. On the other hand, the $'N'$ represents the response of the nonlinear dynamic analysis and $f(t)$ is the external load vector. The energy conservation during the crash time is expressed as equation (2) as follows [22]:

$$SE = KE + IE - W + VD + FD \tag{2}$$

where SE is the summation of all important energies at any time during the crash analysis, KE is the kinetic energy, IE is the internal energy, W is the external work, VD is the viscous dissipation and FD is friction damping. In addition, the sum of energies is approximately equal to the kinetic energy before the crash. Figure 6 illustrates the energy transfer between the impactor and the crash box models during the dynamic crash analysis. The kinetic energy from the impactor is the highest at the beginning of the crushing, then, as it collapse with the crash box; the kinetic energy was reduced and transformed into internal energy. From the figure, there are no noticeable differences in terms of energy transfer for all three crash box models. The energy change occurs almost at the similar time. The value of external work, viscous damping and friction damping is relatively small and can be summed up as energy lost during the collision. Therefore, the energy transfer curve that is shown in Figure 6 is reasonable with the equation (2).

Figure 7 and Figure 8 shows the force curve and the deformation shapes for each crash box models respectively. All of crash box models are capable of stopping the impactor from continuously crashing the crash box. The impact was stopped after t=0.003 s and the impactor was bounce back from the crash box at t=0.004 s. As the dynamic crash analysis are not showing the full deformation or buckling behaviour of the crash box models, the analysis of nonlinear buckling was performed in order to futher study the buckling deformation of each crash box models.

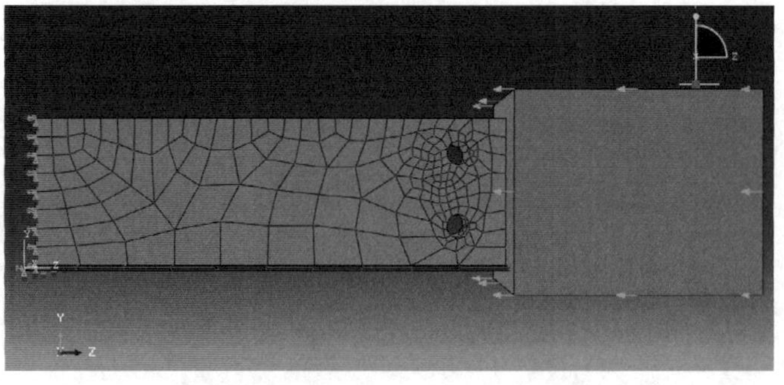

Figure 5. Crash box simulation set up

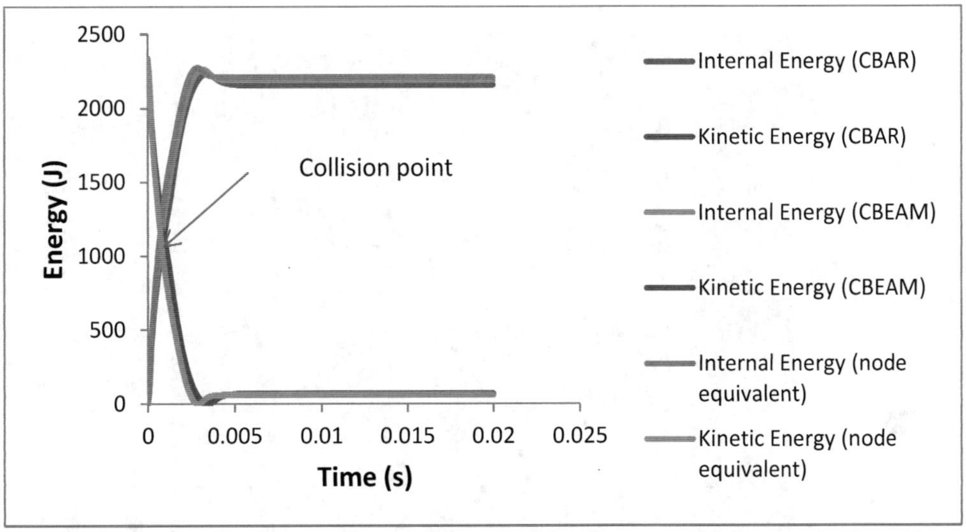

Figure 6. Energy transfer diagram for crash box models of respected joint type in dynamic crash analysis

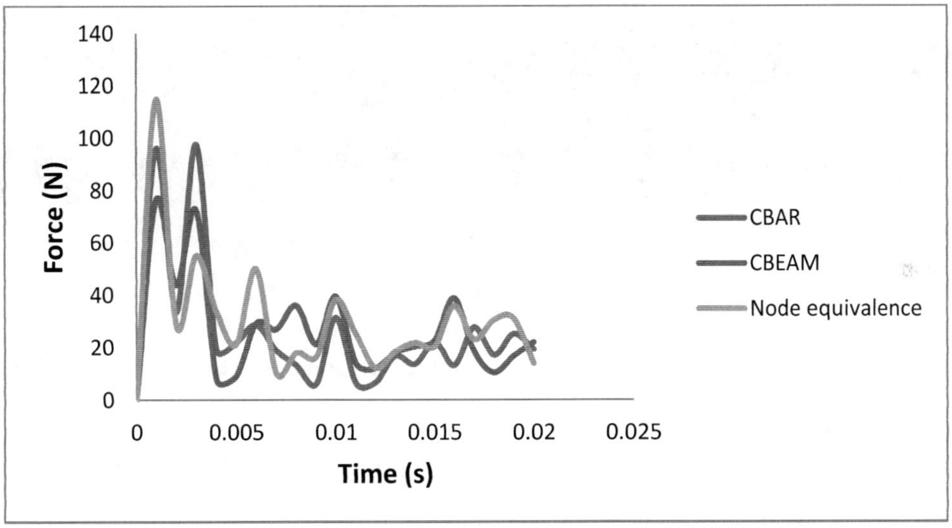

Figure 7. Force diagram for crash box models of respected joint type in dynamic crash analysis

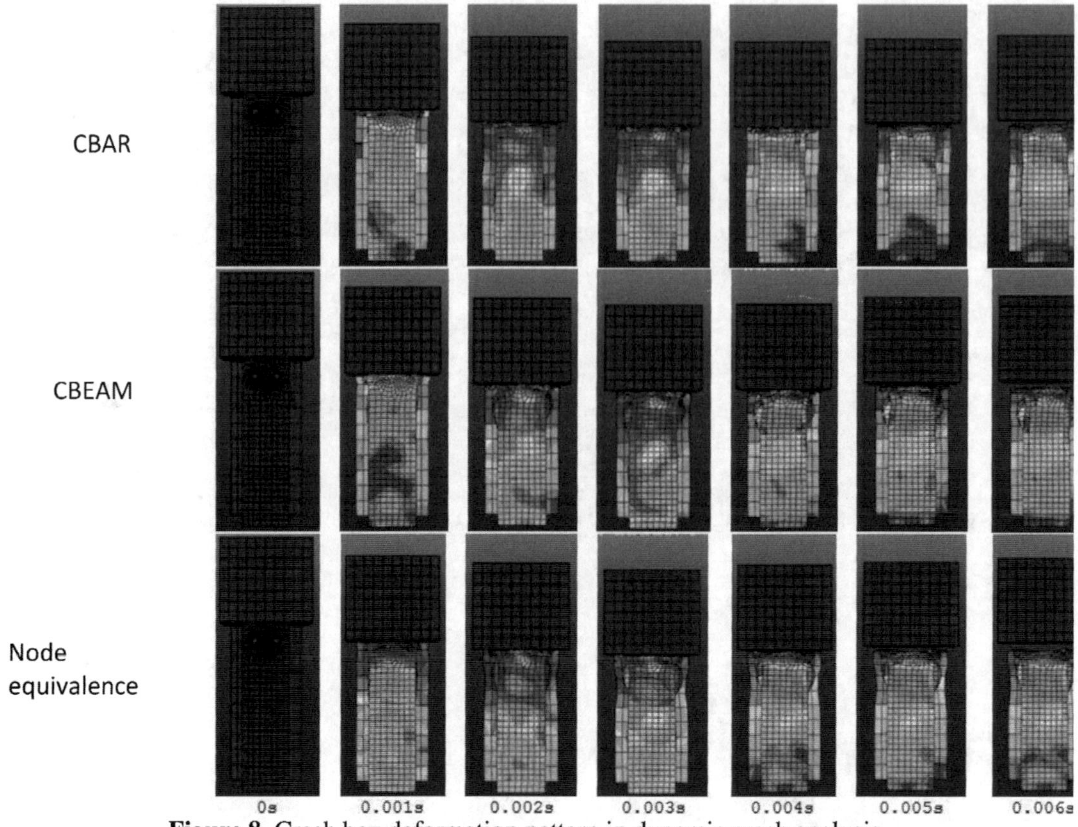

Figure 8. Crash box deformation pattern in dynamic crash analysis

5. Nonlinear buckling analysis on crash box model

Nonlinear buckling analysis was then performed on all crash box models with respective joining type by using same model used in dynamic crash analysis. In this analysis, the crash box models were subjected to mass loading axially in z-axis direction. Instead of rigid impactor with mass and initial velocity, the crash box models were crushed by rigid barrier at a given time and crushing distance. The crash boxes were set up to deform up until half of its original length (100 mm) and the low crushing speed were placed in order to observe the buckling behaviour of the crash box. Contact property remained unchanged with the friction coefficient was set as 0.3.

In nonlinear buckling analysis, among the parameters that are used to evaluate crash box crushing characteristics is the peak crushing force ($\boldsymbol{P_{max}}$). It is the highest force in the load against displacement curve that can indicate the required load to initiate collapse. On the hand, the mean crushing force ($\boldsymbol{P_m}$) is the average of load from zero to total displacement ($\boldsymbol{\delta}$) as shown in equation (**3**) below [18].

$$P_m = \frac{1}{\delta}\int_0^{\delta} P(x)\,dx \tag{3}$$

Based on the analysis, the value of $\boldsymbol{P_{max}}$ that each crash box model can support before it collapses is shown in Figure 9. Crash box with node equivalence joining has the highest crushing force of 70819 N, which is the point where the column of crash box will start buckling and deform. On the other hand, the crash box with CBEAM joint has the lowest value of crushing force which is 54279 N. The crushing force for crash box with CBAR joint is 67992 N. The deformation modes for each of crash models are as shown in Figure 10. The deformation patterns are shown in the interval of 0.005 s. The deformation ended at 0.02 s where the crashing reach its crushing distance that was set up as its boundary condition. Deformation modes for crash box with CBAR and CBEAM are both started the folding at the middle part while the crash box with node equivalence start folding from the impacted region or the top region. However, the folding behaviour is more stable for the crash box with CBAR elements as compared to the crash box with CBEAM elements.

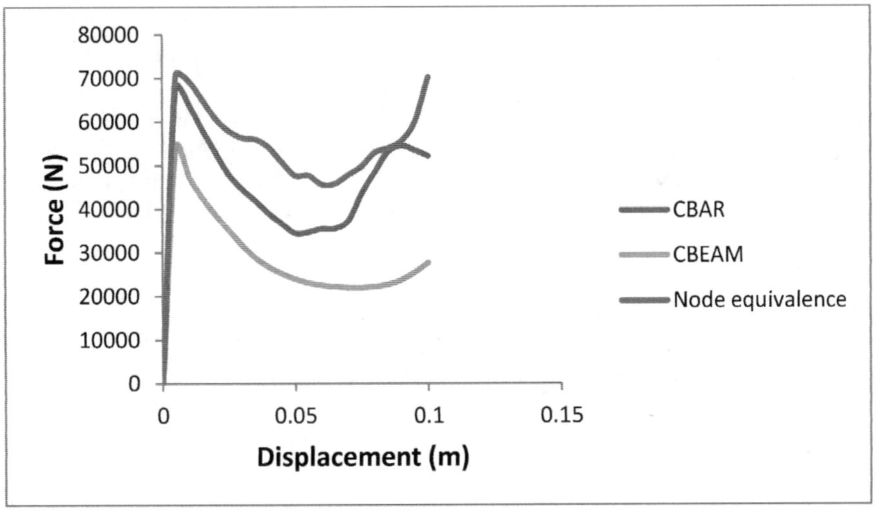

Figure 9. Force diagram for crash box models of respected joint type in nonlinear buckling analysis

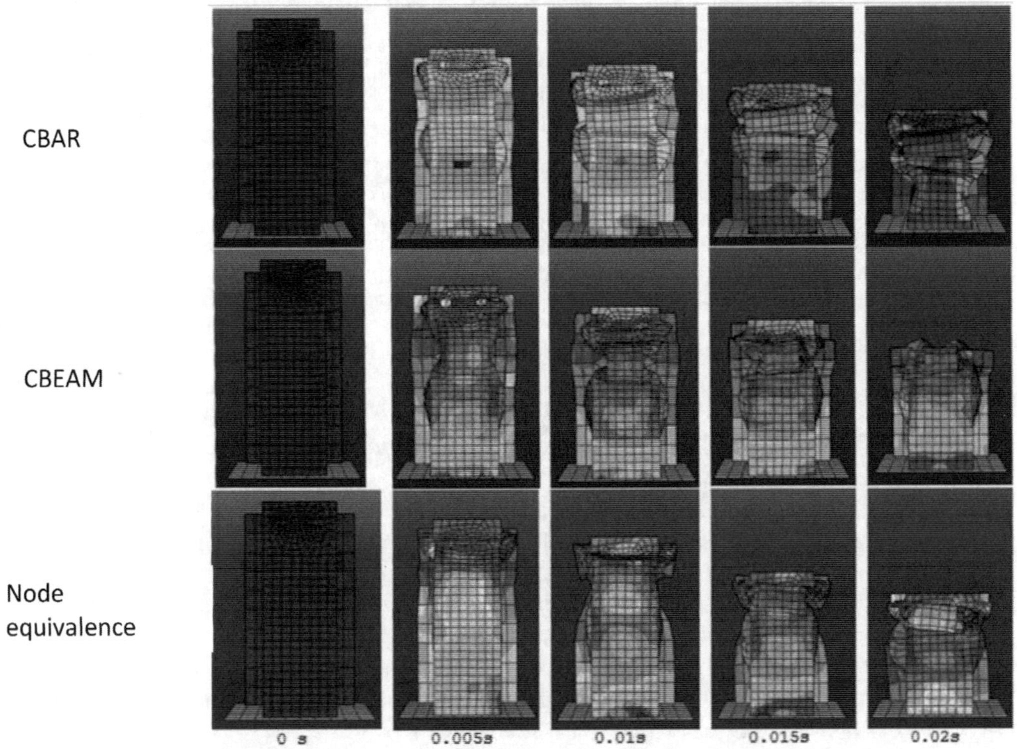

Figure 10. Crash box deformation pattern in nonlinear buckling analysis

6. Conclusion

In the present study, the modal properties and crash behaviour of finite element model of top-hat crash box with trigger and welded joint is investigated. Three different models that were created to model the actual crash box and different joining modelling strategies were used. Based on the output, the crash box modelled with node equivalent method to represent the welded joint shows higher stiffness at the joining area and therefore shows higher value of natural frequencies that the model with CBEAM and CBAR elements as their joining elements. In term of crash behaviour, even if the modal properties of crash box models with CBEAM and CBAR elements shows similar modal behaviour, the model with CBAR element shows higher reaction force compared to the CBEAM. This proves that different type of joint modelling strategies behaves differently. Good correlation with the models that replicate the actual structure behaviour is always needed to ensure the accuracy of computational analysis. Therefore, it is important to have the accurate joining modelling strategy. In future, the application of more joining element such as CWELD, CFAST, CELAS and others in crash and buckling analysis should be investigated.

Acknowledgement

The authors of this paper would like to acknowledge a great support and encouragement by focus group of Advanced Structural Integrity of Vibration Research (ASIVR), Universiti Malaysia Pahang (UMP) for providing all the equipment used for this work Fundamental Research Grant Scheme (FRGS/1/2017/TK03/UMP/02-19) – RDU 170123 and PRGS1903150

References

[1] Yahaya Rashid A S, Ramli R, Mohamed Haris S, Alias A, Rashid A S Y, Ramli R, Haris S M and Alias A 2014 *Sci. World J.* **2014** 1–11
[2] Griškevičius P and Žiliukas A 2003 *Transport* **18(2)** 97-101
[3] Tarlochan F, Samer F, Hamouda A M S, Ramesh S and Khalid K 2013 D *Thin-Walled Struct.* **71** 7–17
[4] Tanlak N, Sonmez F O and Senaltun M 2015 *Eng. Struct.* **95** 49–60
[5] Estrada Q, Szwedowicz D, Rodriguez-Mendez A, Elías-Espinosa M, Silva-Aceves J, Bedolla-Hernández J and Gómez-Vargas O A 2019 *Thin-Walled Struct.* **140** 43–59
[6] Kumar Ȧ As, Ȧ Gh, Raman Ȧ Ms, Kumar Reddy Ḃ Kv, Kumar A S, Kumar Ȧ As, Ȧ Gh, Raman Ȧ Ms and Kumar Reddy Ḃ Kv 2013 *Int. J. Curr. Eng. Technol.* **2** 670-676
[7] Yusof N S B, Sapuan S M, Sultan M T H, Jawaid M and Maleque M A 2017 *Cienc. e Tecnol. dos Mater.* **29** 129–44
[8] Fang J, Sun G, Qiu N, Kim N H and Li Q 2017 *Struct. Multidiscip. Optim.* **55** 1091–119
[9] Abdullah N A Z, Sani M S M, Salwani M S and Husain N A 2020 A review on crashworthiness studies of crash box structure *Thin-Walled Struct.* 153 1-16
[10] Abdullah N A Z, Sani M S M, Husain N A, Rahman M M, and Zaman I 2017 *International Journal of Automotive and Mechanical Engineering (IJAME)* **14** 3887-3897
[11] Peroni L, Avalle M and Belingardi G 2007 *20th International Technical Conference on the Enhanced Safety of Vehicles* 1–13
[12] Peroni L, Avalle M and Belingardi G 2009 *Int. J. Impact Eng.* **36** 498–511
[13] Silva D F M, Silva C M A, Bragança I M F, Nielsen C V., Alves L M and Martins P A F 2018 *Materials (Basel).* **11(7)** 1-15
[14] Back P-E 2010 *Deformation Behaviour of an Energy Absorber* (Luleå University of Tech.)
[15] Hickey A and Xiao S 2017 *Int. J. Mod. Stud. Mech. Eng.* **3** 1–5
[16] Zahari S N, Sani M S M, Husain N A, Ishak M and Zaman I 2016 *J. Teknol.* **78** 57-65
[17] Husain N, Khodaparast H H, Snaylam, James S, Dearden G and Ouyang H 2010 *Proc. Inst. Mech. Eng. Part C J. Mech. Eng. Sci.* **224** 851–61
[18] Dimas A, Dirgantara T, Gunawan L, Jusuf A and Putra I S 2014 *Appl. Mech. Mater.* **660** 578-582

[19] Siemens 2014 Element Library Reference
[20] Fouzi M S M, Jelani K M, Nazri N A and Sani M S M 2018 *Int. J. Automot. Mech. Eng.* **15** 5874–89
[21] Fouzi M S M, Sani M S M and Muchlis Y 2019 *IOP Conf. Ser. Mater. Sci. Eng.* **469** 012099
[22] Zahran M S, Xue P, Esa M S and Abdelwahab M M 2018 *Thin-Walled Struct.* **122** 64–82
[23] Davoudi M and Kim C 2019 *Eng. Optim.* **51** 775–95

A Study of Soil Characteristics of Road damages in Jeram, Kuala Selangor

Intan Najiha A S and Doh S I

College of Engineering, Universiti Malaysi Pahang, Lebuhraya Tun Razak, 26300 Gambang, Kuantan, Malaysia

*Corresponding e-mail: intannajiha58@gmail.com

Abstract. The main purpose of this research work is to identify the classification and strength of soil does it affected the road damages in Jeram, Kuala Selangor. The most factors influencing road damages in Jeram is an overloading vehicle using the main roads. To achieve the objectives of this paper soil laboratory testing has been done in justify the content of peat soil. Location of study can be categorizing as a plantation soil as there are a majority of land has been use in agricultural sector which are oil palm tree plantation. Besides, oil palm plantation suitable for peat soil land which has less soil strength. Sample of soil were collected in Jeram and bring it to soil mechanics and geotechnical laboratory in UMP. Few tests were being soil classification, particle size distribution, moisture content, density, specific gravity, Atterberg limit and direct shear test. Soil sample from three different roads were chosen which are Jalan Pantai Jeram, Jalan Simpang 3 Jeram and Jalan Tambak Tengah. The road has different criteria from the surface damages which no damages occur, less damage and most damages on road. From the data collected, Jalan Simpang 3 Jeram results the most damages road and the less strength of soil due to high content of peat soil. The soils criteria are located near trench and in the area of oil palm plantation. It also is the main road were used to collect the plantation yield to the factory.

Keywords: soil characteristic, road damage, geotechnical, palm plantation

1. Introduction

Malaysia is the country rich with plantation sources due to the suitable weather hot and humid all year long. Malaysia has a vital role to play in achieving the rising global need for oils and fats, as Malaysia is one of the bigger producer and exporter countries of palm oil plantation and palm oil products. In 2015, oil palm plantation in Malaysia turns into huge industry for an overwhelming contribution to world's palm oil production and export which is 39% and 44%, respectively [1]. The location of study in Jeram, can be observed that majority of the land use for agriculture such as oil palm plantation and the type of soil are peat soil which are very suitable for oil palm tree. Generally suitable peat for oil palm cultivation [2]. A soil sample were taken from the site and brought to soil mechanics laboratory to justify the soil types and condition.

Peatland with an area totaling approximately 2.6 million ha is considered a problematic soil in Malaysia, but it has potential for oil palm cultivation. The poor inherent physical properties of peat such as low bulk density, high water-table and rapid subsidence rate make its development for oil palm cultivation difficult and costly as compared to mineral soils. A study has stated that, peat contains high organic content more than 75% [3].

Peat composed of organic material and the component makes peat very spongy, highly compressible and combustible in characteristic according to [4]. Peat is defined as an organic soil with a thickness of

IOP Conf. Series: Materials Science and Engineering 1062 (2021) 012009 doi:10.1088/1757-899X/1062/1/012009

organic matter of more than 50 cm, mineral content not exceeding 35%, and covering an area of more than 1ha. [5]. The objectives of this study are to identify the soil characteristics of road damages in Jeram, Kuala Selangor. In Peninsular Malaysia, Selangor is the 2nd largest area of peat soil and oil palm plantation as shown in Table 1.

Table 1. Distribution of peat in peninsular Malaysia [6].

State	Approximate area under peat (ha)
Johor	216,000
Pahang	285,000
Selangor	182,000
Perak	69,000
Terengganu	46,000
Kelantan	11,000
Negeri Sembilan	4,000
Total	813,000

2. Methodology

The purpose of research methodology is to provide a sound platform for the researcher to achieve the aim and objectives of the study. This chapter highlights the chronological order of research methodology, discusses on how the research methodology will be conducted in order to achieve the objectives of this research. It is also important in obtaining relevant primary data from the selected group. From there, an analysis is conducted to study the data obtained from respondent and finally, based on the results obtained, a conclusion is derived.

In the location of study, three roads and point were chosen as the soil sample in this research. The selected area study is around Jeram, Kuala Selangor is because of the location of oil palm plantation, industrial of oil palm factory and building, accommodation and other developments. The road namely Jalan Tambak Tengah, Jalan Simpang 3 Jeram and Jalan Pantai Jeram. The details of road chosen are as listed in the Table 2 below. To determine the strength of soil test has been done are soil moisture content, particle size distribution, density, atterberg limits and direct shear test. The figure 1 shows the characteristics of soil sample taken if three roads.

Table2. Road details in Jeram.

Soil mark	Road name	Road damages	Road category	Color of soil
Soil A	Jalan Simpang 3 Jeram	Road cracking, settlement	Federal roads	Black brown
Soil B	Jalan Pantai Jeram	Potholes, settlement		Grey
Soil C	Jalan Tambak Tengah	Potholes		Brown

Figure 1. Soil sample in location of study.

The study area has been used for development purposes such as construction of residential area, oil palm factory and near to beach. In Malaysia, oil palm plantations are mainly established on old agricultural land or previously logged-over forest land. Coincides with the developments, there is increasing number of traffic volume in one time of period. Other than that, the study area is the main road for daily activities and the movements of the heavy vehicles such as lorry and heavy trucks make the damage road becomes worse. Classification of soil is needed as it can relate to identify the material can be used for construction of roads and other things.

3. Result and Discussion

3.1. Laboratory Soil Test

3.1.1. Test 1: Organic content

The value of moisture content in soil A, B and C is still categorized as a peat soil but with low natural moisture content. This is because the location of peat soil taken is far away from trench area and also during hot weather season and it has low moisture content.

The result and data analysis of moisture content test for soil sample A, B and C were shown in Table 3. Reference [7] stated that the natural water content of peat in West Malaysia ranges from 200% to 500% and with the organic content in the range of 50% to 95%. The average moisture content obtained for soil A is 41.83%, however for soil B and C are 13.33% and 13.90% itself.

The value of moisture content in soil A, B and C is still categorized as a peat soil but with low natural moisture content. This is because the location of peat soil taken is far away from trench area and also during hot weather season and it has low moisture content. The natural moisture content of peat soil was determined using equations below:

$$w = \frac{M_{SL} - M_{DSL}}{M_{SDL} - M_{DL}} \times 100 \tag{1}$$

$$w = \frac{M_S}{M_D} \times 100 \tag{2}$$

where w is moisture content (%), M_S is the moisture loss and M_D is dry soil.

Table3. Moisture content of soil.

Sample Soil		A	B	C
Container, g	M_{DL}	21.18	14.77	14.20
Wet soil and container, g	M_{SL}	32.35	32.95	32.99
Dry soil and container, g	M_{DSL}	26.97	30.78	30.70
Dry soil, g	M_D	12.85	16.01	16.49
Moisture loss, g	M_S	5.38	2.17	2.29
Average moisture content, %	W	41.83	13.33	13.90
Average moisture, %			23.02	

3.1.2. Test 2: Particle Density

Peat soils are considered as extremely soft, wet and unconsolidated deposits [8]. Particle density is the density of the solid particles that collectively make up a soil sample [9]. The value is commonly expressed in milligrams per meter cubic. The common range among soils is 2.55 to 2.70 mg/m ^{3}. A value of 2.65 were used generally except if some test need a great accuracy of particle density or soils are known to depart from it common range. However, from the test it results for soil sample A with 2.22, sample B 2.26 and sample C 2.28 mg/m^3.

Table 4. Determination of particle density soil.

Soil Sample (specimen reference)		A	B	C
Mass of Bottle		27.30	26.71	26.76
Mass of Bottle + Stopper, m_1	g	32.42	31.67	31.84
Mass of Bottle + Stopper + Dry Soil, m_2	g	42.78	41.97	42.12
Mass of Bottle + Stopper + Soil + Water, m_3	g	138.11	138.32	137.76
Mass of Bottle + Stopper + Water, m_4	g	132.41	132.57	131.99
Mass of Dry Soil, $(m_2 - m_1)$	g	10.37	10.31	10.28
Mass of Water In Full Bottle, $(m_4 - m_1)$	g	99.99	100.90	100.15
Mass of Water Used, $(m_3 - m_1)$	g	95.32	96.35	95.64
Particle Density, ρ_s $\rho_s = \frac{m_2 - m_1}{(m_4 - m_1) - (m_3 - m_1)}$	Mg/m^3	2.22	2.26	2.28
Average Particle Density,	Mg/m^3		2.25	

3.1.3. Test 3: Plasticity Index

There are three limits within Atterberg Limits which are Plastic Limit (PL), Liquid Limit (LL) and Shrinkage Limit (SL). All limits are tested on the three sample soil and repeated three times in order to get accurate description of the soil.

Plastic Limit is the boundary between non-plastic and plastic state of the soil sample. It is the limit in which it starts to act like plastic. This is done by rolling a thread of the soil sample. The moisture content can be simply adjusted by doing so and when it breaks about 1/8 inch in diameter which marks the Plastic Limit. The plasticity index result shows Soil A and Soil C has medium plasticity.

IOP Conf. Series: Materials Science and Engineering 1062 (2021) 012009 doi:10.1088/1757-899X/1062/1/012009

Table 5. Plasticity index result.

PI	Description	Soil A	Soil B	Soil C
0	Non-plastic			
1 – 5	Slightly plastic			
5 – 10	Low plasticity		✓	
10 – 20	Medium plasticity	✓		✓
20 – 40	High plasticity			
> 40	Very high plasticity			

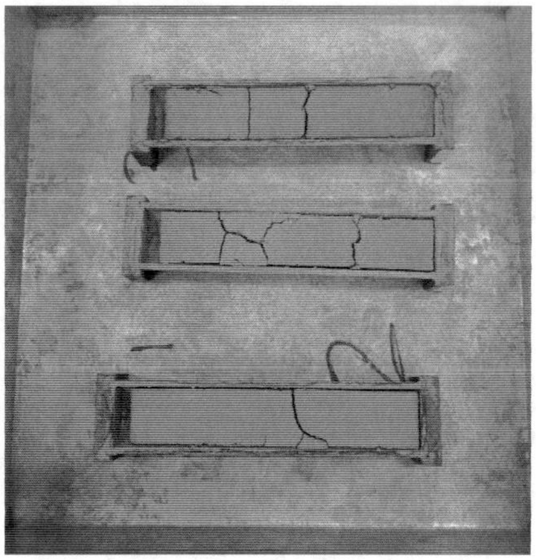

Figure 2. Shrinkage limit test.

Figure 3. Liquid limit.

3.2. Relative Important Index

The horizontal axis represents the RII value and the vertical axis represents the factors influencing road damages. In this study, it involves eight types of factors such as bad drainage system, climatic change, traffic volume, specification of road, overloading vehicles, conditions, sign, traffic, warning and the road markers and safety. From the bar chart it can be seen that the first ranking of factor influencing road damages is overloading vehicle with 0.78 RII value and following by bad drainage system and safety with 0.77 RII value.

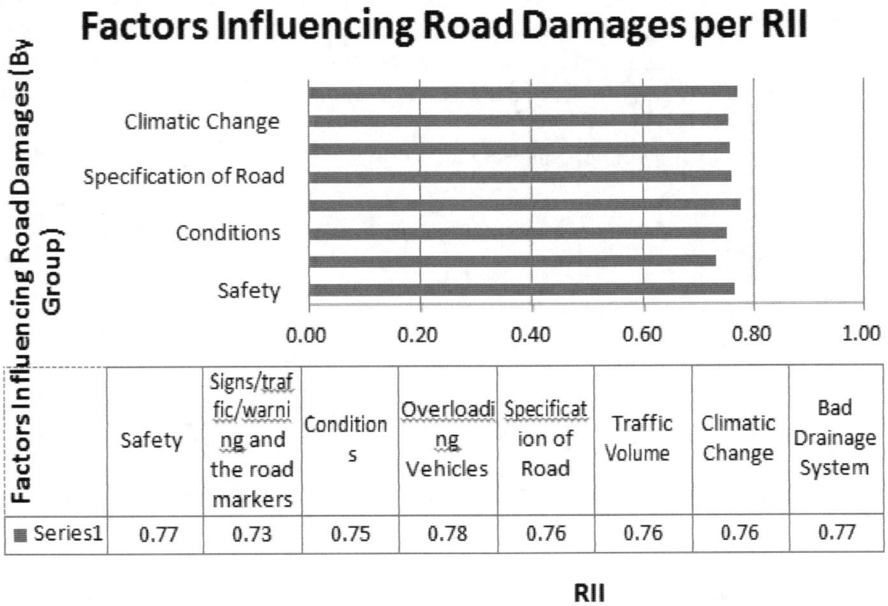

Figure 4. Factors influencing road damages per RII.

4. Conclusion

From user's perspective it can be concluded that eight factors influencing the road damages case study in Jeram, Kuala Selangor. The eight factors are safety, signs/traffic/warning and the road markers, conditions, overloading vehicles, specification of road, traffic volume, climatic change and bad drainage system.

Based on the RII ranking, the factor contributing the most to road damages in Jeram are overloading vehicles, bad drainage system and safety. From table 4.3, it has been recorded that overloading vehicles have the first ranking of RII of 0.778, bad drainage system 0.775 and in the third ranking is safety with RII od 0.771.

With high level of moisture content, high organic content and the behavior of the soil samples used for this research can be classified as the peat soil. This research also proved that the location study affected the properties of the peat soil as they exposed to different environment.

Acknowledgments

This research is supported by Universiti Malaysia Pahang under grant number PGRS190315 entitled A Study of Soil Characteristics of Road Damages in Jeram, Kuala Selangor.

References

[1] Alam A S A F, Er A C and Begum H 2015 Malaysian oil palm industry: Prospect and problem *J. Food, Agri. & Envi.* **13(2)** 143-148
[2] Lim K H, Lim S S, Parish F and Suharto 2013 Pomanualon Best Management Practices (BMPs) Forexisting Oil Palm Cultivation On Peat, RSPO Secretariat Sdn. Bhd.
[3] Kazemian S, Huat B B K, Prasad A and Barghchi M 2011 *International Journal of the Physical Sciences* **6(8)** 1974-1981
[4] Hunt R E 2004 Geotechnical Engineering Investigation Handbook, Second Edition CRC Press
[5] Othman H, Darus F M and Mohammed A T 2009 Experiences in peat development for oil palm planting in the MPOB Research Station at Sessang, Sarawak *Oil Palm Bulletin* **58** 1-13

IOP Conf. Series: Materials Science and Engineering 1062 (2021) 012009 doi:10.1088/1757-899X/1062/1/012009

[6] Melling L, Chua K H and Lim K H 2008 Managing Peat Soils Under Oil Palm *ACT Seminar*
[7] Huat B B K 2004 Organic and Peat Soil Engineering. 1st edn. University Putra Malaysia Press, Serdang, Malaysia.
[8] Kazemian S, Prasad A, Huat B B, Mohammad T A and Aziz F N A 2013 *Journal of the Chinese Institute of Engineers* **36(1)** 114-128
[9] Blake G R 2016 Encyclopedia of Soil Science Springer

IOP Conf. Series: Materials Science and Engineering 1062 (2021) 012010 doi:10.1088/1757-899X/1062/1/012010

Spatial Patterns of Peak Ground Acceleration (PGA) Earthquake on Final Disposal Sites Plan in Samosir Regency

Della Ayu Lestari[1*], Dewi Susiloningtyas[2], Muchamad Hartanto[3], Taufik Ejaz Ahmad[1], Dhea Rahma Azhari[1], Amien Rais[1]

[1] Marine Information System, Universitas Pendidikan Indonesia, Dr. Setiabudi Street No.229, Bandung, 40154, Indonesia
[2] Geography Department, Faculty of Mathematics and Natural Science, Universitas Indonesia, Depok, 16424, Indonesia
[3] Marine Logistics, Universitas Pendidikan Indonesia, Dr. Setiabudi Street No.229, Bandung, 40154, Indonesia

*Corresponding e-mail: della.ayu@upi.edu

Abstract. The strategic location of Samosir Regency, which located in the middle of Lake Toba and enriched with nature tourism, history and culture made tourism were a pre-eminent sector. However, there were several threats that can occur, the threat of earthquake and the waste problems that pollute Lake Toba due to the final disposal sites has not been made. This study aims to analyze the potential of earthquake risk to the development plan of Samosir Regency. The potential for earthquake risk is obtained by the calculation of Peak Ground Acceleration (PGA). The results showed the highest value of PGA was located at the western part of Samosir Regency and the value decreased as it came up to eastern. The highest PGA value is 237,36 SR and the lowest PGA value is 115, 91 gals with range VII to VIII MMI scale. Spatial patterns of low population density were in high PGA areas and otherwise. Based on 250 m buffer analysis on both sides of Semangko Fault, the three planned final disposal sites location (Batu Napal, Simpang PT EJS and Pamittoran) were located outside the active fault zone but these sites located in earthquake prone areas.

Keywords: earthquake, peak ground, acceleration, spatial pattern

1. Introduction

Samosir Regency is one of the areas in the northwest of Sumatra Island were well known because of tourism sector. The strategic location of Samosir Regency, located in the middle of Lake Toba and enriched by natural attractions, history and culture in it, was the main attraction for tourists [1]. However, behind the wealth and success there are several disaster threats that can occur in Samosir Regency, one of which is an earthquake [2]. Samosir Regency affected by the Indo-Australian Plate and the Eurasian Plate. This area is also traversed by the Sumatra Fault (Semangko Fault) and Mentawai Fault. The Mentawai fault serves as a holding zone (locking zone) against the insistence of the sloping subduction of the Australian plate to the south, with a holding capacity of close to 100% or having the ability to store high energy [3]. Local faults also affect the seismic activity in this area. Based on its geological conditions, the formation of Samosir Regency was caused by volcano-tectonic activity in North Sumatra, which was reflected in minor faults on the northwest side of Samosir Island.

The geological conditions make Samosir Regency dominated by old volcanic sediment areas and a small portion to the west of Samosir Island is a young volcanic sediment area with quaternary sediments that have experienced unconsolidated weathering which is very susceptible to earthquake shocks [4]. In general, the level of damage that might occur depends on the strength and quality of the building construction, geological conditions and Peak Ground Acceleration (PGA) at the built-up area from the tremors caused by the earthquake [5- 8]. In this study, the source of earthquake damage parameters is stated in PGA, so PGA data due to earthquake vibrations is important to describe the level of earthquake risk in a particular location.

Data from Development Planning Agency at Sub-National Level of Samosir Regency 2018 showed that Samosir Regency's 2017 local revenue growth was 81%, and the tourism sector is the leading sector. With the rapid development of Samosir Regency, there has been an increase in the economy and population which together with that the production of waste has also increased. Based on Law of the Republic of Indonesia Number 18 of 2008 concerning Waste Management, there must be a hierarchy of waste disposal from Temporary Disposal Sites (TPS) to Final Disposal Sites (TPA), if it is not in accordance with the hierarchy, there will be a problem of garbage accumulation [9]. The garbage accumulation problem occurs in Lake Toba because Samosir Regency does not yet have a waste disposal facility (TPA) so Lake Toba is used as a place to dispose of garbage and waste by the company and the surrounding community. A total of 262 tons of garbage per day flows to Lake Toba [1]. Therefore, the Samosir Regency Government has planned to make a final disposal sites in the spatial plan of Samosir Regency but the location of the final disposal sites plan needs to be studied further because it is close to the Semangko Fault (Development Planning Agency at Sub-National Level of Samosir Regency, 2018). Based on the above background, the purpose of this study is to analyze the potential risk of earthquake disasters in Samosir Regency by using Peak Ground Acceleration (PGA) analysis which is a source of earthquake damage parameters to the population density of each sub-district and analyze the planning of the final disposal sites (TPA) against the threat of earthquake hazards from the Semangko Fault and the distribution of the PGA that has been analyzed. The results of this study are expected to be used as a reference in disaster mitigation efforts and disaster risk management plans in regional development.

2. Material and Method

2.1. Variables and Data

This study used earthquake data from North Sumatra Province because earthquakes emit wave energy that will spread in all directions so that even though the earthquake point is outside Samosir Regency with the energy that spreads it, Samosir Regency will still be affected by the earthquake. The magnitude of the effect of the earthquake point on Samosir Regency was analyzed by looking for maximum ground acceleration or Peak Ground Acceleration (PGA). The PGA data which is the parameter of the source of damage caused by the earthquake, was obtained through the functions of magnitude, epicenter distance, depth of earthquake source and earthquake intensity [8], [10], [11]. Then the maximum PGA value and the average PGA value will produce the spatial pattern of the Peak Ground Acceleration (PGA) of destructive earthquakes in Samosir Regency.

The variables used in this study refer to PGA data with earthquake data used referring to destructive earthquakes which are $\geq$ 5.0 SR and include shallow earthquakes (depth $\leq$ 30 km). PGA data consists of magnitude, epicenter distance, earthquake source depth and earthquake intensity. Data collection in this study was carried out by library studies or literature studies, institutional surveys and field surveys. Data of earthquake from 1900 to 2018 was obtained from Meteorological, Climatological and Geophysical Agency (BMKG) of Bengkulu Province and United States Geological Survey (USGS) [12]. Spatial data of administrative and the plan for the construction of a final disposal sites (TPA) was obtained from Development Planning Agency at Sub-National Level (BAPPEDA) of Samosir Regency. Data of Semangko fault and local faults was obtained from Ministry of Energy and

Mineral Resources of the Republic of Indonesia 2015. Then, data of density was obtained from Statistics Indonesia (BPS) of Samosir Regency.

The Peak Ground Acceleration (PGA) was calculated using Richter's formula and combined with the intensity attenuation formula by Subardjo-Prih Harjadi, which is a function of magnitude, epicenter distance, and intensity of earthquakes. Earthquake's epicenter that was taken in this study was the earthquake's point that occurred in the Province of North Sumatra with boundary coordinates of 4°18 '6" North Latitude to 38°18'05" South Latitude and 97°3'33" East Longitude to 100°39'47" East Longitude. There were 118 earthquake's epicenters with the total of 116 references based on weight point in each village in Samosir Regency. The PGA values were calculated from the earthquake data using the formula below [8], [11]:

a. Measurement of the distance from the epicenter to certain location X:

$$\text{Cos}\Delta = \text{CosLE CosLX} + \text{Sin LE Sin LX Cos}(\lambda E - \lambda X) \quad (1)$$

where:
LX = Latitude of X
λX = Longitude of X
LE = Latitude of Epicenter
λE = Longitude of Epicenter
Δ = Distance between Epicenter and X in degrees
To convert Δ from degrees to kilometers, the value of Δ must be multiplied with a value of 111.11 for every 1o. (This method is used only for the equator area)

b. Calculation of intensity in hypocenter using Richter Empirical formula:

$$Io = 1.5\,(M - 0.5) \quad (2)$$

where:
Io = Source Intensity;
M = Magnitude/body wave (Richter Scale).

c. Calculation of Intensity in location X using attenuation constant formula:

$$I = (Io.\ \exp - b.\Delta) \quad (3)$$

where:
I = Surface Intensity;
Io = Source Intensity
Δ = Distance between Epicenter and X
B = Attenuation Constant (0.00217)
exp = Natural Number (2.786).

d. Calculation of PGA in Location X using Richter Empirical formula:

$$\text{Log}\,\alpha = (I / 3) - 0.5 \quad (4)$$

where:
α = PGA in gals (cm/second2);
I = Surface Intensity (MMI);
3 and 0.5 = constant

2.2. Analysis
All spatial data is processed using mapping software for further analysis. The analysis used in this study is descriptive and comparative analysis. The descriptive analysis used maps to describe patterns and characteristics of destructive earthquakes in the period 1900-2018 based on PGA values in the study area [13]. Based on the results of the spatial distribution of PGA it is known how much the risk level for earthquakes in each region in Samosir Regency was. The PGA region in Samosir Regency was divided into classes that referred form [14] shown in Table 1 below:

Table 1. Earthquake Risk Level

No	Risk Level	Value of Acceleration (gal)	Intensity (MMI)
1	Very low risk	<25	<VI
2	Low risk	25-50	VI-VII
3	Moderate risk level 1	50-75	VII-VIII
4	Moderate risk level 2	75-100	VII-VIII
5	Moderate risk level 3	100-125	VII-VIII
6	High risk level 1	125-150	VIII-IX
7	High risk level 2	150-200	VIII-IX
8	High risk level 3	200-300	VIII-IX
9	Very high-risk level 1	300-600	IX-X
10	Very high-risk level 2	>600	>X

Then for the feasibility analysis of the final disposal sites carried out based on the Government Regulation of the Republic of Indonesia Number 26 of 2008 concerning National Regional Spatial Planning, Article 61 paragraph 4 that the area located in the active fault zone is determined by the criteria of a width of at least 250 (two hundred and fifty) from the edge of the active fault line using buffer analysis in mapping software [15]. The final disposal sites were also analyzed by local faults conditions and PGA area.

3. Results and Discussion

3.1. Distribution of Earthquake and Peak Ground Acceleration (PGA) to the density of population in Samosir Regency
The frequency of earthquakes in the Province of North Sumatra is quite high starting from a small earthquake that only feels ground vibration but has no impact until an earthquake has a damaging effect. A destructive earthquake is the earthquakes with a strength of $\geq$ 5.0 Ricther Scale (SR) and is included in a shallow earthquake (depth $\leq$ 30 km). In the last 118 years, the period from 1900 to 2018 there were 181 destructive earthquakes in the North Sumatra and surrounding areas which could have an effect on Samosir Regency in Figure 1. Earthquakes with magnitude 5.0 to 5.9 SR occur 158 times, earthquakes with magnitude 6.0 to 6.9 SR occur 19 times and earthquakes with magnitudes $\geq$ 7.0 SR occur 4 times each respectively at 7.0 SR, 7.2 SR, 7.6 SR and 8.6 SR.

Peak Ground Acceleration (PGA) is the largest ground acceleration value that has ever occurred somewhere due to an earthquake. In each village in Samosir Regency one coordinate of the center of gravity was taken as a reference point (116 points) so that the PGA value could be known in each village. From the PGA values, the maximum PGA value was sought at each point. The distribution of the PGA region in Samosir Regency is shown in Figure 2 with the lowest PGA value of 115, 91 gals and the highest PGA value of 237.36 SR. The PGA from the earthquake on March 28, 2005 which was centered in Nias, North Sumatra with a magnitude of 8.6 SR had a very significant influence on each village in Samosir Regency. Based on information from the Destructive Earthquake Catalog from

IOP Conf. Series: Materials Science and Engineering 1062 (2021) 012010 doi:10.1088/1757-899X/1062/1/012010

the Central Geological Agency of Volcanology and Geological Disaster Mitigation, the affected areas include Gunungsitoli (VIII MMI), Teluk Dalem, (VII MMI), Banda Aceh (VI MMI), Medan, Padang, Palembang (V MMI), Jambi (MMI IV), Bengkulu (III MMI), Malaysian West Coast (VI MMI), Bangkok (IV MMI), Thailand (III MMI), Singapore and Maldives (III MMI), Andaman and Nicobar Island, India and Sri Lanka. The earthquake also caused Tsunamis at Colombo 40 cm, Sri Lanka 25 cm, Hanimadu 40 cm and Maldives 18 cm. A tsunami with a height of 3 meters damaged the port and airport in Simeulue. The height of the tsunami run-up was recorded at 2 meters on the west coast of Nias, and 1 meter on Singkil and Meulaboh. The earthquake also claimed as many as 1,300 lives while around 350 people were injured and approximately 300 buildings were damaged [16], [17].

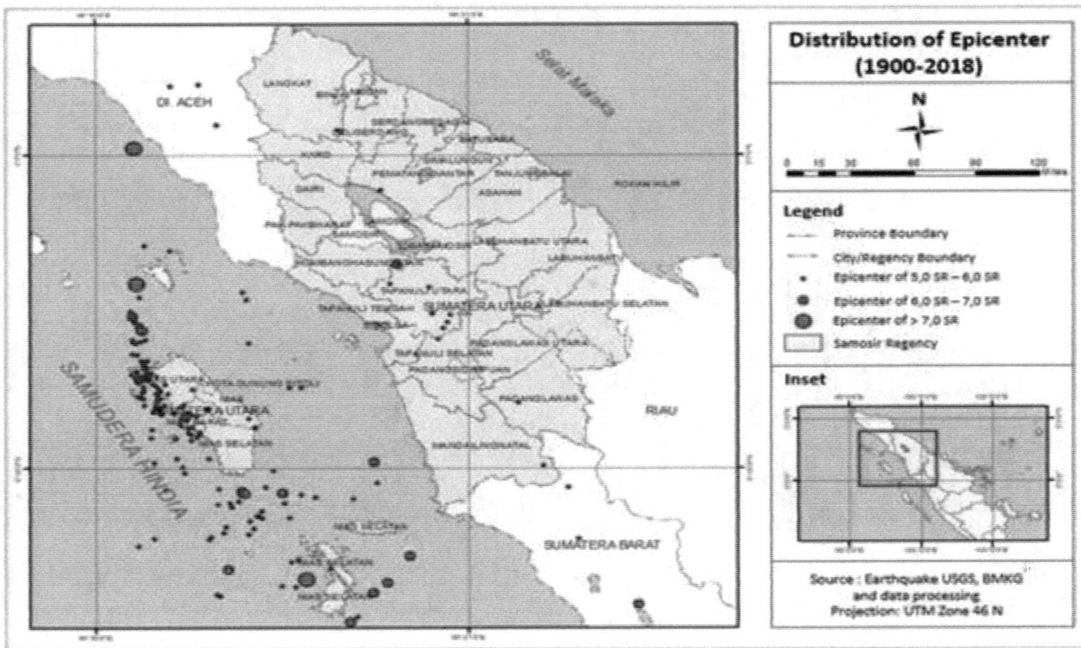

Figure 1. Epicenter Distribution for the Period of 1900-2018 in North Sumatra and Surroundings (Data Processing, 2018)

Based on an interview about the impact of the earthquake in Samosir Regency with the Head of the Regional Disaster Management Agency (BPBD) of Samosir Regency [2], he stated that at the time of the earthquake there was an area which suffered the heaviest damage in the coastal area in Pangururan District. Some of the damages recorded included the collapse of Public Junior High School 1 Pangururan and the need to be rebuilt, a restaurant belonging to Rapidin Simbolon, the Regent of Samosir Regency and Jalan Danau Toba in front of the regent's official residence, which was severely damaged and torn apart.

The PGA distribution in Samosir Regency has a pattern with the highest PGA in the west and the eastward the PGA value decreases in Figure 2. Based on the classification by Fauzi (2000), the PGA of Samosir Regency is divided into 4 classes, Moderate Risk Level 3 (100-125 gals), High Risk Level 1 (125-150 gals), High Risk Level 2 (150-200 gals) and High-Risk Level 3 (200-300 gals). First, class of Moderate Risk Level 3 was located in the southeastern part of Samosir Regency with the lowest PGA value in this class was 115, 91 gals and the highest PGA value was 123,26 gals. Second, High Risk Level 1 was located in the eastern part of Samosir Regency with the lowest PGA value in this class was 125,25 gals and the highest PGA value was 149,99 gals. Third, class of High-Risk Level 2 was located in the central and western part of Samosir Regency with the lowest PGA value in this class was 150,20 gals and the highest PGA value was 179,89 gals. Fourth, High Risk Level 3 was located in the western part of Samosir Regency with the lowest PGA value in this class was 202,40 gals and the highest PGA value was 237,35 gals.

Based on the PGA value, the Samosir Regency area was in the range VII and VIII on the MMI scale. At scale VII MMI identification of events can be in the form of each person leaving the house, minor damage to houses with good buildings and construction, while in buildings with poor construction cracks and even destruction, the chimney is broken and felt by people who ride a vehicle. Whereas on the scale of VIII MMI identification of events can be in the form of minor damage to buildings with strong construction, cracks in buildings with poor construction, walls can escape from the frame of the house, factory chimneys and monuments collapse and water becomes cloudy [5], [18].

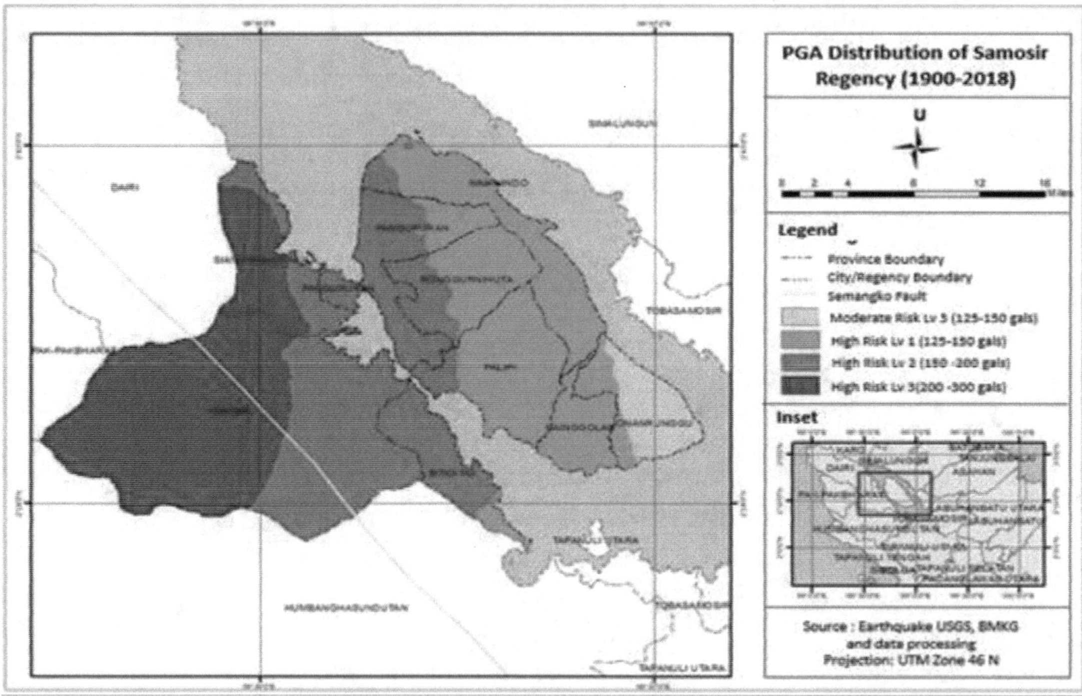

Figure 2. The PGA Distribution of Samosir Regency (1900-2018) in Samosir Regency (Data Processing, 2018)

Figure 3 is the distribution of population density and built-up area in each sub-district in Samosir Regency. Population density is related to built-up density in each region. The higher the population density in an area, the more exposed it will be. This happened because generally areas with high population density are central to the facilities and pre facilities of a region [7]. Based on data from Statisctic Indonesia of Samosir Regency in 2017, the highest population density was in Pangururan Sub-district with a population density of 252,39 people/km^2 while the lowest population density was in Harian Sub-district with population density only 14.56 people/ km^2 [19].

By using the equal method the interval of population density in Samosir Regency was divided into five classes were very low, low, medium, high and very high. The very high population density was located Pangururan Sub-district, which was in the central part of Samosir Regency. The high population density was located in Sitiotio Subdistrict, Palipi Sub-district, Nainggolan Sub-district and Onanrunggu Sub-district which were generally located in the southern part of Samosir Regency. The medium population density was located in Ronggunihuta Sub-district and Simanindo Sub-district in the eastern part of Samosir Regency. The low population density was in Sianjur Sub-district, which was located in the northwestern part of Samosir Regency and the verly low population density was in Harian Sub-district, located the western part of Samosir Regency.

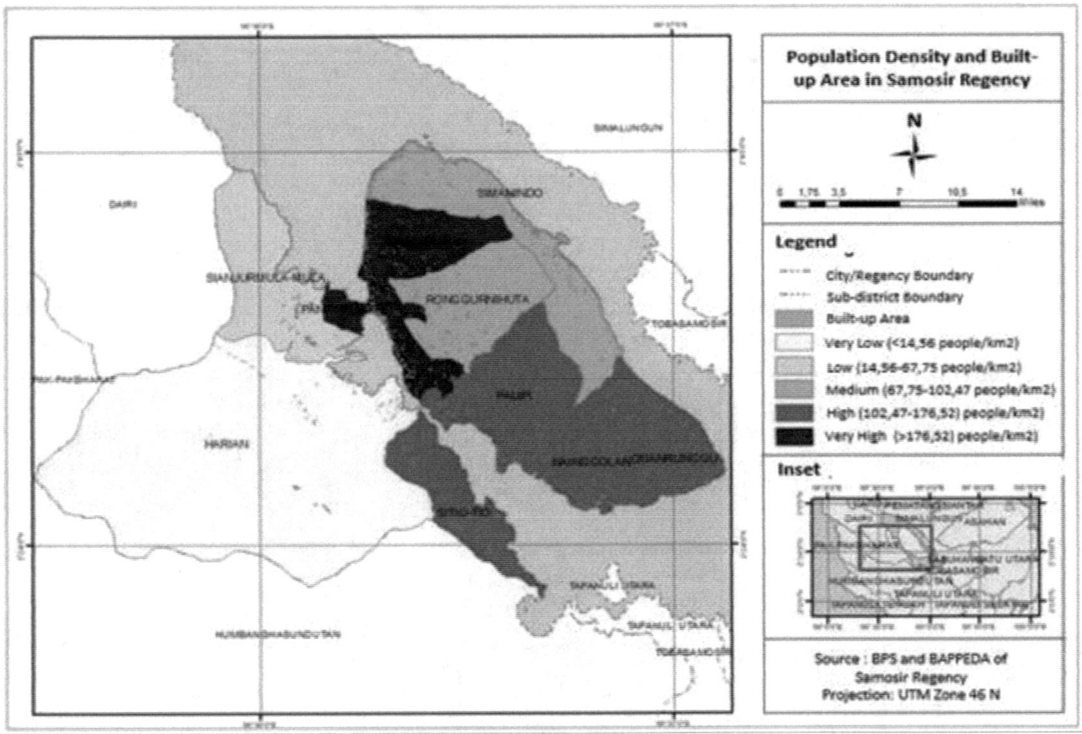

Figure 3. The Distribution of Population Density and Built-up Area in Samosir Regency (Data Processing, 2018)

Based on the spatial pattern of the PGA, regions with very high population density were in the High Risk Level 2 of PGA class, high population density were in the Moderate Risk Level 3 to High Risk Level 2 of PGA class, medium population density were in High Risk Level 1 to High Risk Level 2 of PGA class, the low population density were in High Risk Level 2 to High Risk Level 1 of PGA class and the very low population density were in the PGA region with the highest class, the High Risk Level 1. Based on the population density and the PGA region, the spatial pattern produced was the low population density were spread in the high PGA region while the high population density was spread in the low PGA region. Harian Sub-district which planned to be location of final disposal sites of Samosir Regency were located in very low population density but in the high level of PGA region.

3.2. Study of Construction of Final Disposal Sites (TPA) towards Potential Earthquakes

According to the Indonesian National Standard (SNI) No. 03-3241-1994 issued by the National Standardization Agency (BSN), the final disposal site (TPA) of waste is a physical facility for the final disposal of garbage [20]. TPA is also a place to get rid of or sequester the municipal waste that is safe. The requirement for the establishment of a final disposal sites is that the selection of sites must follow legal requirements, legislation regarding environmental management, analysis of environmental impacts, public order, municipal / environmental cleanliness, local regulations on waste management and urban spatial planning implementation rules. In the Samosir District Medium-Term Development Plan 2016-2021, waste management is one of the strategic issues because the final disposal sites that meets the requirements of Law Number 18 of 2008 that Final Disposal Site (TPA) must use a sanitary landfill does not yet exist [9], [21]. Then in the Draft Regulations of the District Spatial Planning, there are three Final Disposal Sites (TPA) to be built in the Hariara Pintu Village, Harian Sub-district which are the Final Disposal Site of Batu Napal, Final Disposal Site of Simpang PT EJS and Final

Disposal Site of Pamittoran. These sites were close to each other and located in the northeast part of Harian Sub-district. These sites were located at an altitude of approximately 1500 meters above sea level, with a slope of approximately 80-200, the surrounding vegetation in the form of forests and plantations and far from built-up area.

Based on the Regulation of the Minister of Public Works of the Republic of Indonesia Number 03 / PRT / M / 2013 on the Implementation of Infrastructure in Waste Handling Household Waste and Waste kind of Household Waste, Article 35 paragraph 2 that one of the criteria aspect in the selection of the final disposal sites is geological aspect, which not in active faults area [7], [22], [23]. Samosir Regency is traversed by active faults which generally cause destructive earthquakes and local faults that tend not to cause destructive earthquakes. One of the active faults that passes through Samosir Regency is the Semangko Fault. The Semangko Fault is a geological formation that stretches across Sumatra Island from north to south, starting from Aceh to Semangka bay in Lampung [4]. These faults form the Barisan Mountains, a series of plateau on the western side of the Sumatra Island. This fault is a horizontal shear or fracture fault, the flange fault including the horizontal fault, the shift is large around 25-30 km which will have a considerable impact [24]. Therefore, the final disposal sites development plan that has been made needs to be studied further based on the Semangko Fault which indicated in Figure 4.

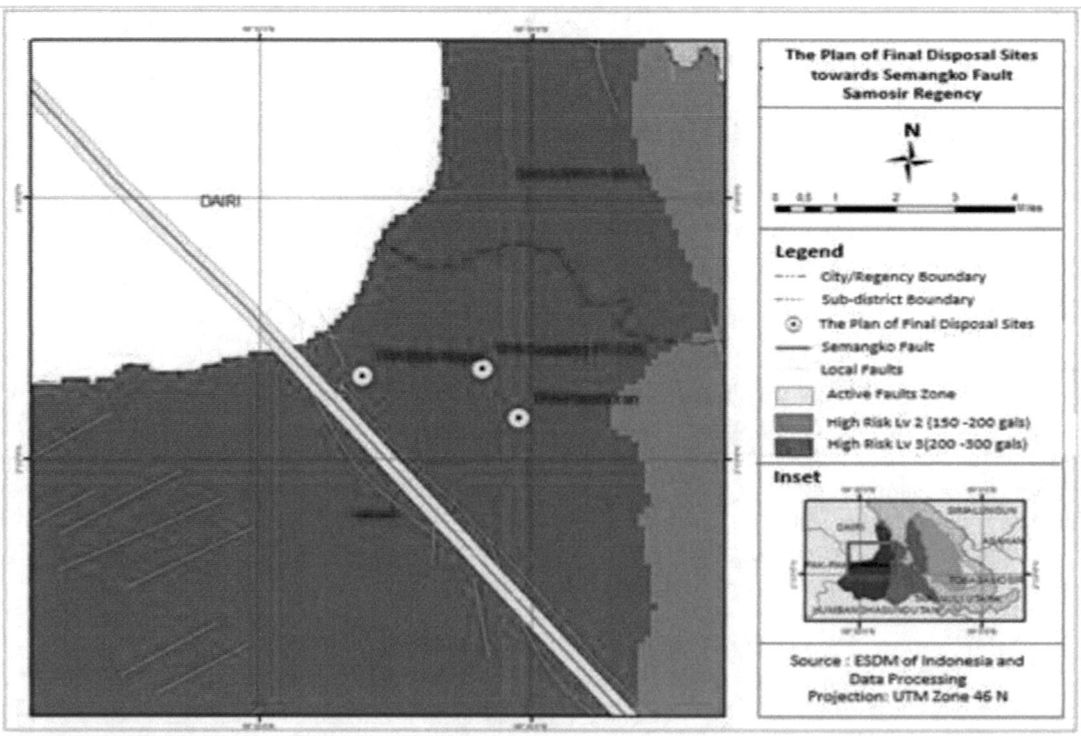

Figure 4. The Plan of Final Disposal Sites towards Semangko Fault in Samosir Regency (Data Processing, 2018)

Based on the Government Regulation of the Republic of Indonesia Number 26 of 2008 concerning the National Spatial Plan, Article 61 paragraph 4 that the area located in the active fault zone is determined by the criteria of a border with a width of at least 250 (two hundred fifty) meters from the edge of the active fault line. Figure 4 shows a buffer is applied to the Semangko Fault with a distance of 250 m on both sides which serves to prevent the effect of faults on the Final Disposal Site construction [15]. Based on the results Semangko Fault buffer towards the planned location of Final Disposal Site, all of sites were outside the buffer of 250 meters. The distance from Semangko Fault to the Final Disposal Site of Simpang PT EJS was 1,014 meters, then 3,788 meters to Final Disposal Site of Batu and 3,578 meters to Final Disposal Site of Pamittoran.

Although the three planned final disposal sites were located outside the active fault zone, these three locations are located in earthquake-prone areas. Based on the Government Regulation of the Republic of Indonesia Number 26 of 2008 concerning National Regional Spatial Planning, Article 61 paragraph 2, Earthquake-prone areas with criteria that have the potential and / or have experienced earthquakes on a scale of VII to XII Modified Mercally Intensity (MMI) [16]. Based on the results of the study, the three Final Disposal Sites locations were located in the High-Risk Level 3 area with the lowest PGA value in this class of 202.40 gals and the highest PGA value of 237.35 gals and in VIII on the MMI scale. Then referring to the Presidential Regulation of the Republic of Indonesia Number 81 of 2013 concerning the Lake Toba and Surrounding Spatial Planning for Article 35 (13) the area located in the active fault zone, one of which was the Harian Sub-district, where the site of the Final Disposal Sites development planned. The existences of local faults in the southwest and east of the final disposal sites also give effect to this earthquake-prone region. Therefore, referring to SNI number 03-3241-1994, the making of this final disposal sites must have further technological input and more technical research on the area to be planned to prevent the failure of investment in environmental facilities and infrastructure.

4. Conclusion

In the development of Samosir Regency there are a number of threats including the threat of earthquakes and the garbage problems that pollute Lake Toba because there was no final disposal site (TPA). The spatial pattern of Peak Ground Acceleration (PGA) showed that the highest value of PGA was located at west part of Samosir Regency and the value decreased as it came closer to eastward. The highest PGA value was 237.36 SR and the lowest PGA value was 115, 91 gals with range VII to VIII Modified Mercally Intensity (MMI) scale. Based on the population density and the PGA region, the spatial pattern produced was the low population density were spread in the high PGA region while the high population density was spread in the low PGA region. Then, based on the 250 m buffer analysis on both sides of the fault, the three plans for the location of final disposal sites (Final Disposal Site of Batu Napal, Final Disposal Site of Simpang PT EJS and Final Disposal Site of Pamittoran) were located outside the active fault zone but this sites also located in an earthquake prone area. The three of final disposal sites plans were located in the High-Risk Level 3 of PGA class with the lowest PGA value was 202.40 gals and the highest PGA value was 237.35 gals which was at VIII on the MMI scale. The existences of local faults in the southwest and east of the final disposal sites also give effect to this earthquake-prone region, therefore this final disposal sites must have further technological input and more technical research.

References

[1] Ministry of Environment and Forestry 2016 *Research, Development and Innovation Agency Center for Social Research and Development, Economics, Policy and Climate Change*. Vol. **10** No.3
[2] Regional Disaster Management Agency of Samosir Regency 2018 Deep Interview Earthquake in Samosir Regency
[3] Natawidjaja D.H 2006 *Indonesian Institute of Sciences* (LIPI)
[4] Salim, Riski and Bagus J S 2014 *Technical Journal of POMITS* Vol. **3** No.2 ISSN: 2337-3539
[5] Meteorological, Climatological and Geophysical Agency (BMKG) 2014 Earthquake Official Website http://www.bmkg.go.id/bmkg
[6] Murti N 2005 *Journal of the Bandung* Institute of Technology
[7] Regulation of the Minister of Public Works of the Republic of Indonesia Number 2013 concerning Implementation of Solid Waste Infrastructure and Facilities in Handling Household Waste and Similar Waste to Household Waste
[8] Supriatna, Samedi J M and Nurmala C 2010 Department of Geography, Faculty of Mathematics

and Natural Science, *Universitas Indonesia*
[9] Law of the Republic of Indonesia 2008 concerning Waste Management
[10] Ibrahim G & Subardjo 2005 Seismology Knowledge. Jakarta: *Meteorology and Geophysics Agency*
[11] Keller E. A and Blodgett R H. 2006 Natural hazards: Earth's Processes as Hazards, Disasters, and Catastrophes Upper Saddle River, NJ: *Pearson/Prentice Hall*
[12] United States Geological Survey 2000 Earthquake, fact & list http://neic.usgs.gov
[13] Hammond, Robert and Patrick McCullagh 1978 Quantitative Techniques in Geography: An Introduction. Second Edition. *Clarendon Press*
[14] Fauzi 2000 Geographic Information System for Natural Disater Map in Indonesia. http://www,reindo,co,id/gempa/Reference/peta-gempa,htm-297k
[15] Republic of Indonesia Government Regulation 2008 concerning National Regional Spatial Planning
[16] Subiyantoro I. 2010 Overview of Disasters *Journal of Disaster Management Dialogue*
[17] Supartoyo and Surono (2008) Ministry of Energy and Mineral Resources, *Central Geological Agency for Volcanology and Geological Disaster Mitigation*
[18] Prihantoro E 2009 Science & Technology 2: Disaster Preparedness Culture Jakarta: *PT Gramedia Main Library*
[19] Statistic Indonesia of Samosir Regency 2017
[20] National Standardization Agency 1994 Guidelines for Utilizing the Area Around the TPA for Waste. (SNI) No. 03-3241-1994 (20)
[21] Republic of Indonesia Presidential Regulation 2013 concerning the Lake Toba Spatial Plan
[22] Law of the Republic of Indonesia 2007 *Jakarta: State Secretariat*
[23] Wibowo M 2008 Hydrological Aspects in Determining Site Locations for Waste Disposal Sites (TPA). *Indonesian Hydrosfir Journal* Vol. **3** No.1 Hal. 1-6. ISSN 1907-1043
[24] Kustanto A R 2015 *Journal of Geophysics*, Padjadjaran University

Monitoring and Optimizing Solar Power Generation of Flat-Fixed and Auto-Tracking Solar Panels with IoT System

Ku Azmie Ku Husin[1], Norfatihah Mohd Adenam[1], Muhamad Yuzaini Azrai Mat Yunin[1], Khairul Nizar Syazwan Wan Salihin Wong[2], Siti Zaiton Mohd Hashim[2] and Hasyiya Karimah Adli[2*]
[1]Faculty of Bioengineering and Technology, Universiti Malaysia Kelantan, 17600, Jeli, Kelantan, Malaysia.
[2]Institute for Artificial Intelligence and Big Data, Universiti Malaysia Kelantan, City Campus, Pengkalan Chepa, 16100 Kota Bharu Kelantan, Malaysia.

*Corresponding e-mail: hasyiya@umk.edu.my

Abstract. In this work, the integration of recent digital technology of Internet of Things (IoT) with solar energy management was made to monitor real-time data of panel voltage (Vpv) and current (Ipv) of the solar panel. The solar panels were installed in Kelantan, Malaysia and this integrated solar power plant with IoT was developed with hardware setup and virtually tested to enhance the tracking behaviour of the proposed system. The comparison of different tilt angle of solar PV shows that solar PV facing the sun perpendicularly (45°) generated higher power generation compared to flat-fixed (0°). Whilst, by comparing fixed angle (45°) and auto tracking solar panel, the power generation recorded can reach up to double performance value (230 W) compared to 170 W generated from 45° fixed angle of solar PV. The IoT-based control of solar power significantly improves the performance of monitoring and maintenance of the power plant parameters with ease monitoring, cost-effectiveness and 24/7 remotely controlled.

Keywords: photovoltaic panel, tilt angles, solar tracker, Internet of Things (IoT)

1. Introduction

Renewable energy promotes significant advantages and benefits compared with conventional energy sources, considering its clean and abundant energy resources [1]. The renewable energies usage is projected to grow massively in coming decades in order to reduce the energy dependence on coal, fuel, oil and other non-renewable energy resources [2-6]. The rising energy demands in every year, uncontrollable environmental pollution and rapid decline of fossil fuels resource have made renewable energy sources, i.e., wind, solar, geothermal, biomass, and tidal energy as better energy options [7-9].

The renewable energy sources in Malaysia have immense potential to generate electricity and these resources play a vital role in resolving the energy crisis. Among the energy resources, solar energy has received greater attention due to its abundantly available, free, inexhaustible and carbon-free energy resources. Solar energy can supply electricity power on a wide scale to satisfy the load requirements, even for rural area use that has limited access to electricity [10-12].

To date, IoT is gaining a prominent position in research worldwide especially in advanced wireless communication and becomes as the foundation for various development applications such as for smart health services [13], smart living [14], smart school education [15] and other sectors [16,17]. IoT uses a wireless sensor network (WSN) for gathering information to monitor and remote-control [18,19]. The hardware consists of end devices fitted with a range of sensors to track various parameters such as temperature, humidity, solar radiation, soil moisture which can be able to communicate the obtained data to other devices [20-25].

Thus, in this work, solar panel is used as primary generation resource with the battery as energy storage system, coupled with AC bus through appropriate converters [20] and DC load is supplied by rectification to AC / DC. Moreover, the Internet of Things (IoT) system is also used to integrate the system, in which this work consists of three objectives:

a) To identify the daily trend and maximum solar power generation in Kelantan, Malaysia.
b) To optimize the maximum power generated from different tilt angle of 45° and 0°.
c) To monitor the solar performance using IoT system by comparing fixed angle (45°) and auto tracking solar panel.

2. Experimental Setup

In this work, 50 W solar photovoltaic (PV) array, charge controller, motors, 12V DC solar battery, DC load, base of PV Array auto moveable, light sensors, inverter and IoT sensors hardware and software were used.

2.1. Solar System Set-up Analysis

This work was conducted from May 2019 until September 2019 at Kampung Pulau Melaka, Kota Bharu, Kelantan, Malaysia. The design of the solar system with IoT-equipped technology was carried out to power water pump for fertigation, as illustrated in Figure 1.

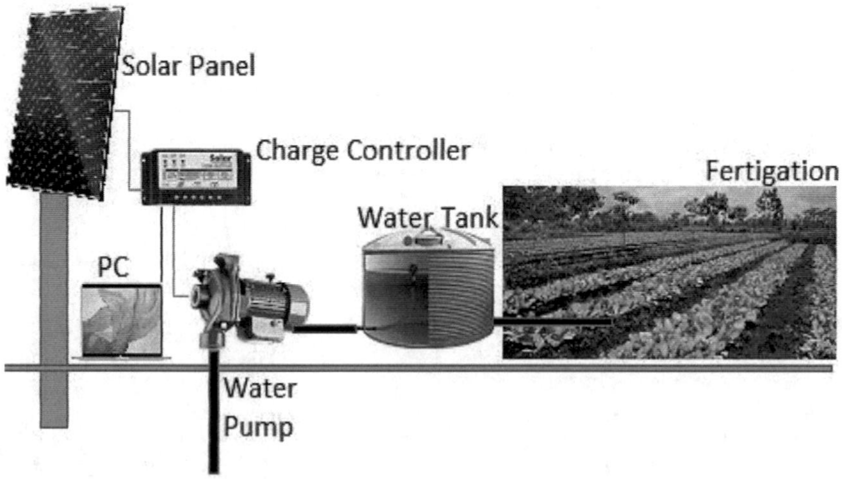

Figure 1. The concept of solar system with IoT-based technology to power water pump for the fertigation

2.2. Hardware Design

The design of the hardware consists of 2 parts, namely mechanical and the control circuit design. The control circuit was equipped with a drive motor, to move the mechanical hardware.

The first step in creating a solar tracker is mechanical design, which the mechanical solar tracker plays a very important role in monitoring the sun. As a part mount and solar tracker frame, PVC pipe was cut and built to support solar panels and has versatile motion feature as required. The built of fixed and auto solar trackers can be seen in Figure 2 as follows:

(a) (b)

Figure 2. (a)Flat-fixed solar panels (b) auto tracker of solar panels

Two LDR driver circuits were designed to place the LDR in the detection of light and transfer to Linear Actuator by connecting with Arduino UNO. Figure 3 shows a schematic description of the LDR driver circuit.

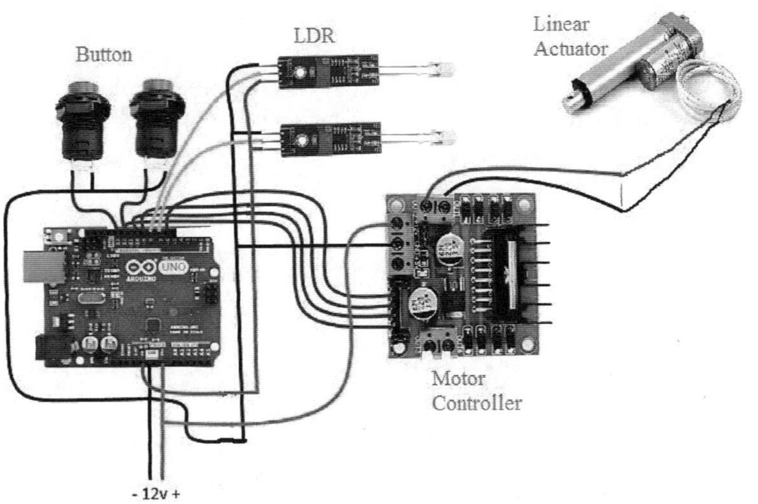

Figure 3. LDR schematic

The linear actuator series was designed as a driving force in light detection, with following specifications:

- Speed: 20mm/s
- Torque: 500N
- Working voltage: 12V DC
- Lever length: 150mm

The electrical architecture consists of sensors as inputs, Arduino UNO kits as signal processors and linear actuators as output.

2.3. Real-time Monitoring with Internet of Things (IoT)

The overall system consists of single axis solar PV system, series of sensors (e.g., current, voltage, temperature), ESP8266 Nodemcu as microcontrollers and Adafruit as platform. The solar PV generates electricity (voltage and current) which were detected by the sensors and data was sent to process by ESP8266 Nodemcu before the user can display and control the data through Adafruit IO platform. IoT was used to connect sensors and actuators and PV power machines to input / output control systems Adafruit application and the Arduino device were connected to Adafruit server via Wi-Fi while Arduino devices use the ESP8266 Wi-Fi module that connected to the local wireless network. Programming Integrated Development Environment (IDE) Arduino UNO to run the Arduino board software was used to write programs in C language that can give commands to the Arduino board and upload programs to a microcontroller. The list of IoT hardware is listed in Table 1.

Table 1. IoT Hardware used in this work

No	Hardware	Number
1	Solar Panel 50 Watt	2
2	30A Charge Controller	1
3	Solar Battery 26 Ah	1
4	Fuse Kit	2
5	INA219 Breakout	2
6	ESP8266 Nodemcu	1
7	Wire of correct gauge for power and distance need to run	As required

3. Result and Discussion

The solar power generated in daily from 0800 hours until 1900 hours was initially investigated in order to identify the solar trend in this region. The data were extracted from the analysis carried out from 26 Jun until 3 July 2019, as shown in Figure 4.

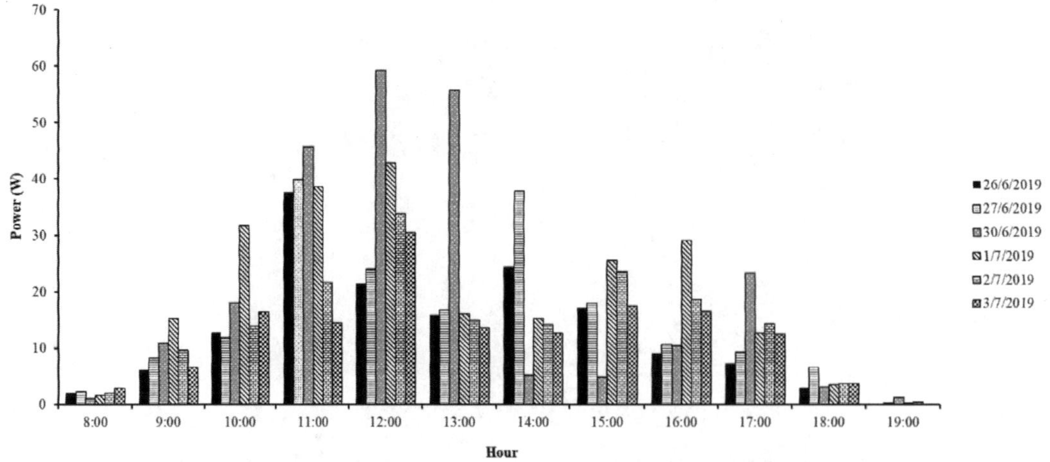

Figure 4. Daily solar power generation from 0800 until 1900

From the results, the highest energy generated was about 60 W recorded at 12 PM showing the maximum sunlight intensity at this hour. On the other hand, the least energy generated at 7 PM only recorded for about 0.14 W, due to sunset. The total solar power generation daily was then recorded everyone hour in order to identify the capacity of solar panel can run the power pump for fertigation. Figure 5 shows the plot of maximum PV watt in daily, exhibited that the total power generated can reach 300 W daily. The value data can be seen varies due to the local weather during analysis was performed.

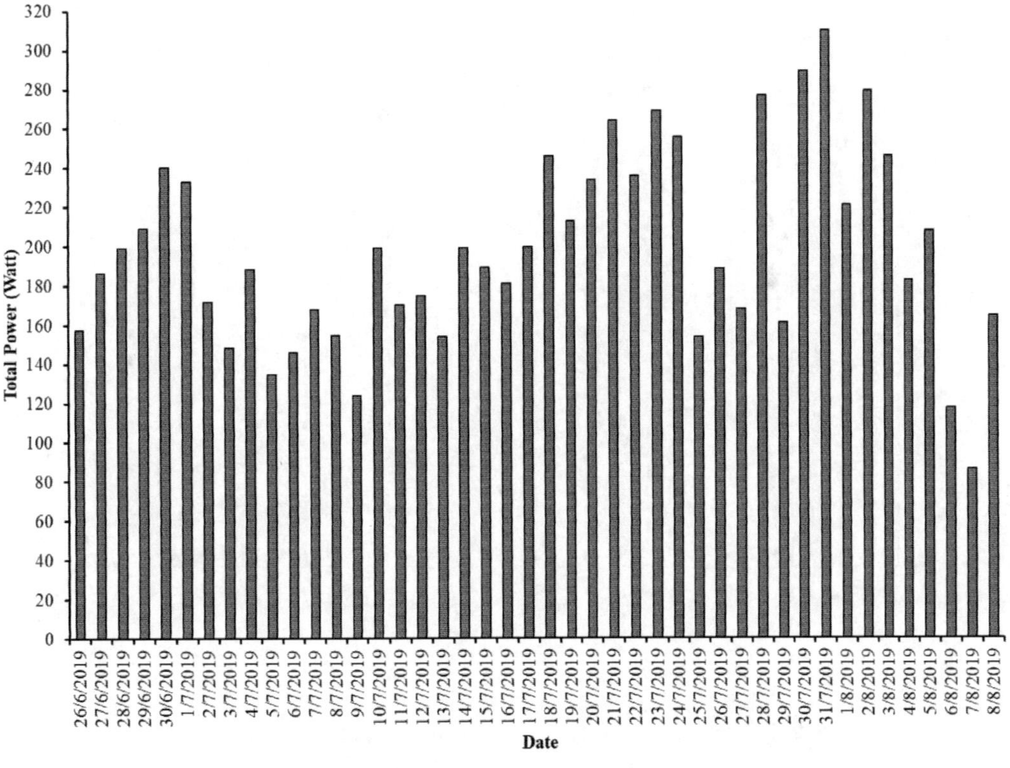

Figure 5. Daily total solar power generated

Meanwhile, the difference of tilt angle analysis of PV panel at 45° and 0° was further evaluated in terms of solar performance. From literature, the rainwater, soil, mud and dust could deposit on the horizontal panel (0°) tilt angle plates which then effect on solar absorption [26]. Figure 6 shows the difference power (W) collected from PV solar panel from manual fitted at 45° and 0°. It can be clearly seen that PV solar panel at 45° generated higher power (W) compared to 0°.

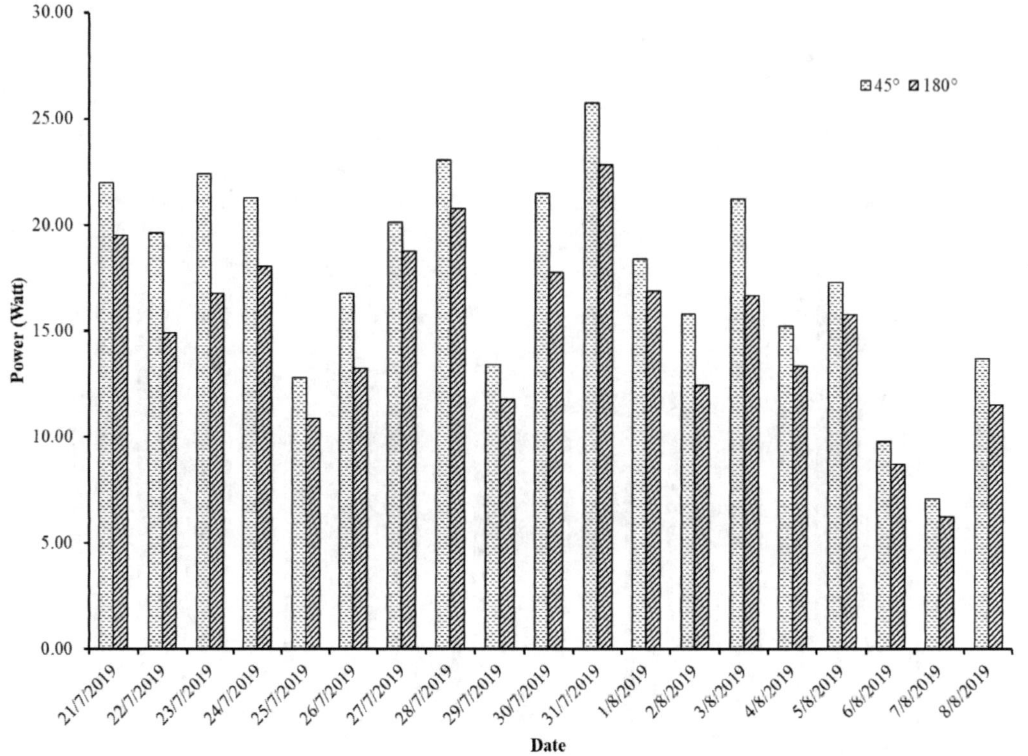

Figure 6. PV solar generation 45° and 180° tilt angles facing the sunlight

As the previous findings showed that better performance of 45° tilt angle solar PV, the analysis continued with the comparison of manual and auto track solar panel. Figure 7 shows the difference of power generation of flat-fixed (45°) and auto track solar panel.

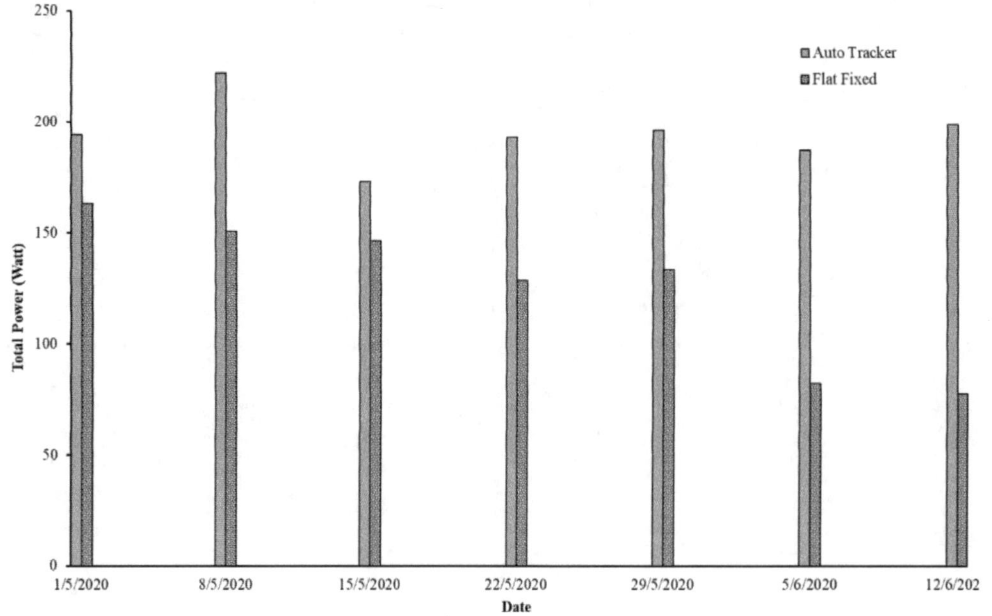

Figure 7. PV solar generation of auto and flat-fixed trackers

From the Figure 7, higher power (W) generation of auto tracker can be explained due to its ability in tracking the sun mode which can maximise more electrical energy can be harvested than flat-fixed position. The sun radiation hits the absorbing surface perpendicularly and it has better conditions of cooling and self-cleaning [26]. The measured values were recorded higher, as the linear actuator rotated the panel perpendicular to the direction of sunlight. Therefore, total solar panel surface areas for solar tracker device for absorbing sunlight are much larger compared to fixed solar panel and produce more performance value.

4. Conclusion

In this study, flat-fixed and auto solar tracker were built with developed IoT system for monitoring the solar generation power generation in Kelantan, Malaysia. The finding shows that solar PV facing the sun perpendicularly (45°) generated maximum power generation compared to flat-fixed (0°). By comparing fixed angle (45°) and auto tracking solar panel, the power generation recorded higher when using auto tracking sunlight. This work highlighted the IoT-based control of solar power significantly improves the performance of monitoring and maintenance of the power with ease monitoring, cost-effectiveness and 24/7 remotely controlled.

References

[1] Alik R, Jusoh A and Sutikno T 2017 *Int. J. Electr. Comput. Eng.* **7** 29–40
[2] Boulmrharj S, Malek Y N, Abdellatif E and Bakhouya M 2018 *Energy Procedia* **153** 56–61
[3] Sreenath S, Sudhakar K, Yusop A F and Solomin E 2020 *Energy Reports* **6** 698–712
[4] Qin J, Hu E and Li X 2020 *Energy Built Environ* **1** 11–26
[5] Shaffer B, Flores R J, Samuelsen S and Anderson M 2018 *Energy Procedia* **149** 25–38
[6] Rizman Z I et al 2018 *J. Fund. Appl. Sci.* 10 689-704
[7] Chandra S and Gaur P 2020 *Procedia Comput. Sci.* **167** 2354–2363
[8] Abrishambaf O, Lezama F, Faria P and Vale Z 2019 *Energy Strateg. Rev.* **26** 1-17
[9] Pavel S O and Nikita V K 2018 *MATEC Web of Conferences* **226** 02023-02026
[10] Munir E, Sipayung F C, Priyani N and Suryanto D 2018 *Earth Environ. Sci.: IOP Conf. Series* **126** 1-7
[11] Soeprijanto A 2018 *IOP Conf Series: Mater. Sci. Eng.* **383** 1-7
[12] López-Vargas A, Fuentes M and Vivar M 2018 *IEEE Sens. J.* **19** 679-691
[13] Solanas A, Patsakis C, Conti M and Vlachos I 2014 *IEEE Commun. Mag.* **52** 74-81
[14] Mahesar S, Baloch M H, Kaloi G S, Kumar M, Soomro A M, Solangi A A and Memon Y A 2018 *Int. J. Adv. Comput. Sci. Appl.* **9** 216–219
[15] Ming T, Hall C, Azman H and Joyes G 2010 *Int. J. Educ. Dev. using ICT* **6** 5-20
[16] Adhya S, Saha D K, Das A, Jana J and Saha H 2016 *2nd. Int. Conf. Control, Instrumentation, Energy, Commun.* **2016** (Kolkata: India) pp 432–436
[17] Kandimalla J 2017 *Int. J. Mod. Trends Sci. Technol.* **04** 16–21
[18] Visconti P 2020 *Int. J. Electron Telecommun.* **66** 59–68
[19] Sabry A H, Hasan W Z W, Kadir M Z A A, Radzi M A M and Safie S 2017 *PLoS One* **12** 1–22
[20] Kyi S and Taparugssanagorn A 2019 *Digit Commun. Networks* **6** 51–57
[21] Witkowski K 2017 *Procedia Eng.* **182** 763–9
[22] Sunhare P 2020 *J. King Saud Univ. – Comput. Inf. Sci.* **2020** 1-22
[23] Ray P P 2016 *Futur. Comput. Informatics J.* **1** 35–46
[24] Reyna A, Martin C, Chen J, Soler E and Diaz M 2018 *Futur. Gener. Comput. Syst.* **88** 173–190
[25] Maghami M R, Hizam H, Gomes C, Radzi M A, Rezadad M I and Hajighorbani S 2016 *Renew. Sust. Energ. Rev.* **59** 1307-1316
[26] Ziemelis I 2017 *Eng. Rural Dev.* **16** 540–545

Effects of Various Alcohol Blends on Performance and Emissions of Small Compression Ignition Engine Fuelled with Diesel-20% Palm Oil Methyl Ester

Helmisyah Ahmad Jalaludin[1], Nik Rosli Abdullah[2], and Muhammad Syafiq Roslee[2]
[1]Faculty of Mechanical Engineering, Universiti Teknologi MARA, 23200 Bukit Besi, Malaysia
[2]Faculty of Mechanical Engineering, Universiti Teknologi MARA, 40450 Shah Alam, Malaysia

*Corresponding email: nikrosli@uitm.edu.my

Abstract. As a promising option of in improving alternative fuels, alcohol such as butanol, methanol, and ethanol can be used in diesel engine. However, there is lack of detailed investigation of the effects of variation addition of butanol, methanol, and ethanol with diesel-palm oil methyl ester (B20) on engine performance and emissions release. This study represents to fully evaluate the potential impact of addition butanol, methanol, and ethanol blended with B20 on engine performance and emissions released by single cylinder compression ignition (CI) engine at engine speeds of 2700, 3100, and 3500 rpm under load of 50%. The results of engine performance and emissions of 10ml ethanol, methanol and, butanol each blended with B20 are compared. Experimental results showed that the B20 + 10ml methanol has an advantage over diesel-biodiesel blend in Brake Specific Fuel Consumption (BSFC) which reduces about 10.42% at every different engine speed. Besides, the B20 + 10ml methanol has reduced the emission of Nitrogen Oxide, Carbon Monoxide, Carbon Dioxide, and Hydrocarbon by 27.84%, 14.28%, 5.19%, and 56.25%. In overall, the addition of 10ml methanol on B20 blend shows the most significance result for the engine performance and emissions at all test condition.

Keywords: fuel, compression ignition, emissions, engine performance

1. Introduction

Extensive use of diesel fuel for transportation all over the world has caused its supplies to continuously decrease and has aggravated environmental pollution. Therefore, the high demand of diesel fuel has stimulated many researchers to investigate alternative fuels such as diesel -biodiesel blend to overcome the crisis. Biodiesel is one of the alternative ways to replace current fuel source that has a good potential to perform up to fill in the demands in case of fuel depletion. Biodiesel can effectively reduce engine-out emissions, particularly unburned HC, CO, and NOx in a typical four-stroke compression-ignition engine. This is crucial in developing an innovative way to substitute or even better, upgrade the current fuel market to be greener without sacrificing performance too much. Higher oxygen content in biodiesel will increase the overall complete combustion in the engine hence higher value of Brake Thermal Efficiency (BTE). This will also help to ensure lower Brake Specific Fuel Consumption (BSFC) which is ideal in finding an alternative fuel [1-6].

A study by Bueno et al. [7] states that both soybean and castor biodiesel increase the CO emissions caused by the biodiesel iodine value. Castor bean or castor oil plant is one of the promising non-edible oil crops since it has low implementation and production costs and relative resistance to hydric stress. However, the result of the study shows that castor oil biodiesel blending into diesel fuel did not advance fuel injection for the engine in use and continuously increased the NOx emissions. Other authors concluded that the effect of biodiesel on NOx emissions depends on the type of engine and its operating conditions. It will be easier to analyse the engine performance results by increasing the biodiesel ratio in the blended fuel because the measurement and evaluation of blended fuel property is an important indicator [8]. An effort to improve popular B20, Asiah et al. [9] investigated the effect of various Tripmexx additive mixed into 80% diesel and 20% biodiesel (B20) with an amount of 0.1 ml (B20-0.1), 0.2ml (B20-0.2), and 0.3ml (B20-0.3) compared to conventional pure diesel. As a result, B20-0.3 gives the best bake specific fuel consumption (BSFC) and higher brake thermal efficiency (BTE) while lower carbon monoxide (CO) and hydrocarbon (HC). Fuel additive is chemical that are used to enhance the properties of the fuel. Currently, numerous chemical additives are used in transportation to improve the quality of biodiesel fuel and diesel fuel in order to convene up the most wanted performance level [10]. Additives may help petroleum to recover its engine combustion, performance, and emission environmental standards. Various studies have proven that adding additives in biodiesel enhancing engine performance and reduce engine emissions.

As a promising option of in improving alternative fuels, alcohol such as butanol, methanol, and ethanol can be used in combustion whether gasoline or diesel [11, 12]. Butanol is produced from through chemical processes based on oxo synthesis, Reppe synthesis, or crotonaldehyde hydrogenation [13]. While methanol is made from coal and natural gas but it can also made from renewable sources, such as wood and papers while ethanol is produced a from the renewable sources that come from fermentation of carbohydrate containing biomass like sugar, grains and tapioca [14, 15]. These trio alcohols may give benefit that could save the environment from pollution by decreasing the toxic emission such as carbon dioxide, carbon monoxide, hydrocarbon, and nitrogen dioxide.

In order to make effect comparison between trio alcohol which blended in 80% diesel blended with 20% palm oil methyl ester (B20), this study is carried out. Thus, experiment of alcohols such as butanol, ethanol, and ethanol blended with B20 on the performance and emissions of a small compression ignition engine. From here, the brake specific fuel consumption (BSFC) and brake thermal efficiency (BTE) of the alcohols blended with B20 biodiesel. Finally, the emission of nitrogen oxides (NOx), carbon monoxide (CO), and carbon dioxide (CO_2) from the combustion may be compared.

2. Experimental Setup

The experiment is started with sample preparation of various types of B20 blended with different but same volume.

2.1. Samples Preparation

Sample of biodiesel fuel blends were prepared through mechanical mixing and blending (based on volume). Briefly, each biodiesel blends are added with constant 10ml alcohol additives. 100% diesel (D100) will be a baseline reference for this research to compare later. The test fuels were B20 + 10ml Methanol, B20 + 10ml Ethanol, and B20 + 10ml Butanol. Each of the mixture was stirred continuously for 30 minutes by using an electric magnetic stirrer and left for 30 minutes to reach an equilibrium at room temperature before they were subjected to any test as recommended by previous research [16]. Sample property tests were carried out to compare different value of Calorific Value (CV), viscosity, and density. The results were differentiated in graphs.

2.2. Engine Test

The experiments are conducted in Automotive Engines Laboratory, Faculty of Mechanical Engineering, Universiti Teknologi MARA (UiTM) Shah Alam. A single cylinder CI engine is used to carry out the experiment as shown in Figure 1. The engine with a specification in Table 1 is operated at different speed of 2700, 3100, and 3500 rpm with load at 50%. The engine speeds are measured using a digital tachometer. Three types of alcohol with 10ml volume are blended with 20% blends of palm oil methyl ester biodiesel (B20) which are the butanol (Bu10), methanol (M10), and ethanol (E10) are fuelled into the engine.

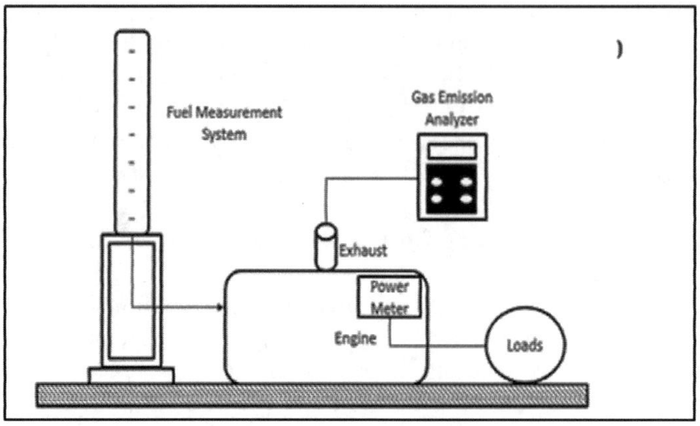

Figure 1. Schematic diagram of the experimental setup.

The engine had been running approximately for ten minutes with diesel fuel after the test then the fuel was changed for the next test to ensure the accuracy and the consistency of the measurement data. Furthermore, the engine fuel tests were repeated three times for each fuel at the specified test conditions in order to have a reliable measurement. Exhaust emissions data was by using a MRU Infrared Gas Analyzer to detect output of CO, CO_2, HC, and NOx. The gas analyser has been calibrated by the company to assure the accuracy of the data collection.

Table 1. Yanmar TF-M engine specifications.

Engine Type	Single-cylinder, 4-stroke air-cooled diesel engine
Fuel Tank Capacity [L]	1.5
No. of Cylinder	1
Displacement [cc]	418
Bore x Stroke [mm]	70x57
Displacement ([L]	0.219
Compression Ratio	8.5
Max Power [hp/rpm]	10/3600
Rated Power [kW]	4.5
Lubrication	Oil SAE10W/30

3. Results and Discussion

Engine performance parameter such as Brake Specific Fuel Consumption (BSFC) and Brake Thermal Efficiency (BTE) are calculated, and the data of exhaust emission such as CO, NOx, CO_2, and HC are justified.

3.1 Properties of Fuel Blends

Figure 2 shows the result comparison between of viscosity for tested fuels (D100, B20, B20 + M10, B20 + E10, B20 + Bu10). Kinematic Viscosity is the ratio of dynamic viscosity to the density of fluid. The results are arranged in the order B20 (0.00494 Pa.s), B20 + Bu10 (0.00434 Pa.s), B20+M10 (0.00416 Pa.s), B20 + E10 (0.00409 Pa.s) and D100 (0.00334 Pa.s). It is stated that B20 has the highest kinematic viscosity compared to other tested fuels. Fuels with high viscosity tend to form larger droplets in the injection resulting in poor fuel atomization, which increases the penetration of the spray tip and decreases the angle of the spray, resulting in poor combustion. This ultimately also increase the exhaust emission engine [17, 18].

However, by adding the alcohol blend with biodiesel, the results seem to be compensated where kinematic viscosity decreases as seen from the result B20 + Bu10 (0.00434 Pa.s), B20 + M10 (0.00416 Pa.s), B20 + E10 (0.00409 Pa.s). Lower viscosity is one of a good consideration for diesel engine in reducing friction losses; hence better fuel consumption [19, 20]. In summary, the percentage of reduction blended fuels compared to B20 is B20 + Bu10 (12.15%), B20 + M10 (15.79%), and B20 + E10 (17.29%). In conclusion, B20 + E10 (0.00409 Pa.s) provides a kinematic viscosity closer to diesel, which makes it more compatible with injection systems used in most of diesel engines. Calorific Value (CV) for D100, B20, B20+M10, B20+E10 and B20 + Bu10 is shown in Figure 3. CV or known as heating value is defined as heat released by the fuel when it is completely burnt and measured at constant volume or constant pressure and the hot gas is cooled back to its initial temperature [21, 22]. It can be classified into low heating value (LHV) and high heating value (HHV). Based on the graph, CVs are obtained as D100 (4.4496), B20 (3.9515), B20 + E10 (3.9266), B20 + M10 (3.8502), and B20 + Bu10 (3.5655). It is proven that the addition of alcohol in biodiesel blend gives a significant lower CV compared to diesel [12]. In summary the percentage reduction of CV for blended fuels when compared with B20 is B20 + E10 (0.63%), B20 + M10 (2.56%), and B20 + Bu10 (9.77%).

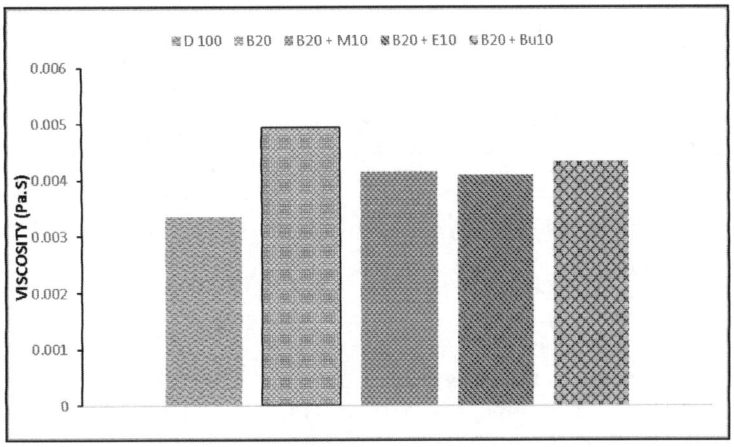

Figure 2. Viscosity of tested fuel

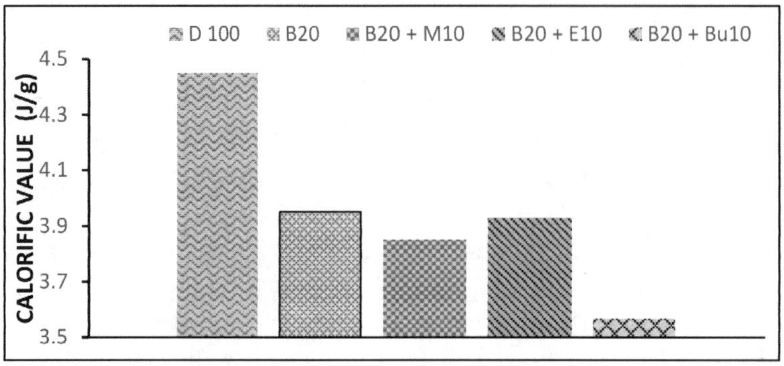

Figure 3. Calorific value of tested fuel.

Figure 4 shows the pattern value of density for all tested fuel (for D100, B20, B20 + M10, B20 + E10, and B20 + Bu10). Density is a mass of fuel per unit volume. The density plays an important role in determining the performances and emissions in internal combustion engine. One important property of diesel fuel is its physical properties such as density [23]. Based on the value shown on bar graph Figure 4, the highest density obtained by B20 + E10 with 1109 kg/m^3 while the lowest density value is obtained by D100 with 922 kg/m^3. If a fuel has higher density than other, it means much more mass is entering into the combustion chamber for the same volume, since new generation diesel injectors which supply fuel into combustion chamber for power generation regulate the amount of fuel by volume, not by mass. Much more fuel entering combustion chamber means an increase of emissions [24]. In conclusion, the sequence of density value for tested fuels is D100 (922kg/m^3), B20 + Bu10 (923kg/m^3), B20 + M10 (950kg/m^3), B20 (958kg/m^3) and B20 + E10 (1109kg/m^3). While the percentage of reduction blended fuels when compared with B20 is B20 + M10 (0.84%) and B20 + Bu10 (3.65%) but increment in B20 + E10 (13.62%).

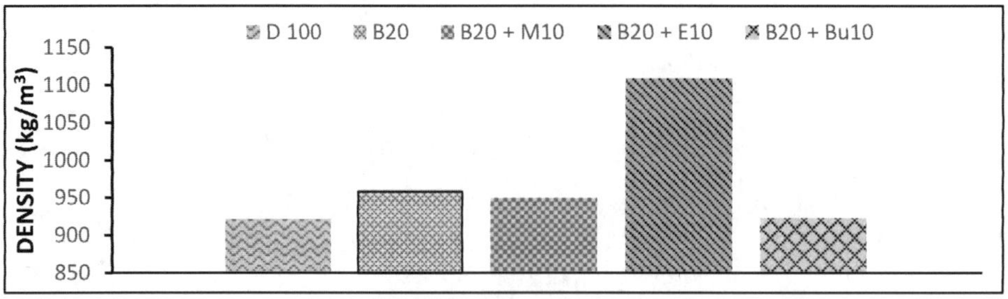

Figure 4. Density value of tested fuel.

3.2 Engine Performance

3.2.1 Brake specific fuel consumption

Figure 5 demonstrate regarding brake specific fuel consumption (BSFC) of all tested full (D100, B20, B20 + M10, B20 + E10 and B20 + Bu10) at different engine speed (2700, 3100, and 3500 RPM) while at constant load (50%). BSFC is known as the relation between mass fuel consumption and brake power. It can be seen from the result that B20 (165, 180 and 210 g/kW.h) blend has the highest value of BSFC as compared to D100 (135, 150 and 170 g/kW.hr) at different engine speed (2700 rpm, 3100 rpm and 3500 rpm). In brief, 18.1% of percentage different is the highest where the BSFC produce by B20 blend is 165 g/kW.h and D100 is 135 g/kW.h at 2700 rpm. The rate of BSFC in B20 blend increase due to the lower calorific value of B20 compared to the D100 [17]. In this case, diesel engine needs to consume more B20 blend to develop the same power as diesel origin.

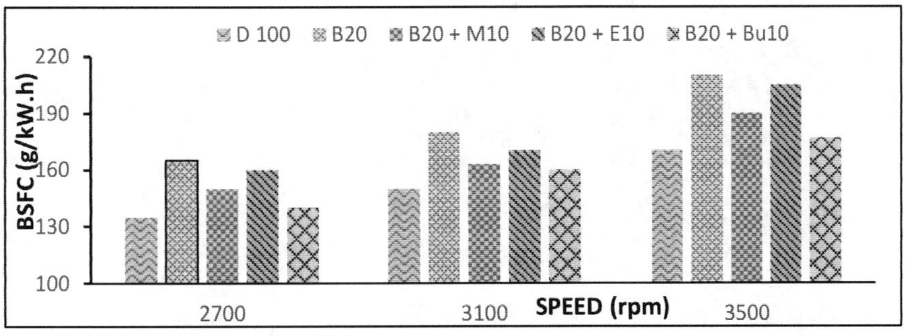

Figure 5. BSFC vs different Engine Speed with constant load

Nevertheless, it could also be seen from the result that BSFC decrease compared to B20 when B20 blend are fuelled with alcohol-diesel blend (M10, E10 and Bu10) at the same volume (10%). The BSFC that produced at 2700 RPM engine speed shows the lowest pattern among three engine speeds as for the example the lowest BSFC for blended fuel is B20 + Bu10, and it is the nearest to diesel origin. It means that the blend of B20 + Bu10 has shown the highest reduction (15.15%) in BSFC when compared to the blend of B20 at constant load of 50%. This concludes the blend of B20 + Bu10 has better BSFC than B20, B20 + M10 and B20 + E10. This case is due to the presence of Bu10 in the blend fuel of B20. The benefit of the presence of Bu10 is butanol has high cetane number, low vapour pressure, and lower miscibility making it more suitable diesel fuel additive than ethanol and methanol [18]. Other than that, butanol has good properties as substitute for ethanol, with higher energy content in addition to enhance mixing properties and has lower autoignition temperature than methanol and ethanol as well [19]. In conclusion the sequence for reduction of BSFC compared to B20 is B20 + E10 (3.03%), B20 + M10 (9.09%) and B20 + Bu10 (15.5%).

3.2.2 Brake Thermal Efficiency

Brake Thermal Efficiency (BTE) results are shown in Figure 6 for diesel fuel and blends of B20, B20 + M10, B20 + E10, and B20 + Bu10. BTE defines the ability of the combustion to accept experimental fuel and offers comparable means of determining how effectively the fuel energy has been converted into mechanical performance. The graph displays the result of BTE for all tested fuels decreasing from low engine speed (2700 rpm) to the highest engine speed (3500 rpm). It may because of the combustion temperature is low, therefore incomplete combustion takes place; on the other hand, at the highest engine speed, the combustion temperature is high and the fuel/air ratio is richer, however there is not enough time for mixing of fuel and air, resulting incomplete combustion and decrease in BTE. It also states the highest BTE among three engine speed is at the lowest engine speed (2700 rpm) where D100 (16.2%), B20 (14.3%), B20 + M10 (17.0%), B20 + E10 (14.0%), and B20 + Bu10 (16.7%) at 50% engine load. Other than that, it can be observed that diesel fuel has the higher BTE compared to B20 blend. It may because of the different number of carbon where D100 contains more carbon atom in longer chains than B20 [20]. Besides, D100 has higher fuel temperature compared to B20 thus increase the BTE value [21].

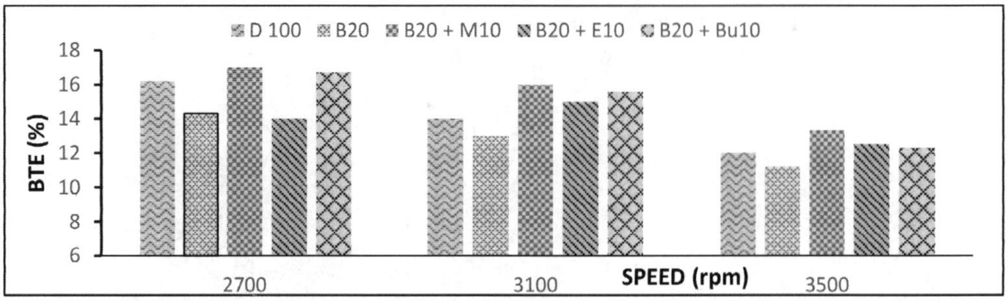

Figure 6. BTE vs different Engine Speed with constant load.

The BTE for blend B20 + M10 at 3 engine speed (2700 rpm, 3100 rpm and 3500 rpm) is the highest (17.0%, 16.0%, and 13.3%) compared to other blend fuels even diesel origin. The increasement of BTE for B20 + M10 is (9.72%) from three different engine speed compared to diesel origin. This is could be attributed to the presence of methanol in B20 blend where methanol has the shortest chain and the lowest molecular weight among all the tested alcohols which leads to easier ignition and better combustion and hence resulting highest BTE [17]. In addition, the lowest boiling point of methanol among the tested alcohols causes reduction in heat losses and hence resulting in higher BTE.

In conclusion, the blends for B20 and B20 + E10 experience reduction at three engine speed where is B20 (8.76%) and B20 + E10 (1.66%) while B20 + M10 and B20 + Bu10 experience increasement three engine speed where is B20 + M10 (9.72%) and B20 + Bu10 (5.38%).

3.2.3 Exhaust Emission

Figure 7 shows the behaviour of concentration carbon monoxide (CO) at different engine speed for tested fuel of D100, B20, B20 + M10, B20 + E10, and B20 + Bu10. All tested fuels are decrease in brake specific carbon monoxide (BSCO) when the engine speed increases. The graph also depicts that all tested blended fuels cause BSCO increase at low engine speed (2700 rpm) compared to diesel fuel. The reason is because at lowest engine speed (2700 rpm) the effect of lower combustion temperature where cooling effect due to higher latent heat of evaporation, dominates the effect of oxygen content of the blended fuel as complete combustion, which suppress the CO oxidation process, resulting in the increase in BSCO emission compared to diesel [18,19]. From the average for three engine speeds, the reduction in BSCO emissions using all alternative fuels are in sequencing B20 + M10 (14.28%), B20 + Bu10 (6.25%), B20 (2.97%), and B20 + E10 (0.14%) comparison with diesel. This results also prove that other studies also obtained reduction in B20 and B20 + Bu10 [25].

However, when the engine speed increase, the effect of combustion temperature becomes lesser and it can be seen at highest engine speed (3500 rpm) which BSCO emissions of all the blended fuels are lower than diesel fuel. From the result, B20+M10 has the highest influence reduction of BSCO (14.28%) that can be attributed to the long duration of combustion, and hence longer time for oxidation of CO to CO_2 [26].

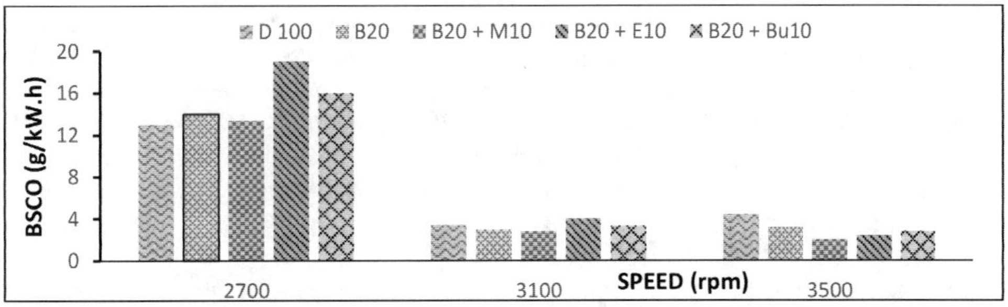

Figure 7. BSCO Vs Engine Speed with constant load.

Figure 8 shows the variation of the brake specific Hydrocarbon (BSHC) of different fuel blends with respect to variation engine speed at constant load (50%). Based on the graph, BSHC for all tested fuels decrease by increase the engine speed. At low engine speed (2700 rpm), BSHC emissions for all tested fuel are high due to the insufficient combustion temperatures to initiate complete combustion. However, at higher engine speed the combustion temperatures are high enough to achieve more complete combustion, leading to decrease in BSHC for all tested fuels. The results from the graph also depicts all the blended fuels (irrespective of B20 + E10 at 2700 RPM) could reduce the BSHC from low engine speed (2700 rpm) to the highest engine speed (3500 rpm).

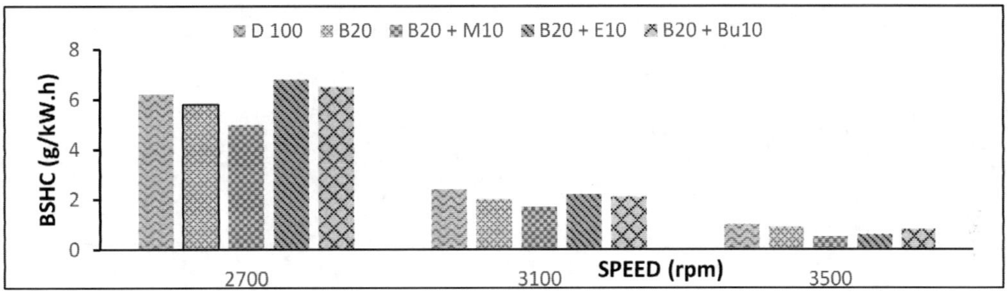

Figure 8. BSHC Vs different Engine Speed with constant load.

The consequence of incomplete combustion contributes to higher latent heat of evaporations of alcohols and biodiesel compared to net diesel is the factor of increments BSHC at low engine speed (2700 rpm). Nevertheless, it is found that BSHC emission for blended fuels is lower than net diesel at high engine speed (3500 rpm) due to the dominant factor of higher oxygen content of the blended fuels that cause more complete combustion and increase the oxidation of unburned hydrocarbons at higher in-cylinder temperatures [27]. There are also proven by some studies that found a rise in THC at low engine speed and reduction at high engine speed with the use of alcohol diesel blends compared to diesel [28, 29]. In addition, other studies also found that the use of B20 + E10 and diesel blended with ethanol or methanol could increase the combustion quality due to more oxygen content in the fuel, and reduce the THC, in comparison with diesel fuel [30-33].

On the average for three engine speeds, the reduction of BSHC are in the order of B20 + M10 (56.25%), B20 (9.38%), B20 + E10 (2.5%), and B20 + Bu10 (1.4%) compared to diesel. From the results, it can be concluded that the effect of using B20 + M10 on reduction of BSHC (56.25%) is higher than other blended fuels which is due to the better combustion.

Figure 9 shows the results for brake specific Nitrogen oxides (BSNOx) emissions with the use of D100, B20, B20 + M10, B20 + E10, and B20 + Bu10. The graph display that the BSNOx emissions approximately decrease with rise in engine speed (RPM). Besides, it also reveals that all the blended fuels (except B20 at all engine speed) causes decrease in BSNOx at all the tested loads compared to diesel fuel. From the results, it can be concluded that on the average reduction for three engine speed, reduction in BSNOx with use of blended fuels are in the order B20 + M10 (27.84%), B20 + E10 (10.79%), B20 + Bu10 (5.81%) while B20 has increasing (6.7%) in BSNOx compared to diesel.

In spites, the huge effect of combustion temperature on formation of NOx can be revealed from the results that length of combustion duration (residence time) also has effect on NOx formation. This is due to the blend of B20 + M10 which has highest decrement of BSNOx (27.84%) among all the tested blended fuels due to the shortest duration of combustion. Meanwhile, the increment of B20 (6.7%) in BSNOx is due to the highest duration combustion.

There are studies found that increase in alcohol branch cause reduction in NOx emission due to the increasing in ignition delay and decrease in adiabatic flame temperature [34]. In addition, there are also studies regarding reduction of BSNOx using ethanol [28, 35] and butanol [36]. It can be concluded from the above result that the effects of higher latent heat of evaporation and lower heating value of alcohol fuels are dominant factors compared to other parameters (such as lower cetane number and higher oxygen content of blends which increase the combustion temperature) that cause reduction in BSNOx for different engine speed.

The behaviour of $BSCO_2$ emissions that fuelled with D100, B20, B20 + M10, B20 + E10 and B20 + Bu10 at three different engine speeds for could be seen in Figure 10. From the results, it is observed that the highest $BSCO_2$ emission is recorded at 2700 rpm engine speed while the lowest $BSCO_2$ emissions is recorded at 3500 rpm. It concludes that higher engine speed cause reduction in $BSCO_2$ due to the more complete combustion.

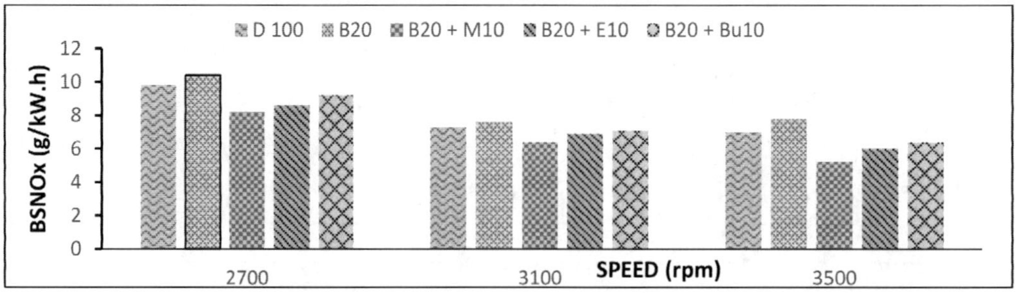

Figure 9. BSNOx Vs different Engine Speed with constant load.

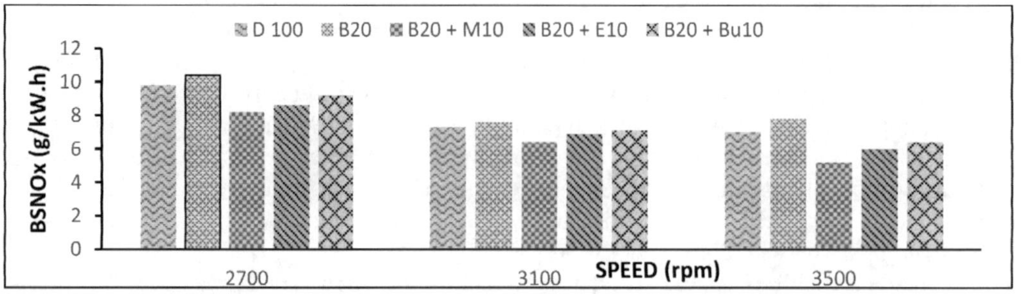

Figure 10. BSCO2 Vs different Engine Speed with constant load.

On the other hand, the graph also shows that all the blended fuels (except B20) could reduce $BSCO_2$ emissions compared to diesel. This is also proven from the studies that reported there are reduction of $BSCO_2$ with ethanol and B20 + E10 [24, 28, 37] and a slight increase in $BSCO_2$ using pure biodiesel

[38]. The reason behind the reduction of $BSCO_2$ in blended fuels compared to diesel is due to the lower carbon-to-hydrogen ratio and the higher oxygen content of the blended fuel [24, 33].

The percentage of reduction in $BSCO_2$ are in the order B20 + M10 (5.19%), B20 + Bu10 (2.42%), and B20 + E10 (0.69%), while increment of $BSCO_2$ emissions in B20 (5.86%) with comparison of diesel. It can be deduced that only B20 + M10 is effective on reducing $BSCO_2$ with reduction of 5.86% among all the blended fuels compared to diesel.

4. Conclusions

From the experiment, B20 has a higher BSFC and lower BTE for engine performances. Oppositely, B20 also shows a disadvantage on exhaust emission where increment of CO, HC, NOx, and CO_2 emission showed up. When the alcohol (butanol, methanol, and ethanol) with 10ml are fuelled into alcohol-diesel-biodiesel blend, the performance characteristics (BSFC, BTE) and emission characteristics shows a better result compared to non-alcohol biodiesel. In overall, the addition of 10ml methanol in B20 shows the most significance result for the engine performance and emissions at all test condition.

Acknowledgement

The authors of this paper would like to express their sincere gratitude to Universiti Teknologi MARA (UiTM) for the opportunity to trigger this research and Fundamental Research Grant Scheme (FRGS) from the Malaysian Higher Education Department (Grant No FRGS/1/2019/TK10/UITM/02/16) for the financial support.

References

[1] Kumar N et al. 2018 *Diesel Gasol. Engines* 1–27 DOI: 10.5772/intechopen.80614
[2] Helmisyah A J et al. 2020 IOP Conf. Series: *Mat. Sci. & Eng.* **8334** 012066
[3] S K and Shukl A 2011*Environ. Impact Biofuels* DOI: 10.5772/20923.
[4] Bae C and Kim J 2017 *Proc. Combust. Inst.* **36** (3) 3389–3413
[5] Owusu P A and Asumadu-Sarkodie S 2016 *Cogent Eng.* **3** (1) 1–14
[6] Hassan M H and Kalam M A 2013 *Procedia Eng.* **56**39–53
[7] Bueno A V et al. 2017 *Applied Thermal Engineering* **125** 559-566
[8] Zaharin M et al. 2017 *Renewable and Sustainable energy reviews* **79** 475-493
[9] Asiah A R et al. 2020IOP Conf. Series: *Materials Science and Engineering* **834** 012051
[10] Yilmaz N et al. 2017 *Fuel* **207** 461-469
[11] Nieuwenhuis P and Wells P 2003 *Automot. Ind. Environ.* 87–99 DOI: 10.1016/b978-1-85573-713-6.50010-x.
[12] Erdiwansyah et al. 2019 *Energy Reports* **5** 467–479
[13] Kolesinska B et al. 2019 *Materials (Basel)* **12** 3
[14] Kowalewicz A and Wojtyniak M 2005 *Proc. Inst. Mech. Eng. Part D J. Automob. Eng.* **219** (1) 103–125
[15] Bušić A et al 2018 *Food Technol. Biotechnol.* **56** (3) 289–311
[16] Lim W et al 2009 *Journal of Oil Palm Research* **21** 683-692
[17] Lapuerta M et al. 2017 *Fuel* **199** 332–338
[18] Knothe G and Steidley K R 2005 *Fuel* **84** (9) 1059–1065
[19] Rejowski E et al. 2016 *XXIV Simpósio Internacional de Engenharia Automotiva* DOI: 10.5151/engpro-simea2016-PAP20
[20] Macian V et al. 2014 *SAE Tech. Pap.* **2014** Oct. DOI: 10.4271/2014-01-2797.
[21] Sharudin H et al 2018 *Jurnal Kejuruteraan UKM* **1** (7) 37–47
[22] Che Mat S et al. 2017 *MATEC Web Conf.* **135** DOI: 10.1051/matecconf/201713500004.
[23] Sarathy S M et al.2018 *Prog. Energy Combust. Sci.***65** 67–108
[24] Tüccar G et al. 2018 *Acad. Platf. J. Eng. Sci.***6** (2) 81–85
[25] Gad M S et al. 2018 *Egypt. J. Pet.* **27** (2) 215–219
[26] Kumar S et al. 2013 *Renew. Sustain. Energy Rev.* **22** 46–72

[27] Altun S et al. 2011 *Ind. Eng. Chem. Res.* **50** (15) 9425–9430
[28] Rahim R et al. 2010 *Natl. Conf. Mech. Eng. Res. Postgrad. Stud.* (2nd NCMER 2010) December 863–870
[29] Mamat R 2009 A thesis submitted to The University of Birmingham for the degree of PhD School of Mech November
[30] Kumar B R and Saravanan S 2016 *Renew. Sustain. Energy Rev.* **60** 84–115
[31] Asokan M A et al. 2018 *Indian J. Chem. Technol.* **25** (3) 225–234
[32] He B Q et al. 2003 *Atmos. Environ* **37** (35) 4965–4971
[33] Shi X et al. 2006 *Atmos. Environ.* **40** (14) 2567–2574
[34] Imdadul H K et al. 2016 *Energy Convers. Manag.* **111** 174–185
[35] Ghadikolaei M A et al. 2018 *Energy* **157** 258–269
[36] Wei L et al 2014 *Energy* **70** 172–180
[37] Tse H and Leung C W 2016 *Adv. Automob. Eng* **01** (1) 1–7 DOI: 10.4172/2167-7670.s1-005
[38] Özener O et al 2014 *Fuel* **115** 875–883

Effect of Fuel Additives (Ethanol) on the Engine Performance Emissions of Single Cylinder Diesel Engine Fueled with Palm Oil Methyl Ester (PME) (B20)

Asiah Ab. Rahim[1], Nik Rosli Abdullah[2*]Mohamad Hafsham Mazlan[2], Hazim Sharudin[3] and Helmisyah Ahmad Jalaludin[1]
[1] Faculty of Mechanical Engineering, Universiti Teknologi MARA, 23200 Bukit Besi, Terengganu, Malaysia
[2] Faculty of Mechanical Engineering, Universiti Teknologi MARA, 40450 Shah Alam Selangor, Malaysia
[3] Faculty of Mechanical Engineering, Universiti Teknologi MARA, 13500 Permatang Pauh, Pulau Pinang, Malaysia

*Corresponding e-mail: nikrosli@uitm.edu.my

Abstract. Transport technology development is a major issue leading to increase the number of vehicles, hence increase emissions and contributing to global warming. In this work the effect of fuel additives (ethanol) on the engine performance and emissions of single cylinder diesel engine fueled with palm oil methyl ester (PME) (B20) were investigated. The tests were performed by varying the engine speed between 2700, 3100 and 3500 rpm with intervals 400 rpm while maintaining the engine load at 50% of rated load. In this study, different fuels which is B0 (100% Pure diesel), B20 (20% PME), E5 (20% PME + 5ml ethanol), E10 (20% PME + 10ml ethanol) and E15 (20% PME + 15ml ethanol) were used. The result indicated that, average brake specific fuel consumption (BSFC) and Brake thermal efficiency (BTE) for usage of E5, E10 and E15 were increased compare with B0 and B20. Meanwhile, average value results of Carbon Monoxide (CO) and Unburned Hydrocarbon (UHC) for E5 was lowered compare to PME (B20). The addition of ethanol with PME (B20) fuel in single cylinder diesel engine can help in controlling exhaust emission and significantly improve engine fuel consumption.

Keywords: fuel, additives, diesel engine, ethanol, palm oil

1. Introduction
Today, energy is one of the most important issues that need to resolve in order to save planet Earth because global energy demand has increased every single day. Currently, 85 % of the world's energy consumption has been fueled by fossil fuels. Limited reserved of existing fossil fuels contribute to seek for alternative sources. [1]. Compression-ignition (CI) engines are widely used in the transportation, power generation, automotive, industrial and agriculture sectors due to their high energy conversion efficiency and ease of operation [2]. However, diesel engine produces high carbon emissions as diesel fuel was derived from crude oil, resulting in high carbon emissions and pollution whenever the incomplete combustion occurs during the combustion process [3]. These pollutants have tremendous effects on human, environment and ozone depletion caused to global warming. Previous studies have proven that emissions from diesel engine can cause respiratory and cardiovascular health problem [4].

In addition, a number of studies have shown that alternative fuel from renewable sources can improve engine performance and reduce exhaust emissions [5].

Biodiesel is an alternative diesel fuel consisting of an alkyl monoester of fatty acids derived from the first-generation feedstock such as vegetable oil or animal fats. Up to date, many studies have reported the use of the biodiesel on diesel engines due to its reproducibility, non-toxicity and Sulphur-free properties [6]. However, the main disadvantage of vegetable oil is its viscosity, which is considerably higher than that of conventional diesel oil. In order to overcome the problems associated with the blend of biodiesel diesel and the suitable alternative part of both diesel biodiesel, ethanol is added as the third liquid fuel along with the blend of diesel-biodiesel. Adding Ethanol into diesel-biodiesel blend can decrease the viscosity of the fuel and at the same time acts as the oxygenating mediator due to ethanol rich in oxygen content. Ethanol become the most encouraging substitute fuel because it can be produced from natural resources compared to fossil fuels produced from non-natural resources [7].

Furthermore, the use of alcohol-based fuels in petroleum fuels has increased due to benefits such as improved combustion and reduced exhaust emissions. Many researchers have researched the effect of mixing different oxygenates with diesel and biodiesel fuel since it was introduced [8]. Mixing oxygenated fuels such as ethanol with baseline fuel (diesel, biodiesel) is considered a potential solution for reducing harmful emissions worldwide as well as meeting the increasing global energy demand.

Therefore, the purpose of the current work is to study the effect of adding ethanol on engine performance and emissions of single-cylinder diesel engine fueled with Palm Oil Methyl Ester (PME) (B20). The research will focus on finding the impact of fuel additives (ethanol) on engine performance and emission of a single-cylinder diesel engine fueled with diesel-palm oil methyl ester (PME) (B20) in response to this issue.

2. Research Method

2.1. Test Fuel

Diesel fuel was supplied by a commercial fuel manufacture. Palm oil methyl ester and ethanol was supplied by local company. The five fuel samples adopted in this study are B0 (pure diesel fuel 100%), B20 [80% diesel + 20% Palm oil methyl ester (PME)], E5 (80% diesel + 20% Palm oil methyl ester (PME) + 5ml ethanol), E10 (80% diesel + 20% Palm oil methyl ester (PME) + 10ml ethanol) and E15 (80% diesel + 20% Palm oil methyl ester (PME) + 15ml ethanol). The fuel properties such as calorific value, viscosity and density were quantified. The calorific values and viscosity values were determined by using Bomb Calorimeter and Viscometer respectively.

2.2. Diesel Test Engine

For the experimental study, the Compression Ignition (CI) diesel engine setup consists of 1.5L engine powered diesel generator was used. The specifications of the engine are shown in Table 1. The engine will be coupled to industrial fan to provide load as shown in Figure 1. Test fuels were tested at 50% of rated constant load condition and with various engine speed condition between 2700 rpm, 3100 rpm and 3500 rpm with 400 rpm intervals. The test bench comprises of a diesel engine, measuring system, an industrial fan as constant load, tachometer for measuring engine speed, and gas analyzer to evaluate the exhaust emissions. Before the test, the engine was warm up with diesel fuel for about 30 minutes until the engine temperature are stabilized.

Table 1. 1.5L Diesel engine specification.

Item	Specification
Engine Series	BENMA 5GF-M Diesel Generator
Engine model	5GF-M
Engine type	4 stroke, single cylinder, air-cooled diesel engine
Frequency (Hz)	50

Displacement Capacity (cc)	418

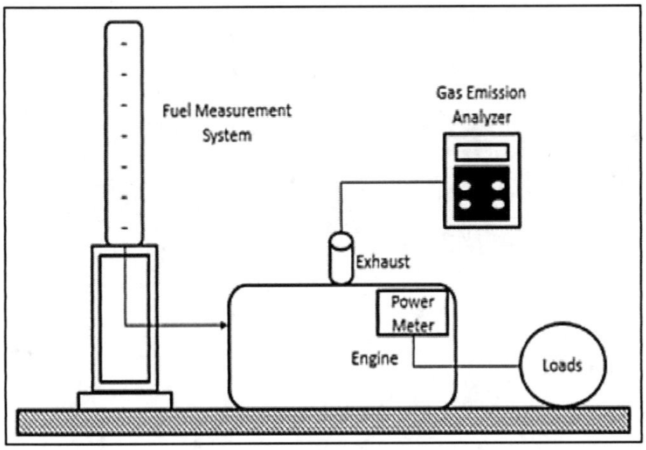

Figure 1. Schematic of an experimental setup.

2.3. Performance and Exhaust Emission Measurement

The most important input parameter in this study is load and speed. The data from the measuring cylinder taken to obtained fuel consumption volume for every 2 minutes for each tested fuel. The performance parameter which are brake specific fuel consumption and brake thermal efficiency calculated.

The output parameters for this experiment include exhaust emission analysis. The exhaust emission that involve in this study is Carbon Dioxide (CO2), Carbon Monoxide (CO), Unburned Hydrocarbon (UHC) and Nitrogen Oxide (NOx). For this experiment, gas analyzer (Portable Kane Automotive gas model 5-1) will be place near to the exhaust system with a nozzle placed in the exhaust. One of the ways to ensure the good data is by running the engine with applying the gas analyzer in 2 minutes and above for every run. Each experiment is run 3 times to obtain an average value of output parameters.

3. Result and Discussion

3.1. Engine Performance

3.1.1. Brake Specific Fuel Consumptions (BSFC). Figure 2 shows the variation of brake specific fuel consumption (BSFC) for all fuel samples with respect to the engine speed. It is seen that the BSFC for fuels with ethanol blend are higher compared to neat diesel fuel for same power output and it is steadily increasing with increase in amount of ethanol in blends. It reveals that the engine uses more ethanol blended fuel than pure diesel fuel due to an improvement in the combustion process by increasing the efficiency of the combustion chamber [4], [5], [9], [10]. The finding shows that BSFC increase is reported due to the ability of the additive as a lubricant to reduce friction in the combustion chamber [7], [11]. Average brake specific fuel consumption for usage of B20, E5, E10, and E15 was 13.86%, 24.61%, 44.61%, and 53.84% respectively higher than that of diesel fuel (B0).

Other than that, the increase in BSFC is primarily attributed to the lower calorific value of biodiesel and ethanol compared to that of pure diesel fuel [7], [12], [10]. Some researchers also have a similar view that higher BSFC for biodiesel blends are due to high viscosity and density values, respectively [3], [13], [14].

Therefore, it can be concluded that ethanol-palm oil methyl ester (E15) produce higher BSFC throughout all engine test conditions as compared with fossil diesel and palm oil methyl ester (PME) (B20). This is because of additive act as lubricant to improve the combustion process in combustion chamber [6], [10], [15].

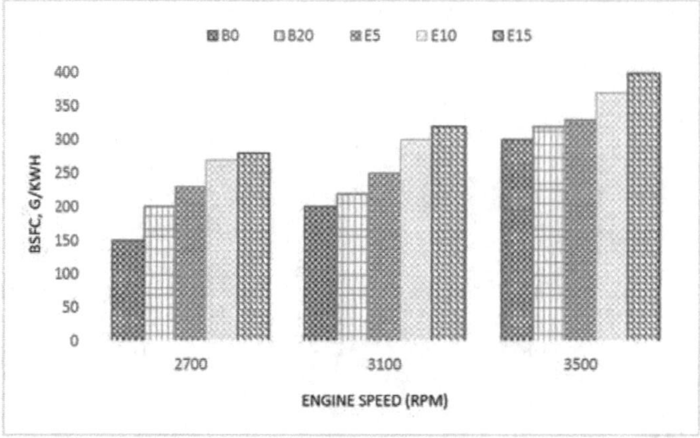

Figure 2. Graph of BSFC (g/kWh) against Engine Speed (rpm).

3.1.2. Brake Thermal Efficiency (BTE). Figure 3 shows the comparison of the BTE with engine speed for the tested fuels. The average of BTE for B0, B20, E5, E10, and E15 is 25.33%, 20 %, 30.33 %, 27.33 %, and 25.33 % respectively. The BTE decrease with an increase in engine speed due to the engine exhibits high friction and heat losses as the engine speed increases [9], [16], [17], [18]. At the lower engine speed 2700 rpm, the BTE value is higher due to the improvement of the combustion process on account of increased oxygen content in the fuels [3], [12]. As a result, the combustion process is improved and thermal efficiency is improved when fuel additive is added [19]. Despite the higher BSFC of ethanol-blends, the higher BTE of the blended fuel is noticed. The justification for BTE improvement is an increase in the compression ratio due to its higher octane number in the case of ethanol blended fuel [5].

Therefore, it can be concluded that ethanol-palm oil methyl ester (E5) will increase the BTE due to the presence of more oxygen and higher latent heat of vaporization as compared to B0 and B20 without additive. It can be said that the most excellent volume of additive is 5ml blended with ethanol-palm oil methyl ester (E5) content in the fuel blends, the higher the value of BTE.

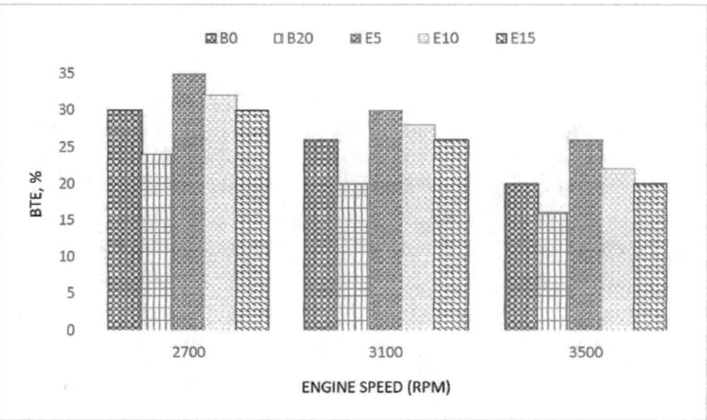

Figure 3. Graph of BTE (%) against Engine Speed (rpm).

3.2. Emission Parameters

3.2.1. Carbon Monoxide (CO). Figure 4 illustrates that the variation in CO emission of the test fuel with diverse engine speed. In general, by increasing the engine speed, the percentage of CO emissions of all tested fuels is increased. As a result, all fuel additive ethanol blended with palm oil methyl ester (B20) at different volumes (5ml, 10ml and 15ml) and palm oil methyl ester (B20) of fuel provide a lower percentage of CO emissions compared to pure diesel. This can be explained by the enrichment of oxygen due to the introduction of ethanol and biodiesel, in which an increase in the proportion of oxygen would facilitate further oxidation of CO during the exhaust cycle of the engine [20], [21]. The significant reduction of CO emissions in biodiesel-ethanol blends is mainly due to the addition of ethanol, a type of oxygenated additives that promotes more complete combustion [22], [17].

Therefore, it can be concluded that CO emissions decreased with increase the fuel additive on ethanol-palm oil methyl ester (5ml, 10ml and 15ml) content. Ethanol-palm oil methyl ester blended with 5 ml (E5) additive emitted the lowest emission of CO compared to other blends. Reduction in CO contributed to lowering the harmful gases be release to the atmosphere.

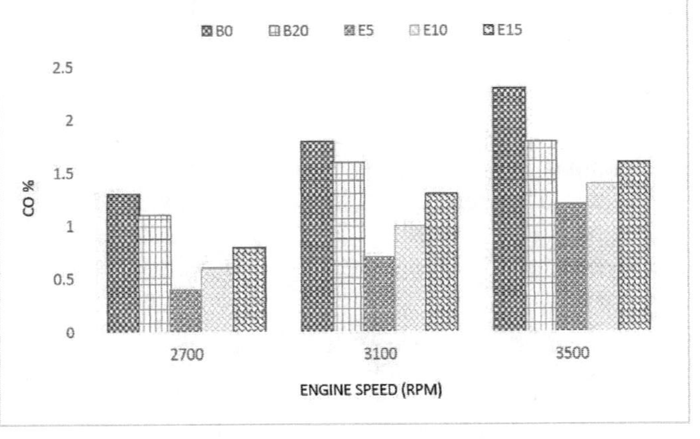

Figure 4. Graph of CO (%) against Engine Speed (rpm).

3.2.2 Carbon Dioxide (CO_2). The Figure 5 shows the trend of CO_2 percentage against various engine speed for various fuel blends B0, B20, E5, E10, and E15 at constant load. As a result, the output of CO_2 for all test blends increases when the engine speed has increased from 2700 rpm to 3500 rpm. It was also shown that Ethanol-palm oil methyl ester blended 5ml (E5) fuel additive demonstrates the most reduced percentage of carbon dioxide emission over engine speed whereas ethanol-palm oil methyl ester blended 15ml (E15) appeared the highest percentage of CO_2 emission if compared with E5 and E10. It is clear that the concentration of CO_2 emission increase as the fuel additive blended on ethanol-palm oil methyl ester substance increase [14], [23]. The higher CO_2 emission of the fuel blend (E10 and E15) with the additive compared to B20 without the additive due to the presence of oxygen content in the biofuel structure. However, the presence of oxygen in fuel improves the process of combustion [13], [19], [16]. In the other hand, high CO_2 formation is due to the complete combustion of the fuel and some previous researchers agreed on these reasons [11], [21], [17].

In conclusion, the result proves that by adding the right amount of fuel additive on palm oil methyl ester and pure diesel fuel content in the fuel blends tends to reduce the percentage of carbon dioxide (CO_2) emission.

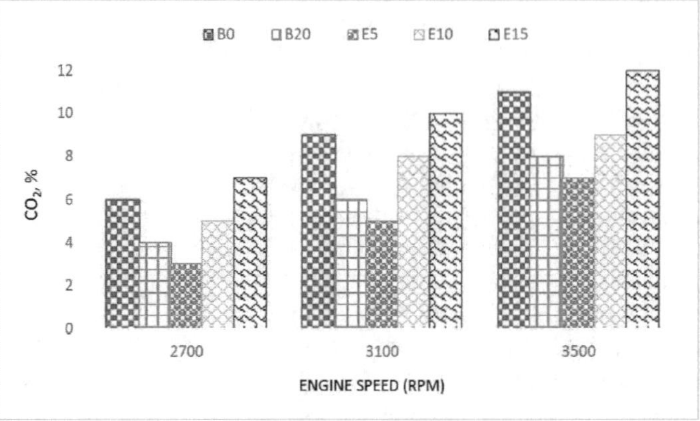

Figure 5. Graph of CO_2 (%) against Engine Speed (rpm).

3.2.3 Unburned Hydrocarbon (UHC). Figure 6 shows the variation of unburned HC (UHC) emissions. In general, UHC emissions increase with an increase in the engine speed. Compared with Pure diesel fuel (B0), biodiesel-palm oil methyl ester (B20) and the Ethanol-palm oil methyl ester (E5, E10, E15) blends give lower UHC emissions. The UHC emissions of biodiesel decrease 50%, 35.71%, and 33.33% at the engine speed of 2700 rpm, 3100 rpm, and 3500 rpm, respectively, in comparison with pure diesel fuel (B0). The higher oxygen content of biodiesel leads to better combustion, resulting in lower UHC [12], [24]. Therefore, decreased UHC emissions in the biodiesel blends are also attributed to their shorter ignition delay [6], [14]. In other words, higher cetane number and oxygen content will reduce the UHC emissions significantly as agreed by many past researchers [21].

In conclusion, ethanol-palm oil methyl ester blended with 5 ml (E5) additive emitted the lowest emission of UHC compared to other blends. Therefore, the result proves that by adding fuel additive content in the fuel blends, it tends to decrease the percentage of unburned hydrocarbon (UHC) emissions.

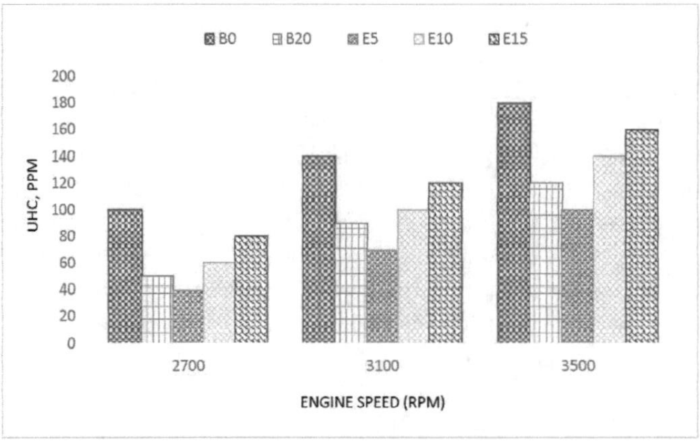

Figure 6. Graph of UHC (ppm) against Engine Speed (rpm).

3.2.4 Nitrogen Oxide (NO_x). The variation of oxides of nitrogen at different engine speeds is shown in Figure 7. The emissions of NO_x are rising with an increase in engine speed. The NO_x emissions were found to be high, 170 ppm and 250 ppm at 3500 rpm and 3100 engine speeds, respectively, for the B0 operated engine. A variety of fuel properties have been shown to have an effect on NO_x emissions. This is because more fuel was injected and combusted in the cylinder as the engine speed increased, which caused higher gas temperatures and resulted in more NO_x formation in the engine cylinder and higher NO_x emissions from the engine [1], [22], [16]. It also shows that one of the most important reasons for NO_x emissions is the combustion temperature of the engine cylinder [3]. Moreover, compared with pure diesel fuel (B0) and biodiesel-palm oil methyl ester (B20), ethanol-palm oil methyl ester (E5, E10, E15) blends give lower NO_x emissions at lower engine speed 2700 rpm. Therefore, adding ethanol to biodiesel as oxygenates can enhance combustion efficiency of the fuel [22], [20].

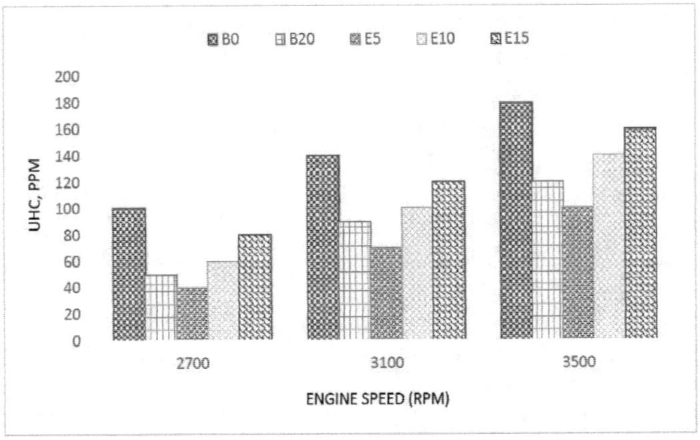

Figure 7. Graph of NO_x (ppm) against Engine Speed (rpm).

4. Conclusions

As a conclusion, the effects of various volume of fuel additive (Ethanol) (5ml, 10ml and 15 ml) on the engine performance emissions of single cylinder diesel engine fueled with Palm Oil Methyl Ester (PME) (B20) were investigated at various engine speed (2700, 3100, 3500) rpm with constant engine load (50% of rated load). Results can draw the following conclusions:

- E5 blends with 5 ml of additive showed increased value of BSFC which is 3.13% than B20 without additive at (3500 rpm, 50% of rated load) due to the properties of additive as a lubricant. E15 blended with 15 ml of additive showed the highest BSFC at all test conditions.
- This study has shown that E5 blended with 5 ml of additive indicates improvement in BTE which is 20% and 30%, B0 and B20 respectively, without additive due to the enrichment of oxygen in the blends. Thus, it promotes better combustion process.
- It was also shown that E5 blended with 5 ml additive shown the lowest emission of CO compare to others blended fuel at all test condition while B0 and B20 blended without additive resulted highest CO and UHC by 35 % and 45% respectively, compared with B0 and B20 without additive.
- The increasing fuel additive volume blended with ethanol showed lower CO, and UHC emissions than the B0 and B20 without additive.

The overall result show engine operates with fuel additive blended with E5 showed promising result in controlling exhaust gas emission and improved BSFC. Therefore, it can be considered that adding the fuel additive will improve combustion behavior inside the combustion. This will be considered as an enhanced strategy in order to improved fuel consumption and diminish exhaust emission. The objectives of this experiment are accomplished, and the ideal volume of additives is found. Hence, the aim for this project is achieved.

Acknowledgments
The authors of this paper would like to express their sincere gratitude to Universiti Teknologi MARA (UiTM) for the opportunity to trigger this research and Fundamental Research Grant Scheme (FRGS) from the Malaysian Higher Education Department (Grant No FRGS/1/2019/TK10/UITM/02/16) for the financial support.

References
[1] Ağbulut Ü, Sarıdemir S and Albayrak S 2019 *J. Brazilian Soc. Mech. Sci. Eng.* **41** pp 389
[2] Pawar R, Jagadale K, Gujar P, Barade V and Solankure B 2018 *Chem. Eng. Trans* **65** pp 45–456
[3] Huang J, Wang Y, Li S, Roskilly A P, Yu H and Li H 2009 *Appl. Therm. Eng.* **29** no 11–12 pp 2484–2490
[4] Zaharin M S M, Abdullah N R, Najafi G, Sharudin H and Yusaf T 2017 *Renew. Sustain. Energy Rev.* **79** pp 475–493
[5] Kumar R and Chaurasia O P 2019 *J. Eur. des Syst. Autom.* **52** no 2 pp 205–214
[6] Park S H, Cha J, and Lee C S, 2012 *Appl. Energy* **99** pp 334–343
[7] El-Araby R, Amin A, El Morsi A K, El-Ibiari N N and El-Diwani G I 2018 *Egypt. J. Pet.* **27** no 2 pp 187–194
[8] Lapuerta M, Armas O and Herreros J M 2008 *Fuel* **87** no 1 pp 25–31
[9] Rakopoulos D C, Rakopoulos C D, Kakaras E C and Giakoumis E G 2008 *Energy Convers. Manag.* **49** no 11 pp 3155–3162
[10] E J *et al* 2017 *Renew. Sustain. Energy Rev.* **80** pp 620–647
[11] Datta A and Mandal B K 2016 *Appl. Therm. Eng.* **98** pp 670–682
[12] Zhu L, Cheung C S, Zhang W G and Huang Z 2011 *Fuel* **90** no 5 pp 1743–1750
[13] Di Y, Cheung C S and Huang Z 2009 *Atmos. Environ.* **43** no 17 pp 2721–2730
[14] Aydin H and Ilkiliç C 2010 *Appl. Therm. Eng.* **30** no 10 pp 1199–1204
[15] Ajav E A, Singh B and Bhattacharya T K 1999 *Biomass and Bioenergy* **17** pp 357–365
[16] Çelebi Y and Aydın H 2019 *Fuel* **236** pp 890–911
[17] Man X J, Cheung C S, Ning Z, Wei L and Huang Z H 2016 *Fuel* **180** pp 41–49
[18] Ban-Weiss G, Gupta R, Chen J Y and Dibble R 2005 *Fall Tech. Meet. West. States Sect. Combust. Inst. 2005 WSS/CI 2005 Fall Meet.* **2** pp 1309–1356
[19] Bilgin A, Durgun O and Şahin Z, 2002 *Energy Sources* **24** no 5 pp 431–440
[20] Rajesh K and Ganesh D B 2019 *Int. J. Innov. Technol. Explor. Eng.* **8** no 9 Special Issue 2 pp 582–587
[21] Banapurmath, N R, Khandal S V, Ranganatha S L and Chandrashekar T K 2015 *Adv. Automob. Eng.* **4** no 01 pp 1–8
[22] Wu-gao Z, Li D, Zhen H and Jian-guang Y 2005 **30** pp 967–976
[23] Ahmed I 2001 *SAE Tech. Pap.* no 2001-01-2475
[24] Imdadul H K *et al RSC Adv.* **5** no 83 pp 67541–67567

Enhancement of the heat transfer in radiator with louvered fin by using Graphene-based hybrid nanofluids

M Sandhya[1,2*], D Ramasamy[1,3*], K Sudhakar[1,3], K Kadirgama[1] and W S W Harun[1]

[1] Faculty of Mechanical and Automotive Engineering Technology, Universiti Malaysia Pahang, 26600 Pekan, Pahang, Malaysia

[2] Department of Mechanical Engineering, KITS College, Warangal, Telangana State, 506015, India

[3] Automotive Engineering Centre, Universiti Malaysia Pahang, 26600, Pekan, Malaysia

*Corresponding e-mail: madderla.sandhya@gmail.com

Abstract. Newly developed hybrid nanofluids can enhance heat transfer performance of various automotive appliances. Improved thermophysical properties of this hybrid nanoparticles when used as coolants will enhance the effectiveness of the car radiators with improved engine cooling system. The present study involves the simulation-based analysis of thermal enhancement of Graphene based Hybrid Nanofluids in a Car radiator. System analysis is performed considering single tube component in the radiator for pattern development. CFD analysis is conducted using ANSYS FLUENT module. Graphene/Crystal Nanocellulose hybrid (Graphene + CNC) nanoparticles are suspended in water & ethylene glycol base fluids (40:60) at 0.01%,0.05%,0.1% &0.2 % volume concentration considering at 3-7 LPM inlet mass flow rate. The effect of various volume concentration, inlet temperature and flow rate on Pressure, Enthalpy, Entropy, Heat transfer coefficient, and Heat transfer rate has been examined by simulation. The results indicate that when using a Louvered fin in a radiator, Maximum enhancement of 60% in heat transfer rate, 53% increase of heat transfer coefficient, 65% increase in pressure are achieved for Graphene with CNC hybrid nanofluids when compared with Graphene based nanofluids. Using this type of hybrid coolants along with louvered fin in radiator can significantly improve the heat transfer performance of radiators.

Keyword: Graphene, Crystal Nano Cellulose (CNC), Hybrid nanofluids, CFD analysis, Heat transfer, Radiator tube

1. Introduction

In current times, vast study has been concentrated on nanofluids having a huge ability for application of heat transfer in a vehicle radiator engine coolant. The particles (metals or metal oxides) suspended in the base fluid with less than 100nm size having higher thermal properties, low concentration with high surface area, corrosion resistant improve the thermal conductivity and heat transfer performance of the radiator. The considerable rise in nanofluids thermal properties has invited scientific consideration in recent years. Nanofluids with smaller size and thermal properties are applied in the radiator to improve the vehicles fuel consumption and better design. Radiator can be designed in a much smaller size and have high heat transfer performance. Conventionally used heat transfer fluids are water, water and ethylene glycol mixture. The laminar & turbulent flow of convective heat transfer nanofluids are experimentally investigated by various authors to predict the thermo-physical properties of the nanofluids. Several investigations are carried out at different mass/volume concentrations to predict the heat transfer coefficients and dynamics of the fluids flowing.

Significant increase in the convective heat transfer of the copper oxide and alumina nanofluid was studied by Vajjha, Das [1] in flat tube of the radiator system. Another set of experiments Leong, Saidur [2] are carried out with copper nanoparticles and concluded that for effective heat transfer application there is a need for higher concentration of the nanoparticles in base fluids. Experiments were carried by pure water with silicon oxide nanoparticles at different temperatures in a horizontal tube by Ferrouillat, Bontemps [3]. Graphene based nanofluids are used by Sarkar, Ghosh [4] to improve the heat transfer performance and pressure drop of the system. High thermal conductivity and thermal stability is the risen attraction in recent research. The thermal conductivity of the graphene nanosheets has been increased at low level temperatures in the study of Sen Gupta, Manoj Siva [5] which is essential to transfer the heat easily. Ahammed, Asirvatham [6] conducted experimental investigations of graphene nanofluid and stated the thermal conductivity enhanced with 37.2% at 50°c. Aravind and Ramaprabhu [7] conducted experiments on thermal conductivity and heat transfer coefficient of graphene and MWCNT and observed there is an increase in 10.5% and 193% improvement. using CFD model developed expected laminar flow heat flux inside cylindrical cross segment tubes. Moraveji, Darabi [8] Used 1%,2%,3% concentrations used nanoparticles size of 45nm and 150nm in single phase simulation. The outcome is 10% higher than experiment value and found that with increase of nanoparticles concentration heat transfer increased.

CFD model developed by Akbarinia and Behzadmehr [9] for Single and Two phase. Obtained results are much nearer to experimental data for two phases rather single-phase approach. Thermal field values are different but similar flow is observed for single- and two-phase approach. Effect of CuO nanofluids is studied by Sharma, Kumar [10] using flat tube. Heat transfer coefficient obtained is 5.1 and increased by 14% compared with water as base fluid while adding 0.1 - 0.5 % vol of CuO nanoparticles. CuO nanofluid performance is also studied by Kumar, Sokhal [11] in 90 flat tube by using 0.1 to 0.4 % vol concentration and found 7-19% increase of heat transfer coefficient than the traditional flat straight tube. Bharadwaja, Ka [12] modeled automobile radiator using siemens NX and CFD analysis has been performed considering only small part of the radiator and defined that there is an increase in effectiveness while using nanofluid coolants and using this carboxyl hybrid nanofluids radiator size can be significantly downsized. Bharadwaj, Mogeraya [13] also studied the hybrid nanofluids behavior in a radiator using CFD analysis for determining the effect of heat transfer performance. Graphene oxide and Carboxyl graphene are used as hybrid nanofluids and results proved that using hybrid nanofluid in radiator the heat transfer performance significantly increased but also more pumping power was required due to friction factor. Using proper concentration, this combination provides better yield effectiveness in an automotive radiator. Pendyala, Ilyas [14] performed CFD analysis using Carbon Nano Tubes and Graphite in a distributor transformer and confirmed that using these nanoparticles the thermal heat transfer performance has improved significantly. Hybrid nanofluids containing platinum and graphene nanoplatelets are used in twisted microchannel by Bahiraei and Mazaheri [15] in their research and studied that higher heat transfer, 1.5x energy efficiency and better flow is observed in twisted microchannel when using hybrid nanofluids.

Hybrid nanofluid containing silver and graphene (Ag/HEG) is used by Zainal, Tan [16] and performed CFD simulation using Ansys fluent in a turbulent circular pipe at 0.1 to 0.9% various volumetric concentrations and results showed that with increase of volume concentrations the heat transfer coefficient and Nusselt number values of Hybrid nanofluid decreased. Along with these studies and few reviews available there is a scope of future work in hybrid nanofluids for better heat transfer capabilities. So, the main objective of the paper is to study the heat transfer performance of the mono and hybrid nanofluid. The heat transfer characteristics of the nanofluids in a turbulent flow conditions are discussed with the CFD simulation. The novelty of the study is to compare the mono and hybrid nanofluid characteristics in the single tube of a car radiator to find the percentage enhancements and heat transfer coefficient estimations along with enthalpy, entropy, total heat transfer rate and Reynolds

IOP Conf. Series: Materials Science and Engineering 1062 (2021) 012014 doi:10.1088/1757-899X/1062/1/012014

number. The CFD simulation is carried out using the Graphene and Graphene +CNC (Crystal Nano Cellulose) nanoparticles at different volume concentrations and flow rates. Comparative study has been made with mono and hybrid nanofluid properties and concluded with better nanoparticles at particular concentration & flow rate using ANSYS FLUENT®. The radiator with lowered strip inserted in the tube is represented in Figure 1

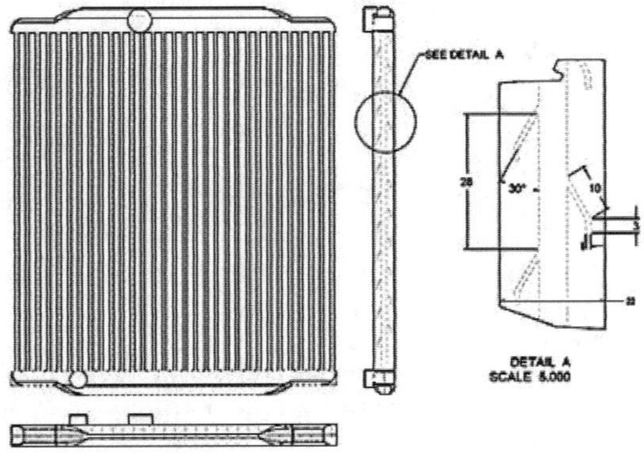

Figure 1. The Lowered strip and radiator details.

2. Mathematical modelling & Governing equations

Nanofluids density is measured based on the following equations (1). The values of density and specific heat of the nanoparticles are considered from vendor. The parameter of concentration of nanofluid is considered the effective density [17].

$$\rho_{nf} = \phi\rho_s + (1 - \phi)\rho_f \tag{1}$$

The specific heat of the nanofluid is measured using equation (2) below. It's a purpose of concentration and base fluid specific heat [18].

$$C_{p,nf} = \frac{\phi(\rho C_p)_s + (1 - \phi)(\rho C_p)_f}{\rho_{nf}} \tag{2}$$

Thermal conductivity has an important role in heat transfer applications. Many researchers focused to improve the thermal conductivity of the fluid experimentally and numerically by many case studies. Maxwell equation [19] as shown in equation (3) is utilized by many researchers to find the thermal conductivity with different volume ratio.

$$k_{eff} = \frac{k_{np} + 2k_{bf} + 2(k_P - k_b)\phi}{k_{np} + 2k_{bf} - (k_{np} - k_{bf})\phi} K_{bf} \tag{3}$$

The dynamic viscosity is calculated by the Einstein [20] proposed in equation (4). Using the flow across one particle, the viscosity of diluted suspension of small particles is calculated. Volume fraction can also be applied to use the Einstein viscosity model. Later, many researchers have improved the viscosity model to find the nature of the viscosity based on its shape and size, most of research studies are focused on finding nanofluids and hybrid nanofluids thermal conductivity as well as viscosity. The viscosity of a fluid is the measure of its resistance to gradual deformation by shear stress or tensile stress [21].

$$\mu_{nf} = \mu(1 + 2.5\phi) \quad (4)$$

The Nusselt number of the nanofluids is given by the equation (5) were h is the convection heat transfer coefficient and T represents the temperature, m stands for the mass flow rate of the nanofluid.

$$Nu = h * d k = m * Cp * (Twall - Tinlet)' a * (Tbulk - Twall) \quad (5)$$

The effectiveness of the nanofluids is given by the equation (6).

$$\varepsilon = T1 - T2/ T1 - T3 \quad (6)$$

2.1. *Boundary conditions*

The Radiator model is developed in CREO, a 3D modelling software by using various tools like Sketch, extrude, pattern and assembly. Dimensions of the Radiator & Louvered strip [22] details are mentioned in Table 1. Once the radiator tube design is completed as shown in Figure 1, the louvered strip is inserted into the radiator. The thermophysical properties of considered base fluid and nanoparticles are listed in Table 2. Using the equations (1) to (4) Thermo physical properties of nanofluids at different concentrations are evaluated and values are listed below in Table 3.

Table 1. The lowered strip and radiator details.

Louvered Strip Details	Radiator
Angle of the strip = 30◦	Length of Radiator tube = 350mm
Thickness = 2mm	Thickness of Radiator = 4mm
Length = 10mm	Width of the Radiator tube = 22mm
Pitch = 30mm	
Number of Tubes = 01	

The main boundary conditions of the simulations applied for the radiator are below:

- Input flow rate = 3,4,5,6 and 7 LPM
- Maximum Inlet temperature = 348K
- Volume concentrations of nanofluids = 0.01, 0.05, 0.1 and 0.2%

Table 2. Thermo-physical properties of Base fluids and Nanoparticles.

	Density (kg/m^3)	Specific heat (J/kg-k)	Thermal conductivity (W/m-k)	Viscosity (kg/ms)
Water	1000	4182	0.6	0.001002
Ethylene Glycol	1100	2430	0.2577	0.0161
Graphene	2267	75	4000	Nil
CNC	1050	2450.324	0.31	0.00495

IOP Conf. Series: Materials Science and Engineering 1062 (2021) 012014 doi:10.1088/1757-899X/1062/1/012014

Table 3. Thermo-physical properties of Nanofluids at different concentrations.

Volume concentration	Nano particles	Density (kg/m^3)	Specific Heat (J/kg-k)	Thermal conductivity(W/m-k)	Viscosity (kg/ms)
0.01	G+H2O+EG	1052.27	1040.267	0.4757	0.02426
	G+CNC+EG	1105.585	2412.4886	1.03	0.06482
0.05	G+H2O+EG	1101.35	1041.39	0.5346	0.0272
	G+CNC+EG	1127.925	2344.17736	1.15787	0.07499
0.1	G+H2O+EG	1162.7	1042.9312	0.6156	0.03132
	G+CNC+EG	1155.85	2262.502	1.3333	0.089466
0.2	G+H2O+EG	1285.4	1046.5756	0.8079	0.04096
	G+CNC+EG	1211.7	2110.444	1.74985	0.125876

3. CFD Governing equations

Computational fluid dynamics (CFD) is the assessment of techniques including liquid flow, heat transfer and related phenomena such as natural reactions by means of computer-based simulation. This technique is influential and has a wide range of engineering and non-industrial application zones such as aerodynamics, turbomachinery, power plant, hydrodynamics, electrical and electronic engineering, chemical procedure engineering, the civil construction areas, marine and environmental engineering, hydrology and oceanology, weather prediction and biomedical engineering. The foundation of CFD is the essential governing equations of fluid dynamics (i)The Continuity, (ii) Momentum, and (iii) Energy equations. These are the mathematical statements of the fundamental physical principles upon which all of fluid dynamics is based on Mass, Force and Energy conservation [23, 24].

CFD approach usages the numerical calculation by solving mass, momentum and energy conservation governing equations (7) to (11) as shown below [25-30].

Continuity equation:

$$\frac{\partial \rho}{\partial t} + \nabla_{.} (\rho U) = 0 \tag{7}$$

Momentum equation:

$$\frac{\partial}{\partial_t}(\rho U) + \nabla_{.} (\rho UU) = -\nabla p + \nabla_{.} \tau + \rho g \tag{8}$$

Energy equation:

$$\frac{\partial}{\partial t}(\rho h) + \nabla_{.}(\rho U C_p T) = \nabla_{.}(k\nabla T) \tag{9}$$

where,

$$\nabla = i\frac{\partial}{\partial x} + j\frac{\partial}{\partial y} + k\frac{\partial}{\partial z} \tag{10}$$

IOP Conf. Series: Materials Science and Engineering 1062 (2021) 012014 doi:10.1088/1757-899X/1062/1/012014

The moment equation for the steady flow:

$$u\frac{\partial u}{\partial x}+v\frac{\partial u}{\partial y}=\left[\frac{1}{\rho}\right]\frac{\partial \tau_{xy}}{\partial y} \tag{11}$$

Here u and v are the x and y velocity components respectively, τ_{xy} is shear stress & ρ is the density of non-Newtonian fluid. The k-epsilon model is given by equation (12) and the dissipation, € is given by equation (13).

$$\frac{\partial\,(\rho k)}{\partial t}+\frac{\partial\,(\rho k u i)}{\partial x i}=\frac{\partial}{\partial x j}\left[(\mu+\frac{\mu t}{\sigma k})\frac{\partial k)}{\partial x j}\right]+Pk + Pb - \sigma€ - YM + Sk \tag{12}$$

$$\frac{\partial\,(\rho\epsilon)}{\partial t}+\frac{\partial\,(\rho\epsilon u i)}{\partial x i}=\frac{\partial}{\partial x j}\left[(\mu+\frac{\mu t}{\sigma€})\frac{\partial\epsilon)}{\partial x j}\right]+C1€\,\frac{\epsilon}{k}\,(Pk + C3€Pb)-\frac{C2€\rho\epsilon^2}{k+S\epsilon} \tag{13}$$

4. Modelling and Procedure

4.1. Modelling

The radiator tube was developed in 3D modelling Creo Parametric software. Extrude tool is used to develop the tube by giving the thickness, width, and length of the radiator along with the louvered fin. Pattern tool was used to fix the number of louvered fins. The direction of the fluid flow is considered as top to bottom during the simulation. The final model is completed in this 3D modelling software and converted into STEP Format and then imported to CFD. The detailed view of Radiator tube with louvered fin is shown in below Figure 2. The cross-sectional view of single radiator tube with louvered fin is shown in below Figure 3 and dimensions are presented in the Table 1.

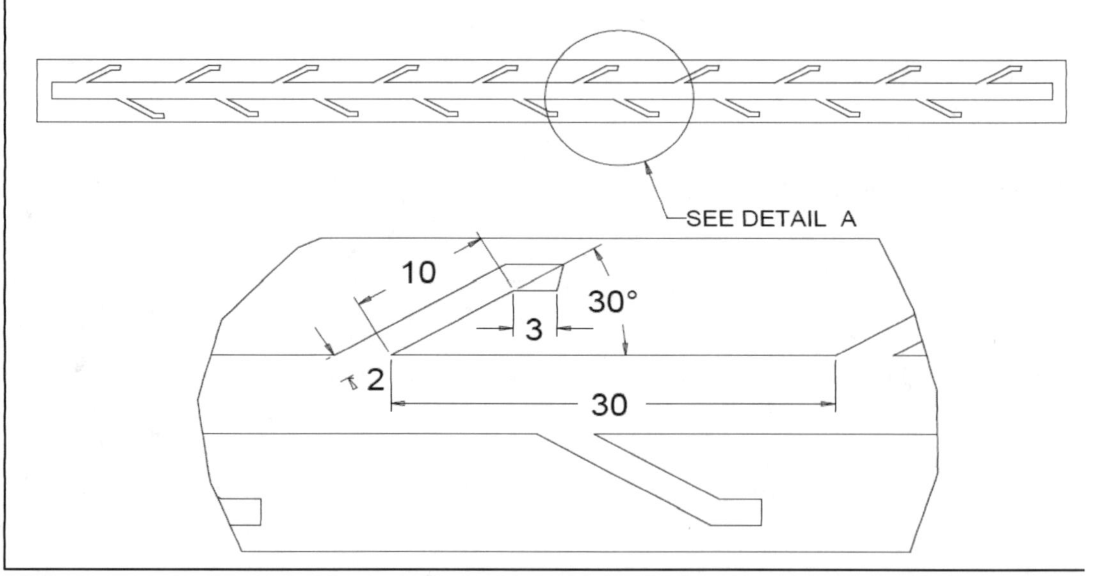

Figure 2. Detail view of radiator tube with louvered fin.

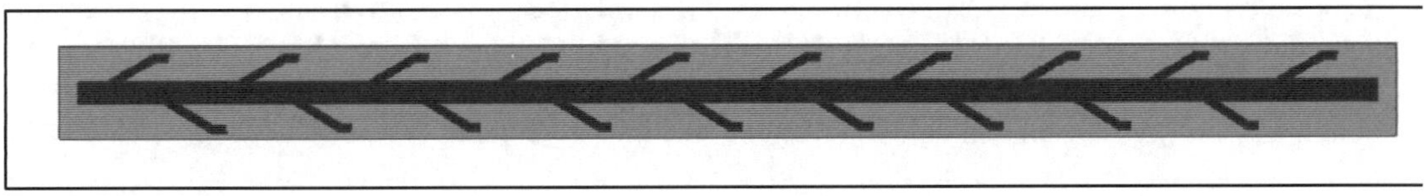

Figure 3. Cross-sectional view of the single radiator tube with louvered fin.

4.2. CFD Simulation

The radiator tube with the louvered fin is imported from the Creo parametric software with the STEP format. The model was undergone with the hexagonal fine mesh. The number of elements and nodes are divided automatically in the software. Total of 328,456 nodes and 282,365 elements were developed during the meshing process as shown in below Figure 4. Energy equation and the turbulence flow is activated. The properties of the hybrid nanofluids are added and applied according to requirement. The inlet flow range is considered as 3,4,5,6 and 7LPM with the temperature of 348K.

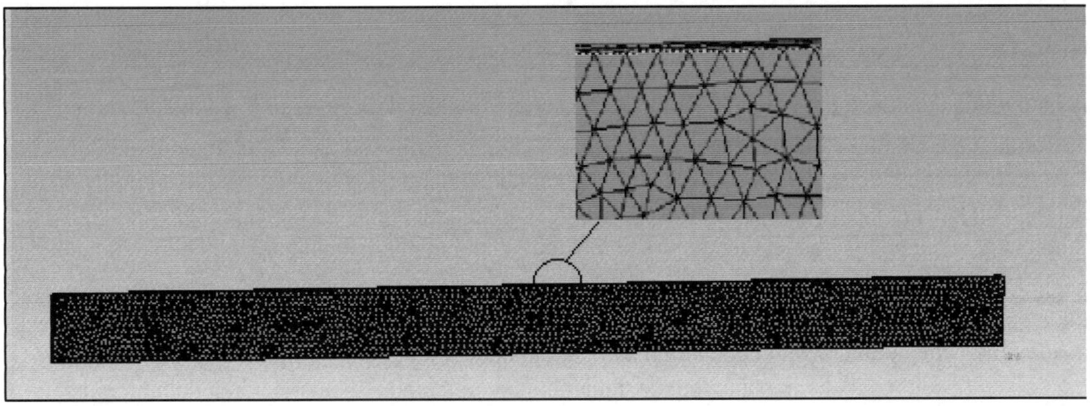

Figure 4. Meshed body.

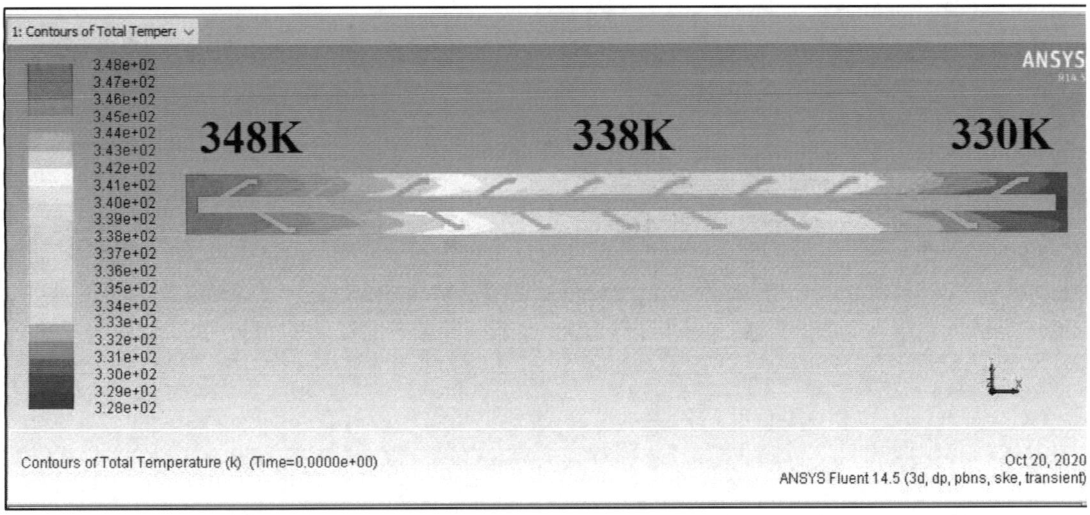

Figure 5. Total temperature of the radiator tube with Louvered Fin.

From the Figure 5 above, it can be observed that the temperature from the inlet point is very high indicated with a red color. The hot fluid is allowed to pass through the tube and blue color from the other end indicates that the hot fluid temperature is decreased. Similar observations of decreasing temperatures are identified for pressure and enthalpy and are shown in Figure 6 and Figure 7 below.

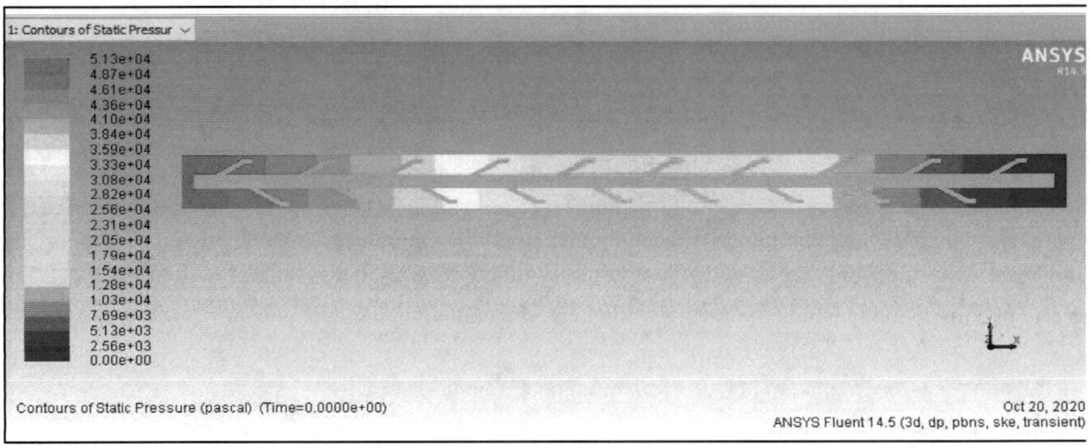

Figure 6. Pressure of the Radiator tube with louvered fin.

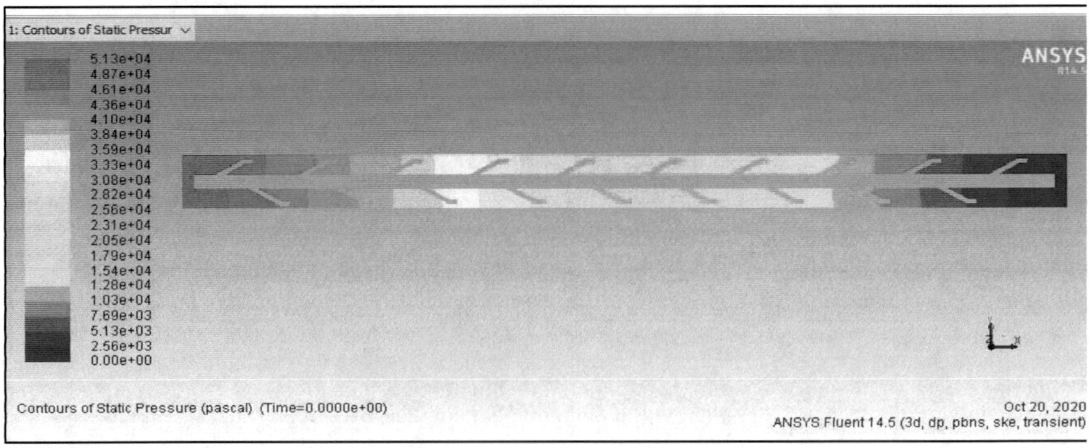

Figure 7. Enthalpy of the radiator tube with louvered fin.

5. Results and Discussion

Results are derived from CFD analysis with respect to different mass flow rates and fixed temperature. From the Figure 8., We observed that the pressure is gradually increasing from 3 LPM to 7 LPM with all volume concentrations (0.01%,0.05%,0.1% and 0.2%) of the nanofluids. The highest pressure was observed for the 0.2 volume concentration at 7 LPM. From the Figure 8(a), it was observed that 2.6 – 9.7% is enhancement from 0.01% to 0.2 % vol conc. Huge increment of pressure was absorbed in Figure 8(b). because of utilizing Graphene+ CNC hybrid nanofluid. Due to thermophysical properties of this hybrid nanofluid, the pressure increase was observed. It was observed with the minimum enhancement at 3 LPM flow rate with 0.01% volume concentration of the nanofluid and maximum at 0.2% volume concentration with 7LPM. The rate of enhancement was observed between 9.33 to 21.39% observed in Figure 8(b). It was noted that almost 55% to 65% enhancement was observed when Graphene+ CNC based hybrid nanofluid as compared with Graphene based nanofluids. This high enhancement happened due to its thermophysical properties of the hybrid nano fluid.

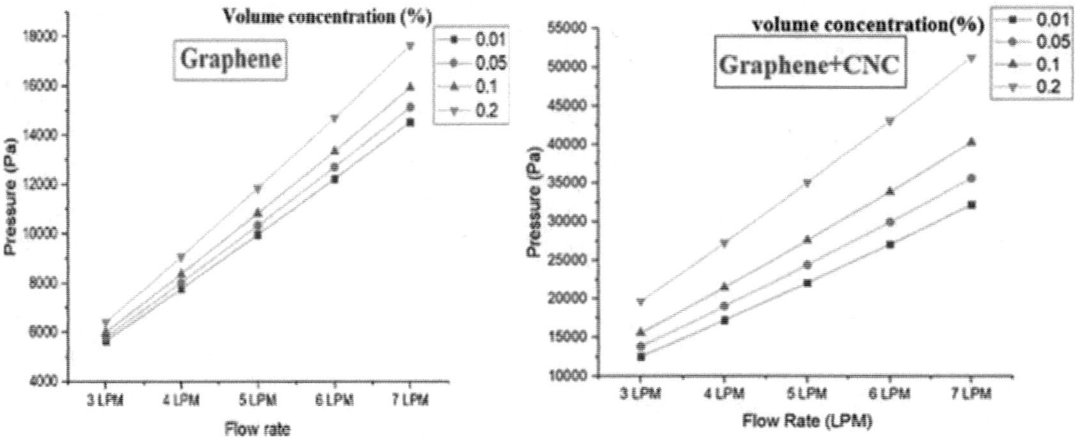

(a) (b)

Figure 8. Pressure (Pa) for different flow rates a) Graphene b) Hybrid Graphene nanofluid.

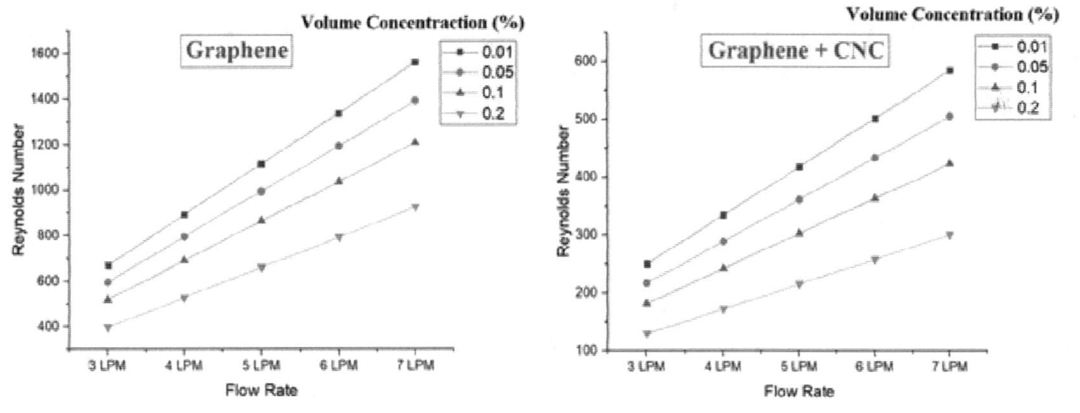

(a) (b)

Figure 9. Reynolds number for different flow rates a) Graphene b) Graphene+ CNC hybrid nanofluid.

The flow rate of the fluid is directly proportional to the Reynolds number and also it is directly proportional to the velocity of the fluid. If the density of the nanofluid (if the volume fraction increases) is increased, then Reynolds number is decreasing. In the Figure 9 it is observed that as the flow rate is increased from 3 LPM – 7 LPM, the Reynolds number value is also increased. Highest Reynolds number observed from both Figure 9 (a) & (b) is at 0.01 % at 7LPM. As the volumetric concentration of the hybrid nanofluid increases, correspondingly the Reynolds number was decreased. This is due to the increase in viscosity of the nanofluid. When compare to 0.01% and 0.2% volume concentration in Graphene nanofluid, nearly 68.83% enhancement observed in 0.01 volume concentration due to low viscosity of the nanofluid. In Figure 9 (b), 93% enhancement was observed for 0.01% volume concentration when compared with 0.2% volume concentration. 167 to 202% enhancement of Reynolds number is obtained for the Graphene nanofluids when compared with Graphene+ CNC based hybrid nanofluid.

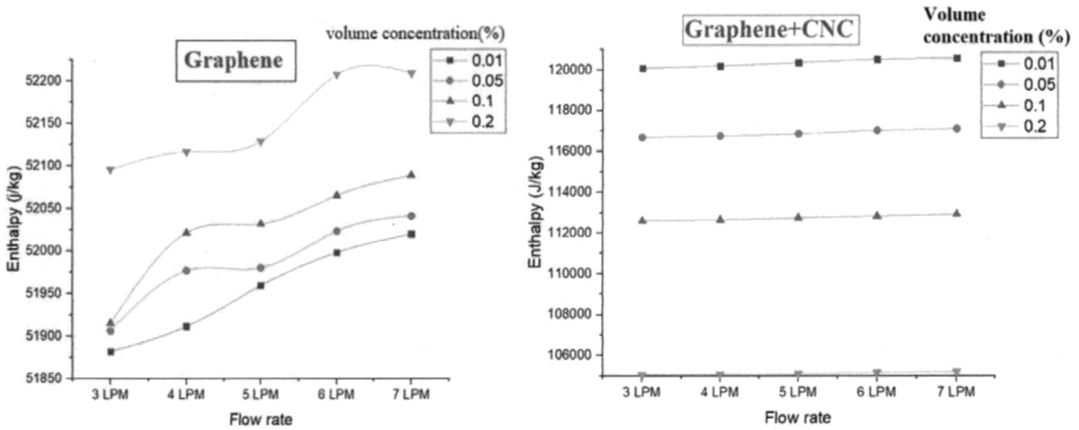

(a) (b)

Figure 10. Enthalpy (J/kg) for different flow rates a) Graphene b) Graphene+ CNC nanofluid.

In the Figure 10(a), The enthalpy is increased immensely as the flow rate increased from 3LPM to 7LPM at volume concentration of 0.01%,0.05%,0.1% and 0.2% because the specific heat of the graphene nanofluid is increasing. The maximum of 0.4% enhanced enthalpy observed when graphene hybrid nanofluid was used when compared with 0.2% with 0.01% of volume concentration of the nanofluid. In the Figure 10(b), the specific heat of the fluid is decreasing when the concentration is increased. So, the enthalpy is very low for the 0.2% of volume concentration when compare with 0.01%. In this condition, value of the enthalpy is enhanced in Graphene hybrid nanofluid and also it raised up to 56% when compared with Graphene nanofluid.

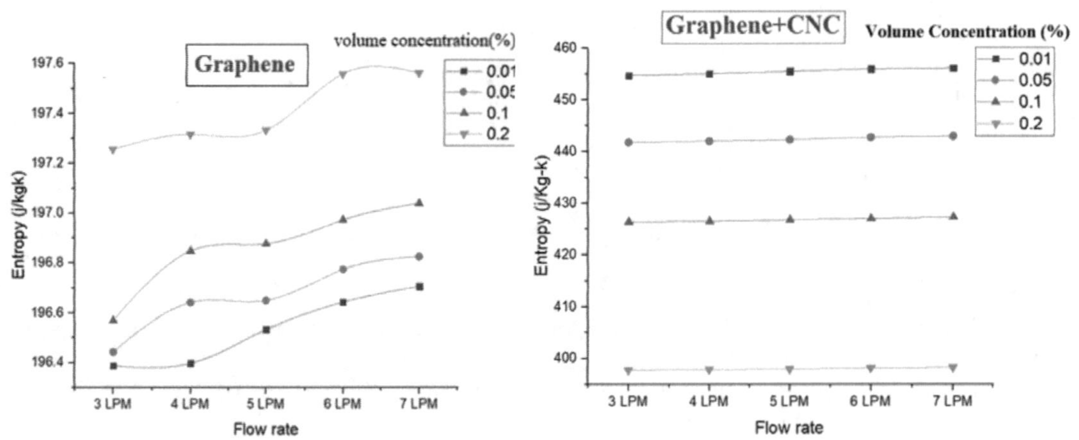

(a) (b)

Figure 11. Entropy (J/kg-K) for different flow rates a) Graphene b) Graphene+ CNC nanofluid.

Entropy in Figure 11 is the amount of a system's thermal energy per unit temperature. In this work, higher temperature reduction is observed in the Graphene+ CNC based hybrid nanofluid than Graphene based nanofluids because of its high thermal conductivity. In the Figure 11 (b), the entropy is very low for the 0.2% of volume concentration when compare with 0.01%. In this condition, value of the entropy is enhanced when compared with Graphene nanofluid Figure 11 (a) and it is raised up to 50-56% in entropy value for Graphene hybrid nanofluid.

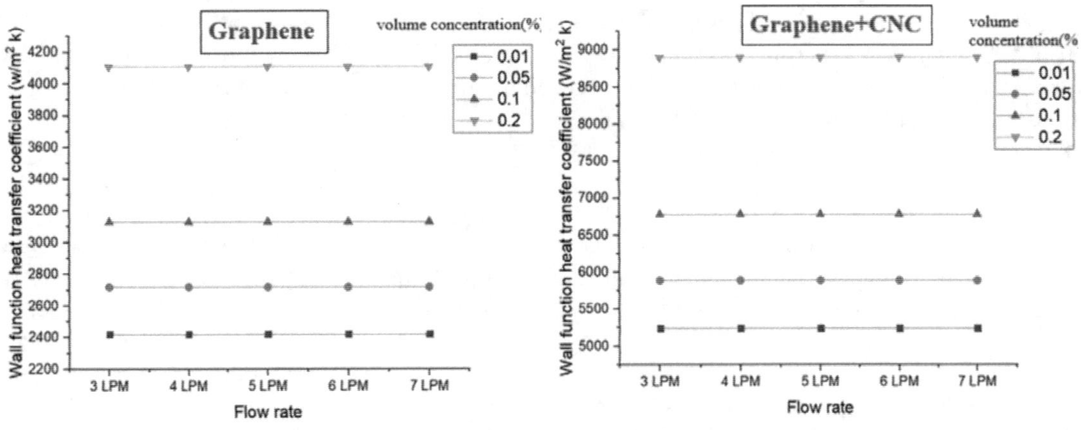

(a) (b)

Figure 12. Wall Function Heat Transfer Rate (W/m2K) for different flow rates a) Graphene b) Graphene+ CNC nanofluid.

The thermal conductivity of the Graphene+ CNC based hybrid nanofluid have high value when compared with graphene-based nanofluid. Based on this, we can say that Graphene+ CNC based hybrid nanofluid have high heat transfer capacity. Same observation was found in the results. From the Figure 12 (a), the heat transfer coefficient is gradually increased when the concentration of the nanofluids was increased. Heat transfer coefficient is enhanced from 11 to 41.11% when graphene-based nanofluid was used. In Figure 12 (b) Same observations are identified for the Graphene+ CNC based hybrid nanofluid utilization. Heat transfer coefficient is 53.83% higher for the Graphene+ CNC based hybrid nanofluid (having high thermal conductivity (Table 3) when compared with Graphene-based nanofluid.

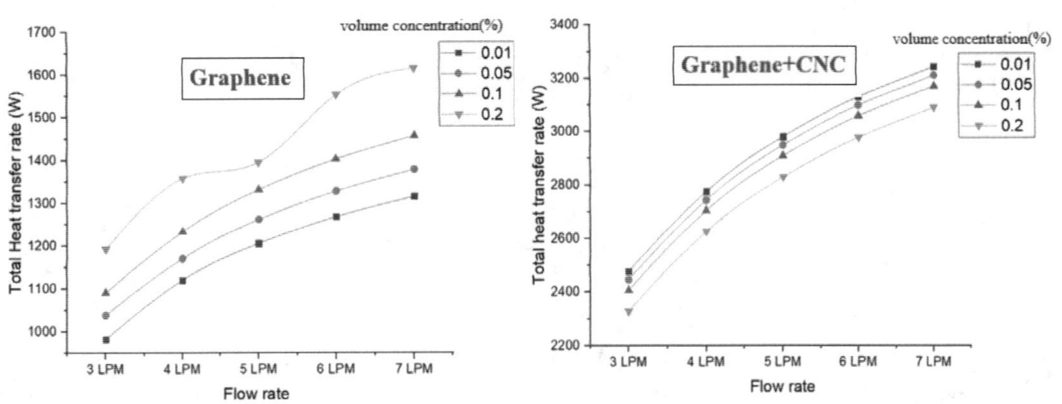

(a) (b)

Figure 13. Total Heat Transfer Rate (W) for different flow rates a) Graphene b) Graphene+ CNC nanofluid.

The heat transfer is always dependent on the thermal conductivity of the fluid. In this work, Graphene+ CNC based hybrid nanofluid have high thermal conductivity at 0.2 % volume concentration. As the flow rate is increased concurrently it is observed that the heat transfer rate is also increased and shown in the Figure 13. In Figure 13 (a), the rate of heat transfer is enhanced from 5.43% to 18.58%. 18.58% enhanced for 0.2% when compared with 0.01 of volume concentration at 7LPM and 17.65 % at 3LPM. In Figure 13 (b), 1 to 3.34% enhancement observed for different volume concentration when Graphene+ CNC nanofluid was utilized. The comparison between Graphene+ CNC based hybrid nanofluid and Graphene based hybrid nanofluid was stated that more than 60% enhancement of heat transfer rate is observed in Graphene+ CNC based hybrid nanofluid utilization.

6. Conclusions

The present study involves the simulation-based analysis of thermal enhancement of Graphene based Hybrid Nanofluids in a Car radiator. System analysis is performed considering single tube component in the radiator for pattern development. CFD analysis is conducted using ANSYS FLUENT module. Graphene/Crystal Nanocellulose hybrid (Graphene + CNC) nanoparticles are suspended in water & ethylene glycol base fluids (40:60) at 0.01%,0.05%,0.1% & 0.2 % volume concentration considering at 3-7 LPM inlet mass flow rate. The effect of various volume concentration, inlet temperature and flow rate on pressure, heat transfer coefficient and heat transfer rate has been examined by simulation.

- More than 60% enhancement of heat transfer rate is observed in Graphene+ CNC based hybrid nanofluid when compared with Graphene based nanofluid.
- 53.83% of heat transfer coefficient enhancement is observed for Graphene+ CNC based hybrid nanofluid when compared with Graphene based nanofluid.
- Entropy is enhanced from 50.01% to 56.0% for the Graphene+ CNC hybrid nanofluid as compared with Graphene based nanofluid.
- 56% Enthalpy enhancement is observed for Graphene+ CNC hybrid nanofluids as compared with Graphene based nanofluid.
- Reynolds number has an increased value of 167-202% for 0.01 %volume concentration for Graphene nanofluid at 7 LPM flow rate.
- Pressure enhanced from 55% to 65% for Graphene+ CNC based hybrid nanofluid as compared with Graphene based nanofluids. This high enhancement happened due to its thermophysical properties of the hybrid nano fluid.

The above results indicate that Graphene+ CNC based hybrid nanofluids will enhance the performance of an automobile radiator when compared with base fluid and Graphene nanofluid.

Acknowledgments

The authors would like to thank University Malaysia Pahang (UMP) for Grant RDU 190194, FRGS/1/2018/TK03/UMP/02/26 and UMP Flagship RDU192204 for financial assistance and facilities provided.

References

[1] Vajjha R S, Das D K and Namburu P K 2010 *Int. J. of Heat & Fluif Flow* **31(4)** 613-621
[2] Leong K, Rahman S, Kazi S N and Mamun A H 2010 *Appl. Therm. Eng.* **30(17-18)** 2685-2692
[3] Ferrouillat S, Bontemps A, Ribeiro J-P, Gruss J A and Soriano O 2011 *Int. J. of Heat & Fluif Flow* **32(2)** 424-439

[4] Sarkar J, Ghosh P and Adil A 2015 *Renew. Sust. Energ. Rev.***43** 164-177
[5] Gupta S S, Siva V, Krishnan S, Sreenivasan S T, Singh P K, Thalappil P and Das S K 2011 *J. Appl. Phys.***110(8)** 084302
[6] Ahammed N, Lazarus G A, Titus J, Bose J and Wongwises S 2016 *Int. Commun. Heat Mass Transf.* **70** 66-74
[7] Aravind S J and Ramaprabhu S 2013 *Rsc. Advances* **3(13)** 4199-4206
[8] Moraveji M K, Darabi M, Haddad S M H and Davarnejad R 2011 *Int. Commun. Heat Mass Transf.* **38(9)** 1291-1295
[9] Akbarinia A and Behzadmehr A 2007 *Appl. Therm. Eng.***27(8-9)** 1327-1337
[10] Sharma P, Kumar V, Sokhal G S, Dasaroju G and Bulasara V K 2020 *Mater. Today* **21** 1800-1808
[11] Kumar D, Sokhal G and Sharma P 2019 *Rev. des Compos. et des Mater. Av.* **29(1)** 63-69
[12] Bhardwaj B R, Karjaal S M, Ponangi B R, Prasad R, Krishna V, Seetharam T R, Shashidhara S, Manjunath D M 2018 *Int. Conf. on Nanotechnology: Ideas, Innov. Innitiatives*
[13] Bharadwaj B R, Mogeraya K S, Manjunath D M, Pinangi B R, Prasad K S R and Krishna V 2018 I*OP Conf. Ser. Mater. Sc. Eng.* **346**
[14] Pendyala R, Ilyas S U, Lim L R and Marneni N 2016 *Procedia eng.* **148** 1162-1169
[15] Bahiraei M and Mazaheri N 2018 *Int. J. Mech. Sci.* **138** 337-349
[16] Zainal S, Tan C, Sian C J and Siang T J 2016 *J. Adv. Res. Appl. Mech.* **23(1)** 20-35
[17] Peyghambarzadeth S M, Hashemabadi H, Hoseini S T and Jamnani M S 2011 *Int. Commun. Heat Mass Transf* **38(9)** 1283-1290
[18] Ponangi B R, Sumanth S, Krishna V, Seetharam T R and Seetharamu K N 2018 I*OP Conf. Ser. Mater. Sc. Eng*
[19] Maxwell J C 1881 *Clarendon Press* **1**
[20] Einstein A 1906 *Ann. Phys.* **19(4)** 371-381
[21] Sreedhar T, Rao B N and Kumar D V 2020 *Emerging Trends in Mechanical Eng.,* 387-397
[22] Mohammed H, Hasan H A and Wahid M 2013 *Int. Commun. Heat Mass Transf* **40** 36-46
[23] Blazek J 2018 *Butterworth-Heinemann*
[24] Wendt J F 2008 *Springer Sc. Business Media*
[25[Anderson J D 1995 *McGraw-Hill*
[26] Sodja J 2007 *Univ. Ljubljana* 1-18
[27] Medvitz R B, Kunz R F, Boger D A, Lindau J W, Yocum A M and Pauley L L 2002 *J. Fluids Eng.* **124(2)** 377-383
[28] Wilcox D C 1998 *DCW Industries La Canada* **2**
[29] Al-Abidi A A, Sohif M, Sopian K, Sulaiman M Y and Abdulrahman M T 2013 *Renew. Sust. Energ. Rev.* **20** 353-363
[30] Arabkoohsar A, Khosravi M and Alsagri A S 2019 *Int. J. Heat Mass Trans.* **141** 432-443

Experimental investigation on noise and vibrations of biodiesel-butanol blends for diesel engine

Erdiwansyah[1*], M.S.M. Sani[2,6], Mahidin[3], M. Zaki[3], Asep Kadarohman[4], Muhibbuddin[5], R.E Sardjono[4], Asri Gani[3]

[1] Faculty of Engineering, Universitas Serambi Mekkah, Banda Aceh 23245, Indonesia
[2] Faculty of Mechanical and Automotive Engineering Technology, Universiti Malaysia Pahang, 26600 Pekan, Pahang, Malaysia
[3] Department of Chemical Engineering, Universitas Syiah Kuala, Banda Aceh 23111, Indonesia
[4] Department of Chemistry, Indonesia University of Education, Bandung 40154, Indonesia
[5] Department of Mechanical and Industrial Engineering, Universitas Syiah Kuala, Banda Aceh 23111, Indonesia
[6] Automotive Engineering Centre, Universiti Malaysia Pahang, 26600 Pekan, Pahang, Malaysia

*Corresponding e-mail: erdi.wansyah@yahoo.co.id

Abstract. Diesel engines have extensively used several developing countries to produce electricity. This is done in various sizes, small and medium, which are driven by millions of diesel engine generators. These engines or generators produce vibrations and noisy and pollute the air. Thus, an alternative to diesel is needed for such applications. Turpentine-diesel fuel mixture is one such alternative. However, there have not many efforts made so far to evaluate noise and vibration characteristics. This study is done to try filling the gap and comprehensively investigate the use of Turpentine-diesel mixed fuels. This experiment uses a single-cylinder-4-stroke diesel engine operated at different rpm speeds and engine loads. Detailed measurements are made for engine noise and vibration. This measurement is used to connect the reciprocity of engine noise and vibration. The relationship of chemical parameters and physical properties to the tested fuel explored. The test results using turpentine-diesel fuel show the vibration and noise characteristics better for the upper-middle load. The correlation between noise and vibration characteristics of the test fuel has a strong relationship. This study has provided a detailed phenomenological explanation of the nature of the fuel. Finally, turpentine blends are a viable alternative to diesel fuel, especially in diesel engines

Keywords: diesel engine, noise, vibration, fuel, generator

1. Introduction

The automotive industry currently uses many compression engines, especially large vehicles and generators. The energy of the fuel has been converted into mechanical and electrical energy in the presence of this machine. In general, generator engines are not like automotive diesel engines; because the nature of the design is simpler, it does not move and runs at a constant speed. In general, this engine is noisy and produces vibrations, which are transmitted to the entire engine structure. In addition, a large

IOP Conf. Series: Materials Science and Engineering 1062 (2021) 012015 doi:10.1088/1757-899X/1062/1/012015

amount of greenhouse gases is donated from this machine. So that alternative fuel is needed for this engine and does not require a significant change in engine design.

Identification of alternative fuels for diesel has been carried out globally in recent years. These alternative fuels can be grouped into two groups including alcohol-diesel and biodiesel-diesel-oil. Research on noise and vibration in diesel engines using alcohol has not been done in recent years. However, the use of biodiesel fuel mixture has been expressed in the characteristics of the noise and vibration of diesel engines. The characteristics of noise and vibration carried out on diesel engines using a mixture of biodiesel fuel have been investigated by [1– 3]. Their analysis and predictions reported that the properties of fuel and engine vibration are related. Engine vibration is reduced for a mixture of biodiesel-diesel fuel because the fuel has oxygen. The investigation of noise and vibration using a mixture of castor-diesel oil for diesel engines has also been discussed by [4– 6]. The results of the tests showed that the engine combustion noise decreased with the amount of noise around 1–3 dBA for a mixture of diesel-fuel oil compared to pure diesel. Among all the fuels tested for the horizontal, lateral and vertical directions, pure diesel shows the highest vibration.

A different study was carried out by [7– 9], where a mixture of soybean biodiesel and canola fuel was carried out to investigate vibrations in diesel engines. The experimental results reported that pure diesel has a lower vibration compared to biodiesel fuel. The correlation between vibration and engine maximum heat velocity for diesel engines has been investigated by [10], [11]. The frequency of engine vibration was reported at 0.3–5 and 1.5–2.5 kHz because it is caused by a piston slap and the engine combustion process. In a different study, investigations were conducted by [12– 14] on vibration for diesel engines using biodiesel fuel with a common rail system that has high pressure. The results of trials conducted by engine vibration experienced a reduction of 13.7% for a mixture of diesel-biodiesel fuel. Engine vibration is reduced because of the low cylinder pressure when using diesel-biodiesel. The use of Karanja biodiesel fuel running on diesel engines was investigated by [6], [15], [16]. The 20% biodiesel fuel mixture can produce maximum heat release and combustion noise among all fuel samples tested. The 20% biodiesel mixture was reported to have the highest vibration in the vertical direction.

The effect of biodiesel fuel for diesel engine noise was also investigated by [17– 20]. The results of the experiments they conducted observed that the physical and chemical properties of the materials used played a very important role in engine noise. They reported that when using biodiesel with low cylinder pressure reduced engine noise compared to pure diesel fuel. Meanwhile, the noise characteristics for direct injection diesel engines triggered by biofuels and synthesis were investigated [1], [20], [21], [22]. The results of their investigation showed that combustion noise decreases with an increasing percentage of biodiesel fuel and the content of the oil synthesized in fuel. Based on the two cases, noise can decrease because of differences in the combustion phases and the speed of fuel injection. Noise of diesel engines when using fuel methyl ester pomace olive oil with diesel has been studied by [23– 25]. The results showed that the biodiesel fuel they used can reduce engine noise by increasing the amount of cetane.

It appears that there is a dearth of work that explores noise and vibration for CI machines. Furthermore, there was little research and knowledge about noise and vibration for CI machines triggered by a mixture of turpentine-diesel blends. Turpentine fuel potential is an alternative to diesel. This study seeks to test and assess the behaviour of CI machines in a comprehensive manner. The work discussed the noise and vibration aspects of the diesel engine for the Turpentine-diesel fuel mixture for different engine speeds.

2. Experimental Setup

This experimental study uses a YANMAR TF120M, 4-stroke, single-cylinder, direct-injection engine. This diesel engine has an output of 7.83 kW and 2400 rpm. The engine specifications for this test are shown in Table 1. The schematic diagram of this experiment can be shown in Figure 1.

Table 1. Detail of engine specifications

Description	Specification
Engine model	YANMAR TF120M
Engine type	Horizontal single-cylinder 4-stroke diesel engine
Fuel injection type	Direct injection
Bore × Stroke (mm)	92 × 96
Displacement (L)	0.638
Injection timing	17° BTDC
Fuel injection pump	Bosch
Injection pressure (kg/cm^2)	200
Compression ratio	17:7
Continuous output (HP)	7.82 kW at 2400 rpm
Rated output (HP)	8.94 kW at 2400 rpm
Cooling system	Water-cooled (radiator type)
Cooling water capacity (L)	2.3

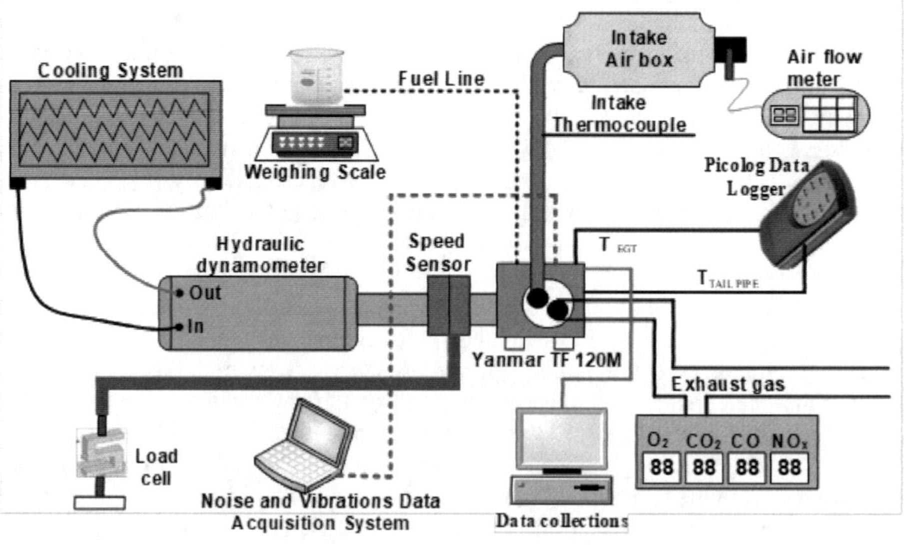

Figure 1. Schematic diagram of the experiment

Noise measurement is taken adjacent to the engine approximately 50 cm to get noise data from each sample of fuel used. Engine vibration is measured in the directions of X, Y, and Z using PCB Piezotronics ICP® Accelerometer (352C22). The specifications of the sensors and the devices used are listed in Table 2.

Table 2. Accuracy of various devices and sensors

Description	Specification
Sensitivity (±15 %)	1.0 mV/(m/s^2)
Measurement Range	±4900 m/s^2 pk
Frequency Range (±5 %)	1.0 to 10000 Hz
Frequency Range (±10 %)	0.7 to 13000 Hz
Frequency Range (±3 dB)	0.3 to 20000 Hz

Resonant Frequency	≥50 kHz
Broadband Resolution (1 to 10000 Hz)	0.04 m/s² rms
Non-Linearity	≤1 %
Transverse Sensitivity	≤5 %

3. Results & Discussion

This experimental study was carried out at five engine speeds at a load of 25% and 50% using turpentine-diesel additive fuel that is, pure diesel with added 0.2% turpentine oil to diesel engines. The vibration measurements result from the use of turpentine additives compared to pure diesel fuel. Experiments were carried out at engine loads of 25% and 50% with engine speeds of 1200 to 2000 rpm at intervals of 200 rpm. Vibration signal data is taken from each load and different engine speeds to see engine vibration levels when using different fuels.

The results of the tests carried out using turpentine oil additives with pure diesel are shown in Fig. 2 and 3 where the use of turpentine oil additives results in lower engine vibrations for all speeds tested at 25% load. The engine vibration also depends on the speed of the engine; the higher the speed of the engine vibration, the more the engine decreases. The use of pure diesel produces the highest vibration at 1200 rpm around 1.1 magnitudes compared to turpentine additive oil which only reaches 0.6 magnitudes. In general, the overall engine speed tested for a 25% additive turpentine load results in less vibration than pure diesel. The reason is that turpentine oil has alcohol in it, so it can produce softness when burning the engine.

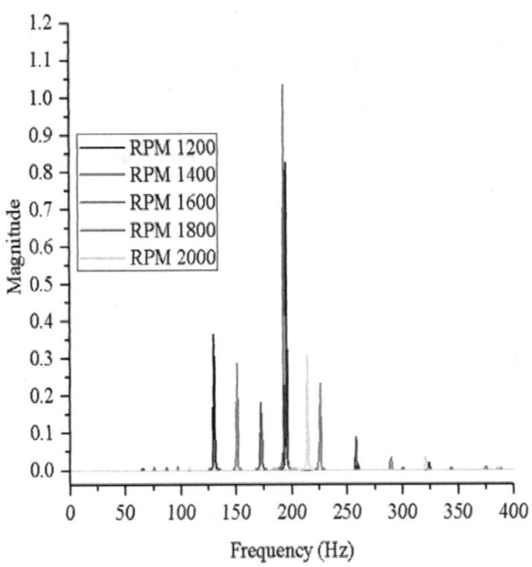

Figure 2. Vibration at load 25% and different speed using diesel fuel

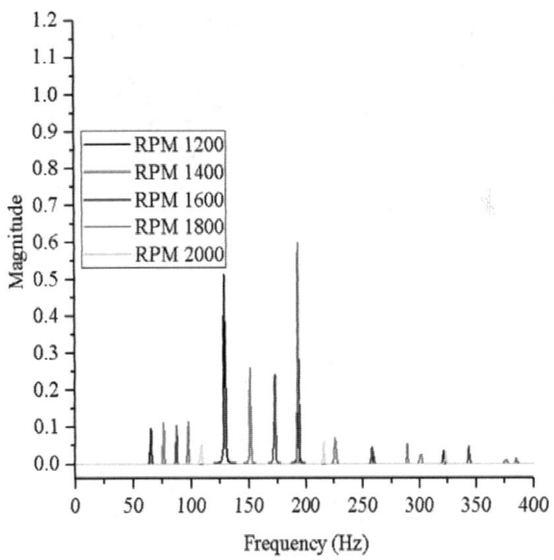

Figure 3. Vibration at load 25% and different speed with additive of turpentine

The next experiment was carried out at 50% engine load with the same speed between 1200 to 2000 rpm using diesel fuel and then compared to turpentine additive oil. From Fig. 5. it can be explained that the use of turpentine additive oil produces less vibration than pure diesel fuel shown in Fig.4. Diesel fuel produces the highest vibrations up to 1.1 more magnitudes for 1200 rpm engine speed whereas when using additive oil for 1200 rpm, it only reaches a magnitude of 0.7. However, there is a slight difference in the engine speed of 1600 rpm diesel fuel as they produced less vibration than the turpentine additive oil. Whereas at 1400 rpm almost has an equation.

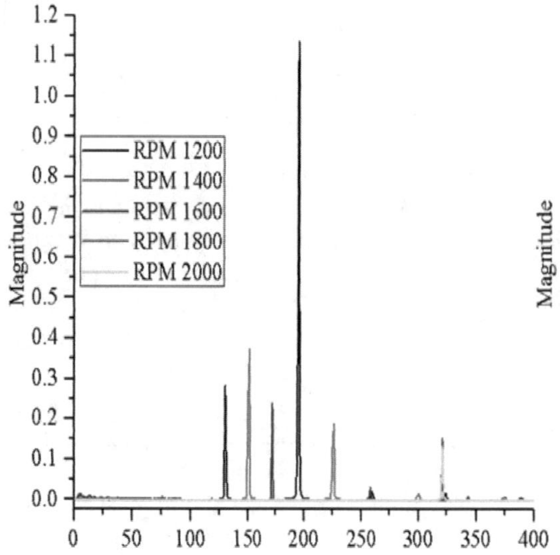

Figure 4. Vibration at load 50% and different speed using diesel fuel

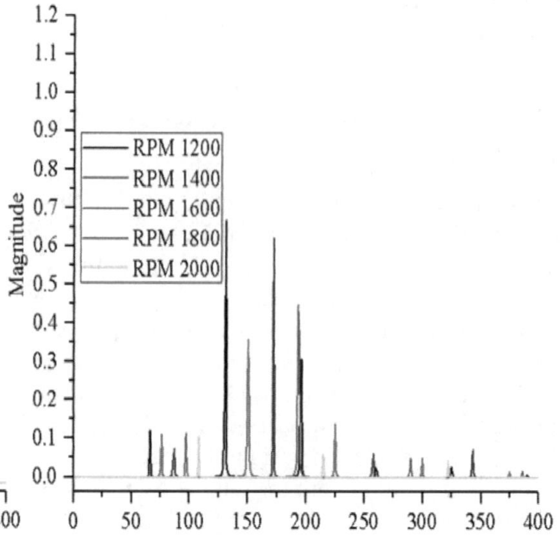

Figure 5. Vibration at load 50% and different speed with additive of turpentine

The results of the experiments that have been done to measure the vibration of the engine using pure diesel fuel with additives show little difference for engine load; where the engine load begins to increase, engine vibration also increases. However, for all speeds tested, turpentine additive oil can generally produce less vibration than when using pure diesel fuel. So it can be concluded that turpentine additive oil is good for use in diesel mixtures. However, turpentine oil is more expensive, therefore, it has not been applied extensively on combustion engines.

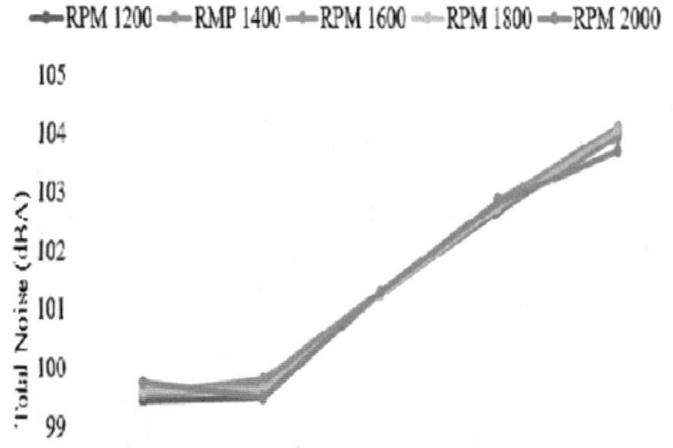

Figure 6. Total noise for different speed at load 25% using diesel fuel

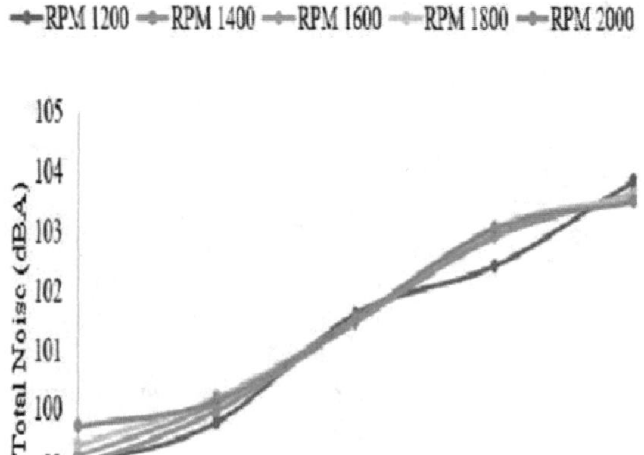

Figure 7. Total noise for different speed at load 25% with additive of turpentine

Figure 6 and 7 show the noise from engine combustion using pure diesel fuel with turpentine additive oil for 25% engine load with different rpm speeds. Noise from combustion using a slightly finer turpentine oil additive are as shown in Fig. 7. The use of diesel fuel for all speeds are higher than that of additive oil shown in Fig. 6. Moreover, the additive oil is slightly more variable compared to the 2000 rpm engine speed, where at 20 seconds it drops slightly while diesel does not have the same trend as additive oil.

Furthermore, combustion engine noise measurements using pure diesel fuel by comparing turpentine additive oil were carried out at 50% engine load with different engine speeds. It is reported that diesel fuel tested as a reference produced higher engine noise than additive oil. The highest noise from diesel fuel is 107 dBA, which is slightly higher than additive oil that reaches 105 dBA. Meanwhile, turpentine additive oil produces the lowest noise at 99 dBA compared to diesel which reaches 100 dBA as shown in Fig. 8 and 9. However, diesel fuel produces more noise trends than oil additives because the use of diesel fuel gets hotter with higher speed and engine load can produce better combustion.

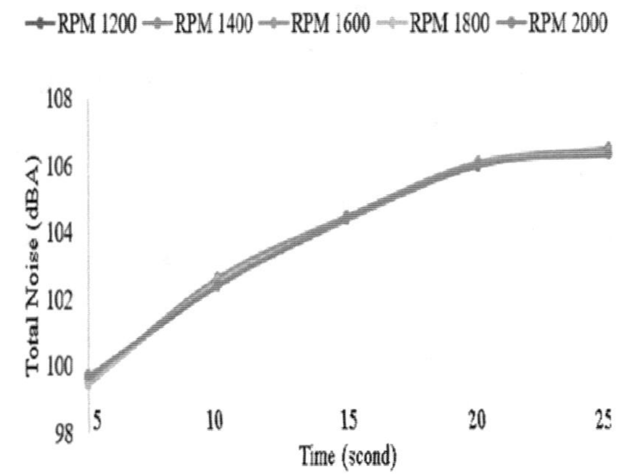

Figure 8. Total noise for different speed at load 50% using diesel fuel

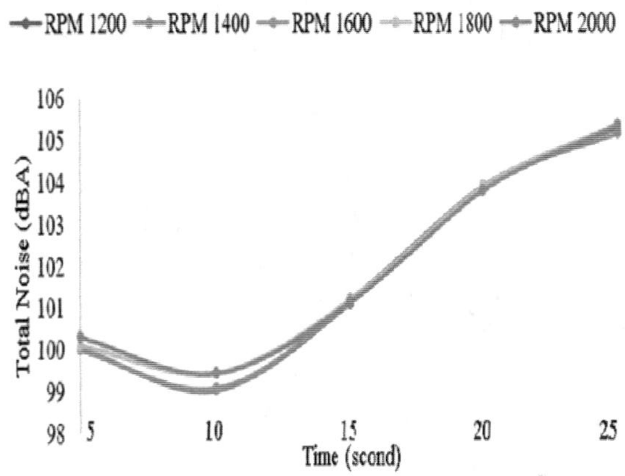

Figure 9. Total noise for different speed at load 50% with additive of turpentine

4. Conclusion

In this work, noise and vibrations of turpentine blends were compared to those for diesel fuel in the engine diesel. Turpentine fuel can produce less noise and vibration, especially when the engine load is higher than when using pure diesel. Moreover, the Turpentine-diesel fuel mixture can reduce considerable exhaust emissions in the engine. For this reason, due to density, the calorific value and viscosity in turpentine fuel are lower than pure diesel. Thus, for the whole sample, turpentine fuel is very suitable as an alternative to diesel fuel, especially for diesel engines. However, this change must be applied with a mixture of fuel and lubricants because additives are very necessary to compensate for the lubrication of the fuel lost due to the presence of turpentine fuel.

Acknowledgements

This research supported by PNBP Universitas Syiah Kuala, Research institutions and community service with the contract number of (32/UN11.2.1/PT.01.03/PNBP/2020).

References

[1] Uludamar E, Tosun E and Aydın K 2016 *Fuel* 177:326–33
[2] Çelebi K, Uludamar E, Tosun E, Yıldızhan Ş, Aydın K and Özcanlı M 2017 *Fuel* 197:159–73
[3] Erdiwansyah, Sani M S M, Mamat R, Zikri J M, Razak N F and Munawir 2019 *J Phys Conf Ser* 1262:012014
[4] Patel C, Tiwari N and Agarwal A K 2015 *SAE Technical Paper*

[5] Satsangi D P, Tiwari N 2018 *Fuel* **221**:44–60.
[6] Erdiwansyah, Sani M S M, Mamat R, Khoerunnisa F, Rajkumar A R, Razak N F D and Sardjono R E 2018 *MATEC Web Conf* 225.
[7] Taghizadeh-Alisaraei A and Rezaei-Asl A 2016 *Fuel* **185**:718–33.
[8] Uyumaz A 2018 *Fuel* **212**:256–67
[9] Javed S, Murthy Y V V S, Ulla R and Rao T N 2016 *J Nat Gas Sci Eng* **33**:217–30
[10] Lee S, Lee Y, Lee S, Song H H, Min K and Choi H 2015 *SAE Technical Paper*
[11] Patel C, Tiwari N and Agarwal A K 2019 *Fuel* **238**:86–97
[12] How H G, Masjuki H H, Kalam M A and Teoh Y H 2018 *Fuel* **213**:106–14
[13] How H G, Masjuki H H, Kalam M A and Teoh Y H 2014 *Energy* **69**:749–59.
[14] Erdiwansyah, Leman A M, Norkhizan A M, Mulyadi L, Khoerunnisa F, Sardjono R E, Maulana M I and Sani M S M 2019 "*IOP Conf. Ser. Mater. Sci. Eng.*, vol. **506**, no. 1, 2019.
[15] Patel C, Agarwal A K, Tiwari N, Lee S, Lee C S and Park S 2016 *Appl Therm Eng* **106**:506–17
[16] Patel C, Lee S, Tiwari N, Kumar A, Sik C and Park S 2016 *Fuel* **185**:410–20
[17] Rizwanul Fattah I M, Masjuki H H, Liaquat A M, Ramli R, Kalam M A and Riazuddin V N.2013 *Renew Sustain Energy Rev* **18**:552–67
[18] Akbarian E and Najafi B 2019 *Fuel* **236**:666–76
[19] Keskin A, Yaşar A, Yıldızhan Ş, Uludamar E, Emen F M and Külcü N 2018 *Fuel* **216**:349–55
[20] Alenezi R A, Erdiwansyah, Mamat R, Norkhizan A M and Najafi G 2020 *Fuel*, vol. **279**, p. 118438
[21] Torregrosa A J, Broatch A, Plá B and Mónico L F 2013 *Biomass and Bioenergy* **52**:22–33.
[22] Erdiwansyah, Leman A M, Khoerunnisa F, Sardjono R E, Munawir, Maulana M I and Sani M S M 2019 *IOP Conf. Ser. Mater. Sci. Eng.*, vol. **506**, no. 1
[23] Mathiarasi R, Mugesh kanna C and Partha N 2017 *J Saudi Chem Soc* **21**:11–7
[24] Redel-Macías M D, Hervás-Martínez C, Pinzi S, Gutiérrez P A, Cubero-Atienza A J and Dorado M P 2012 *Fuel* **98**:280–7
[25] Redel-Macías M D, Pinzi S, Leiva D, Cubero-Atienza A J and Dorado M P 2012 *Fuel* **95**:615–21.

IOP Conf. Series: Materials Science and Engineering 1062 (2021) 012016 doi:10.1088/1757-899X/1062/1/012016

Engine Emissions Analysis of Emulsified Fuel of Different Blend Ratios

Hazman Abu Hassan[1*], Ftwi Y Hagos[1, 2], Asep Kadarohman[3], R E Sardjono[3], Fitri Khoerunnisa[3], Zhang Bo[4] and X X Jiang[4]

[1]College of Engineering, Universiti Malaysia Pahang, Pahang, Malaysia

[2]Automotive Engineering Centre, Universiti Malaysia Pahang, Pahang, Malaysia

[3]Department of Chemistry, Indonesia University of Education, Bandung, Indonesia

[4]School of Mechanical Engineering, Ningxia University, China

*Corresponding e-mail: hazmanptsn@gmail.com

Abstract. Diesel engines are widely used because of their fuel efficiency in producing a higher mechanical energy ratio compared to gasoline engines. The current study is an extension of these efforts where diesel and biodiesel blend are modified to emulsify and oxygenate through water and alcohol to reduce NO_x emissions from the exhaust. Biodiesel on the other hand, has the advantage of being sulfur-free, which reduces the emission of smoke and particulate matters but unfortunately increases NO_x emissions. Meanwhile, emulsion fuel is found to lower the temperature of the cylinder during combustion and thus, reduce the release of NO_x gas. The oxygenated fuels are tested for their emission behaviors in a four-stroke, single cylinder, water cooled, and direct fuel injection engine evaluated at a speed of 1200 rpm until 2400 rpm with the engine load varied at 0%, 50% and 100%. The method of fuel oxygenation and their effect on the concentration of NO_x are discussed. The results have demonstrated a significant reduction in the NO_x emission with an increase in the fraction ratio of palm oil biodiesel. The competition to produce high-performance and environmentally friendly fuels has made emulsion fuel and biodiesel renewable fuels believed to be new alternative fuels that emit very little NO_x emissions into the atmosphere.

Keywords: diesel engine, emissions, biodiesel, emulsion fuel,

1. Introduction

Over the years, the diesel engine has undergone many evolutions leading to a more superior engine performance. Despite the simplicity of the diesel engine design, the problem of NO_x can cause health problems especially to large−scale residents of industries and vehicles using diesel engines. Diesel engine is a notably important contributor to the advancement of the global economy and wellbeing of society while not neglecting its impact on pollutants consequent from the combustion of diesel fuel [1]. The emissions by the fossil-based fuels have shown to damage all the treasures of nature. Naturally, the diesel fuel emits PM (particulate matter), THC (total gaseous hydrocarbon), SO_x (sulphur oxides), CO_2 (carbon dioxides), NO_x (nitrogen dioxide) and CO (carbon monoxide). Moreover, the diesel also emits smoke that harms the air by turning into acid rain and global warming [2]. The cause for the major emissions mainly NO_x, PM and THC are the deficiency of oxygen in the combustion process and the presence of aromatics in diesel fuel [3]. Besides, the combustion process in diesel engine which is best represented as heterogenous where the fuel reaches the oxidizer by diffusion. This causes the flame to travel slowly unlike to the premixed combustion.

Numerous studies have focused on solving issues related to pollution through producing various alternative fuels, engine design modifiers and treatment after engine combustion by applying technological advancements [4, 5]. In recent years, there has been increased attention on the oxygenated fuels because oxygen helps in deep burning and reduces NO_x emissions [6]. The method based on fuel study containing oxygen has become popular because it does not require a change in the diesel engine hardware. Challenges and pollution threats caused by diesel fuels are anticipated to be handled by producing alternative fuels that contain oxygen and that can facilitate the diffusion process [7-9]. As such, oxygenated fuel is a fuel that contains oxygen in its chemical structure [10, 11], where it can reduce carbon monoxide and soot due to complete combustion [12, 13]. Mostly, oxygenated fuels have positive effects on properties such as viscosity, density and cetane number by stabilizing temperature and reducing NO_x product [14, 15].

Fuel modification through blending has been subject to research since the time researchers found the limitation with biofuels to serve as standalone energy source in diesel engine. Blending of two or more fuels is the process of mixing diesel fuel with other naturally miscible liquid fuels to produce new fuel with a different fuel property to its parent fuels and that can provide better engine performance, reduce pollution and cost-efficient effect [16-19]. The method based on fuel study containing oxygen has become popular because it does not require a change in the diesel engine hardware. The demand for biodiesel and biofuels has been growing rapidly since the year 2000. There have been many researches on the blend of diesel with biodiesel [16-18], diesel with alcohols [8, 20-22] and three blends of diesel, biodiesel and alcohols [23, 24]. Biodiesel has many advantages in which the plant does not contain sulphur and the presence of high oxygen content [25]. The oxygen content in biodiesel promotes combustion, and this will further enhance the efficiency of the engine. Biodiesel also has higher cetane number than diesel, which causes the biodiesel to be more combustible and reduce the ignition delay time. Biodiesel is also readily available in Asian countries, especially in Malaysia and Indonesia. Similarly, alcohols are known with their higher oxygen content originally bio-based fuels. Besides, alcohols are known with their higher latent heat of vaporization. However, attaining a stable blend is the challenge.

Several researchers have demonstrated that the experiments on the blended fuel for investigating the characteristics of fuel extraction obtained from ethanol-aviation kerosene revealed the positive outcome. Mainly spray tip penetration, spray cone angle and spray area injection increased the extra performance [22]. Another study has shown that the blended fuel had better performance compared to a triple mixture of diesel/butanol/biodiesel blend, where there was an improvement in a flame tube boiler [26]. On the other hand, Ali Keskin et al. have indicated that mixing of diesel fuel-biodiesel blends with palladium and acetylferrocene has resulted in the reduction of pollution such as particulate matter [27]. Physical properties of fuel are crucial in influencing the combustion, performance and emissions of the engine. It is essential to look into the physical properties of different fuel blends and their impacts on the behavior of engine combustion. Despite the improvement on the performance and emissions of CO and PM, biodiesel have negative impact on the NO_x emissions due to an increase in the combustion temperature in the combustion chamber. Similarly, alcohols have specific only improving NO_x emission by reducing the combustion temperature. This is at the expense of the performance of the engine.

There have been recent efforts to extend the intervention of fuel blends in tackling all the emissions while not affecting the performance of the engine. This was further intensified through the recent discovery of micro−explosion phenomenon in the combustion chamber through the emulsification of different boiling temperature liquid fuels [28, 29]. An addition of alcohols or water into diesel and biodiesel blends has shown through an emulsification technique has shown improvements on emissions of NO_x while improving the performance of the engine [30-32]. Despite the availability of research findings on the double impact of the emulsified fuels, there is limited research on the effect of the emulsified fuel mainly on the NO_x reduction in quantity and its effect on the environmental pollution. Therefore, the primary aim of this study is to investigate the NO_x of diesel, biodiesel and water in biodiesel. The goal is to investigate four new alternative fuels that can produce the minimum hazardous smoke.

2. Experimental Detail

2.1. Material and Method of Blend Preparation

A single cylinder diesel engine is utilized in this investigation. The engine test rig platform and sampling system of this four-stroke engine with water–cooled and direct fuel injection is illustrated in Figure 1. An eddy current dynamometer with a power of 15 kW is used in this study. Meanwhile, a dynamometer controller is used to control engine speed and torque. Table 1 shows the specifications for the diesel engine that was used. The operating temperature of the air intake is at 2.5$\pm$ 0.5 °C with the temperature of the lab controlled by the fan. There are two main types of fuel used in the study, namely diesel and biodiesel. Diesel fuel is obtained from Universiti Malaysia Pahang's lab and biodiesel fuel obtained from Mission Biotechnologies Sdn. Bhd. All fuels have been certified to comply with EN 14214 standards. The engine oil lubricant used is the Delta Super HD40 German Technology.

Figure 1. Engine test rig

Table 1. Specification of YANMAR TF120M.

Engine parameters	Specifications
Engine type	Diesel 1-cylinder
Stroke type	Four – stroke
Cylinder bore x stroke (mm)	92 x 96
Injection timing (deg.)	btdc 17°
Compression ratio	17.7
Displacement (L)	0.638
Maximum torque (kgf.m / rpm)	4.42 / 1800
Specific fuel consumption (g /kWh)	169
Fuel injection pump	Bosch Type
Injection pressure (Nm)	200

2.2. Test Fuels and Experimental Procedures

The preparation of fuel requires carefulness and sensitivity as it can affect the new fuel properties. The right tools are used to mix fuel and test the properties of diesel-based fuels comprising B5M10, B10M10, B5M10E3 and B10M10E3. Fuel test is performed to determine the following properties such as: cetane number, density, viscosity, boiling point calorific value, flash point and moisture content. All test equipment and methods of property analysis will be explained in this section.

The biodiesel used in the current study is palm oil-based biodiesel also known as POME, supplied by a local industrial supplier from Selangor, Malaysia. A commercial fuel supplier provided the diesel fuel which is 100% pure diesel unlike the 7% biodiesel blended fuels available in stations. Methanol, Span 80 and Tween 80 were obtained from the chemistry laboratory of Universiti Malaysia Pahang. Biodiesel blends can be produced by mixing diesel fuel and biodiesel fuel with the help of a magnetic stirrer. The mix was done in a closed beaker for 20 minutes at speed of 600−620 rpm and left at room temperature for 30 minutes. The emulsification process was conducted by an ultrasonic emulsifier in the presence of a surfactant. In addition to the diesel fuel (DIESEL), a methanol emulsion of 10% by volume was added into both palm oil biodiesel and mineral diesel fuel, which correspond to B5M10 and B10M10 fuels, respectively.

Besides that, two emulsions of water were prepared which corresponded to B5M10E3 and B10M10E3 fuels. The emulsion mixtures require the use of a mechanical stirrer and ultrasonic horn agitators to obtain the naturally immiscible liquid by using a surfactant. This method produces stable fuel and micro−emulsion. When a micro-emulsion mixture is formed, it is expected that a micro−explosion phenomenon will occur in the combustion. Consequently, there will be an increase in the performance and reduction in emission. [33]. Figure 2 exhibits the contents of five bottles containing five types of fuels to be tested.

Figure 2. Five bottles containing five types of fuel to be tested.

B5M10E3 is also known as an emulsion fuel as it contains 5% biodiesel oil, 10% methanol, 3% water, 80% diesel and 2% surfactant. The involvement of surfactants is intended to help weaken the liquid surface tension, so that the liquid that attenuates the surface allows other liquids to enter its molecule [34]. The surfactants are required to undergo HLB balance processing (hydrophilic−lipophilic balance) before the use. There were non-ionic surfactants for Tween 80 with HLB 15 and Span 80 with HLB 4.3 being the most suitable option for this study. American Society Testing Material (ASTM) suggests that HLB 0.5%−16% is suitable for fuel, hence 2% of surfactant is considered as an option [17, 35]. The direct mixing of oil and water will deteriorate the function of the engine. This is due to the nature of the water that interacts with the iron, which will cause rust on the engine portion of the iron component [17]. This problem can be solved by ensuring the water is encapsulated in the oil employing an emulsification process called water in oil known as W/O and oil in water and water in oil called O/W/O. Oil in water and water in oil (O/W/O) has shown to be a more stable emulsion and suitable for micro−explosion phenomenon in the diesel engine. The first stage of oil in water (O/W) is a process where oil is blended into the water with a surfactant of HLB 13 to stabilise the emulsion as expressed in Equation (1) and (2) as follows:

$$\% (A) = \frac{100(x - HLB_{(B)})}{HLB_{(A)} - HLB_{(B)}} \quad (1)$$

$$\% (B) = 100 - \% (A) \quad (2)$$

Based on Equations (1) and (2), the percentage of mixtures of Tween 80 and Span 80 surfactant were included in B5M10E3 fuel and mixed with a mechanical stirrer for 20 minutes at 25 °C. The time interval for mixing was between 25 to 30 minutes. According to the ASTM standard, the ideal room temperature for the coolant material is at 25 °C. To ensure that all data obtained is true and can be taken into account, control over pressure, temperature and relative humidity is always given attention to in all the experiments that are carried out, in order to minimize the impact of environmental error. The mechanical stirrer with a power of 80 to 100 hertz and with speed of 600 to 620 rpm was employed for the first stage surfactant to produce O/W. The power of 80 hertz is essential for creating a strong force for the molecular mixture, while 600 rpm was ideal to cause the stirring of the water in a simple state according to the size and volume of the 500 ml jar. The second stage process involved using HLB 6 O/W, which was formed to be used with O/W/O to obtain better fuel emulsion stability.

The mixture of the surfactant percentage was added into B5M10E3 fuel gradually during the fuel recovery process using an ultrasonic emulsifier machine. It should be noted that the ultrasonic emulsifier machine had a power of up to 24 kHz to break the surface tension molecules to ensure optimum mixing of water and fuel. Cycle 0.6 was selected as it provides the appropriate mix of impact. Moreover, amplitude of 70 was selected due to the medium strength of the medium−sized power 500 ml jar, which incorporated O/W into O/W/O with 10 minutes of operation time. The temperature of the fuel sample increases with the increase in the duration of the ultrasonic emulsification process. Furthermore, the recommended loading time was between 5−20 minutes depending on the size of the jar and the requirements for the sample rate. After the emulsification process, the fuel was processed slowly under the stirrer mechanical gauze for 10 minutes for gentle mixing of the molecular composition. It should be noted that B10M10E3 emulsion fuel requires similar processing to B5M10E3.

3. Result and Discussion

3.1. Properties of the test fuels

It has been noted from different research that the performance and emissions improvement with different alternative fuels is directly linked to the change on the physicochemical properties of the modified fuels [36, 37]. Table 2 demonstrates the results of fuel properties obtained from different experiments. The properties considered in the current study are cetane number, calorific value, the density at 25°C, kinematic viscosity at 40°C, flashpoint and boiling point. There are five fuels as a test sample and diesel is taken to be as reference base fuel. The experiment was conducted three times on average, and the tabulated result is the average. Each fuel characteristic affects the combustion behaviour in the combustion chamber, which consequently influences NO_x emission and performance.

Table 2. The fuel properties of five different fuels

Properties		Fuel				
	Unit	Diesel	B5M10	B10M10	B5M10E3	B10M10E3
Cetane Number	-	50	51.6	51.9	52.3	53.9
Calorific Value	MJ/kg	42.7	34.22	35.27	41.69	42.74
Density at 25°C	g/cm³	0.84	0.86	0.84	0.92	0.83

Kinematic Viscosity at 40°C	mm²/s	5	5.69	5.71	5.25	5.25
Flashpoint	°C	80	100-105	95-100	90-95	105-110
Boiling Point	°C	149	208.2	191.7	105.2	73

3.2. Analysis of the NO_x Concentration

NO_x is composed of Nitric oxide (NO) and nitrogen dioxide (NO_2), and its formation mainly depends on the peak temperature, engine design, stoichiometry, ignition delay and operating conditions. The trend for NO_x is low at 1200 rpm followed by a dramatic increase until it reached 1800 rpm due to an increase in thermal efficiency in the combustion chamber. After that, NO_x decreased steadily from 1800 rpm to 2400 rpm due to incomplete combustion which lowers the temperature in the combustion chamber. Figure 3 shows that at 2400 rpm and low load condition, NO_x for B5M10, B10M10, B5M10E3 and B10M10E3 fuels is low compared to diesel with percentages of 7%, 3%, 12.28% and 8.77% respectively. NO_x range varied between 42 ppm and 73 ppm with B5M10E3 being the lowest due to low biodegradability. B5M10E3 and B10M10E3 have lower NO_x emissions compared to B5M10 and B10M10 due to micro-explosion phenomenon which helps burn fuel perfectly and reduce the temperature in the combustion chamber. Similarly, evaporation of droplets will also reduce the temperature in the combustion chamber and assist in the reduction of NO_x emissions [38, 39]. The presence of many oxygen molecules in the combustion chamber will help increase combustion activity and increase temperature thereby causing high NO_x formation [40]. NO_x is largely formed in palm oil biodiesel compared to diesel at constant speeds and different loads.

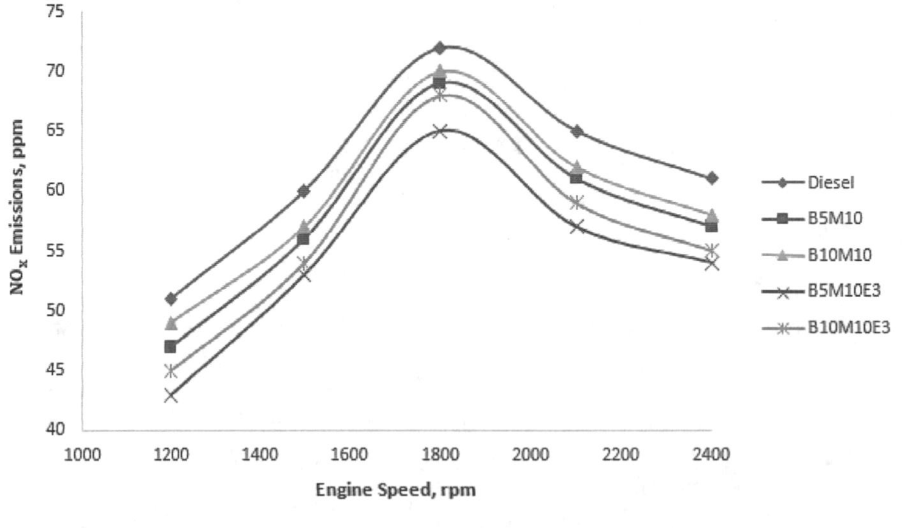

Figure 3. Measured NO_x emissions at low load and various engine speeds using diesel, B5M10, B10M10, B5M10E3 and B10M10E3 and the deviations in NO_x emissions compared to the diesel engine data

At partial load condition and engine speed at 1200 rpm, NO_x slightly increased until it reached 1800 rpm due to increased combustion efficiency. Then NO_x emissions slightly decreased from 1800 rpm to 2400 rpm due to incomplete combustion. Figure 4 shows that emissions for B5M10, B10M10, B5M10E3 and B10M10E3 fuels are low compared to diesel with percentages of 8%, 5.33%, 13.33% and 10.67% respectively. The range for NO_x varied between 57 ppm and 88 ppm with B5M10E3 being the lowest. B5M10 and B10M10 have lower NO_x emissions compared to diesel due to the addition of 10% methanol, which reduces the temperature. It has also been proven by many researchers developing biodiesel alternative fuels that NO_x emissions increase with the increase in biodiesel use [39, 41, 42]. B5M10E3 and B10M10E3 have lower NO_x compared to diesel, B5M10 and B10M10 due to the effect of water evaporation in the cylinder which helps to reduce the temperature in the combustion chamber.

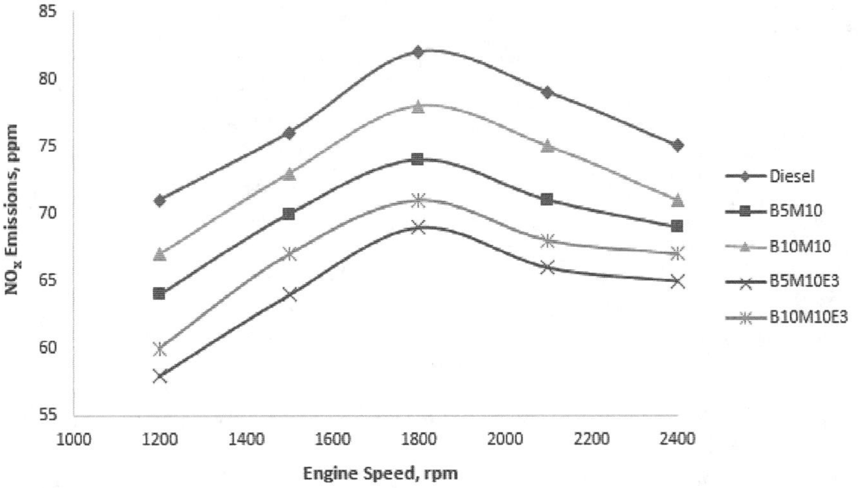

Figure 4. Measured NO_x emissions at partial load and various engine speeds using diesel, B5M10, B10M10, B5M10E3 and B10M10E3 and the deviations in NO_x emissions compared to the diesel engine data

Trend at high load found that NO_x emissions followed a steep upward trajectory from 1200 rpm until it reached 1800 rpm which is the ideal speed for maximum engine efficiency. Then NO_x emissions slowed down noticeably from 1800 rpm to 2400 rpm due to incomplete combustion. Figure 5 shows that at full load and engine speed of 1800 rpm, NO_x emissions for B5M10, B10M10, B5M10E3 and B10M10E3 are lower compared to diesel with percentages of 5.49%, 4.27%, 12.80% and 9.15% respectively. B5M10E3 and B10M10E3 fuels were found to be more able to reduce NO_x emissions compared to B5M10 and B10M 10, approximately 5.4−15% reduction. Reduction of NO_x emissions by B5M10E3 and B10M10E3 fuels is influenced by the temperature factor in the engine combustion chamber. If the temperature in the cylinder increases, NO_x emissions will also increase. Likewise, if the temperature in the cylinder decreases, NO_x emissions also decreases. B10M10 biodiesel blends produced higher NO_x emissions than the B5M10 biodiesel blends due to the fact that B10M10 has a higher oxygen content which enables a more complete combustion. To counteract high NO_x emissions, emulsion fuel is used to compensate for engine performance and lower NO_x emissions due to its 3% water content in disperse phase. Thus, there is a potential for emulsion fuels to compensate for combustion cylinder temperature and at the same time reduce the emission of NO_x [43].

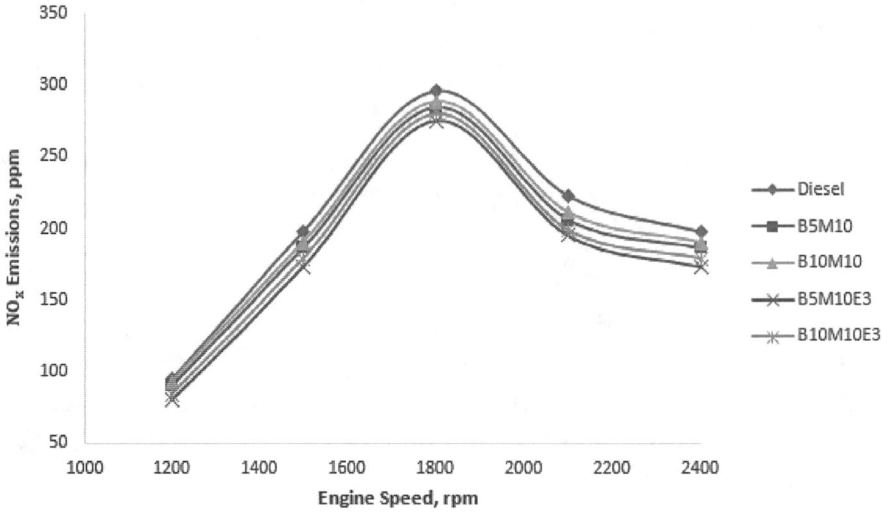

Figure 5. Measured NO_x emissions at full load and various engine speeds using diesel, B5M10, B10M10, B5M10E3 and B10M10E3 and the deviations in NO_x emissions compared to the diesel engine data.

4. Conclusions

Many countries are scrutinizing the use of the diesel engines in the transportation sector due to their negative impact in the environment. Fuel borne emissions from diesel engine which are severely compromising the environment are nitrogen oxide (NO_x). Emulsion fuel derived from palm oil showed superior engine performance and emitted low NO_xat all engine operating condition, compared to biodiesel blends and baseline mineral diesel. The water function caused B5M10E3 and B10M10E3 of collected NO_xlower than diesel, B5M10 and B10M10 at low load conditions, partial load and high load. The B5M10E3 and B10M10E3 as emulsion fuel promote to achieve an optimal temperature of the engine performance with better fuel properties which reduce NO_x.

Acknowledgments

The author would like to acknowledge the financial support by Universiti Malaysia Pahang through grant number RDU1803164 and RDU172204.

References

[1] United States Environmental Protection Agency 2000 *Clean Diesel Trucks, Buses, and Fuel: Heavy-Duty Engine and Vehicle Standards and Highway Diesel Fuel Sulfur Control Requirements*
[2] Hossain F M, Nabi M N, Rainey T J, Bodisco T, Rahman M M, Suara K, Rahman S M A, Chu V T, Ristovski Z and Brown R J 2017 *Energy Convers. Manag.* **152** 186-200
[3] Nabi M N, Zare A, Hossain F M, Ristovski Z D and Brown J D 2017 *J. Clean. Prod.* **166** 860-868
[4] Abdalla I E 2018 *Fuel* **212(x)** 638-655
[5] Eckart S, Penke C, Voss S and Krause H 2017 *Energy Procedia* **120** 149-156
[6] Nair J N, Kaviti A K and Daram A K 2017 *Egypt. J. Pet.* **26(4)** 927-931
[7] Zhang Z H and Balasubramanian R 2018 *Fuel* **215** 161-170
[8] Imdadul H K, ,Masjuki H H, Kalam M A, Zulkifli N W M, Kamruzzaman M, Shahin M M and Rashed M M 2017 *J. Clean. Prod.* **141** 928-939
[9] Prakash R., Singh R K and Murugan S 2015 *J. Ene. Ins.* **88(1)** 64-75

[10] Stepan E, Enascuta C E, Oprescu E E, Radu E, Vasilievici G, Radu A, Stoica R, Velea S, Nicolescu A and Lavric V 2018 *Ind. Crops Prod.* **113**288-297
[11] Park W, Park S, Reitz R D and Kurtz E 2017 *Combust. Flame* **180** 276-283
[12] Alptekin E 2017 *Energy* **119** 44-52
[13] Song H, Peng Z, Quinton K S, Zhao H and Ladommatos N 2015 *Energy Procedia* **66** 17-20
[14] An H, Yang W M, Li J and Zhou D Z 2015 *Ene. Conv.Man.* 2015 **90** 261- 271
[15] Meng X and Wu J 2013 *Fuel* **107** 309-314
[16] Ali O M, Mamat R, Abdullah N R and Abdullah A A 2015 *Renew. Energy* 2015 **86** 59-67
[17] How H G, Masjuki H H, Kalam M A and Teoh Y H 2018 *Fuel* **213** 106-114
[18] Caliskan H 2017 *J. Clean. Prod.* **154** 125-129
[19] Hashim H, Narayanasamy M, Yunus N A, Shiun L J, Muis Z A and Ho W S 2017 *J. Clean. Prod.* **146** 208-217
[20] Phoon LY, Mustaffa A A, Hashim H, Mat R, Manan Z A and Yunus N A 2017 *J.Clean. Prod.* **161** 1192-1202
[21] Suhaimi H, Adam A, Mrwan A G, Abdullah Z, Othman M F, Kamaruzzaman M K and Hagos F Y 2018 *Fuel* **220** 682-691
[22] Song L, Liu T, Fu W and Lin Q 2018 *J. Ene. Ins.* **91(2)** 203-213
[23] Mahmudul H M, Hagos F Y, N A Mukhtar M, Mamat R and Abdullah A A 2017 *IOP Conference Series: Materials Science and Engineering* **318**
[24] Mahmudul H M, Hagos F Y, Mamat R, Abdullah A A and Awad O I 2016 *IOP Conference Series: Materials Science and Engineering* **160**
[25] Sivalakshmi S and T Balusamy 2013 *Fuel* **106** 106-110
[26] Kilic G, Sungur B, Topaloglu B and Ozcan H 2018 *App. Therm. Eng.* **130** 195-202
[27] Keskin A, Yasar A, Yildizhan S, Uludamar E, Emen F M and Kulcu N 2018 *Fuel* **216** 349-355
[28] Mura E, Calabria R, Califano V, Massoli P and Bellettre J 2014 *Exp. Therm. Flu. Sci.* **56** 69-74
[29] Hagos F Y,Aziz A A R and Tan I M 2011 *IEEE 3rd Int. Conf.* (Bali: Indonesia) pp 314-318
[30] Khan Y M, Karim Z A A, Hagos F Y and Aziz A A R 2014 *The Scientific World Journal* **2014(17)** 1-15
[31] Emiroğlu A O and Şen M 2018 *App. Therm. Eng.* **133** 371-380
[32] Mazlan N A, Yahya W J, Ithnin A M, Hasanuddin A K, Ramlan N A, Sugeng D A, Adib A R M, Koga T, Mamat R and Sidik A C 2018 *J. Clean. Prod.* **179(2018)** 559-566
[33] Watanabe H Shoji Y, Yamagaki T, Hayashi J, Akamatsu F and Okazaki K 2016 *Fuel* **182** 259-265
[34] Lee J H, CHoi J G, Kim Y S, Kim K R, Kim S W and Oh D K 2012 *J. Biosci. Bioeng.* **113(4)** 461-466.
[35] Fayyad S M, Ein S A, Al-Marahleh G and Momani W 2010 *J. Eng. Technol.* **2(3)** 268-273.
[36] Zare A, Nabi M N, Bodisco T A, Hossain F M, Rahman M M, Van T C, Ristovski Z D and Brown R J 2017 *J. Clean. Prod.* **162** 997-1008
[37] Nakakita K, Takasu S, Ban H and Ogawa T 1998 *SAE Technical Paper*
[38] Khalife E, Kazerooni H, Miraslim M, Shojaei T R, Mohammadi P, Salleh A M, Najafi B and Tabatabaei M 2017 *Energy* **121** 331-340
[39] Elsanusi O A, Roy M M and Sidhu M S 2017 *Applied Energy* **203** 582-593
[40] Vigneswaran R., Annamalai K, Dhinesh B and Krishnamoorthy R 2018 *Energy Convers. Manag.* **172** 370-380
[41] Ali O M, Mamat R, Abdullah N R and Abdullah A A 2015 *Renew. Energy* **86** 59-67
[42] Koc A B and Abdullah M 2013 *Fuel Process. Technol.* **109** 70-77
[43] Mondal P K and Mandal B K 2019 *Fuel* **237** 937-960

Secondary Metabolites from Colletotrichum phyllanthi, an Endophytic Fungi from Stem of Physalis peruviana and Evaluation of Media Culture

Amelinda Pratiwi[1*], Elvira Hermawati[2], and Yana Maolana Syah[2]
[1] Universiti Pendidikan Indonesia, Bandung, Indonesia
[2] Institut Teknologi Bandung, Bandung, Indonesia

*Corresponding e-mail: amel.pratiwi@upi.edu

Abstract. Endophytic fungi is a microorganism that grows in plant tissue without giving any negative effect directly against its host. Phytochemistry studies from endophytic fungi have been widely reported and showed that the metabolites have excellent activities. The exploration of secondary metabolites from endophytic fungi of *Physalis peruviana* well known as a "cecendet" in Indonesia has never been reported before. Therefore, this research aims to isolate the secondary metabolites of the endophytic fungi from stem of *P. peruviana*. Five known compounds were isolated from MeOH and EtOAc extract of endophytic fungi *Colletotrichum phyllanthi* resulted two steroid compounds, ergosterol (1), and ergosterol peroxide (2), from MeOH extract of *C. phyllanthi* mycelia, and from EtOAc extract of *C. phyllanthi*. The metabolites production of *C. phyllanthi* was influenced with composition of media cultures. In this research endophytic fungi was cultivated using Potato Dextrose Broth (PDB) and Tryptic Soy Broth (TSB) media. Two metabolites including tyrosol (3), and uridine (4) were produced by *C. phyllanthi* through PDB, and another compound that is 1-chloronaphtalene-2-ol (5) was isolated from endophytic fungi while treated by fermentation in TSB. Structure elucidation was achieved using spectroscopic data including 1D, 2D NMR and comparison with reported data. Our results provide a scientific rationale for further explorations into metabolites from endophytic fungi of *P. peruviana*.

Keywords: *Colletotrichum phyllanthi*, endophytic fungi, media culture, *Physalis peruviana*

1. Introduction

Endophytic fungi is microorganism that live in plant tissues without giving a negative effect during the life cycle of host [1, 2]. The role of endophytic fungi, among others, can increase the tolerance of host plants to environmental stresses both biotic and abiotic, increase growth ability, decrease pathogen attack and predators to host plants [3]. One of the role of endophytic fungi, for its host, is production of various secondary metabolites with a variety of bioactivities. It is not surprisingly that endophytic fungi can also be a biological source in the search for bioactive compounds.

Some examples of successful endophytic fungi as a source of bioactive compounds can be seen in the following reports. One of them is the discovery of Paclitaxel (or taxol) from the endophytic fungi of medicinal plants *Paraconiothyrium variabile* [4]. Another case such as a polyphenol derivative, cholletotrate, which is a secondary metabolite of the *Colletotrichum gloeosporioides*, an endophytic fungi of medicinal plants and has antibacterial properties [5]. In addition, environmental factors of fungal fermentation of endophytes have a major effect on the production of secondary metabolites produced [6, 7]. Based on those reports indicate that the endophytic fungi is one source of secondary metabolites that have the potential to be developed and one of them is derived from medicinal plants. Based on this background, research on secondary metabolite isolation, and evaluation of endophytic fungal culture media from *Physalis* genus plants need to be done. The exploration of secondary metabolites from endophytic fungi of *Physalis peruviana* well known as a "cecendet" in Indonesia has never been reported before. Therefore, this research aims to isolate the secondary metabolites of the endophytic fungi from stem of *P. peruviana*

2. Experimental Setup

2.1. General Experimental Procedure
Optical rotations were carried out on an Autopol IV Rudolph Research Analytical polarimeter. All NMR experiments were performed on an Agilent, operating at 500 MHz (^{1}H) and 125 MHz (^{13}C) together with 2D NMR including DEPT135, HSQC, HMQC, HMBC, zTOCSY1D, and ^{1}H-^{1}H COSY with TMS as an internal standard. For the deuterium eluent using $CDCl_3$, acetone-d_6, and CD_3OD, all solvent used in chromatography were technical grade. Polyamide and Silica Gel Merck 60 GF_{254} used for column chromatography, Merck Silica Gel 7749 used for radial chromatograpy, Silica Gel 60 (35 – 70 mesh ASTM) used for chromatography and TLC analysis using Alumunium Silica Gel Merck 60 F254: 0.25 mm (20x20 cm). Spot were visualized using UV light and by spraying with $Ce(SO_4)_2$. Endophyte media culture using Potato Dextrose Agar (PDA), Potato Dextrose Broth (PDB), and Tryptic Soy Broth (TSB).

2.2. Plat Material
Samples of the stem of *P. peruviana* were taken from Bumi Herbal Dago in July 2017, Bandung, West Java Indonesia. The best stem tissue was chosen for further sterilization and inoculation process to isolate endophytic fungi (Figure 1).

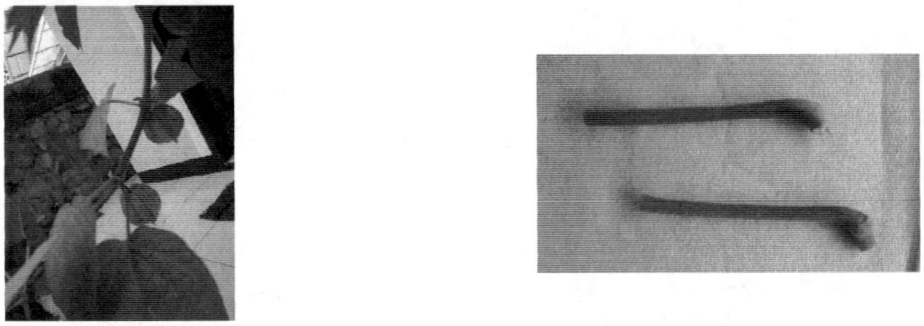

(a) (b)

Figure 1. (a) *P. peruviana* plant; (b) Best stem tissue from *P. peruviana*.

2.3. Fungi Material
Colletorichum phyllanthi was isolated from stem of *P. peruviana* through some steps of cultivation. First of all, selected stem tissue was sterilized in 70% EtOH, 0.5% NaOCl and 70% EtOH respectively for 30 seconds immersed in every step. After the stem tissue was sterilized, the tissue cut to the size of 1 x 1 cm using a sterile scalpel, then it was grown in PDA for seven days. Endophyte fungi with different morphology were repeated subculture until a single potential isolate was obtained. The single strain of endophyte fungi was identified through molecular identification based on Internal Transcribed Spacer (ITS) in Indonesian Culture Collection (InaCC), Biology Research Center-LIPI, Cibinong. Then the isolated endophytic fungi were known as a *Colletotrichum phyllanthi*.

2.4. Extraction of Culture Media
C. phyllanthi was cultivated in two different broth media including PDB and TSB to determine the effect of environment for production of metabolites. The MeOH and EtOAc extract from each culture media were evaluated using TLC to obtain different spot of metabolites.

2.5. Extraction and Isolation
Mycelia was maceration in MeOH for 2x24 h, and obtain 6.54 g MeOH extract. The liquid medium was extracted using EtOAc through liquid-liquid extraction, gained 1.36 g and 1.65 g EtOAc extract from PDB and TSB medium respectively. Then all of the MeOH and EtOAc extract were fractionated using several chromatography methods.

2.5.1 Fractionation MeOH extract of C. phyllanthi cultivated with PDB
The methanol extract of C. phyllanthi (6.54 g) was sequentially fractionated through gravity column chromatography (polyamide as a stationary media, EtOAc, and MeOH in order of increasing polarity) to give five major fraction (Fr. A – E). The first major fraction (Fr. A 358.9 mg) was purified by recrystallization using MeOH, and precipitate 8.3 mg ergosterol (**1**) was obtained as a white needle crystalline compound. The filtrate from recrystallization of Fr. A was further subjected to radial chromatography (Silica Gel 7749, hexane, $CHCl_3$-EtOAc, and MeOH in order of increasing polarity). Six major fractions were ultimately obtained on combining the eluates on the basis of TLC. The fifth major fraction was further subjected to gravity column chromatography and successfully isolated ergosterol peroxide **(2)** (3 mg) was obtained as a shaped white powder.

Ergosterol (a): Shaped white needle crystal.$[\alpha]_D^{25} = -132°$ (c 0,0012 g/ml, CHCl3). Spectrum 1H NMR (500 MHz, CDCl3) δ (ppm): 1,30 (1H, m, H-1a), 1,89 (1H, m, H-1b), 1,90 (1H, m, H-2a), 1,49 (1H, m, H-2b), 3,64 (1H, m, H-3), 2,45 (1H, m, H-4a), 2,28 (1H, m, H-4b), 5,57 (1H, dd, J = 2,6 Hz, 5,6 Hz, H-6), 5,38 (1H, t, J = 2,9 Hz, H-7), 1,97 (1H, m, H-9), 1,73 (1H, m, H-11a), 1,60 (1H, m, H-11b), 2,07 (1H, m, H-12a), 1,26 (1H, m, H-12b), 1,88 (1H, m, H-14), 1,60 (1H, m, H-15a), 1,36 (1H, m, H-15b), 1,77 (1H, m, H-16a), 1,31 (1H, m, H-16b), 1,27 (1H, m, H-17), 0,63 (3H, s, H-18), 0,95 (3H, s, H-19), 2,05 (1H, m, H-20), 1,04 (3H, d, J =6,6 Hz, H-21), 5,20 (1H, dd, J = 15,3 Hz, 7,1 Hz, H-22), 5,19 (1H, d, J = 15,3 Hz, 7,7 Hz, H-23), 1,85 (1H, m, H-24), 1,47 (1H, m, H-25), 0,82 (3H, d, J = 7,2 Hz, H-26), 0,84 (3H, d, J = 7,2 Hz, H-27), 0,91 (3H, d, J = 6,9 Hz, H-28). Spectrum 13C NMR (125 MHz, CDCl3) δ (ppm): 38,3 (C-1), 31,8 (C-2), 70,4 (C-3), 40,7 (C-4), 139,7 (C-5), 119,4 (C-6), 116,2 (C-7), 141,3 (C-8), 46,2 (C-9), 46,9 (C-10), 21,1 (C-11), 38,9 (C-12), 42,7 (C-13), 54,5 (C-14), 22,8 (C-15), 28,1 (C-16), 55,6 (C-17), 11,9 (C-18), 16,2 (C-19), 40,3 (C-20), 21,0 (C-21), 131,8 (C-22), 135,4 (C-23), 42,7 (C-24), 32,9 (C-25), 19,5 (C-26), 19,9 (C-27), 17,5 (C-28).

Ergosterol peroxide (b): Shaped white powder. $[\alpha]_D^{25} = -28°$ (c 0,0008 g/ml, $CHCl_3$). Spectrum ^{1}H NMR (500 MHz, $CDCl_3$) δ (ppm): 1,70 (1H, *m*, H-1a), 1,94 (1H, *m*, H-1b), 1,83 (1H, *m*, H-2a), 1,53 (1H, *m*, H-2b), 3,97 (1H, *m*, H-3), 2,11 (1H, *m*, H-4a), 1,91 (1H, *m*, H-4b), 6,24 (1H, *d*, *J* = 8,6 Hz, H-6), 6,50 (1H, *m*, *J* = 8,5 Hz, H-7), 1,49 (1H, *m*, H-9), 1,51 (1H, *m*, H-11a), 1,23 (1H, *m*, H-11b), 1,95 (1H, *m*, H-12a), 1,24 (1H, *m*, H-12b), 1,56 (1H, *m*, H-14), 1,58 (1H, *m*,H-15a), 1,39 (1H, *m*, H-15b), 1,75 (1H, *m*, H-16a), 1,36 (1H, *m*, H-16b), 1,22 (1H, *m*, H-17), 0,81 (3H, *s*, H-18), 0,88 (3H, *s*, H-19), 2,01 (1H, *m*, H-20), 1,00 (3H, *d*, *J* = 6,7 Hz, H-21), 5,14 (1H, *dd*, *J* = 8,3 Hz, 15,3 Hz, H-22), 5,22 (1H, *dd*, *J* = 7,5 Hz, *J* = 15,3 Hz, H-23), 1,85 (1H, *m*, H-24), 1,46 (1H, *m*, H-25), 0,84 (3H, *d*, *J* = 7,2 Hz , H-26), 0,81 (3H, *d*, *J*= 7,2 Hz, H-27), 0,91 (3H, *d*, *J*= 6,9 Hz, H-28). Spectrum ^{13}C NMR (125 MHz, $CDCl_3$) δ (ppm): 34,5 (C-1), 30,0 (C-2), 66,4 (C-3), 36,8 (C-4), 82,0 (C-5). 135,3 (C-6), 130,6 (C-7), 79,3 (C-8), 51,0 (C-9), 36,9 (C-10), 23,3 (C-11), 39,2 (C-12), 44,5 (C-13), 51,6 (C-14), 20,5 (C-15), 28,6 (C-16), 56,1 (C-17), 12,8 (C-18), 18,1 (C-19), 39,6 (C-20), 20,8 (C-21), 135,1 (C-22), 132,2 (C-23), 42,7 (C-24), 33,0 (C-25), 19,9 (C-26), 19,6 (C-27), 17,5 (C-18).

2.5.2 Fractionation EtOAc extract of C. phyllanthi cultivated with PDB
The EtOAc extract of *C. phyllanthi* that cultivated in PDB (1.36 g) was sequentially fractionated using vacuum chromatography (Silica Gel, hexane, hexane-EtOAc, ETOAc-MeOH, MeOH in order of

increasing polarity) obtained nine major fractions (Fr. A- Fr.I) on combining the eluates on the basis of TLC. The fourth major fraction (Fr. D) (44.8 mg) was further fractionated by radial chromatography (Silica Gel 7749, $CHCl_3$, $CHCl_3$-EtOAc, EtOAc-MeOH, MeOH in order of increasing polarity) and gained tyrosol **(c)** (12 mg) was obtained as a shaped yellowish-coloured compound. While, from the last major fraction (208.6 mg) was further fractionated using radial chromatography (Silica Gel 7749, $CHCl_3$, $CHCl_3$-EtOAc, EtOAc-MeOH, MeOH in order for increasing polarity) and successfully isolated uridine **(d)** (4 mg) was obtained as a Shaped white crystals compound.

Tyrosol (c): Shaped yellowish-coloured. Spectrum ^{1}HNMR (500 MHz, $CDCl_3$) δ (ppm): 6,66 (1H, *d*, *J* = 8,4 Hz, H-2, H-6), 6,97 (1H, *d*, *J* = 8,4 Hz, H-3, H-5), 2,62 (2H, *t*, *J* = 7,2 Hz, H-1'), 3,59 (2H, *t*, *J* = 7,2 Hz, H-2'), 8,03 (1H, *s*, -OH). Spectrum ^{13}C NMR (125 MHz, $CDCl_3$) δ (ppm): 156,6 (C-1), 115,9 (C-2), 130,7 (C-3), 131,1 (C-4), 130,7 (C-5), 115,9 (C-6), 39,5 (C-1'), 64,3 (C-2').

Uridine (d): Dark yellow gum of. Spectrum ^{1}H NMR (500 MHz, aseton-d_6) δ (ppm): 5,70 (1H, *d*, *J* = 8,05 Hz, H-5), 8,01 (1H, *d*, *J* = 8,1 Hz, H-6), 5,90 (1H, *d*, *J* = 4,6 Hz, H-1'), 4,18 (1H, *t*, *J* = 4,85 Hz, H-2'), 4,15 (1H, *t*, *J* = 4,85 Hz, H-3'), 4,01 (1H, *m*, H-4'), 3,84 (1H, *dd*, *J* = 3,1 Hz, 12,25 Hz, H-5'), 3,73 (1H, *dd*, *J* = 3,1 Hz, 12,25 Hz, H-5'). Spectrum ^{13}C NMR (125 MHz, $CDCl_3$) δ (ppm): 152,5 (C-2), 166,2 (C-4), 102,6 (C-5), 142,7 (C-6), 90,7 (C-1'), 75,7 (C-2'), 71,3 (C-3'), 86,3 (C-4'), 62,3 (C-5').

2.5.3 Fractionation EtOAc extract of C. phyllanthi cultivated with TSB

The EtOAc extract of *C. phyllanthi* that cultivated in PDB (1.65 g) was sequentially fractionated using gravity column chromatography (Silica Gel, hexane, EtOAc, ETOAc-MeOH, MeOH in order of increasing polarity) obtained four main fractions (Fr. A- Fr.D) on combining the eluates on the basis of TLC. The third major fraction (Fr. C) (22.7 mg) was further fractionated by gravity column chromatography (Silica Gel 7749, hexane, EtOAc:hexane, EtOAc in order of increasing polarity) gained two major fraction (Fr. C1 – Fr. C2). After Fr. C2 was crystallized gained 1-choloronaphtalene-2-ol **(5)** (3.8 mg) was obtained as a shaped white crystal compound.

1-chloronaphtalene-2-ol (e): Shaped white crystals. Spectrum ^{1}H NMR 500 MHz, $CDCl_3$) δ (ppm): 7,39 (1H, *d*, *J* = 8,9 Hz) H-3, 7,98 (1H, *d*, *J* = 8,9 Hz) H-4, 7,16 (1H, *d*, *J* = 8,6 Hz) H-5, 7,31 (1H, *td*, *J* = 1,4 Hz, 8,3 Hz, 7,6 Hz) H-6, 7,61 (1H, *td*, *J* = 1,3 Hz, 8,1 Hz, 7,5 Hz) H-7, 7,89 (1H, *d*, *J* = 8,1 Hz) H-8, 5,07 (1H, *s*). Spektrum ^{13}C NMR (125 MHz, $CDCl_3$) δ (ppm): 110,7 (C-1), 152,7 (C-2), 117,6 (C-3), 131,3 (C-4), 129,4 (C-4a), 128,3 (C-5), 123,9 (C-6), 124,1 (C-7), 127,3 (C-8), 133,3 (C-8a).

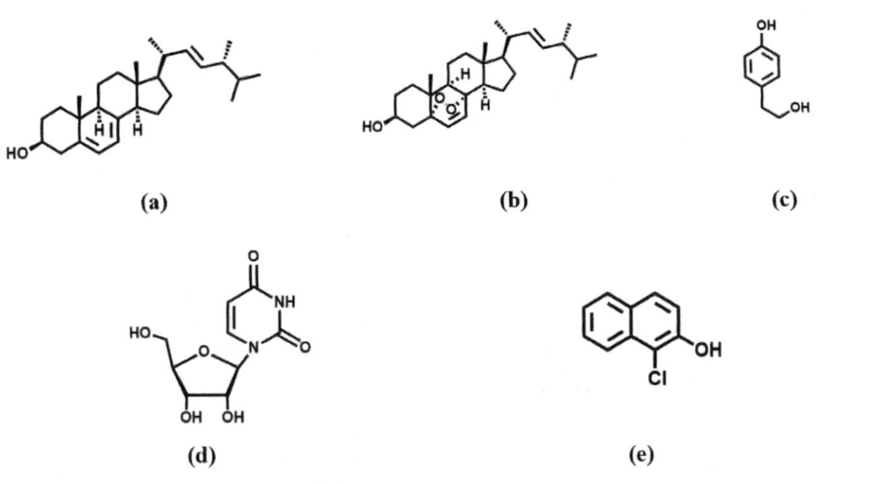

Figure 2. Skeleton of (a) Ergosterol (b) Ergosterol peroxide (c) Tyrosol (d) Uridine (e) 1-chloronaphtalene-2-ol

3. Results and Discussion

3.1. Compound Elucidation

Compound **(1)** was obtained as a white needle crystals with mass 8.3 mg and was optically active $[\alpha]_D^{25} = -132°$ (c 0,0012 g/ml, $CHCl_3$). Based on the ^{13}C-NMR spectrum together with the spectrum of DEPT135 and HSQC, revealed 28 carbon signals consisting of six carbon peaks of olefin (δ_C 131.8, 135.4, 139.7, 119.4, 141.3, 116, 2 ppm) and 22 sp^3 carbon signals was comprising a methynoxy carbon signal (δ_C 70.4 ppm), six methyl carbon signals (δ_C 11.9, 16.2, 21.0, 19.5, 19.9, 17, 5 ppm), seven methylene carbon signals (δ_C 38.3, 31.8, 40.7, 21.1, 38.9, 22.8, 28.1 ppm), six methyl signals (δ_C 46.2, 54 , 5, 55.6, 40.3, 42.7, 32.9 ppm) and two quaternary carbon signals (δ_C 36.9, 42.7 ppm). Considering the amount of sp^3 carbon (in total 22 carbon) and the three alkene groups (DBE 7), it was suspected that compound **(1)** was a tetracyclic steroid derivative with a C-28 ergostand ($C_{28}H_{44}O$) skeleton (Figure 3).

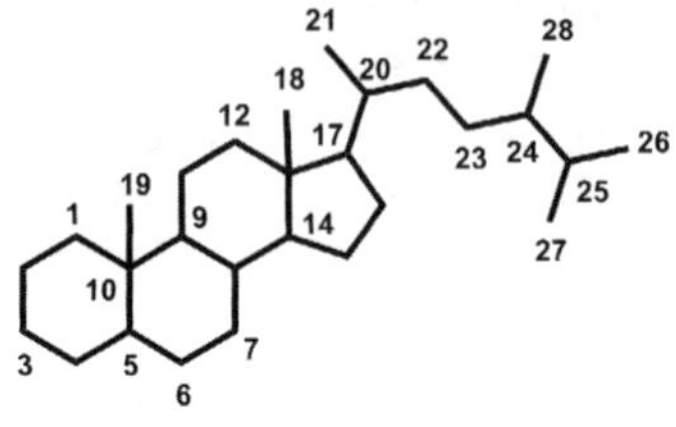

Figure 3. Ergostand skeleton.

The ^1H-NMR spectra also indicated the presence of signals δ_H 5.20 and 5.19 ppm which were characteristic of proton olefins that having a multiplicity of *dd*. These two protons had a large coupling value (J = 15.3 Hz) which showed trans-position and have a small coupling value (J = 7.2 Hz) which informed both of the proton in a vicinal position. The certainty of the location of the alkene system was confirmed by the correlation of HMBC (Figure 4). The spectral data of zTOCSY1D which showed the proton-proton correlation on (H-21), (H-22), (H-23), (H-24), (H-25), (H-26), (H-27) and (H-28) exhibit the side chains of ergostand skeleton (Fig.2). The position of the hydroxy group on the skeleton of compound **(1)** lied on C-3, this corresponded to the biogenesis of the steroid group compounds [8]. These signals from the ^1H, ^{13}C NMR, HMBC, and zTOCSY1D suggested that compound **(1)** was ergosterol **(1)**, similar to the NMR data of ergosterol that found in *Chaetomium globosum l18*, an endophytic fungi in the medical plant *Curcuma wenyujin* [9].

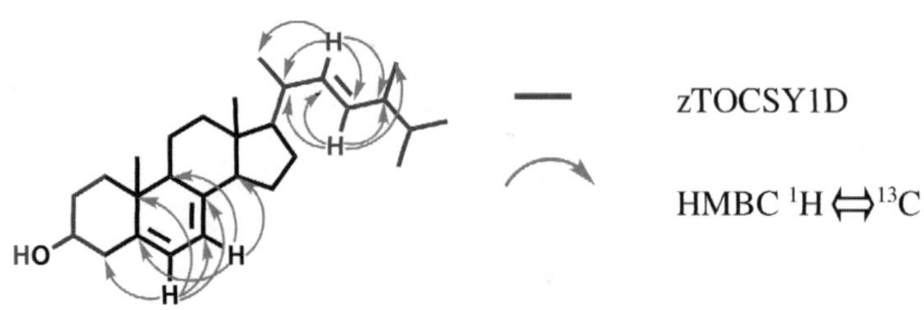

Figure 4. HMBC and zTOCSY1D correlation of ergosterol **(1)**.

Compound **(2)** was obtained as a white amorphous solid and it was optically active $[\alpha]_D^{25} = -28°$ (c 0,0008 g/ml, $CHCl_3$). Based on ^1H and ^{13}C-NMR data assisted with HSQC and HMBC data showed that compound **(2)** had similarities with ergosterol **(1)**, so it was presumed to have the same basic skeleton with ergosterol **(1)**. The differences between these compounds was the compound **(2)** only had two alkene groups at δ_C 135.3, 130.6, 135.1 and 132.2 ppm. There were also two C-quaternary peaks which bind the O atoms at carbon δ_C 82.0 and 79.3 ppm, this data suspected there were substituted peroxide groups in C-5 and C-8. This is confirmed by the DBE value of compound **(2)** was 7, which denotes 2 for two alkene units, 4 for ergostand cyclic and 1 for cyclic of peroxide bridge. In addtition, the ^{1}H-NMR shift also indicated that the proton at δ_H 6,24 ppm and 6,25 ppm with value J = 8,6 Hz showing cis-position. This provided information that an alkene system presented in C-6 and C-7. The assurance of second position of proton methyn olefin could be seen from HMBC correlation. Proton δ_H 6.24 (H-6) correlated with three C-quaternary (δ_C 36.9 (C-10), 79.3 (C-8), 82.0 (C-5) ppm) and proton δ_H 6,50 (H-7) ppm correlated with two C-quaternary (δ_C 79.3 (C-8) and 82.0 (C-5) ppm) and one carbon methylene (δ_C 51.0 (C-9) ppm). Based on data ^{1}H, ^{13}C-NMR, HSQC and HMBC (Figure 5) it can be suggested that compound **(2)** is ergosterol peroxide **(2)**. These data also compared with reported data of ergosterol peroxide isolated from *Chaetomium globosum* 118, an endophytic fungi in the medical plant *Curcuma wenyujin* [9].

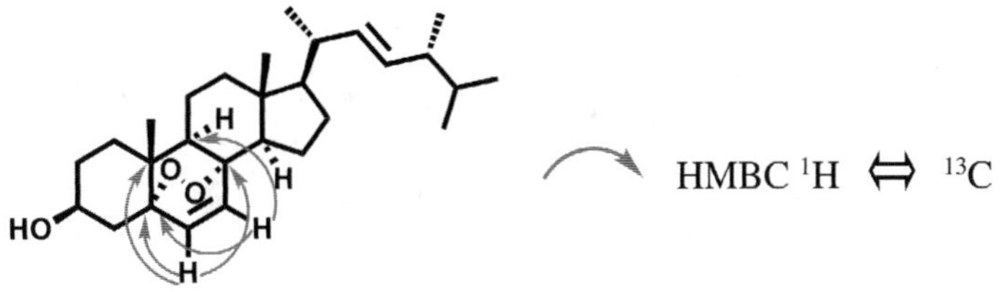

Figure 5. HMBC correlations of ergosterol peroxide **(2).**

Compound **(3)** was obtained in the form of a yellowish solid. Based on HMQC measurement, DBE value of compound **(3)** was 4. Based on the ^{13}C-NMR spectrum of compound **(3)** shows the presence of six carbon signal peaks consisting of two methylene carbon signals (δ_C 39.5 and 64.3 ppm) of two C-sp^2 methyl. Two carbon signals representing four symmetric aromatic methyn (δ_C 115, 9 and 130.7 ppm) and there are two quaternary C- sp^2 carbon signals (δ_C 131.0 and 156.6 ppm) suggesting that the compound **(3)** had a substituted aromatic ring structure in the para position. The presence of a substituted aromatic ring with the para position of compound **(3)** is amplified by the presence of two aromatic proton signals, each representing 2H (δ_H 6.66 and 6.97 ppm) with ortho-coupling multiplicity (J = 8.4 Hz). The presence of a highly downfield carbon signal (δ_C 156.6 ppm) indicates the presence of oxy-aryl carbon. Another substituent in the form of an alkyl chain group is characterized by the presence of one quaternary aromatic carbon with a carbon shift (δ_C 131.0 ppm). The spectrum analysis of zTOCSY1D shows the presence of ^{1}H-^{1}H correlation in one unit. There are three correlated proton signals (δ_H 2.62, 3.59 and 8.03 ppm) with these multiplicities were *t*, *m* and *s* respectively. This data confirmed that the side groups for aromatic systems are hydroxyethyl groups. The proton and carbon position certainty was seen from the environmental influences on the structure of compound **(3)**, in which proton carbon shifts in C-3 and C-5 will be more downfield compared to C-2 and C-6 carbon because of ortho positions with electron donating group (-OH, C-1). On the side chain, methylene-oxy carbon has a more downfield carbon and proton shift compared to the methylene group due to the electronegativity of the -OH group. Thus, based on the spectra ^{1}H, ^{13}C-NMR, HMQC and zTOCSY1D show that compound **(3)** was known to be a tyrosine amino acid derived that was tyrosol **(3)**. These NMR data confirmed with previous research [10, 11].

Compound **(4)** was obtained as a dark yellow. Based on the ^{13}C-NMR spectrum shown the presence of nine signals consisting of five C-methyn signals located in the C-sp^3 shift region (δ_C 62.3, 71.3, 75.7, 86.4 and 90.7 ppm) which were typical shift for ribofuranoside. In addition, the presence of a carbon signal at δ_C 90.7 ppm which was a typical shift for anomeric carbon. Downfield carbon (δ_C 166.2 ppm) was carbonyl carbon with a system of α, β unsaturated carbonyl. In addition, the presence two proton signals (δ_H 5.70, 8.01 ppm) with multiplicity *d* (J = 8.05 and 8.01 Hz) that was indicated the coupling cis. Another carbonyl carbon has δ_C 152.5 ppm which was characteristic for the carbonyl group that was bonded by two N atoms in C-2. The four carbon signals indicate the carbon shift for the pyrimidine base group of nitrogen compounds such as uracil. Based on the 1D and 2D NMR (Figure 6) spectrum, it was found that compound **(4)** had a structure consisting of uracil and arabinose sugar which was to be a nucleotides uridine. This was reinforced by the HMBC correlation between proton methyn olefin (δ_H 8.01 ppm (C-6)) correlated with C-anomeric (δ_C 90.7 ppm (C-1 ')), in the arabinose sugar system indicating a bond β-N1 glycosides which were characteristic of the nucleoside compound. So based on data spectrum ^{1}H, ^{13}C-NMR, HMBC and ^{1}H-^{1}H COSY it was known that compound **(4)** was uridine **(4)**.

(a)

(b)

Figure 6. (a) HMBC correlations (b) ^{1}H-^{1}H COSY correlations of uridine **(4).**

Compound **(5)** was isolated as a white solid. Based on the ^{13}C-NMR spectrum data, there were 10 carbon signals located in the C-sp^2 shift region. The 10 carbon signals consist of four carbon quaternary signals (δ_C 110.7, 152.7, 129.4 and 133.3 ppm) and six carbon methyl olefins (δ_C 117.6, 131.3, 128.3, 123, 9, 124.1 and 127.3 ppm). Based on the value of the carbon shift it can be seen that compound **(5)** has a structure in the aromatic ring system where one of carbon quaternary was oxyaril carbon (δ_C 152,7 ppm). In addition, the proton shifts with a singlet-wide multiplicity at δ_H 5.07 ppm which was characteristic of -OH freely protons. One other carbon quaternary is thought to be a carbon olefin with a substituted chlorine atom. This is seen from the shifting of highly aromatic quaternary carbon shifts at δ_C 110.7 ppm and this quite similar with the data oh 1-chloronaphthalene-2-ol from previous research that shown in Table 1 [12]. The ^{1}H-NMR spectral data together with zTOCSY1D show that compound 5 has a naphthalene base framework. These data suggest that four signals in the aromatic proton region are an early feature of the substituted aromatic system (δ_H 7,16, 7.31, 7.38 and 7.89 ppm). Certainty of the substituent position is confirmed from the spectral data of zTOCSY1D showing the correlation of the four protons, so that both substituents are ortho positions. Based on ^{1}H-NMR spectrum data shown that two other protons are δ_H 7.39 and 7.98 ppm with coupling $J = 8.9$ Hz indicating that both protons were in the ortho positions. This was also confirmed by the correlation of zTOCSY1D on both protons. Based on ^{1}H, ^{13}C-NMR spectrum data as well as zTOCSY1D and comparing with reported data compound **(5)** is proposed as 1-chloronaphthalene-2-ol **(5)** [12].

3.2. Evaluation of Media Culture

The results of qualitative evaluation on EtOAc extract TLC profile of PDB and TSB media and three pure isolated compounds (Figure 7) showed that composition of media culture could influence *C. phyllanthi* in producing secondary metabolite. The difference composition of PDB and TSB was the nitrogen sources, where nitrogen source in PDB medium was amino acid mean while nitrogen source in TSB medium were *Enzymatic Digest of Casein and Enzymatic Digest of Soybean Meal*. Interaction

IOP Conf. Series: Materials Science and Engineering 1062 (2021) 012017 doi:10.1088/1757-899X/1062/1/012017

between endophytic fungi and fermentation condition could affect the production of secondary metabolite [13].

Table 1. ^{1}H dan ^{13}C-NMR data of 1-kloronaftalen-2-ol (**5**).

No. C	5*		5#	
	δ_H (*mult.*, *J* in Hz)	δ_C	δ_H (*mult.*, *J* in Hz)	δ_C
1	-	110,7	-	113,3
2	-	152,7	-	149,3
3	7,39 (*d*, 8,9)	117,6	7,59 (*t*, 8,8)	117,2
4	7,98 (*d*, 8,9)	131,3	8,07 (*d*, 8,6)	129,4
4a	-	129,4	-	128,4
5	7,16 (*d*, 8,6)	128,3	7,27 (*s*)	128,1
6	7,31 (*td*, 1,4, 7,6)	123,9	7,42 (*t*, 7,9)	122,7
7	7,61 (*td*, 1,3, 7,5)	124,1	7,73 (*d*, 8,9)	124,1
8	7,89 (*d*, 8,1)	127,3	7,81 (*d*, 8,1)	127,5
8a	-	133,3	-	131,0
-OH	5,07 (*s*)	-	5,90 (*s*)	

*measured in $CDCl_3$, 500 MHz (^{1}H) and 125 MHz (^{13}C)
#measured in $CDCl_3$, 500 MHz (^{1}H) and 126 MHz (^{13}C) [12]

Ergosterol **(1)** and ergosterol peroxide **(2)** were found in *C. phyllanthi* MeOH extract from both media culture, because these compound were the major compound that consist in fungi cell membrane [14]. In other side, Tyrosol **(3)** was produced more by *C. phyllanthi* in EtOAc extract from cultivation medium with PDB, while production of tyrosol **(3)** decreased when *C. phyllanthi* was cultivated with TSB medium. This is demonstrated by the TLC profile under 254 nm UV light suspected of tyrosol **(3)** concentration in the EtOAc extract from TSB culture is decreased. The decline concentration of tyrosol **(3)** was also suspected because there were other secondary metabolites on TSB media which was not produced by *C. phyllanthi* when cultivated on PDB media such as 1-chloronaphthalene-2-ol **(5)**. The presence of 1-chloronaphthalene-2-ol **(5)** could affect the concentrations of other compounds that found in PDB media. Based on the results of research on media culture evaluation proven that the state of fermentation environment could affect the production of secondary metabolites in endophytic fungi *C. phyllanthi*.

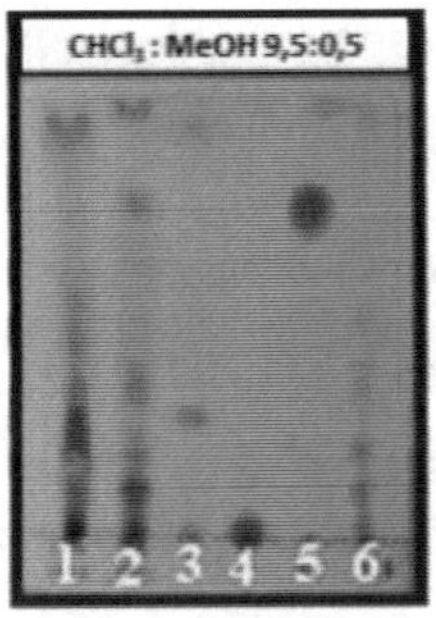

(a)

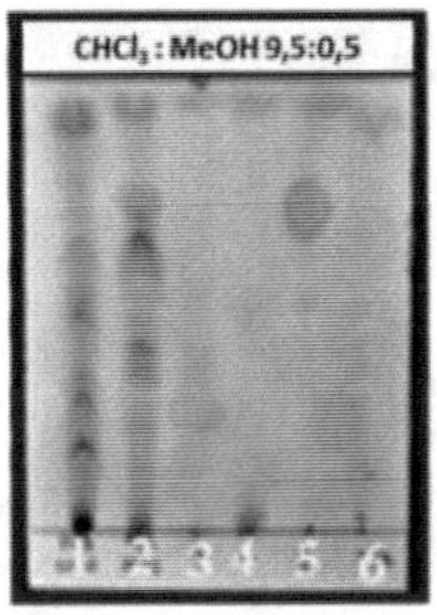

(b)

Figure 7. TLC Profile comparison of EtOAc *C. phyllanthi* extract from cultivation medium with PDB and TSB and three pure isolated isolates ie tyrosol (3), uridine (4) and 1-chloronaphthalene-2-ol (5).

1: EtOAc extract of cultivation with PDB media, 2: EtOAc extract cultivated with TSB medium, 3: tyrosol (**3**), 4: uridine (**4**), 5: 1-chloronafatlen-2-ol (**5**), 6: extract media TSB

4. Conclusion

Based on the present research that has been done, it can be concluded that it has been successfully isolated single isolate of the endophytic fungi *Colletotrichum phyllanthi* from stem's *Physalis peruviana*. Successfully isolated five known compounds, two steroid compound that is ergosterol **(1)** and ergosterol peroxide **(2)** from the MeOH extract of mycelia, a tyrosine amino acid derivative that is tyrosol **(3)** and uridine which is nucleoside compound **(4)** from EtOAc extract cultivated with PDB culture medium from endophytic fungi *C. phyllanthi*. Another compound that is 1-chloronaphthalene-2-ol **(5)** isolated from EtOAc extract from cultivation with TSB medium. Composition of cultural media can affect the growth capability of endophytic fungi as well as secondary metabolites produced by endophytic fungi *C phyllanthi*. *C. phyllanthi* can produce more tyrosol **(3)** when cultivated on PDB media compared with TSB medium while 1-chloronaphthalene-2-ol **(5)** is produced only when endophytic fungi *C. phyllanthi* is cultivated on TSB media.

References

[1] Gouda, S, Das, G, Sen, S K, Shin, H –S, Patra, J K 2016 *Frontier in Microbiology*, **7**, 1538
[2] Kusari, S, Hertweck, C, Spiteller, M 2012 *Chemistry and Biology Perspective*, **19**, 792 – 798
[3] Yan, J F, Broughton, S J, Yang, S L, Gange, A C 2015 *Fungal Ecology*, **13**, 53 – 59
[4] Somjaipeng, S, Medina, A, Mahan, N 2016 *Enzyme and Microbial Technology*, **90**, 69 – 75
[5] Zou, W X, Meng, J C, Lu, H, Chen, G X, Shi, X, Zhang, T Y, Tan, R X J 2000 *Nat. Prod.*, **63**, 1529 – 1530
[6] Calvo, A M, Wilson, R A, Bok, J W, Keller, N P 2002 *Microbiology and Molecular Biology Reviews*, **66** (3), 447 – 459
[7] VanderMolen, K M, Raja, H A, El-elimat, T, Oberlies, N H 2013 *AMB Express*, **3**, 71
[8] Dewick, P M 2009 *Medicinal natural products: a biosynthetic approach* (Wiley)
[9] Wang, Y, Xu, L, Ren, W, Zhao, D, Zhu, Y, Wu, X, B 2012 *Phyto* **19**, 364 – 368
[10] Cimmino, A, Cinelli, T, Masi, M, Reveglia, P, Silva, M A, Mugnai, L, Michereff, S J, Surico, G, Evidente, A 2017 *J. Arg. Food. Chem.,* **65**, 1102 – 1107
[11] Capasso, R, Cristinzio, G, Evidente, A, Scognamiglio, F 1992 *Phytochemistry*, **31** (12), 4125 – 4128
[12] Nahide, P D, Ramadoss, V, Juárez-Ornelas, K A, Satkar, Y, Alvarado, R O, Villanueva, J M C, Castro, A J A, Morales, J R Z, Morales, M A. R, Padilla, A J R, Álvarez, M A D, Alvarado, C R S 2018 *Eur. J. Org. Chem*, 485 – 493
[13] Kusari, S., Hertweck, C, Spiteller, M 2012 *Chemistry and Biology Perspective*, **19**, 792 – 798
[14] Carvalho, J M, Paixão, L K O, Dolabela, M F, Marinho, P S B, Marinho, A M R 2016 *Acta Amazonica*, **46** (1), 69 – 72.

Susceptibility of Sesbagrandiflorain B against Chlorination: A DFT Study

Mita Nurhayati[1*], Hafiz Aji Aziz[1], Muhamad Abdulkadir Martoprawiro[2]
[1] Universitas Pendidikan Indonesia, Bandung, Indonesia
[2] Institut Teknologi Bandung, Bandung, Indonesia

*Corresponding e-mail: mita@upi.edu

Abstract. Sesbagrandiflorain B is a novel aryl benzofuran derivative isolated from Sesbania grandiflora. Due to its novelty, there is almost no study on the reactivity and properties on this compound. In this study, we perform calculation to determine the reactivity of sesbagrandiflorain B against chlorination reaction based on fukui indices and energy calculation of reactant, intermediate, and product structures. The properties of the system are calculated using DFT B3LYP/6-311++g(d), and to simulate the effect of solvent we employ polarizable continuum model (PCM) on GAMESS-US 2018 package software. From the structure of sesbagrandiflorain B, there are five possible positions that susceptible to chlorination. The energy of reaction in those positions are significantly different from each other with energy of the intermediate formations are 446.48 kJ, 428.06 kJ, 428.63 kJ, 364.08 kJ, and 388.86 kJ, indicating that the selectivity of sesbagrandiflorain B against chlorination reaction. The intermediate formation of sesbagrandiflorain B chlorination that is simulated without solvent tend to be endothermic while the implementation of PCM to the system significantly reduce the intermediate energy, suggesting that the solvent promote the reaction to be thermodynamically driven. For some position, the solvent also stabilizes the product of chlorination reaction.

Keywords: benzofuran, sesbania grandiflora, chlorination, sesbagrandiflorain

1. Introduction

Heterocyclic compounds have important roles in discover and synthesis novel molecules which possess physiological and pharmacological activity. One of heterocyclic compound that widely used is benzofuran. Benzofuran derivatives are very useful for developing new potential therapeutic agents. It is the important intermediate on synthesis a lot of new and useful heterocyclic compound that is difficult to obtain without involving benzofuran as intermediate [1]. It also has a good bioactivity such as antihyperglycemic, analgesic, antiparasitic, antimicrobial, antitumor, and kinase inhibitor. Substituted benzofuran also applied in fluorescence sensor, oxidant, bleaching agent, and so on [2].

Many benzofuran derivatives compound isolated from natural products. Recently, the isolation of novel benzofuran derivatives compound, 6-methoxy-2-(2′,3′-dihydroxy-5′-methoxyphenyl)-1-benzofuran-3-carbaldehyde (sesbagrandiflorain A) and 6-hydroxy-2-(2′,3′-dihydroxy-5′-methoxyphenyl)-1- benzofuran-3-carbaldehyde (sesbagrandiflorain B) from the stem bark of Sesbania grandiflora has been reported [3]. Antibacterial activity and cytotoxicity of EtOAc extract of the stem bark of S. grandiflora including sesbagrandiflorain A and B were also conducted [4]. S. grandiflora is a flowering plant that is native to tropical Asia. Different parts of this plant have been used as herbal medicines to treat anemia, bronchitis, fever, and tumors [5-7]. Earlier studies on the leaves, seeds, and roots of S. grandiflora showed the presence of various secondary metabolites, e.g., α-5-methyl-5-

pentacosanol, galactomannan, and flavonoids, some of which exhibited antituberculosis activity against Mycobacterium tuberculosis H37Rv [8-11]

Based on the structure and the abundance of S. grandiflora, sesbagrandiflorain A and B is expected to be a beneficial compound and potential starting material of various benzofuran derivatives compound that is hard to be synthesized from another starting material. However, the study of sesbagrandiflorain's chemical reactivity is very rarely conducted. Chemical reactivity of sesbagrandiflorain can be studied theoretically through computational calculation. Density Functional Theory (DFT) is widely used and effective to obtain various information of a molecule [12]. Recently, we have reported theoretical study of chemical reactivity descriptor, such as ionization energy, electron affinity, hardness and electronegativity of sesbagrandiflorain A [13]. Another descriptor is Fukui indices, that predicts the susceptibility of molecule against electrophilic, nucleophilic and radical reactants [14].

Electrophilic substitution plays the important roles in reaction mechanism of many bioactive molecule syntheses. One of the important electrophilic substitution is chlorination reaction. The presence of chlorine plays the pivotal roles in a large number of natural products. It has been found that the introduction of a chlorine atom into one or more specific positions of a biologically active molecule may improve the intrinsic biological activity [15,16]. The application of chlorine in medicinal chemistry is one of the fastest growing hot areas in chemistry as its fascinating and instructive role of halogens distribution in the field of drug development. Interestingly, among four halogens, chlorine is the one which is more frequently found in drugs than others, even fluorine [17,18]. In this study, we perform Fukui indices calculation of sesbagrandiflorain B. Furthermore, susceptibility of sesbagrandiflorain B against electrophilic reagents is studied through chlorination reaction. Hopefully it can provide information to develop sesbagrandiflorain into other useful compounds.

2. Experimental Setup

All of computational calculation were conducted on high performance computer of ITB and public server XSEDE [19,20]. Quantum mechanics calculation was performed using GAMESS-US 2018 package software [21].

2.1. Geometry Optimization and Single Point Calculation

Geometry optimization were conducted to obtain the combination of bonds length, bonds angle and dihedral angle of the molecular structure which has the lowest energy [12]. Geometry optimization is performed on stable structures (products and reactants) as well as intermediate structures using Hartree-Fock (HF) method with basis set of 6-31G. The HF method was chosen due to its quite accurate results of structure prediction, but shorter calculation time compared with higher level methods. All of geometry optimization were conducted without solvent and with solvent, using Polarizable Continuum Model (PCM) solvent. PCM is one of methods that represent the solvated molecule model based on quantum mechanical calculation [22]. The solute, that influenced by interacting solvent, was treated by quantum mechanics, while the solvent was modeled as unlimited polarizable continuum [23]. In this experiment, we simulated the calculation without solvent, PCM water and PCM ethanol.

Higher level methods were needed to calculate the energy of molecule since the energy that obtained from HF calculation is quite overestimate [24]. In this experiment, single point energy calculation is performed using various levels of theory and basis set, such as B3LYP/6-31G, B3LYP/6-31++G, B3LYP/6-311++g, B3LYP/6-311++g(d) and M06/6-31G. Those result were used to determine the electronic population distribution in the sesbagrandiflorain molecule. Furthermore, electronic population distribution was used to calculate Fukui Indices, that determine the susceptibility of molecule against electrophilic, nucleophilic and radical attack [14].

2.2. Fukui Indices

Local reactivity of molecule can be predicted from Fukui Indices based on electronic population [25]. In this experiment, we utilized Lowdin population that obtained from single point calculation to calculate Fukui Indices, based on equation (1), (2) and (3).

$$fk^{+} = qk(N+1) - qk(N) \tag{1}$$

$$fk^{-} = qk(N) - qk(N-1) \tag{2}$$

$$fk^{0} = \frac{qk(N+1) - qk(N-1)}{2} \tag{3}$$

Which qk is Lowdin population, N is the number of neutral molecules and fk^{+}, fk^{-}, and fk^{0} correspond to Fukui Indices that in a good agreement with the reactivity towards nucleophilic, electrophilic and radical attack, respectively. In this study, we perform the fk^{-} to determine the reactivity towards electrophilic, especially chlorine species that is involved in chlorination reaction.

2.3. Chlorination Reaction Energy

In this research, we performed chlorination reaction energy calculation on several possible position based on Fukui indices results. Chlorination reaction energy was calculated from the differences between product and reactant energy. Intermediate structure of each possible mechanism also calculated. All of the structures mentioned were optimized using HF/6-31g and single point energy calculation was calculated using B3LYP/6-311++g(d). The results in various solvents were then compared.

3. Results and Discussion

3.1. Geometry Optimization and Single Point Calculation

The result of geometry optimization without solvent shows that benzofuran moiety of sesbagrandiflorain is planar, while the aryl moiety that connected to sesbagrandiflorain B has different plane with benzofuran (Figure 1). There is no significant difference between structure of sesbagrandiflorain B that optimized with or without various solvent. This structure was then used as input for single point calculation to determine the susceptibility electrophilic attacks.

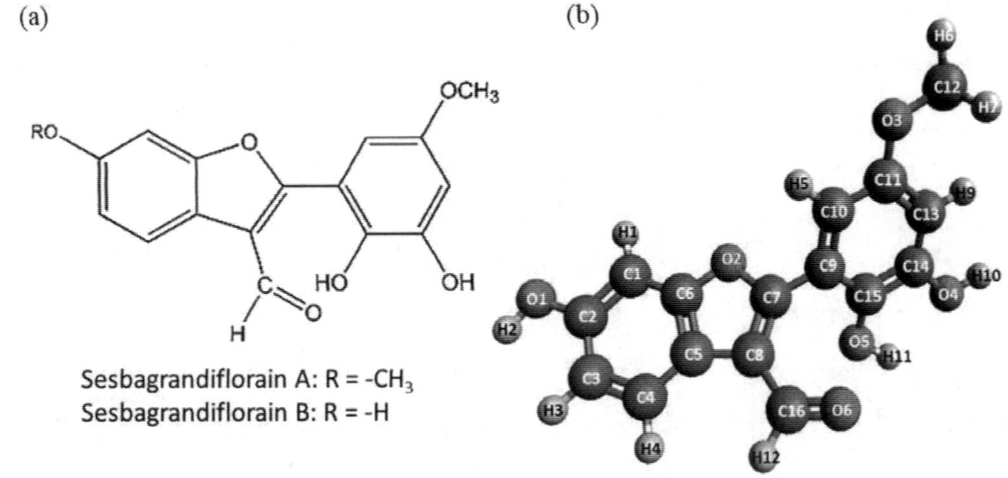

Figure 1. (a) Molecular structure of sesbagrandiflorain A and (b) optimized structure of sesbagrandiflorain B

IOP Conf. Series: Materials Science and Engineering 1062 (2021) 012018 doi:10.1088/1757-899X/1062/1/012018

3.2. Fukui Indices

Fukui Indices was then calculated with various theory. The order of reactivity towards electrophilic attack can be seen in Table 1, 2 and 3, that demonstrate almost similar tendency when it simulated in non-solvated or solvated (PCM water and methanol) model. However, the molecule tends to undergo electrophilic attack on the water as its solvent, due to the higher Fukui Indices.

Table 1. Reactivity order of atoms in sesbagrandiflorain B molecule toward electrophilic attack (no solvent).

No	HF/ 6-31g	B3LYP/ 6-31g	B3LYP/ 6-31++g	B3LYP/ 6-311++g	B3LYP/ 6-311++g(d)
1	O5	O5	O5	O5	O4
2	C15	C15	C15	C15	O5
3	C8	O3	O4	O4	C15
4	O6	O4	O3	O3	C14
5	O4	C11	C11	C11	C11

Table 2. Reactivity order of atoms in sesbagrandiflorain B molecule toward electrophilic attack (PCM Water).

No	HF/ 6-31g	B3LYP/ 6-31g	B3LYP/ 6-31++g	B3LYP/ 6-311++g	B3LYP/ 6-311++g(d)
1	O5	O5	O5	O5	O4
2	C15	C15	C15	C15	O5
3	C8	O3	O4	O4	C15
4	O6	O4	O3	O3	C14
5	O4	C11	C11	C11	C11

Table 3. Reactivity order of atoms in sesbagrandiflorain B molecule toward electrophilic attack (PCM Ethanol).

No	HF/ 6-31g	B3LYP/ 6-31g	B3LYP/ 6-31++g	B3LYP/ 6-311++g	B3LYP/ 6-311++g(d)
1	O5	O5	O5	O5	O4
2	C15	C15	C15	C15	O5
3	C8	O3	O4	O4	C15
4	O6	O4	O3	O3	C14
5	O4	C11	C11	C11	C11

Based on Table 1, 2, and 3, we can be clearly observed that some atoms in the sesbagrandiflorain molecules tend to be more reactive (higher fk^{-}) in solvated condition rather than non-solvated, such as O1, C8 and C2. This is the finding that in order to carry out a reaction involving a certain electrophilic, a suitable solvent is necessarily needed. From the structure of sesbagrandiflorain B, it is known that there are five points that allow electrophilic substitution reaction to occur due to the substitutable proton availability (Figure 2), labelled as C1, C3, C4, C10 and C13 as well as written on Figure 1.

IOP Conf. Series: Materials Science and Engineering 1062 (2021) 012018 doi:10.1088/1757-899X/1062/1/012018

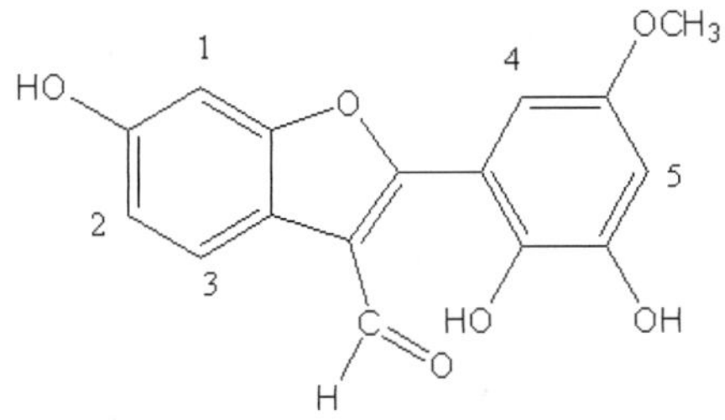

Figure 2. Five plausible of chlorination reaction take place

The order of fk^- for some atoms that plausible to undergo electrophilic attack that derived from calculation with various method and condition is served in Table 4. The order of reactivity is quite similar for the same system that calculated with various methods, but exhibit the different result regarding to the order of reactivity. It confirms that each solvent is affecting the regioselectivity.

Table 4. Order of group reactivity toward electrophilic attack

Methods	Order of reactivity		
	No solvent	PCM water	PCM methanol
HF/6-31g	C10>C13>C3>C1>C4	C3>C13>C10>C4>C1	C3>C13>C10>C4>C1
B3LYP/ 6-31g	C10>C3>C13>C4>C1	C13>C4>C3>C1>C10	C13>C4>C3>C1>C10
B3LYP/ 6-31++g	C10>C3>C13>C4>C1	C13>C3>C4>C1>C10	C13>C3>C4>C1>C10
B3LYP/ 6-311++g	C10>C3>C13>C4>C1	C13>C3>C4>C1>C10	C13>C3>C4>C1>C10
B3LYP/ 6-311++g(d)	C3>C10>C13>C1>C4	C13>C3>C4>C1>C10	C13>C3>C4>C1>C10

3.3. Chlorination Reaction Energy

Sesbagrandiflorain B molecules can undergo reactions that involve either an electrophilic or a nucleophilic attack. One plausible reaction is an electrophilic substitution reaction. In this research, we calculated the species involved in the chlorination reaction of the sesbagrandiflorain molecule, which is an electrophilic substitution reaction. Generally, the mechanism of electrophilic substitution of an aromatic compound occurs in two stages; attack by an electrophile to produce an intermediate cation and releasing a proton from the cation to restore aromatization [26]. The energy calculation was carried out on the reactant, intermediate and product in different condition (non-solvated and PCM water) to investigate the susceptibility of molecule towards chlorination reagent.

Chlorination reaction tends to occur exothermically [27]. We calculate the energy of reactant, intermediate and product to investigate the susceptibility of sesbagrandiflorain B towards chlorination reaction thermodynamically. The energy calculation of chlorination reaction in position 1-5 were shown in Figure 3-7. Total reaction that occur in all possible position is exothermic, both in the solvated model and non-solvent model. There is no significant different between the energy of total reaction. Interestingly, the intermediate formation reaction energy was significantly different. In the non-solvent model, stage 1 of the chlorination reaction was endothermic, while other was exothermic. It suggests that use of solvent stabilize the intermediate, thus making it easier for molecules to undergo reactions thermodynamically since the intermediate formation is rate determining step of chlorination reaction.

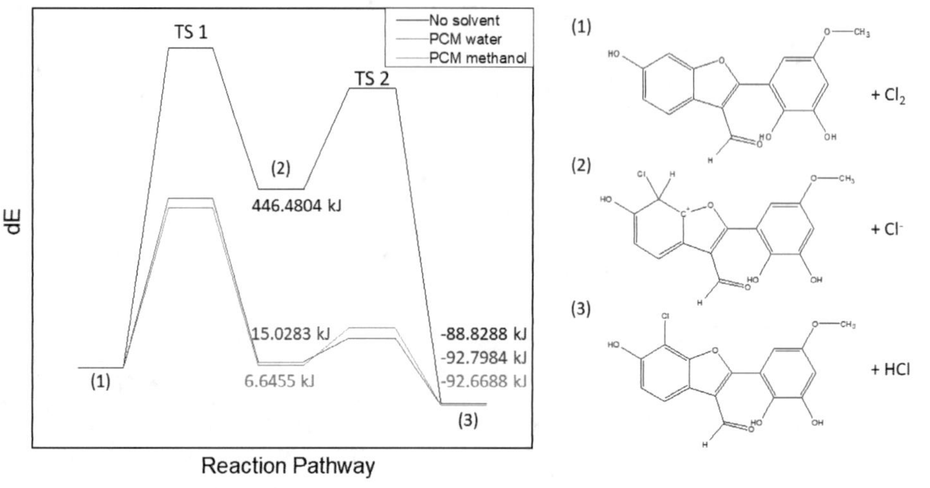

Figure 3. Energy diagram of the chlorination at position C1

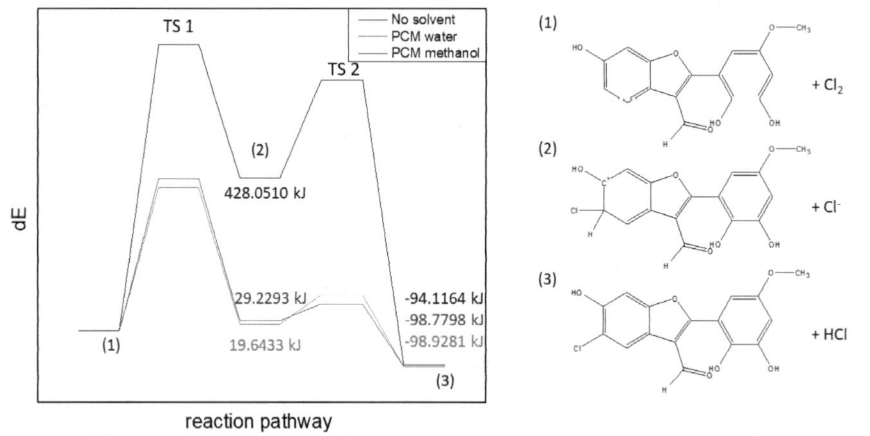

Figure 4. Energy diagram of the chlorination at position C3

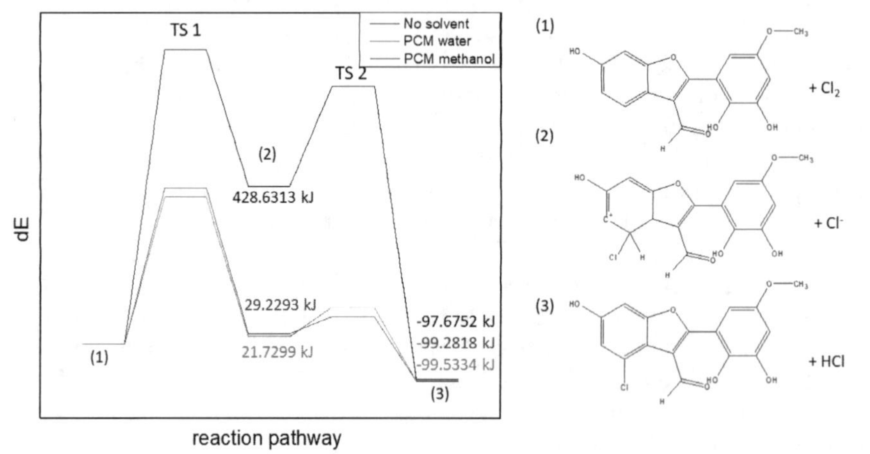

Figure 5. Energy diagram of the chlorination at position C4

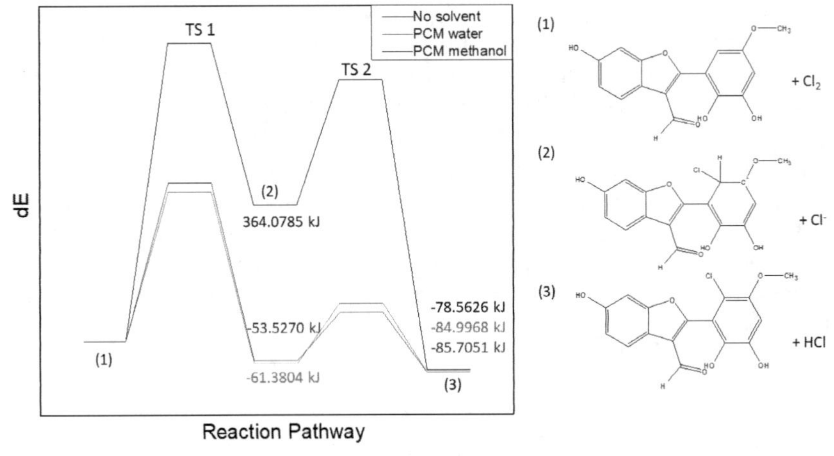

Figure 6. Energy diagram of the chlorination at position C10

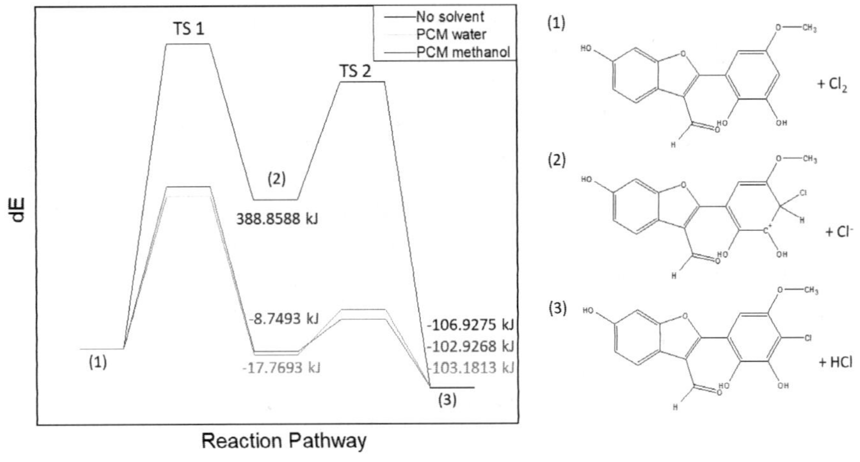

Figure 7. Energy diagram of the chlorination at position C11

There is a difference between the energy of the intermediate formation, especially at position C1, C3, C10 and C13, indicating that the selectivity of sesbagrandiflorain B against chlorination reaction. The lowest intermediate formation reaction energy of the system that calculated in non-solvent model is C3, indicating that this position is the most reactive moieties toward chlorination reaction (based on thermodynamic point of view), in line with the reactivity prediction from Fukui indices (Table 4). Fukui indices prediction also in a good agreement with the intermediate formation reaction energy of solvated model, demonstrate that the lowest energy of reaction is reached by the chlorination at position C13. It indicates that the most reactive moieties toward chlorination reaction in the solvated model is C13. The intermediate formation of chlorination reaction at position C10 is the lowest, but the product energy is not significantly different and the product is not stabilized thus chlorination reaction is not thermodynamically favourable at that position. The energy of chlorination reaction in PCM water have a tendency to be slightly lower than the reaction that modelled in PCM methanol. It confirms that water stabilize the intermediate and the product better than that of methanol. Thus, the reaction will be more favourable to occur in the more polar solvent.

4. Conclusion

The chlorination reaction of sesbagrandiflorain B is exothermic reaction, while the intermediate formation of some position is endothermic. The energy of reaction in those positions are significantly different from each other with energy of the intermediate formations are 446.48 kJ, 428.06 kJ, 428.63 kJ, 364.08 kJ, and 388.86 kJ based on non-solvent model; 6.64 kJ, 19.64 kJ, 21.73 kJ, -61.38 kJ, and -17.77 kJ for PCM water; 16.03 kJ, 29.23 kJ, 29,22 kJ, -53.53 kJ and -8.75 kJ for PCM methanol; indicating that the selectivity of sesbagrandiflorain B against chlorination reaction. The intermediate formation of sesbagrandiflorain B chlorination that is simulated without solvent tend to be endothermic while the implementation of PCM to the system significantly reduce the intermediate energy, suggesting that the solvent promote the reaction to be thermodynamically driven. For some position, the solvent also stabilizes the product of chlorination reaction.

References

[1] Abuhashem A A et al 2014 *Synthetic Communication* **44** 2899 – 2920.
[2] Shamsuzzaman H K 2015 *European Journal of Medicinal Chemistry* **97** 483-504
[3] Noviany Net al 2018 *Natural Product Research* 1-7
[4] Noviany N et al 2020 *Phytochemistry Letters* **35** 211-215
[5] Wagh V D et al 2009 *Journal of Pharmacy Research* **2** 889–892
[6] Laladhas K P et al 2010 *J. Cell. Mol. Med.* **14** 636-646
[7] Powell R G et al 1984 *Phytochemistry* **23** 2789-2796
[8] Tiwari R D and Bajpai R K 1964 *Proc. Nat. Acad. Sci. India Sect A* **34** 239-244
[9] Pollard MA et al 2011 *Carbohyd. Polym.* **84** 550-559
[10] Hasan N et al 2012 *Pharmaceuticalas* **5** 882-889
[11] Noviany N et al 2012 *J. Basic & Appl. Sci* **8** 253-256
[12] Cramer C J 2004 *Essentials of Computational Chemistry Theories and Models 2nd Edition* (England, John Wiley & Sons Ltd)
[13] Nurhayati M and Martoprawiro M A 2019 *Prosiding Seminar Nasional Kimia UIN Sunan Gunung Djati Bandung* 53-61
[14] Kolandaivel P et al 2005 *J. Chem. Sci.* **117** 591-598
[15] Naumann K 2000 *Pest Management Science: Formerly Pesticide Science* **56**(1) 3-21
[16] Fang et al 2019 *European Journal of Medicinal Chemistry* **173** 117-153
[17] Kosjek T et al 2012 *Halogenated Heterocycles* **27** 219-246
[18] Parsini E et al 2011 *Chem. Soc. Rev.* **40** 2267-2278
[19] Perri M J et al 2014 *Journal of Chemical Education* **91**(12) 2206-2208
[20] John T 2014 *Computing in Science & Engineering* **16**(5) 62-74
[21] Barca G M J et al 2020 *J. Chem. Phys.* **152** 154102
[22] Mennucci B 2012 *WIREs Comput Mol Sci* **2** 386-404
[23] Cossi M et al 1998 *Chemical Physics Letters* **286** 253-260
[24] Martinez J P 2017 *Annalen der Physik* **529**(1-2) 1600328
[25] Yang W and Mortier W J 1986 *J. Am. Chem. Soc.* **108** 5708-5711
[26] Clayden J et al 2012 *Organic Chemistry 2nd edition* (United States: Oxford University Press)
[27] Altarawneh M et al 2009 *Progress in Energy and Combustion Science* **35**(3) 245-274

Experimental Determination of Water, Water/ Ethylene Glycol and TiO_2-SiO_2 Nanofluids mixture with Water/Ethylene Glycol to Three Square Multilayer Absorber Collector on Solar Water Heating System: A Comparative Investigation

M A Fikri[1,4*], W M Faizal[1], H K Adli[1], Z Bo[2], X X Jiang[2], A I Ramadhan[3]
[1]Faculty of Bioengineering and Technology, Universiti Malaysia Kelantan (Jeli Campus), Malaysia
[2]School of Mechanical Engineering, Ningxia University, China
[3]Faculty of Engineering, Universitas Muhammadiyah Jakarta, Jakarta, Indonesia
[4] Engineering Division, Jabatan Kesihatan Negeri Kelantan, Kementerian Kesihatan Malaysia, Malaysia

*Corresponding e-mail: moamfik.selehor@gmail.com

Abstract. This paper investigates three square multilayer absorber solar collector in Solar Water Heating System (SWHS) experimentally. The main aim was to study the output temperature of the solar absorber collector based on distilled water (DW), water/ ethylene glycol (W/ EG) and TiO_2-SiO_2 nanofluids. An experimental apparatus for testing absorber solar collector was designed and built at Universiti Malaysia Kelantan (Jeli Campus). By using the two-step method, TiO_2-SiO_2 nanofluids with ratio 30:70 and volume concentration at 1.0%. For W/ EG the ratio is 60:40. The effect working fluid was studied experimentally on the output temperature of three square multilayer absorber solar collector in SWHS with Angle of Sunlight (AoS) at 45°, volume flow rate at 3 litres per minute (LPM) and varies Intensity of Light (IoL) at 300, 500 and 700 W/m^2. The output temperature is determined through the experimental results by using this apparatus. The result reveals that the output temperature of the solar absorber collector is TiO_2-SiO_2 nanofluids with 1.0% of volume concentration as a best working fluid at each intensity of light.

Keywords: solar collector, water heating, nanofluids, inrensity

1. Introduction

Renewable energy comes from continuously replenishing natural processes [1, 2]. It derives directly from the sun or heat generated deep within the earth in its different forms. The definition includes electricity and heat from solar, wind, sea, hydropower, biomass, geothermal resources, and renewable biofuels and hydrogen [3]. Renewable sources of energy are classified into three categories: direct solar energy, indirect solar energy and non-solar renewable energy [4].

Most renewable sources of energy derived from solar radiation. Natural solar energy refers to the conversion of solar thermal and photovoltaic solar energy [5]. Indirect solar energy comprises wind, wave, and biofuels [6]. Non-solar renewable energy sources are those that do not rely on solar radiation. Two non-solar renewable sources of energy are tidal and geothermal. Renewable energy has become a critical engineering sector, as fossil fuel resources have diminished, and environmental concerns have increased. There is no doubt that a large proportion of the energy the human race consumes would come from several renewable sources in the future.

A solar water heater is a device capable of heating water by sunlight. It not only allows for substantial energy savings as, contrary to natural gas or fuel oil, but solar power is also free. Besides, it is a way of producing hot water for year-round sanitation without emitting any CO_2. Although solar water heating systems (SWHS) have been in decades, SWH technologies can be run efficiently and affordably in any environment with today's technological advances. The plans are specifically designed for different climatic and geographical areas of the country, depending on the heating needs SWHS can be classified into five primary types: direct circulation systems, indirect water heating systems, drain-back systems, air systems and pool heaters [7]. If some form of heat pump arrangement is used, the solar heating system will provide all the heat needed throughout the year. The heat pump can therefore be considered an efficient and relatively inexpensive auxiliary heat source. Naturally, the economy will be improved if the same heat pump system is used for both winter heating and summer cooling.

A solar collector is a flat box, consisting of three main parts, a transparent cover, tubes carrying a refrigerant, and an insulated back plate. The solar collector works on the principle of the greenhouse effect; incident solar radiation is transmitted through this surface to the transparent cover of the solar collector [8]. Usually the inside of the solar collector is evacuated; the energy contained in the solar collector is trapped and heats the coolant contained in the tubes. Typically the pipes are made of copper, and black is painted on the back plate to help absorb solar radiation [9]. The solar collector is usually insulated to avoid heat loss, the. Solar collector characterization is based on experimental techniques in addition to the validation of similar models. For different complexities, both methods can be implemented [10].

A nanofluid is a mixture in the base fluid of nanoparticles suspended. It is an advanced heat transfer fluid with superior heat transfer properties [11, 12]. Recent nanotechnology advances produce fluids that have better thermal properties than conventional fluids [13, 14]. The frequency at which a substance transfers heat is thermal conductivity. It is a significant factor in increasing the efficiency of nanofluid in heat transfer, and it has been extensively studied by researchers [15, 16]. The rate of heat transfer by solids is much higher than that by liquids and gasses, which is why nanofluids have higher thermal conductivity values compared to their base fluids [15, 17, 18]. Viscosity is the critical parameter for determining the coefficient of convective heat transfer. Understanding the nanofluid of viscosity mechanisms is worrying due to the lack of a general mathematical model predicting viscosity activity in nanofluids. Several mathematical models have been developed to predict the viscosity of nanofluids. Nanofluids can be used in engine cooling, solar water heating, cooling devices for heat exchange, improving chillers heat transfer efficiency, and so on.

Nanofluids performed well and showed satisfactory results which prompted the researchers to think about the suspension in the base fluid of various combinations of nanoparticles that were later developed and called hybrid nanofluids [15]. Thermal conductivity is the most glittering nanofluid property that has required researchers attention most. These are because of the genuinely complex and important thermophysical property, but unfortunately, experimental studies and theories have not yet explained the hybrid nanofluids of increased thermal conductivity. Research has revealed the following important factors that have influenced hybrid nanofluids of thermal conductivity. In recent years, there have been many advancements in this field, but the most significant was its progress towards hybrid nanofluids [19-21]. The experts made a step forward at that time and started analyzing advanced nanofluids.

A result based on the effect of the working fluid is used in the solar collector is still new, especially by using nanofluids in the system. However, we still need to research to make sure that the use of nanofluid in solar thermal engineering devices improves their efficiency. Hence, this study aims to study the effect of solar water heater based on the difference of working fluid used. In this experiment, it is attempted to determine the output temperature of the three square multilayer absorber solar collectors at different working fluids between distilled water (DW), water/ethylene glycol (W/EG) and TiO_2-SiO_2 nanofluids with 1.0% of volume concentration.

2. Materials and Method

2.1. Nanoparticles Materials and Base Dilution

The average diameters for TiO_2-SiO_2 nanoparticles obtained from US Research Nanomaterials, Inc. were 50 and 22 nm with weight concentrations of 40 and 25 wt. %. Table 1 shows the properties of the selected nanofluids. However, distilled water and ethylene glycol were used as liquid-based, and the mixture ratio used was in the ratio (60:40). Table 2 shows the characteristics of the ethylene glycol.

Table 1. Nanoparticle Properties [21]

Type of nanoparticle	Diameter, (nm)	Weight concentration, ω (%)	Volume concentration, ø (%)	Specific heat, C (J/Kg.K)	Density, ρ (kg/m^3)
TiO_2	50	40	13.62	692	4230
SiO_2	30	25	13.06	745	2220

Table 2. Physical and Chemical Properties of Ethylene Glycol [22]

Parameter	Value
pH	5.5-7.5
Melting Point (°C)	-13
Boiling Point (°C)	197.6
Vapor pressure (mmHg)	0.12
Density (kg/m)	1113
Molar Mass (g/mol)	62.07

2.2. Thermo-physical Properties Measurement

Figure 1 shows the KD2 Pro is used as a property analyst to measure the thermal conductivity of a sample from Decagon Devices, Inc., USA. After two hours through the sonication process, data were collected for a temperature range of 30 to 70 °C.

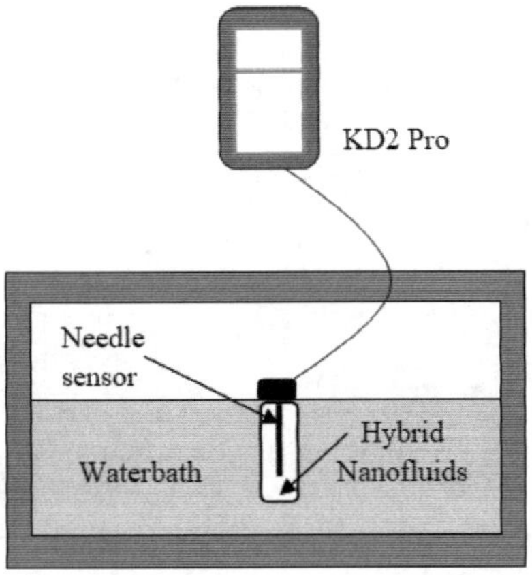

Figure 1. Apparatus and Instrument used in thermal conductivity measurement.

Figure 2 shows the Brookfield LVDV-III Ultra Rheometer is used to measure the viscosity of nanofluids. This method is used in viscosity measurement by several researchers [21, 23, 24]. At temperatures between 30 °C and 70 °C, viscosity was measured, and 30 min was allowed to stabilize the temperature, and its value was recorded at stable conditions.

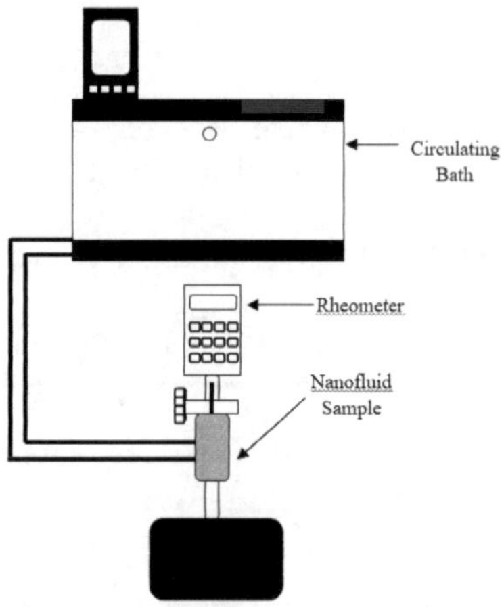

Figure 2. LVDV III Ultra Rheometer used in viscosity measurement.

2.3. Experimental Setup

Figure 3 shows a simple schematic overview of components for solar absorber collector which consists of a few essential elements such as collecting tank, solar submersible pump, collector, control panel, 12 pieces of the spotlight, interface of Arduino Uno, K-type thermocouple and data logger. A solar submersible pump is connected to a collecting tank to circulate the working fluid through the system. The collector consists of three square multilayers (0.23 × 0.23 m; 0.33 × 0.33 m; 0.43 × 0.43 m) and insulated by spraying with the flat black spray. K-type thermocouples are used and fixed at the inlet, outlet and three square multilayer absorber solar collector of solar water heating system.

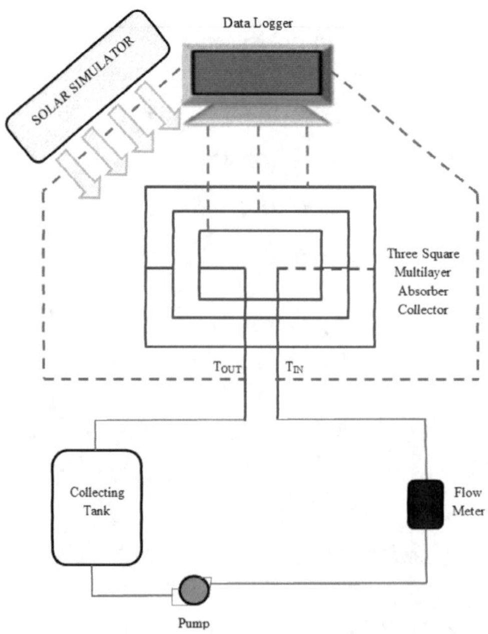

Figure 3. Schematic diagram of the heat transfer test setup.

2.4. Experimental Procedure

The experiment started with distilled water being filled into the supply tank, followed by W/ EG and then TiO_2-SiO_2 nanofluids. Solar simulators replace sunlight as these experiments are conducted in a lab. By exposing the solar collector absorber to the sunlight for 15 minutes of heating, data of solar collector is collected. Then, the light on the solar simulator is turned off for the cooling process for 15 minutes. The data obtained is recorded by the data logger every 2 minutes for up to 30 minutes. The data taken are as follows:

a. Input temperature (T_{in})
b. Output temperature (T_{out})
c. Ambient temperature (T_{at})
d. Absorber temperature (T_{ab})

Table 3. Detail description of the solar radiation test rig

No.	Description	Specification
1	Absorber solar collector	Triple square multilayer
2	Thermocouple	K-type
3	Pyranometer	Apogcc Logan UT SP-110
4	Spotlight	Halogen 500 W (12 pieces)
5	Data logger	ADAMView Advantech Data Acquisition
6	Solar submersible pump	Max 7LPM
7	Dimmer	Setting up Intensity Light (300, 500 & 700 W/m^2)
8	Interface Arduino Uno	Coding system
9	Arduino Uno	Reading up-flow meter (LPM)
10	Supply tank	5L
11	Controller	Controlling the speed of the pump (by Voltage and Current)

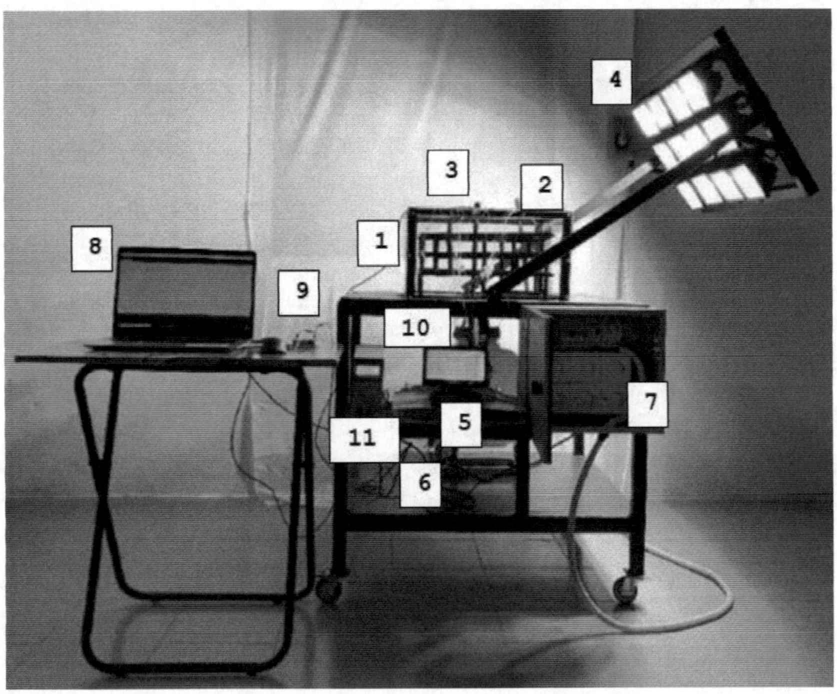

Figure 4. Experimental setup at Angle of Sunlight (AoS) 45°.

3. Results and Discussion

Experiments were performed to compare the response of the three square multilayer absorber solar collector to the fluid difference between distilled water, water/ ethylene glycol (W/EG) and TiO_2-SiO_2

nanofluids with 1.0% of volume concentration. The intensity of Light (IoL) varies starting by 300, 500 and 700 W/m^2. All data obtained is through data logger with a k-type thermocouple, which is attached to absorber collector for getting output temperature.

3.1. Effect of working fluid

The results are shown in Figure 5 the output temperature in the working fluid using TiO_2-SiO_2 nanofluids with 1.0% of volume concentration higher compared to W/ EG and DW. The output temperature at the beginning of the cooling process obtained on working fluid TiO_2-SiO_2 nanofluids with 1.0% of volume concentration is 34.9 °C. It maintains the temperature rise to 35.3 °C and always continues until the end of collecting data.

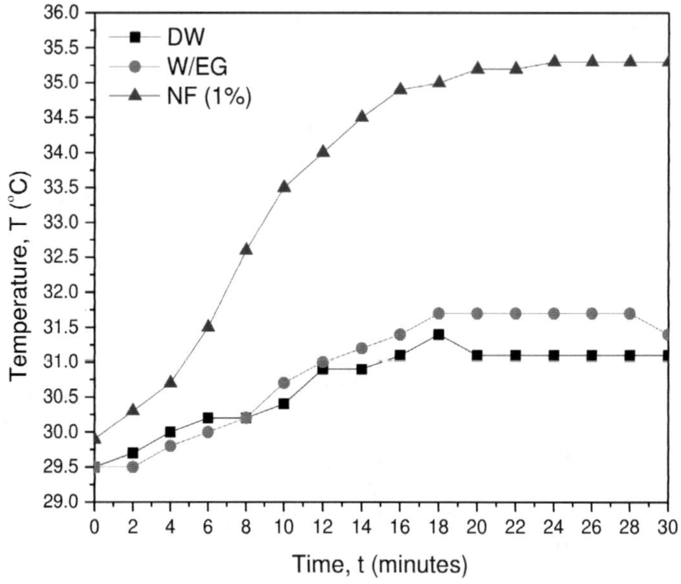

Figure 5. Temperature output of the absorber with solar radiation 300 W/m^2.

The volume flow rate of 3 LPM is observed consistently on a three square multilayer solar collector absorber through a tank in a closed-loop system assisted by a pump to complete the working fluid cycle. Figure 6 shows the output temperature in working fluid distilled water is lower compared to W/ EG and TiO_2-SiO_2 nanofluids with a volume concentration of 1.0%. However, the onset of cooling process output temperature for distilled water and W/EG did not show much difference of 31.2 and 32.5 °C respectively compared to the output temperature of TiO_2-SiO_2 nanofluids with a volume concentration of 1.0% of 34.6 °C.

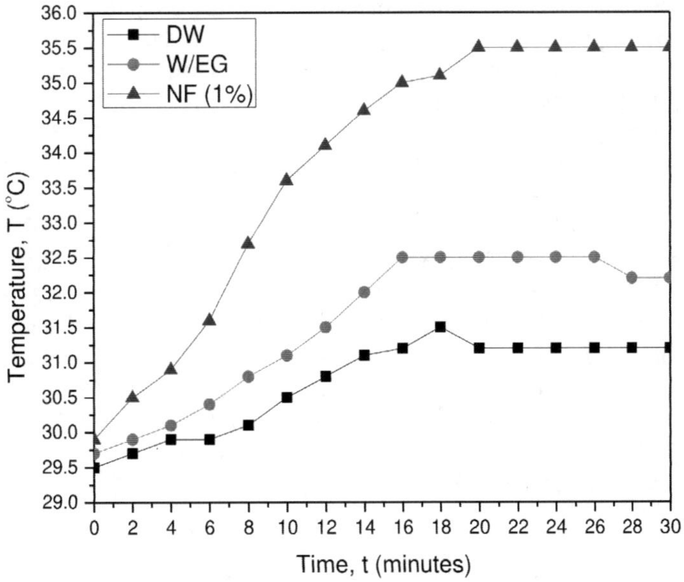

Figure 6. Temperature output of the absorber with solar radiation 500 W/m^2.

The output temperature obtained is based on the angle of sunlight (AoS) 45° lighting to the solar collector absorber. The average result got shows that TiO_2-SiO_2 with a volume concentration of 1.0% is the highest. Similarly, Figure 7 shows that the output temperature for TiO_2-SiO_2 with a volume concentration of 1.0% is 34.9 °C after the start of the cooling process. Followed by 33.6 °C by W/EG and the last one is DW, which is 32.4 °C. Then the output temperature will show constant value until the end of the cooling process at 30 minutes. The output temperature at 30 minutes is 32.4, 33.1 and 35.5 °C for DW, W/EG and TiO_2-SiO_2 with a volume concentration of 1.0% respectively.

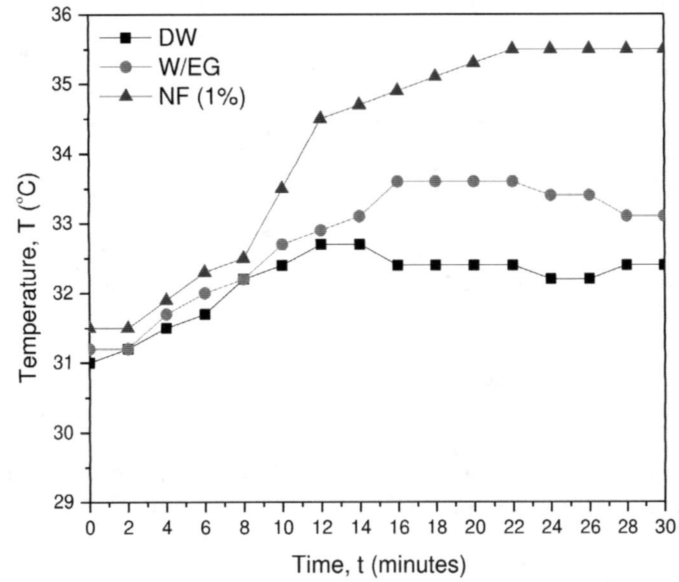

Figure 7. Temperature output of the absorber with solar radiation 700 W/m^2.

Therefore, it can be concluded that the working fluid on the three square multilayer absorber collector affects the data for the output temperature. The mass flow rate in the solar collector absorber provides an increase in temperature output rapidly based on the type of working fluid used. Therefore, copper pipes can transfer heat to the working fluid through a convection heat transfer and thus increase its temperature. M. Norhafana et al. [25] in the study of solar collector absorber using working fluid distilled water has shown an increase in heat transfer also found in the study conducted.

4. Conclusions

In this study, experimental determination to study the comparison of distilled water (DW), water/ethylene glycol (W/EG) and TiO_2-SiO_2 nanofluids as working fluid in three square multilayer absorber solar collector of Solar Water Heating System. In conclusion, the solar collector of three square multilayers of solar absorber collector is proven to play a vital role when it comes to adapting in multi-directional solar radiation with different radiation intensity of 300, 500 and 700 W/m2. The heat transfer performance by TiO2-SiO2 nanofluids with 1.0% of volume concentration showed the highest temperature output obtained, followed by water/ ethylene glycol (W/EG). The lowest heat transfer performance is seen in distilled water (DW).

References

[1] Ab Kadir M Z A, Rafeeu Y and Adam N M 2010 Prospective scenarios for the full solar energy development in Malaysia *Renewable and Sustainable Energy Reviews* **14** 3023-31
[2] Vakulchuk R, Overland I and Scholten D 2020 Renewable energy and geopolitics: A review *Renewable and Sustainable Energy Reviews* **122** 109547
[3] Hosseini S E and Wahid M A 2016 Hydrogen production from renewable and sustainable energy resources: promising green energy carrier for clean development *Renewable and Sustainable Energy Reviews* **57** 850-66
[4] Kalogirou S A 2013 *Solar energy engineering: processes and systems*: Academic Press)
[5] Tritt T M, Böttner H and Chen L 2008 Thermoelectrics: Direct solar thermal energy conversion *MRS bulletin* **33** 366-8
[6] Stephens E, Ross I L, Mussgnug J H, Wagner L D, Borowitzka M A, Posten C, Kruse O and Hankamer B 2010 Future prospects of microalgal biofuel production systems *Trends in plant science* **15** 554-64
[7] Abdelrazik A S, Al-Sulaiman F A, Saidur R and Ben-Mansour R 2018 A review on recent development for the design and packaging of hybrid photovoltaic/thermal (PV/T) solar systems *Renewable and Sustainable Energy Reviews* **95** 110-29
[8] Fikri M A, Asri F F, Faizal W M, Adli H K, Mamat R, Azmi W H, Najafi G and Yusaf T 2020 TiO2-SiO2 nanofluid characterization: Towards efficient with water/ethylene glycol mixture for solar application *IOP Conference Series: Materials Science and Engineering* **863** 012055
[9] Amiruddin Fikri M, Fatihah Asri F, Mohd Faizal W, Karimah Adli H, Mamat R, Azmi W H, Ramadhan A I and Yusaf T 2020 Effects of heat transfer based water for three square multilayer absorber solar collector *IOP Conference Series: Materials Science and Engineering* **788** 012078
[10] Jamar A, Majid Z A A, Azmi W H, Norhafana M and Razak A A 2016 A review of water heating system for solar energy applications *International Communications in Heat and Mass Transfer* **76** 178-87
[11] Ramadhan A I, Azmi W H, Mamat R and Mazlan M 2020 A new correlation of thermal-properties of tri-hybrid nanoparticles in water-ethylene glycol mixture *Technology Reports of Kansai University* **62** 1151-60
[12] Ramadhan A I, Azmi W H and Mamat R 2020 Heat transfer characteristics of car radiator using tri-hybrid nanocoolant. In: *IOP Conference Series: Materials Science and Engineering*: IOP Publishing) p 012054

[13] Ramadhan A I, Azmi W and Mamat R 2020 Experimental investigation of thermo-physical properties of tri-hybrid nanoparticles in water-ethylene glycol mixture *Walailak Journal of Science and Technology (WJST)*

[14] Kamajaya K and Umar E 2015 The empirical correlations for natural convection heat transfer Al_2O_3 and ZrO_2 nanofluid in vertical sub-channel. In: *IOP Conference Series: Materials Science and Engineering*: IOP Publishing) p 012053

[15] Fikri M A, Faizal W M, Adli H K and Majid Z A A 2019 TiO2-SiO2 nanofluids in water/ethylene glycol mixture for collector of solar water heating system: An experimental study for thermo physical properties in different ratios *NaSMiNT 2019 ORGANIZING COMMITTEE* 1

[16] Ramadhan A I, Azmi W H, Mamat R and Hamid K A 2020 Experimental and numerical study of heat transfer and friction factor of plain tube with hybrid nanofluids *Case Studies in Thermal Engineering* **22** 100782

[17] Fikri M A, Faizal W M, Adli H K and Amran N A 2018 Preparation of hybrid nanofluids for multilayer absorber solar collector. In: *Proceeding of National Conference on Mechanical Engineer (NCME)* (Politeknik Kota Bharu pp 68-75

[18] Umar E, Kamajaya K and Tandian N P 2015 Experimental Study of Natural Convective Heat Transfer of Water-ZrO2 Nanofluids in Vertical Sub Channel *Contemporary Engineering Sciences* **8** 1593-605

[19] Abdul Hamid K, Azmi W H, Mamat R and Usri N A 2014 Thermal conductivity enhancement of aluminium oxide nanofluid in ethylene glycol. In: *Applied Mechanics and Materials*: Trans Tech Publ) pp 730-4

[20] Azmi W, Zainon S, Hamid K and Mamat R 2019 A review on thermo-physical properties and heat transfer applications of single and hybrid metal oxide nanofluids *Journal of Mechanical Engineering and Sciences* **13** 5182-211

[21] Nabil M, Azmi W, Hamid K A, Mamat R and Hagos F Y 2017 An experimental study on the thermal conductivity and dynamic viscosity of TiO2-SiO2 nanofluids in water: ethylene glycol mixture *International Communications in Heat and Mass Transfer* **86** 181-9

[22] Hamid K A, Azmi W H, Nabil M F, Mamat R and Sharma K V 2018 Experimental investigation of thermal conductivity and dynamic viscosity on nanoparticle mixture ratios of TiO_2-SiO_2 nanofluids *International Journal of Heat and Mass Transfer* **116** 1143-52

[23] Ramadhan A I, Azmi W H, Mamat R, Hamid K A and Norsakinah S 2019 Investigation on stability of tri-hybrid nanofluids in water-ethylene glycol mixture. In: *IOP Conference Series: Materials Science and Engineering*: IOP Publishing) p 012068

[24] Azmi W H, Hamid K A, Usri N A, Mamat R and Mohamad M S 2016 Heat transfer and friction factor of water and ethylene glycol mixture based TiO_2 and Al_2O_3 nanofluids under turbulent flow *International Communications in Heat and Mass Transfer* **76** 24-32

[25] Mohamed N, Ismail A F, Majid A and Azran Z 2015 Performance evaluation of solar collectors using a solar simulator *IIUM Engineering Journal* **6** 79-90

Investigation on stability of TiO_2-SiO_2 nanofluids with ratio (70:30) in W/EG mixture (60:40)

M A Fikri[1,4*], W M Faizal[1], H K Adli[1], Z Bo[2], X X Jiang[2], A I Ramadhan[3]

[1] Faculty of Bioengineering and Technology, Universiti Malaysia Kelantan (Jeli Campus), Malaysia
[2] School of Mechanical Engineering, Ningxia University, China
[3] Faculty of Engineering, Universitas Muhammadiyah Jakarta, Jakarta, Indonesia
[4] Engineering Division, Jabatan Kesihatan Negeri Kelantan, Kementerian Kesihatan Malaysia, Malaysia

*Corresponding e-mail: moamfik.selehor@gmail.com

Abstract. Nowadays, thermal energy is essential to the protection of the renewable technology ecosystem in the world. Most researchers study nanofluid related to increased demand for thermal energy, which includes the heating and cooling process. The goal of this analysis was therefore to determine the stability of TiO_2-SiO_2 nanofluids (70:30) suspended in W/EG (60:40). This study will help researchers gain insight into the impact of nanoparticle stability on hybrid nanofluids, which will enable researchers to add nanoparticles to solar collector. TiO2-SiO2 nanofluids with ratio 70:30 was prepared using a one-step process at a volume concentration of 0.3, 0.5, 0.7, and 1.0%. Two of nanoparticles are used, namely TiO_2 and SiO_2, and distributed in the base W/EG mixture (60:40). The stability investigation of hybrid nanofluids in this research is performed by UV-Vis, visual sedimentation. The findings from this investigation is stability analysis of hybrid nanofluids by UV-Vis process, which is stable up to 10 days after preparation with 2-hour sonication time. For concentration ratio of 1.0% of the volume concentration of hybrid nanofluids. The optimal absorbance value is 2.3 after 24 hours, which is the ideal consistency of the fluid. Concentrations of 0.3, 0.5 and 0.7% begin to decrease after 24 h and continue to decrease until 10 days (240 h).

Keywords: nanofluids, renewable energy, thermal, particles

1. Introduction

Nowadays, thermal energy is essential to the protection of the renewable technology ecosystem in the world. Most researchers study nanofluid related to increased demand for thermal energy, which includes the heating and cooling process. This investigation shows that it relates to the equipment industry, such as cooling equipment, lubrication, solar collector, heat exchanger and automotive cooling systems [1-7].

The nanofluid preparation method is important in order to minimize the agglomeration of nanoparticles and thus improve stability. The most popular technique used in the processing of nanofluids is one-step and two-step processes [8- 10]. The one-step approach is the mechanism of synthesizing and dispersing nanoparticles simultaneously in a base liquid. However, this approach is not feasible for the industry, and refers only to low vapor pressure liquids. Another nanofluid preparation process known as the two-step method. There are two processes in this system, namely (i) the synthesis of nanoparticles in powder form (ii) the distribution of nanoparticles in the base liquid to form a stable and homogeneous solution [11].

Nanofluid stability and size of nanoclusters impact thermal conductivity parameters [12]. Nanofluid stability is known as the resistance of nanoparticles to aggregation. Factors such as Van der Waals attractions induce aggregation, which results in the creation of nanoclusters in nanofluids [13], [14]. The structure of nanoclusters depends on the scale of the nanofluid particles. The stability of nanofluid is very important for its application; a small analysis has been performed to estimate the stability of the suspension. UV – Vis spectrophotometric tests have been used to quantitatively describe the colloidal stability of the dispersion [15]. It can be extended to all baseline fluids, while zeta potential analysis restricts the viscosity of baseline fluid [16], [17].

The mixture of two or more separate nanoparticles in the liquid base, according to [18], is intended for hybrid nanofluids. Higher thermal conductivity and lower viscosity than microfluidics are the benefits of using nanofluid in heat transfer applications. A variety of research and analyses has been released on hybrid nanofluids for possible investigations [19], 20]. Their papers contribute to the planning, performance and implementation of hybrid nanofluid. The presence of strong particles results in curious features of the basic thermo-physical properties of nanofluids. A variety of researchers have studied thermal conductivity, viscosity, density and stability in recent years [10], [15], [16], [21].

The goal of this analysis was therefore to determine the stability of TiO_2-SiO_2 nanofluids (70:30) suspended in W/EG (60:40). This study will help researchers gain insight into the impact of nanoparticle stability on hybrid nanofluids, which will enable researchers to add nanoparticles to solar collector.

2. Materials and Method

2.1. Material

The preparation of hybrid nanofluids involved three distinct forms of single nanofluids, namely TiO_2 and SiO_2 blended together and distributed in the base Water/EG mixture. Both single nanofluids were collected from US Research Nanomaterials, Inc. The corresponding nanoparticle sizes for TiO_2 and SiO_2 are 50 and 2 nm with a purity of 99% and 99.99% respectively, with weight concentrations of 40 and 25 wt. %. The properties of each nanoparticle are shown in Table 1. The base fluid used in this analysis was a mixture of water and EG at a ratio of 60:40 (vol.%). The properties of ethylene glycol are shown in Table 2. Nanoparticle scale characterization of the hybrid nanofluid by field electron microscope (FESEM) scanning technique. The definition of FESEM for nanoparticles shown in Figure 1 (a) and Figure 1 (b).

Table 1. Nanoparticle Properties [16]

Properties	TiO_2	SiO_2
Molecular mass, g mol^{-1}	79.86	60.08
Average particle diameter, nm	50	30
Density, kg m^{-3}	4230	2220
Thermal conductivity, W m^{-1} K^{-1}	8.4	1.4
Specific heat, J kg^{-1} K^{-1}	692	745

Table 2. Properties of Ethylene Glycol (EG) [10]

Properties	EG
Vapour pressure, mmHg at 20 °C	0.08
Boiling point, °C	195–198
Melting point, °C	-13
Density, g ml^{-1} at 25 °C	1.113

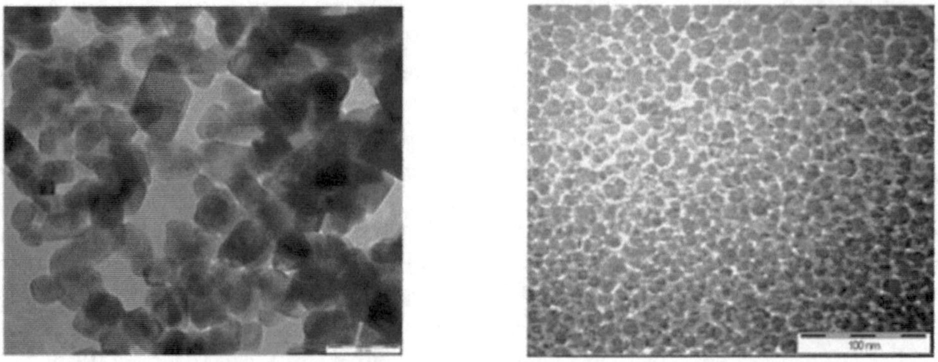

(a) TiO_2 (b) SiO_2

Figure 1. Transmission Electron Microscope (TEM) image for each nanoparticles [22]

2.2. Preparation of TiO_2-SiO_2 nanofluids with ratio (70:30)

The one-step process is used for the preparation of TiO_2-SiO_2 nanofluids with ratio 70:30. TiO_2-SiO_2 nanofluids were prepared by mixing together all two single nanofluids (TiO_2 and SiO_2) and undergoing a mixing and sonication phase. Initially, the preparation of nanofluids began with the measurement of the necessary volume by concentration. In this analysis, hybrid nanofluids were prepared at a concentration of 0.3, 0.5, 0.7 and 1.0%. Nanofluids were first prepared at the maximum concentration, 1.0%, and then reduced to the lowest concentration. TiO_2 and SiO_2 single nanofluids were supplied in a water suspension with a weight concentration of 40, 25% for TiO_2 and SiO_2 respectively. Eq. (1) [9] is used to translate the concentration of weight to the concentration of volume. The dilution from a higher concentration of volume to a lower concentration of volume used the Eq. (2) [10] by introducing the essential fluid (ΔV).

$$\phi = \frac{\omega \rho_w}{\frac{\omega}{100}\rho_w + \left(1 - \frac{\omega}{100}\right)\rho_p} \tag{1}$$

$$\Delta V = (V_2 - V_1) = V_1\left(\frac{\phi_1}{\phi_2} - 1\right) \tag{2}$$

All single nanofluids were mixed together at a volume ratio of 70:30 to form TiO_2-SiO_2 nanofluids. Complete volumes of 100 mL have been prepared for each concentration of hybrid nanofluid. The blended solution of the two single TiO_2 and SiO_2 nanofluids was mixed together using a magnetic stirrer for 120 minutes. The solution was then subjected to a sonication phase using an ultrasonic bath to improve stability.

2.3. Stability of TiO_2-SiO_2 nanofluids with ratio (70:30)

The stability investigation of TiO2-SiO2 nanofluids with ratio (70:30) in this research is performed by visual inspection, calculation of UV-Vis spectrophotometer and zeta potential. Sedimentation by visual examination was done for up to 60 days. Nanofluid shall be called stable when the concentration is steady. Previously, name [23] have used the same process for visual sedimentation of [18], prepared nanofluids. The UV-Vis was done for 10 days (250 h) based on the length of the sonication. Following the analysis by [10] the wavelength of the UV-Vis spectrophotometer is set at 900 nm. UV-Vis measures

IOP Conf. Series: Materials Science and Engineering 1062 (2021) 012020 doi:10.1088/1757-899X/1062/1/012020

the absorption and light strength of the nanofluid dispersion by matching the intensity level with the base fluid. The absorbance ratio of the sonication time is different during the sedimentation time at a constant wavelength of 900 nm. Previous studies [9] have used UV-Vis stability measurement. Zeta potential calculations were done using Zetasizer Nano ZS (Malvern Instruments Ltd., GB).

3. Results and Discussion

3.1. Visualization effect of TiO_2-SiO_2 nanofluids

Stability by Hwang et al, 2007 is one of the variables that can influence the efficiency of nanofluids. Decreasing thermal conductivity of nanofluid is the solution and blocking of microchannels formed by the aggregation of nanoparticles. Consequently, the properties that affect nanofluid are also stable. It is one of the big problems that can alter the essence of nanofluid by either studying influences that influence the distribution of nanofluid stability or nanofluid stability that are specifically related to its electro-kinetic properties. The balance of the Al_2O_3-Cu nanofluids hybrid was studied at various volume levels [24] and occurred when the delay in freezing triggered their possible lack of heat transfer. The stability of the assessment and prosecution should also not be overlooked. As heat transfer is applied, the thermal properties of the hybrid nanofluid thermal, if it can be decreased, then good stability can alter and deliver efficient results.

(a) First day (b) After 60 days

Figure 2. Sedimentation observation of TiO_2-SiO_2 nanofluids samples for (a) first day; (b) after 60 days

Figure 2 (a) presented that nanofluids prepared using a one-step process do not exhibit deposition, and this proves that the nanofluid is in a good shape. On the basis of this visual analysis, many researchers are doing the same thing (Ramadhan, et al, 2020) and are studying their stability assessments at various temperatures. The nanofluid stability of TiO_2-SiO_2 is determined by absorption as seen in Figure 2 (b). The sample was left at the return temperature for 60 days. The ideal absorption ratio would be one (100 percent) which reflects excellent stability during the deposition phase. The nearer rate is one with a rise in deposition period to assess the balance of the sample.

In order to determine the consistency of the concentration of the TiO_2-SiO_2 nanofluid hybrid volume, distributed in W: EG with a mixing ratio of 60:40, simple indoor deposition monitoring. The parameters taken into account during the sedimentation process are based on the percentage of velocity progress in time. A quicker phase happens when the setting is higher than seen in Figure 3.

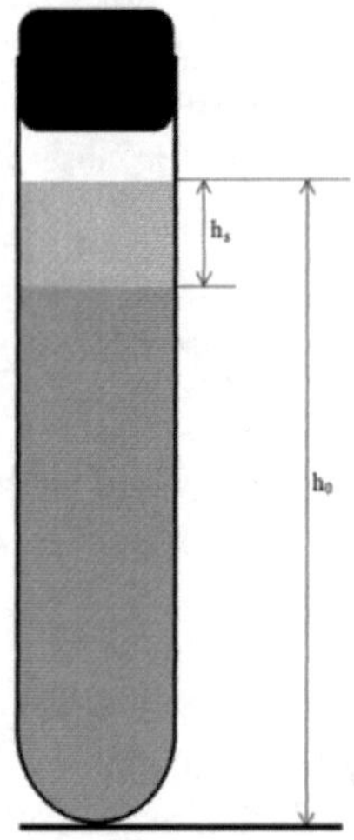

Figure 3. Illustration of the nanofluid sample on the sedimentation process

3.2. Stability of TiO2-SiO2 nanofluids using UV-Vis spectrophotometer

The same study of nanofluid stability using the UV-Vis spectrophotometer was previously suggested and studied by researchers [25]. The absorbance of nanofluids is linearly increasing with the increase of volume concentration. This trend is in agreement with the Beer-Lambert Law, which is the value of the absorbance is equivalent to concentration [4, 9]. TiO_2-SiO_2 Nanofluid of absorption and sedimentation time was commonly observed for 240 hours for a different wavelength of 850 nm as seen in Figure 4.

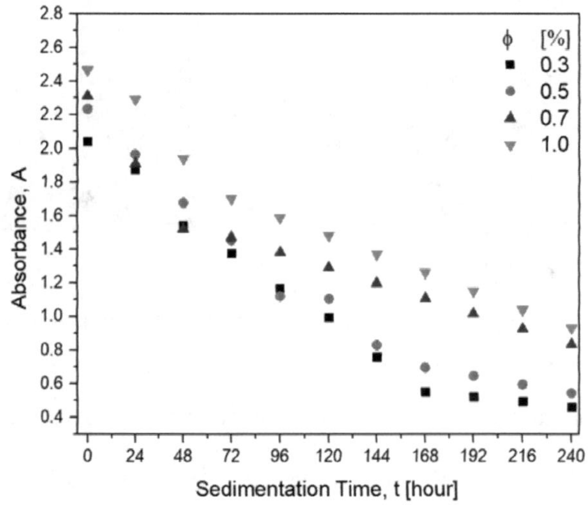

Figure 4. Absorbance of TiO_2-SiO_2 nanofluids with sedimentation time for different concentration

Figure 4 indicates a concentration ratio of 1.0% of the volume concentration of hybrid nanofluids. The optimal absorbance value is 2.3 after 24 hours, which is the ideal consistency of the fluid. Concentrations of 0.3, 0.5 and 0.7% begin to decrease after 24 h and continue to decrease until 10 days (240 h). Thus, it can be inferred that nanofluid has a greater capacity for agglomeration and accelerated deposition at lower concentrations. In terms of diffusion, there is also a high concentration of more suspended nanoparticles.

4. Conclusions

In this study, TiO_2-SiO_2 nanofluids for ratio (70:30) preparation can be successfully prepared using a one-step process. Stability analysis of hybrid nanofluids by UV-Vis process, which is stable up to 10 days after preparation with 2-hour sonication time. For concentration ratio of 1.0% of the volume concentration of hybrid nanofluids. The optimal absorbance value is 2.3 after 24 hours, which is the ideal consistency of the fluid. Concentrations of 0.3, 0.5 and 0.7% begin to decrease after 24 h and continue to decrease until 10 days (240 h).

Acknowledgements

The facilities of a laboratory to the authors by the Advanced Automotive Liquid Laboratory, Faculty of Mechanical Engineering, Universiti Malaysia Pahang is gratefully acknowledged.

References

[1] Azmi W H, Zainon S N M, Hamid K A and Mamat R 2019 *Journal of Mechanical Engineering and Sciences* **13** 5182-211
[2] Ramadhan A I, Azmi W H and Mamat R 2020 In: *IOP Conference Series: Materials Science and Engineering*: IOP Publishing) p 012054
[3] Fikri M A, Asri F F, Faizal W M, Adli H K, Mamat R, Azmi W H, Ramadhan A I and Yusaf T 2020 In: *IOP Conference Series: Materials Science and Engineering*: IOP Publishing) p 012078
[4] Sharif M, Azmi W, Redhwan A and Mamat R 2016 *international journal of refrigeration* **70** 93-102
[5] Redhwan A, Azmi W, Sharif M and Mamat R 2016 *International Communications in Heat and Mass Transfer* **76** 285-93
[6] Kamajaya K and Umar E 2015 In: *IOP Conference Series: Materials Science and Engineering*: IOP Publishing) p 012053
[7] Ramadhan A I, Diniardi E and Dermawan E 2016 In: *AIP Conference Proceedings*: AIP Publishing LLC) p 050003
[8] Said Z, Sabiha M, Saidur R, Hepbasli A, Rahim N A, Mekhilef S and Ward T 2015 *Journal of Cleaner Production* **92** 343-53
[9] Ramadhan A I, Azmi W H, Mamat R, Hamid K A and Norsakinah S 2019 In: *IOP Conference Series: Materials Science and Engineering*: IOP Publishing) p 012068
[10] Nabil M, Azmi W, Hamid K A, Mamat R and Hagos F Y 2017 *International Communications in Heat and Mass Transfer* **86** 181-9
[11] Bahiraei M, Mazaheri N and Rizehvandi A 2019 *Applied Thermal Engineering* **149** 588-601
[12] Prasher R, Phelan P E and Bhattacharya P 2006 *Nano letters* **6** 1529-34
[13] Hong J and Kim D 2012 *Thermochimica acta* **542** 28-32
[14] Umar E, Kamajaya K and Tandian N P 2015 *Contemporary Engineering Sciences* **8** 1593-605
[15] Ramadhan A I, Azmi W H, Mamat R and Mazlan M 2020 *Technology Reports of Kansai University* **62** 1151-60
[16] Hamid K A, Azmi W H, Nabil M F, Mamat R and Sharma K V 2018 *International Journal of Heat and Mass Transfer* **116** 1143-52
[17] Ramadhan A I, Azmi W H, Mamat R and Hamid K A 2020 *Case Studies in Thermal Engineering* **22** 1-9
[18] Azmi W H, Hamid K A, Usri N A, Mamat R and Mohamad M S 2016 *International Communications in Heat and Mass Transfer* **76** 24-32
[19] Hatwar A S and Kriplani V 2014 *Int. J. Adv. Res. Sci. Eng* **3** 175-83
[20] Sarkar J, Ghosh P and Adil A 2015 *Renewable and Sustainable Energy Reviews* **43** 164-77

[21] Ramadhan A I, Azmi W H and Mamat R 2020 *Walailak Journal of Science and Technology (WJST)*
[22] Azmi W H, Sharma K V, Sarma P K, Mamat R and Najafi G 2014 *International Communications in Heat and Mass Transfer* **59** 30-8
[23] Abdul Hamid K, Azmi W H, Mamat R and Usri N A 2014 In: *Applied Mechanics and Materials*: Trans Tech Publ) pp 730-4
[24] Suresh S, Venkitaraj K and Selvakumar P 2011 In: *Advanced Materials Research*: Trans Tech Publ) pp 1560-7
[25] Hwang Y-j, Lee J, Lee C, Jung Y, Cheong S, Lee C, Ku B and Jang S 2007 *Thermochimica Acta* **455** 70-4

Effect of n-Butanol Blends on Engine Performance and Exhaust Emission of Compression Ignition Engine Fuelled with Diesel-Palm Oil Methyl Ester (B20)

Mohd Azrul Ahmad[1, 2], Nik Rosli Abdullah[1]*, Zaky Zuhairi Zawawi[1], Asiah Ab Rahim[3] and Hasannuddin Abdul Kadir[4]
[1]Universiti Teknologi MARA, Shah Alam, Malaysia
[2]Malaysia Japan Higher Education Program, Beranang, Malaysia
[3] Faculty of Mechanical Engineering, Universiti Teknologi MARA, 23200, Bukit Besi, Terengganu, Malaysia)
[4] Faculty of Mechanical Engineering, Universiti Teknologi MARA Cawangan Johor, Jalan Purnama, Bandar Seri Alam, 81750 Masai, Johor, Malaysia

*Corresponding author Email: nikrosli@uitm.edu.my

Abstract. The study aims to evaluate the performance and emission characteristic in a compression ignition engine of conventional diesel fuel-biodiesel blends with different percentage of fuel additive n-butanol. The experiment was operated at a constant engine load (50% throttle condition) with different engine speed (2700, 3100, 3500 rpm). A blend of biodiesel and diesel fuel known as B20 (20% palm oil methyl ester and 80% diesel in volume) was prepared, and then n-butanol was added to B20 at a volume of 5ml, 10ml and 15ml (denote as B20+Bu5, B20+Bu10 and B20+Bu15, respectively) and the tested fuel samples were compared with diesel fuel and diesel-biodiesel (B20). The experimental results show that when the proportion of n-butanol was increased in B20 blends, kinematic viscosity was larger while calorific value was smaller than those of the neat diesel. Although n-butanol have some negative impacts on engine performance parameters, its generally positively affect exhaust emission parameters compared to diesel fuel. According to engine performance and exhaust emission test result of n-butanol fuel blends with B20 blends, average values of brake thermal efficiency (10.19%, 7.58% and 4.29%), carbon monoxide (21.75%, 17.06% and 11.28%), hydrocarbon (18.51%, 15.68% and 12.13%) are lower, while brake specific fuel consumption (BSFC) (27.48%, 45.37% and 59.20%) are higher and carbon dioxide (CO2) and oxides of nitrogen (NOx) are comparable than those of diesel fuel.

Keywords: renewable energy, biodiesel, n-butanol, palm oil methyl ester

1. Introduction

The compression ignition engine, also known as diesel engine, is the most commonly used in industries and automobile due to its capability for heavy duty application [1,2] and because of their high fuel conversion efficiencies and durability. It has a higher brake thermal efficiency than spark ignition (SI) engines [3]. This type of engine is one of the main users of diesel fuel. The demand of diesel fuel has been growing constantly because the diesel engines are more efficient, use relatively less fuel per engine effective power unit and have a less impact on the environment compared to petrol engines. However, the limited amount of fossil fuels and pollutants emission resulting from diesel combustion have negative effects on both human health and the environment. This has led researchers to investigate a new alternative and renewable energy sources that suitable for the diesel engine to operate [4].

Alternative fuel is a fuel for internal combustion engines that is derived partly or wholly from a source other than petroleum and that is less damaging to the environment than conventional fuels (fossil fuels). Some examples of alternative fuels are natural gas, biodiesel, methanol, ethanol, vegetable oil, biomass and others [5, 6] are potential candidates to reduce fossil fuel consumption, exhaust emission, and global warming in the future. Currently, biodiesel is one of the alternative fuels which is widely used biofuel for diesel engines which have been examined by previous researchers [7, 8]. Biodiesel is generally derived from a variety of vegetable oils and animal fats. It is a diesel fuel substitute that can be used in diesel engines without modifying the engines. Biodiesel is cleaner, produces fewer emissions and is a viable alternative energy source. Moreover, biodiesel is less toxic fuel, biodegradable, renewable, non-explosive and emit less air pollutant compared to mineral diesel [9]. This alternative fuel designed to extend the usefulness of petroleum and diesel engine longevity and cleanliness [10, 11]. However, biodiesel produce lower engine power because it has less energy per unit volume than diesel fuel. The fuel efficiency tends to be slightly lower when using biodiesel due to the lower energy content of the fuel [10],[12-14]. Biodiesel also slightly increase in nitrogen oxide emission compare than diesel fuels. Therefore, fuel additives such as alcohols are used to improve the fuel properties [10, 11],[5].

Fuel additives are chemical formulated compounds to improve the quality and efficiency of fuels in motor vehicles. The fuel additives increase a fuel's cetane rating or act as corrosion inhibitors or lubricants, thus allowing the use of higher compression ratios for greater efficiency and power. For biodiesel, the biodiesel fuel additive which have higher oxygen content are blends in biodiesel to enhance the performance of engine and reduced the brake specific fuel consumption (BSFC). One of the examples of fuel additives are is alcohol n-butanol. Butanol ($C_4H_{10}O$), which is known as butyl alcohol is an alcohol that comprised of a four-carbon structure [17]. It is a clear and colourless liquid with a four-carbon straight-chain structures. The demand for butanol nowadays is due to its usable as biofuel. Butanol have been used as blending component in diesel fuels due to its higher miscibility [16]. the lower viscosity and higher volatility in alcohols would be improved with the addition of biodiesel. Furthermore, the increased oxygen content of the blended fuel can improve the combustion process. Butanol can be used as biodiesel additive due to high oxygen content in it [8],[11],[16],[18]. It has a high cetane number compare to some alcohols [10], good solubility and higher heat content appeal many researchers to consider butanol as an additive to biodiesel blends [18],[19]. However, the use of butanol as fuel additive has found increase the brake thermal efficiency (BTE) and the BSFC of the engine which is stated by previous researchers [10],[18]. Additional of n-butanol in fuel blends also found reducing the exhaust emissions such as nitrogen oxide (NOx), carbon monoxide (CO) and particulate matter (PM) except for hydrocarbon (HC) emissions [20].

Hence, the aim of this research is to study the engine performance and emissions profile of compression ignition engine fuel with palm oil methyl ester (B20) blends with alcohol butanol with various of proportions. The fuel blends were tested for some important properties such as density, kinematic viscosity and calorific value. Engine performance parameters such as brake thermal efficiencies and specific fuel consumption were investigated. The emission of CO, CO_2, HC and NOx were also considered in this study. Thus, the objective of this study was to evaluate the potential of diesel-palm oil methyl ester-n-butanol ternary blends as an alternative to diesel fuel. In addition, the effect of n-butanol proportion on combustion and emissions was investigated. The current study adds importance to the topic of ternary blends and may support the idea of using ternary fuel blends of diesel-biodiesel-n-butanol in CI diesel engine in the future.

2. Experimental Setup

2.1. Sample Preparation
There are five samples that were used in this experiment which are 100% diesel fuel (D100) which will be a baseline reference for this research to compare with other fuel blends, B20 (80% diesel fuel + 20% Palm Oil Methyl Ester biodiesel), B20+Bu5, B20+Bu10 and B20+Bu15 (B20 palm oil methyl ester + 5ml, 10ml, 15ml n-butanol respectively). The biodiesel Palm Oil Methyl Ester was obtained from the Automotive Lab at Faculty of Mechanical Engineering (FKM) whereas alcohol n-butanol was obtained from chemical lab at Faculty of Chemical Engineering (FKK) in UiTM Shah Alam. The preparation of fuel properties for the fuel blend was examined at the Instrumentation Lab in Faculty of Chemical Engineering, UiTM Shah Alam. Calorific value, viscosity and density of the fuel blends are the properties that have been identified at the lab which will be discussed further in results of the experiment. The density for the fuel blends were calculated by obtaining the mass of 1mg each of the fuel blends via analytical balance, then calculate by using density formula. After obtaining the densities, calorific value of the fuel blends was evaluated via bomb calorimeter machine. For the kinematic viscosity, those values were obtained by dividing the dynamic viscosity of the fuels with its density. The dynamic viscosity of the fuel blends was obtained via physical rheometer at chemical lab, FKK. The fuel properties data that obtained from FKK are tabulated in table 1 below.

Table 1. Properties of test fuel

Fuel	D100	PME B20	B20 + 5 ml n-Butanol	B20 + 10 ml n-Butanol	B20 + 15ml n-Butanol
Density (kg/m^3)	0.9220	0.9590	0.8782	0.9229	0.9536
Calorific Value (MJ/kg)	44.496	39.515	39.328	39.114	38.946
Kinematic Viscosity (m^2/s)	0.00362	0.00515	0.00512	0.00476	0.00435

2.2. Engine Test Performance
The experiment was conducted at Automotive Workshop, Faculty of Mechanical Engineering, University Teknologi Mara (UiTM) Shah Alam. The experiments were performed by using a single cylinder, four stroke, air cooled, compression ignition engine and the schematic of the test setup is presented in Figure 1. The engine with a specification in Table 2 was operated with constant load at 50% with a different speed of 2700, 3100 and 3500 rpm. The engine speeds are measured using a digital tachometer. Three different volume of n-butanol with 5ml, 10ml and 15 ml were blended with B20 diesel-palm oil methyl ester blends which are B20+Bu5, B20+Bu10 and B20+Bu15 are fuelled into the engine.

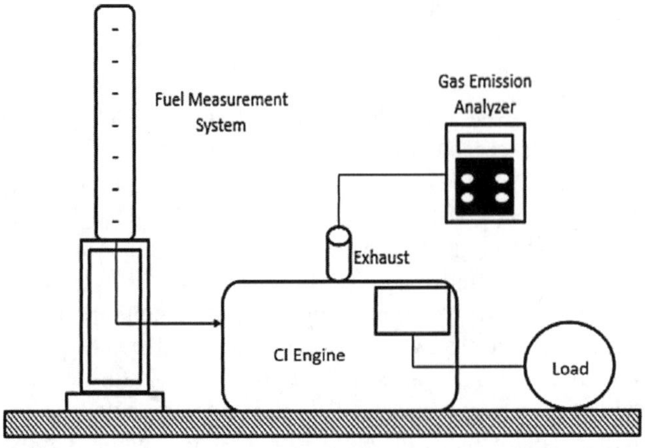

Figure 1. Schematic diagram of the experimental setup

To ensure the accuracy and consistency of the measurement data the engine had been running approximately for ten minutes with diesel fuel after each test. Also, each test was repeated for 3 times in order to have a consistent and reliable measurement. The average values were used for drawing the graphs. Exhaust emissions data was by using a MRU Infrared Gas Analyzer to detect output of CO, CO2, HC, and NOx. The gas analyser has been calibrated by the company to assure the accuracy of the data collection.

Table 2. Engine specification.

Model	BENMA 5GF-ME
Engine Type	Single cylinder, 4-stroke engine, air cooled
Fuel Tank Capacity	1.5 L
Fuel	Diesel
Displacement	418cm^3
Starting System	Recoil Start
Rated AC Power (Net)	4.5 kW / 3600 rpm
Max AC Power (Net)	5.0 kW / 3600 rpm
Max.Output	10 HP/ 3600 rpm
AC Output Voltage	240 V
Net Weight/Gross Weight (kg)	100/104
Dim (L x W x H)	727 x 495 x 597
Lubrication oil	SAE10W/30
Frequency	50 Hz

3. Result and Discussion

The results and discussion are presented in this chapter based on the outcome of the experiment that has been conducted and by referring the literature, reviews, journal, internet and books to get a clear point of view on this study. Besides, the engine performance parameters that are evaluated are BSFC and BTE. For engine emissions parameters, CO_2, CO, NOx, and HC emissions are evaluated. These parameters were compared between diesel fuels and four sample blended fuels. In addition, the engine was tested under three different speeds which are 2700, 3100 and 3500 rpm with a constant load of 50%. The results of the parameters of engine performance and emissions are plotted in bar chart.

3.1. Properties of fuel blends

3.1.1. Density of fuels

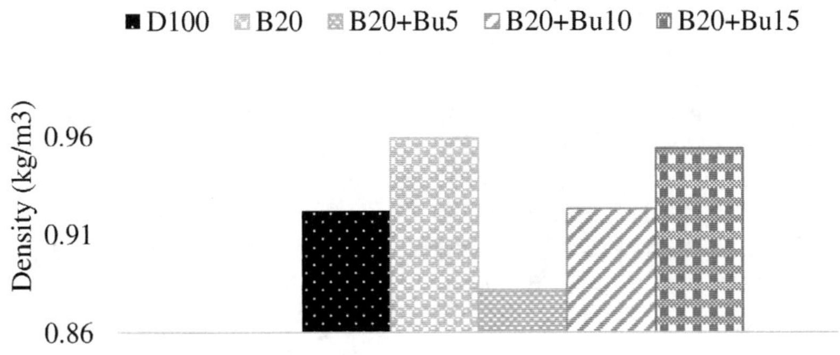

Figure 2. Density of tested fuel blends

Figure 2 shows the graph of density comparison among tested fuel blends which are diesel (D100), B20, B20+Bu5, B20+Bu10 and B20+Bu15. Density is known as the mass per unit volume of a substance [21]. Previous study also state that the density is one of the fuel properties that affect the performance and emission of an engine [18]. From the graph, B20 shows the highest density value with increment of 4.01% (0.959 kg/m^3) compare to diesel fuel. However, the n-butanol fuel blend B20+Bu5 has the lowest density with reduction of 4.34% (0.8882 kg/m^3). The density of B20+Bu10 and B20+Bu15 increases as more n-butanol content is added into the fuel blends with increment of 0.01% (0.9229 kg/m^3) and 3.43% (0.9536 kg/m^3) respectively compare to baseline diesel (0.922 kg/m^3). When the density of fuel is higher than another, it indicates more mass of fuels with same amount of volume entering the combustion chamber, as new generation diesel injectors that supply fuel to the combustion chamber for power generation control the amount of fuel by volume rather than mass. The more fuel entering the cylinder resulting more emissions such as the increase of CO_2 and NOx emission by biodiesel fuel [22]. It can be concluded that density have affect the efficiency of the combustion process in diesel engine.

3.1.2. Viscosity of fuels

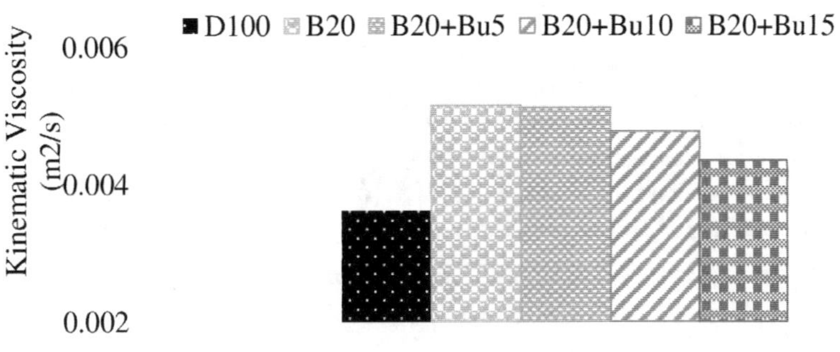

Figure 3. Kinematic viscosity of tested fuel blends

The comparison of kinematic viscosity between diesel fuel, D100 and biodiesel-n-butanol fuel blends (B20, B20+Bu5, B20+Bu10 and B20+Bu15) are shown in figure 3. Kinematic viscosity is the ratio of dynamic viscosity to density of a substance. It is a crucial property which affects the fuel fluidity [23]. From the graph, B20 has the highest kinematic viscosity value compare to diesel (0.00362 m^2/s) with increment by 42.27% (0.00515 m^2/s). Previous study also found that biodiesel has higher viscosity compare to diesel and n-butanol fuel blends [8],[10],[24]. For n-butanol blends, the kinematic viscosity decreased as the n-butanol content in the fuel blend increases. The average increment of kinematic viscosity values for B20+Bu5, B20+Bu10 and B20+Bu15 are 41.44% (0.00512 m^2/s), 31.49% (0.00476 m^2/s) and 20.17% (0.00435) respectively compare with diesel fuel. The higher viscosity of B20 can lead to deterioration of spray properties and resulted in poor atomization in cylinder [24]. Higher viscosity and fuel density result in reduced evaporation due to larger fuel droplets that cannot be sufficiently atomised in the cylinder [17]. To overcome this problem, adding alcohol such as n-butanol into biodiesel can reduced the density and viscosity of the fuel blends due to its fuel characteristics which is agree by previous researchers [8],[18],[24]. The higher oxygen content of n-butanol, lower viscosity and density which improve the atomization for the fuel blends [18]. Thus, it can be concluded that adding n-butanol into biodiesel fuel blends improve the kinematic viscosity of the fuel.

3.1.3. Calorific values of fuels

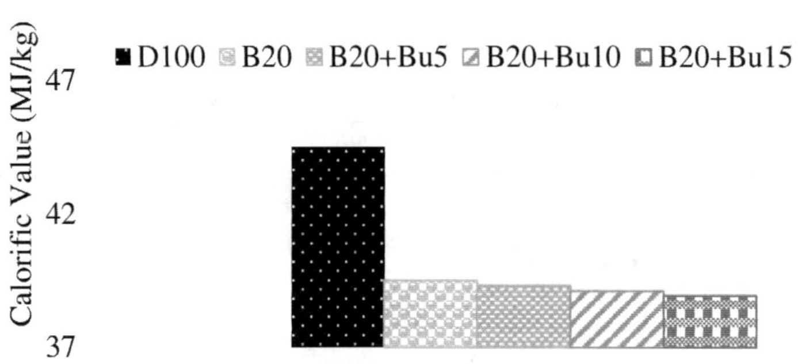

Figure 4. Calorific values of tested fuel blends

The figure 4 shows a graph of comparison of calorific values between diesel and other fuel blends. Calorific value is important parameter in determining the fuel properties. Calorific values can be defined as the amount of heat that a fuel produced when it is fully burnt and the combustion products are cooled to standard temperature and pressure (STP) [25, 26]. From the graph, diesel has the highest calorific value at 44.496 MJ/kg compare to biodiesel B20 and other n-butanol fuel blends. The percentage reduction of calorific value for B20, B20+Bu5, B20+Bu10 and B20+Bu15 compare to diesel are 11.19% (39.515 MJ/kg), 11.61% (39.328 MJ/kg), 12.10% (39.114 MJ/kg) and 12.47% (38.946 MJ/kg) respectively. From the results, it is found that the calorific value decreases as more alcohol content such as n-butanol is added into the fuel blends which also can be found in previous study [12], [18]. Both biodiesel and butanol have lower energy content than diesel fuel due to their oxygen content that will causes lead to increase in BSFC [6, 18, 27]. It can conclude that adding more alcohol into fuel blends can lowering the calorific value of the fuel blend which is agree by previous researcher [14], [18].

3.2. Engine Performance

3.2.1. Brake Specific Fuel Consumption (BSFC)

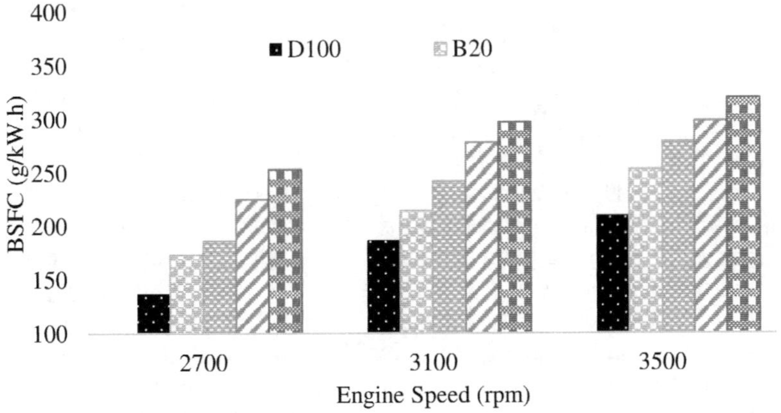

Figure 5. BSFC of fuel blends at different engine speeds

Brake specific fuel consumption (BSFC) off test fuel under constant engine half load and at various engine speeds are presented in Figure 5. It can be seen that BSFC values of B20 are higher on average by 17.24% than those of Diesel fuel. An increasing volume of n-butanol to B20 will further increase in BSFC values for all engine speed test and the average increments of BSFC compare to baseline Diesel fuel are 27.28%, 45.37% and 59.2% for B20+Bu5, B20+Bu10 and B20+Bu15 respectively. The highest BSFC values for the fuel blends at engine speed of 3500 rpm is B20+Bu15 which is 304 g/kW.hr, following B20+Bu10 (288 g/kW.hr), B20+Bu5 (262 g/kW.hr), B20 (241 g/kW.hr) and Diesel fuel which has the lowest of BSFC values (210 g/kW.hr) among the others. It is obvious that increasing the n-butanol volume will increased BSFC compared to Diesel fuel and B20. The finding is consistent with findings of past studies [10,11],[18] which both B20 and n-butanol have lower energy content than Diesel fuel, due to their oxygen content. The lower calorific value is the reason for the higher BSFC for the same power output, hence consumption energy per power is also increased which was consistent with the finding of İ. Örs *et al.* [18]. Furthermore, the ignition efficiency of n-butanol is lower compared to diesel. This lead to partial combustion of fuel and increase on the BSFC value for the fuel blends as stated by other researchers [8],[28].

3.2.2. Brake Thermal Efficiency (BTE)

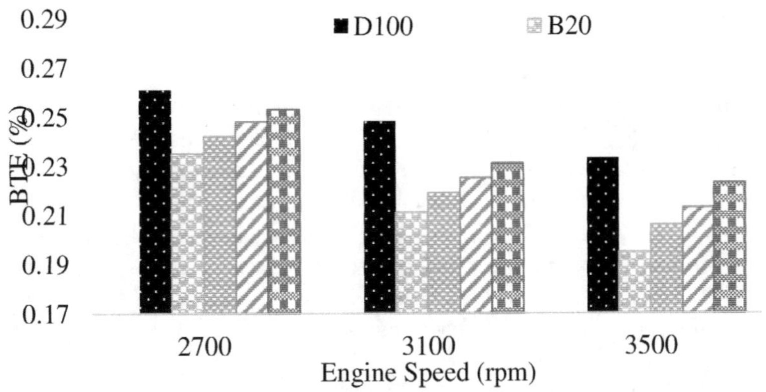

Figure 6. BTE of fuel blends at different engine speeds

The result for brake thermal efficiency (BTE) of all test fuel under constant engine load at 50% and different engine speeds (2700rpm, 3100rpm and 3500rpm) are shown in Figure 6. It can be seen that BTE values of B20 are lower on average by 13.73% than those of Diesel fuel. This is due to higher viscosity and lower energy content of B20 reflects the findings of G. Goga *et al.* [8]. However, with increasing n-butanol volume to B20, BTE increased for all engine speed test compared to B20 but still lower by an average of 10.19%, 7.58% and 4.29% for B20+Bu5, B20+Bu10 and B20+Bu15 respectively compared to Diesel fuel. The BTE values of butanol fuel blends are slightly higher than B20 due to the increased of n-butanol fraction resulting increase in oxygen content in the fuel blends [4],[8],[10],[29]. Oxygen content of butanol promote in complete combustion during the diffusion combustion phase while the lower cetane number of the blends prolong the ignition delay, burning larger fraction of the fuel during the premixed mode, thus increase the temperature in the cylinder, resulting in increasing of BTE [4],[7]. However, the butanol blends have lower BTE compare to baseline diesel fuel because butanol has high latent heat of evaporation that reducing the combustion temperature in the cylinder that will lead to impartial combustion of fuel especially at high engine speed, thus reducing BTE [4],[8]. On the other hand, lower BTE is observed for ternary blends containing B20 and the overall viscosity and lower calorific value of the blend are the proposed to be the contributing factors for lower combustion efficiency [32].

3.3. Exhaust Emissions

3.3.1. Carbon Monoxide (CO) Emission

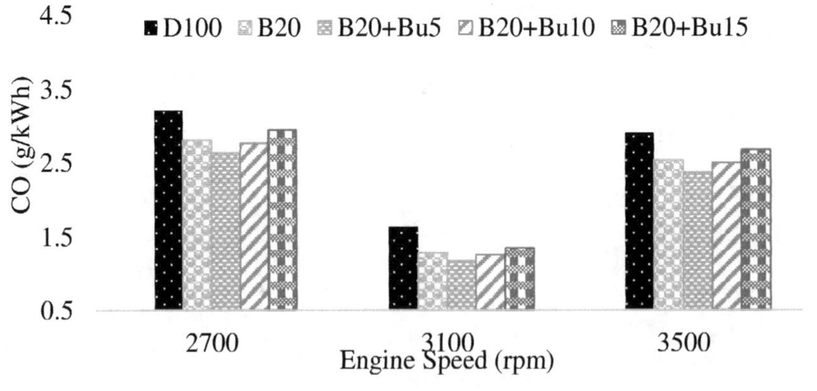

Figure 7. CO emission of fuel blends at different engine speeds

Changes in CO emission of all test fuel under constant engine load at 50% and different engine speeds (2700rpm, 3100rpm and 3500rpm) are shown in Figure 7. It can be observed that maximum CO emission value for all test fuel were obtained at 2700rpm which is due to lack of oxygen content in the air-mixture at lower engine speed, causing partial combustion to occur [8],[28]. The lowest CO emissions for all test fuel were obtained at 3100rpm. The use of biodiesel reduced the CO emission values compared to diesel at all engine speed tested with average CO emission reduction of 15.78% compared to diesel fuel. The reduction of CO emission is due to lower carbon contents and higher oxygen content of the biodiesel, thus lead to more complete combustion in the combustion chamber [8],[10],[18]. Higher oxygen particles presence in the fuel caused the carbon atoms to react with oxygen atoms to convert into CO2 gas as many as possible which led to complete combustion which was consistent with the finding of G. Goga *et al.* [8] and İ. Örs *et al.* [18]. In addition, the average reductions of CO emission for B20+Bu5, B20+Bu10, and B20+Bu15 compare to baseline diesel are 21.75%, 17.06% and 11.28% respectively. Similar results were obtained by İ. Örs *et al.* [18], who concluded that by using biodiesel and n-butanol decreases CO emission due to presence additional oxygen in the additives improve the combustion process compared to diesel fuel. However, there are slight increases of CO emission as the number of n-butanol increased in the fuel blend due to high latent heat of vaporization of n-butanol. This cause the cooling effect in the combustion cylinder, reducing the cylinder temperature which lead to incomplete combustion, causes CO formation to occur [18], [29]. Although there are slight increase of CO emission at higher n-butanol blend, the results will not go any higher than baseline diesel [18].

3.3.2. Carbon Dioxide (CO_2) Emission

The CO_2 emission results for all test fuels under constant engine speed at 50% load with different engine speeds are shown in Figure 8. The minimum CO_2 emissions value were observed at 2700rpm for all test fuels due to lack of oxygen content in the air-mixture at lower engine speed, causing partial combustion to occur [8],[27]. However, higher engine speed increases the engine volumetric performance and fluid mixing process which leads to more complete combustion consequently increase of CO_2 emission value for all test fuel [30]. From the results, B20 has higher CO_2 emission values

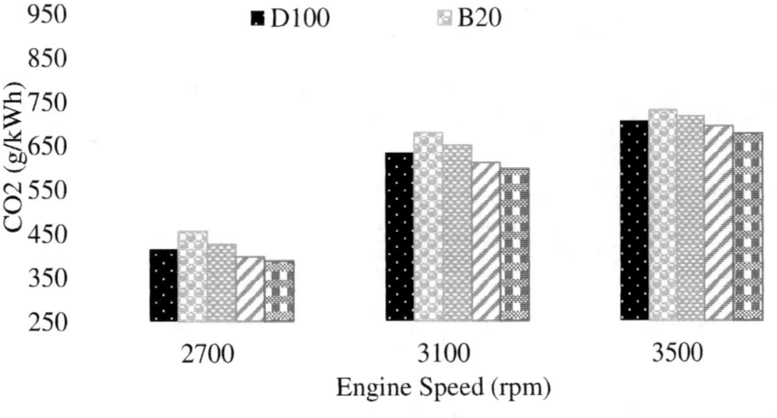

Figure 8. CO_2 emissions of fuel blends at different engine speeds

Compared to diesel at all engine speed tested with average increase CO_2 emission of 6.71% due to higher oxygen contents in biodiesel. The abundant of oxygen atoms in the fuel causes carbon atoms to fuse with adequate of oxygen atoms to form into CO_2 gas, thus lead to more complete combustion in the combustion chamber [18]. From the result, increasing n-butanol volume to B20 blends decrease CO_2 emission value for all engine speed condition. The average increase of CO_2 emission for B20+Bu5 is 2.12 %, whereas the average reduction of B20+Bu10, and B20+Bu15 compare to baseline diesel are 3.35 % and 5.75 % respectively. Previous studies also reported that increase the n-butanol proportion in the diesel-biodiesel blend also reduced the CO_2 emission [15],[31]. The reduction of CO_2 emission for n-butanol fuel blends is due to high latent heat of vaporization of n-butanol which lead to the cooling effect in the combustion cylinder, reducing the cylinder temperature causing the combustion to be incomplete. Thus, lead to CO formation to occur and reducing CO_2 emission for the fuel blends [18, 29]. The chilling effect of n-butanol causing ineffective oxidation of CO to CO_2 has also been documented by previous researchers [29]. Also, this could be explained by the balance between CO and CO_2 formation during the combustion process. The formation of CO_2 decreased as the emission of CO increase [18]. Hence, the CO_2 emission of n-butanol fuel blend decrease as the content of n-butanol in the fuel blend increases.

3.3.3. *Hydrocarbon (HC) Emission*

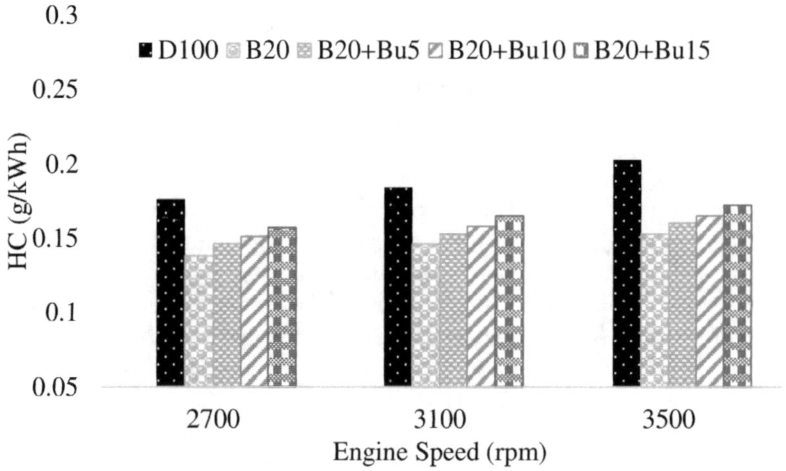

Figure 9. HC emissions of fuel blends at different engine speeds

Figure 9 shows the result of hydrocarbons (HC) emission of the internal combustion engine fuelled with diesel (D100), B20, B20+Bu5, B20+Bu10, and B20+Bu15 under the engine condition constant engine load of 50% with different engine speeds of 2700, 3100 and 3500 rpm. The lowest HC emission values of diesel, B20 and n-butanol blends can be observed at lowest engine speed which is 2700rpm while the maximum HC emission values occurred at 3500rpm for all test fuel. The average reductions of HC emission for B20, B20+Bu5, B20+Bu10, and B20+Bu15 compare to baseline diesel fuel are 22.29%, 18.51%, 15.68% and 12.13% respectively. The highest reduction of B20 is due to the higher oxygen content in biodiesel improved combustion and increase of the exhaust gas temperature [10],[14],[18]. Besides, compared to diesel fuel, biodiesel with higher cetane number promote premixed combustion by shorten the ignition delay period hence reducing the HC emissions [8],[14],[18],[29]. The lower density and viscosity values of n-butanol causing the combustion to be improved better than diesel fuel, thus reducing the HC emission value [18],[23]. However, increasing the volume of n-butanol in the fuel blends slightly increase the HC emissions compared to B20. The low cetane number of n-butanol causing longer ignition delay period which consume more time for the evaporation of the fuel blend in combustion, thus lead partial combustion to occur [8],[11]. Besides, higher heat of evaporation of n-butanol causes the fuel blends evaporating at slower rate, thus increase the HC emission for the n-butanol blend [11]. Therefore, the increasing of n-butanol content in the fuel blends lead to increase in HC emission compare to biodiesel B20.

3.3.4. *Nitrogen Oxides (NOx) emissions*

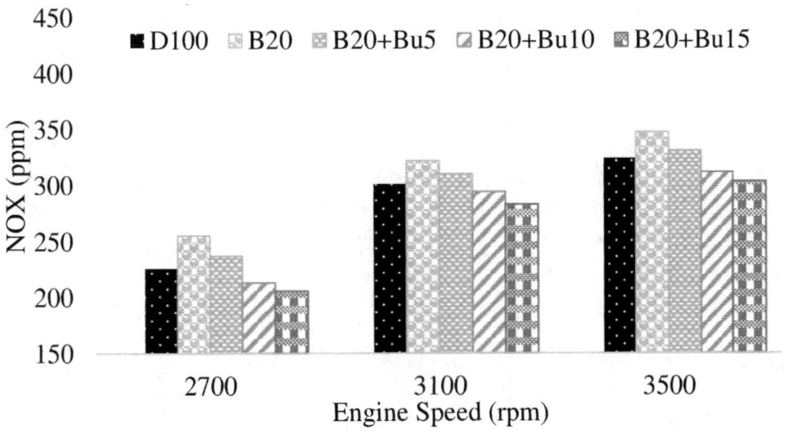

Figure 10. NOx emissions of fuel blends at different engine speeds

Formation of NOx emission substantially depends on the temperature of the combustion flame. Generally, low cetane number will affected ignition delay by increasing the end of combustion temperature for the quantity of premixed combustion. Figure 10 illustrates NOx emission results for all test fuel under a constant half load 50% and at a different engine speed (2700, 3100, 3500rpm). The maximum NOx values were obtained for all test fuels at 3500rpm engine speed, while the minimum values were recorded at 2700rpm. The average increases in NOx emission value for B20 and B20+Bu5 fuels compared to diesel fuel were 8.97% and 3.34% respectively. In contrast, the average reduction in emission values of B20+Bu10 and B20+Bu15 fuel blends were 3.34% and 7.10% respectively. From the result, B20 blend has been found to increase NOx emissions as the oxygen content of B20 is higher than that of diesel fuel which improves more combustion, thus increase the combustion temperature in the cylinder [8],[12],[18]. Previous study also reported that the high oxygen particles presence in the biodiesel caused the nitrogen atoms to react with adequate amount of oxygen atoms to form into NOx gas in the cylinder [18]. As the cylinder temperature increase due to improved combustion, the nitrogen atoms reacts with oxygen atoms at higher rate, generates more formation of NOx gas, hence increase the NOx emission for biodiesel fuels [12]. In contrast, increasing n-butanol in the blends has positively decreased NOx emissions. Higher heat of evaporation and lower calorific value of butanol induces the cooling effect in the cylinder and reduces NOx emissions as stated by previous researchers [8],[11],[18]. Also, further reduction on NOx emission for all different engine speeds has been recorded when higher butanol concentration added on the blends. This is due to lower cetane number and longer ignition delay reduces in cylinder temperature during the premixed combustion phase. Hence, minimizing the formation of NOx emission [8],[18].

4. Conclusion

In this study, the effect of diesel-biodiesel blend with fuel additives has been explored as an alternative fuel for future. The performance of the engine and the exhaust emissions in a single cylinder compression ignition engine was fully evaluated at constant engine load of 50% and three different engine speed fuelled with diesel, biodiesel B20 and the n-butanol – biodiesel fuel blends. The n-butanol fuel blends were compared with diesel as a base fuel. From the result, it can be concluded that biodiesel B20 has a disadvantage on engine performance such as higher BSFC and lower BTE compared to diesel fuel. Oppositely, B20 has an advantage on exhaust emission such as lower CO and HC but slightly higher for CO_2 and NOx. However, addition of n-butanol to biodiesel significantly decreases some essential thermo-physical parameters such as density, viscosity, and calorific value, in addition to enhancing the properties of combustion. Furthermore, n-butanol shows a better result on engine performance for BTE and exhaust emission for CO_2 and NOx when blends with B20. This is due to higher oxygen and latent heat of vaporization of n-butanol promoting better combustion hence increasing BTE on engine performance and reducing CO_2 and NOx on exhaust emission. In overall, the addition of 5ml n-butanol to biodiesel B20-diesel fuel blends show the most significance result for the engine performance and exhaust emissions at all test condition.

Acknowledgement
The author of this paper would like to express their sincere gratitude to Universiti Teknologi MARA (UiTM) for the opportunity to trigger this research and Fundamental Research Grant Scheme (FRGS) from the Malaysian Higher Education Department (Grant No FRGS/1/2019/TK10/UITM/02/16) for the financial support.

References
[1] Geng P, Cao E, Tan Q and Wei L 2016 *Renew. Sustain. Energy Rev.* **71** 523–534
[2] Resitoglu I A, Altinisik K and Keskin A 2014 *Clean Technologies and Environmental Policy* **17(1)** 5–27
[3] Ashok B and Nanthagopal K 2019 *Eco friendly biofuels for CI engine applications* (In Woodhead Publishing Series in Energy) p 407-440
[4] Zhang Z and Balasubramanian R 2014 *Appl. Energy* **119** 530–536
[5] Shirneshan A 2013 *Procedia - Soc. Behav. Sci.* **75(x)** 292–297
[6] Yilmaz N, Vigil F M, Benalil K, Davis S M and Calva A 2014 *Fuel* **135** 46–50
[7] Yoshimoto Y, Kinoshita E, Shanbu L and Ohmura T 2013 *Energy* **61** 44–51
[8] Goga G, Singh B, Kumar S and Muk H 2019 *Energy Reports* **5** 78–83
[9] Huang J, Wang Y, Li S, Roskilly A P,Yu H and Li H 2009 *Appl. Therm. Eng.* **29** 2484–2490
[10] Altun Ş, Öner C, Yaşar F and Adin H 2011 *Ind. Eng. Chem. Res.* **50(15)** 9425–9430
[11] Dogan O 2011 *Fuel* **90(2011)** 2467-2472
[12] Çelebi Y and Aydın H 2018 *Fuel* **222(2017)** 385–393
[13] Norhafana M, Noor M M, Hairuddin A A, Hagos F Y, Ayob S I, Kadirgama K and Ramasamy D 2019 *AIP Conf. Proc.* **2059(2019)** 1-10
[14] Radhakrishnan S, Devarajan Y, Mahalingam A and Nagappan B 2017 *J. Oil Palm Res.* **29(3)** 380–386.
[15] Akar M A 2016 *Adv. Mech. Eng.* **8(2)** 1–7, 2016
[16] Zaharin M S M, Abdullah N R, Najafi G, Sharudin H and Yusaf T 2017 *Renewable and Sustainable Energy Reviews* **79** 475-493
[17] Yoruklu H C, Koroglu E O and Demir A 2019 *The Electromotive-Induced Regulation of Anaerobic Fermentation : Electrofermentation.* (Elsevier B V) p 739-756
[18] Örs I, Sarıkoç S, Atabani A E and Ünalan S, 2020 *Biofuels* **11(2)** 121–134
[19] Madiwale S, Karthikeyan A and Bhojwani V 2017 *IOP Conf. Series: Materials Science and Engineering* **197(2017)** 012015
[20] Rahman M M, Rasul M G, Hassan N M S, Azad A K and Uddin M N 2017 *Environ. Sci. Pollut. Res.*
[21] Adams C W M 1966 *Nature* **213(5080)** 974–974
[22] Tüccar G, Tosun E and Uludamar E 2018 *Acad. Platf. J. Eng. Sci.* **6(2)** 81–85
[23] Stournas S, Lois E and Serdari A 1995 *J. American Oil Chem. Soc.***72** 433-437
[24] Yesilyurt M K, Aydin M, Yilbasi Z and Arslan M 2020 *Fuel* **269(Jan)** 117455
[25] Sarkar D K 2015 *Fuels and Combustion*
[26] Thangarasu V and Anand R 2019 *Advances in Eco-Fuels for a Sustainable Environment* (Woodhead Publishing Series in Energy) p 309-336
[27] Atmanli A 2016 *Fuel* **176** 209–215
[28] Satsangi D P and Tiwari N 2018 *Fuel* **221(Feb)** 44–60

[29] Sharon H, Ram P J S, Fernando K J, Murali S and Muthusamy R 2013 *Energy Convers. Manag.*, **73(X)** 95–105
[30] Shahad H A K and Wabdan S K 2015 *J. Renew. Energy* **2015** 1–7
[31] Lujaji F, Kristóf L, Bereczky A and Mbarawa M 2011 *Fuel* **90(2)** 505–510
[32] Atmanlı A, Yüksel B and Ileri E. 2013 *Fuel* **109(1)** 503-511

Experimental Investigation on the Optical and Stability of Aqueous Ethylene Glycol/Mxene as a Promising Nanofluid for Solar Energy Harvesting

K Kadirgama[1*], L Samylingam[2], Navid Aslfattahi[3], M Samykano[4], D Ramasamy[4] and R Saidur[2]

[1] Faculty of Mechanical and Automotive Engineering Technology, Universiti Malaysia Pahang, 26600, Pekan, Pahang, Malaysia
[2] Research Center for Nano-Materials and Energy Technology (RCNMET), School of Science and Technology, Sunway University, Bandar Sunway, Petaling Jaya, 47500,Selangor Darul Ehsan, Malaysia
[3] Department of Mechanical Engineering, Faculty of Engineering, University of Malaya, 50603, Kuala Lumpur, Malaysia
[4] College of Engineering, Mechanical Department, Universiti Malaysia Pahang

*Corresponding e-mail: kumaran@ump.edu.my

Abstract. In this research work, emerged nanomaterial (MXene) and Ethylene Glycol (EG) as base fluid is used to formulate a homogenous mixture of nanofluid with promising optical properties. Amphiphilic structure of ethylene glycol allows the dispersion of water and MXene nanoflakes appropriately. The studied nanofluids are prepared with a mixture of 70 % of EG and 30% of deionized water. The prepared aqueous EG is mixed with Ti_3C_2 nanoflakes with three different loading concentrations of 0.02, 0.05, and 0.1 wt.%. The Fourier Transform Infrared spectrum (FTIR) of the aqueous EG/MXene nanofluids is measured using a Perkin Elmer Spectrum Two-Universal ATR. Particle analyzer is used to measure the stability of the prepared nanofluids. The experimentally achieved data represents the highest stability with mean zeta potential value of -92.6 mV for aqueous EG/MXene with a loading concentration of 0.02 wt.%. Optical absorbance measurements are performed using Ultraviolet-visible (UV-Vis) spectroscopy. Perkin Elmer Lambda 750 is used to acquire spectra. The result of the observation indicates that the solution of nanofluid with a loading concentration of 0.1 wt. % of Ti_3C_2 exhibits the highest peak magnitude of light absorbance among the rest of the three samples. This is because of the more mass concentration value of that nanofluid solution, which is proportional to light absorption. Due to the hydrophilic nature of the utilized nanomaterial and base fluid, there is an excellent probability for them to form a potential homogenous mixture of nanofluid, which might accomplish high optical performance in terms of applying at the solar system.

Keywords: nanofluids, optical absorbance, solar energy, harvesting

1. Introduction

Due to the inevitable demand for high energy consumption, it is essential to exploit renewable energy as the source of generating power to fulfil the request. Since the cost of fossil fuels or petroleum energy are expensive and also sources of environmental pollution, so that from the past year's renewable energy is being utilized as the alternative source of producing electrical power such as by the wind, solar, geothermal, ocean, tidal-wave energy and so on [1]. Among those energy sources, the contribution of solar power is increasing drastically. Owing to having the privileges of being sustainable, inexhaustible, and its non-polluting qualities, at present solar energy has been facilitated tremendously as a coherent, prominent, and convenient form of energy source. For this reason, the demand for solar energy is spreading for different purposes. As a result, solar plants are being implemented in many places around the world. The solar system contains some required accessories and elements where nanofluids are considered as one of the significant integral segments of the system. Because the adequate performance of nanofluids can improve the system efficiency of the solar plant [2]. Generally, nanofluids are formed by homogeneous mixing of nanoparticles along with a base fluid. Characteristics of nanofluids hold some significant properties such as thermal, optoelectronics, optical to maintain the energy efficiency of the solar plant. Energy efficiency is highly dependent mainly on the performance of optical properties from nanofluids because, based on optical properties, suitable nanofluids are selected [3]. Since solar energy contributes a vital role in generating power, but the system efficiency is low. Because of this issue, it is not adequate to meet the demand yet. Researchers noticed this problem, and still, they are conducting their experiments upon different potential nanofluids to enhance the efficiency of the solar plant by exploiting the waste heat along with improving the optical properties of nanofluids.

From the past decades, the scientist has elucidated plenty of experiments to improve the potential performance of optical properties of nanofluids to yield more energy from the solar system that is heading to achieve high efficiency. Light absorption, transmittance, scattering, and coefficient of extinction are the significant aspects involved in the optical properties of nanofluids. The intake of sunlight is reliable on the volume fraction of nanoparticles mixed with nanofluids. At a high-volume fraction of nanoparticles, desirable light can be captured from the sun. There are some intrinsic characteristics of nanofluids that have significant contributions to developing effective optical performance where an increasing amount of viscosity is less [4]. Like selecting suitable pH, adding a surfactant to nanofluid also has a positive effect on improving the system efficiency [3].

Moreover, various types of solar collectors depend on the performance of different nanofluids are being manufactured to accumulate more solar energy. The optical properties of nanofluids have a potential impact on the performance of several types of solar collectors [5]. Khullar et al. have investigated the effect of using nanofluid based concentrating solar collectors to harvest solar energy. They concluded that while maintaining the external conditions, nanofluid based focusing solar collector exhibits 5-10% higher energy efficiency compared to the conventional parabolic solar collector [6]. Otanciar et al. have proposed a theory of enhancing the efficiency of solar collectors. They claimed that direct absorption of the solar collector has a unique advantage over conventional methods that tends to improve the effectiveness of the solar plant up to 5% [7]. Besides, optical and all other properties of nanofluids are size-dependent, which indicates that any desired properties of the fluid can be acquired by the proper size and shape of the nanoparticles [8]. These are also very sensitive in terms of geometrical configurations [9]. Even though theoretically, nanoparticles are adapted to enhance the efficiency of the solar plant, but in case of application, it causes some crucial issues like low specific heat, little thermal stability, altering thermophysical properties concerning temperature [10]. All these issues raised the curiosities of researchers and encouraged them to continue their investigation for the possible acceptable solution.

To overcome the drawbacks above, a new family of two-dimensional nanomaterial from metal carbides or nitrides called MXene has been emerged recently and swiftly gained the attention for the application of better energy conversion and energy storage [11, 12]. The new type of nanomaterials (MXene) is a two-dimensional transition nanocomposite material. The structure of MXene (Ti_3C_2) nanomaterial contains a lot of nanosheets filled composite that can display effective performance in the optical field like absorbing microwaves [13]. Optical properties are considered as the most crucial aspects in terms of metal band structure and transition of electronics in them [14]. Depending on the electronics structure of Ti_3C_2, it is related to the molar absorptivity, which indicates that absorption in optical properties is a sequel between the intra and interband transition [15]. In nanocomposite materials like MXenes, the dimension of the sizes of particles are so petty that it makes it perfect to yield a high absorption rate by receiving more sunlight [16]. This small size of particles forms metallic bonds that tend to increase the band gap energy. This phenomenon eventually leads to having more high-frequency light absorption from solar energy. Because of possessing some nomadic configuration like the feature of overlapping thin film flakes, Ti_3C_2 can be applicable in many useful purposes such as capturing harmful gases and adsorb storage of natural gases [17] and designing metasurface that possesses broadband absorption quality [18].

Optical properties of base fluids have to be proficient enough to ensure to get high performance by the formed nanofluids. Ethylene Glycol is a crucial organic compound that is considered as one of the most efficient base fluid for synthesizing nanofluids. Ethylene glycol exhibits good qualities of optical properties that make it useful enough to work as a base fluid. Optical properties like absorption are highly dependent on the activities of Ethylene Glycol. Band gap energy of the nanofluids ordains the quality of the optical properties of it. A tremendous band gap energy can increase the performance of optical properties in Ethylene Glycol based nanofluid [19]. Gawel et al. have explained the optical properties of EG based Silicon Nitride nanofluid. They stated that in the ultraviolet region of light, with the increase of volume fraction of particles, the capacity of absorbance increases [20]. Optical properties such as the coefficient of extinction can also be enhanced with a high mass concentration of nanoparticles but only in the visible range [21]. In terms of application, Ethylene Glycol can be exploited in different sectors. This chemical intermediate can be used in the field of heating or cooling plastic, automobiles, energy, chemicals, industrial process, and solar thermal energy system [22, 23].

In terms of making a competent homogenous mixture of nanofluid, both nanoparticles and base fluid must be hydrophilic. Hydroxyl (OH) or Oxygen terminated surface can be found in the molecular structure of Titanium Carbide (Ti_3C_2) that makes it hydrophilic enough. These surfaces ended groups such as F, and OH has a strong influence on the optical properties, especially in transmitting light of Ti_3C_2. Light transmittance in the visible range can occur upon the termination of F or OH [24]. Apart from that, the molecular structure of Ethylene Glycol ($C_2H_6O_2$), it is noticeable that two pairs of OH bonds are presented that tend to mix with water at an arbitrary position [25]. These characteristics make Ethylene Glycol hydrophilic enough.

In the present study, a novel highly stable EG/Water/MXene (Ti_3C_2) nanofluid has been prepared successfully. The studied EG with a molecular weight of 62.07, represented an excellent dispersion with DI water due to the presence of hydroxyl groups and hydrophilic nature. To the best of authors' knowledge, this is the first report of the utilization of MXene nanoparticles as filler for aqueous EG-based nanofluids with promising optical and stability properties. Interestingly the unique balance and optical performance of this promising EG/Water/Ti_3C_2 nanofluid were revealed according to the experimental achieved data. This highly promising EG/Water/Ti_3C_2 nanofluid might contribute to the development of the next-generation of optical-based nanofluids for efficient solar energy harvesting.

2. Materials and Method

2.1. Materials

Ethylene glycol ($C_2H_6O_2$) with a molecular weight of 62.07 and CAS number of 107-21-1 was procured from R&M Chemicals and used as received without further purification. The aqueous solution was obtained in a predefined ratio (70% EG+30% water). MAX phase material (Ti_3AlC_2) was procured from Y-Carbon Ltd.

2.2. Synthesis of MXene (Ti_3C_2) Nanomaterials

In the synthesis of MXene (Ti_3C_2,) the following materials were used without any further purification: MAX Phase material (Ti_3AlC_2) from Y-Carbon Ltd., lithium fluoride (325 mesh powder, 98.5% purity, Alfa Aesar), hydrochloric acid (37% wt., Fisher chemicals), and sodium hydroxide (97% purity, pellets, Sigma Aldrich). Firstly, 15 ml of DI water was added to a 50 ml volume beaker followed by adding 15 ml of HCL to obtain 30 ml of HCL (6M). Then 3 g of LiF was added to the HCL solution with stirring at 300 rpm for 30 minutes until dissolved. This etching process was continued with adding 3 g of MAX phase material (Ti_3AlC_2) to the solution slowly (within 15 minutes) to avoid overheating (exothermic reaction), and the resultant solution is left to stir at 40 °C for 48 h. After the etching process, a dilute solution of NaOH was added slowly until the pH of the solution reached 6 and was filtered and rinsed several times with deionized water. The washing process was conducted using an ultrahigh centrifuge (Sorvall LYNX 6000, Thermo Scientific) for four times (each time of 10 minutes) at 3500 rpm. The achieved multilayered MXene (m-Ti_3C_2) was sonicated for one h using an ultrasonic probe sonicator (FS1200N) to obtain delaminated flakes of the MXene (d-Ti_3C_2). The synthesized delaminated flakes of MXene nanomaterial was dried in a vacuum oven (VO 500, MEMMERT Germany) overnight.

2.3. Preparation of EG/Water/Ti3C2 Nanofluid

The aqueous solution of EG/Water was prepared with ratio of 70% EG and 30% DI water. The total volume of each sample was 50 ml were comprised 35 ml of Ethylene Glycol and 15 ml of deionized water. Mixing the Ethylene Glycol with deionized water is required prior to becoming a suitable base fluid. Because at the time of mixing, deionized water opens the chain structures of the molecules of Ethylene Glycol so that it becomes less viscose. The viscosity of Ethylene Glycol must be less to make potential homogenous mixing with Ti_3C_2 nanoparticles.

Firstly, 35 ml of EG was added to the beaker followed by adding 15 ml of DI water. Acquired 50 ml of the aqueous EG/Water was weighed using a microbalance (TX323L, UNIBLOC) to calculate the amount of MXene in terms of weight percentage. The weight of the solution was measured in 100.98 g. Secondly, for preparing three different loading concentrations of MXene nanoparticles, including 0.02, 0.05, and 0.1 wt.%, the amount of the required Ti3C2 nanoflakes was calculated in 20.2, 50.95 and 100.1 mg respectively. The amount of 50 ml of aqueous solution (EG/Water) was stirred at 1000 rpm for 10 minutes to obtain homogenous dispersion. The required amount of Ti_3C_2 for loading concentration of 0.02 wt.% was weighed using a microbalance (EX224, OHAUS), then it was added to the solution followed by stirring for another 60 minutes at 400 rpm. The resultant EG/Water/Mxene solution was sonicated using an ultrasonic probe sonicator (FS-1200N) for one h with the power of 60%. This protocol was conducted precisely for preparing a loading concentration of 0.05 and 0.1 wt.% as well. Figure 1 represents a schematic of the sample preparation method.

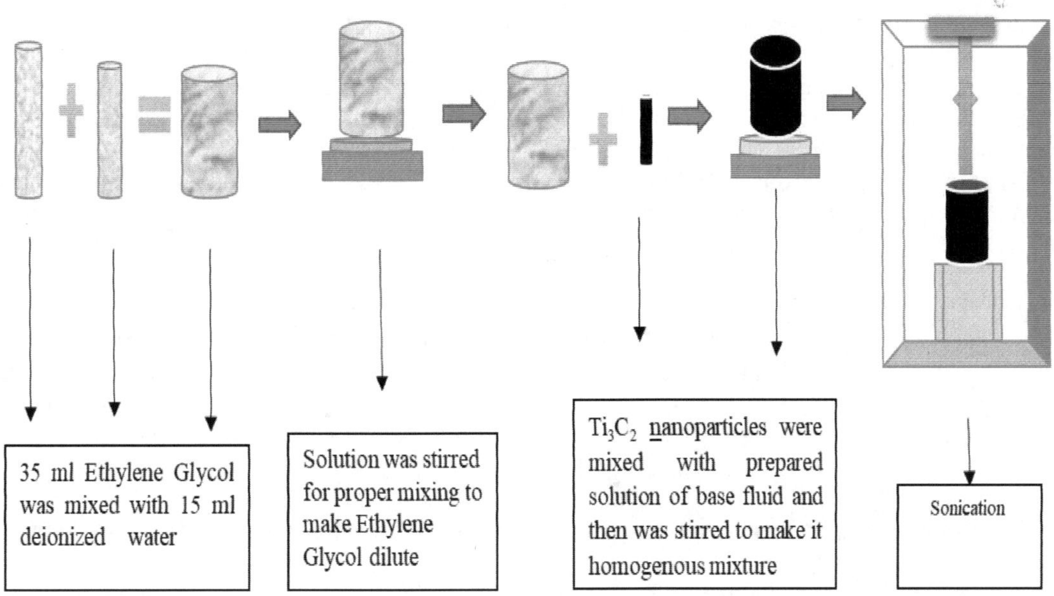

Figure 1. Schematic representation of the samples prepared during the experiment.

2.4. Fourier Transform Infrared Spectroscopy (FTIR) Test

The Fourier Transform Infrared spectrum (FTIR) of the EG/MXene nanofluids was measured using a Perkin Elmer Spectrum Two-Universal ATR. The integrated detector of MIR TGS was used to detect the spectra. Scanning speed was 0.2 cm/s within the optimum scan range of 350-4000 cm^{-1}. The number of scanning was adjusted to 16 scans. The FTIR measurements for all prepared samples was followed with a unique protocol to assure the accuracy of the experiments.

2.5. Stability Analysis of Aqueous EG/MXene Nanofluids.

To measure the stability of the prepared aqueous EG/MXene nanofluids, particle analyzer (LITESIZER 500, ANTON PAAR) device was used. This type of measurement is called Zeta Potential. While observing this experiment, solutions of Ti_3C_2 based nanofluid and 70% of Ethylene Glycol base fluid from the four samples were filled inside the Omega-cuvette and inserted those inside the device. Prior to achieving practical results, apparatus such as cuvettes must be clean enough to get rid of contamination related issues. During that stage of the experiment, some exigent parameters were required. Such as the voltage was set 200V, the temperature was 25°C (room temperature), the number of the run was 1000, and the refractive index, relative permittivity, and viscosity were also inputted in the device accordingly.

2.6. UV-Vis Ultraviolet (UV-Vis) Spectroscopy

Optical absorbance measurements were accomplished using Ultraviolet visible (UV-Vis) spectroscopy. Perkin Elmer Lambda 750 was used to achieve spectra. Correction curve measurements was conducted initially. Then the four solutions of different mass concentrations of Ti3C2 based Nano fluids and 70% Ethylene Glycol base fluid-filled cuvettes were inserted inside the machine to commence the investigation. At room temperature, experimental data was collected within the wavelength range from 800 to 200 nm. The adjusted scan speed was 266.75nm.

3. Results and Discussion

3.1. Chemical Structural Analysis using FTIR

Based on the observation of the experiment, the chemical structures of the molecules of the three different mass concentration based nanofluid and 70% Ethylene Glycol based fluid were investigated by using the FTIR spectroscopy analysis. The result of the acquired FTIR spectrum was found out in a graphical exposure, and it is presented in Figure 2. Distinctive mass concentrations of Ti_3C_2 like 0.02, 0.05 and 0.1 wt. % involved in nanofluid and 70% Ethylene Glycol based fluid are shown separately in one single graph for better comparison. Both MXene and base fluid (Ethylene Glycol) contains common surface terminated groups such as O-H. Moreover, characteristics peaks for EG/MXene nanofluid and 70 % Ethylene Glycol based fluid revealed the presence of O-H, C-H, C-O, and C-O-C groups in the samples. The peak at 2878.89-2939.73 cm^{-1} displays the skeletal vibrations from C-H, O-H groups in the chemical structure of the nanofluid. Due to O-H extending vibrations, the presence of O2-functionalities was detected in peaks at 1411.72 cm^{-1} [24]. In that experiment, it is clearly observed that the wavenumber (cm^{-1}) characteristics peaks of the four different samples of nanofluid are similar. They are following a similar trend line in the graph. It explicates that negligible chemical changes occurred in that prepared nanofluids due to mixing a low amount of wt. % of the Ti_3C_2 with the base fluid (Ethylene Glycol). During that experiment, no new chemical bonding was formed. The molecules involved inside the mixture of nanofluid of the samples are chemically stable. The dispersion was occurred due to physical reaction. A chemical reaction is not required here. On the contrary, chemically unstable molecules are not suitable for homogenous mixture because it shows less hydrophilic quality. The result of this investigation reveals that the performance of Ti_3C_2 is potential. This quality makes the MXene more hydrophilic, which tends to mix with Ethylene Glycol since it is also hydrophilic and thus exhibited effective performance during the experiment by forming an influential homogenous mixture.

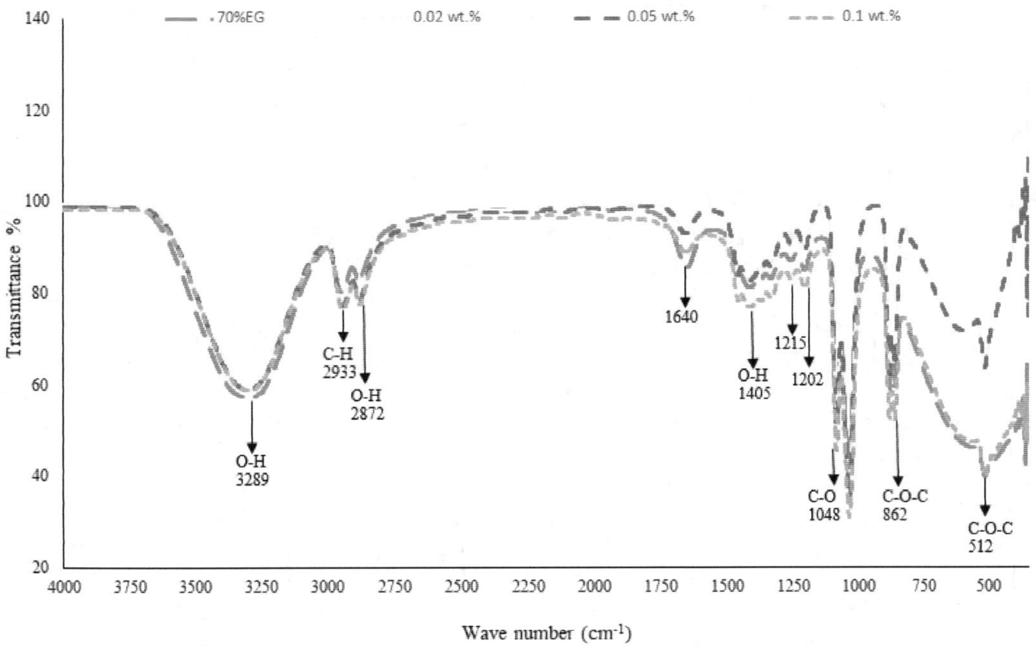

Figure 2. Full range (4000-350) cm^{-1} FTIR spectra of Ti_3C_2 and 70% Ethylene Glycol (base fluid) based Nano fluid with varying MXene loading mass fraction.

3.2. Stability Characterization of Aqueous EG/MXene Nanofluids.

The potential difference between the stationary layer of fluid aligned to the dispersed diminutive particles and the dispersion medium is known as zeta potential. One of the crucial factors that affect zeta potential is the size of the molecules or particles. The size of particles must be small enough to achieve high molecular stability of the solution of nanofluids, and this aspect also resists the phenomena of getting agglomeration or sedimentation during the experiment. Nanoparticles inside the nanofluid follow the Brownian motion. This Brownian motion suppresses the instability condition of the tiny particles. Moreover, zeta potential is the electrostatic repulsive force between the base fluid and nanoparticles, which denotes that this high repulsive force causes high stability of the molecules of nanofluids [26]. Generally, the magnitude of mV may be defined as the estimated stability value of the nanofluid, whereas the excellent stability of the molecules of nanofluid may exceed the extent of 60 mV [27, 28].

Table 1. Experimentally achieved zeta potential values from different samples of nanofluid.

Samples	Mean zeta potential (mV)	Standard deviation (mV)	Transmittance (%)
70% Ethylene Glycol	-34.5	123.5	76.6
0.02 wt% of Ti_3C_2	-92.6	2.4	67.8
0.05 wt% of Ti_3C_2	-80.5	1.7	86.4
0.1 wt% of Ti_3C_2	-69.5	1.5	86.3

According to the magnitudes above of Table 1, the experimental value of mean zeta potential for 70% Ethylene base fluid is found -34.5 mV. For this type of base fluid, the suitable range for potential molecular stability is around -35 mV to +35 mV. From this result, it is noticeable that the molecules

IOP Conf. Series: Materials Science and Engineering 1062 (2021) 012022 doi:10.1088/1757-899X/1062/1/012022

inside the 70% Ethylene base fluid are in a stable range. The rest of the three samples where different amounts of mass concentration of MXene were mixed with 70% Ethylene Glycol also displays excellent particle stability. For example, the mean potential values of 0.02, 0.05 and 0.1 wt. % of Ti_3C_2 based nanofluids are subsequently -92.6, -80.5, and -69.5 mV. From these illustrations, it can be clarified that nanoparticles of all these samples are in stable condition. This experimental result reveals that the molecules of three different values of mass concentration of MXene maintain excellent stability inside the samples of nanofluid because the mean potential magnitudes of these samples are in the range of 60 mV. Since molecules of the nanoparticles are in stable conditions so that those tiny particles were able to arrange chronological sequences of their position inside the solutions of nanofluid for different load concentrations. The distance of each of the particles was synced to each other, and incoming light got enough space for transmitting. Because of this reason, the magnitude of the percentages of light transmittance through the mass concentration mostly for 0.05 and 0.1 wt. % is high in this experiment, such as 86.4 and 86.3% subsequently. Surface charges of Ti_3C_2 are negatively surface charge because it contains surfaces terminated groups such as OH^-, F^- and -O, which are also negative charges. In this experiment, Ethylene glycol was used as the base fluid, which surface charges is also negative. Because it contains two strong hydrogen bonds that are aligned with OH^- group, which possesses negative charge, since both nanoparticles and base fluid exhibit negative charges inside the solution of nanofluid, there is a massive repulsive force between the particles and base fluid was occurred during the experiment and thus achieved excellent particle stability condition. High stability of particles inside the solution indicates a high surface area of the thin nanosheets of Ti_3C_2, which makes the solution into suitable nanofluid to achieve potential optical performance.

3.3. UV-Vis Ultraviolet (UV-Vis) Spectroscopy Characterization.

The conversion of solar energy to any other form of energy or directly to generate power is a significant aspect. Increasing the performance of optical properties such as absorption, transmittance, scattering, and coefficient of extinction can lead to gain desired output results. The measurement of absorption of sunlight by nanofluid can be executed by using Ultraviolet-visible (UV-Vis) spectrometry. In this investigation, the typical optical absorption spectra of Ti_3C_2 for different mass concentrations like 0.02, 0.05, and 0.1 wt. % in the nanofluid were determined and shown in Fig 3. From the result of the figure, pure Ethylene Glycol indicates almost the same trend throughout the whole range of wavelength did not express any peak point of absorbance. Nanofluid contains 0.02 wt. % of MXene nanoparticles exhibit only one peak point of light absorbance at 280 nm based on the conjugation π system of the electronic transition of the MXene particles. According to the beer lambert law, it is known that absorbance of light increases with the number of wavelengths and with the quantity of molar concentration of the fluid. From Figure 3, it is found that there are five peak points for 0.05 wt. % of EG/Ti_3C_2 nanofluids. The highest peak is observed at a wavelength of 371 nm that was happened because of the electronic transition ($n \rightarrow \pi$) of the particles—the nanofluid solution of 0.01 wt. % of Ti_3C_2 nanoparticles has the maximum number of peak points. The graphical expression of this nanofluid follows the beer lambert law, and at the wavelength of 781 nm it shows maximum light absorbance. Electrons of the atoms involved in the molecules of MXene nanoparticles gained energy from light and increased their strength and transited through lower bands to higher-level bands. As a result, their band gap energy enhanced, which accelerated those nanoparticles to absorb more incoming lights. In this sample of nanofluid solution, the mass concentration of MXene nanoparticles is high that made the mixture darker, which led to capture more light to meet the objectives of this experiment.

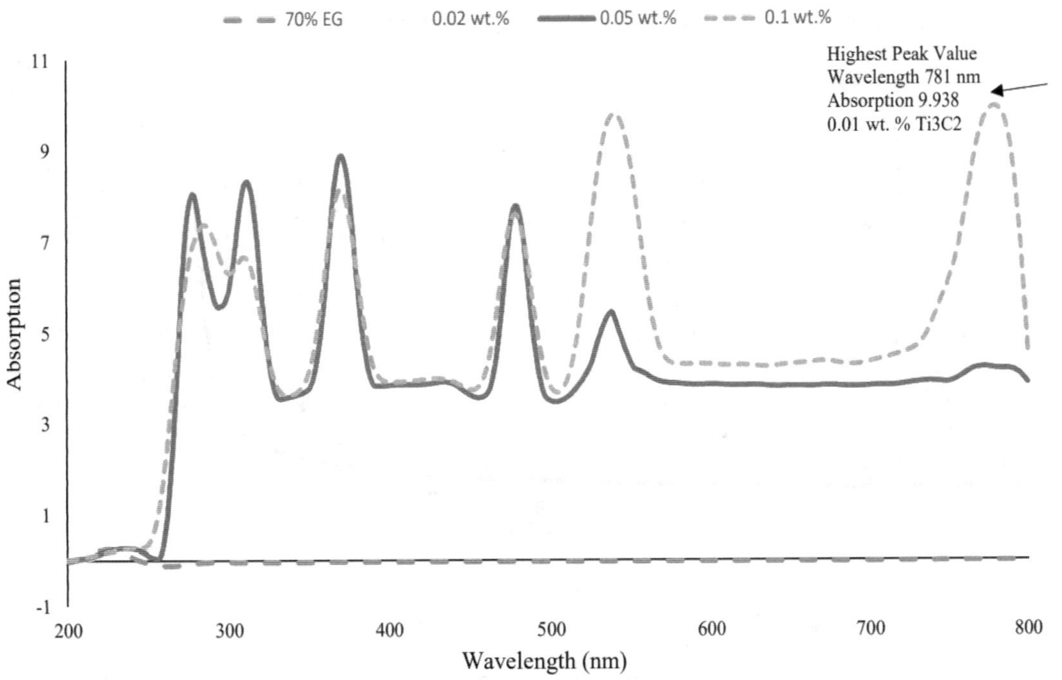

Figure 3. Light absorption characteristics for different loading concentration of Ti_3C_2 into EG/MXene nanofluids.

4. Conclusions

In this study, optical properties of different mass concentrations of prepared EG/MXene nanofluids were determined. Absorbance, transmittance, extinction coefficient, and scattering are the crucial parameters of the optical property of a nanofluid. Nanoparticles of Ti_3C_2 were mixed with the base fluid (Ethylene Glycol) to form the homogenous mixture of nanofluid. Base fluid (Ethylene Glycol) contains strong hydrogen bonds that assist in withstanding against high temperature during the experiment. For this reason, at the beginning of the investigation, it was mixed with deionized water to reduce its viscosity. Mixing with deionized water enables to open the chain structure of the molecules of Ethylene Glycol and makes it dilute enough to get mixed with the MXene nanoparticles. 0.02, 0.5 and 0.1 wt. % of the mass concentration of Ti_3C_2 nanoparticles were selected for comparison. These particles were mixed with 70% Ethylene Glycol to comprise a new form of nanofluid. During the experiment, it is found that the pattern of the graphs of each sample is displaying a similar trend that indicates that the molecules inside the prepared samples are chemically stable. The physical reaction occurred in the dispersion of the particles. No new chemical bonds were formed. It was also noticed that particles inside the solution of different mass concentrations of Ti_3C_2 expressed excellent stability performance by showing the zeta potential values in the range of 60 mV. While conducting the observation, the concentration of 0.01 wt. % of Ti_3C_2 nanoparticles solution reflects a darker color, which led this mixture to absorb more incoming light, and the result was collected from the graph. In the graph, it is found that light absorption is the highest with an increasing number of wavelengths for 0.1 wt. % mass concentration quantity of Ti_3C_2 compared to the other values of the samples. Based on this investigation, both nanoparticles and base fluid show hydrophilic property. Because of this reason, during the experiment, those were able to make the new type of homogenous mixture of nanofluid. After the observation of the experiment, it can be clarified that this novel type of nanofluid displays potential optical properties and can be applied to enhance the system efficiency of the solar plant. The future work is to investigate the optical properties such as absorption, transmittance by using this novel type of nanofluid that includes a different amount of weight percentages of Ti_3C_2 nanoparticles because it is a promising nanofluid that can meet the future demand.

Acknowledgments
"R. Saidur would like to acknowledge the financial support provided by the Sunway University through the project no# STR-RCTR-RCNMET-001-2019". The author would like to thank Malaysia Higher Education Minister for providing financial under the Fundamental Research Grant Scheme FRGS/1/2019/TK07/UMP/02/3 and Universiti Malaysia Pahang no. RDU 192207.

References

[1] Krishna Y, Saidur R, Faizal M, Aslfattahi N, Ng K, Arifutzzaman A and Karinka *S* 2019 *J I J N M*
[2] Taylor R A, Phelan P E, Otanicar T P, Walker C A, Nguyen M, Trimble S, Prasher R 2011 *J J o R and Energy S* **3** 023104
[3] Raj P, Subudhi S J R and Reviews S E 2018 **84** 54-74
[4] Taylor R A, Phelan P E, Otanicar T P, Adrian R and Prasher R 2011 *J N r l* **6** 225
[5] Chen M, He Y, Zhu J, Shuai Y, Jiang B and Huang Y J S E 2015 **115** 85-94
[6] Khullar V, Tyagi H, Phelan P E, Otanicar T P, Singh H, Taylor R A 2012 *J J o N i E and Medicine* **3**
[7] Otanicar T P, Phelan P E, Prasher R S, Rosengarten G, Taylor R A 2010 *J J o r and energy s* **2** 033102
[8] Verma S K, Tiwari A K J E C and Management 2015 **100** 324-46
[9] Barcaro G, Sementa L, Fortunelli A and Stener M 2015 J P C C P **17** 27952-67
[10] Saidur R, Meng T, Said Z, Hasanuzzaman M, Kamyar A 2012 *J I J o H and Transfer M* **55** 5899-907
[11] Chaudhari N K, Jin H, Kim B, San Baek D, Joo S H and Lee K 2017 *J J o M C A* **5** 24564-79
[12] Aslfattahi N, Saidur R, Arifutzzaman A, Sadri R, Bimbo N, Sabri M F M, Maughan P A, Bouscarrat L, Dawson R J and Said S M 2020 *J J o E S* **27** 101115
[13] Qing Y, Zhou W, Luo F and Zhu D 2016 *J C I* **42** 16412-6
[14] Zhang R-Z, Cui H-L and Li X-H 2019 *J P B C M* **561** 90-6
[15] Lashgari H, Abolhassani M, Boochani A, Elahi S and Khodadadi J 2014 *J S s c* **195** 61-9
[16] Shaviv E, Schubert O, Alves-Santos M, Goldoni G, Di Felice R, Vallée F, Del Fatti N, Banin U and Sönnichsen C 2011 *J A N* **5** 4712-9
[17] Liu F, Zhou A, Chen J, Jia J, Zhou W, Wang L and Hu Q 2017 *J A S S* **416** 781-9
[18] Chaudhuri K, Alhabeb M, Wang Z, Shalaev V M, Gogotsi Y and Boltasseva A 2018 *J A P* **5** 1115-22
[19] Zhu Y, Chen Y and Liu L 2011 *J J o c g* **328** 70-3
[20] Żyła G J 2020 *J o M L* **297** 111732
[21] Di Rosa D, Wanic M, Fal J, Żyła G, Mercatelli L and Sani E 2019 *J P T* **356** 508-16
[22] Yue H, Zhao Y, Ma X and Gong J 2012 *J C S R* **41** 4218-44
[23] Sani E and Dell'Oro A 2014 *J O M* **37** 36-41
[24] Xue Q, Zhang H, Zhu M, Pei Z, Li H, Wang Z, Huang Y, Huang Y, Deng Q and Zhou J 2017 J A M **29** 1604847
[25] Kao M, Tien D, Jwo C and Tsung T 2005 In: *Journal of Physics: Conference Series: IOP Publishing)* p 442
[26] Xian H W, Sidik N A C, Saidur R J 2020 I *C i H and Transfer M* **110** 104389
[27] Kumar D D, Arasu A V 2018 *J R and Reviews S E* **81** 1669-89
[28] Yu F, Chen Y, Liang X, Xu J, Lee C, Liang Q, Tao P and Deng T 2017 *J P i n s M I* **27** 531-42

Optimization of Performance and Exhaust Emissions of a Spark-Ignition (SI) Engine Fueled with Bioethanol-Gasoline Blends using TOPSIS Methodology

I Baratian[1], Muhammad Yasar[2,3*], G Najafi[1], B Ghobadian[1], Bo Zhang[4], X X Jiang[4]
[1] Tarbiat Modares University, Tehran, Iran
[2] Department of Agricultural Engineering, Faculty of Agriculture, Universitas Syiah Kuala, Indonesia
[3] Politeknik Aceh Selatan, Tapak Tuan, Aceh, Indonesia
[4] School of Mechanical Engineering, Ningxia University, China

*Corresponding e-mail: yasar@unsyiah.ac.id

Abstract. Energy crises and environment impact forced governments and countries to develop utilizing renewable energy in different forms. One of the first forms of alternative energy which was in use is bioethanol which is using in SI (spark ignition) engines as pure fuel or after blending with gasoline. The selection of proper blending of bioethanol plays an important role in the production of alternative energy. In this investigation, the use of TOPSIS (Technique for Order Performance by Similarity to Ideal Solution) to select the optimal value of the engine performance and exhaust pollutants which run at various bio-ethanol-gasoline fuel blends, namely E5, E10, E15 and E20. Obtained results indicate that performance parameters (i.e. torque, power) for different bioethanol- gasoline blends improved. It is evident from results that there is an increase in NOX as well Carbon dioxide (CO2) emission while UBHC and CO decreased. TOPSIS methods indicate that a mixture ratio equal to 20% (E20) and an engine speed equal to 3000 rpm were chosen to be optimal working condition.

Keywords: energy, exhaust emissions, fuel, performance, pollutants

1. Introduction

Energy saving is one of the most effective issues for strategic planning in the industrialized world [1–3]. On the other hand, the demanding energy is clearly a driving force of the industry and world economy. [4] Among different types of energy, the fossil energy is the largest source of world energy which is currently used in internal combustion engines [5]. One of the most profitable options for governments to replace conventional fossil fuels in is the use of renewable fuels including biodiesel, bioethanol, bio methanol, biogas, etc. [6].

Bioethanol, which is extracted from sugar and starch feedstock, can be used alone or blend with gasoline in SI engines without the need to any major modification in the engine [6–8]. Using bioethanol – gasoline blends in spark ignition engines improves engine performance and reduces emissions [9–12]. Due to the complexity of issues associated with internal combustion engines, it is difficult to model using one decision modeling [13–17]. Therefore, the use of modeling techniques in combination with experimental methods provides a lot of benefits for understanding these phenomena [3,14,18]. There are many research works deal with predicting engine parameters fueled with bio-based additive [19].

In a number of studies, artificial neural networks (ANNs) have been used to predict engine performance and emissions. For instance, Najafi et al. [20] analyzed the engine performance and pollutants of a four-cylinder SI engine operating on ethanol–gasoline blends of 5%, 10%, 15% and 20% with the aid of ANN model. An ANN model was developed to predict engine performance and exhaust emissions. Their results showed that with increasing brake thermal efficiency and the volumetric efficiency (VE), the brake specific fuel consumption (BSFC) was declined. Although, the condensation of CO2 and NOx were increased with increase the amount of bioethanol. In an experimental work, Kiani et al. [21] used artificial neural network (ANN) modeling for predicting of performance and exhaust emissions in a spark ignition engine fueled with different ethanol- gasoline fuel blends. They used experimental results for training and validating. Their results proved that ANN demonstrated the best accuracy in predicting the emission characteristics, engine brake power (BP) and torque.

Najafi et al. [22] studied the influence of five different blends of ethanol- gasoline on the performance and exhaust emission parameters of a four-cylinder spark ignition engine using response surface methodology (RSM). Their results proved that design of experiments (DOE) based on RSM has many advantages in modeling of engine performance. A mixture consists of 10 % ethanol (E10) were identified as best blend with highest desirability. A number of investigators have developed novel techniques for modeling of spark ignition (SI) engines. For instance, Najafi et al. [23] proposed a comparison of both ANFYS and support vector machines (SVM) model for prediction of engine performance and gaseous exhaust emissions of a SI engine with various bioethanol ratios and validates models with experimental test results. Their results demonstrated that the machine learning based on ANFIS –SVM techniques has great advantages for modeling engine application and emissions using alternative fuels. They also mentioned that the ANFIS technique is much efficient than SVM method.

In an interesting study, Cay et al. [24] strived to model the performance (i.e. BSFC, torque, pe, APe) and emissions (i.e. CO, HC) characteristics of a four-stroke spark-ignition engine with blend of methanol-gasoline fuel by exerting ANN method. Their finding showed that logsig (logistic sigmoid) algorithm provide the best accuracy statistics (R) for the designed network. TOPSIS (Technique for Order of Preference by Similarity to Ideal Solution) is based on multi-criteria decision analysis which was introduced by Hwang and Yoon in 1981[25]. This method is a relatively simple and quick method for analyzing and monitoring the best strategy [15,26]. In an isolated study, ANN modeling followed by multi-objective optimization (MOP) was used to finding the best solution of the engine performance and emissions of the CI engine fueled with conventional petro diesel and castor oil biodiesel blends. They explain that, applying a hybrid method of NSGA-II integrated with TOPSIS design provides some trade-off optimum points on the non-dominated solutions which can be used for achieving high performance [25]. Dehghani Soufi et al. [27] conducted experiment on the use of TOPSIS techniques for choosing the best performance with more desirability reported that the palm bio-lubricant illustrated better performance in comparison with the mineral two-stroke lubricants.

It is needless to mention that, there is rarely study on the application of TOPSIS to select optimal ethanol blend based on the engine working condition and emission characteristics using TOPSIS technique. Hence, in this paper tried to evaluate the best appropriate blend and engine working conditions by using TOPSIS techniques. To implement the experimental tests was used to evaluate engine parameters and then TOPSIS analyzing method was applied to optimize and sort for all test results based on the improvement of engine parameters. In this paper, TOPSIS is proposed as the most effective method to evaluate the optimal blend and engine working condition.

2. Material and Methods

2.1. Description of the experimental setup

In this experimental work, four-cylinder naturally aspirated, water-cooled SI gasoline engine KIA 1.3 SOHC was used. The technical characteristics of the engine are shown in Table 1.

Table 1. Technical specifications of the test engine.

Type of engine	SI engine,4 cylinder, water cooled
Bore×Stroke(mm)	71.0×83.6
Engine displacement volume (cc)	1323
Compression ratio	17.5
Maximum Torque (Nm-rpm)	410-1300
Maximum Power (kW-rpm)	47-2300

Exhaust gas emissions were measured by means of AVL DIGAS exhaust gas analyzer. Spark advance was set to 25 before TDH (top death center). Engine performance parameters and gas emissions have been determined for gasoline at the compression ratios (CR) of 9.7. An eddy current dynamometer which is a SCHENCK-WT190 brand is capable of absorbing 190 kW at 10000 rpm engine speed. Dynamometer was equipped with a barometer to determine the pressure inside the cylinder chamber. The data collected from the pressure sensor is recorded in the dynamometer control unite. AVL Flow-sonix air flow meter was used to determine air consumption. The consumption rate of fuel was quantified with a Pierburg laminar-type flow meter. A Distributor-Less Ignition (DLI) was employed for ignition apparatus. An Ignition system was Semi-static distributor less ignition (DLI) was used as. The schematic view of test setup is depicted in Figure 1. Instruments accuracies and parameter uncertainties are illustrated in Table 2.

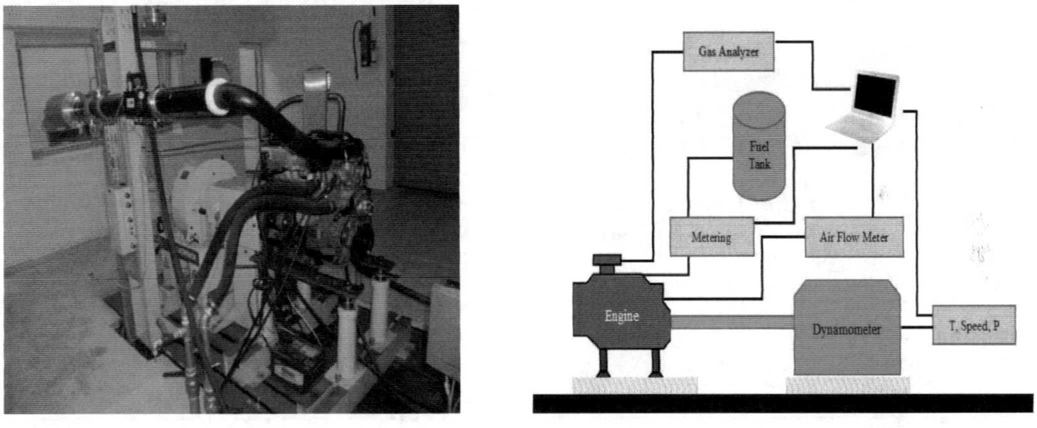

(a) (b)

Figure 1. System configuration for engine testing (a) substantive and (b) schematic.

Table 2. The accuracies of the measurements in the calculated parameters.

Instrument variables	Accuracy (±)
Engine speed (rpm)	2.00
Engine load (N)	0.1
Fuel Consumption (g)	0.1
Temperature (℃)	0.1
Time	0.5
Emissions parameters	Uncertainty (%)
UHC	1.80
CO2	2.40

IOP Conf. Series: Materials Science and Engineering 1062 (2021) 012023 doi:10.1088/1757-899X/1062/1/012023

CO	2.40
NOX	1.4
Performance variables	Uncertainty (%)
Brake power (B.P)	1.03
Torque	1.10
BSFC	1.10

In this study, a continuous solid fermentation setup based on fermentation of sugar has been employed for producing bioethanol from inedible potato biomass. The specifications of potato-based bioethanol are cited in Table 3:

Table 3. Characteristics of ethanol-gasoline blended fuels.

Specification	Method	Ethanol
Density (kg/m3)	ASTM D 4052	785
Viscosity (cSt)	ASTN D 88	1.1
Calorific value (kJ/kg)	ASTM D 240	27000
Octane number	ASTM D 2699	108.6
Pour point (℃)	ASTM D 97	-50<
Flash point (℃)	ASTM D 93	-14

In this investigation, potato biomass has been considered as a bioethanol source. Various blends of ethanol and gasoline including E5, E10, E15, and E20 were exanimated. Since engine could not run normally at higher ethanol ratios, therefore, experimental data obtained up to E20 (80% gasoline - 20% ethanol) not presented. The characteristics of blended fuels are presented in Table 4.

Table 4. characteristics of ethanol-gasoline blended fuels.

Fuel Specification	Tested fuel				Standard
	E5	E10	E15	E20	
Vapor Pressure (kPa)	55.16	55.16	55.16	55.16	ASTM-D323
Octane number	89.7	93.8	95.6	99.4	ASTM-D2,699
GHC (MJ/kg)	44.15	40.15	38.6	40.51	ASTM-D340

Experimental study done for various blends in various days, but for all tests is done in one day for each blend. Correction factors are calculated and used to correct results for comparing together. Test variables and conditions were controlled. Test pattern for all blends is the same. Engine speed is varying from 1000 to 5000 rpm in 500 rpm steps. In each rpm level, output parameters including torque (Nm), power (kW), CO (%Vol.), CO2 (% Vol.), UBHC (ppm), NOX (ppm), specific fuel consumption (SFC) and volumetric efficiency (VE) are recorded.

2.2. Optimization based on TOPSIS.

2.2.1. Normalization matrix. In this experiment data obtained in different scale and order. In addition, the recorded data must be homogenized so that they can be compared and use together. After that, the data have been normalized in the form of matrix using Equation (1):

$$n_{ij} = \frac{r_{ij}}{\sqrt{\sum_{i=1}^{m} r_{ij}^2}} \tag{1}$$

In this equation r_{ij} refer to evaluation matrix. Other parameter (i.e. i, j) are matrix index.

2.2.2. Weighted normalized decision matrix. Weighted matrix is used to show the importance of parameters. Weigh of each parameter is determined between 0 and 1. The sum of all of weigh is equal to1. In this study we used ENTROPY method for weight calculation. The corresponding value is defined as follows in equation (2) to (6). Where p_{ij}, K, E_j, d_j and W_j are:

$$p_{ij} = \frac{r_{ij}}{\sum_{i=1}^{m} r_{ij}} \;\; ; \;\; j = 1, \cdots, n \quad \forall_{ij} \tag{2}$$

$$K = \frac{1}{Ln\, m} \tag{3}$$

$$E_j = -K \sum_{i=1}^{m} p_{ij} \ln p_{ij} \quad \forall_j \tag{4}$$

$$d_j = 1 - E_j \tag{5}$$

$$W_j = \frac{d_j}{\sum_{j=1}^{m} d_j} \quad \forall_j \tag{6}$$

2.2.3. Positive and negative ideal solution. The positive ideal solution (PIS) shows the most preferable option and the negative ideal solution (NIS) shows the least preferable option. PIS is maximum option for parameters which in bigger case is better. And minimum for which in lower case is better. PIS is an alternative and that is the minimum option for parameters which in bigger case is better. And maximum for which in lower case is better. Using equation (7) and (8), calculate the separation from PIS and NIS for each parameter. The relative closeness to the ideal solution for each parameter is calculated using equation (9). Using the relative closeness to the ideal solution, increasing of the relative closeness to the ideal solution is more preferable.

$$S_{i*} = \sqrt{\sum_{j=1}^{n} \left(v_{ij} - v_j^*\right)^2} \tag{7}$$

$$S_{i-} = \sqrt{\sum_{j=1}^{n} \left(v_{ij} - v_j^-\right)^2} \tag{8}$$

$$C_{1*} = \frac{S_{1-}}{S_{1*} + S_{1-}} \tag{9}$$

3. Results and Discussion

In this section, the data have been analyzed and sorted using TOPSIS methodology. Parameters and test points were arranged based on benefit and favor methodology. These data are sorted based on the ethanol- gasoline fuel blend, engine speed and output parameters including torque, power, specific fuel consumption, volume efficiency and emission i.e. CO, CO_2, O_2, UBHC and NOX. Experimental results are shown at different engine speed for various fuel blend in Figures 3 to 9. Figure 2 indicates the effect of engine speed on the engine torque. It can be observed that engine torque increases with increase in engine speed for all fuel blend ratios. According to Figure 2, increasing in engine speed lead to less torque. The decline in peak torque at high speeds was because of a decrease in time for filling the cylinder chamber. Hence, cylinder pressure decreases slightly when reducing the volumetric efficiency (VE) [22,28–30]. Figure 3 shows the variation of engine brake power (BP) with engine speed for various fuel blend ratios. Figure 3 clearly shows that increase in the percentage of biethanol (i.e. less proportion of gasoline) increase drastically the brake power of the engine at all tested fuel proportions. Additionally, an increase in engine power corresponds to the increment in the indicated mean effective pressure (IMEP). This can be attributed to the fact that by increasing proportion of bioethanol, volumetric efficiency increase due to high density of the mixture. One of the distinct advantages of ethanol is promote complete combustion, therewith increasing the engine brake power and torque. [31,32]

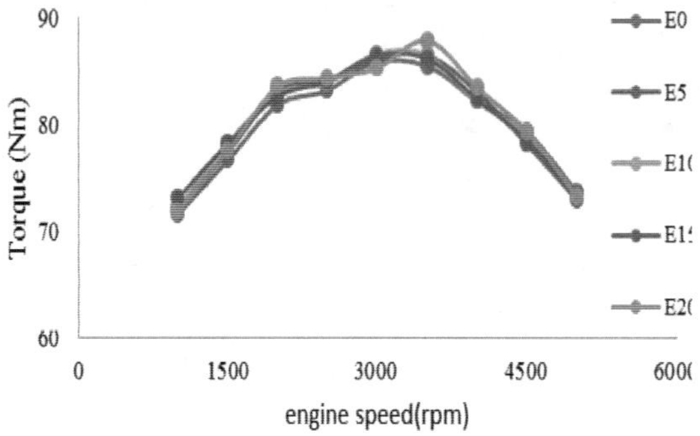

Figure 2. Variation of torque with engine speed for different test fuel blends.

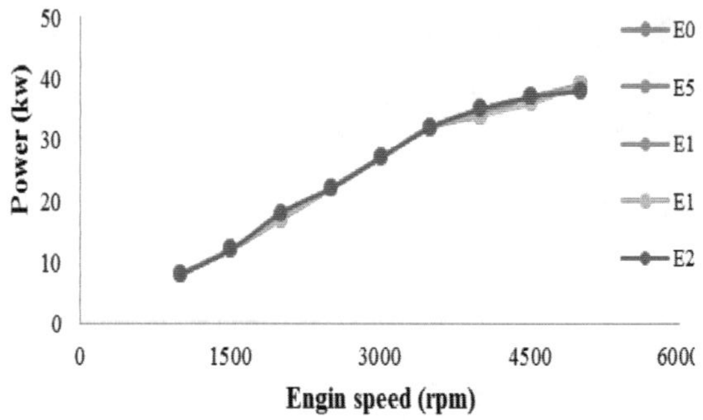

Figure 3. Variation of engine power with engine speed for different test fuels.

Carbon monoxide (CO) emission as a function of engine speed for different bioethanol fuel blends is shown in Figure 4. As seen in this figure, a decrease in CO emission was observed when the percentage of bioethanol in blend increase. The variation of the CO_2 emission versus engine speed for different blends is presented in Figure 5. It is clearly understood from Figure 5 that CO_2 formation increases while the bioethanol ratio increases as well as engine speed. Generally, CO_2 forming related with both stoichiometric ratio and CO emission concentration. On the other hand, increasing bioethanol ratio causes a lean burning and improvement in the combustion process, resulting in increase of the CO_2 emissions.[33,34]

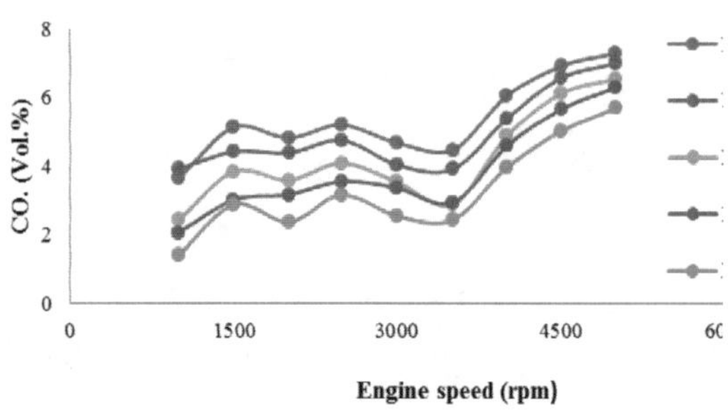

Figure 4. Effect of ethanol blend on CO emission at various engine speeds.

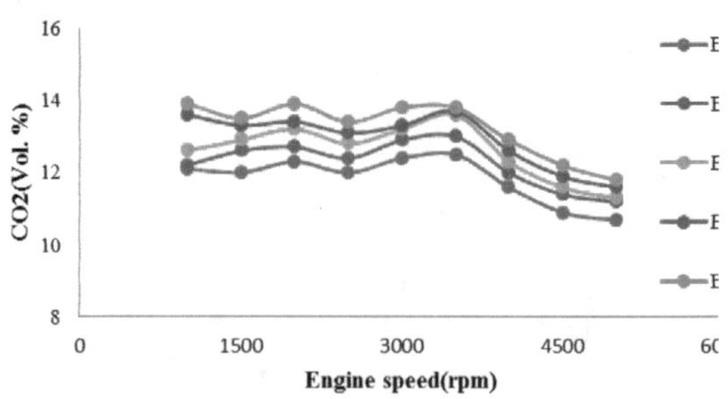

Figure 5. CO_2 emission vs. engine speed for various test fuels.

Figure 6 depicts the variation of UHC emissions for each engine speed and bioethanol concentration. UHC emissions decreased using bioethanol in blend of gasoline due to lower in-cylinder temperatures. As shown in this figure, when bioethanol blends were increased UHC emissions was decreased. The formation of UHC as unburned fuel component increases with increase in air-fuel ration. The reason for reduction of UHC emission at high bioethanol blends would be attributed to fuel oxygen content promoting complete combustion, the same as the CO concentration [35,36]. Figure 7 illustrate the relationship between the NOx concentration and engine speed for various bioethanol ratios. As shown in test results, increasing ethanol, results increasing NOx which is because of increasing in temperature. [37,38]. This effect is the only negative effect of ethanol increasing in blend. However, the higher oxygen content in the mixture and lower cetan number promote the NOx formation. Increasing engine speed, improves most of parameters including power, CO_2, NOx, and has negative effect on fuel consumption and CO. For this reason, positive effect of engine speed is more than negative effect and increasing in engine speed increase improvement. In fact, ethanol and engine speed have a super position effect and improvement with increasing in these factors is predictable. Of course, if choose another method in weight matrix calculation, and using different weighted matrix can change this results totally. Based on the goals and purpose of users, parameter's weight can change and affect weighted matrix and final results definitely will be different.

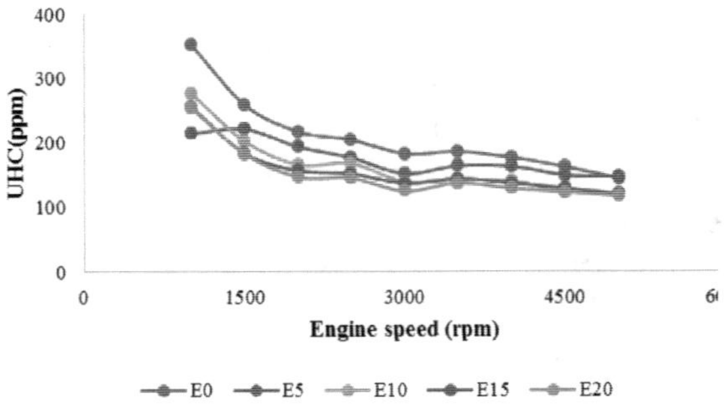

Figure 6. Effect of engine speed on UHC emission for various fuel blends.

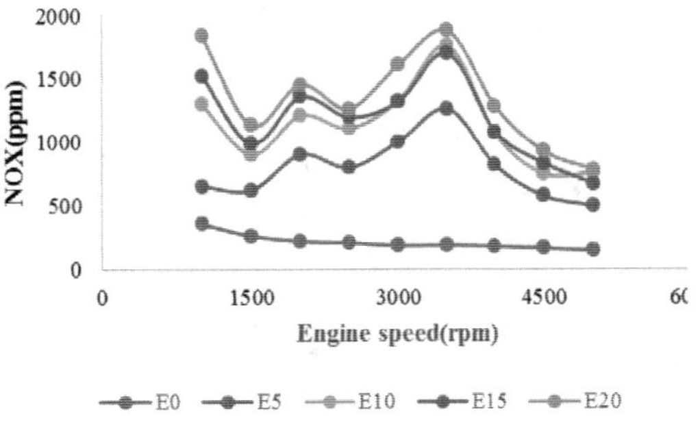

Figure 7. NOx emission vs. engine speed for different test fuels.

Figure 8 shows that specific fuel consumption (SFC) increase by ethanol increasing in blend, however, this amount is not high. Energy content of ethanol is lower than gasoline; normally we use more ethanol for going same distance regarding gasoline.

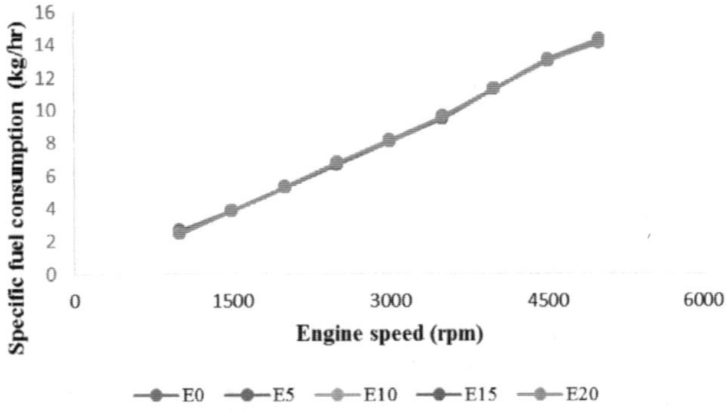

Figure 8. Variation of specific fuel consumption w.r.t engine speed for bioethanol blends.

3.1. Optimization based on TOPSIS.

The TOPSIS methodology is used for determining the best ethanol- gasoline blend. The decision matrix which is built by test results data, analyzed by using TOPSIS method and related equations. The first TOPSIS algorithm step is calculating normalized matrix which results are shown in table 5 for all test point and data. PIS and NIS index are calculated for test parameters using weighted matrix and shown in table 5. In next step, S+ and S- value are calculated for all parameters and test point.

Table 5. PIS and NIS index.

	S (rpm)	T	P	F	A	CO	CO_2	HC
PIS	0.00016	0.009027	3.79365E-06	0.003633	0.01016	3.2E-06	0.000492	0.001289
NIS	0	0	0.000135623	0.120465	0.225358	4.52E-06	3.73E-05	0.144414

The relative closeness at different engine speed for various fuel blends are shown in Figure 9. According to Figure 9, increasing ethanol in blends is definitely preferable and increase engine performance. Figure 10 shows the relative closeness versus fuel blend for different engine speed. It is revealed from the results that engine performance increases with increasing engine speed and ethanol blends.

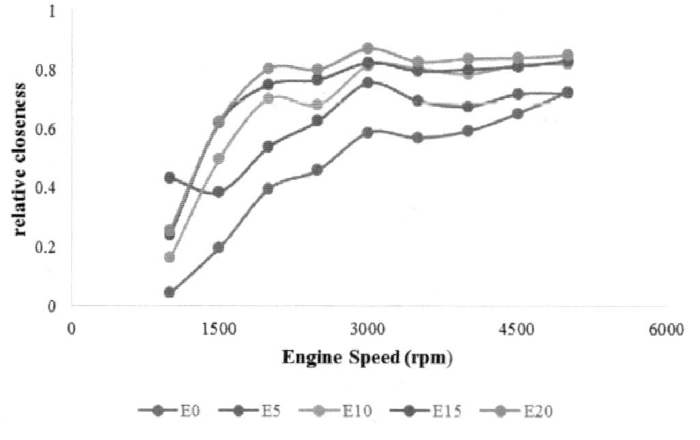

Figure 9. The relative closeness at different engine speed for various fuel blends.

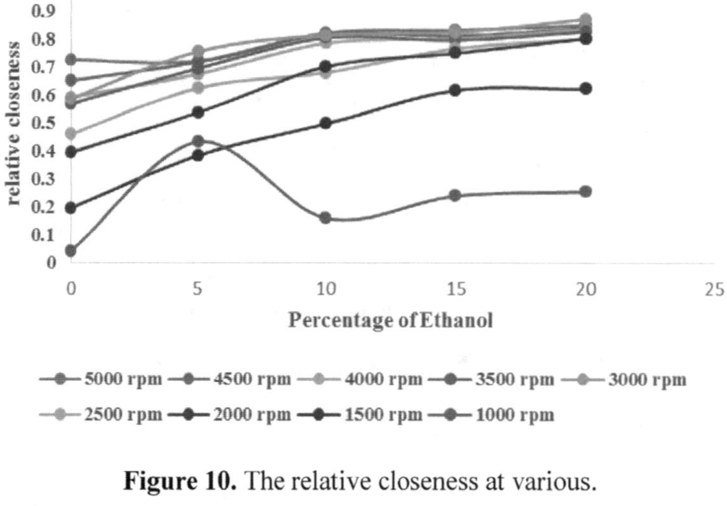

Figure 10. The relative closeness at various.

The distance to the good point is higher and relative closeness is very low particularly in lower engine speed. 3D schematic of the relative closeness is shown in Figures 11 and 12. Preferable points on the test are located in the top of the graph. With increasing ethanol blends exhaust gas temperature and engine power increases while CO_2, CO and UBHC decreased. Based on TOPSIS methods results, optimal value of engine speed and ethanol-gasoline ratio were obtained 3000 rpm and 20% ethanol in blend fuel (E20).

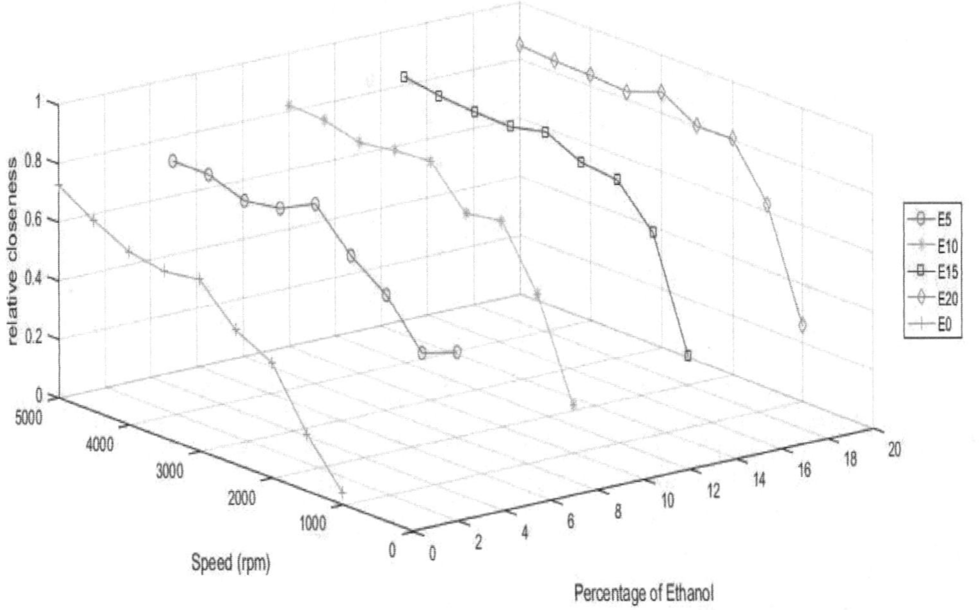

Figure 11. 3D schematic of the relative closeness at various fuel blends for different engine speed.

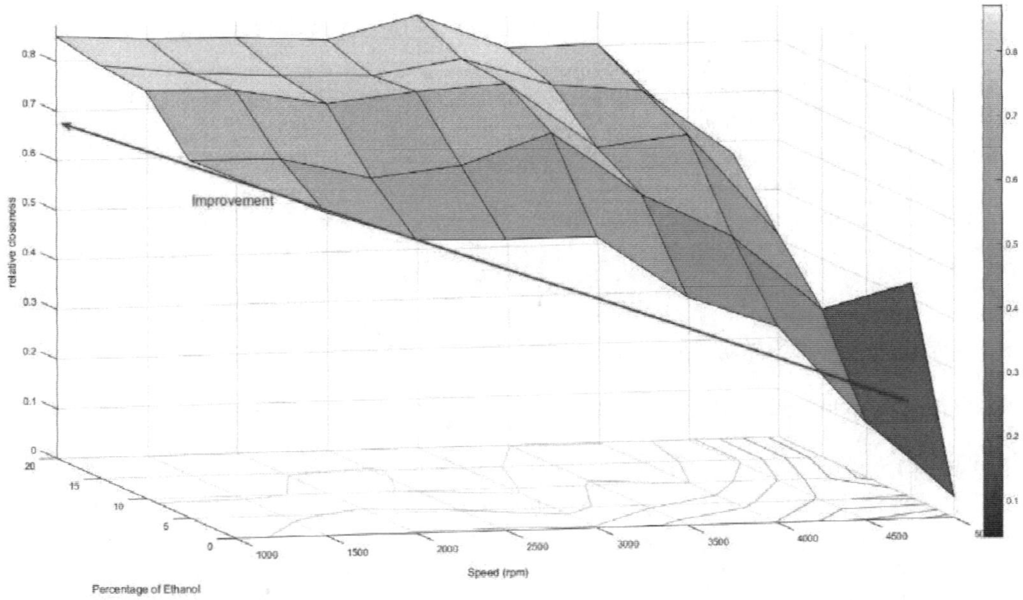

Figure 12. 3D schematic and contour of the relative closeness at all test points.

4. Conclusions

In this investigation, experimental studies of using gasoline- ethanol blends including E0, E5, E10, E15 and E20 and comparison between various blends are carried out. Using TOPSIS analyzing method, the results sorted and ranked to improve performance of the engine. The obtained results are summarized as below:

- With increasing bioethanol ratio, torque, pressure and NOx increased.
- Carbon monoxide, HC, and carbon dioxide reduced as bioethanol in blend increase,
- TOPSIS methods indicate that with increasing ethanol in blend and engine speed, engine performance will approve.
- TOPSIS methods indicate that engine rotational speed and ethanol percentage are 3000 rpm and 20% (E20), respectively, for optimal working conditions.

References

[1] Ardebili M S, Ghobadian B, Najafi G and Chegeni A 2011 *Renew. Sustain. Energy Rev.* **15** 3041–3044
[2] Samuel O D, Emovon I 2018 *Int. J. Integr. Eng.* **10** 176–181
[3] Azizan N H, Kisim N F M, Saedudin R D R, Abidin Z A Z, Ashaari M M and Kasim S 2018, *Int. J. Integr. Eng.* **10** 133–138
[4] Xiaohua Wand Jingfei L 2005 *Renew. Sustain. Energy Rev.* **9** 229–236
[5] Solmaz H 2015 *Fuel Process. Technol.* **133** 20–28
[6] Renzi M, Bietresato M and Mazzetto F 2016 *Energy* **115** 1069–1080
[7] Jambo S A, Abdulla R, Mohd Azhar S H, Marbawi H, Gansau J A and Ravindra P (2016) *Renew. Sustain. Energy Rev.* **65** 756–769
[8] Zabed H, Sahu J N N, Suely A, Boyce A N N and Faruq G (2017) *Renew. Sustain. Energy Rev.* **71** 475–501
[9] Park S H, Cha J and Lee C S 2012 *Appl. Energy* **99** 334–343
[10] Rastogi M and Shrivastava S 2017 *Renew. Sustain. Energy Rev.* **80** 330–340
[11] Sebayang A H, Masjuki H H, Ong H C, Dharma S, Silitonga A S, Mahlia T M I and Aditiya H B 2016 *RSC Adv.* **6** 14964–14992
[12] Tan Y H, Abdullah M O, Nolasco-Hipolito C, Zauzi N S A and Abdullah G W 2017 *Energy Convers. Manag.* **132** 54–64
[13] Ming N K, Philip N and Sahlan S 2019 *Int. J. Integr. Eng.* **11** 119–130
[14] Raheem A, Hassan M Y and Shakoor R 2014 *Int. J. Integr. Eng.* **6** 1–8
[15] Iradiratu Diah P K, Dewantara B Y, Abduh M, Sudirman S and Mulyo Utomo W 2019 *Int. J. Integr. Eng.* **11** 181–193
[16] Ambak K, Hadipramana J, Abas N A, Mokhatar S N and Jaini Z 2018 *Int. J. Integr. Eng.* **10** 151–155
[17] Abdal-Kadhim A M 2019 *Int. J. of I. Engineering* **11** 19–26
[18] Hamid A H F A, Chang K W, Rashid R A, Mohd A, Abdullah M S, Sarijari M A and Abbas M 2019 *Int. J. Integr. Eng.* **11** 149–158
[19] Emovon I and Mgbemena C O 2018 *Int. J. Integr. Eng.* **10** 169–175
[20] Najafi G, Ghobadian B, Tavakoli T, Buttsworth D R, Yusaf T F and Faizollahnejad M 2009 *Appl. Energy* **86** 630–639
[21] Kiani M K D, Ghobadian B, Tavakoli T, Nikbakht A M and Najafi G 2010 *Energy* **35** 65–69
[22] Najafi G, Ghobadian B, Yusaf T, Safieddin Ardebili S M and Mamat R 2015 *Energy* **90** 1815–1829
[23] Najafi G, Ghobadian B, Moosavian A, Yusaf T, Mamat R, Kettner M and Azmi W H 2016 *Appl. Therm. Eng.* **95** 186–203
[24] Çay Y, Korkmaz I, Çiçek A and Kara F 2013 *Energy* **50** 177–186
[25] Etghani M M, Shojaeefard M H, Khalkhali A and Akbari M 2013 *Appl. Therm. Eng.* **59** 309–315

[26] Roslee D, Sabudin S, Ariffin N A N, Jamari F N and Batcha M F M 2018 *Int. J. Integr. Eng.* **10** 148–155
[27] Soufi M D, Ghobadian B, Najafi G, Sabzimaleki M R and Yusaf T 2015 *Energies* **8** 13960–13970
[28] Awad O I, Mamat R, Ibrahim T K, Ali O M, Kadirgama K and Leman A 2017 *Appl. Therm. Eng.* **123** 1374-1385
[29] Solmaz H *Fuel Process. Technol.* **133** 2015 20–28
[30] Ismail M M, Fawzi M, Zulkifli F H and Osman S A 2018 *Int. J. Integr. Eng.* **10** 88–93
[31] Wang C, Zeraati-Rezaei S, Xiang L and Xu H 2017 *Appl. Energy* **191** 603–619
[32] Singh S B, Dhar A and Agarwal A K 2015 *Renew. Energy* **76** 706–716
[33] Catapano F, Di Iorio S, Magno A, Sementa P and Vaglieco BM 2015 *Energy* **88** 101–110
[34] Tutak W 2014 *Energy Convers. Manag.* **86** 39–48
[35] Solmaz H 2015 *Fuel Process. Technol.* **133** 20–28
[36] Elfasakhany A 2017 *Renew. Sustain. Energy Rev.* **71** 404–413
[37] Doğan B, Erol D. Yaman H and Kodanli E 2017 *Appl. Therm. Eng.* **120** 433–443
[38] Li Y, Gong J, Deng Y, Yuan W, Fu J and Zhang B 2017 *Appl. Therm. Eng.* **115** 53–63

Investigation on the Fouling and Heat Transfer Characteristics of Novel EGR Cooler (Semi-spiral) for Diesel Engine Fueled with Biodiesel

F. Jamali[1], Mustaqimah[2,*], B. Ghobadian[1], G. Najafi[1], Bo Zhang[3], X.X Jiang[3]
[1]Department of Mechanical & Biosystems Engineering, Tarbiat Modares University, Tehran, Iran
[2]Department of Agricultural Engineering, Faculty of Agriculture, Universitas Syiah Kuala, Indonesia
[3]School of Mechanical Engineering, Ningxia University, China

*Corresponding author: mustaqimah@unsyiah.ac.id

Abstract. In this study, heat transfer characteristic and fouling of Novel EGR cooler (Semi-spiral) system were investigated. In this study, waste vegetable oil (Waste Cooking Oil) was used as feedstock for biodiesel production. The experiment was conducted in different loads (25%, 50% and 75%) and different speeds (2100rpm and 2400rpm). Four fuel blends were prepared and used (B0, B5, B10, and B15). The new EGR cooler has 6 tubes in length of 19 cm and overall heat transfer coefficient of 35.63 W/m^2 K which is acceptable with compared to overall heat transfer coefficient of shell and tube heat exchanger. Also, the total resistance of fouling (Rft) derived was 0.002712 m^2 K/W. The maximum heat transfer efficiency is 94% and the minimum heat transfer efficiency is 68%. Also, the results showed that when B0 is used, the deposition was 1.37 gr that amounts to 0.23gr/hr. When B15 is used, the deposition was 1.21gr that amounts to 0.20gr/hr.

Keywords: heat transfer, diesel engine, vegetable oil, heat exchanger, pressure drop

1. Introduction

Diesel engines have good potential to reduce greenhouse gases. However, some of the pollutants of diesel engines play an important role in polluting the environment [1]. For example, one of the important and significant emissions of diesel engines is NOx emissions. NOx emissions easily pollute the environment and are a carcinogenic pollutant [2, 3]. NO and NO_2, which are called NOx emissions, combined with hemoglobin in blood and moisture content in liver and creates nitric acid. Due to low concentrations of acid produced with NOx, their effects will be less. But through time they accumulate in the body and cause respiratory diseases. The cause of NOx in diesel engines is the high-temperature combustion [4]. Many strategies have been used to reduce NOx emissions in diesel engines including the use of after-treats like technology and combustion management [5-7]. Another way to reduce NOx is through exhaust gas Recirculation (EGR). EGR is the Recirculating amount of exhaust gas to inlet [8]. Exhaust gas Recirculation causes dilution of oxygen at the inlet, increasing thermal capacity and thus reducing the maximum temperature of combustion. Therefore, by using EGR, NOx emissions can be significantly reduced. EGR has a high temperature; hence there are no significant impacts on reducing the maximum temperature combustion. Therefore, to enhance the performance of the EGR, cooled EGR is used. In which case the heat capacity can be significantly increased [9, 11]. To cool the EGR, heat exchangers (EGR cooler) is used. Until now, various heat exchangers and EGR cooler has been evaluated by researchers. Jang et al. [10] examined two types of shell- tube EGR cooler and a stack EGR cooler. In this study, heat transfer and pressure drop in three types of EGR coolers were experimentally investigated. The results indicated that temperature of EGR significantly decreases in the stack type; therefore when stack type EGR cooler is used, the amount of NOx is reduced. Liu et al. [12] investigated the characteristics of flow and heat transfer in EGR cooler with Helical Baffled Cooler with Spiral Corrugated Tubes (EGR-HBCSCT).The results indicated that the tube side heat transfer in the EGR-HBCSCT is 60 to 130 percent more than the tube side heat transfer of EGR-SBCST as shell side resistance in the EGR-HBCSCT is about 2.8 to 5.9 percent less than the shell side resistance of EGR-SBCST. The results showed that the pressure drop in the EGR-HBCSCT was less than EGR-SBCST. Liu et al. investigated the turbulent flow and heat transfer characteristics in the 7 types of cooling pipes of Internal Longitudinal Plate-Rectangle Fins EGR Cooler. The results showed that the friction factor increases with the number of fins, and the friction factor decreases with increase in internal diameter and width of the fin. Liu et al. [13] investigated the flow and heat transfer performances of an EGR cooler with internal longitudinal fins. This study was conducted experimentally and through three-dimension computation. In this study, the effects of fin width and fine height on heat transfer and flow were investigated. The results showed that in both internally finned tubes the heat transfer significantly increases. It was also revealed that the heat transfer in the pipes with blocked structure is better while the pressure drop in the tube without blocked structure is lower. Shabgard et al. [14] investigated the effects of several types of nano-fluid as a coolant on the performance of heat transfer and fluid flow characteristics of EGR cooler. The results showed that using nano-fluid as coolant, the heat transfer increases significantly. Fouling (Deposit) of EGR cooler is of great importance. Deposit formation can be defined as the accumulation of undesirable substances on a surface. In general, Deposit formation refers to the accumulation of unwanted substances that lead to undesired operations on the surface. During the operation of the EGR cooler, the heat transfer surface is deposited. This leads to increase in thermal resistance and pressure drop [15]. Park et al. [16] investigated the effects of deposited PM on the heat transfer in the Fin-Type EGR Cooler. In this study, six different EGR coolers with various pitches and grooves were used. The results showed that the lowest temperature and lowest pressure drop of exhaust gas occurred in the P3.6/36 EGR cooler. Abd-Elhady et al. [17] investigated the effects of velocity gas passing through the EGR cooler on the deposition of PM. In this experiment, the shell–tube EGR cooler was used. The flow rate through the EGR cooler was considered at three levels: 30, 70 and 120 m/s. The results showed that when the flow speed is 120 m/s, the least amount of soot is formed because fewer deposits were formed in the EGR cooler. Hong et al. [18] investigated the size and SOF soot particles deposited in the one type EGR cooler. Effect of EGR at three temperatures of 150, 250 and 350 °C were investigated. The coolant temperature was 80 °C. The results showed that as a whole, by increasing the temperature of EGR, the size of particulate matter (PM) decreased but the amount of deposited formed in the 10 hours of testing increased.

With regards to the review of the above results, it was found that the geometry of the EGR cooler has a significant impact on the flow and heat transfer performance of EGR cooler. It was also revealed that the deposit of EGR cooler pipes has a significant impact on performance of EGR cooler. So far, little research has been done on the effects of the pipe deposition of EGR cooler performance. In this study, a new EGR cooler was designed and built. After fabrication, the EGR cooler was mounted on a single-cylinder diesel engine. Then the performance parameters such as heat transfer efficiency, Nusselt number, and overall heat transfer coefficient and pressure drop along the way were evaluated. Also, the fouling of EGR cooler tubes for different working times and the effects of EGR cooler fouling on heat transfer tubes were investigated.

2. Material and method

2.1. Geometry of EGR cooler

Using the design criteria stated in the previous parts, the EGR cooler was designed and built. Semi spiral-semi smooth tube n was applied to Recirculation of the exhaust gas. The built EGR cooler is shown in Figure 1. The smooth semi spiral-semi smooth tubes used in the EGR cooler is shown in Figure 2.

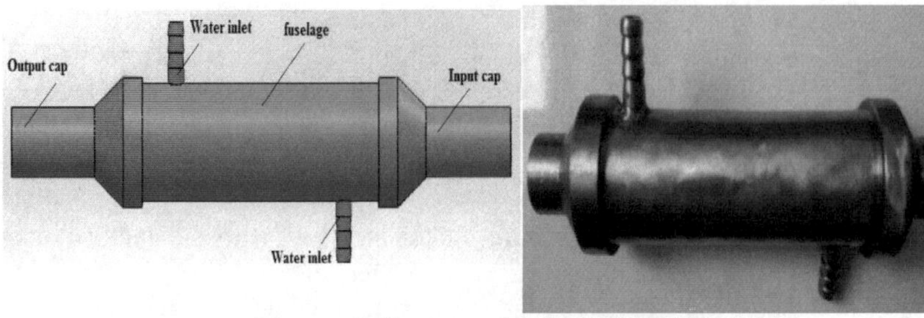

Figure 1. Photograph of EGR cooler

Figure 2. Semi spiral-semi smooth tubes

2.2. Engine test and experimental procedure

2.2.1. Experimental set-up

The schematic of experimental set-up is shown in Figure 3. In this study, air-cooled single-cylinder engine (Lombardini DIESEL 3LD 510) was used and the details of its engine is shown in Table 1.

Table 1. Specifications for the test engine

Description	Specification
Engine type	Lombardini- Diesel 3LD510
Cylinder number	1
Displacement(cm3)	510
Induction type	Non- turbocharger
Bore * stoke (mm)	85*90
Maximum torque (Nm-RPM)	32.8-1800
Compression ratio	18

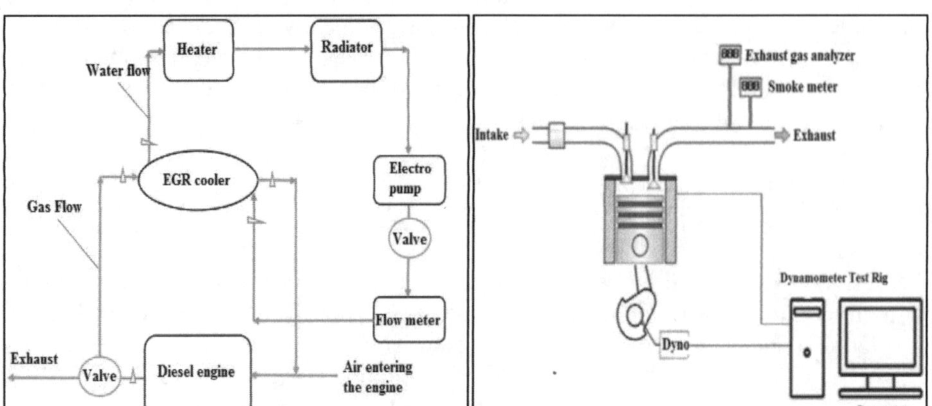

Figure 3. Engine test set-up with EGR cooler.

The single-cylinder engine used in this study was coupled with the dynamometer (Eddy Current Type DC Dynamometer). In order to measure temperature, 5 Thermocouples (K type) were used. The thermocouple measurement range was -200 to 1200 °C. In order to monitor the temperature measured by the thermocouple, TM 925 thermometer was used.

2.2.2 Fuel properties

In this study, waste vegetable oil (Waste Cooking Oil) was used as feedstock for biodiesel production. After producing biodiesel with the trans-esterification method, some important properties such as kinematic viscosity, flash point, cloud point, pour point, the amount of water and sediment, the amount of free glycerine, and corrosion were measured and the results were compared with the international standard (ASTM D-6751). In Table 2, the properties of biodiesel with the relevant standard acceptable ranges were presented.

Table 2. Properties of biodiesel with ASTM standard.

Properties	**Standard**	**Exposure Limits**	**Biodiesel**	**Unit**
flash point	ASTM D-92	≤130	176	°C
kinematic viscosity	ASTM D-445	1.9-6	4.73	mm^2/s
Cloud point	ASTM D-2500	-	-1	°C
pour point	ASTM D-97	-	-4	°C
corrosion	ASTM D-130	3 ≤	1a	-
free glycerin	ASTM D-6584	0.02≤	0.016	% mass
water and sediment	ASTM D-2709	0.05≤	0.05	% Vol
Density	-	-	0.88	g/cm^3

2.2.3. Experimental procedure

Given that the best performance of EGR occurs in the part load, so the EGR Cooler Performance Test were done in 25%, 50% and 75 % of load. Experiment was conducted in 2100 rpm and 2400 rpm. Four fuel blends were prepared and used. These blends were, B5, B10, and B15 in each of the speeds and blends tested at three levels of EGR (10%, 20% and 30%). A matrix experiment is shown in Table 3.

Table 3. A matrix experiment

Variable	**Levels**
EGR rate (%)	10, 20, 30
Load (%)	25, 50, 75
Fuel blends	B0,B5,B10,B15
Engine speed (rpm)	2100, 2400
Coolant inlet temperature (°C)	85

2.2.4 Fouling test

In order to measure the weight of sediment accumulated in the EGR cooler tubes, weighing scales with 0.01gr of precision were used.

3. Result and discussion

3.1 Inlet temperature of EGR cooler

Inlet gas temperatures of EGR cooler at different speeds and loads to different EGR rates for the blends (B0, B5, B10, and B15) is shown in Figures 4-9. According to Figures 4(a-f), the increase of the EGR rate causes the EGR cooler inlet temperature to also increase. On average, at 2100 rpm and 25% load, with 20% increase of EGR, the inlet temperature of EGR cooler is increased by 14.75%.

The increase of the inlet temperature of EGR cooler also occurred for four blends with similar trends. At 2100 rpm and 50% load, with 20% increases of EGR on average, the inlet temperature of EGR cooler increased by 20%, Also considering Figures 4(a-f), when using biodiesel fuel blends, the inlet temperature of EGR cooler is enhanced. For example, at 2100rpm and 50% load, when using a B10 blend, the inlet temperature of EGR cooler showed 8-14 % increase compared to when the B0 blend was used, According to Figure 4-9, it is clear that the highest inlet temperature of EGR cooler is obtained in the following conditions: 2400 rpm, 75% load andB15 blend.

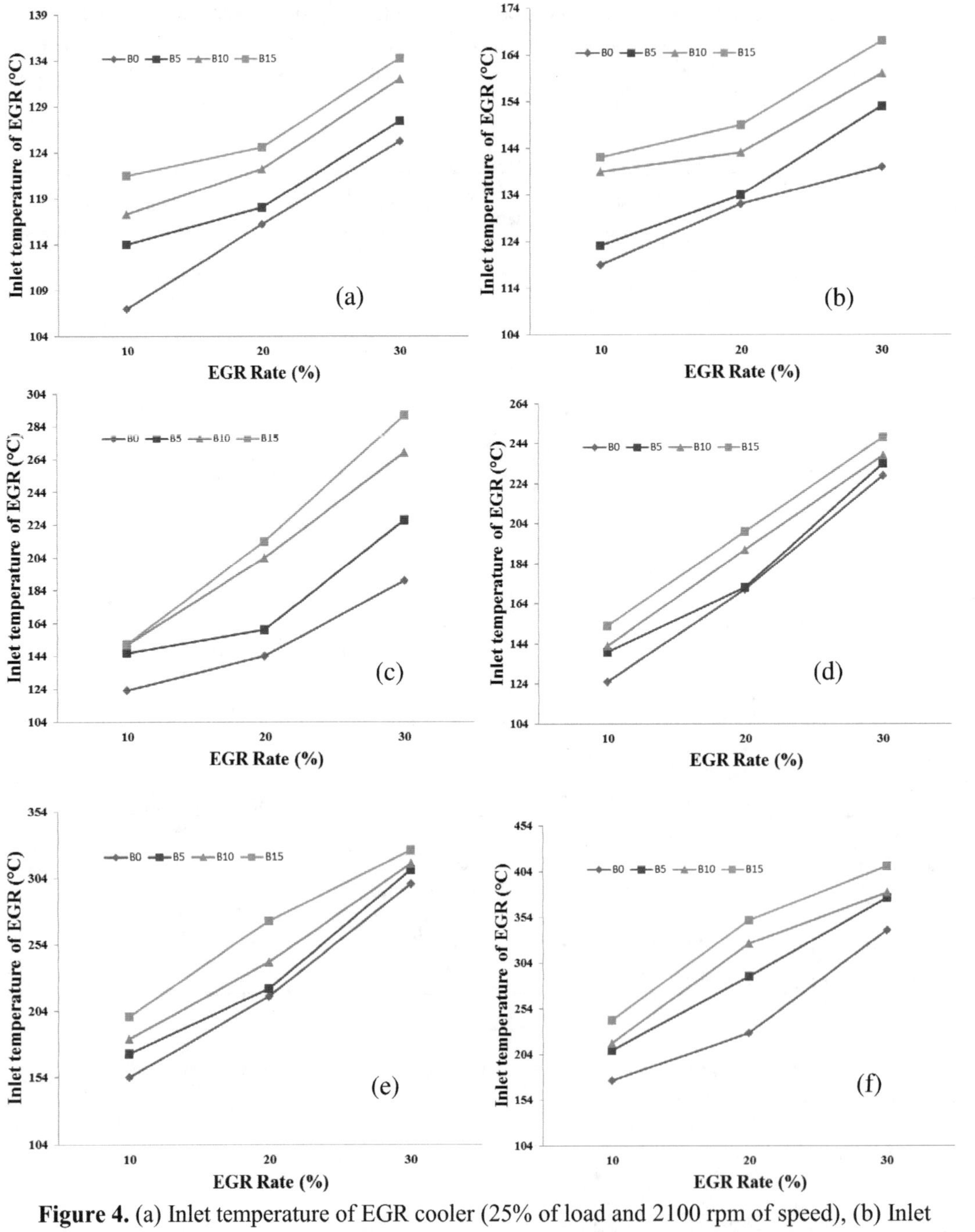

Figure 4. (a) Inlet temperature of EGR cooler (25% of load and 2100 rpm of speed), (b) Inlet temperature of EGR cooler (50% of load and 2100 rpm of speed), (c) Inlet temperature of EGR cooler (75% of load and 2100 rpm), (d) Inlet temperature of EGR cooler (25% of load and 2400 rpm of speed), (e) Inlet temperature of EGR cooler (25% of load and 2100 rpm of speed), (f) Inlet temperature of EGR cooler (25% of load and 2100 rpm)

IOP Conf. Series: Materials Science and Engineering 1062 (2021) 012024 doi:10.1088/1757-899X/1062/1/012024

3.2 Temperature difference of EGR cooler

In Figure 5a, and 5b, temperature difference of EGR coolers between inlet and outlet of tube side for two different engine speeds is shown at 25% load. According to these data, with 2100 rpm and 25% load, by increasing the EGR rate, the temperature difference of EGR cooler between inlet and outlet of tube side increased. With the increase of EGR rate to 20%, the temperature difference between inlet and outlet of tube side increased by 66.4%. Also, with 2400 rpm and 25% load, by increasing the EGR rate, the temperature difference between inlet and outlet of tube side of EGR cooler increased. According to results, it is clear that when B0 fuel was used for all EGR rates, the temperature difference of EGR cooler between inlet and outlet of tube side is greater than when B5, B10 and B15 blends were used.

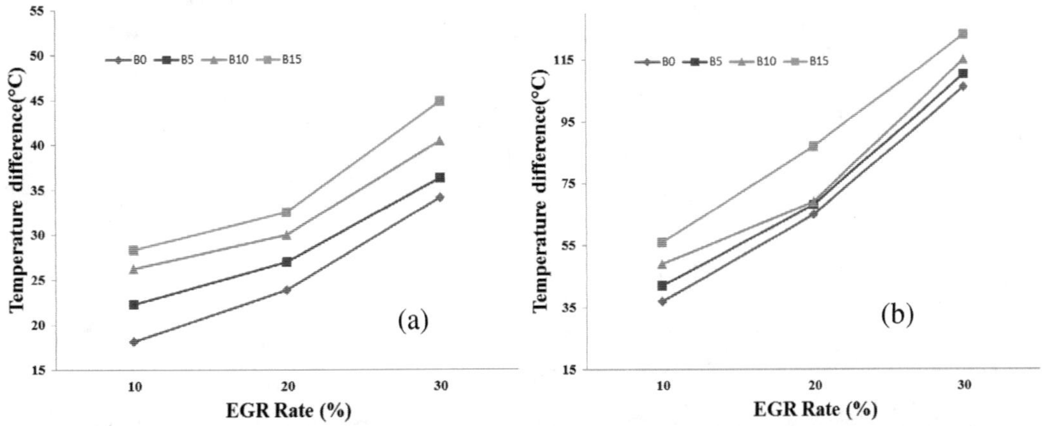

Figure 5. Temperature changes between input and output: (a) 25% of load and 2100 rpm, (b) Temperature changes between input and output (25% of load and 2400 rpm).

3.3. Heat transfer efficiency of EGR cooler

The heat transfer efficiency of EGR cooler at different speeds and loads to different EGR rates for the fuel blends (B0, B5, B10, B15) are shown in Figures 6 (a-f). It is clear that by increasing the EGR rate, the heat transfer efficiency at 2100 rpm and 25% load did not decrease much. The heat transfer efficiency of EGR cooler with B0 blend compared to other blends (B5, B10 and B15) is more. The heat transfer efficiency of EGR cooler at 2100 rpm and 50% load, where the heat transfer efficiency of EGR cooler with B0 blend compared to other blends (B5, B10 and B15) is more. At 2100 rpm and 75% load, with increase of EGR rate, the heat transfer efficiency of EGR cooler is decreased by 5.75%.

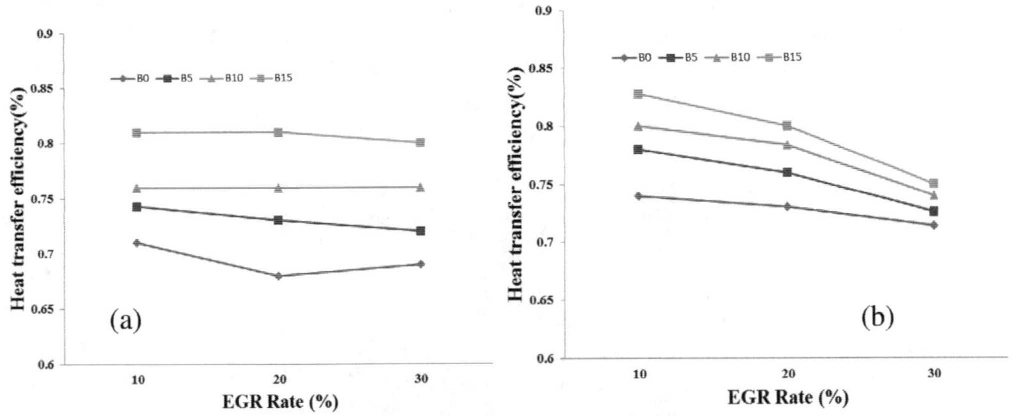

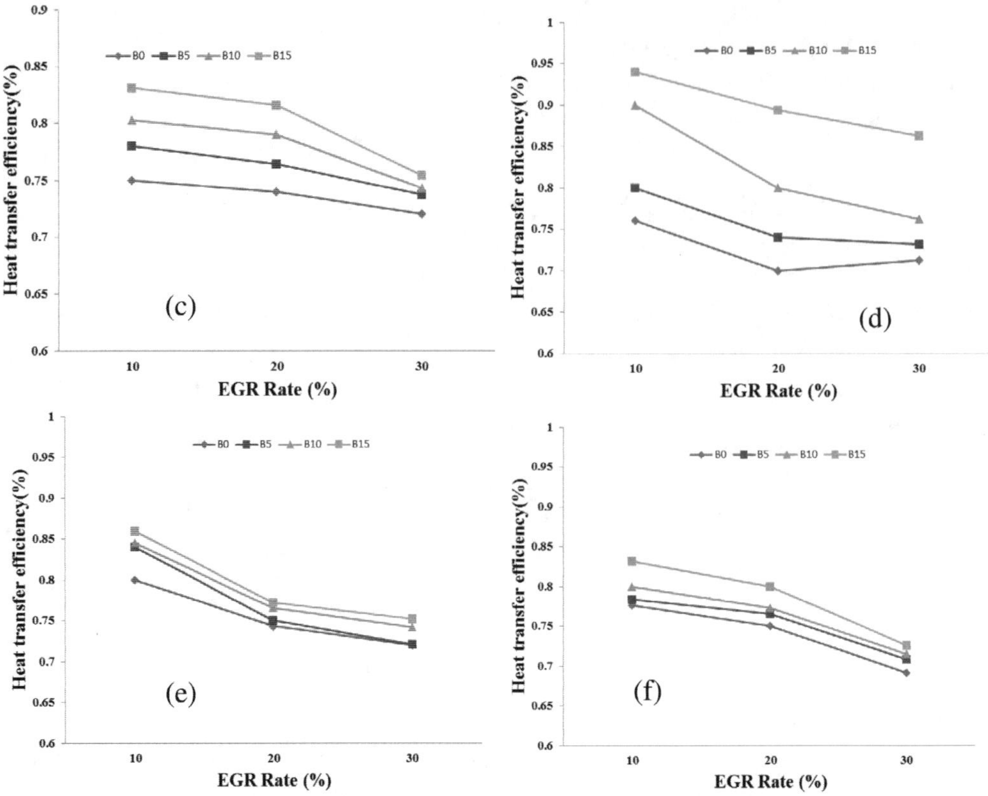

Figure 6. (a) Heat transfer efficiency: 25% of load and 2100 rpm of speed, (b) Heat transfer efficiency: 50% of load and 2100 rpm of speed, (c) Heat transfer efficiency: 75% of load and 2100 rpm, (d) Heat transfer efficiency: 25% of load and 2400 rpm. (e) Heat transfer efficiency: 50% of load and 2400 rpm, (f) Heat transfer efficiency: 75% of load and 2400 rpm.

Heat transfer efficiency of EGR cooler at 2400 rpm and 25% load is shown in Figure 15. A 20% increase of EGR rate decreases the heat transfer efficiency of EGR cooler by 14%. Heat transfer efficiency of EGR cooler at 2400 rpm and 50% load is shown in Figure 16. A 20% increase of EGR rate decreases the heat transfer efficiency of EGR cooler by 10.5%. Heat transfer efficiency of EGR cooler at 2400 rpm and 75% load is shown in Figure 17. A 20% increase of EGR rate decreases the heat transfer efficiency of EGR cooler by 9%.

3.4 Fouling of EGR cooler

To measure the sediment inside the EGR cooler pipes, the standards shown in the table were used. To measure the deposition, two blends (B0, B15) were used. When B0 is used, the deposition was 1.37gr that amounts to 0.23gr/hr. When B15 is used, the deposition was 1.21gr that amounts to 0.20gr/hr. The results show that in B15, the Fouling decreases by 13% when compared to B0. Cross section views of EGR cooler (when B0 is used) after deposition is shown in Figure 7.

Figure 7. Cross section views of EGR cooler (when B0 is used) after deposition.

The causes of sedimentation when biodiesel blend is used are as follows:

- Gasoline often contains 20 to 40 volume percent aromatics which increased emissions such as soot and particulate matter whereas biodiesel is substantially devoid of aromatics.
- Biodiesel is free from sulphur whereas in pure diesel fuels, sulphur is present so the products of the pure diesel combustion are sulphur dioxide and sulphur trioxide.

4. Conclusion

Conclusions of this study are summarized as follows:

- The overall heat transfer coefficient of 35.63 W/m^2 K is acceptable compared to overall heat transfer coefficient of shell and tubes heat exchanger because the desirable and standard quantity of heat transfer coefficient for these kinds of coolers which use both water and gas is between 10 and 200 W/m^2 K.
- The maximum thermal efficiency obtained was 94 percent and the minimum thermal efficiency obtained was 68 percent. The average of efficiency in speeds was obtained at 2100 and 2400 rpm as 77.2 percent, which is considerable in comparison with other similar studies.
- When B0 is used, the deposition was 1.37gr that amounts to 0.23gr/hr whereas when B15 is used, the deposition was 1.21gr that amounts to 0.20gr/hr. As a result, the use of fuels containing biodiesel in terms of fouling is more appropriate compared to diesel fuel.

References

[1] Shao J, Tao Y and Hansen K K 2016 *Electrochemistry Communications* **72** 36-40

[2] Zheng M, Reader G T and Hawley J G 2004. *Energy Conversion and Management* **45** 883-900

[3] Geng P, Tan Q, Zhang C, Wei L, He X, Cao E and Jiang K 2016 *Science of The Total Environment* **572** 467-475

[4] Dich A, Grimsley K, Koufos D, Sarasin T and Rodas M 2012 Alternative and renewable energy

[5] Gunnarsson, F, Pihl J A, Toops T J, Skoglundh M and Härelind H 2017 *Applied Catalysis B: Environmental* **202** 42-50.

[6] Wang J, Zhao H, Haller G and Li Y 2017 *Applied Catalysis B: Environmental* **202** 346-354.

[7] Feng X, Cao Y, Lan L, Lin C, Li Y, Xu H, Gong M and Chen Y 2016 *Chemical Engineering Journal* **302** 697-706

[8] Liu L, Li Z, Liu S and Shen B 2017 *Mechanical Systems and Signal Processing* **87** 195-213

[9] Abu-Hamdeh N H 2003 *Energy Conversion and Management* **44** 3113-3124

[10] Jang S H, Hwang S J, Park S K, Choi K S and Kim H M 2011 *Heat and Mass Transfer* **48** 1081-1087

[11] Liu L, Ling X and Peng H 2013 *Experimental Thermal and Fluid Science* **44** 275-284

[12] Liu L, Ling X and Peng H 2013 *Applied Thermal Engineering* **54** 145-152

[13] Liu L, Ling X and Peng H 2015 *Heat and Mass Transfer* **51** 1017-1027

[14] Shabgard H, Kheradmand S, Farzaneh H and Bae C 2017 *Applied Thermal Engineering* **110** 244-252

[15] Kakac S and Liu H 2002 Heat exchangers - selection, rating and thermal design *CRC Press*

[16] Park S, Choi K, Kim H and Lee K 2010 *Heat and Mass Transfer* **46** 1221-1227

[17] Abd-Elhady M S, Zornek T, Malayeri M R, Balestrino S, Szymkowicz P G and Müller-Steinhagen H. 2011 *International Journal of Heat and Mass Transfer* **54** 838-846

[18] Hong K S, Lee K S, Song S, Chun K M, Chung D and Min S 2011 *Atmospheric Environment* **45** 5677-5683

[19] Özçelik Y 2007 *Applied Thermal Engineering* **27** 1849–1856

[20] Adrian B and Allan D K 2003 Heat transfer handbook *John Willey and Sons*

[21] Shah R K and Sekulic D P 2003 Fundamentals of heat exchanger design *John Willey and Sons*

[22] Patel V K and Rao R V 2010 *Applied Thermal Engineering* **30** 1417-1425

[23] Babu B V and Munawar S A 2007 *Chemical Engineering Science* **62** 3720-3739

[24] Naphon P, Nuchjapo M and Kurujareon J 2006 *Energy Conversion and Management* **47** 3031-3044

[25] Ponce-Ortega J M and Serna-González, M.; Jiménez-Gutiérrez, A. 2009 *Applied Thermal Engineering* **29** 203-209.

[26] Bell K J 1980 *Conf. NATO Adv. Study Inst. Heat Exchangers: Thermal-Hydraulic Fund. & Design* 559-580

[27] Bell K, Kakac S, Bergles A and Mayinger F 1981 Delaware method for shell side design, heat exchanger thermal-hydraulic fundamentals and design *Taylor and Francis*

[28] Stachura V J 1988 *Standards of the tubular exchanger manufacturers association.* Tubular Exchanger Manufacturers Association

[29] Kakac S, Liu H and Pramuanjaroenkij A 2012 Heat exchangers: Selection, rating, and thermal design *CRC Press*

[30] Caputo A C, Pelagagge P M and Salini P 2008 *Applied Thermal Engineering* **28** 1151-1159

[31] Kakac S and Liu,H 2002 Heat exchangers: Selection. Rating and Thermal Design *CRC Press* 88-89

Implementation of the UTAUT Model to Understand the Technology Adoption of MOOC at Public Universities

Hafiza Haron[1*], Supyan Hussin[2], Ahmad Rizal Mohd Yusof[3], Hafiza Samad[4] and Hafidzan Yusof[5]
[1,4,5] Faculty of Computing and Multimedia, Kolej Universiti Poly Tech Mara, 56100 Cheras Kuala Lumpur, Malaysia.
[2,3] Universiti Kebangsaan Malaysia, 43600 UKM Bangi, Selangor, Malaysia.

*Corresponding e-mail: afieza@kuptm.edu.my

Abstract. MOOC is a national agenda in the institutions as it is addressed in the 11th Malaysian Plan, the National Economic Model, Economic Transformation Programme and the Malaysian Education Blueprint. Recently, have witnessed the development of massive open online courses (MOOCs) as a learning trend in the field of open distance education and online learning. As it is considering new technology for us, a research towards its acceptance is essential. This study examines the adoption of MOOC deployment among students at Universiti Kebangsaan Malaysia (UKM). The survey was used in this research where data were collected from 400 respondents and then analysed using Structural Equation Modelling (PLS-SEM). The findings were obtained based on Unified Theory of Acceptance and Use of Technology (UTAUT) model. The result shows that performance expectancy, effort expectancy, social influence and facilitating condition have influenced the respondents to use MOOC in their learning. This research exposed that the UTAUT factors implemented in this study have a significant effect to behavioural intention in order students use MOOC technology. It illustrates the positive impact of these technologies and beneficial to understand the MOOC adoption for all students, especially at public universities and in the area of online learning perspective.

Keywords: MOOC, Technology Adoption, UTAUT, PLS-SEM, Online Learning

1. Introduction

A Massive Open Online Course (MOOC) has been frequently a free course that can facilitate learning material which can be accessed through a web site. The course contains all related material such as text in presentation slide and video content which user can download it on their own. It also contains a task such as quizzes and assignments which to assess the student's ability during the courses as well as social interaction through online forum. MOOCs are accessible to people who plan to take a course or to be schooled in a particular subject area [1]. Nowadays, online learning has become one of the main channels for online teaching and learning between lecturers and students, especially in colleges and universities. MOOC is the course provided to learners in the way of at distance. These scenarios have viewed the appearance of MOOC as learning trends in the field of open distance education. It is an emerging trend practice in e-learning. The culture of teaching and learning is often challenging and changing by the passage of time. MOOC is an e-learning concept that is open for any interested participants attended and access courses with materials that are free of charge. It brings thousands of participants to register for MOOC courses every day. Although these concepts of learning are still growing, it is important to inspect the technology adoption of MOOC especially at a higher level of the institution. The Malaysian government is very supportive of the use of MOOC and sees it as a platform to integrate learning technology, lifelong learning and concurrently lead the way towards a new direction in teaching methodologies for undergraduate programmes [2]. In addition, the Malaysia MOOCs program which is based in Open Learning platform entails more than one million participants in more than 100 courses offered [3]. In order to make MOOC beneficial to use for a long journey ahead, the acceptance of these technologies is crucial. Effective implementation of any information technology (IT) or information system (IS) depends on user acceptance [4]. Thus, the objective of this research is to observe and understand the technology adoption of MOOC by using a Theory of Acceptance or UTAUT at selected public university. By using a survey method, the result illustrates the positive impact of MOOC adoption among students where all variables of UTAUT have a significant effect to behavioural intentions to use MOOC technology in their teaching and learning.

2. MOOC

Currently, MOOCs adoption in Malaysia is developing in a cycle with a few important plans, which is 11th Malaysia Plan (2016-2020), the National Economic Model, Economic Transformation Programme and the anticipated Malaysian Education Blueprint for Higher Education. The last of which has specifically addressed MOOCs in its preliminary discussion document. Online learning, as an essential component of the delivery mechanism in MOOCs, is also addressed in the soon-to-be-released Blueprint [5]. MOOCs in Malaysia are likely to see various developments in the next several years, as we can anticipate greater involvement from higher education institutions in response to the Malaysian government's recent statements that have revealed several national objectives for MOOCs in the next few years [6]. In addition to the apparent advantage of broader access to potentially high-quality instruction and instructional material, some believe a MOOC can help revolutionise higher education pedagogy [7]. Nevertheless, the effectiveness of MOOC whether it can facilitate bottomless and significant to learning environment for students was still arise. Abeer and Miri [8] asserted that the more autonomous, diverse and open the MOOC is, the more the potential for students' learning to be limited by the lack of structure, support, and moderation which are normally associated with a regular course.

MOOCs popularity is growing rapidly despite its novelty and age. Large numbers of participants are enrolling continuously in massive number of courses. MOOCs attracted attention quickly and acquired the interest of academics [9]. If the Malaysian Government and local higher education institutions intend to adopt MOOCs on a large scale, this will inevitably have significant repercussions on the entire national higher education landscape, especially if they are made a part of the delivery approach in higher education institutions (as currently explored by public universities), as a means for branding and internationalisation, or even as part of the advancement of online learning and ODL [5]. Surprisingly, even though MOOC was new at Higher Learning Institution, especially Malaysia, several scholars found that there are an issue and challenges that came across for MOOC as global. This learning system attracts many diverse learners, so there is need to tailor the course instruction to meet the students' individual needs [10].

In general, MOOC can be divided into two categories. cMOOCs and the xMOOCs are two different types of MOOCs [11]. The cMOOC (CCK08 course) was created based on the learning theory of Connectivism, a concept that has principles developed by George Siemens [1] . Meanwhile the second category was xMOOCs. xMOOCs are online versions of traditional learning formats applying a knowledge transmission model using video recordings of classroom lectures or custom-produced mini-lectures [12]. Its look alike the structure of courses at MOOC, will fully setup by the lecturer or instructor. Besides, the xMOOCs comes with a clearly specifies syllabus of recorded lectures and self-test problems. They also employ the original elements of MOOC, but it affects the branded IT platforms which offer content distribution partnerships to institutions [13]. Thus, xMOOCs was a course designed by instructor like a traditional classroom converted into online class [14].

IOP Conf. Series: Materials Science and Engineering 1062 (2021) 012025 doi:10.1088/1757-899X/1062/1/012025

2.1. Unified Theory of Acceptance and Use of Technology (UTAUT)

Previous studies have found that there are many acceptance studies that apply the UTAUT model in various fields. However, studies applying the UTAUT model in Education in Malaysia are limited [15]. It is because the use of Information Communication and Technology (ICT) in the Malaysian education system is relatively modest [16]. Thus, research on individual acceptance and use of information technology (IT) is one of the most established and mature streams of information systems (IS) research [17]. The UTAUT model which aims to explain technology acceptance, is based on eight technology acceptance theories or model [18]. In particular, the UTAUT draws on the Theory of Reasoned Action (TRA), the Technology Acceptance Model (TAM), the Motivational Model, the Theory of Planned Behaviour (TPB), the combined TAM and TPB, the model of Personal Computer Utilization, the Innovation Diffusion Theory and the Social Cognitive Theory [16]. Figure 1 illustrate the UTAUT Model.

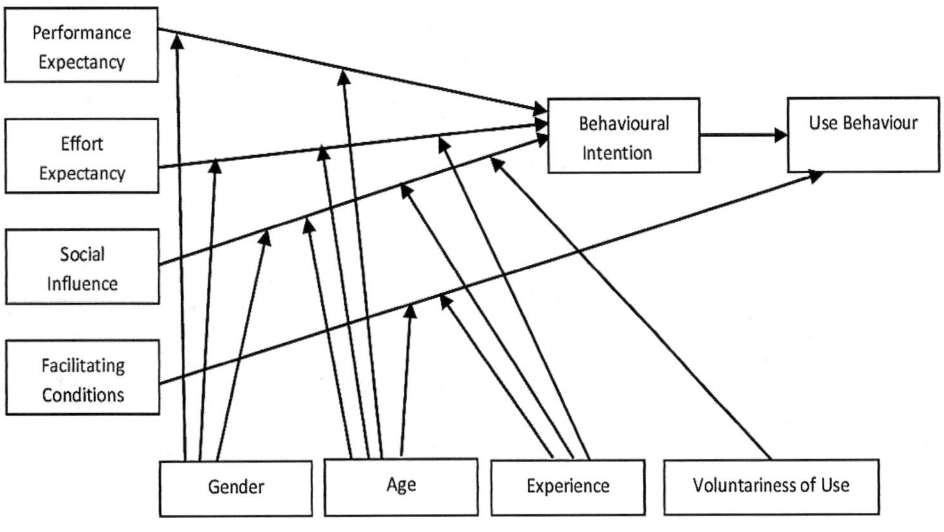

Figure 1. UTAUT model (Venkatesh et al. 2003)

The Unified Theory of Acceptance and Use of Technology (UTAUT) as illustrated in Figure 1 was developed by Venkatesh, Morris, Davis and Davis in 2003 to address the limitations of the Technology Acceptance Model (TAM) and other popular models used in the study of information systems adoption [1]. According to Venkatesh et al. [19] the variance in intention to use explained by the contributing models ranged from 17 to 53%. The UTAUT model was found to perform better in terms of variance in intention to use compared to any of the other eight models. Table 1 below describes the main construct and variables used in this model.

Table 1: Four main variables of UTAUT

Construct/Variable	Description
Performance Expectancy	Performance expectancy is the degree to which an individual trust that using the system will help him or her in the work performed.
Effort Expectancy	Effort expectancy is the level of convenience associated with the use of the system.
Social Influence	Social influence is the degree to which an individual sees the significance of others believes he or she should practice the new system.
Facilitating Condition	Facilitating condition is the amount to which individuals believe that organisational and technical infrastructure occurs to support the use of the system.

Previous research that using the UTAUT model has shown that all four constructs have significant effects on behavioural intention to use e-learning systems. As in this study, MOOC is also one of the e-learning systems. Therefore, the hypothesis of this research is stated as follows:

Hypothesis 1: *Performance expectancy has a significant effect on students' Behavioural Intention to use MOOC.*

Performance expectancy: Performance expectancy is the degree to which an individual trust that using the system will help him or her in work performed.

Hypothesis 2: *Effort expectancy has a significant effect on students' Behavioural Intention to use MOOC.*

Effort expectancy: Effort expectancy is the level of convenience associated with the use of the system.

Hypothesis 3: *Social influence has a significant effect on students' Behavioural Intention to use MOOC.*

Social influence: Social influence is the degree to which an individual sees the significance of others believes he or she should practice the new system.

Hypothesis 4: *Facilitating condition has a significant effect on students' Behavioural Intention to use MOOC.*

Facilitating condition: Facilitating condition is the amount to which individuals believe that organisational and technical infrastructure occurs to support the use of the system.

Hypothesis 5: *Behavioural Intention has a significant effect on students' Usage Behaviour.*

Behavioural Intention and **Use Behaviour**: Behavioural Intention refers to the pre-determined decision (*antecedent*) of a student to use the MOOC soon. Behavioural intention is theorised to result in Use Behaviour. A few concepts and models have suggested a direct impact of behavioural Intention on Use Behaviour, such as Technology Acceptance Model (TAM), Theory Plan Behaviour (TPB), UTAUT and UTAUT2.

3. Methodology

The quantitative method was conducted during the research period where online questionnaires have been distributed to students. This is in line with the majority of UTAUT studies that have used survey research method [20]. The respondents were students at Universiti Kebangsaan Malaysia (UKM) who had involved in one or more MOOC courses. After cleansing, a total of 400 usable responses were obtained. Since MOOC has been deployed at the university starting 2014, the MOOC was implemented as a blended learning where the method of face-to-face combined with the online material setup by instructor or lecturer. The theory of acceptance or UTAUT is used in this research where a few variables act as independent variables and dependent variables [17]. Thus, the performance expectancy, effort expectancy, social influence and facilitating condition as independent variables and behavioural intention and use behaviour as dependent variables. The framework of this study take place as follow:

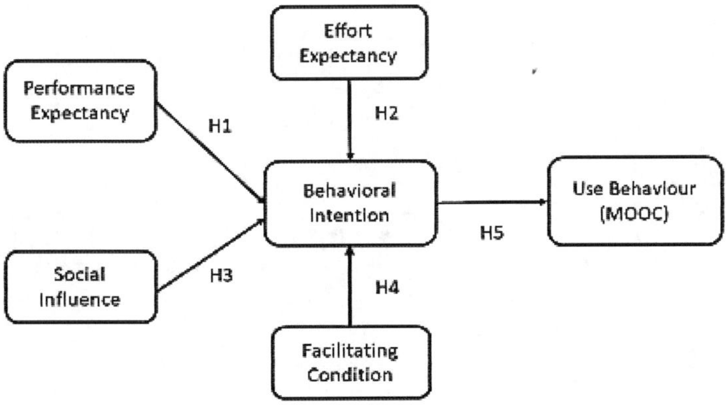

Figure 2. Research framework

4. Results and Discussion

The partial least squares approach which is structural equation modelling (PLS-SEM) on Smart PLS version 3 was carried out to analyse the collected data. Structural equation modelling is a second-generation statistical technique that enables researchers to examine causal relationships between latent variables [1]. The results are discussed according to the UTAUT factors which is all four-construct act as the independent variables. Meanwhile behavioural intention as antecedent and use behaviour as dependent variable. Figure 3 illustrated the diagram of the structural model or paths between the variables in the research framework. This was done using Smart PLS software as well as to evaluate the reliability and validity of the construct in this study. The measurement model was assessed based on reliability, discriminant validity and convergent validity. Reliability of the constructs was assessed using Alpha Cronbach. Table 2 illustrated that the value of Alpha Cronbach is in the range of 0.7 until 0.9. It can be determined that the measurement model shows the good reliability of the construct used in the study.

Table 2: Composite and reliability statistics

Construct	Number of Item	Composite Reliability	Alpha Cronbach
Performance Expectancy	3	0.942	0.907
Effort Expectancy	4	0.905	0.860
Social Influence	3	0.936	0.809
Facilitating Condition	3	0.892	0.818
Behavioral Intention	3	0.944	0.910
Usage Behaviour (MOOC)	3	0.933	0.902

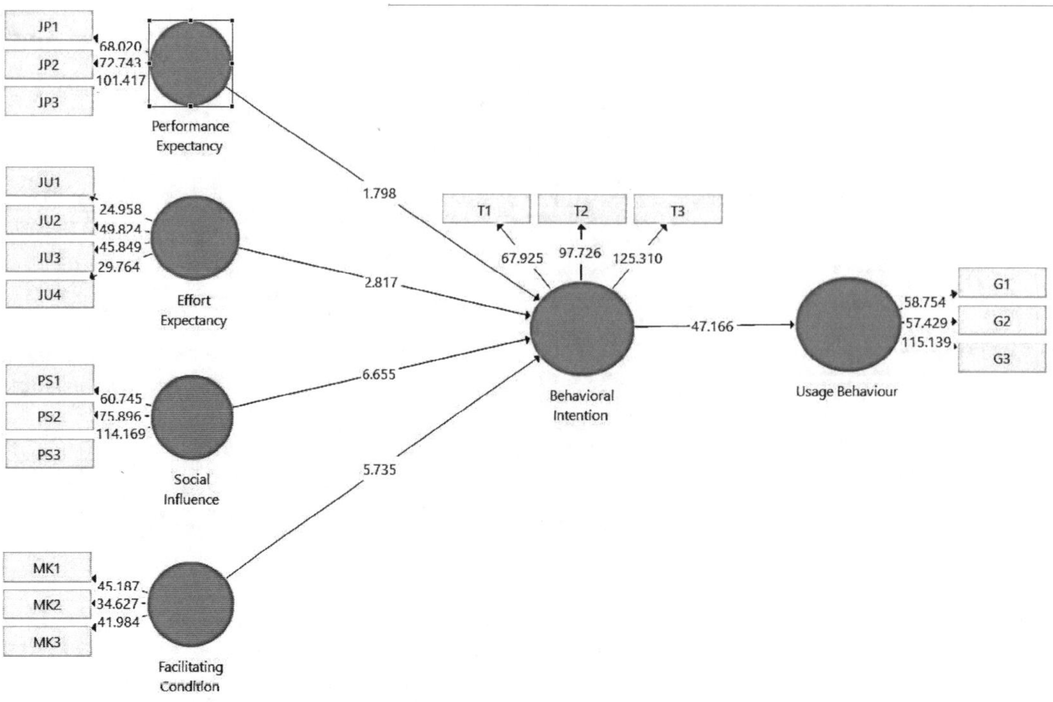

Figure 3. Measurement model

Convergent validity of the measurement model was also assessed based on recommendations by Henseler et al. [21] that the average variance extracted (AVE) for each latent construct should be greater than 0.5. Table 3 shows that the AVE values for all constructs are higher than the 0.5 thresholds. Meanwhile, the Fornell-Larker criterion was used in order to assess a discriminant validity. The method states that the AVE of each latent construct should be greater than the highest squared correlations between any other construct. Thus, it can also be seen from Table 3 that the square root of the AVEs for each construct is greater than the cross-correlation with other constructs. From the results, it shows that the measurement model exposes good validity of the construct developed.

Table 3. Discriminant validity using Fornell-Larker Criterion

Construct	AVE	EE	FC	BI	PE	SI	UB
Effort Expectancy	0.703	*0.839*					
Facilitating Condition	0.703	0.720	*0.856*				
Behavioral Intention	0.848	0.738	0.734	*0.921*			
Performance Expectancy	0.843	0.786	0.668	0.719	*0.685*		
Social Influence	0.830	0.757	0.683	0.785	0.743	*0.911*	
Usage Behavior	0.837	0.743	0.713	0.847	0.706	0.784	*0.915*

Note: Square roots of average variances extracted (AVEs) shown on diagonal in italic.

After that, the structural model was measured which is to confirm the validity and reliability of the measurement model. The structural model was measured based on the sign, magnitude, and significance of path coefficients of each hypothesised path. The bootstrapping technique was conducted to determine the significance of each estimated path. Then, the coefficient of determination which is R^2 was assessed to determine the detailed description of the research model. In other words, the assessment was to predict the endogenous constructs that used in the model. Table 4 shows the detailed results of assessment.

Table 4: Hypothesis Testing

Hypothesis Path	Path Coefficient (B	T Statistics	P Values	Results
Behavioral Intention → Usage Behaviour	0.847	46.498	0.000	Supported
Effort Expectancy → Behavioral Intention	0.161	2.897	0.002	Supported
Facilitating Condition → Behavioral Intention	0.303	5.617	0.000	Supported
Performance Expectancy → Behavioral Intention	0.120	1.815	0.035	Supported
Social Influence → Behavioral Intention	0.363	6.608	0.000	Supported
R square (Behavioral Intention) = 0.720				
R square (Usage Behaviour) = 0.717				

The first variable which is Effort expectancy was found to have a significant effect on behavioural intention where ($\beta = 0.161$, $p = 0.002$). It has provided support for H1. The second variable which is performance expectancy also was found to have a significant effect on behavioural intention. It is because the value of ($\beta = 0.120$, $p = 0.035$). It has also provided support for H2. And also facilitating condition was found to have a significant effect on behavioural intention where ($\beta = 0.303$, $p = 0.000$). Again, it has also provided support for H3. Furthermore, social influence was found to have a significant effect on behavioural intention ($\beta = 0.363$, $p = 0.000$). It has also provided support for H4. As estimated, behavioural intention was found to have a significant effect on usage behaviour where ($\beta = 0.847$, $p = 0.000$). Thus, it has also provided support for H5. Generally, the suggested research framework accounted for 71.7% of the variance in usage behaviour, and 72% of the variance in behavioural intention (R^2 of 0.720 and 0.717 respectively).

5. Conclusion

This research framework and data collection have appointed to explore on technology adoption of MOOC using the UTAUT model. From our point of view, the findings of our study show that the behavioural intention to use MOOC is influenced by four variables or constructs suggested in the research framework. The constructs were performance and effort expectancy, social influence and facilitating condition. All five hypotheses identified were sustained, which, to a huge scope supports the exploration of the study. The study hypothesised the impact of variables or construct used in the acceptance model which is consistent with UTAUT. From this finding, it indicates that the main construct of UTAUT Model was in-line with the literature where social influence is found to be the most prominent predictor. Another three predictor or factors were also found to be significant in term of MOOC adoption.

Facilitating conditions were found to be the second prominence in this study, possibly due to individuals believe that appropriate technical infrastructure in the organization takes place to provide students at the university in term of MOOC technology. Furthermore, the respondents believed that those related to the technology are essential to sustenance the usage of MOOC in learning due to the relationship between these factors. Availability of resources such as good internet bandwidth and appropriate devices to access MOOC can't be taken for granted for all potential users of MOOC. Thus, it would be a determining factor for MOOC adoption. Performance and effort expectancy were also found to have an influencing role in the adoption of MOOC. Even though, the value of T corresponding of these two factors considers lower than other prominence, this finding indicates that learners expect a good point of effortlessness related with the use of MOOC. They were also trust that using the system will help him or her in work performed.

The constraint of the study is that the research conducted at selected public university only. For further research it could apply to all the public universities in Malaysia in order to understand the MOOC adoption and their usage as well. Furthermore, the UTAUT also can be extended with other variables or construct in order to provide a better understanding of MOOC continuance in teaching and learning.

Therefore, improvement and solution can be proposed for the long-term benefit of the technology used especially in the Industrial Revolution 4.0 era.

References

[1] Fianu E, Blewett C, Ampong G and Ofori K *Edu. Sci.* **8(2)** 70
[2] Saha R and Halder U K 2018 *Shinjan* **IV(7)** 527– 539
[3] Azami H H R and Ibrahim R 2018 *Open I. J. Info.* **6(2)** 45–51
[4] Chao C M 2019 *Frontiers in Psychology* **10(July)** 1–14
[5] Fadzil M, Abdol L, Tengku L and Munira A 2016 *E-ASEM Forum Renewing the Lifelong Learning Agenda for the Future* 1–17
[6] Gamage D, Fernando S and Perera I 2015 *I. Conf. Ubi-Media Com.* 224–229
[7] Canbek N G 2015 *GLOKALde* **1(1)** 19–32
[8] Abeer W and Miri B 2014 *Procedia - Social and Behavioral Sciences* **152** 318–323
[9] Abu-Shanab E A and Musleh S 2018 *International Journal of Web-Based Learning and Teaching Technologies* **13(4)** 62–76
[10] Alzaghoul A and Tovar E 2016 *Proc.13th Int. Conf. on Remote Engineering and Virtual Instrumentation* **2016** 127–132
[11] Haron H, Hussin S, Samad H, Adnan R and Tahir W A 2019 *Int. J. Man. App. Sci.* **5(6)** 38–41
[12] Kocdar S, Recep Okur M and Bozkurt A 2017 *Turkish Online Journal of Distance Education* **18(4)** 52–65
[13] S A Al-shami Aziz H and Rashid N 2018 *J. of Fund. App. Sci.* **10(6S)** 2634–2654
[14] Haron H, Hussin S, Yusof A R and Yusof H 2019 *J. Info. Sys. Tech. Man.* **4(14)** 26–33
[15] Mohamad M, Hamdan N A, Shaharuddin S and Khalid F 2015 *Turkish Online Journal of Educational Technology* **2015(1)** 301–310
[16] Hamdan A, Din R, Manaf S Z A, Salleh N S M and Kamsin I F 2015 *J. Adv. Rev. Sci. Res.* **5(1)** 10–29
[17] Venkatesh V, Thong J Y L and Xu X 2016 *J. Assoc. Info. Sys.***17(5)** 328–376
[18] Khalid R, Raman A, Hussin F, Ghani M, Don Y and Omar M S 2014 *Asian Social Science* **10(11)** 71–85
[19] Venkatesh V, Morris M G, Davis G B and Davis F D 2003 *MIS Quarterly Management Information Systems* **27(3)** 425–478
[20] Mulik S, Srivastava M and Yajnik N 2018 *Nmims Management Review* **36(2)** 26–44
[21] Henseler J, M R Cristian and Sarstedt M 2009 *Partial Least Squares Equation Model*

Implementation of Anti-Profanity Words in Mobile Application Platform

F Razali[1*], Z Jusoh[1], A Salleh@Omar[1] and N Azizan[1]
[1] Faculty of Computing and Multimedia, Kolej-Universiti Poly-Tech MARA, Kuala Lumpur, Malaysia

*Corresponding e-mail: nurfarahbalqishrazali@gmail.com

Abstract. Communication technology increases day by day in our daily life due to the rapid growth of variety mobile application platforms. For instance, mobile application users are using chat rooms for sharing data, pictures, videos and documents. However, the users are exposed to receive something inappropriate such as bad words, bad photos and sensitive data in their social media platforms. The developed chat profanity filtering application can be used to detect and filter the bad words before the sender send any messages via the application. One way to detect and filter profanity words is by using the *PurgoMalum* web service. Therefore, the recipients are unable to view any bad words sent to the chat application. This research has shown that the developed chat profanity filtering application will minimize the potential of users abusing the chat platform in daily communication. Result from this study had successfully demonstrated that the developed chat profanity filtering application could minimize the potentials for users to abuse the chat platforms in communicating various forms of obscenity. Therefore, chat profanity filtering application was developed to minimize the abusing of social media platform in daily communication.

Keywords: mobile apps, social media, chatting, profanity, *PurgoMalum*

1. Introduction

Social media platform using mobile application play prominent role in our daily life due to the advancement of the mobile phone technologies and network accessibility. The popularity of social media platform is supported by the rapid growth of internet technology and social media platform. Commonly, the most popular social media platforms such as Facebook, WhatsApp, WeChat and Twitter provide facilities to communicate with other people by the sharing data such as pictures, videos and documents. In contrast, the negative effects of the social media to the people such as their perception to society and the way they view themselves and to the other people. By using the social media, the users tend to send any kind of content without considering whether the content or text is physically threatening or humiliating and not just offensive to the recipients. On top of that, the offensive contents might be affected other people feelings when the recipient received an inappropriate text or messages. Consequently, they might get anxiety and bad feeling which could influent to their behaviour and daily life.

In order to encounter the discourteous communication such as abusive words in text or media content between lecturers and students in Kolej University Poly-Tech MARA (KUPTM), a mobile social media platform was developed to curb the problem. This platform implements a chat profanity filtering application that able to detect and filter the bad words before the sender send any messages to the recipient. The application is integrated with a *PurgoMalum* web service that being used to detect and filter the profane words. Moreover, the platform offers the easy and secure ways of communication between the lecturers and students by using smartphones. A secure environment is very important to ensure only authorized users able to use and access this application.

1.1. Problem Statement

Today more than 3 billion people worldwide use a smartphone [1]. They use smartphone not only for communication but also can be used to surf website, take photos, record videos and play games. There are many online applications being used for communication such as Facebook, WhatsApp, WeChat and Twitter. These applications are free and can be used by everyone. The applications offer the fastest way for people to communicate with each other. However, all these applications cannot filter any offensive contents or words from the sender to the recipient. According to Prof Timothy, the professor of psychology at Massachusetts College of Liberal Arts and his assistant Kristin, people who received the abusive words in text or verbal might have the significant impact both emotionally and physically, for instance, they will get anxiety or sleep disorder [1]. Understandably, some of the social media applications do not provide any authentication methods to authenticate the authorized users. However, this feature is very important to ensure the communication is in a secure environment and to prevent from anyone might access the platform and read the confidential message.

1.2. Objective

There are two objectives of this research, namely 1) Develop a mobile social media platform for KUPTM lecturers and students to communicate in a healthy way with filtering profanity. This application provides a feature will be used to ensure the recipient will not get any bad words from the sender. If the sender intends or wants to send offensive contents or bad words, he/she will get the warning notification because the system detects the bad words in the message. 2) Create a unique username and high secure password for authentication purposes before the user is able to use the application.

2. Literature Review

Social Media is a platform that gives individuals the opportunity to interact, using two way communication; meaning, that anyone who has online accounts can share their opinions with other social media users [2]. Hartshorn defines that Social Networking is "the act of engagement," while social media is the tool used to communicate with mass audiences [3]. In social media, people chat with each other without necessary face-to-face interaction, hence promoting a sense of being liberal without breaking the cultural norms, resulting in an increased in the use of profanity in cyber space. Although most social media platforms have developed profane filters to censor profanity, but the cyber bullies have continuously improved on their profanity techniques to make the existing filters less and less effective. This has led to a major problem in social media, resulting in potentially serious adverse effects on young users if undetected [4].

A study conducted by [5], has found that an online community are often plagued with negative content user-generated content that is negative in tone, hurtful in intent, mean, profane, and/or insulting. However, according to [6], most research works have encountered several challenges in solving the profanity problem in social media. The three main challenges affecting profanity detection are namely; category of language spoken from region to region, types of profanity filters, and techniques used for detecting profanity. Meanwhile, according to [7], the categorization of commercialized profane filters is classified into three categories: blacklist filtering (profane words), free form whitelist filtering (non-profane words), and restricted entry whitelist filtering (text prediction of non-profane words).

Abelardo Pardo believes that technology offers a platform for innovation, and allows its users to express their opinions about how they feel towards the information being published. He adds that, social media is also a platform that allows students to interact with one another, with their teachers and communities that share their same education. Moreover, Pardo also stated that these types of interaction are "an essential part of how humans learn" [8]. Mobile Application Development (MDA) refers to the process of making application software for handheld devices such as mobile phones and Personal Digital Assistants (PDA). Through the usage of mobile apps, the user is provided with various features that will enable him to fulfil all his needs and much more. Apps should be interactive to the users. Apps can be downloaded from various platforms such as Google Play Store and iOS App Store [9].

3. Methodology

Systems development life cycle (SDLC) describes a process of planning, analysis, design, implementation, testing & integration and maintenance as shown in Figure 1. This model is easy to understand and has a clear view of an entire project, the tasks involved, estimated costs and project timelines. Furthermore, the advantage of using this model makes the development of the project more efficient in order to achieve the objectives and standards.

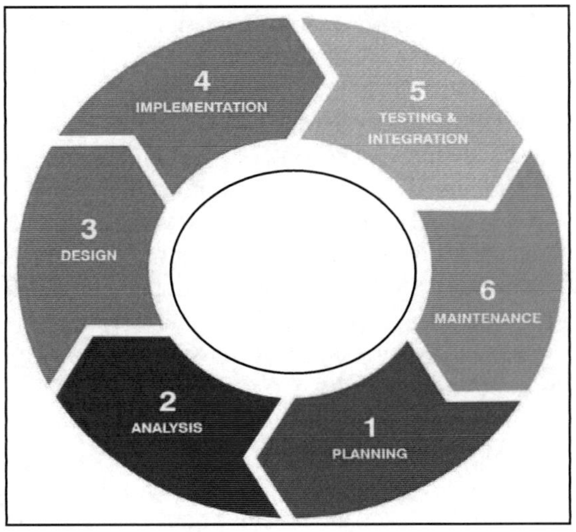

Figure 1. System Development Life Cycle (SDLC) model.

3.1. Planning

The planning phase involves the process required to define the problem statements, objectives, costs needed and the duration of the development process. The sufficient time needs to allocate included conducting extensive interviews with the target users and distribute questionnaires to them. It is better to structure all the incoming data and analyse them. Not only that, all technical limitations that may arise on the user's side also need to be clarified and come out with a ready-to-follow specification to meet the user requirements.

3.2. Analysis

The analysis phase involves the process to analyse all data gathered from dissemination of questionnaires to the KUPTM lecturers and students. The user requirements such as the features of the apps needs to be a focus on this phase. Hence, all the requirements will be finalized before proceed to the design phase.

3.3. Design

The design phase involves the process to transform the analyse data to system design. Usually the ideas come from the sketches, including the user interface design, database design, system modelling, system components and security level. The whole system will model by using use-case diagram and flowchart diagram. The advantages of using system modelling are assisting business analysts, project leader, the assistant project leader, the tester, or related organization and also the user that uses the system. Moreover, the model must fulfil all the user requirements and the application design should be user-friendly.

3.4. Implementation

The implementation phase involves the process to write and compile programming codes by using a software Visual Studio Code. JavaScript is a client-side programming language and used as the front-end of the apps. *PurgoMalum* is a simple, free, RESTful web service for filtering and removing content of profanity, obscenity and other unwanted text. *PurgoMalum* interface input several parameters for customization and can return results in plain text, XML and JSON [10].

This application will require a unique username and high secure password to log in. Besides, this application will work to detect the bad words when the sender tries to send it. Only registered users can log in into the application. The users need their username and password to log in into the system. An unauthorized person is not allowed to log in into the system.

3.5. Testing and Integration

The testing and integration phase involves the testing process to the whole system where there will be a system checking on the specification and also the user requirements. During this phase, bugs and glitches will be detected so that it can be corrected to ensure the system runs smoothly, able to perform all the functional tasks and meets the user requirements. The system will be tested by the specific users which are KUPTM lecturers and students. They are required to give any feedback and comment after completed the test. The testing phase includes functional and non-functional requirements. After the system passes each testing phase, the system is deployed to the user environment. However, this paper only shows the results of non-functional testing.

3.6. Maintenance

The maintenance phase involves the activities of modifying the system module after it is delivered to the target user, fixed any bugs or coding error and making changes to improve the system's performance. In reality, it is inevitable there are some defects or issues come up, especially for a new system which leads to modifications to the system in improving the performance and efficiency. Furthermore, regular maintenance and support will be provided to the user.

4. System Design

Data and system modelling is a process of creating a data model for an information system by applying certain formal techniques. Besides, it is also a process of organizing and structuring data to be implemented in software or system development. Data modelling is a process used to illustrate and analyses the data requirements needed to support the processes within the scope of the corresponding information system in an organization. System model represents the view of the whole system in a visual diagram and clearly shows the functionality of the whole system.

4.1. Use-Case Diagram

The use-case diagram can be defined as a representation of a user's interaction with the system. For chat profanity filtering application, there are two actors involved, which are lecturer, student and system.

Based on Figure 2, it shows that lecturers and students are able to sign up and log in to the system. Furthermore, they can view messages, compose messages and send messages. The system is able to detect any type of bad words and display a warning.

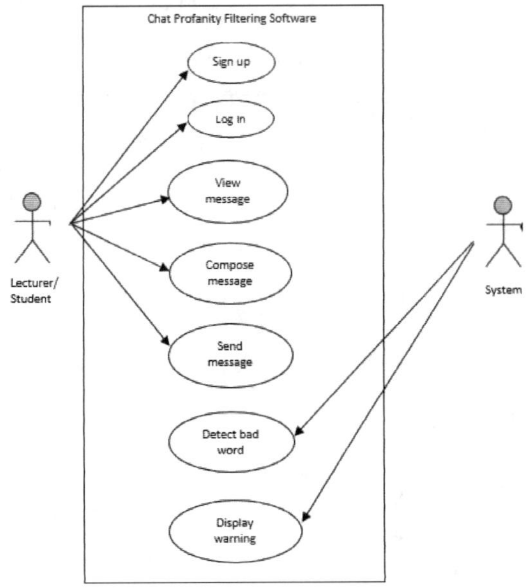

Figure 2. Use-case diagram for chat profanity filtering application.

4.2. Flowchart Diagram

Flowchart is a type of diagram that represents a workflow or process in sequential order. It can be used to standardize a process for effective analysis, efficiency and to ensure the procedures are being followed.

Figure 3 depicts the flowchart for chat profanity filtering application. The process flow begins with start symbol to indicate that the process is started, then it's connected to the first task which is a lecturer and student need to sign up to become members of the chat group. Then, they can log into the chat group. After that, they can view messages and compose messages. Before messages being sent to the recipient, the system will detect any bad words. If any bad words are detected in a message, the warning message will be displayed and send button will be disabled automatically. Then, the user has to delete the bad word and again the system will check the message. If there is no bad word detected, so user will able to send the message to the recipient.

5. Results and Discussion

This section discusses the findings and the result of data gathering from dissemination of questionnaires to KUPTM lecturers and students. The questionnaires were distributed to the respondents through WhatsApp and it was made by using Google Form. The feedback is important to thoroughly understand the needs to develop a chat profanity filtering application.

5.1. Result from Questionnaire

The objective of the questionnaire is to obtain new ideas and information about the development of profanity filter application for KUPTM lecturers and students. Two types of questionnaires have been constructed, one for the KUPTM lecturers and another one for KUPTM students as shown in Table 1and Table 2. There are 30 lecturers and 40 students were responded to the questionnaires.

Based on the survey, both lecturers and students preferred to use smartphone rather than using laptop to communicate with each other. Communication using a smartphone is much more convenient and easier. In addition, they preferred to use mobile application to communicate rather than a website. Both of the target users were also agreed to filter bad words on their social media platform, hence to avoid from inconvenience relationship.

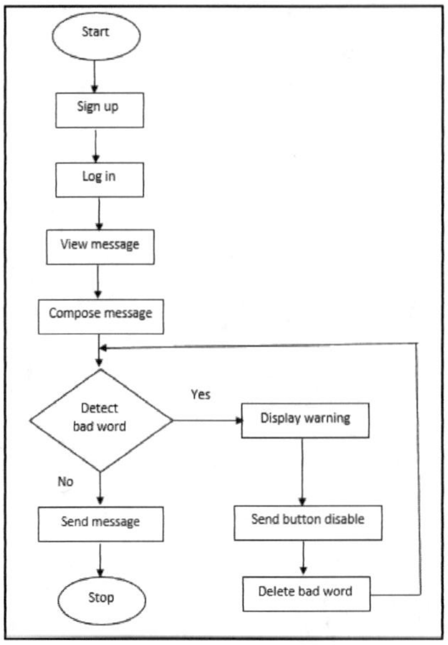

Figure 3. Flowchart for chat profanity filtering application.

Table 1. Response from the lecturers.

Questions	Response	
1. Do you think using a mobile social media platform is the fastest way to communicate with students?	Yes 100%	No 0%
2. Which device do you prefer to use when communicate with students?	Smartphone 97%	Laptop/Computer 3%
3. Which platform do you prefer to communicate with your students?	Mobile application 100%	Website 0%
4. Have you ever received harmful/bad words when communicate with students through online social media platform?	Yes 71%	No 29%
5. Sometimes, in group chat/personal chat, students accidentally sent bad words. Do you agree chat application should be able to filter the bad words?	Yes 100%	No 0%

Table 2. Response from the students.

Questions	Response	
1. As a student, how do you prefer to communicate with your lecturer?	Social Media 80%	Face to face 20%
2. Do you think using a mobile social media platform is the fastest way to communicate with lecturers?	Yes 99%	No 1%
3. Which device do you prefer to use when communicate with lecturers?	Mobile application 100%	Website 0%
4. Sometimes, in group chat/personal chat, students accidentally sent bad words. Do you agree chat application should be able to filter the bad words?	Yes 48%	No 52%

6. Conclusions

This research has shown that the developed chat profanity filtering application will minimize the abusing of social media platform in daily communication. Furthermore, result of this study has successfully demonstrated that the developed chat profanity filtering application could minimize the potentials for users to abuse the social media platforms in communicating various forms of obscenity.

Acknowledgments

The authors gratefully acknowledge the support provided by the supervisor, Puan Noraliza Azizan, the Senior Lecturer of Faculty of Computing and Multimedia (FCOM), Kolej Universiti Poly-Tech MARA (KUPTM). The authors would like to thank all the lecturers from the Faculty of Computing and Multimedia (FCOM) for their contribution, especially Puan Zurina Jusoh and Puan Nor Azura Salleh@Omar.

References

[1] Timothy J and Kristin 2012 *The Science of Swearing*
[2] Tarek A. El-Badawy and Yasmin Hashem 2015 *International Journal of Business Administration*
[3] Hartshorn S 2010 Social Media Today *5*
[4] Abdulrehman A Mohamed, Dr George O Okeyo and Dr Michael W Kimwele 2018 International Journal of Scientific & Engineering Research **9** (10) 2229-5518
[5] Sood S O, Antin J and Churchill E 2012 *Conference on Human Factors in Computing Systems ACM* 978-1-4503-1015
[6] Benedictus L 2014
[7] King, M. 2013 *Types of Profanity Filters for Online Safety*
[8] Pardo A 2013 *Int. J. Social Media and Interactive Learning Environments*, **1**(1), 43-58.
[9] Kishore Baktha 2017 *International Journal of Computer Science and Mobile Computing* **6** (9) 15-20
[10] Purgomalum.com. 2019 *PurgoMalum*
[11] Harish Yenala, Ashish Jhanwar, Manoj K Chinnakotla, Jay Goyal 2017 *Deep Learning for Detecting Inappropriate Content in Text*
[12] K R Subramanian 2017 *International Journal of Scientific Progress and Research (IJSPR)* **38** (109) ISSN: 2349-4689
[13] W Akram, R Kumar 2017 *International Journal of Computer Sciences and Engineering* **5**(10) E-ISSN: 2347-2693

Hole Introduction Effect on The E-Glass WR 185 Fibre Reinforcement Polyester

Ilhamdi[1*], D Hidayat[2], D E Sepriyani[1]
[1] Mechanical Engineering Department, Faculty of Engineering, Andalas University, 25163 Padang, Indonesia
[2] Indonesian National Institute of Aeronautics and Space, Bogor 16350, Indonesia

*Corresponding e-mail: ilhamdi@eng.unand.ac.id

Abstract. Challenge to use a light structural material for aero vehicle could be answered by fibre reinforced polymer composite. This composite has high ratio of strength to weight. In application this composite should be in mechanical assembling such as bolt. So that, investigation of hole introduction on the composite integrity and behaviours are needed. In this study, E-glass WR 185 fibres used as reinforcement in polyester BQIN EX 157 matrix. The composite made by hand lay-up methods; in which 13 layers of E-glass fibres were laminated one by one layer with polyester resin. After dried up, composite was cut off to made specimen for density, tensile and compression test. Some of tensile and compression specimens were drilled (marked as "open hole specimen") in their centre. The result showed density of 1.12 g/cm3, maximum strength of 178 MPa (tensile) and 130 MPa (compressive). Introducing an open hole reduced strength to 119 MPa (tensile, decreased ~33%), and to 112 MPa (compressive, ~14%). Reducing strength is well confirmed with change of failure type; from lateral at grip top (LAT) to lateral gage middle (LGM). LGM of the open hole specimen means crack starting from the hole to both of lateral direction.

Keywords: polymer, composite, reinforcement, fibre, polyester resin

1. Introduction

Global warming due to change in earth atmosphere especially in ozone. The ozone breakdown is triggered by glass house gasses likes CO and CO_2. There are many sources of the gasses in the atmosphere likes forest firing, coal and gasoline vehicles [1]. Environment saving articles demand as low as gas emission from the vehicle. That could be achieved through higher efficiency of petroleum, and weight reducing of the vehicles. Mechanically, weight reducing should be with no or less sacrifice on mechanical properties of vehicle material. Introducing lightweight metal or composite material might satisfy the requirement. Among them, composite seemed more flexible in physical and mechanical properties [2] and kind of green or recyclable material [3]. So that composite be more familiar in some structural purposes' material like in aero plane vehicles.

CFRP, a carbon fibre polymer composite can be found in aeroplane body or in unmanned aircraft vehicle, UAV. The lighter body of vehicle means longer travel time and more covered area can be achieved. UAV or drone is useful to surveillance purposes likes mapping of forest, farming and disaster mitigation [4]. In polymer matrix composite, lightweight refer to polymer-based matrix, in which mechanical strength source from fibre material and bounding between fibre and matrix [2]. Fibre material usually made from carbon, Kevlar, or ceramic material such as fibre glass. E-glass fibre is widely used because high mechanical strength, and good insulator. Polyester has good dimension stability, and high modulus that comparable to epoxy. Both of these two materials are easily to find in our national market. In this study, e-glass fibre namely WR 185 introduced into matrix of polyester resin. In our previous study, 40% volume of WR185 in polyester matrix provides best mechanicals performance in the composite. This work tries to understand the composite characteristics when open hole presence. That may be useful information prior to applied for floating material of small amphibian aircraft vehicle.

2. Methodology

2.1. Material and specimen

Composite was made from polyester resin and e-glass fibre. A woven fabric e-glass fibre WR 185 with thickness of 0.11 mm was cut off in size of 400 x 450 mm. ASTM standard demand composite thickness of 3.5 mm. Fibre volume fraction of 40% means 0.4 of total composite thickness is thickness of the fibres (1.4 mm). So that, one composite needs 13 plies of e-glass fibre in total. The other part of composite was occupied by unsaturated polyester resin Yukalac BQIN-EX 157. In order to accelerated curing of resin, certain volume of catalyst was added, in ratio of 1:100, then manual stirred for about 45 minutes or till homogenous mix.

Making of composite was started with introduced wax on bottom surface (glass material) of die, to prevent resin sticking. Followed by introducing a layer of resin with brush, and one sheet e-glass fibre was carefully put on resin, then flattened with touch up method to ensure resin filled the pores of the fibre. It is one set of resin and e-glass fibre. For the next set, a layer of resin followed with a sheet fibre again. In total, one composite panel has 13 layers of fibre and 14 layers of resin. Last layer of fibre was covered with a last thin layer of resin as well as first layer of resin. Fully dried up composite had achieved after 24 hours. Then composite surface was mechanically smoothed with sand paper to remove some surface defects and to achieved thickness of 3.5 mm.

Specimen for density, tensile and compressive test was machined cut off from the composite plate. Density testing specimen has dimension of 10x10 mm^2 with composite thickness. Density test was repeated for 10 times. Tensile testing specimen was made according ASTM D3039; with total length of 250 mm, width of 25 mm and thickness of accordingly to composite. The compressive testing specimen made according to ASTM D6641; total length of 140 mm, and width of 13 mm. Both of testing and compressive specimen is rectangular shape material.

Half of tensile and compressive specimens were drilled to get an open hole. As above explanation, joining composite by bolt will results an open hole on composite. This study investigated effect of an open hole on the composite. According ASTM, ratio of width-to-diameter should be equal to 6. So that, tensile and compressive test specimen would have hole diameter of ~4 mm (25/6) and ~2 mm (13/6), respectively.

2.2. Experimental

Measuring density was conducting 10 times using densitometer machine. Specimen was weighed before and during immersion in water that put on the plate of densitometer. Condition for density test are at room temperature 24.5°C and relative humidity of 38.8%. Tensile testing conducted on UTM Tensilon RTF-2410 with condition of travel speed of 2 mm/min, at room temperature of 22 ± 2°C and relative humidity of 47 ± 10 %. Compressive testing also on UTM Tensilon with condition equal to tensile testing. Setting for compressive test is shown in Figure 1. All the mechanical tests were done for 7 times.

Figure 1. Setting for a compressive test.

3. Results and Discussion

Densitometer showed density of the composite on its monitor at the end of testing. Density of specimen varied from 0.94–1.62 g/cm^3, with average of 1.12 g/cm^3. This value is less than GFRP WR 185 laminates with lycal resin that reported by K Abdurohman et al [5]. This value is also less than density of lightweight metal that widely use in aero industry such as aluminium alloy (~2.7 g/cm^3) or magnesium alloy (~1.7 g/cm^3). However, variety of value density implies that hand lay-up technique is not quite good to result homogenous composite.

Tensile strength of composite decreased with introduction of hole at centre of specimen as shown in Figure 2. With no hole composite has tensile strength in average of 178 MPa, meanwhile with open hole specimen only provided tensile strength of 119 MPa. These value is lower than reported by Ilhamdi et al with stacking sequence 0°, 90°, 0° [6], but higher than reported by S.M. Kumar et al [7]. By number, introducing hole at the centre of specimen had weakened composite about 33%, lost strength about 59 MPa.

Other tensile parameter is strain at failure that showed in Figure 3. Tensile strain of no hole composite varied from 3.48% to 4.79%, while open hole composite has 2.68% to 3.64%. Different among 7 repetitions is quite closed each other, implies that composite has good homogeneity in tensile strain. No hole composites seemed have higher strain compare to open hole ones. However, the different between these two kind composite is about 1% in average, 4.2% of no hole composite and 3.2% of open hole composite.

Figure 2. Tensile strength of the composite.

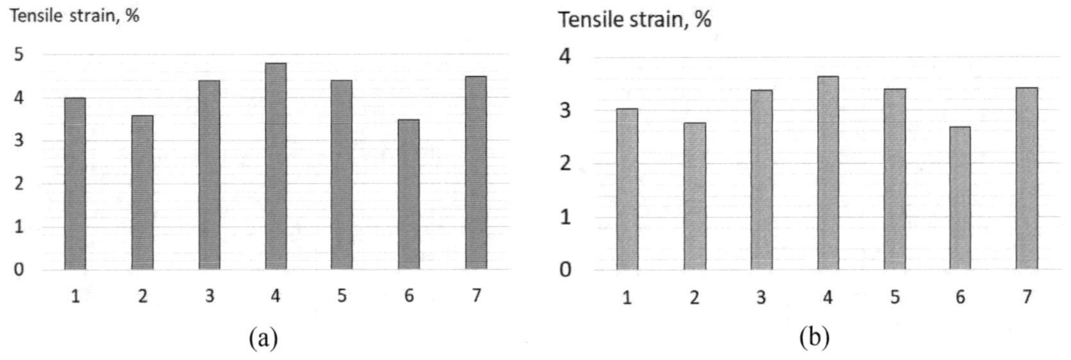

Figure 3. Tensile strain of composite for 7 repetition test; (a) no hole, (b) open hole.
Note: Horizontal dash line is average value.

Compressive strength of composite also decreased with introduction of open hole at centre of specimen as shown in Figure 4. In average, compressive strength reduced from 136 MPa to 119 MPa. These value is much lower than reported by Ilhamdi et. al. [6] and M Tanoglu et. al. [8] that used ply-lay up in plane direction technique in making composite. Introducing hole weakened composite about 12%, lost compressive strength about 17 MPa. Deviation in compressive strength is larger than tensile strength. But, coincidentally minimum compressive strength both of two condition composite is rather equal at 106 MPa. Other impact of open hole introduction on composite is enabling composite achieved more deformation during compressive test i.e. compressive strain increased almost two times from 0.75% to 1.13%, as shown in Figure 5. Empty area in hole might allow composite deformed larger before broken.

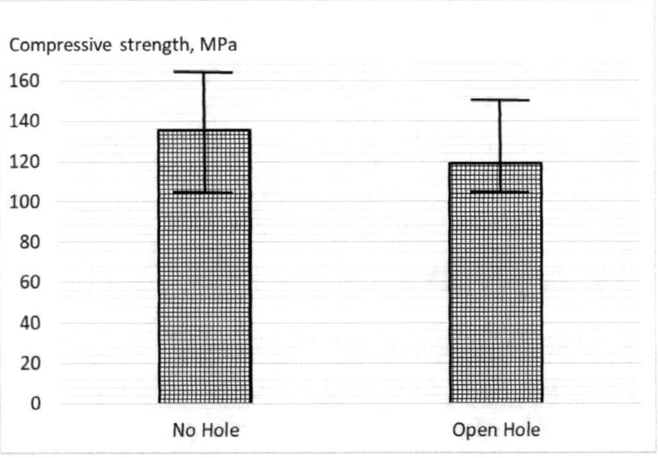

Figure 4. Compressive strength of the composite.

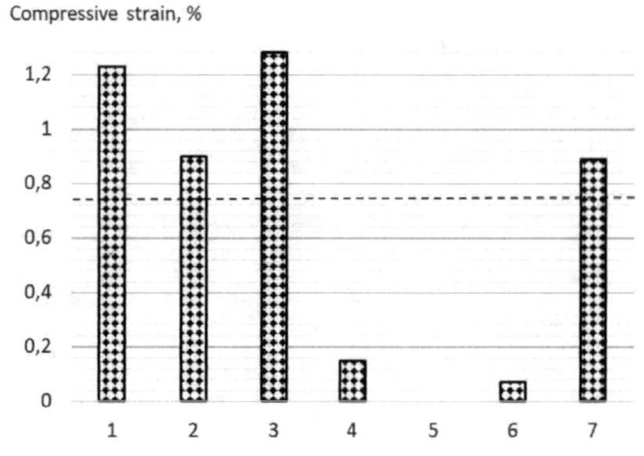

(a)

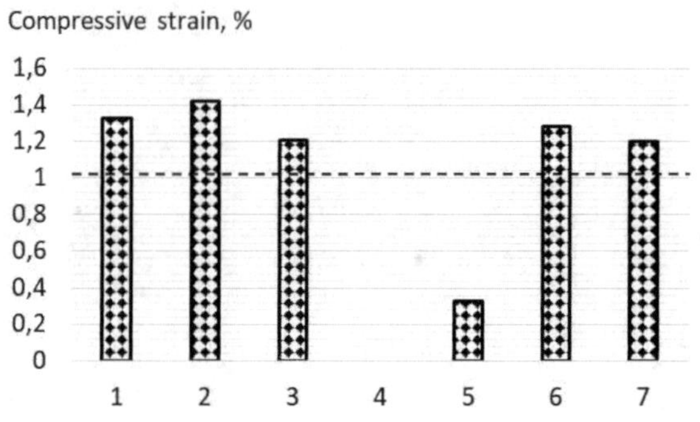

(b)

Figure 5. Compressive strain of the composite for 7 repetition test; (a) no hole, (b) open hole. Note: specimen number 5 of (a)and 4 of (b)are failed. Horizontal dash line is average value

Introduction of hole on the composite also change failure mode tensile testing specimen, as shown in Figure 6. According to ASTM, no hole specimen revealed lateral at grip top (LAT) failure mode, where fracture happened at the end of examination area that closed to grip. Total different failure mode presence on open hole specimen in which fracture lines cross the hole geometry. It is known as lateral gage middle (LGM). Both of specimen condition showed pull out fibres from their matrix.

The same failure modes also revealed on compressive testing specimens with identical condition as shown in Figure 7. No hole specimen of compressive test showed LAT mode while open hole compressive test had LGM mode. The only difference compressive specimen from tensile specimen is no pull out fibre was exposed after testing. Fracture line on compressive test was marked as blur-like grey line in examination area.

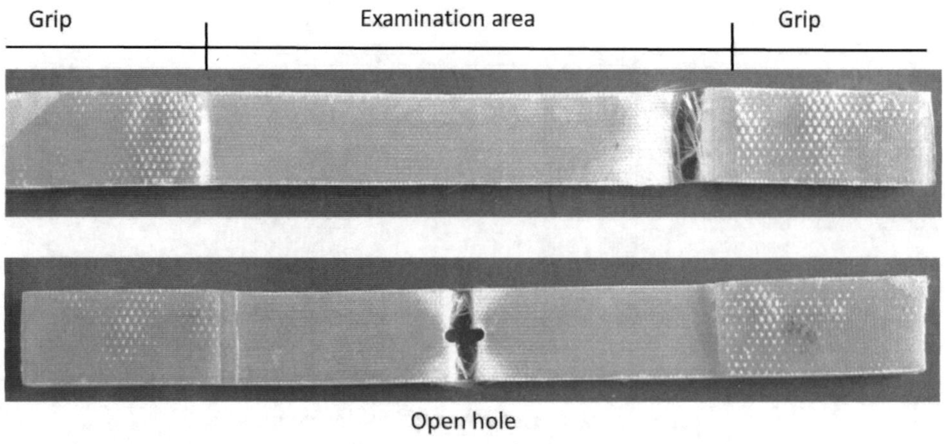

Figure 6. Fracture appearance of tensile test specimen; (above) no hole, (below) open hole.

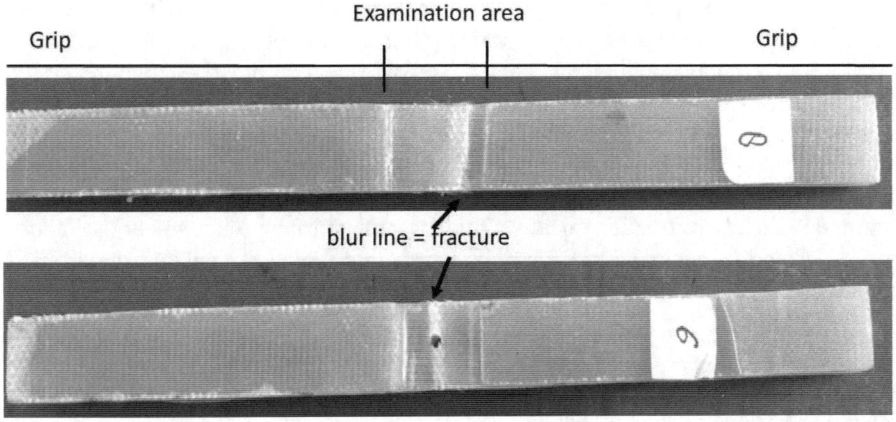

Figure 7. Fracture appearance of compressive test specimen; (above) no hole, (below) open hole.

Effect of open hole introduction at centre of composite could be resumed in Table 1. Open hole change fracture behaviour of composite from lateral at top grip into lateral gage middle. This change might be enabled by fact that hole reduces actual solid cross section area of the specimen. So that, when equal applied force with no hole specimen occurred, specimen will experience higher stress. However, results showed that tensile stress of open hole specimen is lower than of no hole specimen. It is should be another reason for this change.

Table 1. Results resume of open hole effect on composite.

	UTS	TS	UCS	CS	TFM	CFM
No hole	178	4.2	136	0.75	LAT	LAT
Open hole	119	3.2	119	1.13	LGM	LGM

Note: UTS = ultimate tensile strength, MPa. TS = tensile strain, %. UCS = ultimate compressive strength, MPa. CS = compressive strain, %. TFM = tensile failure mode. CFM = compressive failure mode

Hole in the composite could be treated as geometry defect. Because hole made by drilling on composite. The material flow could be said break off by that cutting process, and no change in mechanical strength by deformation. Fortunately, circle shape of hole made no stress concentration around. However, when external tensile force applied on composite, tensile stress will raise. When stress conquer a local yield strength, a little plastic deformation will be promoted at two ends of hole. This deformation made hole change from defect to crack starter, that also induced stress concentration increasing [6]. Further applied stress will open crack at the two ends of hole, then crack propagated perpendicular to applied stress direction, as indicated in Figure 6. It is seemed that crack that started from open hole is easier to nucleated than specimen with no hole. It is reasonable for failure mode change. It became clear that not smaller area that made hole dangerous for structure integrity, but when hole experienced small plastic formation does.

4. Conclusions

Polyester resin e-glass fibre (40% vol.) reinforced composite is a kind of lightweight material with density of 1.12 g/cm3. This value is much smaller than lightweight metal alloys such as aluminium or magnesium base. Introducing open hole has negative effect on mechanical properties of composite; reduced tensile strength, tensile strain and compressive strength, but increased compressive strain. Tensile strength significantly dropped from 178 MPa to 119 MPa, weakened about 33%. Tensile strain is only reduced about 1%, dropped from 4.2% to 3.2 %. Compressive strength dropped from 136 MPa to 119 MPa, weakened about 12%. Existence of hole enables a larger compressive strain, up to 1.31%.

Open hole existence made a total change in nature of composite fracture. When no hole happened on the specimen of tensile and compressive, LAT mode failure always occurred. But, open hole would have resulted LGM failure mode on these two tests. Tensile and compressive fracture could be easily distinguished by exposed pull out fibre on tensile and grey blur line fracture on compressive, respectively.

Acknowledgments
Author's gratitude addressed to Mechanical Engineering Department of Engineering Faculty, Andalas University who provide financial fund for publication this work through Hibah Publikasi contract number 059/UN16.09.D/PL/2020.

References
[1] Ijagbemi C O, Oladapo B I and Campbell H M 2016 *Pro. Eng.* **159** 124–132
[2] Callister W D 2007 *Materials Science and Engineering: An Introduction* (John Wiley & Sons, Inc.)
[3] William J 2003 *Acta Materialia,* **51(19)** 5775-5799
[4] Budiyantara A S, Pandoyo T, Hidayat D and Prabowo G S 2013 *The 2013 Asia-Pacific International Symposium on Aerospace Technology*
[5] Abdurohman K, Satrio T and Muzayadah N L 2018 *J Phys: Conf. series* **1130(1)** 012018
[6] Ilhamdi et al 2019 *J. Metal* **2** 94-100
[7] Kumar S M et al 2018 *Mat. Today Pro.* **5-5-2**, 13,367-13,374
[8] Tanoglu M and Seyhan A T 2003 *Mat. Sci. Eng. A* **363** 335-344

Physical and Mechanicals Characteristic of the Nano Particulate TiO_2-Epoxy Composite

Ilhamdi[1*], A Nugroho[2], D Hidayat[2] and R A Nur[1]
[1] Mechanical Engineering Department, Faculty of Engineering, Andalas University, Padang, Indonesia 25163
[2] Indonesian National Institute of Aeronautics and Space, Bogor, Indonesia 16350

*Corresponding e-mail: ilhamdi@eng.unand.ac.id

Abstract. Using nanoparticle as filler in polymer composite material is not widely reported yet. In this study, nano particle of TiO_2 (4%, 8% vol.) was added in epoxy resin matrix through casting method; a liquid homogenized mix of epoxy- TiO_2 was carefully poured into a simple die for air dried-up. The composite is characterized by physical test of density, and mechanical test such as tensile, compression, flexural and shear test. The results show that actual density of the composites is 1.111, 1.219 and 1.309 g/cm^3 with addition of 0%, 4% and 8% vol. TiO_2, respectively. These values are a less than theoretic density. The difference is related to the estimated porosity. Each of mechanical properties has its own trend regard to content of TiO_2. Maximum tensile strength is 27.8 MPa on 4% Ti specimen. Further increasing of TiO_2 made tensile strength and modulus elasticity decreased. Compressive stress is gradually decreased with increasing content of TiO_2 ; decreased from 60 to 52 MPa. The addition of TiO_2 by 4% decreased flexural strength, but further addition of TiO_2 by 8% had increased the flexural strength. Specimen with no TiO_2 had provided maximum flexural strength of 40 MPa. Shear stress has same trend as compressive stress; gradually decreased from 14.4 to 9.2 MPa.

Keywords: nanoparticle, composite material, porosity, tensile, compression

1. Introduction

Gasoline vehicles innovation affected by two main issues i.e. global warming and natural resources deposit. The global warming issue is closed related to CO_2 gas emission coming from gasoline vehicles [1]. A great reducing of petroleum-based deposit natural resource issue becomes popular since crisis of 1970's [2]. These issues indirectly demand a more efficient use of petroleum for the vehicles. The engineers had been developed a less weight vehicles to achieve this demand. Ideally, reducing weight of the vehicles should be with no or less sacrifice on mechanical aspect of the vehicles material. It is seemed that lightweight metal or composite material can satisfies this purpose. A composite has more flexibility in physical and mechanical properties than metal [3]. Composite is also could be categorized as green material [4].

With a comparable strength to metal, and less density, composite become a promising material for structural purposes material such as in automotive and aero vehicle. In archipelago country that have some narrow area, small island, forestry area and watery area, a light aero vehicle is preferable than a common aeroplane. One of the light aero vehicles is unmanned aircraft vehicle, UAV. This UAV is useful to surveillance purposes of forest, farming and disaster mitigation [5]. So that, UAV that capable for a longer travel time and wider covered area is preferable.

The light term in the composite is come from polymer-based matrix material that naturally light material. Polymer nanocomposite is made from nanoparticle powder (as filler) and polymer matrix. Commonly, nanocomposite has better properties than conventional composite. This superiority is depending on structure, composition and properties of filler and matrix [6].

In this study, nano powder TiO_2 of 0, 4 and 8% volume percentage are introduced into epoxy resin. In the further research this powder will be used as a filler on the fibre reinforced composite, so its content should be not in a large number, namely less than 10%. TiO_2 is widely used because of its mechanical, optical, dielectric, anti-corrosion and biocompatibility. Epoxy has good dimension stability, high modulus, thermal and chemical resistance that better than other polymers such as polyester or vinyl ester, but relatively brittle material and expensive [7]. It would be interesting to know this composite characteristic, before to developed as floating material of the amphibian aircraft vehicle.

2. Methodology

2.1. Material and specimen

Composite was made of from epoxy resin and TiO_2 powder. Volume fraction of TiO_2 is 0%, 4% and 8%. TiO_2 powder has particle size of 30 - 50 nm, with density about 4.25 g/cm^3. Epoxy and TiO_2 powder were mechanical stirring mixed for 60 minutes at 600 rpm, meanwhile an open simple die was rub with wax. Then a homogenized mix was poured into die. The complete dried up composite was achieved after 3 days. The boundary between composite and die that formed with wax, enable easy take off composite from die. Then composite surface was mechanically smoothed with sand paper.

Density testing specimen has area dimension of 10x10 mm2, and the thickness refer to thickness of the composite. Testing was repetition for 10 times. Tensile testing specimen was made according ASTM D638; with total length of 165 mm, gage length of 50 mm, gage width of 13 mm and thickness of 7 mm as shown in Figure 1. The compressive testing specimen was made according to ASTM D6641; total length of 140 mm, width of 13 mm and thickness of composite of 7 mm. Flexure test of three points bending specimen was made according to ASTM D790; width length of 112 mm and width of 27 mm. The shear testing specimen was made according to ASTMN D5379; length of 76 mm, width of 20 mm, thickness of 7 mm, at half of length has V-notch of 90° at both side of wide.

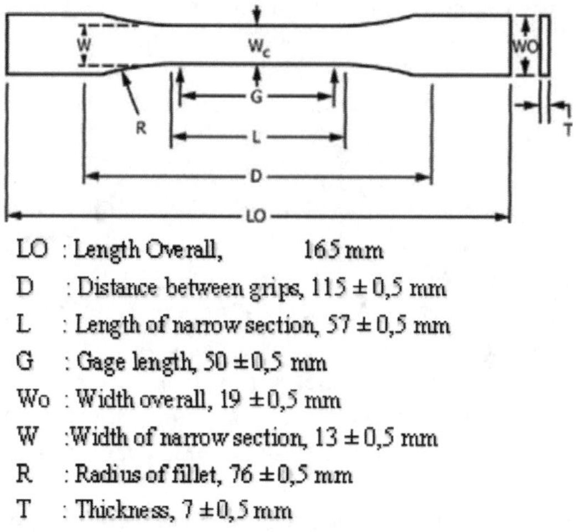

Figure 1. Tensile testing specimen dimension.

2.2. Experimental

Measuring density was conducting 10 times using densitometer machine. Specimen was weighed before and during immersion in water that put on the plate of densitometer. Mathematically actual density of specimen calculated as weight before immersed multiplied with water density divide by difference between weight before to during immerse, as shown as equation 1. Condition for density test are at room temperature 24.5oC and relative humidity of 38.8%.

$$\rho_{actual} = \frac{w_{before}}{w_{before} - w_{during}} \rho_{air} \quad (1)$$

Other, theoretical density is calculated by fraction volume of epoxy and TiO2 multiplied with their density as common composite equation. Difference between theoretical and actual density to theoretical density marked as porosity that existed in the composite.

Tensile testing conducted with condition; initial specimen length distance of 115 mm, travel speed of 5 mm/min, at room temperature of 22 ± 2 °C and relative humidity of 47 ± 10 %. Compressive testing was conducted with condition; initial sample length distance of 12.7 mm, travel speed of 2 mm/min, at room temperature of 23± 3 °C and relative humidity of 50 ± 10 %. Condition for flexural testing; initial sample length distance of 121 mm, travel speed of 5 mm/min, and the same temperature - relative humidity with tensile testing. In shear testing, a specimen was properly stapled on the grip by jig to avoid slip during the test as shown in Figure 2. All the mechanical tests were done on UTM Tensilon, and repeated for 7 times.

Figure 2. Setting for Shear Test.

3. Results and Discussion

3.1. Density test

Densitometer showed the actual density of the composite on its monitor at the end of testing. Complete result can be seen in the Table 1. It is shown that increasing content of TiO_2 made composite become more heavier, and the existence of porosity also increased. It implies that more defect introduced during the addition of TiO_2, but also resulting denser composite.

Table 1. Density measurement results (density in g/cm3, porosity in %).

Content of TiO2	Actual density	Theoretical density	Porosity
0%	1.111 ± 0.005	1.17	5.04
4%	1.219 ± 0.006	1.29	5.74
8%	1.309 ± 0.004	1.42	7.58

3.2. Tensile test

Tensile strength of composite increased with addition of 4% TiO_2, but further addition TiO_2 (8%) had decreased composite strength, as shown in Figure 3. Composite with no TiO_2 (epoxy only) has tensile strength, modulus elasticity and elongation at break in the range of 21.5 – 27.4 MPa, 1,900 – 2,200 MPa and 1.3 – 1.6 %, respectively. Their average value is 25.7 MPa, 2,100 MPa and 1.5%, respectively. Their low in elongation showed a typical brittle material characteristic. Addition 4% volume of TiO_2 increased tensile strength up to 27.8 MPa, modulus elasticity up to 2,400 MPa, but decreased elongation at break down to 0.95%, all in average.

Compare to epoxy only material, their change in all parameters showed moving behaviour from brittle to more ductility. However, in addition of 8% TiO_2 tensile strength and modulus elasticity bounce back to 20.5 MPa and 2,300 MPa, respectively. Other hand, elongation is the same grade with previous addition, about 0.97%. Over all, highest in tensile strength and modulus elasticity was found in the specimen with 4% TiO_2.

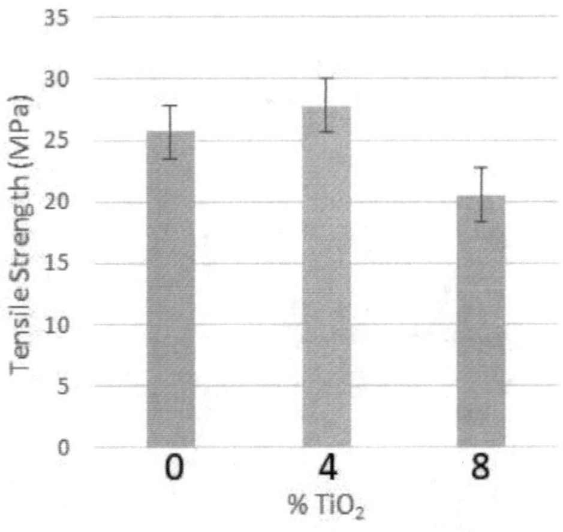

Figure 3. Tensile strength of the composite.

3.3. Compressive test

Compressive strength of composite was gradually decreased with addition of TiO_2, as shown in Figure 4. Epoxy only material has compressive strength, modulus elasticity and elongation at break in the range of 60 – 63.5 MPa, 500 -1,000 MPa and 3.6 – 4.2 %, respectively. Their average value is 61.4 MPa, 700 MPa and 3.8%, respectively. Addition 4% volume of TiO_2 decreased compressive strength to 57.6 MPa, but increased modulus elasticity up to 1,200 MPa, and decrease elongation at break to 3.4%, all in average.

Further addition of TiO_2 (8%) resulted continue decreasing of tensile strength to 51.9 MPa, also decreased modulus elasticity to 900 MPa. On other hand, elongation was in the same grade with initial material, about 3.9%. Over all, highest compressive strength was found in the epoxy only specimen.

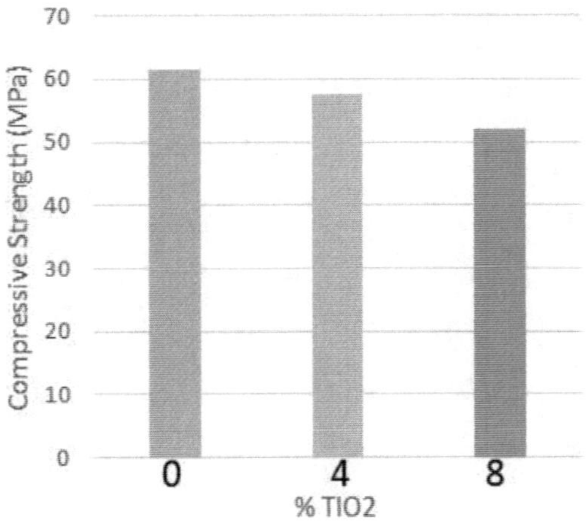

Figure 4. Compressive strength of the composite.

3.4. Flexural test

The highest flexural modulus of the composite of 2,100 MPa was found in the specimen with addition of 4% TiO_2. This composite has relatively same strength with addition of 8%; 29 compare to 32 MPa MPa. High flexural strength means low flexural modulus. It is seemed that the optimum is resulted by composite 4% TiO_2. The complete result can be seen in Figure 5.

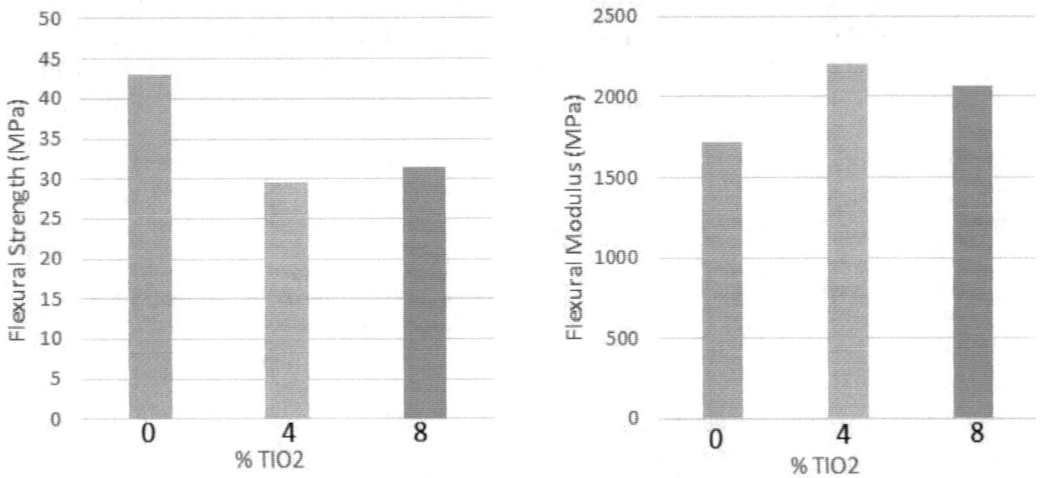

(a) (b)

Figure 5. Flexural properties of the composite; (a) flexural strength.(b) flexural modulus.

3.5. Shear test

The same behaviour with compressive strength, shear strength was also gradually decreased with more addition of TiO_2, namely decrease from 14.4 to 7.3 MPa, as shown in Figure 6. The same trend was found in break strain from 1.14 to 0.43%. But, modulus elasticity of the composite was irregular changed; increased from 600 to 900, then dropped back to 600 GPa. Compilation of the results indicated that the best condition in composite 4% TiO_2; shear strength of 13.1 MPa, that comparable with epoxy only, highest modulus elasticity of 900 GPa, and moderate break strain of 0.894%.

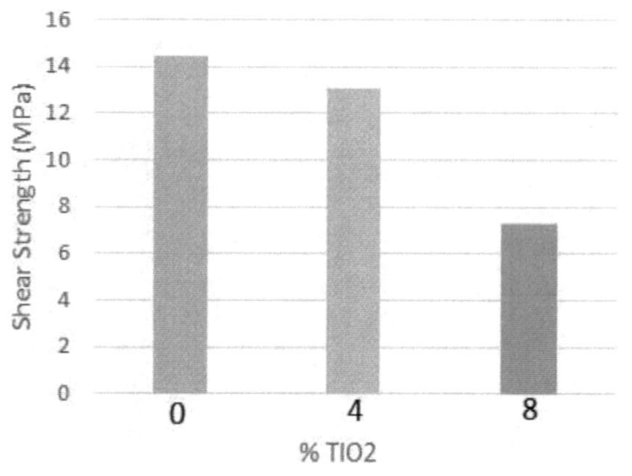

Figure 6. Shear strength of the composite.

3.6. Discussion

The 4% vol. TiO_2 is seemed provide the best combination of physical and mechanical properties. This composition provide moderate density of 1.2 g/cm^3, higher than epoxy only material but still in the range of lightweight material. This value is even much less than magnesium and its alloys, of 1.7 g/cm^3. Magnesium alloys are popular lightweight metal. Composite with 4% TiO_2 is kind of lightweight material with 5.7% porosity. A more proper preparation and/or better production technique of composite is needed to reduce the porosity. Remarkable properties of this composite are tensile strength of 27.8 MPa, compressive strength of 57.6 MPa, Flexural modulus of 2.2 GPa, and shear strength of 13.1 MPa.

Increase in tensile strength in this kind of composite is also reported by A Chaterjee et al which completing with thermal and viscoelastic [8], A. Turaif et al [9], Mirmohseni et al [10]. But, no confirmation about compressive strength and shear strength of this composite from other researcher. In the term of flexural properties, decrease in flexural modulus might related to reaction of TiO_2 nano particle that may have agglomerated inside composite. This will have reduced interface area between matrix and particle that reduces load transfer from matrix epoxy to the filler [11].

4. Conclusions

Addition TiO_2 as filler in epoxy matrix composite is not always results a higher mechanical performance. Among the variation of TiO_2 addition in the composite, the 4% vol. TiO_2 is seemed provide the best combination of physical and mechanical properties i.e. moderate density of 1.2 g/cm^3, highest tensile strength of 27.8 MPa, highest modulus elasticity of 2,400 MPa, moderate compressive strength of 57.6 MPa, moderate flexural strength of 29 MPa, highest flexural modulus of 2,200 MPa. and high shear strength of 13.1 MPa. Value of 5.7% of porosity in 4%vol. In the future, porosity might be decreased with a better production technique.

Acknowledgments

Author's gratitude addressed to Faculty of Engineering Andalas University whom provide financial support for publication through Hibah Publikasi Prosiding contract No. 040/UN.16.09.D/PL/2020.

References

[1] C O Ijagbemi, Oladapo B I, Campbell H M and Ijagbemi C O 2016 *Pro. Eng.* **159** 124 – 132
[2] A N Oumer, Hasan M M, Baheta A T, Mamat R and Abdullah A A. 2018 *Renew. and Sust. En. Reviews* **88** 82-98
[3] W D Callister 2007 *Mat. Sci. and Eng. : An intro.* John Wiley & Sons, Inc.
[4] J William 2003 *Acta Mat. Prog. In Struc. Mat. for Aerospace S*
[5] A S Budiyantara et al. 2013. *APISAT-2013*
[6] Hader et al, 2018 *Ultra. - Sonochem.* **2** 861-873
[7] D Pinto, Bernardo L, Amaro A and Lopes S 2015 *Construc. and Build. Mat.* **1** 506-524
[8] A Chattterjee and Islam M S 2008 *Mat. Sci. Eng. A* **487** 574-585
[9] H A Al-Turaif 2010 *Prog. in Org. Coat.* **69** 241-246
[10] A Mirmohseni and Zavareh S 2010 Mat. and Des. **31** 2,699-2,706
[11] V Prasad et al 2018 *Mat. Today Pro.* **1** 24,862-24,870

Slowloris DoS Attack Based Simulation

Shima Sabri[1*], Noraini Ismail[2] and Amir Hazzim[1]
[1]Faculty of Computing and Multimedia, Kole Universiti Poly-Tech Mara, Jalan 6/91, Taman Shamelin Perkasa, Cheras, 56100 Kuala Lumpur
[2]Faculty of Ocean Engineering Technology and Informatics, Universiti Malaysia Terengganu, 21030 Kuala Nerus, Terengganu Darul Iman, Malaysia

*Corresponding e-mail: shimasabri@kuptm.edu.my

Abstract. Denial of Service (DoS) attack is a type of cybercrime when an internet site is unavailable to be accessed by the user. The DoS attack is one of the popular attacks which can be launched by using a single machine and could take down many web servers by sending a lot of request to the server until it is disconnected. The objective of this paper is to simulate the launch and prevention against the DoS attack towards the website. The simulation of DoS attack is implemented by using ActivePerl Language and tested by using Slowloris DoS Attack. Both software shows a good combination of simulation program because it supports each other to exhaust the available connections on the servers. The result shows that the website connection is successfully unavailable to be accessed by legitimate user. In addition, the website still could not be served even clearing the cache of the browser. Finally, this paper discusses about future solution of DoS attack and recommendation for secure environment architecture.

Keywords: DoS Attack, Apache server, Slowloris DoS attack, simulation

1. Introduction

People are using website as one of communication technologies to manage life. Websites allows user to carry out online activities such as e-payment, shopping, reservation, booking as well as storing and retrieving information. To do this, a website must use apache web server to have the capability to manage and organize an online process. The problem arises when the websites that are using the Apache server becomes a main target by attackers to launch Denial of Service (DoS) attack [1]. DoS attack is a kind of cyber-attack in which the perpetrator seeks to make a machine or network resource unavailable to its intended users by temporarily or indefinitely disrupting services of a host connected to the Internet [1]. Services affected may include email, websites, online accounts (e.g., banking), or other services that rely on the affected computer or network. Authorized user will find that the website fails to provide the page in a short time and may lead to system coming down for a duration of time and then ultimately causing huge amount of revenue loss. Denial of Service has the most devastating effect, putting tremendous pressure on security experts to bring out effective defence solutions on a computer or a network [11]. These attacks could be implemented with a variety of tools and codes. However, there is a lack of solution for DoS attack has took place over the internet for nearly a decade. Hence, it becomes essential to carry out these attacks in small test bed environments in order to extrapolate the patterns of known threats, detecting and blocking them efficiently. These real time attacks are measured and analysed using the network traffic monitors. Also, this simulation presents defence strategies that could be executed in order to mitigate these attacks [1]. As shown in Figure 1, the attacker could inject multiple packets into the channels. As a result, the communication between the networks components are blocked, which is seen as the DoS attack [2]. On top of that, the hacker aims to attack the system in order to degrade the estimation performance by injecting interference packets into the channel to block the packets from the victim [2].

In this paper, we point out traditional approach of Slowloris DoS attack based on statistical method, the signature methods and the cluster analysis. According the authors in [3], they stated that this method can recognize the Dos attacks of transport and network layer by filling the channel capacity.

This paper focuses on the web framework that uses the Apache server to become the key target for attackers to initiate Denial of Service (DoS) attack. To solve this problem, we simulate and launch a DoS attack using ActivePerl Language and test it using Slowloris DoS Attack. Results of the simulation reveals that the website is unreachable even after clearing the cache of a browser used by an attacker.

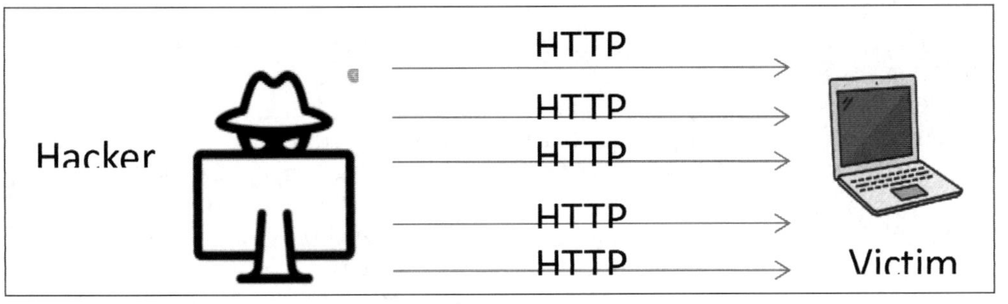

Figure 1. DoS attack.

2. An Online Instruction Detection Approach to Identify Low-Rate DoS Attack

The main purpose of a DoS attack is to consume the resources of a targeted victim. A victim may be a server, a router or any other kind of peripheral device that is connected to a network. This simulation will concentrate on HTTP server. In the past years, the popular DoS attack that has been introduced is Low-Rate Denial of Service attack that uses tiny amount of bandwidth [4].

There are several types of attack that is categorized as Low-Rate DoS Attacks that perform slow DoS attack. Although most of the slow rate DoS attack target the HTTP, they also can be adapted to attack other protocol. Table 1 summarises the comparison for DoS attack technique.

Slowloris attack is the most popular. It exploit the HTTP protocol by establishing a large amount of pending request with the targeted web server [5]. Additionally, the SlowLoris is underpinning the evolution of variety attacks such like SlowComm [5], Slow Next attack [4] , Slow Read Attack [6], and SlowReq Attack [7].

All of the attacks above are known as low and slow attack. Slowloris attack [8] is one of the most known slow rate Denial of Service attacks. Slowloris works by sends legitimate HTTP request to the web server with a catch, as it is incomplete HTTP. Due to this, the attack forces the targeted web server to wait for the end of the request, which will never be send by the attacker because the HTTP request is incomplete. In addition, Slowloris sends to the server a first part of incomplete HTTP request, therefore the wait timeout is triggered. After the tie wait timeout expired, an additional HTTP parameter is sent to the server, and the timeout is triggered again.

In relation to the above explanation, the objective of this paper is to attack the server by using Slowloris that allow perpetrator to overwhelm a targeted server by opening and maintaining many simultaneous HTTP connections between the attacker and the victim server. Further, by implementing DoS attack, the countermeasure can be recommended, thus enhancing the security level of the server to perform the task.

Table 1. Comparison for DoS attack technique.

Name of the Malicious	Malicious Behaviour	Parameter	Evaluation Parameter
DoS attack on Wireless Sensor Network	Disrupt privacy security data	Sensor nodes	Limited sensor node energy
SlowComm	Long request DoS attack	Internet service	Saturating the available connection
Resilient load frequency control	Time delay system model	Uncertainty of the system	Sleep internal and attack internal
R-U-Dead-Yet (RUDY / Slow HTTP POST RUDY)	Exploits the body of POST request.	Opens up desired number of concurrent connections	POST request
Slow Next attack	Exploiting the protocol persistency	Permits client to send multiple request in the same connection	Catching a specific amount of connection
Slow Read Attack	Sending legitimate HTTP request to the server at regular rate	Delay and will slow down the response of the server	Identifying a small client side reception buffer in the SYN packets sent to the victim during the connection establishments
SlowReq Attack	Makes the server to wait to an never-ending wait	Incomplete HTTP request is sent to the server	Each packet is composed by a single character
Slowloris	Opening connections to a targeted Web server and then keeping those connections open as long as it can	Utilizing partial HTTP requests	Incomplete HTTP request

3. Methodology

The detection and mitigation mechanism designed here are effective for small network topologies and can be extended to analogous large domains. Slowloris DoS Attack tool can be executed using multiple programming language such as Phyton, Perl, Go, Java, and Shell. The most common programming language used to run the attack is Perl and Phyton. This simulation will use Perl as a programming language purposely to extract information from a text file and printing out a report or for converting a text file into another form.

3.1. Active Perl

Perl is a software known as "Practical Extraction and Reporting Language". As the Slowloris script is using Perl programming language, Perl is needed to run the script and to complete the DoS attack to the targeted website. As the attack will be executed just using the "Command Prompt" Perl is needed to have a functioning DoS attack.

IOP Conf. Series: Materials Science and Engineering 1062 (2021) 012029 doi:10.1088/1757-899X/1062/1/012029

3.2. Oracle VM VirtualBox
Oracle VM VirtualBox is used to install Linux Server. This is purposely to create two different environments in a VirtualBox and at the same time to implement a DoS attack protection onto the server.

The purpose of creating two servers is to simulate the attack and to demonstrate how the attack would be executed by a hacker when the hacker is trying to perform DoS attack on Apache web server. As VirtualBox can run multiple Operating systems by using single host in this way, this simulation attack will be run in emulated environment.

The simulation will be using two machines. Ubuntu will be used to run the Apache2 and to host the website and another machine observe on how DoS attack affects the normal users and normal traffic.

4. DoS Attack Architecture
The Slowloris attack was design to flood the webserver by sending multiple requests under the same IP just make sure that the webserver crashed or make it operate ineffectively. Based on Figure 2, the architecture of the attack starts from the attacker and down to the user. The attacker first launches the attack by sending multiple HTTP GET request to the website. The website then send to the webserver for processing in order to complete the request. Due to Slowloris sending too much request for the server to handle, it will be slowing the server down by overwhelming all meaningless request. Slowloris will begin to open connections. As a new one free up from someone else using the website, they will open until they got all the connection. As a result, there is no room for legitimate users to request that particular website.

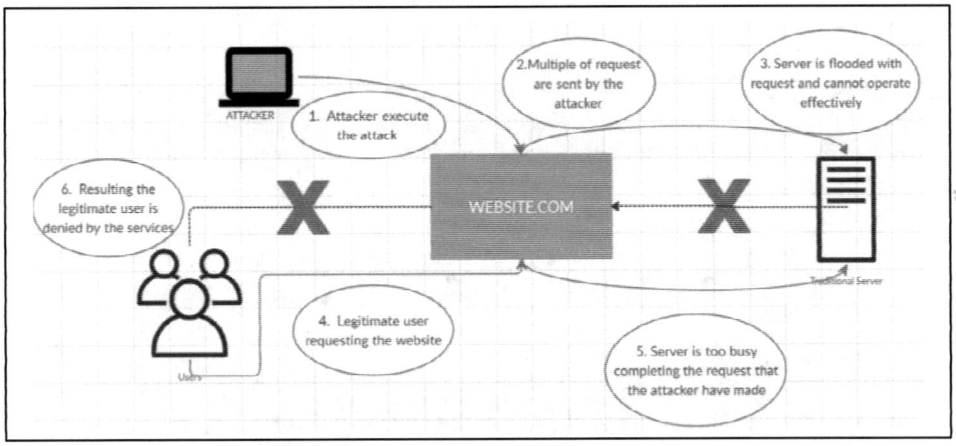

Figure 2. DoS attack architecture.

The command line to execute the DoS attack is shown in Figure 3 below: C:\Perl64>perl slowloris.pl –dns 192.168.56.1

IOP Conf. Series: Materials Science and Engineering 1062 (2021) 012029 doi:10.1088/1757-899X/1062/1/012029

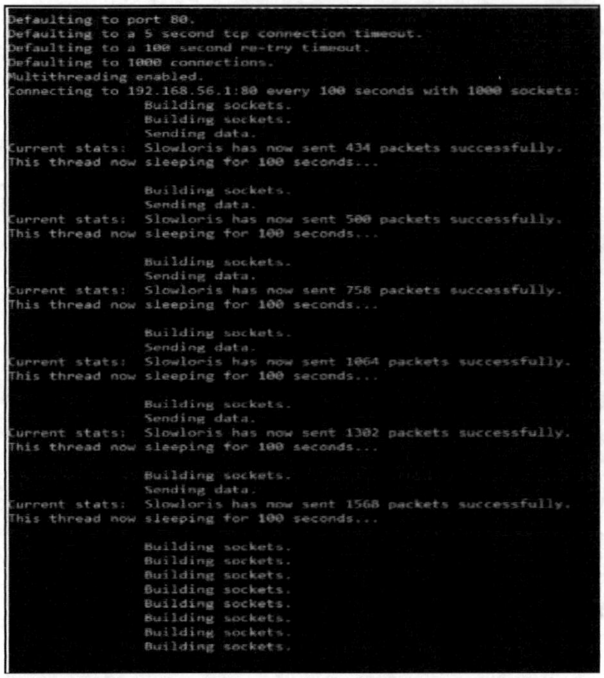

Figure 2. Command Prompt interface to execute the attack.

5. Results and Discussion

The execution of DoS attack by using Slowloris DoS Attack tool is presented and the result is shown in Figure 4 and Figure 5. To check if the website is still working, the webpage needs to be refreshed. As a result, the output would be the webpage is unable to load.

Figure 4. Interface of testing webpage.

Based on Figure 6, the attack is successfully running and flooding the server with multiple request. When the user or the victim browse through the internet and visit the sites, the browser saves several contents and data in temporary storage. This temporary storage is called "cache". To get rid of the fluff, the cache needs to be cleared from time to time. But in this DoS attack case, the website is not served even after clearing the cache of the browser.

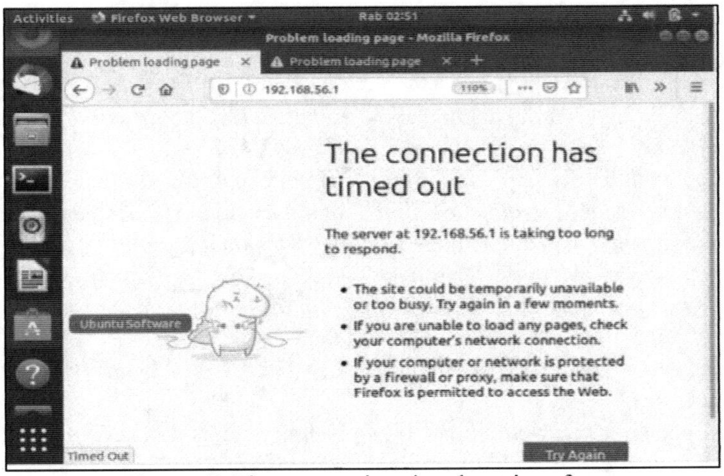

Figure 5. The connection timed out interface.

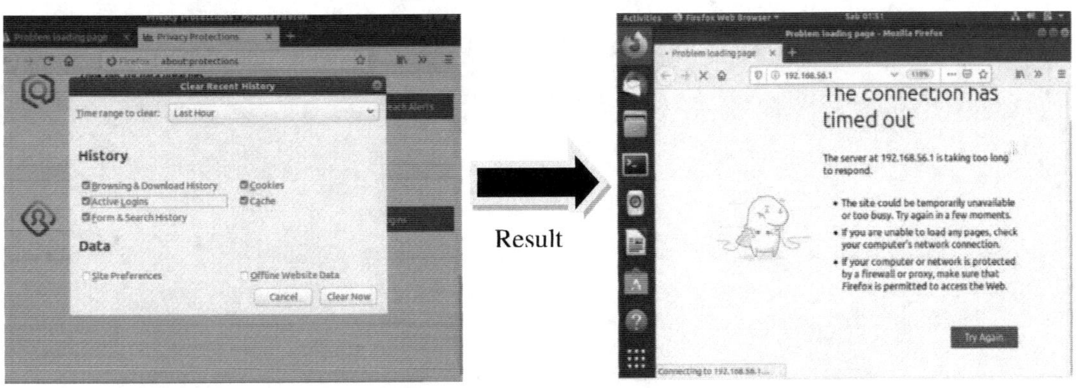

Figure 6. Clearing the cached of the browser and the connection timed out.

6. Conclusions

In order to learn the potential behaviour and help protect the system, designing a type of DoS attack may help to reveal the technique as well as the algorithm. Users may take a measurement of a process and strategies to plan for suitable prevention actions. The invader intends to block the communication channels by DoS attack in order to degrade the estimation performance. Therefore, in this paper, we studied the reliable channel case. In real applications, the communication is likely to be wireless and may be unreliable. Therefore, the simulation is the first step while the unreliable channel case can be left for future work. We do not directly consider the defence and protection of the system in this paper, but aim at pointing out the way DoS attack is implemented, which pave the right direction in providing the protection.

Acknowledgments
We gratefully thank the Faculty of Computing and Multimedia, Kolej Universiti Poly-Tech MARA for providing the opportunity to join ICCEM 2020, Research Management Center (RMC) KUPTM for Micro Grant Support and Dr. Adam Shukri Ali for his proofreading service.

References
[1] H H Jazi, H Gonzalez, N Stakhanova and A A Ghorbani 2017 *Computer Networks* **03** 018
[2] M A Shruthi C V 2018. *International Journal of Scientific Research in Computer Science, Engineering and Information Technology* 175-180
[3] C Yang, W Yang and H Shi 2018 *IET Control Theory & Applications* 1244-1253
[4] A L a A C I Duravkin 2014 *Problems of Infocommunications Science and Technology* 102-106
[5] E Cambiaso, G Papaleo, G Chiola and M Aiello 2015 *Computational Intelligence in Security for Information Systems Conference* (Burgos)
[6] E Cambiaso, G Papaleo and M Aiello 2017 *Journal of Information Security and Applications* 23-31
[7] S Tayama and H Tanaka 2017 *International Conference on Mobile and Wireless Technology* (Kuala Lumpur)
[8] M Aiello, G Papaleo and E Cambiaso 2014 *International Joint Conference SOCO'13-CISIS'13-ICEUTE'13 Advances in Intelligent Systems and Computing* **239**
[9] C C I M d D E C Luis Campo Giralte 2013 *Computers and Electrical Engineering* **39** (7) 2252-2262
[10] Z Cheng, D Yuea, S Hua, C Huanga, Chunxia Doub and Lei Chena 2019 *Electrical Power and Energy System* 105496
[11] M Abushwereb, M Mustafa, M Al-kasassbeh and M Qasaimeh 2020 7

Secure Database on Attendance Verification System by using Data Masking Technique

Siti Faizah Miserom[1*], Shima Sabri[1,2], Fatin Farina Aida[1] and Noraini Ismail[2]
[1] Faculty of Computing and Multimedia, Kolej Univesiti Poly-Tech Mara, Jalan 6/91, Taman Shamelin Perkasa, Cheras, 56100 Kuala Lumpur
[2] Faculty of Ocean Engineering Technology and Informatics, Universiti Malaysia Terengganu, 21030 Kuala Nerus, Terengganu Darul Iman, Malaysia

*Corresponding e-mail: faizah@kuptm.uitm.edu.my

Abstract. Data Masking Technique is a system designed to secure sensitive data by replacing the front data but not the original data. For instance, the data masking techniques capable to hide any data from being viewed by any unauthorized users. However, fewer studies about data masking techniques in the attendance system are implemented. Previous studies have primarily concentrated on defining sensitive data exposure and the technique of securing sensitive data using data masking. Therefore, the aim of this paper is to integrate the use of the data masking technique towards the attendance verification system. Thus, it is sufficient to stop unauthorized users from accessing confidential data. This study also applied the iterative model as a software development methodology as well as integrated with the hashing concept. Therefore, the attendance verification system can be used to perform the functionalities of easily manage records. By using this technique, the attendance of a student was added with a security element. The result shows that data masking and hashing techniques ensure the confidentiality of student data from being viewed by unauthorized parties. The authorized person is the ability to see the masked of sensitive student data. The significance of this study is capable to implement the security element on the attendance system and verify the user before entering the system

Keywords: Data masking; Security; Database; Attendance verification system

1. Introduction

Data Masking Technique is a system designed to secure sensitive data by replacing the front data but not the original data. For instance, the data masking techniques capable to hide any data from being viewed by any unauthorized users. However, fewer studies about data masking techniques in the attendance system are implemented. Previous studies have primarily concentrated on defining sensitive data exposure and the technique of securing sensitive data using data masking [1-2]. Therefore, the aim of this paper is to integrate the use of the data masking technique towards the attendance verification system. Thus it is sufficient to stop unauthorized users from accessing confidential data.

This study also applied the iterative model as a software development methodology as well as integrated with the hashing concept. Therefore, the attendance verification system can be used to perform the functionalities of easily manage records. By using this technique, the attendance of a student was added with a security element. The result shows that data masking and hashing techniques ensure the confidentiality of student data from being viewed by unauthorized parties. The authorized person is the ability to see the masked of sensitive student data.

The significance of this study is capable to implement the security element on the attendance system and verify the user before entering the system. With technological evolution, the world has changed drastically. At every age that human beings have gone through various challenges, and need to acquire new inputs. Due to the dean award ceremony previously had no computerized system and the lecturer on duty only use Microsoft Excel as a platform to store student data. Previously, attendance verification goes through Google Form which students are required to fill out the form provided if they want to verify their attendance.

An unauthorized user can read or steal sensitive information [3]. They have no monitoring database system. Sensitive data exposure will occur. There are some security concerns. The leaks of the information can cause a significant problem to the victim, such as loss of institution reputation and credibility, and worse can cause the institution to specific law acts regarding sensitive data protection [2]. Uncertified students can fill out the provided form. Google Form has no restricted access and if they have a connection, anyone can access it. In KUPTM, MPP will provide Google Form links on multiple platforms such as WhatsApp and Twitter for students to verify their attendance to attend the student appreciation award.

Sensitive information is characterized as data that protects against unauthorized disclosure. Confidential data covers all data containing personal information, records of student education, cardholder information, and more [2]. Some examples of data exposure when a person is accidentally uploaded somewhere, poor crypto, meaning an intruder can read the information if they break into the target of the system successfully [3].

The data masking technique is a technique that replaces the realistic data but not the original data [4]. The primary objective of data masking is to ensure that the actual data not exposed from the outside of the environment [2-3]. In general, data masking ensures that only an authorized party can see the original data. Data masking is used to protect multiple types of data, including personal data, protected health data, and financial data, such as payment card information. Therefore, this project is proposed to develop an attendance verification system for the KUPTM student appreciation award with a secure database by using the data masking technique.

This system helps Majlis Perwakilan Pelajar (MPP) easily to manage the record of student attendance verification for student appreciation award. In this system, the student needs to verify their identity first by their ID number. MPP members will register the certified student and only certified students in the list of student appreciation awards can access the system. After that, a student can verify their attendance to attend the student appreciation award by filling the form provided. The MPP can see the student's data. The problem here is that the MPP is also a student, and they should not see the details of the student. With the privileges that colleges give to the MPP, they can misuse their power.

2. Methodology

As a cyclical process, the iterative model is best reflected. Design modifications are developed in every iteration, and new functional capabilities are introduced. The basic plan after this method is to develop the system through an iterative cycle and in an incremental part [5]. Improvements can quickly be identified and completed throughout each iteration, moving the next iteration to be at least slightly better than the last [6]. The six stages included in the iterative model are as follows:

2.1. Initial Planning

Attendance Verification System for KUPTM Student Appreciation Award with Secure Database by Using Data Masking Technique and will implement the rule-based algorithm. These algorithms are easy to comprehend and expressive algorithms that will extract information from the classification model in the form of rules.

2.2. Planning Phase
Through several studies and observations, the problem statements have been identified. It is possible to identify the key discussion of an issue based on an unauthorized person being able to access confidential information.

2.3. Requirement Phase
The existing method is analysed and all the requirements that are needed to develop a computerized system are identified. Gathering the requirement of the project with the questionnaire which helps the developer to initiate the development process and at the same time create a mutual understanding of the project objective.

2.4. Analysis and Design Phase
Use Case Diagram and Flowchart are designed based on the system requirement.

2.5. Implementation and Testing Phase
This system is developed by using PHP, JavaScript, and SQL Server. Three modules were developed which are Administrator module, MPP module, and student module. The module was tested to ensure the system is fully working and meets the requirement.

2.6. Evaluation Process
The system is evaluated in the admin section to prove that the student's sensitive data will be secure or not, especially in the part of the data masking technique. Evaluate the system whether MPP can easily handle a list of student attendance verification and students can verify their attendance to attend the student appreciation award without no problem and meets or not in user requirement

2.7. Deployment Phase
Once the evaluation is completed, this system is deployed during the student appreciation award announced by the college, so students who are nominated for the award can use the system. The system will be released to the admin to implement whether the system is working properly or not.

As shown in Figure 1, users can test the system and can provide input, which is the involvement of the user more intense positive impact on development. With a prototype, errors and omissions in development can be immediately known. The results can also be received promptly [7]. Furthermore, the model is easier to manage risk, especially the high-risk part. Usually, the high-risk part is done first. Moreover, it also supports changing requirements and supports user feedback.

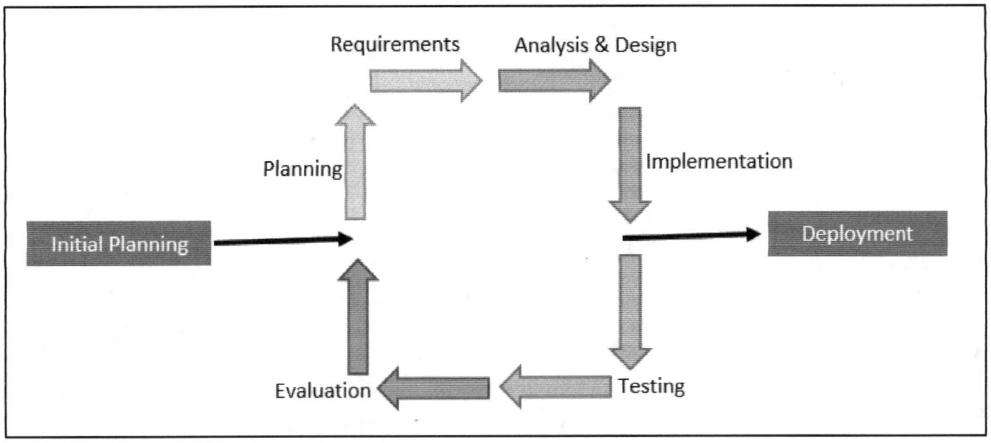

Figure 1. Iterative model phases.

3. System Design

This section explains the process of the system by using two methods which use case diagram and flowchart. These methods possible to classify accurately the requirements of the designed system and determine users with functions they meet within the system [7].

3.1. Use Case Design

The use case diagram shown in Figure 2 shows both interactions between system admin, MPP, and student with Attendance Verification System for KUPTM Student Appreciation Award with Secure Database by Using Data Masking Technique shows the relationship between the admin and users and the different use cases where both parties involved. There are two modules namely the admin site and user site.

In site admin, admin roles areas to login into the system. Admin can register MPP members that have a request to use this system. The admin will also provide a list of certified student names that are eligible for the student appreciation award. So, only the admin system can view the details of the student and only those who can update student data.

The second actor's character is the MPP who is responsible for managing the list of student attendance verification. If the MPP members are assigned a Person in charge (PIC) in managing student attendance verification, they will need to request ID to the admin system. Finally, the student's roles towards the system are that they need to verify their identity using the student ID. This is the reason why MPPs need to register a list of students they got from admins. If identity verification is successful, students will need to fill out the form and verify their attendance to attend the student appreciation award. Only admin and MPP actors can logout as they only go through the system login process.

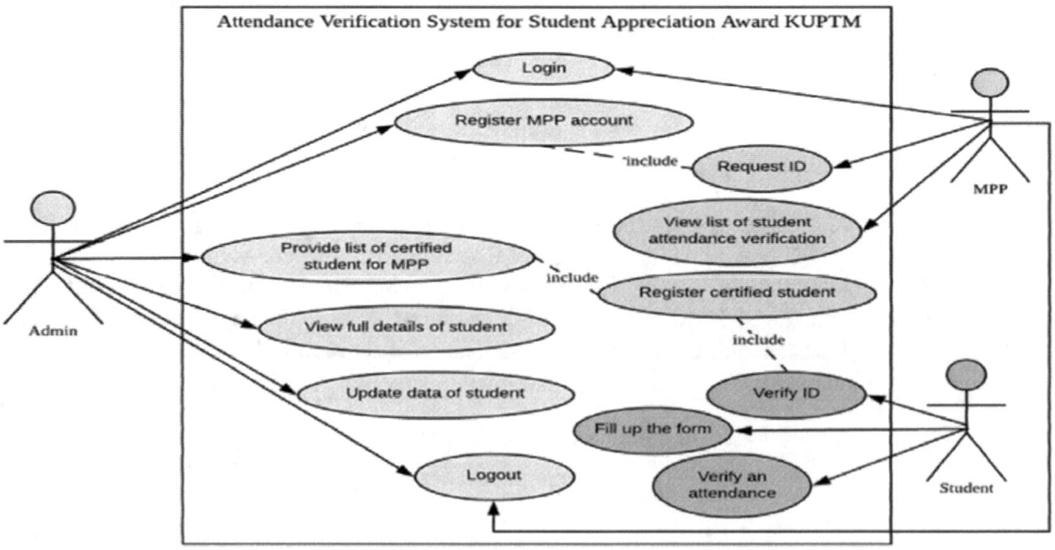

Figure 2. Use case diagram of attendance verification system.

3.2. Flowchart Design

The flow chart below shows the formalized graphic representation of the system and the development process. It supports the developer to gain a basic language and a reference point when the developer with the project.

Figure 3 present a Flowchart of the Attendance Verification System for KUPTM Student Appreciation Award with Secure Database by Using Data Masking Technique for the admin site. In the site admin, the system flow is to start with the admin will be logged into the system. After that, the admin can make an option where the admin can select the menu. There are two sections: register account for MPP or view details of the student. If MPP members request an account for them, the admin will register for the MPP. Admins can also view student details if they want and update student data for specific reasons. After finishing the business in the system, admins can go directly to process logout.

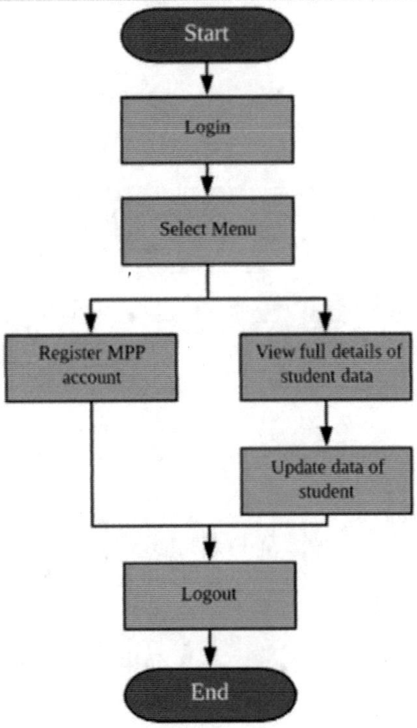

Figure 3. Flowchart of admin site.

Figure 4 demonstrates a flowchart of the system of the MPP site who is appointed as a person in charge (PIC) manages the student attendance verification records. PIC does not have an account, they can request the system admin manually to create their account. If the MPP already has its account, they can log into the system. They can also make choices by selecting the menu provided. These include register an account for certified students or view the list of students who have verified attendance. MPP should register a certified student in the list of student appreciation awards that they receive from the admin system. This is to verify student identity later. The admin will provide the list of student names nominated in the student appreciation award list where the MPP receives only the student's name, student ID, and program code. MPP is not allowed to view the full details of students. Only admins can view the details of the student. After the MPP is done in the system, the MPP can log out of the system.

The flowchart illustrated in Figure 5 shows students do not need to have their account for access to the system. Students only need to verify their identity by entering the student ID into the system. If the identity is not verified, the process will end where they cannot access the system. If their identity is verified, meaning that their names have been registered as students who attend the student appreciation nominated for the award. After that, students will go through a process where they have to fill out the form provided. After completion of the form, the student will have to verify the presence of the award and the process for students to end.

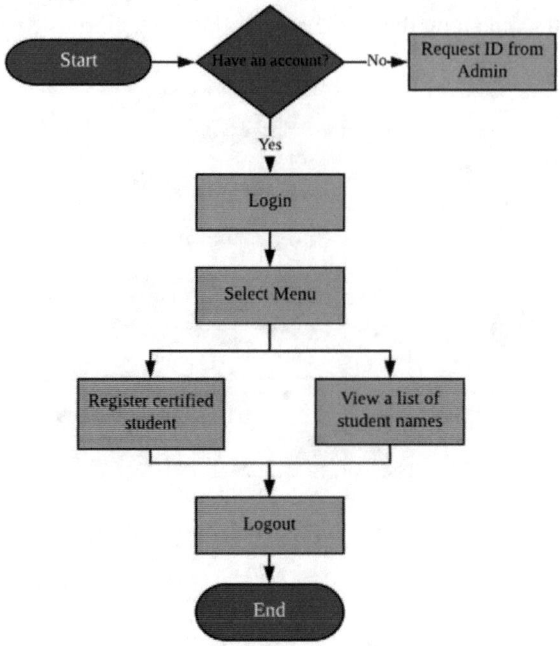

Figure 4. Flowchart of the system.

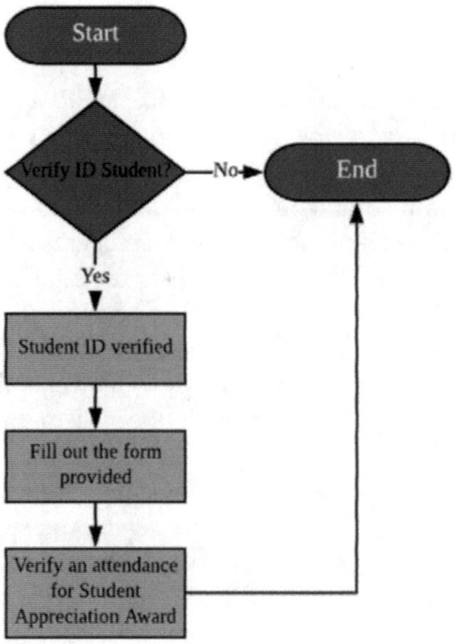

Figure 5. Flowchart of the system for student module.

4. The Proposed System

The objectives of the proposed system are to secure sensitive student data from being read and stolen by unauthorized parties. Using the Data Masking Technique also helps Majlis Perwakilan Pelajar (MPP) manage the student attendance verification record because the system automatically filters out an uncertified student in the list of student appreciation awards from accessing the system. MPP members may only register students with data provided by the chosen lecturer assigned to handle the event.

All the sensitive student data that has been collected will be masked by using the data masking technique. Only the system admin can see the details of the student. The process of changing or hiding certain data elements within a set of data so that the data structure remains the same while the contents are changed will be applied to the proposed system. This is done to protect sensitive information such as real names, addresses, identity card (IC) numbers, ID students, and so on.

Figure 6 is the main page that will pop-up which includes full information such as date, time, and location of the event. Students need key-in their ID number to verify join the event.

Continue from Figure 6 its will go to interface Figure 7. This form to verify attendance is shown in Figure 7. Students need to enter their IC, CGPA, Email, and Telephone Number. Students are required to answer the three questions details industrial training, attend the event, and transportation options. The last requirement, press the "Submit" button to verify the attendance of the KUPTM Student Appreciation Award.

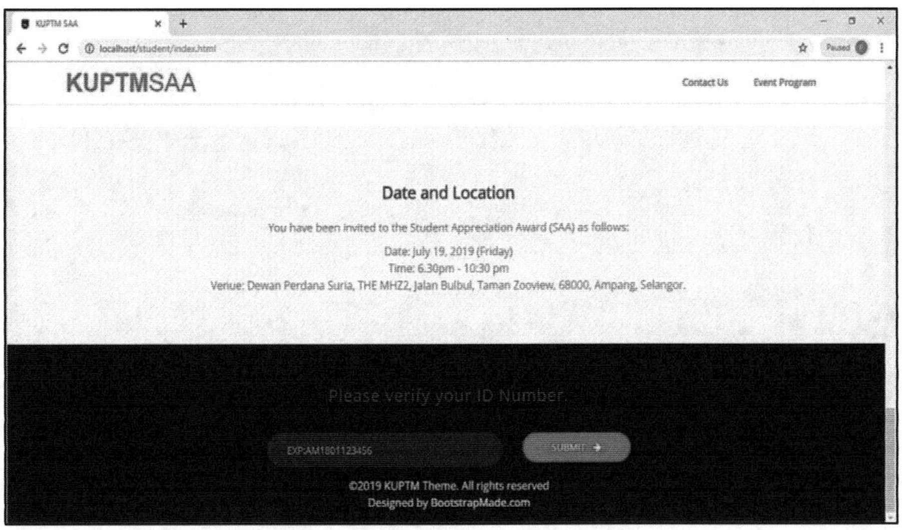

Figure 6. Main page.

As shown in Figure 8, Majlis Perwakilan Pelajar (MPP) manage the student attendance verification record student appreciation award from accessing the system. MPP members may only register students with data provided. MPP will enter Student ID, Full Name of student are qualified and Program Code.

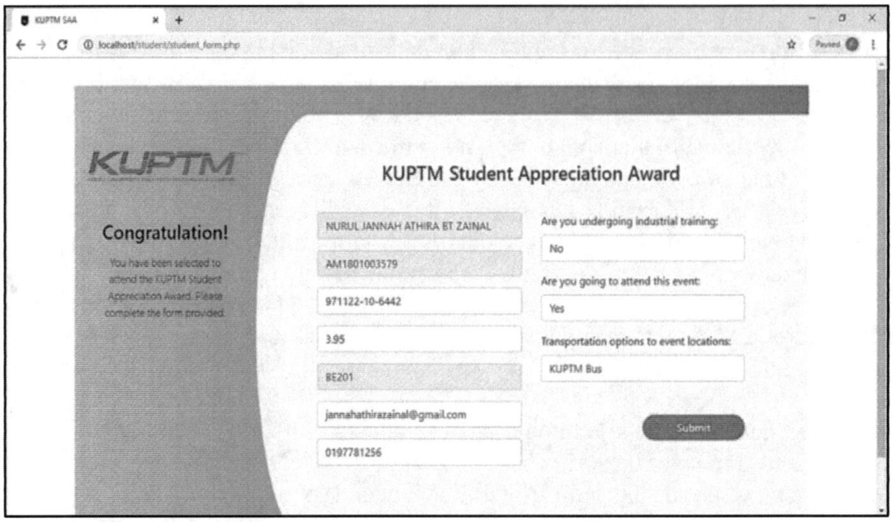

Figure 7. Verify and update information.

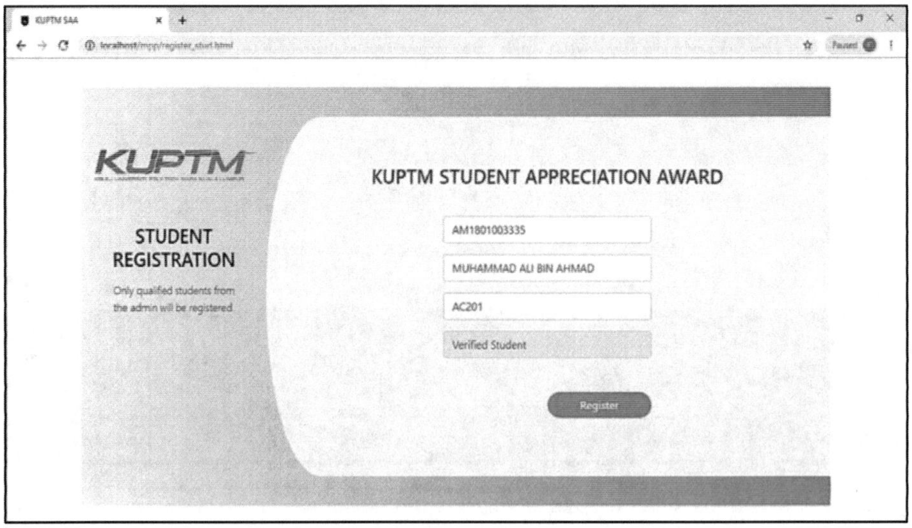

Figure 8. Register form by admin.

5. Conclusion and Recommendation

Attendance Verification System for the KUPTM Student Appreciation Award with the Secure Database by Using Data Masking Technique do has the limitation and other elements that need to be enhanced and implemented to gain and have the perfect system project. The system would be able to reduce the workload of the KUPTM Student Appreciation Award committee and hasten the process of managing the attendance verification for the event. By using this system, integrate the use of the data masking technique towards the attendance verification system is successful. Based on the result, data masking and hashing techniques can ensure the confidentiality of student data from being viewed by unauthorized parties. The features and elements that will be improving in the future might be as stated below.

The Student Appreciation Award system automatically imported the database of a student who nominated to attend the Student Appreciation Award by the student's CGPA. Also, the system can filter all students by the previous semester so easy to admin handle the KUPTM Student Appreciation Award through Campus Management System (CMS) and MPP System.

Account verification is essential whenever a new user who is MPP who assigned as Person in Charge (PIC) registered on the admin system. Whenever a new user registers a verification code with the link is sent on the email, and if the user is genuine, MPP clicks on a link with the code sent on their email, and then they verified themselves for the website. Email is more effective, fraud prevention is improved and the ability to protect sender reputation increases.

Compatible mobile view for KUPTM Student Appreciation Award System. Students prefer a simple and user-friendly system. In this way, the attendance confirmation record is quickly available to the MPP. MPP will also do their job efficiently.

References

[1] Ajayi O O, and Adebiyi T O 2014 *IOSR Journal of Engineering* **4(2)** 13-21
[2] Ravikumar G K, Justus R B, Manjunath T N, Ravindra S H and Archana R A 2011 *International Journal of Engineering Science and Technology (IJEST)* **3(6)** 5150-5159
[3] Santos R J, Bernardino J and Vieira M 2011 *IDEAS '11: Proceedings of the 15th Symposium on International Database Engineering & Applications* 61–69
[4] Manjunath T N, Ravindra S H and Ravikumar G K 2010 *International Journal of Computer Science and Information Technologies* **2(1)** 477-485
[5] Domingo-Ferrer J and Mateo-Sanz J M 2002 *IEEE Transactions on Knowledge and Data Engineering* **14(1)** 189- 201
[6] Adam N R and Wortmann J C 1989 *ACM Computing Surveys* **21(4)** 515 –556
[7] Goyal C 2015 *International Research Journal of Management Science & Technology* **6(5)** 221-229

Adhesion Strength of Hydroxyapatite Coating on Titanium Alloy (Ti-6Al-4V ELI) for Biomedical Application

Gunawarman[1*], Jon Affi[1], Agus Sutanto[1], Dian Mustika Putri[1], Dian Juliadmi[2], Nuzul Ficky Nuswantoro[2], Hidayatul Fajri[2], Djong Hon Tjong[3], Menkher Manjas[2]

[1] Mechanical Engineering Department., Andalas University, Padang, Indonesia
[2] Biomedical Science, Medicine Faculty, Andalas University, Padang, Indonesia
[3] Biology Department., Mathematics and Science Faculty, Andalas University, Padang

*Corresponding author: gunawarman@eng.unand.ac.id

Abstract. Implantation using titanium Ti-6Al-4V ELI still has disadvantages, in particular, low osseointegration cause of less bioactive of the material. Hydroxyapatite was then applied as a coating material to increase the bioactivity and the osseonintegration as well. The Ti-6AL-4V ELI samples were coated with the hydroxyapatite using the electrophoretic deposition (EPD) method, with voltage variations of 5, 8 and 11 volts for 5 minutes and variations in deposition time for 5,8 and 11 minutes with voltage of 8 volt. Surface properties (coarseness, coating mass, thickness, and surface coverage) were then determined by using related equipments. Adhesion strength the HA layer was tested a cross-cut test method. The results showed that the EPD could be used properly to coat the Ti-6Al-4V ELI material with hydroxyapatite. Voltage and deposition time influence coating layer properties that is related to adhesion bonding of the layer on the metal surface. Applying the voltage of 8 volt for 5 minutes resulted in optimal surface properties with coating roughness of hydroxyapatite layer (Ra) as much as 0.88. The removed area after the cross-cut tape test is 2.25%, which means adhesion strength between the coating layer and the subtrate is high enough. A relatively high hydroxyapatite mass (0.3 mg), is covering the material surface with a surface coverage of 82.1%, and the average coating thickness is 73.3 μm. Such kind of coating layer characteristics could improve bioactivity, biocompatibility and thus enhancing osseonintegration capacity of prosthetic medical implants.

Keywords: hydroxyapatite, coating, titanium, bioactivity, biocompatibility

1. Introduction

Titanium $\alpha + \beta$ type Ti-6Al-4V ELI, has been used in orthopaedic applications because it has biocompatibility and relatively low modulus of elasticity [1] which can reduce the effects of stress shielding when implantation [2]. The titanium also has excellent physical and mechanical properties such as tensile and fatigue strength, low density and good corrosion resistance; those are the advantages of titanium as an implant material [3,4]. Moreover, implantation of titanium material results in a low inflammatory, hypersensitivity and allergic response when in contact with the biological environment [5].

However, titanium is bioinert or less bioactive biomaterial that produces less integration between the material with bone tissue, and it can trigger implantation failure during application in human body [6]. Titanium alloys can also corrode and even cause toxic effects on the body. Therefore, the degradation of material particles must be hindered by conducting surface modification by utilizing hydroxyapatite

(Ca10(PO4)6(OH)2) which has a chemical structure similar to the bone components [7] and bioactive [8,9]. Such coating can initiate osseointegration and involved in the molecular biological mechanism of the body and accelerating the healing process of bone damage during implantation [10]. However, coating with hydroxyapatite still shows the disadvantage that there is a particle releasing due to the low bonding of coating layer to titanium substrate [5], [11]. Bond strength of coating includes adhesion strength (coating with substrate) and cohesion strength (between the hydroxyapatite particles) [12]. Furthermore, delamination and disintegration of the coatings can disrupt fixation of material to bone tissue which also trigger implantation failure [13]. Thus, the adhesion strength between hydroxyapatite layer and substrate should be high enough. The way to improve the layer adhesion can be done by the heating process. However, high temperatures result in phase transformations tends to reduce the physical properties of the coating. Hydroxyapatite coating can be carried out using the electrophoretic deposition (EPD) as an alternative method due to the efficiency of the coating process, which utilizes electric current and can be applied at low temperatures [14]. The characteristics of the coating can be controlled by treating the voltage or deposition time of the EPD coating process and followed by sintering to improve the coating adhesion. This study was conducted to determine the effect of hydroxyapatite coating by EPD method to the adhesion strength of coating layer to the Ti-6Al-4V ELI surface.

2. Materials and Methods

The Ti-6Al-4V ELI material with a size of 10mm x 4mm was used as a substrate of the study. The material was then heat-treated to reduce residual stress effect after sample preparation and also to improve the mechanical properties of the material. Solution treatment was conducted at temperature 950°C for 1 hour with a heating rate of 10°C/minute using a vacuum furnace (High-Temperature Vacuum Tube Furnace GSL-1100). The rough surface of the material was reduced by grinding process using sandpaper with mesh sizes of #120, #500, and #800. The sample was then cleaned by soaking it using ethanol solution for 15 minutes and acetone for 15 minutes in an ultrasonic bath. The sample was then immersed in the nitric acid solution for 30 minutes and NaOH solution for 1 hour and finally dried at 50°C for about 5 minutes.

Hydroxyapatite (Sigma Aldrich) with a particle size of 10 μm was dissolved in ethanol (4gram/100mL). The suspension was homogenized with a stirring hot plate for 1 hour. Subsequenly, nitric acid was added to reduce pH up to 4. The electrophoretic deposition was applied with the materials acts as the cathode and carbon as the anode with a distance of 1cm. The coating process was carried out by utilizing a DC digital power supply with voltage variations of 5, 8, and 11 volts for 5 minutes, and also variations in deposition time for 5,8, and 11 minutes with voltage of 8 volt. The coated material was dried for 24 hours at room temperature and then heated at 800°C for 10 minutes with a heating rate of 10°C/ minutes using a vacuum furnace. Microstructure observations were conducted using an optical microscope (Olympus LG-PS2). Thickness measurements were made using the Sanfix thickness gauge series GF-280, and surface roughness measurements were carried out with a surface roughness tester. The strength of the coating adhesion on the surface of the material was determined by a cross-cut tape test method based on ASTM D3359 standard.

3. Results and Discussion

Hydroxyapatite coating layer, in general, has been formed on the surface of all titanium Ti-6Al-4V ELI samples with variation of voltage and deposition time as shown in Figure. 1, while the weight of hydroxyapatite coating layer on material surfaces were indicated in Figure 2. However, there is differences in the coating morphology as a function of the voltage or deposition time. In the voltage variations, it can be seen that the treatment with 8 volt shows the best appearance of coating layer, although in some parts there is an agglomeration of hydroxyapatite particles. The treatment with a lower voltage (5 volt) shows that there are some black spots of the uncoated surface. While at a high voltage (11 volt) shows a fairly thick coating layer.

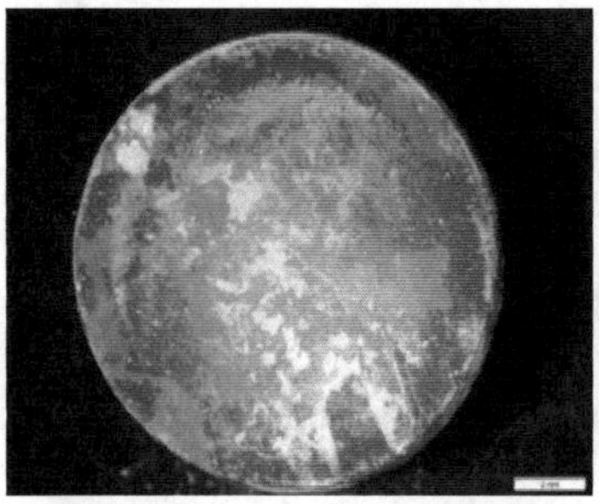

Figure 1 (a). 5 volt for 5 minutes

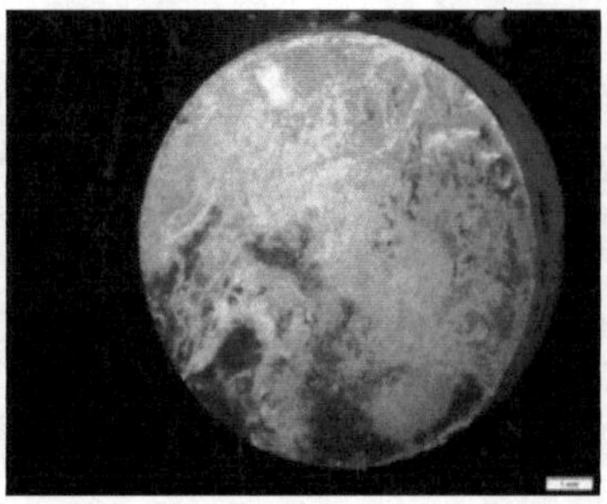

Figure 1 (b). 8 volt for 5 minutes

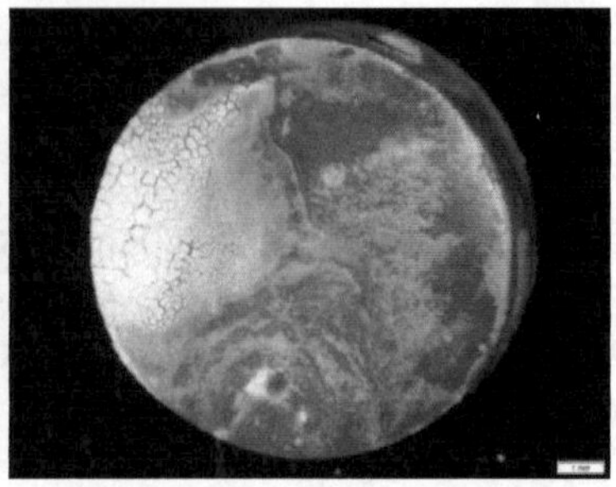

Figure 1 (c). 11 volt for 5 minutes

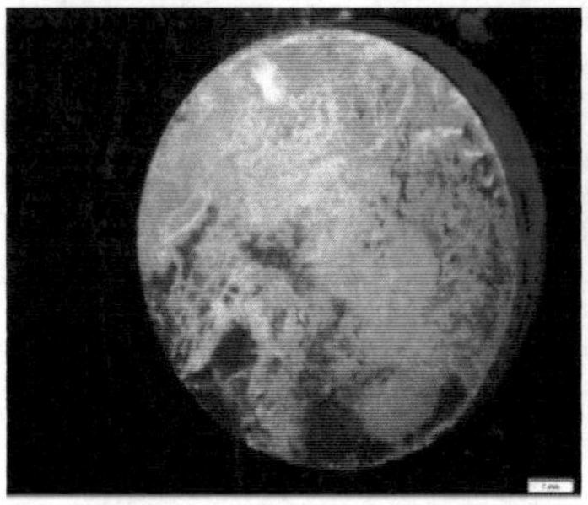

Figure 1 (d). 5 minutes at the voltage of 8V

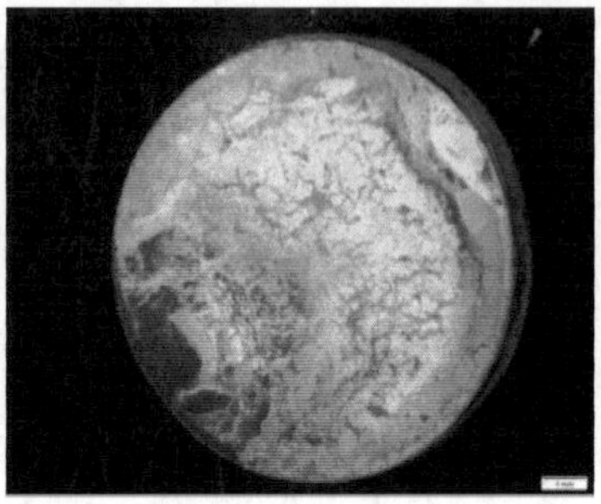

Figure 1 (e). 8 minutes at the voltage of 8V

Figure 1 (f). 11 minutes at the voltage of 8V

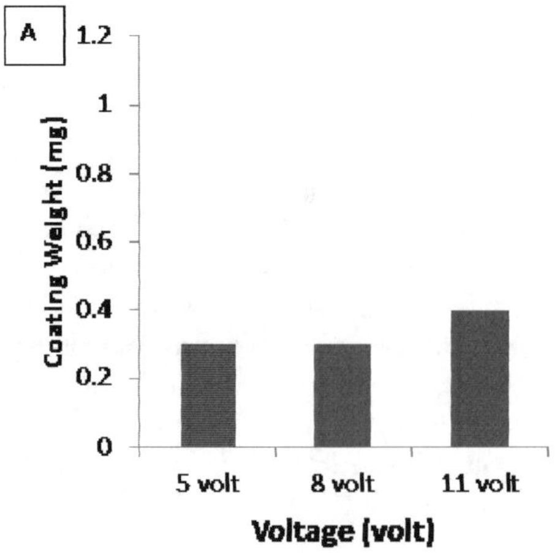

Figure 2 (a). Weight of hydroxyapatite coating layer by voltage variations (5,8, and 11 volts) for 5 minutes

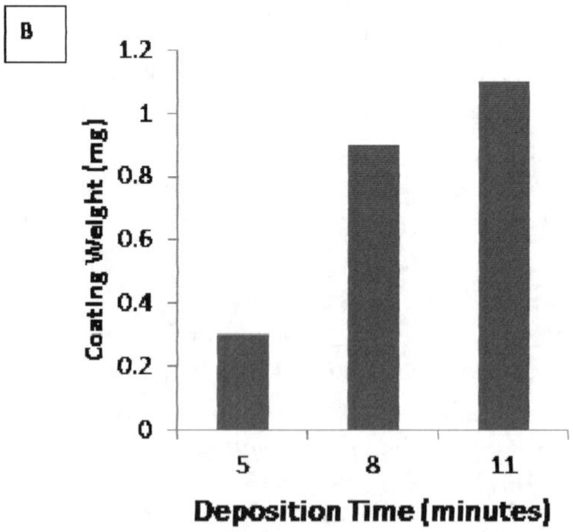

Figure 2 (b). Weight of hydroxyapatite coating layer by variations in deposition time (5,8, and 11 minutes) at the voltage of 8 volt

Increasing deposition of particles on the material surface appears to be influenced by the increase of deposition time. Optimum particle deposition can be obtained with a relative lower deposition time treatment. The results indicate that deposition for 5 minutes is sufficient to produce a good enough coating layer. However, cracks appear in some parts of the layer which tends to be thicker after sintering process. The cracked coating layer can be caused during heating process is due to differences in thermal expansion coefficients and large volume reduction between titanium material and hydroxyapatite [15].

Coating cracks can be avoided by coatings of bilayers using nano/micro-sized hydroxyapatite particles with sintering temperature of 800°C [16]. Heating process with this temperature does not cause phase transformation because the phase transition α to β occurs between 883°C to 960°C [17].

The amount of deposition particles on the surface is influenced by the voltage and deposition time. The increase of the voltage tends to increase of slight deposited particles on the material surface in Figure 2a. The increase in deposition time increases significantly the weight of coating layer in Figure 2b. However, if the applied voltage is too high in a short time, it can result in agglomeration of particles revealing un-uniform particle distribution on the material surface [18], [19]. Therefore, the coating process carried out with low energy in a relatively short time produces a better particle deposition as can be seen by EPD treatment at the voltage of 8 volt for 5 minutes.

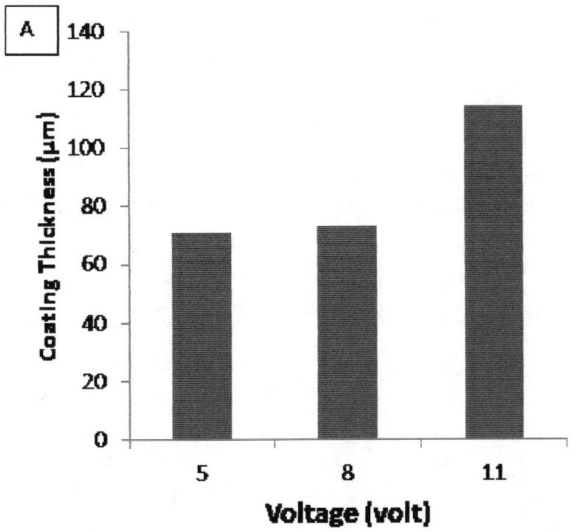

Figure 3 (a). Thickness of hydroxyapatite coating by voltage variations (5,8, and 11 volts) for 5 minutes

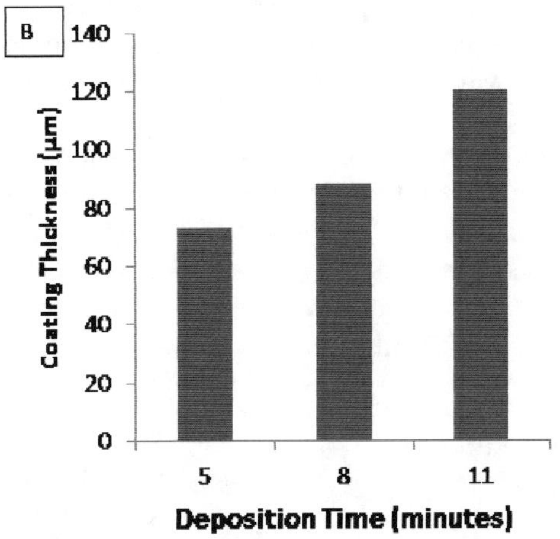

Figure 3 (b). Thickness of hydroxyapatite coating by variations in deposition time (5,8, and 11 minutes) by voltage 8V

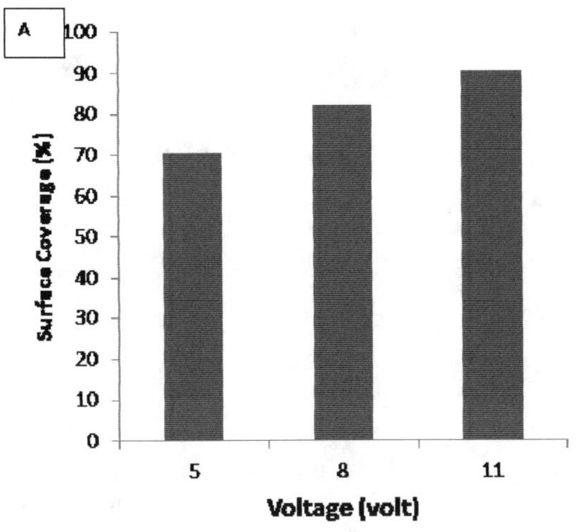

Figure 4 (a). Surface coverage of hydroxyapatite coating by voltage variations (5,8, and 11 volts) for 5 minutes

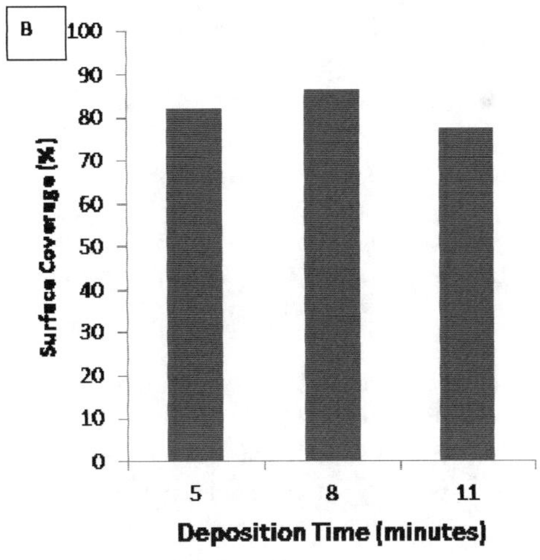

Figure 4 (b). Surface coverage of hydroxyapatite coating by variations in deposition time (5,8, and 11 minutes) by voltage 8 volt

The voltage and deposition time treatments give an effect to the thickness of hydroxyapatite coating on the material surfaces in Figure 3. The result revealed that the highest thickness by 11V in Figure 3a have morphology with cracks in Figure 1. In contrast, the treatment with 8V for 5 minutes showed a preferred coating layer (thickness 73.3 μm). This value is in the range of thickness requirements standard for biomedical applications, that is around 50-100 μm [20- 22]. The layer can undergo resorption faster with thicknesses below 75 μm but will be delamination with thicknesses beyond 200 μm [23], [24]. Hydroxyapatite coating with a thickness of 50 μm obtained by the plasma sprayed method has an amorphous structure, but it is easily resorbed in physiological environments [25]. The coating obtained in this study visually has a dense structure because it might be affected by sintering after coating. The dense structure will prevent the dissolution of the layer and avoid corrosion of material constituents to the biological environment [26].

Surface covered by hydroxyapatite coating is more affected by voltage rather than deposition time in Figure 4a. The optimum coating properties are not found on high voltage or deposition time. Despite having widest area (surface coverage of 90.5%), deposition with a voltage of 11 volts for 5 minutes produces a thick layer and has a crack in Figure 1. For long deposition time (11 minutes), the hydroxyapatite coating has a too high thickness (120.1 μm) but a low surface area coverage (77.4%). Some accumulation of particles is found on certain parts of the material surface (Fig. 1). However, using low voltage, material covered with relative wide in Figure 4 and thin hydroxyapatite coating (Figure 3a) thus result in even distribution on the material surface in Figure 1. Therefore, coating with a voltage of 8 volts for 5 minutes has a better hydroxyapatite coating with surface area coverage of 82.1% and a thickness of 73.3 μm without cracks and agglomeration.

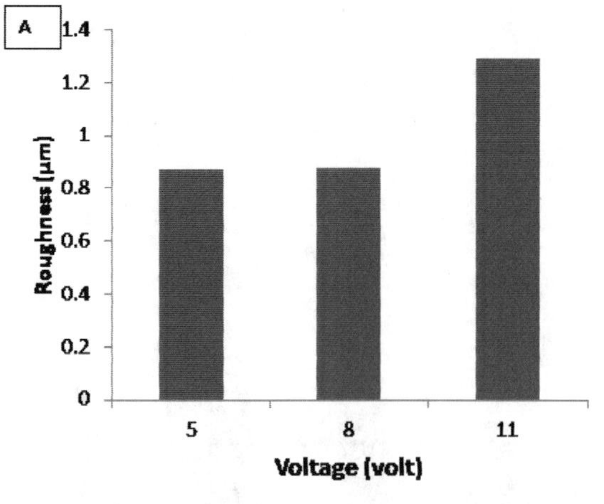

Figure 5 (a). Roughness of hydroxyapatite coating by voltage variations (5,8, and 11 volts) for 5 minutes

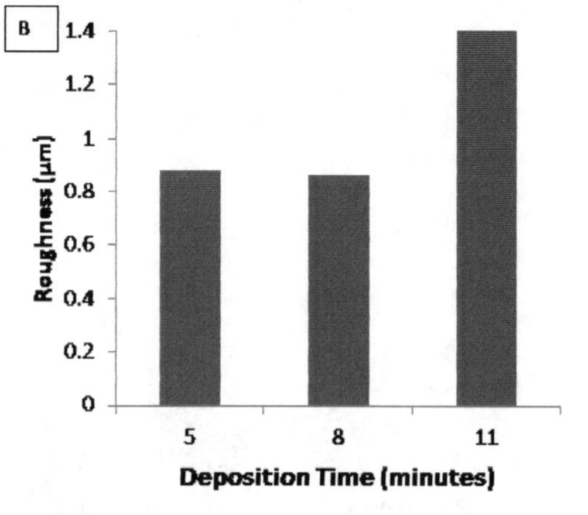

Figure 5 (b). Roughness of hydroxyapatite coating by variations in deposition time (5,8, and 11 minutes) by voltage 8 volts

The distribution of hydroxyapatite particles on the surface of the material influences the roughness (Ra) of material surfaces. The higher voltage or deposition time tends to produce the higher roughness value in Figure 5. The Ra of surface for voltage of 5 and 8V is almost the same, 0.86 μm and 0.88 μm, respectively. High roughness allows an increase in the area that will be in contact with the biological environment such as proteins, growth factors or interactions with cells around the material [27]. The rough surface of the hydroxyapatite layer has a direct impact on cellular responses by increasing the adhesion and proliferation of cells involved in the process of repairing bone damage [28]. Thus, roughness by hydroxyapatite has good biocompatibility and triggers osteoblast cell activity on the surface of the material [29]. However, Holthaus et al [29] found hydroxyapatite coatings do not increase osteoblast cell proliferation within 7 days of incubation. Different levels of roughness may have different effects on biological responses. Osteoblasts have a better preference for the surface of the rougher titanium material with a value of Ra = 0.60 μm to Ra = 7.50 μm [30,31]. The value of Ra = 0.86 μm to Ra = 1.61 in this research shows that the coating produced by the EPD method is predicted to have much better biocompatibility and bioactivity. However, an *in vivo* test using animal trial is necessary to prove this predictions. This is our next-step works, and it will be reported later.

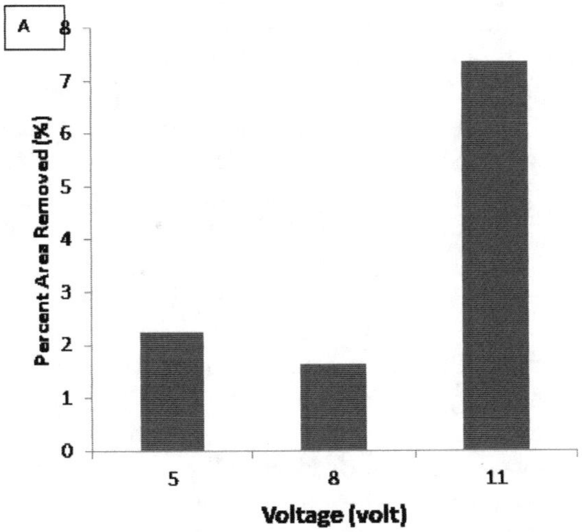

Figure 6 (a). Percentage of removed area by cross-cut tape test over hydroxyapatite coating by voltage variations (5,8, and 11 volt) for 5 minutes

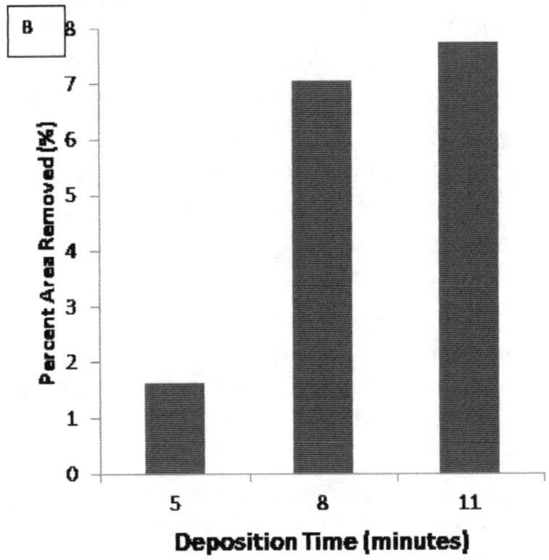

Figure 6 (b). Percentage of removed area by cross-cut tape test over hydroxyapatite coating by variations in deposition time (5,8, and 11 minutes) by voltage 8 volt

In the other hand, too high surface roughness can trigger dissolution of coating layers, but the optimum roughness value of hydroxyapatite coating is still unknown [32]. The positive impact of roughness can be lost when the value is very high [12]. In addition, the roughness of the hydroxyapatite coating can decrease due to an increase in sintering temperature due to decreasing in pore size and densification [17]. Thus, optimum roughness was needed for biomedical application encouraging biocompatibility of materials.

The adhesion strength of the coating layer to the material is important characteristic for biomedical application. Applying high voltage or long deposition time seems resulting in the formation of coating layer with poor adhesion to the material surface in Figure 6. Applying high energy produces irregular particle deposition, and long-term deposition makes the charged particles become saturated, so it cannot form a good bond on the material [33- 35]. At low voltages with short deposition times in Figure 6, the hydroxyapatite coating tends to bind strongly to the material surface. The coating with the voltage of 8V for 5 minutes shows the lowest percentage of exfoliating area, meaning that the hydroxyapatite layer has a good bond on the material surface. Hydroxyapatite coating which is firmly attached to the material surface, is needed to maintain the structural and functional role of the coating during implantation. Structurally, the hydroxyapatite layer acts as a barrier against the release of toxic constituents from the material into the biological environment and prevents corrosion when in contact with body fluids [36]. The functional role of the hydroxyapatite coating is involved in the body metabolism which initiates bone tissue growth. The hydroxyapatite coating forms a direct bond with bone tissue without fibrous tissue intervention in interfacial tissue and implant material. Encapsulation of fibrous tissue that isolates implant material can trigger implant looseness [37], [38].

Applying high voltages (11V) and long deposition times (11 minutes) produces high thickness, roughness and surface area coverage, but do not result in better coating adhesion as compared to the low voltage and time. Hydroxyapatite coating with impure phase acts as source of crack initiation and transversal cracks over the coating thickness. Iy will be the site of triggering the dissolution of the layer.

The cracks can also reduce the strength of the coating adhesion after immersion in simulated body fluid solution [39]. By using the same material, coating using plasma-sprayed by Oh et al [40] produced coating adhesion strength of 4.5-6.8 MPa. Rojaee et al [41] obtained the adhesion strength of the hydroxyapatite coating on magnesium alloys is 4.2±0.3 MPa using sol-gel method.

A stable hydroxyapatite coating is needed to avoid rapid and dangerous dissolution to the body [23]. Debris from coating exfoliation is a risk problem due to it can trigger implantation failure [42]. The adhesion of hydroxyapatite coating to the material surface is also related to the sintering after the coating process [43- 45]. Sintering at high temperatures increases adhesion strength due to the increase of particle diffusion in interfacial parts and triggers diffusion of the bond between the phosphorus element and the material. In addition, adhesion of the coating also can be affected by material preparation before the coating (pre-treatment). Soaking the material with NaOH solution without heating before the coating can increase the coating adhesion and reduce the delamination associated with the increase of osseointegration [46].

4. Conclusions

The electrophoretic deposition method was applied to produce a hydroxyapatite coating on the surface of the Ti-6Al-4V ELI material in order to enhance bioactivity of the alloy, and thus induces osseonintegration. Differences in voltage and deposition time influence the surface characteristics including layer thickness, surface coverage, layer roughness and the adhesion strength of the coating layer. The EPD process with a voltage of 8V for 5 minutes resulted in the best peformances with a coating roughness of hydroxyapatite layer (Ra) as much as 0.88. The removed area after the cross-cut tape test is 2.25%, which means adhesion strength between the coating and the materials is high enough. A relatively high hydroxyapatite mass (0.3 mg), is covering the material surface with a surface coverage of 82.1%, and the average coating thickness is 73.3 μm. The important things is almost no cracks are found in the surface layer. The thickness coating layer meets to implants requreiment between 50-100 μm. Such kind of coating layer characteristics are strongly believed to increase bioactivity, biocompatibility and thus enhancing osseonintegration capacity of prosthetic medical implants. However, an animal trial is necessary to carry out for proving this prediction. This is our next-step works, and it will be reported later.

Acknowledgements

The corresponding author thanks the Directorate General Higher Education, Ministry of Education and Culture, the Republic of Indonesia for the Basic Research Grant under contract no. T/5/UN.16.17/PT.01.03/PT-MM/2019 and Mechanical Engineering Dept. of Universitas Andalas for publication grant support under contract no. 057/UN.16.09.D/PL/2020.

References

[1] Surmeneva M.A, Vladescu A, Surmenev R. A, Pantilimon C. M, Braic M and Cotrut C.M 2016 *RSC Advances* **6**, 87665–87674.
[2] Hsieh, M F, Perng L H and Chin T S 2002 *Mater chemical and physic* **74**, 245-50.
[3] Bauer S, Schmuki P, Mark K v d and Park J 2012 *Prog Mater Sci* **58**, 261–326.
[4] Oldani C and Dominguez A 2012 *Recent Advances in Arthroplasty*
[5] Mohseni E, Zalnezhad E and Bushroa A R 2014 *International Journal of Adhesion and Adhesives* **48**. 238–257.
[6] Fujibayashi S, Nakamura T, Nishiguchi S, Tamura J, Uchida M, Kim H M and Kokubo T 2001 *Journal of Biomedical Materials Research*, **56**, 562–570.
[7] Kattimani V S, Kondaka S and Lingamaneni K P 2016 *Bone and Tissue Regeneration Insights* **7**, 9-19
[8] Zahrani E M, Fathi M H and Alfantazi A 2011 *Metallurgical and Materials Transactions A* **42**(11):3291-3309
[9] Gadow R, Kilinger A and Stiegler N 2010 *Surface and Coatings Technology* **205**, 1157–1164.

[10] Ripamonti U, Roden L C and Renton L F 2012 *Biomaterials* **33**, 3813-3823.
[11] Sun L, Berndt C C, Gross K A and Kucuk A 2001 *J Biomed Mater Res* **58**, 570 –592.
[12] Yang Y C and Yang C Y 2013 *Ceramics International* **39**, 6509–6516.
[13] Gineste L, Gineste M, Ranz X, Ellefterion A, Guilhem A, Rouquet N and Frayssinet P 1999 *Journal of Biomedical Materials Research* **48**, 224–234.
[14] Gunawarman, Nuswantoro N F, Juliadmi D, Fajri H, Budiman A, Tjong D H and Manjas M 2019. IOP Conf. Series: *Materials Science and Engineering* **602** 012071, pp 1-11.
[15] Zhang Z, Dunn M F, Xiao T D, Tomsia A P and Saiz E 2001 MRS Proceedings, *Cambridge Univ Press.*
[16] Juliadmi D, Harlendri, Tjong D H, Manjas M and Gunawarman 2019 IOP Conf. Series *Mater Sci and Eng* **547**, 1-7.
[17] Jafari S, Taheri M M and Idris J 2012 *Adv Mater Res* **341-342**, 48-52.
[18] Kazek-Kęsik A, Łastówka D, Donesz-Sikorska A, Dercz G and Simka W 2015 *Journal of the Electrochemical Society* **162**, 589-597.
[19] Nuswantoro N F, Budiman I, Septiawarman A, Tjong D H, Manjas M and Gunawarman 2019 IOP Conf. Series: *Materials Science and Engineering* **547** 012004, pp.1-11.
[20] Herrera A, Mateo J, Gil-Albarova J, Lobo-Escolar A, Ibarz E, Gabarre S, Más Y and Gracia L 2015 *Biomed Research International* 386461, 1-14.
[21] Geesink R G, Groot K de and Klein C P 1988 *The Journal of Bone and Joint Surgery* Vol. **70**-B, No. 1 17-23.
[22] Juliadmi D, Fauzi V R, Gunawarman and Hadi N 2017 *International Journal on Advanced Science Engineering and Information Technology* **7**(6):2152
[23] Hung K Y, Lo S C, Shih C S, Yang Y C, Feng H P and Lin Y C 2013 *Surface and Coatings Technology* **231**, 337–345.
[24] Yang C Y, Wang B C, Lee T M, Chang E and Chang G L 1997 *Journal of Biomedical Materials Research* **36**, 39–48.
[25] Svehla M, Morberg P, Bruce W, Zicat B and Walsh W R 2002 *The Journal of Arthroplasty* **17**, 304-312.
[26] Ohtsu N, Takahara T, Hirano M and Arai H 2014 *Surface and Coatings Technology* **239**, 185–190.
[27] Anselme K 2000 *Biomaterials* **21**, 667–681.
[28] Deligianni D D, Katsala N D, Koutsoukos P G and Missirlis Y F 2000 *Biomaterials* **22**, 87–96.
[29] Holthaus M G, Treccani L and Rezwan K 2013 *Journal of Biomaterial Application*, **27**(7):791-800
[30] Kunzler T P, Drobek T, Schuler M and Spencer N D 2007 *Biomaterials* **28**, 2175–2182.
[31] Schneider G B, Perinpanayagam H, Clegg M, Zaharias R, Seabold D, Keller J and Stanford C 2003 *J Dent Res* **82**, 372–376.
[32] Dudek A and Adamczyk L 2013 *Optica Applicata* **43**, 143-151.
[33] Maleki-Ghaleh H, Khalili V, Khalil-Allafi J and Javidi M 2012 *Surface and Coatings Technology* **208**, 57–63.
[34] Hussain R, Tabassum S, Gilani M A, Ahmed E, Sharif A, Manzoor F, Shah A T, Asif A, Sharif F, Iqbal F and Siddiqi S A 2016 *Applied Surface Science* **364**, 117-123.
[35] Iqbal N, Nazir R, Asif A, Chaudhry A A, Akram M, Fan G Y, Akram A, Amin R, Park S H and Hussain R 2012 *Current Applied Physics* **12**, 755–759.
[36] Cavalli M, Gnappi G, Montenero A, Bersani D, Lottici P P, Kaciulis S, Mattogno G and Fini M 2001 *Journal of Materials Science*. **36**, 3253-3260.
[37] Kokubo T and H Takadama 2006 *Biomaterials* **27**, 2907–2915.
[38] Suchanek K, Bartkowiak A, Gdowik A, Perzanowski M, Kąc S, Szaraniec B, Suchanek M and Marszałek M 2015 *Materials Science and Engineering: C* **51**, 57–63.
[39] Zhang S, Wang Y S, Zeng X T, Khor K A, Weng W and Sun D E 2008 *Thin Solid Films* **516**, 5162–5167.

[40] Oh I H, Nomura N, Chiba A, Murayama Y, Masahashi N, Lee B T and Hanada S 2005 *Journal of Materials Science: Materials in Medicine* volume **16**, pages 635–640
[41] Rojaee R, Fathi M H and Raeissi K 2013 *Materials Science and Engineering: C* **33**, 3817–3825.
[42] Baltag I, Watanabe K, Kusakari H, Taguchi N, Miyakawa O, Kobayashi M and Ito N 2000 Journal of Biomedical Materials Research, **53** (1), 76-85.
[43] Mhaede M, Ahmed A, Wollman M and Wagner L 2015 *Materials Science and Engineeing: C* **50**, 24–30.
[44] Vechietti F A, Muniz N O, Rezwan K, Schumacher M, Gelinsky M and Loureiro dos Santos L A 2018 *Material Research Express*, **6** (2)
[45] Filiaggi M J, Pilliar R M and Coombs N A 1993 Journal of Biomedical Materials Research **27**, 191–198.
[46] Lakstein D, Kopelovitch W, Barkay Z, Bahaa M, Hendel D and Eliaz N 2009 *Acta Biomaterialia* **5**, 2258–69.

The Effect of Wing Spar Cross-Sectional Profile Variation on the Unmanned Aerial Vehicle (UAV) Natural Frequency

Lovely Son[1], Arif Rahman[1] and Mulyadi Bur[1]
[1] Structural Dynamic Laboratory, Mechanical Engineering Department, Andalas University, Padang, West Sumatera, Indonesia

*Corresponding e-mail: lovelyson@eng.unand.ac.id

Abstract. Natural frequency is a critical dynamic parameter in designing the unmanned aerial vehicle (UAV) wing structure. The low-value natural frequency causes the aircraft wing to deflect easily while flying. Furthermore, the low-value natural frequency can causes resonance when this natural frequency is closest to the wind frequency. An increase in the UAV natural frequency can be conducted by increasing the wing structural stiffness. However, increased stiffness by enlarging dimensions is often followed by an increase in mass that is avoided in the design of aircraft structures. One technique that can be used to increase UAV natural frequency is by increasing the cross-sectional moment of inertia of the wing spar. In this study, an analysis of the effect of the wing spar cross-sectional inertia variations on the UAV natural frequency is conducted. Several wings spar cross-sectional profiles with different cross-sectional inertia are evaluated, such as U profile, circular, T profile, square, L profile, I profile, and Z profile. The results of the investigation showed that the wing spar using T profile is better than other profiles in increasing the wing structure natural frequency. This profile could increase the 1st elastic natural frequency by 22.9%, with only a 3.4 % increase in mass.

Keywords: wing spar, UAV, natural frequency, aircraft structure

1. Introduction

Unmanned Aerial Vehicle (UAV) is a pilotless aircraft that is controlled using a remote control or computer [1]. UAV can come in a variety of different types, that depends on operational purposes. Fixed-wing is a kind of UAV that is specifically designed to meet the needs of aerial mapping and monitoring [2]. AFRG 006 is a fixed-wing type UAV produced by Mechanical Engineering Department, Andalas University. This aircraft has participated in the national level competition on Indonesia Flying Robot Contest (KRTI). During the flying test, the aircraft wing experienced significant deformation and flapping. The wing deformation becomes large as the aircraft speed increases. This phenomenon indicates that the wing stiffness, as well as its natural frequency, is low so that it deforms easily and resonates at low excitation frequencies. The large vibration response occurred during resonance can damage the sensitive electronic components inside UAV body [3] and under certain conditions; it can lead to aircraft failure.

Figure 1a shows the initial design of AFRG 006 structural components. Two fuselage spars are positioned on the UAV wing. The wing structure uses the beam spar for increasing the structural stiffness. The spars are mounted transversely against the aircraft fuselage. During the flight test, a large deflection occurs on the UAV wing because the structural strength and stiffness is low. Several studies have been proposed by many researchers to increase the strength and stiffness of the wing structure. Ernnie et al. [4] proposed a technique by modifying the orientation of ply combinations in composite

IOP Conf. Series: Materials Science and Engineering 1062 (2021) 012032 doi:10.1088/1757-899X/1062/1/012032

material which contributed to increasing the strength and stiffness of UAV wing structure. A method for increasing the wing's torsional stiffness by optimizing the rib spacing was proposed by Tolga et al.[5]. Adam et al.[6] developed an evolutionary algorithm to obtain an optimal design of UAV wing structure utilizing three different functional functions. In this study, a method for increasing the first body bending frequency of UAV structure is conducted by modifying the wing spar cross-sectional profile.

Figure 1b shows the modified design of AFRG 006 structure for increasing the wing structural stiffness. An additional spar is added to the wing structure. The location of the additional spar is coinciding with the center gravity of the aircraft. The high stiffness wing structure can be obtained using more rigid spar material. However, an increase in spar rigidity is often followed by an increase in wing mass. In this research, increasing the wing stiffness is performed by varying the moment of inertia of the spar cross-sectional area. Several spar cross-sectional profiles are numerically simulated using commercially finite element software to find the optimum condition for the spar profile that is resulting in a high stiffness and low mass wing structure.

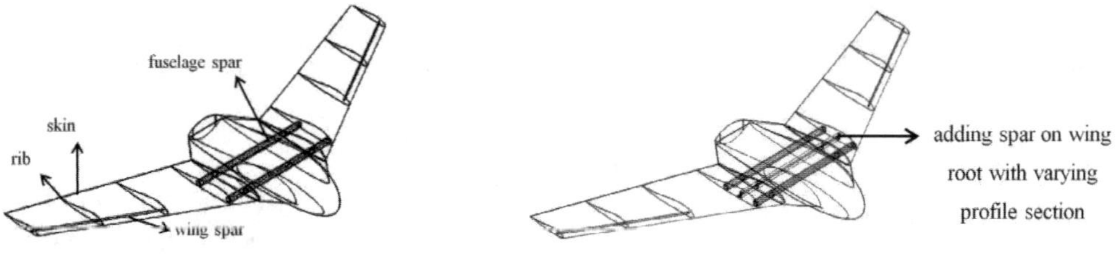

a. Initial design b. Modified design

Figure 1. UAV structure.

2. Theoretical Background

2.1. Finite Element Modelling of UAV

The finite element method (FEM) is used to analyze the structural dynamics of UAV model. The numerical simulation using FEM model is performed using commercially finite element software. The modal analysis technique is utilized to calculate the UAV natural frequencies and mode shapes. Calculation of the UAV natural frequencies and mode shapes is conducted using several wing spar cross-sectional profiles such as U profile, circular, the T profile, square, L profile, I profile, and Z profile. A free-free support condition is used in the simulation FE model. The UAV structural model was discretized using three-dimensional elements into 268490 nodes and 100976 elements. The normal mode analysis is conducted in the frequency range between 0 to 400 Hz. Figure 2 shows the finite element meshing of UAV structure.

Figure 2. Finite element meshing of UAV structure.

2.2. Simple Model of Aircraft Wing Vibration

The aircraft wing vibration can simply be modeled using the cantilever beam model, as shown in Figure 3. The beam vibration natural frequency can be calculated as:

$$f = \frac{1.758}{\pi \ell^2}\sqrt{\frac{EI}{\rho A}} \tag{1}$$

Equation (1) can be used to analyze the wing vibration at its first elastic vibration mode. Equation (1) shows that the wing natural frequency is proportional relating to the square root of the cross-sectional moment of inertia and inverse relating to the square root of the wing structure mass.

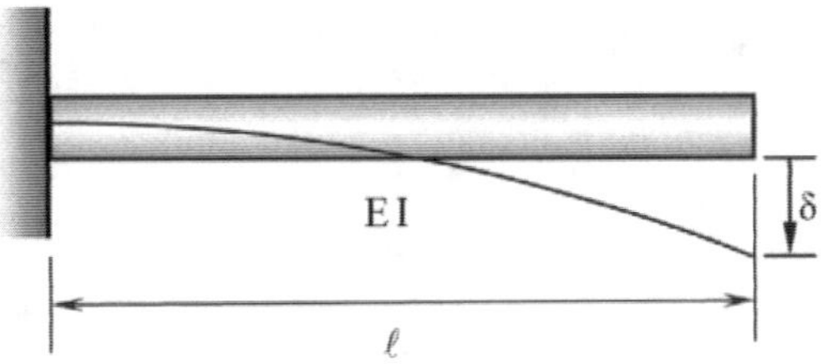

Figure 3. Wing vibration modelled as a cantilever beam.

3. Research Method

The research method is shown schematically in Figure 4. In the first step, AFRG 006 UAV structure produced by the Mechanical Engineering Department, Andalas University is used as the initial design. The commercially FEM software is utilized to calculate the modal parameters of the UAV model. The experimental modal analysis [7-8] is conducted to validate the simulation model. The simulation is performed until the error between the theoretical and the experimental modal parameters is small. The next step is the modal analysis of UAV structure by adding one spar coincides with the center gravity of the UAV body, as shown in Figure 1b. The analysis is conducted to evaluate the effect of wing spar cross-sectional profile variation on the natural frequencies of UAV structure. Several spar cross-sectional profiles are evaluated, such as U-shape, circular-shape, T-shape, rectangular shape, L-shape, I-shape, and Z-shape.

IOP Conf. Series: Materials Science and Engineering 1062 (2021) 012032 doi:10.1088/1757-899X/1062/1/012032

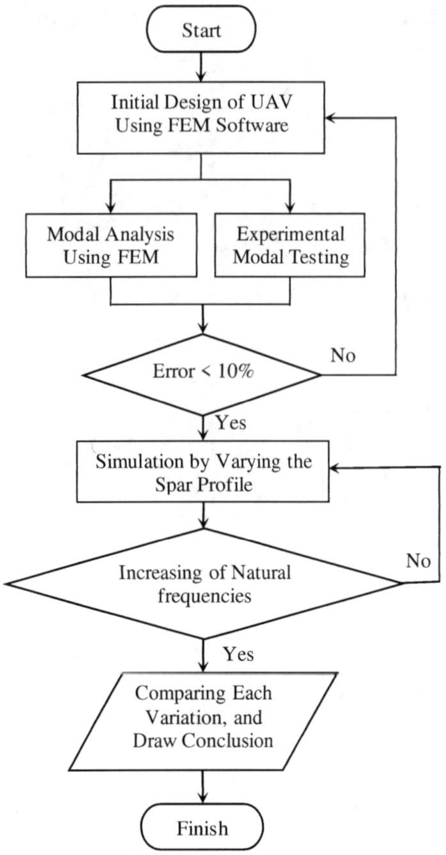

Figure 4. Research method.

UAV structure is consisting of several components i.e. spar, rib, skin, and fuselage. The material properties of UAV structure are depicted in Table 1 [9-11]. As shown in Table 1, the wing spar, rib, and skin are made of aluminum alloy, balsa wood, and GRFP composite, respectively. The elastic modulus of wing spar is larger than those of other components. The wing bending deflection due to the wind load is mainly affected by the spar stiffness [12].

Table 1. Material properties of UAV structure.

Structure	Material	Properties		
		ρ (kg/m^3)	E (GPa)	Poisson's Ratio
Spar	Aluminum alloy	2770	71	0.33
Rib	Balsa wood	120	3	0.38
Skin	Composite of GFRP and resin	2000	51	0.29
Fuselage	Composite of balsa, resin, and GFRP	1060	45	0.35

Figure 5 shows the experimental modal testing set-up of the UAV structure. A free support condition is realized by hanging the UAV body using a flexible rod. The impact hammer is used to apply an impulsive load to the UAV structure. The acceleration response is measured using a PCB Piezotronics accelerometer. The acceleration signal is collected using the National Instrument Data Acquisition System type NI USB-6221 and processed using Lab-View software.

The experiment is performed by applying the impulsive excitation load on the UAV fuselage. The acceleration responses of UAV are measured at some sensor locations. The sensors are located on the UAV fuselage and wing. The frequency response function (FRF) is calculated as the ratio between the acceleration response and the excitation force in the frequency domain. To obtain a good FRF signal, the impulsive excitation and the position of the sensor are applied on the wing stiffener (rib) for avoiding the wing local vibration response.

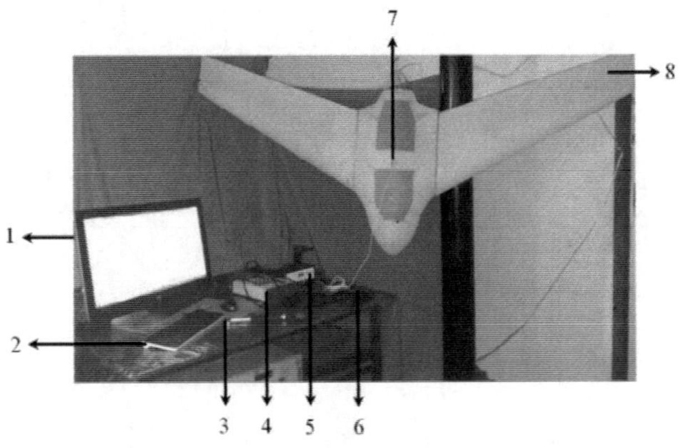

1. Personal computer (PC) for processing and displaying the data
2. Keyboard
3. Impact hammer
4. National instrument type NI USB-6221
5. Amplifier to amply the signal
6. BNC cable
7. Accelerometer
8. UAV wing

Figure 5. Experimental set-up.

4. Results and Discussion

Figure 6 shows the position of spar on the wing structure. The initial design consists of two spars located on the upper and lower position of the aircraft center of gravity. For the modified design, an additional spar with varying cross-sectional profiles is positioned to coincide with the aircraft center of gravity. The variation of the spar cross-sectional profile used in the research is depicted in Table 2.

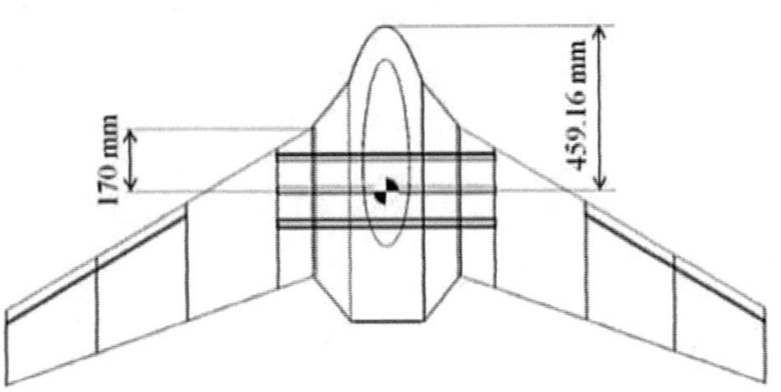

Figure 6. Spar position on wing structure.

Table 2. Spar profile.

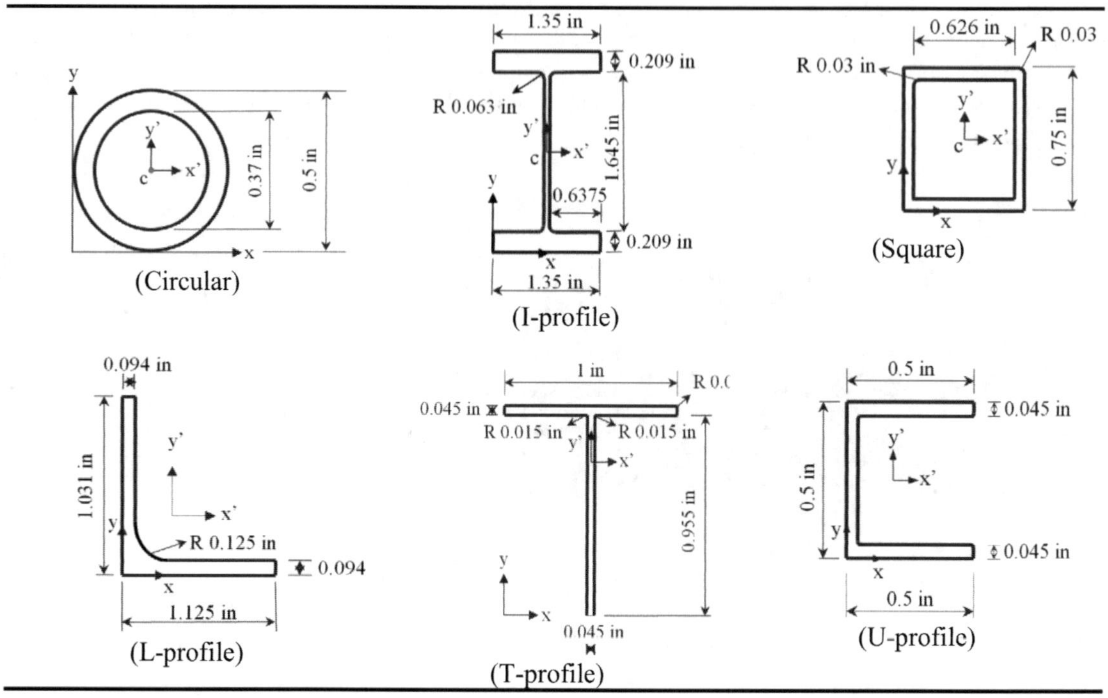

Figure 7 shows the experimental modal testing scheme used in the research. The UAV body is hanging on an elastic rod. An impulsive force from the impact hammer is applied at point B, and the acceleration response is measured at point A. The measured signals are conditioned using the conditioning amplifier. The National Instrument Acquisition System type NI USB-6221 is used to acquire the signal. The acquired signal is sent to the personal computer (PC) for post-processing. To analyze two lower mode shapes of the UAV structure, the frequency range for the experiment is selected between 0 to 400 Hz.. The reason for investigating only the two lowest modes is due to the lower mode vibration response are larger and more dangerous than those of higher-order vibration modes. In addition, according to the UAV operational speed, the wind excitation frequency relating to this speed is lower than 100 Hz. Figure 8 shows the FRF and coherence signal obtained from the experimental study. As shown in Figure 8, there are two natural frequencies of UAV structure located at 102 Hz and 131 Hz. These frequencies are relating to the two lowest elastic vibration modes of the wing structure.

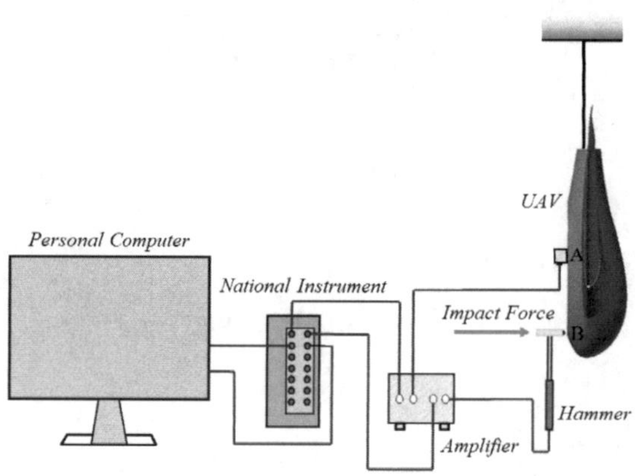

Figure 7. Experimental modal analysis.

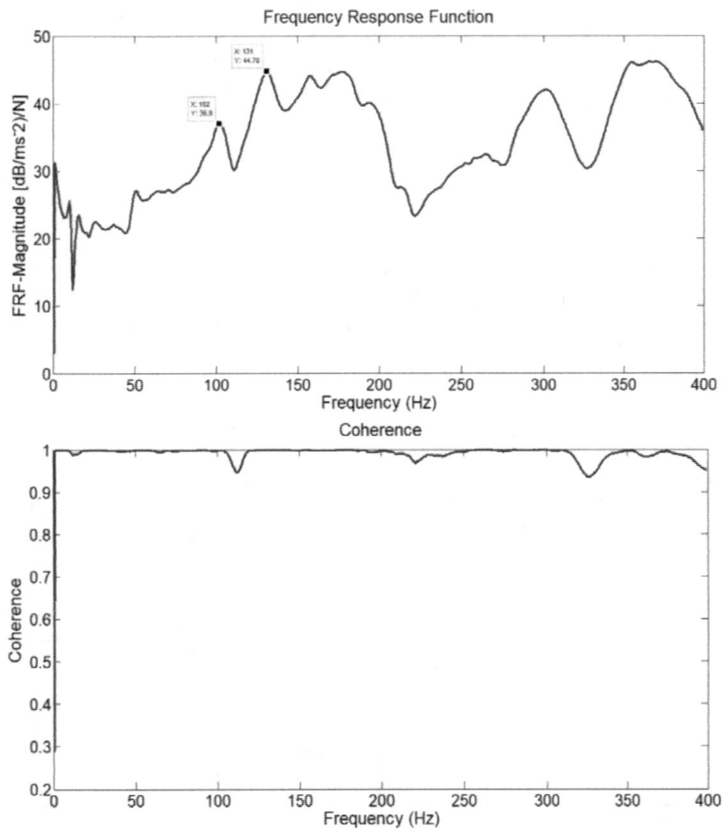

Figure 8. Experimental FRF and coherence.

Figure 9 shows the simulated natural frequencies and mode shapes for the 1st and the 2nd elastic modes. As depicted in Figure 9, the simulated natural frequencies for the 1st and the 2nd elastic modes are 104.18 and 119.74, respectively. Comparison of the UAV natural frequencies obtained from the simulation and experiment are depicted in Table 3. It can be shown from Table 3 that the differences between the experimental and the simulation results (error) are relatively small. Furthermore, the error obtained for the 1st natural frequency is 2.14 %.

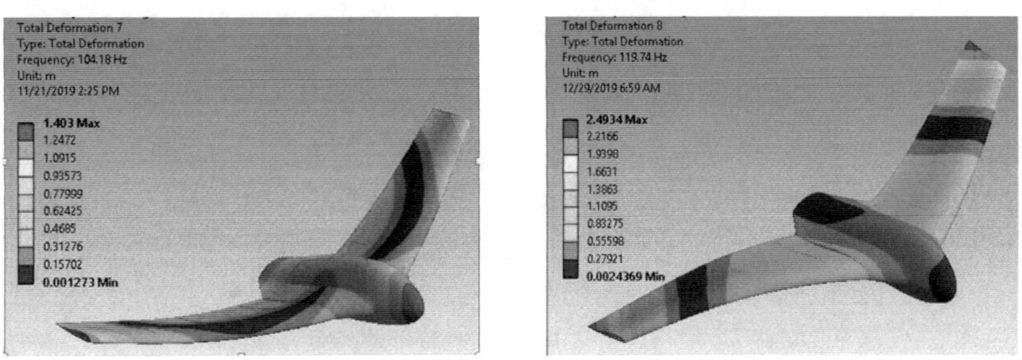

1. (a) 1st mode (b) 2nd mode

Figure 9. The simulated natural frequencies and mode shapes.

Table 3. Natural frequencies of the UAV structure.

Natural Frequency	Experiment	Simulation	Error
1st elastic mode	102 Hz	104.18 Hz	2.14 %
2nd elastic mode	131 Hz	119.74 Hz	8.60%

Table 4 shows a comparison of the 1st and the 2nd elastic mode natural frequencies obtained from the simulation and the experimental study. The initial profile in Table 4 is a condition of UAV wing structure without an additional spar located coincides with the UAV center of gravity. Variation of spar cross-sectional profile causes the variation of the 1st and the 2nd elastic mode natural frequencies of UAV structure as depicted in Table 4.

Table 4. Natural frequency variation.

Profile.	Total Mass (kg)	1st natural freq. (Hz)	2nd natural freq. (Hz)
Initial	2.74	104.18	119.74
U-shape	2.81	121.32	123.21
Cir-shape.	2.84	122.11	123.84
T-shape	2.84	128.09	131.35
Rec.-shape	2.92	118.27	123.31
L-shape	2.96	119.03	123.08
I -shape	3.24	117.85	138.25
Z -shape	3.34	118.91	122.77

Figure 10 shows the relation between spar cross-sectional profile variation to the mass and the UAV natural frequency variation. It can be observed from Figure 10 that T-profile results in the highest 1st natural frequency of UAV structure in comparison with other profiles. Furthermore, the increase of UAV mass for T-profile is relatively low compare to those of other profiles.

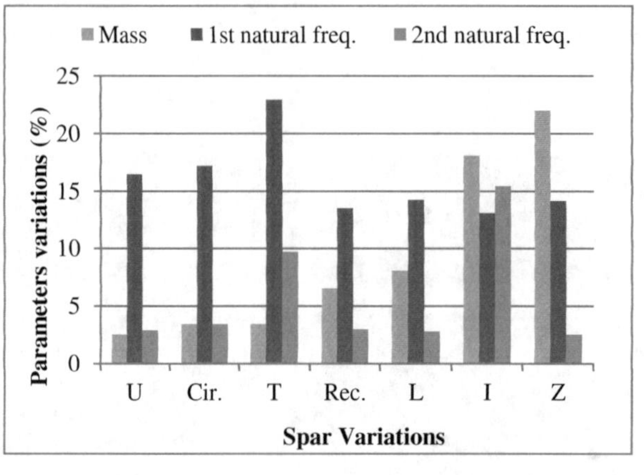

Figure 10. Relation of spar cross-sectional profile variation to mass and natural frequency.

5. Conclusion

The effect of spar cross-sectional profile variation on the natural frequency of UAV structure is evaluated. It is found that variation of the spar cross-sectional profile significantly affects the dynamics characteristic of UAV structure. A simulation study using theoretical modal analysis has shown that UAV spar using T-shape cross-sectional profile is better than other profiles in increasing UAV structure 1st natural frequency without significantly increase the UAV mass. The experimental modal analysis is conducted to validate the simulation results. It is found from the experiment that the UAV natural frequencies are in good agreement with those obtained from the simulation study.

Acknowledgments

The corresponding author thanks the Faculty of Engineering Andalas University for financial support from research publication grand under contract number 061/UN16.09.D/PL/2020.

References

[1] Son L, Bur M and Rusli M 2018 *J. Vib.and Cont.* **24** 8 173–214
[2] Valavanis K P 2015 *Handbook of Unmanned Aerial Vehicles* (Netherlands: Springer)
[3] Verbeke J and Debruyne S 2016 *Int. Conf. on Noise and Vibration Eng. (ISMA 2016)* 2329-2338
[4] Ernnie I B, Mohammed T H S, Faizal M, Basri A A, Abas M F, Majid M S A, Mandeep J S and Ahmad K A 2019 *J. of Mat. Research Tech.* **8** 5 3822–3834
[5] Tolga V and Ceyhun T 2019 *4th International Symposium on Innovative Approaches in Engineering and Natural Sciences* 108–111
[6] Adam D and Wiktor K 2018 *AIP Conf. Proc.* **1922** 120009
[7] Lemler K J and Semke W H 2012 *Proc. 30th Int. Modal Anal.Conf. (IMAC 2012)* (Florida: Engineering Media Technology) 515-523
[8] Abdullah N A Z, Sani M S M, Husain N A, Rahman M M, and Zaman I 2017 *International Journal of Automotive and Mechanical Engineering (IJAME)* **14** 3887-3897
[9] Hibbeler R C 2011 *Mechanics of Materials* (Prentice Hall)
[10] Andre D S and Stelios K 2007 *Solids and Struct.* **44** 8685–8717
[10] Fazle E, Hossain M M, Afrin S and Khan M A 2014 *International Conference on Mechanical, Industrial and Energy Engineering* pp ICMIEE-PI-140304-3
[11] Megson T H G 2007 *Aircraft Structures for Engineering Student Four Edition.* (Oxford: Elsevier)

Numerical Study on Various Position of Multiple Water Tanks toward Earthquake on High-rise Building

Izzul Syazwan Ishak*[1], Norhayati Abdul Hamid[2], Norliyati Mohd Amin[3]
[1]Faculty of Civil Engineering, Universiti Teknologi MARA, 40450 Shah Alam, Selangor, Malaysia.
[2]Institute for Infrastructure Engineering and Sustainable Management(IIESM), Faculty of Civil Engineering, Universiti Teknologi MARA, 40450 Shah Alam, Selangor, Malaysia.
[3]Faculty of Civil Engineering, Universiti Teknologi MARA, 40450 Shah Alam, Selangor, Malaysia.

*Corresponding e-mail:izzul.syazwan88@gmail.com

Abstract. Earthquake can cause many problems to the structures, which lead to building collapse and may takes humans life. One of the effort to reduce the structural response is by introduce the damping system to the buildings where the energy of the system is slowly reduced until the vibration of the system is totally eliminated and the system is brought to rest. Several techniques are available nowadays, however passive control system has advantage in term of cost especially in Malaysia where the earthquake is not the major threat like in Japan but it is good to tune existing water tank to became a passive damper where it is there went situation needed. The objective of the present paper is to analyse the practicality of implementing multiple water tank as passive Multiple Tuned Liquid Damper and finding the best position of water tanks, which would reduce peak response of the structure subjected to seismic forces using SAP2000. 15 multi-storey concrete structure attached with five water tanks with 3 different situation of water tanks location under consideration. In order to study the depth ratio parameter, the water level was varies for each situation such as ¼ tank for all water tanks, half tanks for all water tanks and ¾ tank for all water tanks. The results shows that Position 2 filled with ¾ water has highest significant reduction in term of structural displacement which is 12.6mm compare to other water tanks positions and water level.

Keywords: earthquake, building, water tank, control system

1. Introduction

The breaking and movement of tectonic plate due to sudden energy release inside the earth cause natural disaster known as earthquake. This phenomenon can cause tsunami, rock falls, landslide ground settling and liquefaction. Furthermore, human life and manmade structures will also affected from this event which contains a very powerful energy which leads to destruction. Malaysia is not located at the active seismic region but surrounded by an active seismic bay and can be considered exceptional as it is in a non-earthquake zone as opposed to Indonesia (Sumatera Island) and the Philippines. However, Malaysia will be affected as well if there is any big magnitude earthquake took place 350 km away from the active fault line. During the earthquake on 22 January 2003 originated near Sumatra, it had affected buildings of 1 to 10 storeys high in Penang and Kuala Lumpur, so water tank as a passive Tuned Liquid Damper (TLD) could be considered as a precaution measurement for future unpredicted earthquake [1].

Nowadays, the advance development of light-weight high strength material has cause the trend in construction industry move to skyscrapers and tall buildings. This is due to over high density of population in the cities in conjunction of land space scarcity that makes tall buildings more economic. These new generation of tall structures is vulnerable to earthquake excitations due to its light-weight, material flexibility and low inherit damping. These factors will increase failure possibility of structural elements. Vibration can be periodic or non-periodic that cause problems to the structures as well as the discomfort to the occupants inside. Vibration control is important for machinery, space shuttle, airplane and ship. With the modernization of engineering design and construction techniques, the vibration mitigation technique has find a way to civil engineering and infrastructure field. In this modern day, the total damping for structures is nearly 5% of the critical [2]. Conventional approach to cater the vibration from horizontal motions especially from earthquake is by strengthening the structural members. Alternatively, isolation systems and supplemental damping systems can be considered.

Isolation systems are installed between the foundation and the base elements of the building to separate the building and the moving ground during earthquake. In order to absorb vibrations from earthquake, this base isolation needs to be designed less amount of lateral stiffness compare to the main structure. Meanwhile, for supplemental damping system, the external device is installed at the top of the building. This supplemental damping system can be categorized into four groups which are active, semi-active, passive and hybrid [3]. Passive systems work without required any external power source that makes this damper system is cost-effective compare to other supplemental damping systems. One of the example of passive damping system is motion-activated damper which absorb structural energy through motion [4]. Tuned mass damper (TMD) and tuned liquid damper (TLD) are examples of this category. The TLD is also known as tuned sloshing damper because it uses water sloshing motion to dissipate energy. Tuned liquid damper is a passive damper which depends on hydrodynamic force to harmonize with natural frequency of the building [5].

TLD is a rigid tank filled with liquid which is typically water is used in practice [6]. TLD shape can be square, rectangular or circular and usually TLD is a water tank placed at the top of building. This is another advantageous of TLD because with proper tuning, the water tank not only use for water storage but it can be used as a passive damper to control and reduce the structural vibration. The sloshing and wave breaking during the seismic excitation is very important criteria to control and reduce the building vibration.

A single water tank as TLD will be placed at the centre of the building so that each center of mass will coincide [7]. In real situation, single tank as TLD cannot contain the necessary liquid mass as a water storage and supplement as well as to achieve the desired structural acceleration reduction so a combination of several water tanks is necessary. This combination of several water tanks known as multiple tuned liquid damper (MTLD) where the sloshing frequencies of each tanks are tuned to a range of the structural frequency [8]. Furthermore, slightly detuned of single TLD can reduce the system efficiency and it is impossible to detuned the TLD frequency during the excitation event since it is passive damper system [9].

Several studied had been conducted to find the practicality of implementing the water storage tank as a passive damper. Shad et al. [10] had conducted experimental and numerical studied for seven structures on one storey single degree of freedom system attached with TLD. In this study, the TLD was actually a single water tank that been divided into three section. The position of this rectangular water tank was maintain at the same location, only the water filled inside these three divided compartment was varying. The highest reduction of structural response reported when all three compartment filled with water. Love & Tait [8] conducted a research to find the performance of structure when installed with one (traditional TLD), two and three water tanks as MTLD that having distributed properties for single degree of freedom structure. It showed that the MTLD had better result compare to single TLD when the tanks have different damping ratio and natural sloshing frequency. Gowda [11] had considered several parameters for TLD-structure and found out that TLD is to be more functional and cost-effective for larger buildings. Tuong et al. [12] concluded that the dynamic response of the 8-story building subjected to harmonic loading with a rigid TLD remarkably reduced compare to the flexible TLD where did not prove to be as efficient as the rigid TLD.

2. Numerical Work

The present work aims to find the best water level with it position at the top of building which give the highest reduction in structure response subjected to seismic load using SAP2000.

2.1. Equation of motion for seismic analysis

The movement of the structure is the response due to the motion of the ground and depends on its mass, rigidity, damping and load or displacement applied. An equation of motion in the system was affected by the earthquake intensity can be written as equation (1) below:

$$u_t = u + u_g \tag{1}$$

where u_t is single mass total displacement; u is the relative displacement between mass and ground; and u_g is the ground motion displacement.

By applying Newton's law and D'Alembert's principle of dynamic equilibrium, it can be shown that

$$f_I + f_D + f_S = 0 \tag{2}$$

where f_I is the single mass inertial force and is due to the acceleration of the mass by $f_I = m\ddot{u}_t$; f_D is the damping force on the mass and connected to the velocity across the viscous damper by $f_D = c\dot{u}$; f_S is the elastic force applied on the mass and related to the relative displacement between the mass and the ground by $f_S = ku$, where m is mass of dynamic system; c is damping ratio ; and k is the stiffness. Fill in these equations for f_I, f_D and f_S into equation (2) gives

$$m\ddot{u}_t + c\dot{u} + ku = 0 \tag{3}$$

Equation (3) is often referred to as the motion equation for the SDOF. New equations can be obtained for a system subject to ground motion by substituting equation (1) into equation (3)

$$m\ddot{u}_t + c\dot{u} + ku = -mu_g \tag{4}$$

The motion equation for the MDOF system is equivalent to the SDOF system. The difference is only the stiffness k, mass m and damping c are in matrices form. The aid of software is used to speed up the analysis process for MDOF system.

2.2. Model

15 multi-story concrete structure was considered for this analysis with the aid of analysis software; SAP2000. The structure was design using space frame with fixed at the bottom. The details are given in Table 1.

Table 1. Details of the model

Parameter	Details
Building type	Reinforced Concrete
Number of floors	15
Number of bay	5 x 3
Bay size (m)	4
Floor height (m)	3
Dimension of column (m)	0.45 x 0.45
Dimension of beam (m)	0.23 x 0.45
Depth of slab (m)	0.15
Concrete strength (N/mm^2)	35
Steel reinforcement (N/mm^2)	460

During the modelling process, for an optimization of the analysis, the structure was design as lumped mass single degree of freedom spring-mass system. The MTLD placed at the rooftop act as a spring mass system. Figure 1 and figure 2 shows the views of the model in SAP2000.

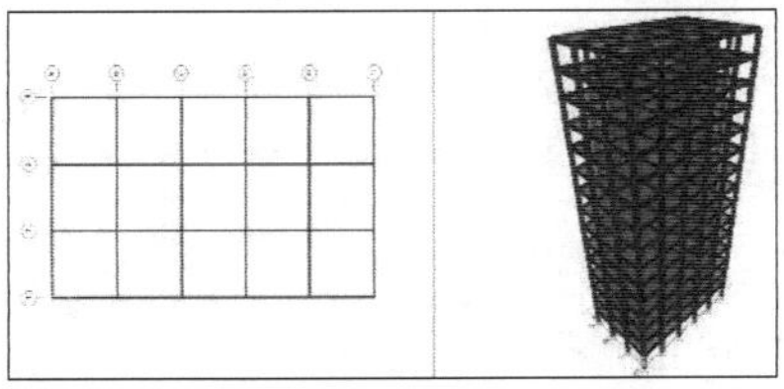

Figure 1. X-Y Plane and 3-D View.

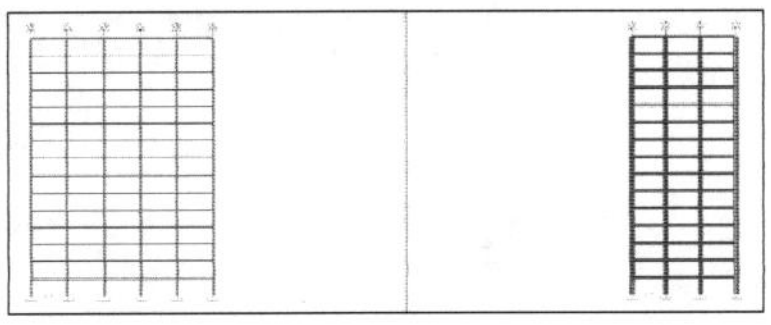

Figure 2. X-Z Plane and Y-Z Plane View.

2.3. Time History Loading
The El-Centro 1940 earthquake was consider in the present study in the direction of global-X towards the structure. Time history analysis was conducted in this study for structures without water tank and with water tank that varies in water level. The summary of time history records used for this analysis was obtained from the website of Pacific Earthquake Engineering Research Center (PEER) as shown in Table 2. Time history of El-Centro is considered with acceleration values at 0.01 sec time interval as shows in Figure 3 below. The reason to consider El-Centro 1940 in the analysis is because this time history was widely used in study and research so it is easy to make comparative study for future research.

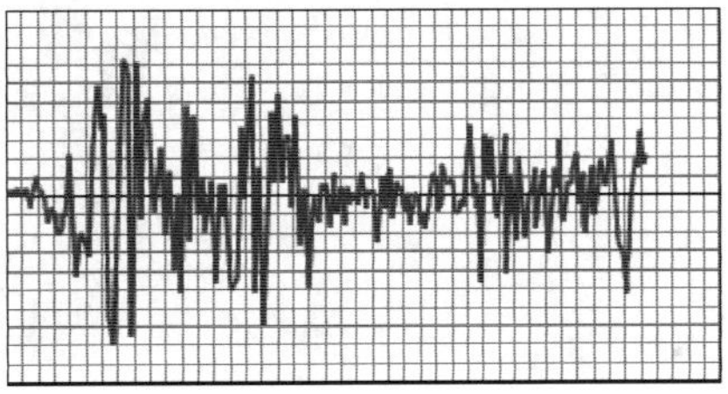

Figure 3 Time history El-Centro 1940.

Table 2. Time history summary

Record	Station/Year	Magnitude	PGA (g)
Imperial Valley	El-Centro 1940	6.9	0.32

2.4. Tuned Liquid Damper
High-rise building will has several water tanks at the top of it for a water storage. During earthquake occurrence, this huge dead load can be utilize as TLD to absorb extra energy exert to the building. In the present study, a rectangular water tank will performing as tuned liquid damper. Five water tanks had been considered during the study. One water tank will be permanently located at the center of the top building so that the center structure mass and the tank mass will coincide [7].

The water tank dimension plan is summarize in Table 3. The water tank was placed over 1m high columns. The water level consider in this study is; (i) ¼ tank for all tanks, (ii) half tanks for all tanks (iii) ¾ tank for all tanks. The water level was varies in each situation in order to study one of the MTLD parameter that is depth ratio (the ratio of water depth to tank length) which give the optimum structural response reduction. The reason full water level not consider in study is because the mechanism of TLD to fully functioning is through water sloshing so a gap is required inside water tank for water to slosh. For optimization, the fluid inside water tank was assume incompressible. The fluid and MTLD was

IOP Conf. Series: Materials Science and Engineering 1062 (2021) 012033 doi:10.1088/1757-899X/1062/1/012033

simplify by modelled it using special joints with added mass and connected these joint to the structure [13].

Table 3. Details of water tank

Dimension (m)	Height (m)	Column height (m)	Water (liters)
4 x 4	4	1	64000

The considered water tanks positions is shown in Figure 4, Figure 5 and Figure 6 below.

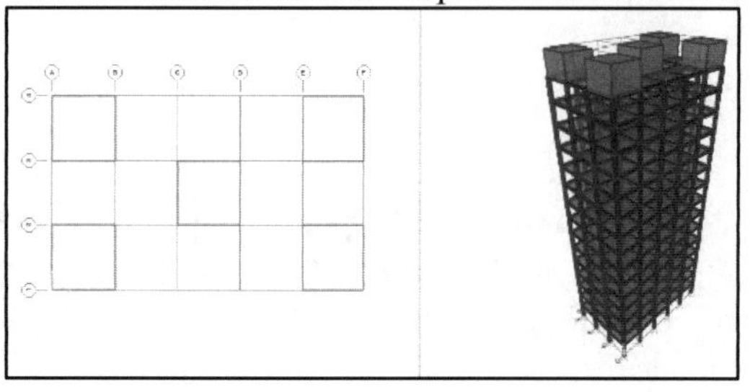

Figure 4. Position 1 of water tanks.

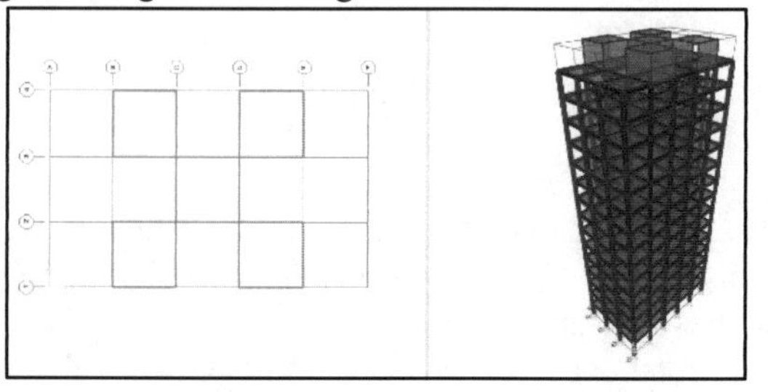

Figure 5. Position 2 of water tanks.

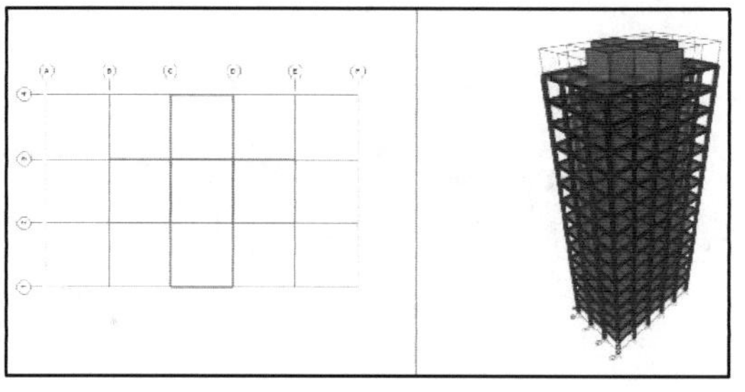

Figure 6. Position 3 of water tanks.

3. Results and Discussion

The objective of the study is to find the effective position of multiple water tanks in reducing the structural response during earthquake event.

3.1. Roof Displacement

Roof displacement of the structure without water tank, Position 1, Position 2 and Position 3 with varies water level were compared as shown in Figure 4,5 and 6. The results was summarize in Table 4.

Table 4. Roof Displacements for EL Centro Earthquake Data

Conditions	Water level	Roof Displacement (mm)
Position 1	¼ tank	15.0
	Half tank	13.6
	¾ tank	13.3
Position 2	¼ tank	14.6
	Half tank	14.4
	¾ tank	12.6
Position 3	¼ tank	14.2
	Half tank	14.9
	¾ tank	13.8

The time history response at the roof level is presented in Table 4. The following conclusion can be made:

- The highest reduction can be seen at Position 2 when it is filled with ¾ tank of water where the roof displacement is 12.6mm.
- Less significant position is Position 1 with water level ¼ tank (15.0mm).
- Base on Figure 7, Position 3 had shown the increase in roof displacement (14.9mm) when the water level is half tank. However the roof displacement reduce once the water level is ¾ tank.

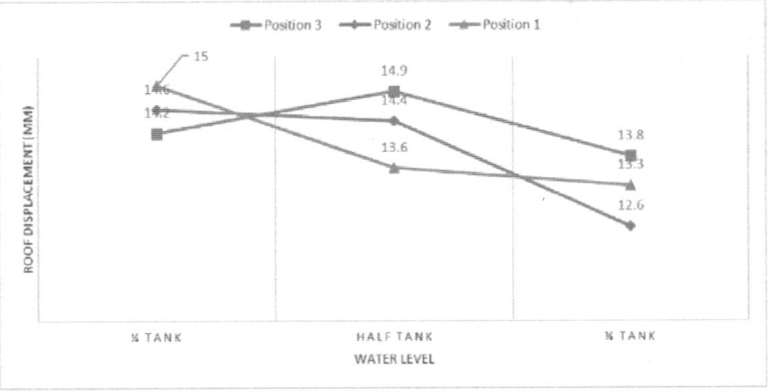

Figure 7. Comparison roof reduction between Position 1,2 and 3.

3.2. *Base shear*

Table 5. Base shear in X-direction summary

	Position 1 Base Shear X (kN)	Position 2 Base Shear X (kN)	Position 3 Base Shear X (kN)
¼ water level	414.77	414.14	407.35
Half tank	395.09	393.20	402.75
¾ water level	373.97	402.75	394.43

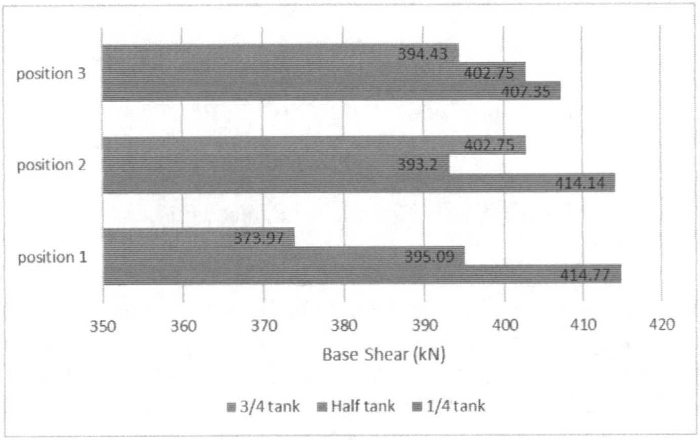

Figure 8. Base Shear in X-direction.

Base shear approximates the maximum predictable lateral force that will happen caused by seismic ground motion at the structure base. Table 5 shows the summary of base shear in X-direction

- Referring to Figure 8, the highest and lowest base shear in X-direction occur at Position 1

- The lowest base shear is 373.97 kN meanwhile the highest is 414.77 kN (both at Position 1)
- For Position 3, the base shear distance is not really big gap compare to Position 1.
- For position 2, the pattern of base shear between all there water condition is different compare to Position 1 and 3, which half tank causes less shear base compare to ¾ tank.

3.3. Acceleration Response

i. It is shows that the ¾ tank of water level has the lower maximum acceleration response regardless MTLD position. Please refer Figure 9.
ii. The highest maximum acceleration response experiences during Position 3 with ¼ tank of water level

Table 6. Maximum acceleration of models of TLD-structure system

Conditions	Water level	Maximum acceleration response (m/s^2)
Position 1	¼ tank	0.527
	Half tank	0.449
	¾ tank	0.407
Position 2	¼ tank	0.548
	Half tank	0.467
	¾ tank	0.429
Position 3	¼ tank	0.568
	Half tank	0.502
	¾ tank	0.447

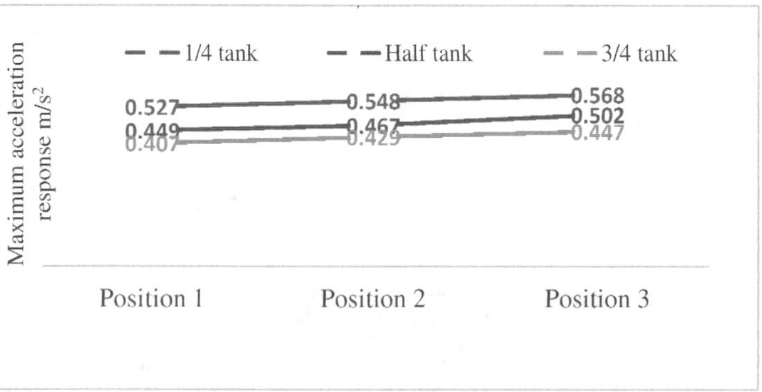

Figure 9. Graph of maximum acceleration.

4. Conclusion

The analysis results demonstrate that multiple tuned liquid damper (water tanks) can be used to reduce structure response due to earthquake forces. It has several advantages over other damping systems such as; low installation and RMO (Running, Maintenance and Operation) cost and less worried about mechanical problems. TLD natural frequency is controllable by adjusting the water depth and container dimension and TLD is applicable for any directional vibration Based on above study, the following conclusions are drawn:

- Base on the overall results, it can be concluded that Position 2 with ¾ water tank reduce most of the structure displacement.
- This position 2 shows the water tanks located around the center of the building has a better structural response reduction. However, when the water tanks pointed at the center like position

3, the efficiency to reduce the structure roof displacement had reduced when the water level increase. This shows water tanks need to be located at the center of the building but in distributing around it not all together near each other.

- It can be concluded from the outcome that the water tanks can be used in designing passive damper by adjusting the water level. Hence, this MTLD can be considered economical especially in the region such as Malaysia where this water tank has an extra purpose rather than just water storage.

References

[1] Adnan A, Marto A, and Hendriyawan 2004 Lesson Learned From the Effect of Recent Far Field Sumatra Earthquakes to Peninsular Malaysia *13th World Conf. Earthq. Eng. Vancouver Canada* **416** 1–10
[2] Kundu P 2012 Vibration Control of Frame Structure Using Multiple Tuned Mass Dampers
[3] Meshram R S and Khante S N 2016 Effectiveness of Water Tank as Passive TMD for RCC Buildings *Int. J. Eng. Res.***5013(5)** 731–736
[4] Rao V K 2013 Experimental and Numerical Studies on Tuned Liquid Damper
[5] Kartha R V and Ritzy R 2015 Tuned Liquid Damper to Control Earthquake Response in a Multi-Storied Building Frame *Int. J. Eng. Res. Appl.* **5(8)** 49–56
[6] Silva M J F and Costa A C 2008 Experimental Studies On The Characteristics Of Tuned Liquid Dampers For Reducing Vibration In Structures *14th World Conf. Earthq. Eng.*
[7] Gulve T S and Murnal P 2013 Feasibility of Implementing Water Tank as Passive Tuned Mass Damper *Int. J. Innov. Technol. Explor. Eng.* **3(3)** 12–19
[8] Love J S and Tait M J 2015 Multiple Tuned Liquid Dampers for Efficient and Robust Structural Control *J. Struct. Eng.* **141(12)** 04015045
[9] Kareem A and Kline S 1995 Performance of Multiple Mass Dampers Under Random Loading
[10] Shad H, Adnan A, and Behbahani H P 2013 Performance evaluation of tuned liquid dampers on response of a SDOF system under earthquake excitation and harmonic load *Res. J. Appl. Sci. Eng. Technol.* **6(16)** 3018–3021
[11] Gowda A P 2020 Seismic Analysis Of High-Rise Building Using Tuned Liquid Damper *Int. J. Futur. Gener. Commun. Netw.***13(3)** 2843–2861
[12] Tuong B P D, Huynh P D, Bui T T, and Sarhosis V 2019 Numerical Analysis of the Dynamic Responses of Multistory Structures Equipped with Tuned Liquid Dampers Considering Fluid-Structure Interactions *Open Constr. Build. Technol. J.***13(1)** 289–300
[13] Khalaf H M K and Kumar K S 2016 Analysis and Design of Multi Storied Building for Vertical and Horizontal Loading With and Without Dampers using SAP2000 *Int. J. Sci. Eng. Technol. Res.***05(10)** 2079–2089

IOP Conf. Series: Materials Science and Engineering 1062 (2021) 012034 doi:10.1088/1757-899X/1062/1/012034

Simulation Modeling The Performance of Ocean Thermal Energy Conversion Power Cycle

N Samsuri[1*], N Sazali[2], A S Jamaludin[3], and M N M Razali[3]
[1] Ocean Thermal Energy Centre (OTEC), Ground Floor, Block Q, Universiti Teknologi Malaysia Kuala Lumpur, Jalan Sultan Yahya Petra, 54100 Kuala Lumpur, Malaysia.
[2] Department of Electrical Engineering, College of Engineering, Universiti Malaysia Pahang, 26300 Gambang, Kuantan, Pahang, Malaysia.
[3]Faculty of Manufacturing and Mechatronic Engineering Technology, Universiti Malaysia Pahang, 26600 Pekan, Pahang, Malaysia

*Corresponding e-mail: norazreen1989@gmail.com

Abstract. Ocean Thermal Energy Conversion (OTEC) is a foundation for an appealing renewable energy technology with regards to its vast and inexhaustible resources of energy, renewability, stability, and sustainable output. The principle of an OTEC power plant is to exploit the energy stored in between the upper layer of warm surface seawater (heat source), and the cold layer of deep seawater (heat sink). The plant operates based on a Rankine cycle to produce electricity between the source and the sink at the minimum temperature difference of approximately 20 K. The main objective of this study is to evaluate the performance of the proposed OTEC closed Rankine cycle using ammonia as the working fluid, to be paralleled with basic OTEC Rankine cycle. Preliminary simulation was performed at the initial stage of the study to validate the simulation model by referring to previous OTEC studies. The same developed model was deployed to test the efficiency of the proposed modified OTEC Rankine cycle, resulting in an enhancement in terms of thermal cycle performance from 3.43% to 7.98%. This study has revealed that the proposed OTEC closed Rankine cycle which introduced an interstage superheating as well as an improved condenser cooling system, augmented the system competence of an OTEC power cycle.

Keywords: ocean, thermal energy, renewable, modelling, Rankine cycle

1. Introduction

Ocean Thermal Energy Conversion (OTEC) has tremendous prospective in deep ocean water area, in which there is an enough high temperature difference between the surface water and a precise depth that is necessary to effectively run an OTEC power plant. In 1881, Arsonval's initial thought specified that the optimum temperature difference needed for the installation of an OTEC plant is larger than 20K [1]. The system will work between the surface seawater at 30°C (known as heat source), and seawater at 1000 m depth with temperature of 4°C (known as heat sink) [2-4]. OTEC power plant technology is developed on a basis of open (OC-OTEC) and closed Rankine cycles (CC-OTEC). Aforementioned research reported that the process must be founded upon the Uehara cycle for optimal power plant output, applying ammonia-water mixture as the working fluid with smaller than 20 K of temperature difference, and at 5-6% thermal efficiency [5].

IOP Conf. Series: Materials Science and Engineering 1062 (2021) 012034 doi:10.1088/1757-899X/1062/1/012034

Malaysia has never previously been on any world map showing areas with the potential for OTEC power generation. Nevertheless, owing to the works done by Sapura-Crest Group's Subsidiary Teknik Lengkap GeoSciences Sdn Bhd, a marine survey was carried out at the South China Sea in 2008 [6]. It proved that the temperature of approximately 4°C is attained at the ground under the North-Borneo Trough, which is also acknowledged as the Sabah Trough (illustrated in Figure 1), as different to the temperature of nearly 29°C on the surface. The discovery described above offered Malaysia a chance to effectively run an OTEC plant. Rendering to data estimates, the Sabah Trough is 60 km in width, 100 km in length and has an average depth of 1000 m. Necessary research is required to reach the maximum output of OTEC while at the same time minimalizing the cost of capital.

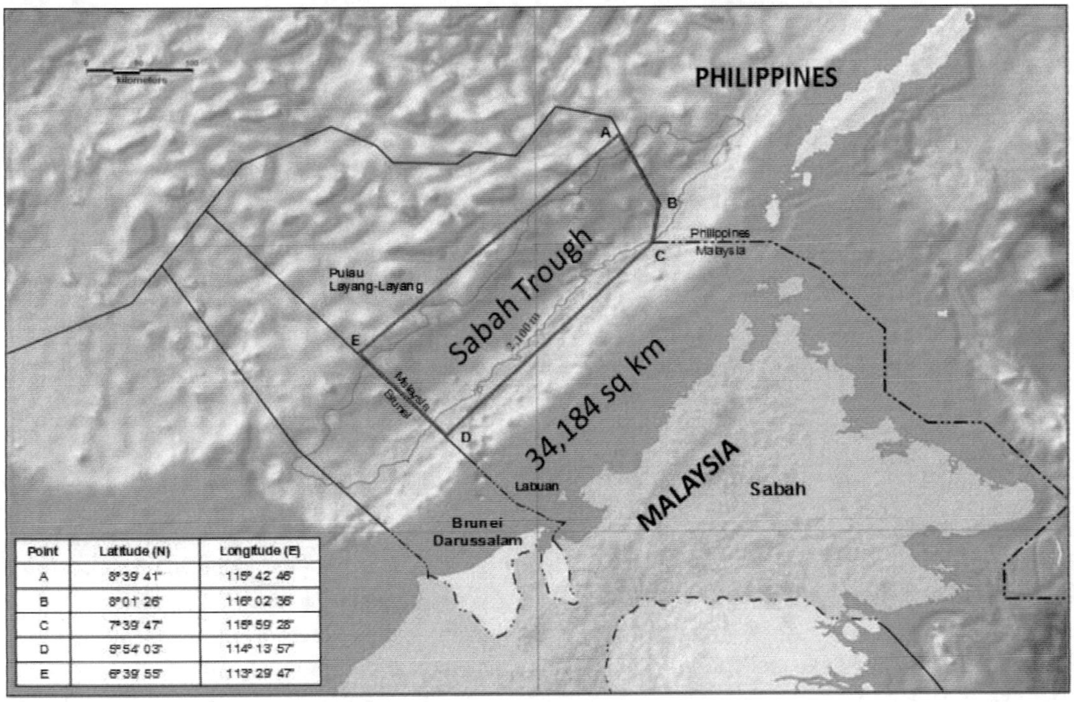

Figure 1. The Sabah Trough's position and region discovered by Sapura-Crest Group's Subsidiary Teknik Lengkap GeoSciences Sdn. Bhd. [6].

The first OTEC system was suggested by D'Arsonval (1881) which associated with the closed Rankine cycle, using a turbine with low-pressure attributed by its low operating temperature, thus formed 3% of thermal efficiency [1]. The introduced closed Rankine cycle consisted of an evaporator, a turbine, a condenser, and three designated pumps for working fluid, surface seawater (SSW) and deep seawater (DSW) as shown in Figure 2. In 1985, Dr. Kalina has come out with the Kalina cycle which was a modification from closed Rankine cycle with the addition of a separator, an evaporator, an absorber, and a decreasing valve. The cycle has developed the OTEC system performance by 4% [5]. Far ahead at Saga University in Japan, the Uehara cycle was introduced, in which the outcome showed 3% higher efficiency compared to a basic closed-cycle OTEC system [7,8]. Uehara cycle was conceived based on the Kalina cycle with the overview of a turbine, a heater, a tank, a pump for working fluid and a diffuser. the Uehara cycle trend showed a slight improvement in plant efficiency with some risk in terms of plant difficulty and production cost.

IOP Conf. Series: Materials Science and Engineering 1062 (2021) 012034 doi:10.1088/1757-899X/1062/1/012034

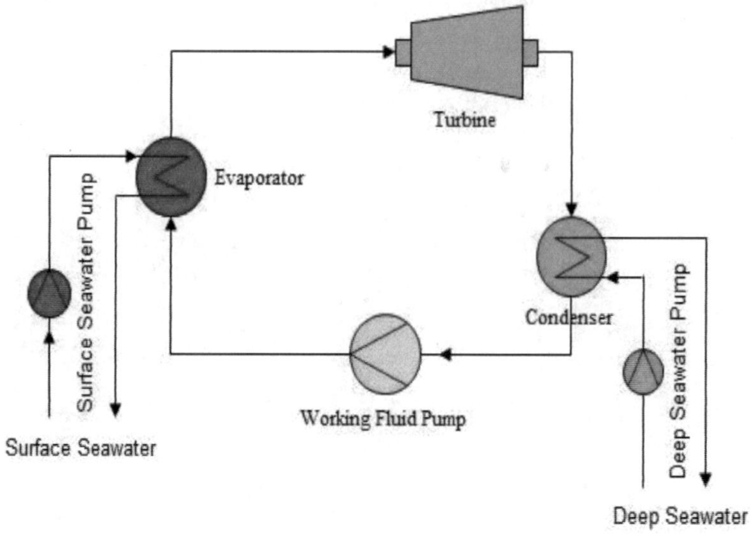

Figure 2. Schematic diagram of OTEC closed Rankine cycle by Arsonval (1881).

Given that there is sufficient energy supply from the SSW, this current study proposes preheating using an external source of heat, as illustrated in Figure 3. To alleviate the performance failure, a multi-stage turbine that uses an exterior source of heat to conduct interstage superheating is suggested. In spite of that, the multi-stage expansion would raise the condenser's cooling demand, and thus increase the flow rate or power needed to pump the DSW. From now, the suggested OTEC closed Rankine cycle (illustrated in Figure 3) utilizes a closed loop condenser to reduce the probable adverse effect of DSW upwelling on the underwater environment. Furthermore, the commercial plant used titanium as pipe material to surmount biofouling. Undeniably, this customization will also decrease the biofouling effect on the inner side of the heat exchanger tube, enabling the use of low-price materials for condenser fabrication. Consequently, in this study, the issue of evaporator and condenser as the most lavish components will be addressed by introducing the use of a closed loop of evaporator and condenser.

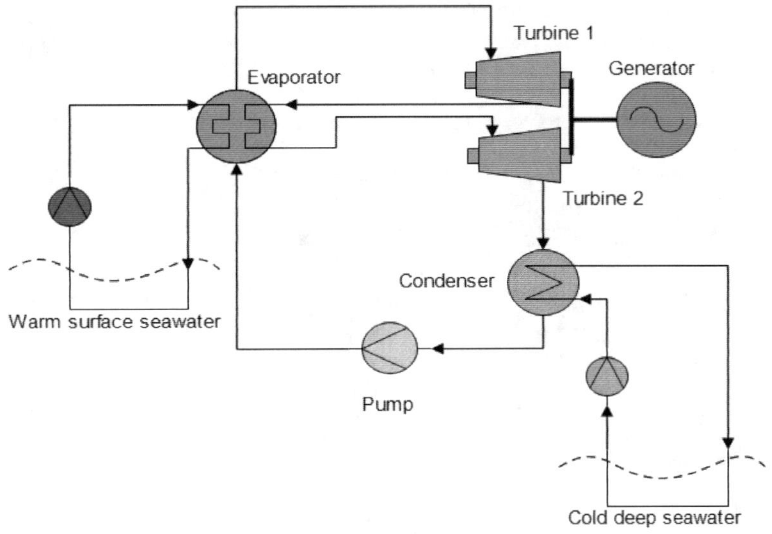

Figure 3. Schematic diagram of the proposed OTEC closed Rankine cycle with modified evaporator heating and condenser cooling systems.

2. Methodology

2.1. Introduction
Laboratory Virtual Instrument Engineering Workbench (LabVIEW) is a platform for framework design created by National Instruments employed as a language for visual programming. Its use in frequent fields of engineering (e.g. aeronautical, mechanical, electrical, etc.) has led to the advancement of many of the world's major and most complex applications to achieve future demands. LabVIEW compromises the users with flexibility through intuitive graphical programming which helps to reduce the time needed for test development. A simulation of LabVIEW is conducted through five stages as indicated in Figure 4. The thermodynamic model has been made in LabVIEW and linked to the working fluid data base in National Institute of Standards and Technologies (NIST) RefProp 9.

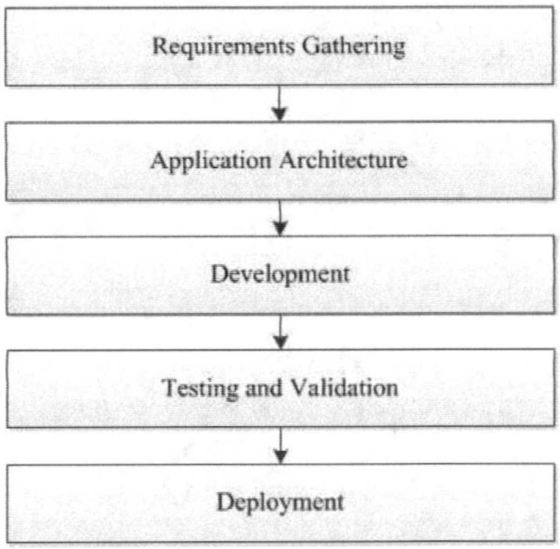

Figure 4. Steps involved in LabVIEW simulation.

2.2. Analytical Techniques of Thermodynamics

The simulation was based upon the thermodynamic analysis of the OTEC Rankine cycle performance. The Rankine cycle contains of four major components, which are condenser, coolant pump, turbine, and evaporator. Numerous expectations were comprised to facilitate the simulation analysis and assessment [13,14] defined as follows:

- Each component is in steady state;
- Any heat loss and pressure drop are disregarded;
- The system is entirely insulated; and
- All pumps and turbines are given isentropic efficiency.

For the steady state energy balance equation, the total energy entering a system is equal to the total energy exiting the system, as stated in equation (1)

$$E_{in} = E_{out} \tag{1}$$

or it can be elaborated as in equation (2)

$$W_{in} + Q_{in} + \sum \dot{m}_{in} = W_{out} + Q_{out} + \sum \dot{m}_{out} \tag{2}$$

where $\dot{Q}$ represents heat transfer rate; $\dot{m}_{in}$ and $\dot{m}_{out}$ is inlet and outlet mass flow rate; whereas W_{in} and W_{out} is work inlet and outlet, respectively. By assuming the system is completely insulated and any heat losses are neglected; which $Q_{in} = Q_{out} = W_{out} = 0$, therefore the energy balance in the pump is expressed as in equation (3):

$$W_{in} + \sum \dot{m}_{in} = \sum \dot{m}_{out} \tag{3}$$

From equation (3), the work supplied is given as in equation (4):

$$W_{in} = \sum \dot{m}_{out} - \sum \dot{m}_{in} \tag{4}$$

Rate of heat supplied to the cycle (evaporator), $\dot{Q}_{\epsilon}$ is expressed as in equation (5) :

$$\dot{Q}_e = \dot{m}_{wf}\, \Delta h_e \tag{5}$$

Rate of heat rejected from the cycle (condenser), $\dot{Q}_c$ is indicated as in equation (6):

$$\dot{Q}_c = \dot{m}_{wf}\, \Delta h_c \tag{6}$$

Rate of heat absorbed from the warm seawater, $\dot{Q}_{e,ws}$ is expressed as in equation (7):

$$\dot{Q}_{e,ws} = \dot{m}_{ws}\, c_p \Delta T_{ws} \tag{7}$$

Rate of heat rejected into the cold seawater, $\dot{Q}_{c,cw}$ is indicated as in equation (8) :

$$\dot{Q}_{c,cw} = \dot{m}_{cs}\, c_p \Delta T_{cs} \tag{8}$$

where $\dot{m}_{ws}$ and $\dot{m}_{cs}$ are the mass flow rate of warm and cold seawater, respectively. c_p is the seawater specific heat capacity at constant pressure. The working fluid pump, $W_{P_{wf}}$ and the turbine work, W_T is written as in equation (9) and equation (10)

$$W_{P_{wf}} = \dot{m}_{wf}\, v(P_2 - P_1) \tag{9}$$

$$W_T = \dot{m}_{wf}\, \Delta h \tag{10}$$

where Δh represents the enthalpy difference in the turbine system.

Referring to Uehara and Ikegami (1990) [14], the working fluid pumping power, is given as in equation (11). The pumping power of warm seawater, is indicated in equation (12); whereas the pumping power of cold seawater, is expressed in equation (13)

$$P_{Pwf} = \frac{\dot{m}_{wf}\, \Delta H_{wf}\, g}{\eta_{wf,p}} \tag{11}$$

$$P_{ws} = \frac{\dot{m}_{ws}\, \Delta H_{ws}\, g}{\eta_{ws,p}} \tag{12}$$

$$P_{cs} = \frac{\dot{m}_{cs}\, \Delta H_{cs}\, g}{\eta_{cs,p}} \tag{13}$$

where ΔH refers to the difference in pressure. The net power output, P_n is indicated as in equation (14) below:

$$P_n = P_G - P_{ws} - P_{cs} - P_{Pwf} \tag{14}$$

2.3. Preliminary Simulation

An ideal working fluid should have the related thermophysical properties corresponding with its application, as well as the capability to withstand its chemical stability within the definite range of temperature. Working fluid selection plays a foremost part on the system in terms of its performance, operating conditions, effects on the environment and economic feasibility. In this section, the parameters for recognizing a suitable working fluid for the cycle system are defined. Sami and co-workers (2012) [15] has itemized the main factors disturbing the properties of thermodynamics and thermophysics of the system, among which are thermal conductivity, chemical stability, specific heat, boiling temperature, latent heat, toxicity, in addition to flash point, as defined in Figure 5.

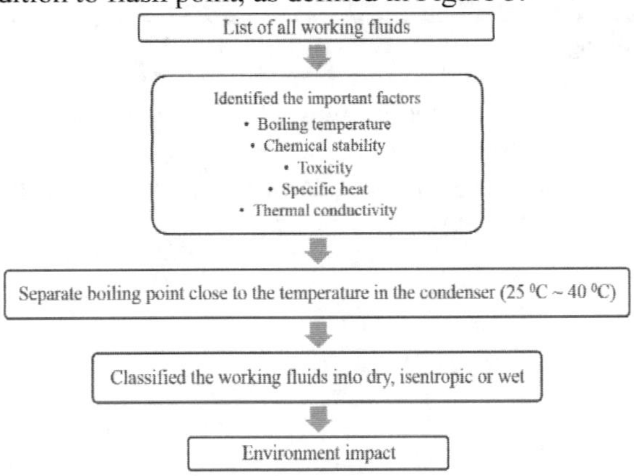

Figure 5. Steps in selecting the working fluids [15].

The OTEC closed Rankine cycle in this study exploited the boiling point of the working fluid near the evaporator operating temperature, which is about 25°C to 40°C [16]. In addition, the fluids were classified as dry, isentropic or wet relative to the saturation curve (dT/ds). A dry or isentropic fluid is proper to be used in OTEC closed Rankine cycle [17]. The purpose of separating the type of fluids is to make sure that the fluids are completely superheated after isentropic expansion, intended to evade the presence of liquid drops on the blades of the turbine.

2.4. Types of Working Fluids

There are two kinds of working fluid, specifically pure fluid (pure compound) and pseudo-pure fluid (a mix of several pure compounds of fluid). Ammonia, propane, R22, R134a, and R143a, are marked as pure fluid, and are not combined with any other compounds. Temporarily, ammonia-water mixture, R404a, R410a, R470c, and R507a are pseudo-pure fluids. Figure 6(a) show that the highest enthalpy change can be exposed in ammonia-water mixture, followed by ammonia, propane and R32. According to Figure 6(b), in contrast with other working fluids, ammonia-water mixture has the highest quantity of heat applied. This situation is instigated by its larger latent heat value, also can be defined as the amount of heat that a liquid absorbs to stay at a constant pressure or temperature throughout the process of vaporization.

2.5 *Preliminary Simulation*

A preliminary design model for simulation of a 1 MWe OTEC closed Rankine cycle was accompanied by means of ammonia as the working fluid. This preliminary simulation is to validate the model developed by [11]. Apart from that, the preliminary design model allows the approximation for 5MWe and 10MWe OTEC closed Rankine cycle.

Table 1. Parameters for three OTEC cycles to be investigated

Parameter	Symbol	Unit	Value
Evaporating temperature	T_E	°C	28
Condensing temperature	T_C	°C	8
Warm seawater inlet temperature	T_{wsw}	°C	30
Cold deep seawater inlet temperature	T_{csw}	°C	5
Working fluid pump efficiency	η_{wf}	%	0.75
Turbine efficiency	η_T	%	0.82
Generator efficiency	η_G	%	0.95
Warm seawater pump efficiency	$\eta_{pump,wsw}$	%	0.80
Cold deep seawater pump efficiency	$\eta_{pump,csw}$	%	0.80

The OTEC closed Rankine cycle simulation based on reference [10] was showed according to the fixed condition parameters as tabulated in Table 1, in which the ammonia is in steady state. The graph that signifies the simulated model is shown in Figure 6(b). When relating the reference case with the preliminary analysis, it was found that ammonia generated the uppermost total work output. Such results are strengthened by the fact that ammonia influenced the maximum as well as the most appropriate value of latent heat for the OTEC cycle system.

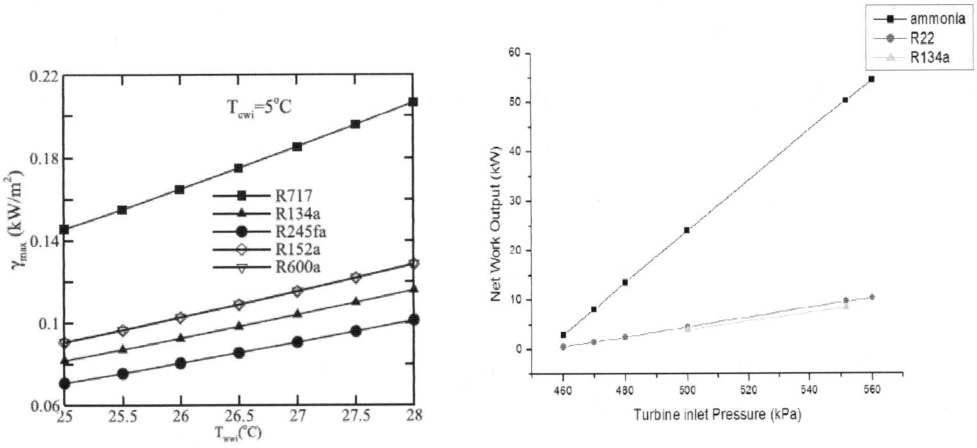

(a) (b)

Figure 6. (a)The network output of closed Rankine cycle using several working fluids[11]; (b)Simulation model using LabVIEW and RefProp.

Table 2. Analysis of OTEC Closed Rankine cycle using ammonia as working fluid

	Unit	1MWe	5MWe	10MWe
Q_{in}	kW	19724.40	81375.50	162751.00
Q_{out}	kW	18685.00	76166.90	152334.00
$W_{p(wf)}$	kW	13.22	54.53	109.06
$W_{p(wsw)}$	kW	96.74	399.12	798.24
$W_{p(cws)}$	kW	118.76	484.11	968.21
$\dot{m}_{wf}$	kg/s	15.82	65.25	130.50
$\dot{m}_{WSW}$	kg/s	1793.01	7397.29	14794.60
$\dot{m}_{CSW}$	kg/s	1587.67	6471.91	12943.80
W_T	kW	905.00	4525.00	9050.00
W_{net}	kW	676.28	3587.24	7174.48

As can be seen from Table 2, the net power output increased expressively after the system was scaled up [12]. The primary study acts as an initiation to the visualization procedure used in the subsequent assessment in Section 3. On the other hand, the preliminary simulation is exposed to elucidate the adequacy of the parameters used in this study.

3. Results and Discussion

The basic OTEC closed Rankine cycle has been optimized to generate more work output by assuming two-stage steam turbines that implement interstage superheating along with revised condenser cold deep seawater cooling system regardless of the cycle's cost effect. The assumptions made regarding the working condition is the critical factor and these considerations arise in order to emphasis the evaluation on the impact of implementing interstage superheating as well as modified condenser cooling system. Both the basic and proposed closed Rankine cycle used the alike considerations and methodology.

Figure 7 shows the T-s diagram of the suggested OTEC closed Rankine cycle. The process from point 1 to 2 corresponded to the constant pressure heat addition in the evaporator. Points 2 to 3 signified the complete steam expansion that was disturbed in the high-pressure turbine, and steam was emitted after partial expansion. Using warm seawater, the reheated steam was complete additionally at point 4 to a turbine with low-pressure for full expansion, and the pressure of steam reaches the condenser pressure value at point 5. The process from point 5 to 6 represented the constant pressure of heat rejection process in the condenser; whereas points 6 to 1 corresponded to the working fluid pump compressing the working fluid to the operating pressure of the evaporator. Additionally, the T-s diagram in Figure 7 presented a slight increase in the region that linked with the implementation of steam reheating. Consequently, the reheating process improved the work output due to the expansion work of the turbine at low-pressure. The primary intention of reheating is to shield the blade of the turbine by removing the moisture subsequent from the steam.

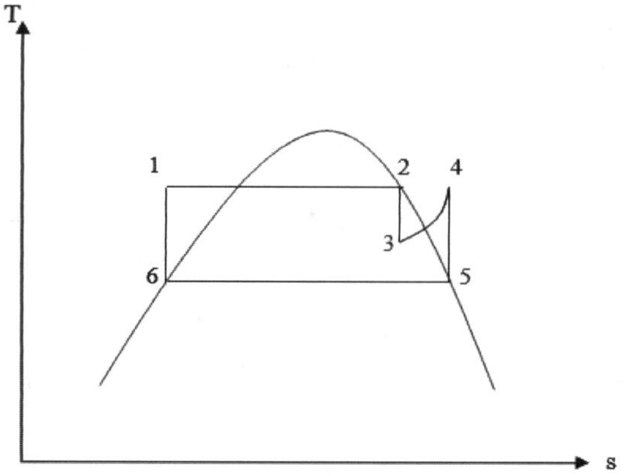

Figure 7. T-s diagram of the proposed OTEC closed Rankine cycle.

Additionally, the use of two turbines has given the advantage of improving the cycle performance [13]. It frequently occurs because when a working fluid at a higher temperature is organized, a greater proportion of heat flows through the cycle. Furthermore, the exhaust steam energy and the heat attained from the warm seawater permitted a significant amount of work to be generated. Two turbines generated much greater work output compared to a single turbine. Nevertheless, the disadvantage of the use of two turbines was the high cost in terms of manufacturing and maintenance.

The preliminary simulation performance of the suggested cycle not only showed a minimal improvement in terms of net power output, but also the cycle efficiency was found to be twice as more associated to a basic OTEC closed Rankine cycle. Clearly, Table 3 revealed that the implementation of interstage superheating contributed to an enhancement of 4.5% to the basic OTEC closed Rankine cycle. In reality, utilizing two or more phases in the Rankine cycle might include significantly smaller turbines that might lead to a performance decline [14,15]. Nonetheless, this study concluded that a higher work output and performance can be attained with the modified OTEC closed Rankine cycle, however it essential supplementary costs due to the use of two turbine stages.

Furthermore, a modified condenser cooling system has also been recognized in this work. The condenser used an open loop for the previous works, but a modification has been made in this recent research where a closed loop was used for the condenser cooling system using ethylene glycol as the chiller. Ethylene glycol was identified as the typical working fluid in the air conditioning sector. Thus in this case, the synergy between air conditioning technology and OTEC technology might be presumed. Ethylene glycol was extensively used for standard cooling applications as an antifreeze fluid; henceforth it was designated as the heat transfer medium for cold seawater in this proposed cycle. Taking into account a 20% of safety factor, the system cooling capacity using ethylene glycol was 9569.88 kW, while the system cooling capacity applying deep seawater was 10471.75 kW. In brief, the use of ethylene glycol in the suggested closed-loop condenser cooling system helped to minimize the cooling capacity required for the condenser. Besides, applying the recommended closed-loop condenser cooling system would mitigate the biofouling effect in the tube of condenser.

Table 3. Calculated results of the proposed OTEC closed Rankine cycle and the basic OTEC closed Rankine cycle using ammonia as working fluid.

	Unit	Evaluation of 1MW OTEC power plant	
		Basic OTEC Closed Rankine Cycle	Proposed OTEC Closed Rankine Cycle
P_E	kPa	1099.30	1099.30
P_C	kPa	573.70	573.70
Q_{in}	kW	19724.40	9237.98
Q_{out}	kW	18685.00	8189.88
$W_{p(wf)}$	kW	13.22	6.17
$W_{p(wsw)}$	kW	96.74	43.80
$W_{p(cws)}$	kW	118.76	52.05
$\dot{m}_{wf}$	kg/s	15.82	7.38
$\dot{m}_{wsw}$	kg/s	1793.01	811.77
$\dot{m}_{csw}$	kg/s	1587.67	695.89
W_T	kW	905.00	756.58
W_{net}	kW	676.28	737.22
η	%	3.43	7.98

4. Conclusion

In summary, a model which incorporated LabVIEW and Refprop softwares was efficaciously established and deployed for a preliminary assessment of the OTEC cycle efficiency. The initial analysis of a test run at a net power output of 1 MW showed a close agreement with that of existing data. The study was further extended to study the competency of a proposed modified OTEC Rankine cycle, in which the evaluation was led using the similarly developed model. The outcome presented an enhancement in thermal cycle efficiency from 3.43% to 7.98%. Therefore, this study has proved that there are enhancements on the performance of changed closed Rankine cycle that introduced interstage superheating as well as modified condenser for cold deep seawater cooling system.

Acknowledgments

The authors express gratitude to the Ocean Thermal Energy Centre (OTEC) lecturers and staffs, Universiti Teknologi Malaysia for their knowledge, laboratory, and equipment supplied. The authors also thanked to Ministry of Higher Education (MOHE) for financially supported this study through The Fundamental Research Grant Scheme for Research Acculturation of Early Career Researchers, FRGS-RACER/1/2019/TK05/UMP//1 - RDU192621.

References

[1] D'Arsonval J 1881 *Reveu Scientifique* 370-372
[2] Nihous G C 2005 An Order-of-Magnitude Estimate of Ocean Thermal Energy Conversion Resources *Journal of Energy Resources Technology* **127(4)** 328
[3] Nihous G C 2007 A Preliminary Assessment of Ocean Thermal Energy Conversion Resources *Journal of Energy Resources Technology* **129(1)** 10
[4] Vega L A 2002 Ocean Thermal Energy Conversion Primer. *Marine Technology Society Journal* **36(4)** 25-35
[5] Zhang X, He M and Zhang Y 2012 A Review of Research on the Kalina Cycle. *Renewable and Sustainable Energy Reviews* **16(7)** 5309-5318

[6] Sapura-Crest Group's Subsidiary TLG GeoSciences Sdn. Bhd. 2008 *Malaysia Marine Survey in the South China Sea (MyMRS)*
[7] Uehara H, Ikegami Y and Nishida T 1994 Performance Analysis of OTEC Using New Cycle with Absorption and Extraction Process *Proc. of Oceanology Int. '94* 1-11
[8] Wang C M, Yee, A A, Krock H and Tay Z Y 2011 Research and Developments on Ocean Thermal Energy Conversion *The IES Journal Part A: Civil & Structural Engineering* **4(1)** 41-52
[9] Wu C and Burke T 1998 Intelligent Computer Aided Optimization on Specific Power of an OTEC Rankine Power Plant *Applied Thermal Engineering* **18(5)** 295-300
[10] Uehara, H., & Ikegami, Y. 1990. Optimization of a Closed-cycle OTEC System. *Transaction of the ASME Journal of Solar Energy Engineering*. 112-4: 247-256.
[11] Yeh R H, Su T Z, and Yang M S 2005 Maximum Output of an OTEC Power Plant. *Ocean Engineering* **32(5-6)** 685-700
[12] Upshaw C R 2012 Thermodynamic and Economic Feasibility Analysis of a 20 MW Ocean Thermal Energy Conversion (OTEC) Power
[13] Sun F, Zhou W, Ikegami Y, Nakagami K and Su X 2014 Energy-exergy Analysis and Optimization of the Solar-boosted Kalina Cycle System 11 (KCS-11) *Renewable Energy* **66** 268-279
[14] Yamada, N, Hoshi A and Ikegami Y 2009 Performance Simulation of Solar-boosted Ccean Thermal Energy Conversion Plant *Renewable Energy* **34(7)** 1752-1758
[15] Kim N J, Ng K C, and Chun W 2009 Using the Condenser Effluent from a Nuclear Power Plant for Ocean Thermal Energy Conversion (OTEC) *International Communications in Heat and Mass Transfer* **36(10)** 1008-1013

Axial Crush Response of Aluminium Square Tube with Origami Patterns

Amir Radzi Ab Ghani[1*], **M Amirul Syafiq Zulfikry**[1], **Ramzyzan Ramly**[1] **and Mohd Zaid Othman**[2]
[1] Faculty of Mechanical Engineering, Universiti Teknologi MARA, Shah Alam, Selangor, Malaysia
[2] Department of Mechanical Engineering, Universiti Pertahanan Nasional Malaysia, Kuala Lumpur, Malaysia

*Corresponding e-mail: amirradzi@gmail.com

Abstract. Thin-walled square tubes are widely used as impact energy absorber in automotive structures due to their ease of fabrication and installation, high energy absorption capacity in terms of progressive plastic deformation and long stroke. However, the main drawback of square tube is its high initial peak force during the initial stage of crushing. An origami pattern on the faces of the tube is proposed to reduce the high initial peak force and increase the energy absorption of the tube. Static and dynamic axial crushing were performed using finite element analysis to determine the initial peak force (IPF), crush force efficiency (CFE) and plastic specific energy absorption (SEA) of tubes with origami pattern. The results of simulations were validated by experimental data. Then, various combinations of origami patterns were studied using finite element simulation only. It was found that the origami pattern significantly enhanced the tubes crush performance. Comparison between plain square tube and tubes with various origami patterns was carried out and it was found that the origami patterns reduced the initial peak force and increased the crush force efficiency for both static and dynamic loading conditions.

Keywords: thin-walled, square tubes, origami pattern, plastic deformation, energy absorber

1. Introduction

Energy absorbers in the form of tubular structures of various shapes are mainly used for impact protection of vehicles during collision. Ductile materials such as steels and aluminium are the preferred choice as they exhibit large plastic deformation before total fracture. These materials absorb the excessive kinetic energy during impact by various mechanisms such as friction, fracture, shearing, bending, tension, crushing and plastic deformation. Widely used geometries such as square and circular tubes, frusta and polygons subjected to axial and lateral loadings has been extensively studied for the past six decades [1-3]. The need to further optimise the energy absorption of structures has prompted researchers to explore new geometries, configurations and material combinations.

Zhang X [4] introduced two types of patterns using the basic pyramid elements on to the surface of conventional thin walled square tubes. Type A pattern was aimed at triggering the extensional collapse mode while Type B was planned to fail in a new mode. The tubes were numerically subjected to quasi-static axial crushing using LS Dyna explicit finite element code. Tubes with Type A pattern failed in extensional collapse mode while tubes with Type B pattern failed in octagonal collapse mode. Both patterns resulted in increased specific energy absorption of the tubes. Song J et al. [5] implemented origami patterns on to thin walled tubes with square, hexagonal and octagonal cross-sections subjected to axial crushing. Numerical results showed that the patterned tubes exhibited lower initial peak force and more uniform crushing load as compared to the plain tubes. Parametric study has shown that the pre folding angle affected the initial peak force and mean crushing load. A prototype of the patterned tube was then fabricated and tested, and the experimental results validated the simulation results.

Song J et al [6] introduced patterned windows on to thin walled square tubes with the aim to reduce weight while maintaining the mechanical property of the original tubes. Experimental results showed that windowed tubes have lower initial peak force and higher specific energy absorption. Parametric study using finite element analysis was conducted to investigate the effect of window size on the tubes collapse characteristics. Three collapse modes were observed where the symmetric and extensional mode gave an increase in specific energy absorption. Zhou C et al. [7] manufactured origami crash boxes using a customised stamping die. Two types of origami patterns were proposed and subjected to axial impact by using a drop hammer test rig. The origami crash boxes performed better in terms of mean crushing force and energy absorption. Three collapse modes were observed where the complete diamond mode produced the highest energy absorption.

Crushing behaviours of circular tubes with shallow and deep corrugations were experimentally and theoretically investigated under axial loading conditions by Eyvazian A et al [8]. Three types of tube designs were tested and their failure modes and failure history were presented and discussed. The experimental results showed that the corrugated metal tubes provide perfect energy absorbing characteristics in terms of uniformity of load-displacement curve, reduction of initial peak force and controlling failure modes. The theoretical model based on experiment and modified simplified super folding element (MSSFE) theory was proposed to predict the mean crush load. Yang K et al. [9] introduced predesigned ellipsoidal dimples on to the surface of circular brass tubes subjected to quasi-static axial crushing. Finite element analysis was used to investigate the influence of various design parameters of the dimples on the tubes' crush response. The dimpled brass tubes were fabricated using 3D printing and experiments were carried out to validate the simulation results. The results showed that the dimpled tubes had substantially lowered the initial peak force and produce more uniform mean force as compared to the plain tube.

Cetin E and Baykasoglu C [10] investigated the crashworthiness performance of graded lattice structure filled tubes (GLSFTs) under multiple impact loadings. Results showed that the graded lattice structure designs provide variable stiffness throughout the length of the tubes and enable more folds to be formed without the tubes experiencing global bending. This consequently led to higher energy absorption under axial and oblique loadings. Significant improvement of energy absorption performance can be obtained by appropriate selection of design parameters. Sadeghi A et al. [11] studied circular tubes with different axial corrugations subjected to axial loading. The corrugated tubes were fabricated using forming process. Comprehensive experimental and numerical analysis have been conducted to investigate the effects of various geometrical parameters on the crushing behaviour of tubes. The corrugated tubes showed more uniform force-displacement curves and higher energy absorption. The corrugated tubes deformed in an inversion mode where the additional frictional force between the tube and inverted surfaces resulted in increase of specific energy absorption, mean force and crush force efficiency.

The crush performance indices of a tubular structure under axial loading is expressed in terms of initial peak force (IPF), crush force efficiency (CFE) and specific energy absorption (SEA). The initial peak force is the maximum force experienced by the structure before it collapses. It is desirable to have a low initial peak force so as to reduce the injury level in an event of crash. The crush force efficiency is the measure of the uniformity of the collapse force. It is expressed in term of percentage and is given by equation (1). An ideal energy absorber would have a crush efficiency of 100% which is very difficult to achieve in reality. Typical values of CFE for tubular structures under axial loading are between 30 to 50%.

IOP Conf. Series: Materials Science and Engineering 1062 (2021) 012035 doi:10.1088/1757-899X/1062/1/012035

$$CFE = \frac{Mean\ Force}{Peak\ Force} \tag{1}$$

Specific Energy Absorption (SEA) is the ratio of energy absorbed divided by the total mass of the structure. It allows a direct comparison of similarly shaped structures made from different materials.

$$SEA = \frac{Energy\ Absorption}{Mass} \tag{2}$$

2. Finite Element Modelling and Simulation

Based on literature review, an origami patterned square tube which has yet to be studied is proposed. The tube geometric configuration is illustrated in Figure 1. The function of the origami pattern is to allow the tube to fail in a predetermined and stable manner. The 3D CAD model of the origami patterned tube was created in Abaqus CAE preprocessor. The tube was modelled as surfaces where it was partitioned into several regions to represent the origami pattern. Also partitioning the surfaces enabled more uniform elements to be created during the subsequent meshing process. Table 1 shows the dimensions of the plain tube and origami patterned tube.

Table 2 shows the detail parameters for the different origami pattern arrangements on the square tube. The tube model was meshed using 4-noded quad S4R deformable elements. Altogether about 13832 elements were used for the origami pattern tube. Prior to that, a mesh sensitivity analysis was carried out to determine the suitable element size for the analysis. The complete assembly model is shown in Figure 2. The top and bottom plates are modelled as surfaces with discrete rigid elements. The bottom plate is fully fixed while the top plate is allowed to move in a vertical direction only. The square tube is made from Aluminium AA6065-T5 and the material mechanical properties is given in Table 3. For the contact and interaction of the multiple bodies, a coefficient of friction of 0.3 was assigned. A point mass of 36.34 kg and impact speed of 6 m/s were assigned to the top plate to simulate the drop hammer property. The values were chosen based on machine capacity.

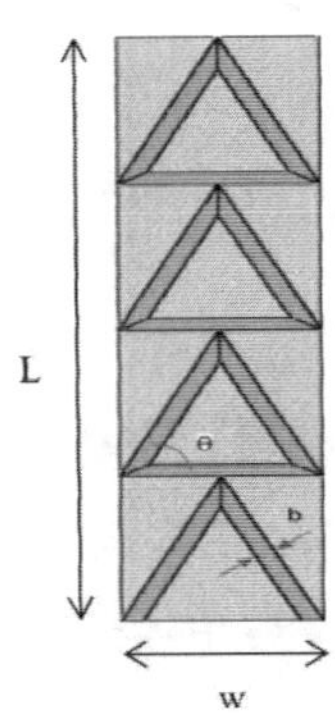

Figure 1. Geometric configuration of origami patterned square tube.

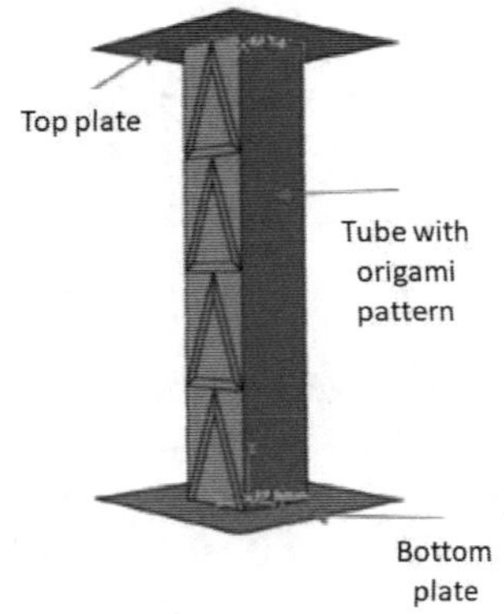

Figure 2. Complete assembly of the bottom plate, square tube and top plate in the finite element simulation.

IOP Conf. Series: Materials Science and Engineering 1062 (2021) 012035 doi:10.1088/1757-899X/1062/1/012035

Table 1. Dimensions for the plain square tube and the origami patterned square tube.

Parameter	Plain square tube	Square tube with origami
Length, L (mm)	200	200
Width, w (mm)	39	39
Thickness, t (mm)	1.5	1.5
Width of groove, b (mm)	-	3,4
Depth of groove, d (mm)	-	0.37, 0.75
Pattern angle, Θ (°)	-	64°, 68.7°

Table 2. Detail parameters for origami pattern arrangement of square tube.

Pattern arrangement		Detail parameters
Pattern 1	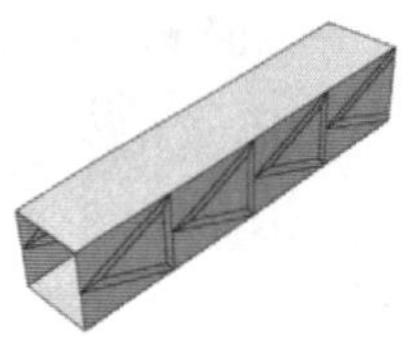	The pattern is employed on two opposing surfaces only Width of groove, b = 4 mm Depth of groove, d = 0.75 mm Pattern angle = 68.70°
Pattern 2	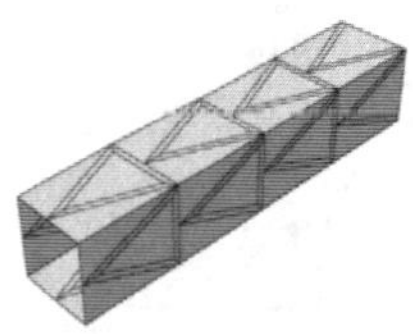	The pattern is employed on all surfaces Width of groove, b = 3mm Depth of groove, d = 0.37 mm Pattern angle = 68.70°
Pattern 3	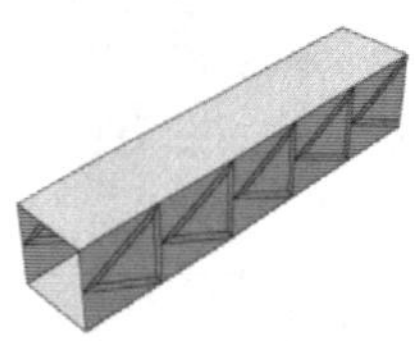	The pattern is employed on two opposing surfaces only Width of groove, b = 3 mm Depth of groove, d = 0.75 mm Pattern angle = 64°
Pattern 4	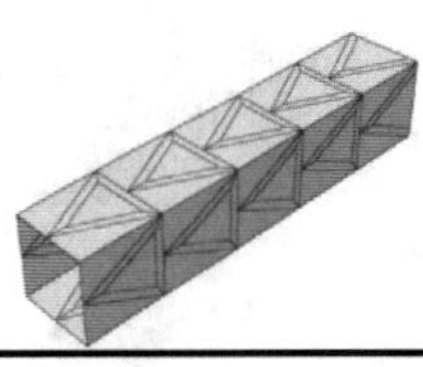	The pattern is employed on all surfaces Width of groove, b = 4 mm Depth of groove, d = 0.37 mm Pattern angle = 64°

Table 3. Mechanical properties of Aluminum AA6063-T5.

Density	2700 kg/m^3
Young's Modulus, *E*	73 GPa
Poisson's Ratio, ν	0.33
Yield strength, σ_y	180 MPa
Ultimate tensile strength, *UTS*	220 MPa
Plastic strain at *UTS*, ε_f	0.1

3. Static and Impact Experiments

The fabrication of origami patterned square tubes begins with the cutting of plain aluminum tubes into specified lengths by using an abrasive disc cutter. In order to ensure both ends of the tube are perfectly parallel to one another, the surfaces were finished using the milling machine. The origami pattern was manually machined using the vertical milling machine. The average time to machine the grooves on each surface of the tube was about 15 minutes, hence the total time to completely machine one origami patterned tube was about 40 minutes to one hour. The faces of the tube must be totally flat to ensure proper placement during the experiment. The spirit level ruler was used to check the straightness and flatness of the top and bottom tube faces on the stable table top.

The experiments were conducted to validate the simulation results. Static axial loading was carried out the INSTRON 3382 universal testing machine as shown in Figure 3. The tubes were crushed between the top and bottom compression platens with a loading rate of 5 mm/ min. The tube was crushed until 70% of its original length to ensure complete crushing of the tube. The force-displacement results were recorded using the Instron data acquisition software Bluehill.

An impact test was performed using INSTRON DYNATUP 8250 drop hammer machine as shown in Figure 4. The impact hammer is pneumatically assisted and the impact speed is controlled by the compressor pressure and drop height. Also the machine comes with a safety feature such as automatic locking sensor and impact resistance enclosure to ensure safety during an experiment. The experiment results as such as force, energy and acceleration were recorded using the Instron data acquisition software Impulse.

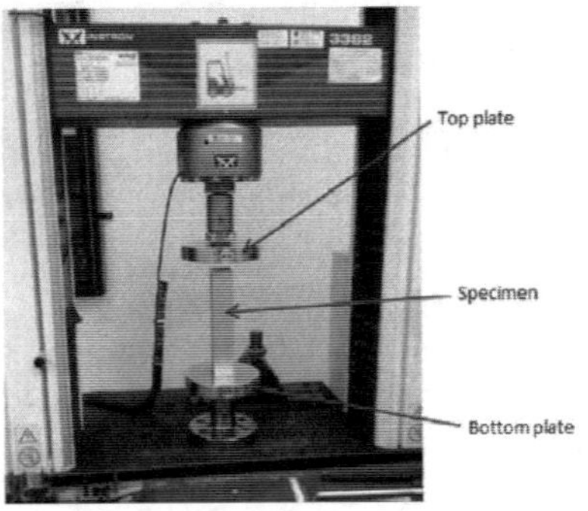

Figure 3. INSTRON 3382 Universal testing machine.

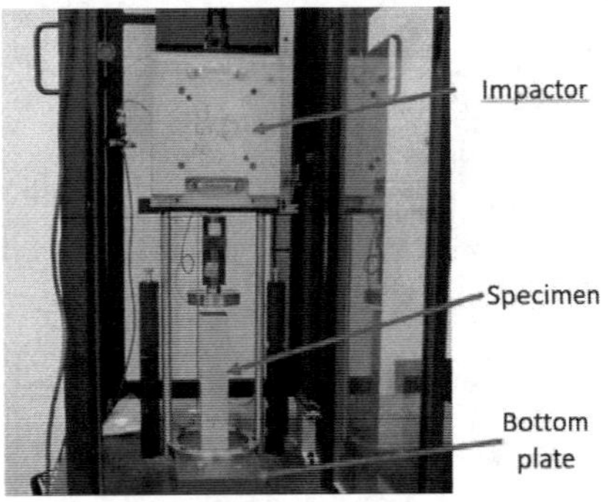

Figure 4. INSTRON Dynatup 8250 drop hammer machine.

4. Results and Discussion

4.1. Validation of Simulation Results

Figure 5 shows the experimental and simulation force-displacement curves of the plain square tube subjected to axial static loading. Altogether 5 samples were tested for the experiment. Simulation result showed good agreement with the experimental result where both curves showed high initial peak forces followed by lower fluctuating mean forces. The experimental IPF value was slightly lower than the simulation value. This could be due to the specimen imperfections in the form of tube geometry and material properties.

Figure 6 shows the experimental and simulation force-displacement curves of the square tube with origami pattern subjected to axial static loading. Simulation results showed good agreement with the experimental results. The experimental IPF and mean force were slightly lower than the simulation values due to the specimen imperfections. In addition, machining work to create the origami pattern may result in stress concentrations and residual stress that may further weaken the tube. Both simulation and experimental results have shown that the tube with origami pattern has lower IPF value compared to the plain tube. The outcome of which demonstrates the effectiveness of the origami pattern in reducing the IPF of the square tube subjected to static axial loading.

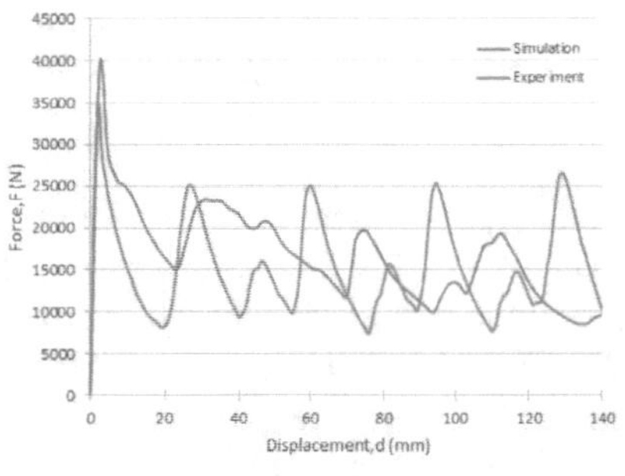

Figure 5. Experimental and simulation force-displacement curves for plain tube under axial static loading.

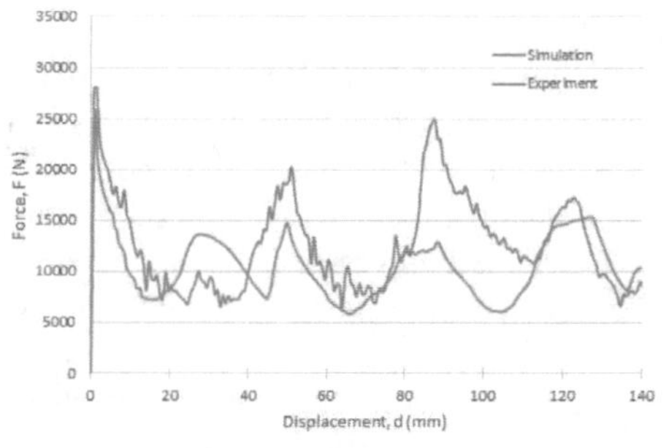

Figure 6. Experimental and simulation force-displacement curves for tube with origami pattern under axial static loading.

Figure 7 shows the experimental and simulation force-displacement curves of the plain square tube subjected to axial impact loading. Figure 8 shows the experimental and simulation force-displacement curves of the tube with origami pattern subjected to axial impact. Simulation results showed a similar trend as the experimental results. The simulation result has higher crushing distance compared to the experimental. This could be due energy lost in the form of heat, sound and vibration in the experimental impact which resulted in the lower crushing distance. Experimental results showed highly fluctuating force which gradually stabilized towards the end of the travel. This could be due to stress wave propagation in the contacting bodies during impact. Tube with origami pattern showed a lower IPF value and longer crushing distance as compared to the plain tube. This indicates that the tube with origami pattern absorbed impact energy in a more gradual and safer manner.

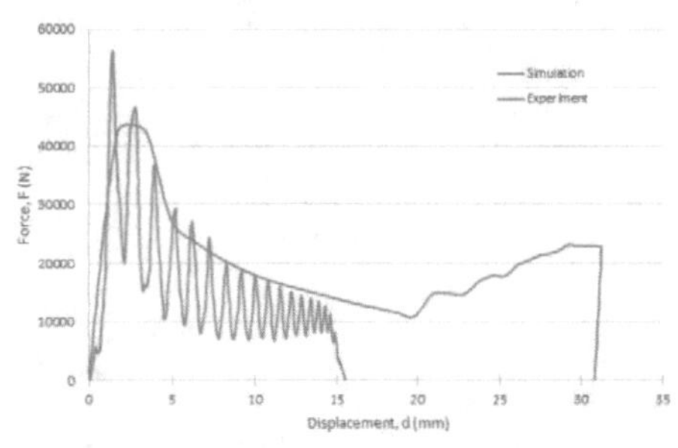

Figure 7. Experimental and simulation force-displacement curves for plain tube under axial impact loading.

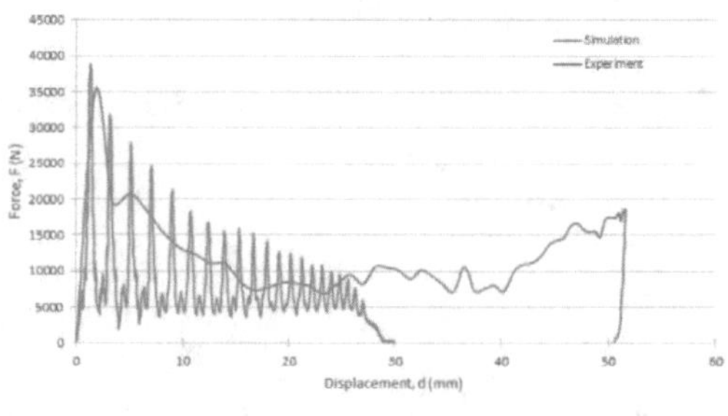

Figure 8. Experimental and simulation force-displacement curves for tube with origami pattern under axial impact loading.

Figure 9 shows the simulation energy-time curves for the tube with origami pattern subjected to axial static loading. Internal energy (IE) is the energy absorbed by the tube by plastic deformation. The artificial energy (AE) indicates the element distortion energy. Since this is a quasi-static loading, kinetic energy is almost negligible as there is no inertia effect. Figure 10 shows the simulation energy-time curves for the tube with origami pattern subjected to axial impact loading. The kinetic energy (KE) is the energy of the drop mass just before impacting the tube. The internal energy is the energy absorbed by the tube by plastic deformation. As the tube plastically deforms, the kinetic energy of the drop hammer is converted to internal energy absorbed by the tube. For perfect impact energy absorption, the kinetic energy is equal to the absorbed internal energy.

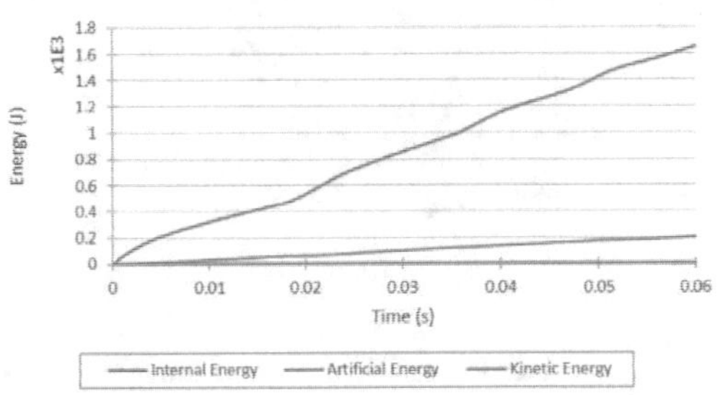

Figure 9. Simulation energy-time curves for tube with origami pattern under axial static loading.

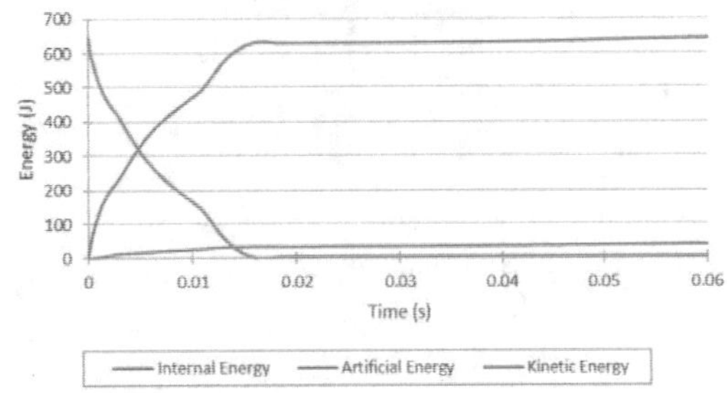

Figure 10. Simulation energy-time curves for tube with origami pattern under axial impact loading.

Figure 11 shows the final experimental and FE deformed shapes of the plain square tube subjected to static axial loading. Figure 12 shows the final experimental and simulation deformed shapes of the tube with origami pattern subjected to static axial loading. Both experimental and simulation results showed similar failure modes for the plain tube and tube with origami pattern. Failure was progressive buckling mode with similar number of folds. The size and number of the fold of the deformed shapes will dictate the fluctuating mean forces of the tubes crushing after the initial peak forces.

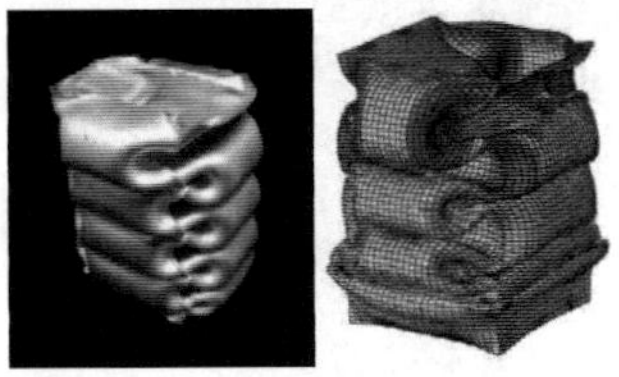

Figure 11. Experimental and simulation deformed shapes of the plain square tube under static axial loading.

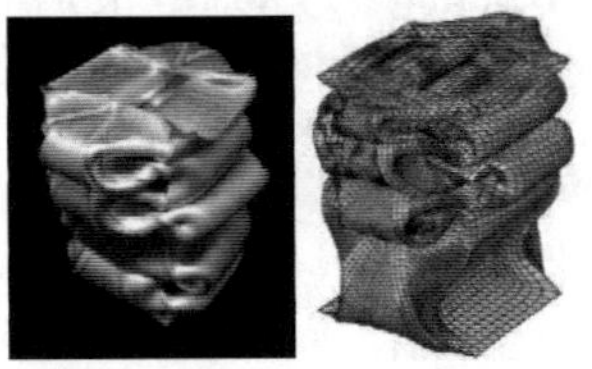

Figure 12. Experimental and simulation deformed shapes of the tube with origami pattern under static axial loading.

Figure 13 shows the final experimental and simulation deformed shapes of the plain square tube subjected to axial impact loading. Figure 14 shows the final experimental and simulation deformed shapes of the tube with origami pattern subjected to axial impact loading. Both experimental and simulation results showed similar failure modes for the plain tube and tube with origami pattern. For the plain tube, progressive buckling failure mode occurred at the bottom of the tube. Due to machine limitations in terms of impact speed and impact mass, only one fold could be achieved which resulted in only one peak recorded in the force-displacement curves. For the tube with origami pattern, the failure mode was similar but at a different location. For the experimental, failure occurred near the bottom of the tube while for the simulation, failure occurred near the top of the tube. This could be due stress wave propagation and stress concentrations on the sample due to machining process of the origami pattern.

Figure 13. Experimental and simulation deformed shapes of the plain square tube under axial impact loading.

Figure 14. Experimental and simulation deformed shapes of the tube with origami pattern under axial impact loading.

4.2. Static and Impact Simulation Results of Tubes with Different Origami Patterns

Figure 15 shows the force-displacement curves for plain square tube and tubes with different origami patterns subjected to axial static loading. Figure 16 shows the failure modes of all tubes with different origami patterns. All tubes failed in progressive buckling mode with different folding size and shape. The plain square tube exhibited the highest IPF while all tubes with origami patterns showed a significant reduction in IPF. From Table 4, it can be seen that Pattern 4 had the lowest IPF, a decrease of 69.1% compared to the plain tube. Pattern 4 also showed the highest CFE with an increase of 134.7% compared to the plain tube. The origami patterns on all four sides of the tubes reduce the resistance to progressive collapse resulting in a lower IPF value and an improved CFE value. In terms of SEA, Pattern 3 showed the highest increase of 27.9% compared to the plain tube. The plain opposing sides of Pattern 3 may result in higher energy absorption due to its higher stiffness. The findings above suggest that the addition of origami pattern on the tube may reduce the IPF and increase the CFE but at the expense of SEA. More parametric study needs to be carried out to optimize the design of origami pattern on the tube for energy absorption under static loading.

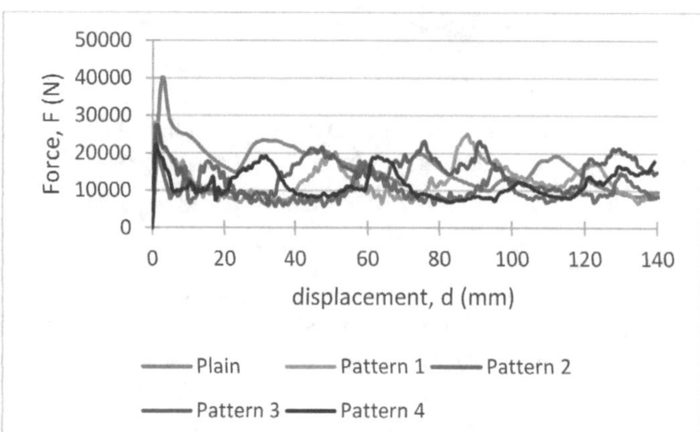

Figure 15. Simulation force-displacement curves of plain square tube and tubes with different origami pattern under axial static loading.

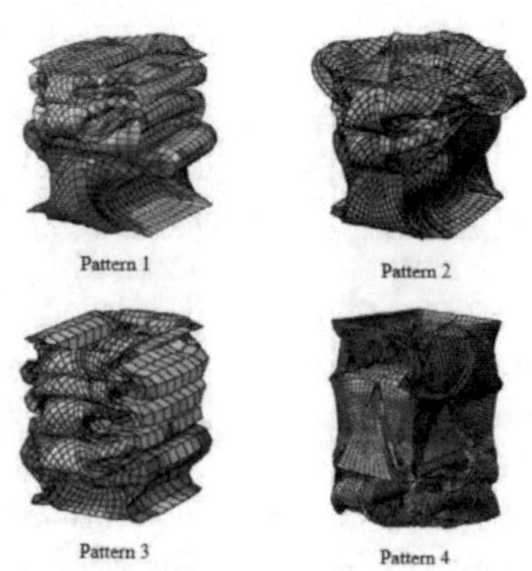

Figure 16. Failure modes of tubes with different origami patterns subjected to axial static loading.

Table 4. Changes in crush performance indices of tube with different origami patterns subjected to axial static loading.

Crush Performance Index	Plain Square Tube	% Change			
		Pattern 1	Pattern 2	Pattern 3	Pattern 4
IPF	39152 N	-28.5	-65.5	-29.8	-69.1
CFE	40.5 %	+6.5	+80.1	+34.1	+134.7
SEA	12692 J/kg	+6.7	-16.5	+27.9	+20.6

Figure 17 shows the force-displacement curves for plain square tube and tubes with different origami patterns subjected to axial impact loading. Figure 18 shows the failure modes of all tubes with different origami patterns. All tubes failed in progressive buckling mode where failures occurred at different parts of the tubes. For Patterns 1 and 2, failures occurred near the top of the tubes. For Patterns 3 and 5, failures occurred at the middle of the tubes. The plain square tube exhibited the highest IPF while all tubes with origami patterns showed a significant reduction in IPF. From Table 5, it can be seen that Pattern 2 had the lowest IPF, a decrease of 73.5% compared to the plain tube. Pattern 2 also showed the highest CFE with an increase of 126.9% compared to the plain tube. The origami patterns on all four sides of the tubes reduce the resistance to progressive collapse resulting in lower IPF and better CFE. In term of SEA, all tubes with origami pattern showed slight reduction compared to the plain tube. Among all tubes with origami pattern, Pattern 3 gave the best SEA with a 3.6% reduction compared to the plain tube. The findings above suggest that the addition of origami pattern on the tube may reduce the IPF and increase the CFE but at the expense of SEA. More parametric study needs to be carried out to optimize the design of origami pattern on the tube for energy absorption under impact loading.

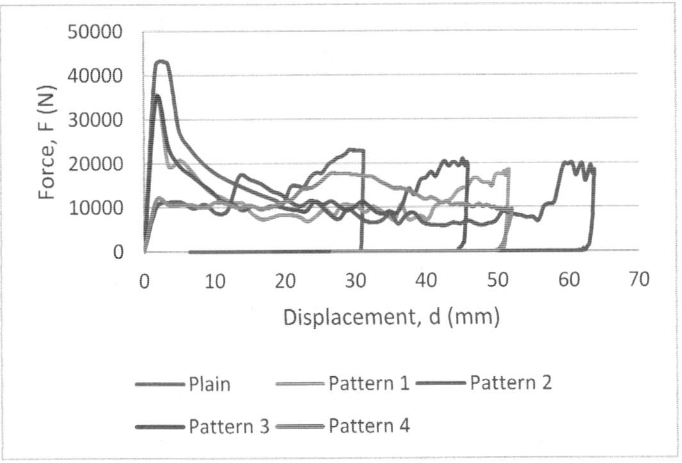

Figure 17. Simulation force-displacement curves of plain square tube and tubes with different origami pattern under axial impact loading.

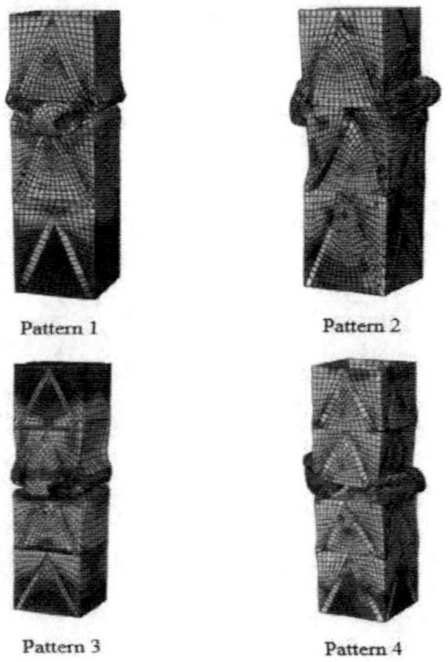

Figure 18. Failure modes of tubes with different origami patterns subjected to axial impact loading.

Table 5. Changes in crush performance indices of tube with different origami patterns subjected to axial impact loading.

Crush Performance Index	Plain Square Tube	% Change			
		Pattern 1	Pattern 2	Pattern 3	Pattern 4
IPF	42390 N	-17.1	-73.5	-18.4	-72.2
CFE	41.1 %	-21.8	+126.9	-11.2	+106.8
SEA	5098 J/kg	-4.1	-4.7	-3.6	-4.2

5. Conclusion

In axial static loading, Pattern 4 showed the lowest IPF which is a reduction of 69.1% compared to the plain square tube. Pattern 4 also gave the highest increase in CFE with an increase of 134.7% compared to the plain square tube. In terms of SEA, Pattern 3 showed the highest increase of 27.9% compared to the plain square tube. In axial impact loading, it can be seen that Pattern 2 shows the lowest IPF, a reduction of 73.5% compared to the plain square tube. Pattern 2 also gave the highest increase in CFE of 126.9% compared to the plain square tube. In terms of SEA, Pattern 3 gave the best performance which is a reduction of only 3.6% compared to the plain square tube.

Overall, the predicted results have shown that the employment of origami pattern on the square tube has reduced the initial peak force (IPF) and increased the crush force efficiency (CFE). The specific energy absorption (SEA) of the origami tubes under static loading is improved while for impact loading, the SEA showed a slight reduction. Based on this study, it can be observed that the geometry and dimension of the origami pattern influence the crush response of the tube. Further detail parametric study needs to be carried out to optimize the crush response of the tube in terms of IPF, CFE and SEA under static and dynamic loadings. The study has shown that the origami pattern can be employed on various tubular structures to improve their crush performance for a wide range of application.

References

[1] Alghamdi AAA 2001 *Thin-walled Structures* **39**(2) 189
[2] Lu GX and Yu TX 2003 *Energy Absorption of Structures and Materials* (Cambridge: Woodhead)
[3] Abramowicz W 2003 *Thin-walled Structures* **41** 91
[4] Zhang X, Cheng GD, You Z and Zhang H 2007 *Thin-walled Structures* **45** 737
[5] Song J, Chen Y and Lu G 2012 *Thin-walled Structures* **54** 65
[6] Song J, Chen Y and Lu G 2013 *International Journal of Mechanical Sciences* **66** 239
[7] Zhou C, Wang B, Ma J and You Z 2016 *International of Mechanical Sciences* **118** 1
[8] Eyvazian A, Tran TN and Hamouda AG 2018 *International Journal of Non-linear Mechanics* **101** 86
[9] Yang K, Xu S, Zhou S, Shen J and Xie XY 2017 *Thin-walled Structures* **112** 31
[10] Cetin E and Baykasoglu C 2020 *Thin-walled Structures* **154** 1
[11] Sadighi A, Eyvazian A, Asgari M and Hamouda AG 2019 *Thin-walled Structures* **145** 1

Vulnerability Assessment of Tsunami-Affected Inundated Area Using Geospatial Analysis Based Tsunami Run-Up Simulation

Rina Febrina[1*], Lusmelia Afriani[2], Rr M I Retno Susilorini[3] and Tri Prastio[1]
[1]Departement of Civil Engineering, Universitas Malahayati, Bandar Lampung 35153, Indonesia
[2]Departement of Civil Engineering, Universitas Lampung, Bandar Lampung 35153, Indonesia
[3]Departement of Infrastucture and Environmental Engineering, Soegijapranata Catholic University, Semarang 5023, Indonesia

*Corresponding e-mail: febrinacivil@yahoo.com

Abstract. A tsunami can cause considerable loss of life, extreme property damage and infrastructure destruction. A tsunami run-up simulation can be used to forecast wave generation caused by an earthquake. This study aims to assess the tsunami-affected inundated area on the west coast of Lampung province using geospatial analysis-based tsunami run-up simulation. The parameters used in the assessment were applied elevation and slope computed from Aster GDEM, coastline distance determined from a vector map and tsunami inundation obtained by Hills and Mader equation and vegetation density calculated from ALOS ANVIR-2 image. The AHP process was applied for weighing the parameter through a pair-wise comparison using four iterations of the normalised matrix. Five classes of vulnerability were defined and analysed for the tsunami inundation mapping. Weighted overlay through spatial analyst in GIS was used to draw the final map of tsunami vulnerability. The assessment result indicates that 11.26 square kilometre of the study area falls under the high tsunami vulnerability zone and 6.6 square kilometres of the building area comes under the inundation area. The results presented in this study may facilitate improvements in basic data for city planning related to evacuation processes and management strategies during disasters.

Keywords: tsunami, property damage, infrastructure, vulnerability

1. Introduction

Tsunami based disasters are natural, highly destructive and unavoidable phenomena. Vulnerability mapping assessments of a tsunami are observed as an effective way of reducing the impact of the disasters and have been successfully applied for disaster mitigation measures [1-4]. The assessments include the mapping of evacuation covering routes and buildings, the construction of tsunami barriers, disaster risk management and regulation and disaster education for the public [5]. Several pieces of formative information such as reliable and accurate spatial and temporal characteristics and impact of a tsunami with damage potential at different magnitude scales are required for the assessments. The availability of the information has created more parameters which are vitally important in developing a realistic description of vulnerability for a spatial and temporal assessment. The probability of risk, exposure, capacity measurement, and vulnerability are parameters used to determine a disaster risk [6-8]. However, it was shown that external pressure imperilled a person, community, or a system. Vulnerability is explained as a process and condition refers to that produced by economic, environmental, physical, and social factors. Moreover, vulnerability enhanced the susceptibility of hazards impact community [9,10].

Mapping the distribution of damage area because of disasters and assessing the vulnerability area via remote sensing and GIS are seen to be a potentially effective way of mitigating the disastrous impact of a tsunami. Geo-reference data were used to make a decision, while GIS analysis was employed for the assessment through spatial multi-criteria. On top of that, geographical references and criterion attributes were needed [11,12]. Previous studies used the GIS method for tsunami and an overlay of official land-use mapping [13]. In addition, while GIS can be used to analyse the risk of tsunami, remote sensing data and analysis of the physical built-up of infrastructures, such as building, contour of inundation area, and highest level of tsunami can be used to predict vulnerability [5],[14].

The aims of this study are to map tsunami vulnerability areas on the west coast of Lampung Province and to evaluate the tsunami impact by mapping possible inundation areas using a geospatial analysis-based tsunami run-up simulation. Vulnerability is calculated based on the run-up, elevation, slope, coastline distance and vegetation density using tsunami run-up simulation generated by software L-2008. Furthermore, the Soil Adjusted Vegetation Index (SAVI) is applied. The vegetation mapping is used as a Normalised Difference Vegetation Index (NDVI). In addition, the analytical hierarchy process is used to estimate the weights of the parameters.

2. Research Methodology

2.1. Data

The ASTER Global Digital Elevation Model (ASTER GDEM) version 2 provided the elevation data. ASTER GDEM was used as a platform for data collection via a space-borne observing optical device, while the 2nd version of ASTER GDEM was used to enhance the resolution of global digital elevation and the accuracy of elevation. In addition, ASTER GDEM was also used in advancing algorithm and reprocessing a total of 1.5 million scene data, including 250,000 additional scenes acquired after the previous release. The data were posted approximately 30 m at the equator. Subsequently, the ALOS ANVIR-2 image was analysed for the vegetation density mapping. Moreover, run-up information was used based on the tsunami run-up simulation (L-2008). The protocols of analysis were started from the data collection itself. It proceeded with the surface analysis of DEM data, vector data processing, and inundation processing which continued with the AHP process and finally the raster overlay processing through the GIS approach as in Figure 1.

2.2. Study area

The study was conducted on the west coast of Lampung Province as in Figure 2, non the island of Sumatra. It is an area which is highly exposed to a wide range of potentially devastating impacts from earthquake and tsunami threats.

2.3. Simulation of tsunami

2.3.1. Source modelling

The initial profile of a tsunami centre can be obtained by considering it as a seabed deformation with fault parameters, determined on the mechanism solution data of an earthquake centre. The parameter setting for the source modelling is shown in Figure 3. A11 (km) and A12 (km) refer to the fault length, and Aw1 is the fault width. The positions x_eq fault, y_eq, and z_eq are defined as the longitude, latitude and depth of the earthquake, respectively. Furthermore, the source modelling was conducted using the L-2008 software. The output of the source modelling is the vertical displacement that occurs in the seabed. Subsequently, this deformation was used in the ocean and run-up modelling.

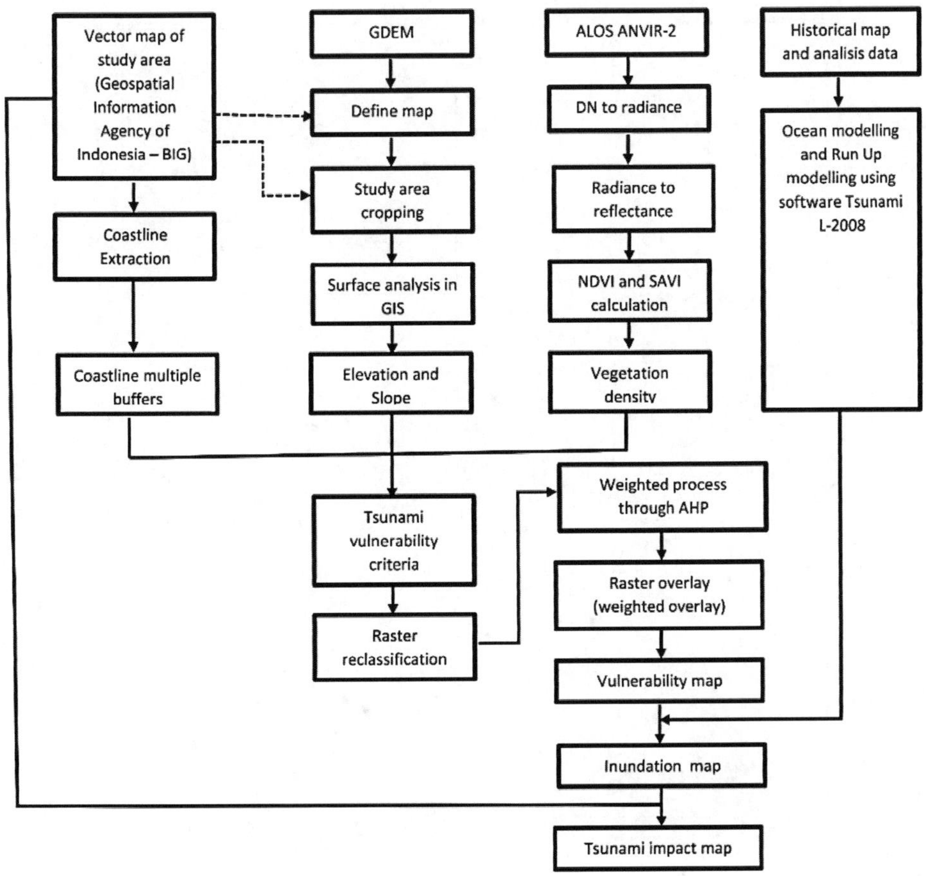

Figure 1. General framework of study

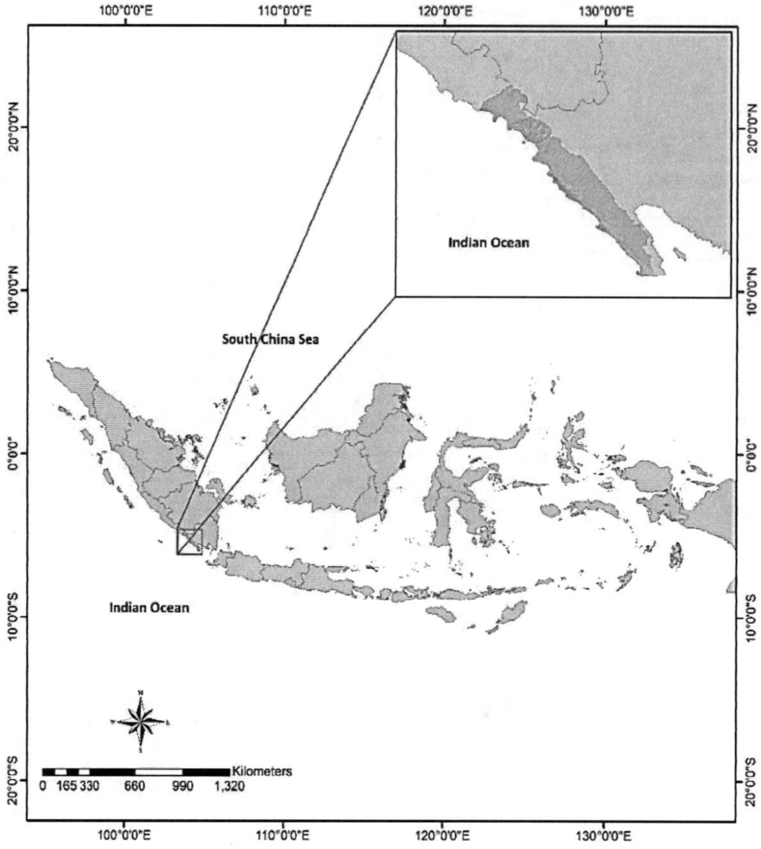

Figure 2. Map of study area

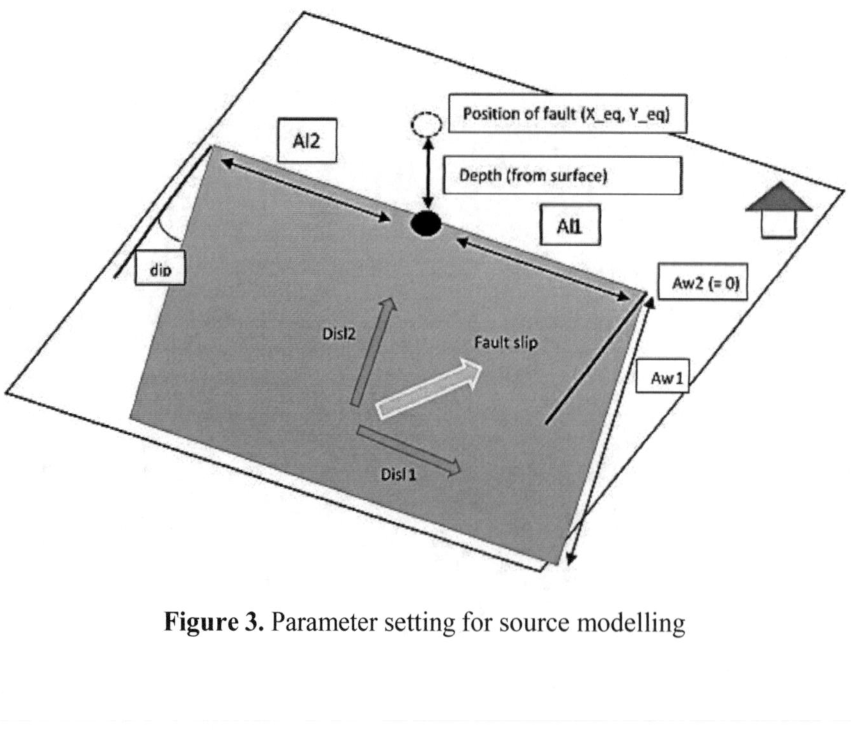

Figure 3. Parameter setting for source modelling

IOP Conf. Series: Materials Science and Engineering 1062 (2021) 012036 doi:10.1088/1757-899X/1062/1/012036

2.3.2. Ocean modelling
Output data of source modelling and bathymetric data were needed for ocean modelling. In addition, the appropriate bathymetric data with a study area was needed to get a good simulation result. In this ocean modelling, the parameter simulation as an input software tsunami L-2008 was maximum time. Maximum time was needed to vine the simulation of 3600s. Subsequently, the data save internal of 25s was added and resulted in 1 output per 25s.

2.3.3. Run-up Modelling
In this study, the height of the tsunami modelling (run-up modelling) was demonstrated by digitising the predetermined area. Digitising was conducted at various points along the coastline of the tsunami-affected area. On observation, every affected area recorded the highest run-up value.

2.4. Spatial analysis

2.4.1. Elevation classification
Elevation data are required to calculate vulnerability and tsunami inundation. A digital elevation model was created using an elevation map. The data from GDEM which were downloaded from DEMNAS website were used to derive a set of physical vulnerability parameters. The elevation was then classified into five vulnerability categories as in Figure 4 based on the height of the surface. A low elevation for example indicated a high tsunami vulnerability wave as in Table 1.

Table 1. Vulnerability class based on elevation, slope and coastline distance [15], [16]

Elevation (m)	Slope (%)	Coastline distance (m)	Vulnerability class
< 5	0 – 2	0-545.65	High
5 - 10	2 – 6	545.65 - 1261.80	Slightly high
10 - 15	6 – 13	1261.80 - 2187.40	Medium
15 - 20	13 – 20	2187.40 - 3154.65	Slightly low
>20	>20	>3154.65	Low

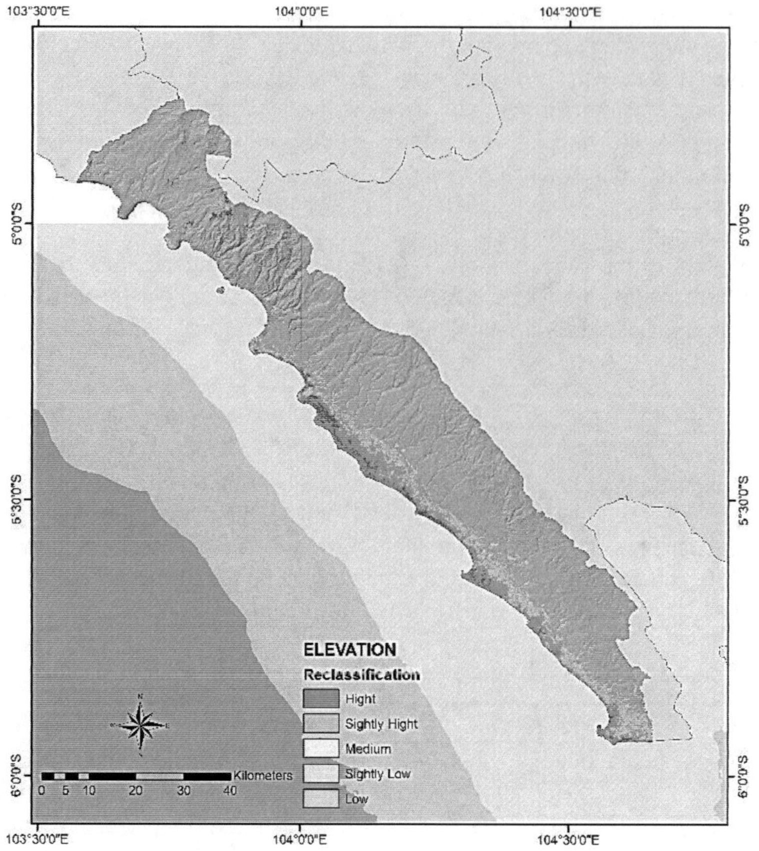

Figure 4. Tsunami vulnerability map based on parameters of elevation

2.4.2. *Slope*

In this study, a slope is identified based on the maximum changing of the z-value rate from each cell of the image. The third-order finite difference method and GDEM were applied to create a slope map. The range of the slope percentage was identified as zero to almost infinity. The slope map was categorised into five categories according to the value of tsunami vulnerability, as described in Table 1. The vulnerability map shown in Figure. 5 was made by applying slope classification, which is based on the tsunami vulnerability class.

2.4.3. *Coastline distance*

Coastal areas were divided into five classes of vulnerability using a multi-buffering calculation. Buffering distance was determined using the possibility range of tsunami when reaching the ground. The maximum tsunami height which was obtained from historical reports was used to calculate the buffering distance below [13],[17]:

$$logX_{max} = \log 100 + \frac{4}{3} log\left(\frac{Y_0}{10}\right) \tag{1}$$

where X_{max} is the maximum reach of the tsunami over land, and Y_o is the tsunami height at the coast. Besides, the buffering distance was determined based on the range of possible tsunami flows flowing

IOP Conf. Series: Materials Science and Engineering 1062 (2021) 012036 doi:10.1088/1757-899X/1062/1/012036

toward the inland. The score of the X_{max} in this study was based on the simulation of a tsunami using software L-2008. Coastline distances in the vector map of Lampung through GIS processing were used to generate a vulnerability map based on coastline distance.

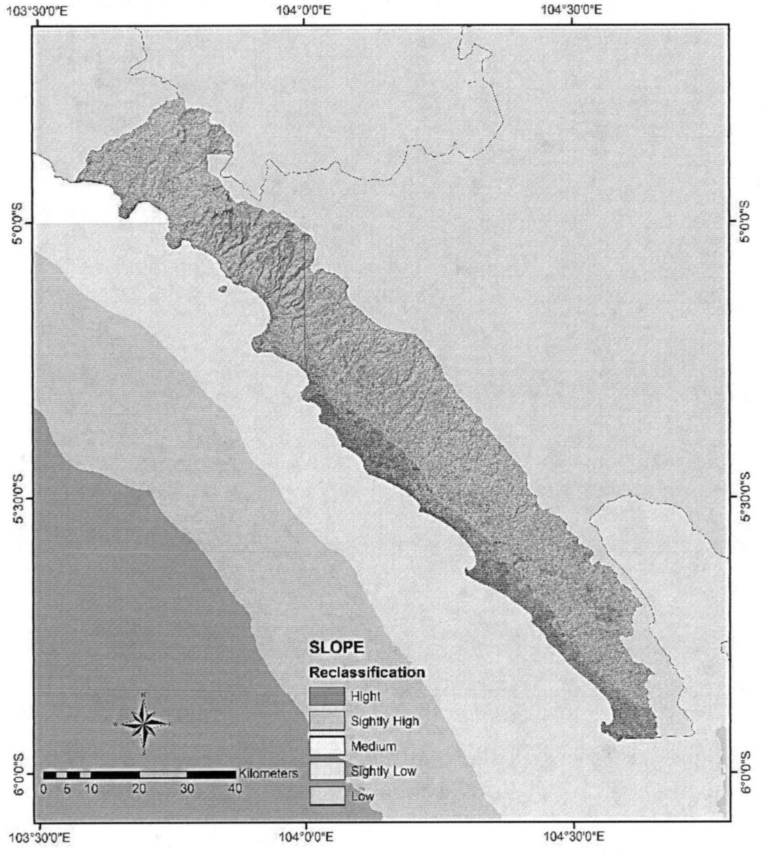

Figure 5. Tsunami vulnerability map based on parameters of slope

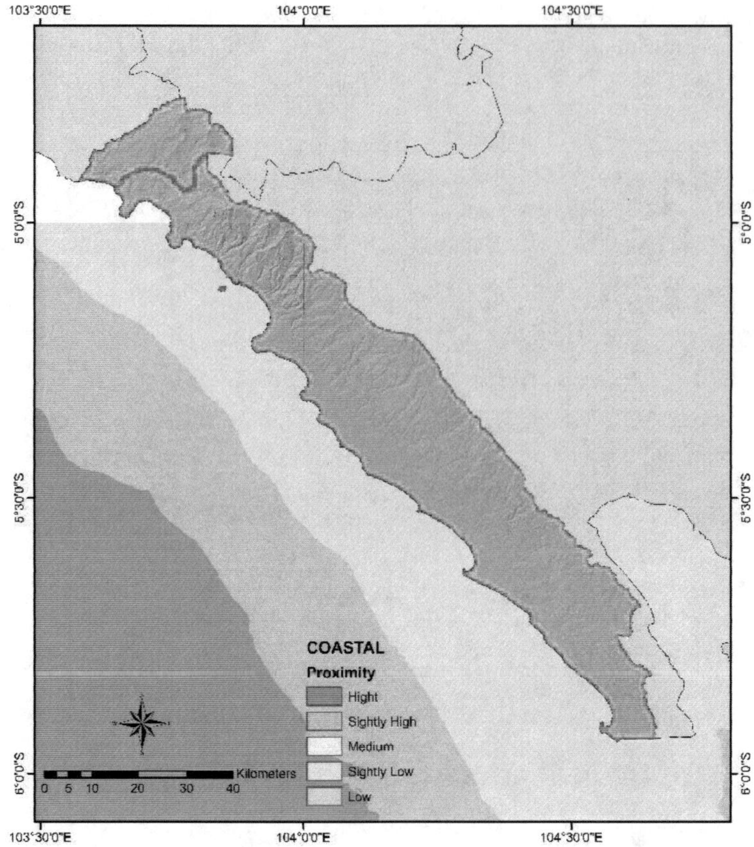

Figure 6. Tsunami vulnerability map based on the parameters of coastline distance

2.4.4. Inundation

The simulation of a tsunami was demonstrated using ArcGIS 10.6 software with the Raster calculator functions [21] as shown below:

$$X_{max} = (Hs)^{1,33} n^{-2}\, k\, Cos\, S \tag{2}$$

where X_{max} is the maximum inundation (m), Hs is the maximum tsunami run-up (m), $k$is constant (0.06), manning coefficient of the land is n, the slope of the land is S, and X_{max} in the vector map of Lampung through GIS processing to generate a vulnerability map based on inundation.

2.5. ALOS AVNIR-2 Image Processing

SAVI was used for mapping vegetation density. Briefly, the digital numbers of ALOS AVNIR-2 were converted to reflectance values before generating the synthetic NDVI and SAVI images [19,20]. A vegetation density map was created as below:

2.5.1. Reflectance Calculation

A radiance conversion from digital number values is required before calculating reflectance. Calculating radiance is a fundamental step in relating image data from multiple sensors and platforms into a common radiometric scale [21]. The general equation for DN to radiance conversion is demonstrated in equation (3) [22].

$$L_\lambda = G_{rescale} \times QCAL + B_{rescale} \tag{3}$$

where L_λ is spectral radiance at the sensor's aperture (W/m2/sr/μm), Grescale is rescaled gain, $QCAL$ is Digital Number (DN), and $B_{rescale}$ is rescaled bias [23]. Conversion of reflectance is obtained using Equation (4),

$$\rho_\lambda = \frac{\pi \times L_\lambda \times d^2}{ESUN_\lambda \times cos\theta_s} \tag{4}$$

where ρ_λ is unitless planetary reflectance and L_λ is spectral radiance at the sensor's aperture. Meanwhile, $d^2$is the earth-sun distance in astronomical units from a nautical handbook, $ESUN_\lambda$ is mean solar Exoatmospheric irradiances, and θ_s is solar zenith angle in degrees.

2.5.2. NDVI Calculation

The concept of NDVI is about the first channel causing considerable absorption of incoming sunlight and the second channel creating considerable reflectance [24-26]. The differences in wavelength are calculated using NDVI. The range of wavelength used is between -1 and 1, in which 0.5 indicates dense vegetation and < 0 indicates no vegetation, including water [27]. NDVI is calculated as

$$NDVI = \frac{(NIR - VIS)}{(NIR + VIS)} \tag{5}$$

where NIR is Near Infra-Red band, and $VIS$is a visible band of the red band. ALOS AVNIR-2 band 3 is red, and band 4 is NIR.

2.5.3. SAVI Calculation

SAVI can be defined as an algorithm developed to generate vegetation indexes by eliminating soil factor [28]. A SAVI based model is capable of characterising the forested and non-forested area [20]. Therefore, the model is created to reduce soil brightness, such as shadow [29]. NDVI is calculated as [30],

$$SAVI = \frac{(NIR - R)}{(NIR + R + L)} x(1 + L) \tag{6}$$

where NIR is Near Infra-red band, R is red band, and L is the soil calibration factor, in which L is equal to 0.5 [31]. The secondary backscattering effect of canopy transmitted is reduced using a correction factor (L).Lvalue of 1 is optimal in semiarid environments. The SAVI generated map is shown in Figure 7(a).

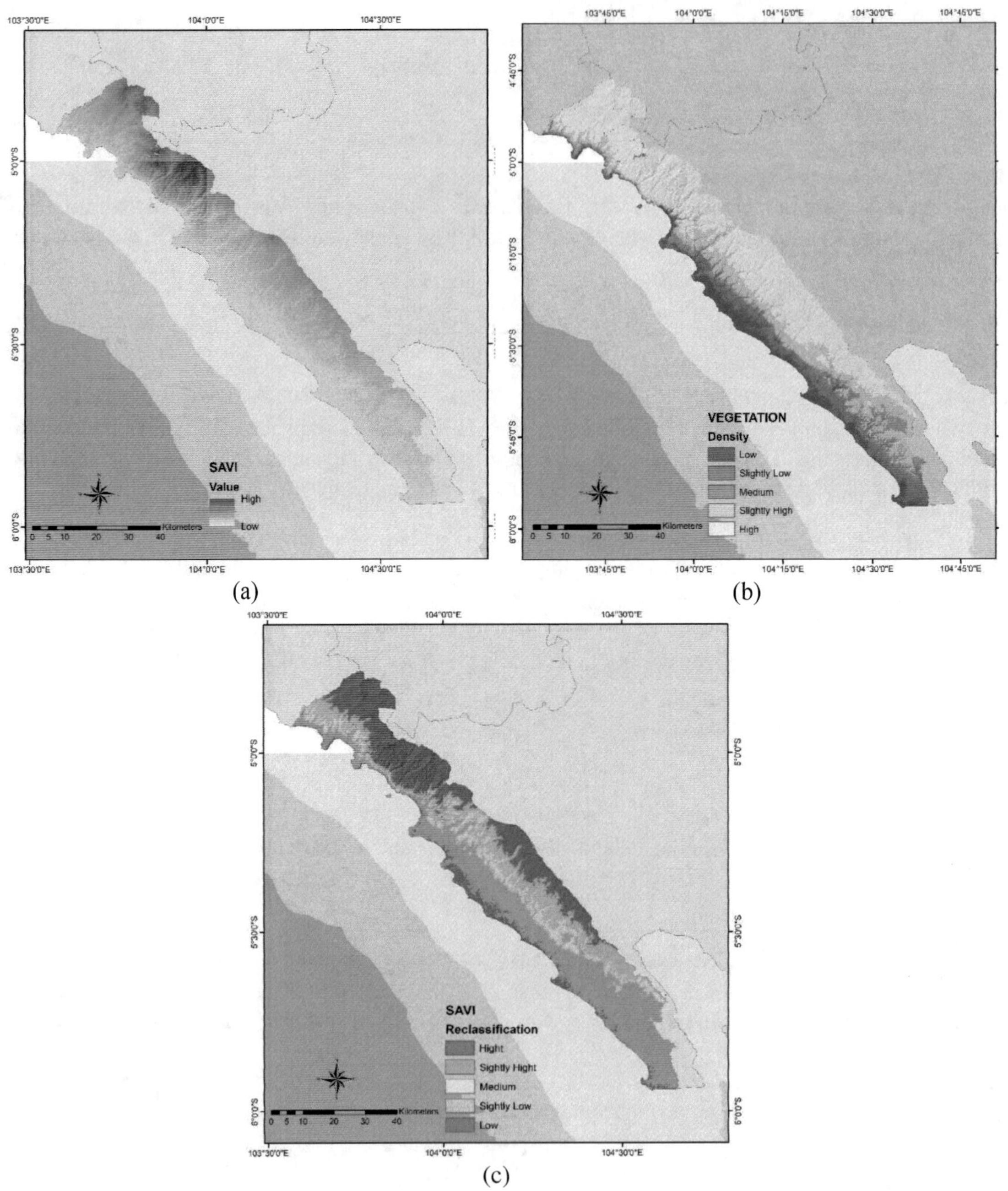

Figure 7. (a) SAVI Map; (b) Vegetation density; and (c) Vulnerability map based on vegetation density

3. Results and discussion

3.1. Tsunami run-up simulation

A scenario simulation was demonstrated in the central earthquake location on the ocean plate of the West coast of Lampung. Tsunami modelling parameters which are length, width, maximum and average displacement were created for measuring time, amplitude, and wave height of the tsunami in one area. The input process of the tsunami model was generated using the bathymetric input data of the earthquake location on the Western region of Sumatera island, and topography data required were used in the ASCII

XYZ format. The results of the input parameters decision for source modelling are shown in Table 2. Subsequently, the mechanism data of earthquake were prepared by deciding on the design modelling, while the grid creation of the bathymetric value, simulation design of tsunami and the wave height of tsunami modelling were then determined. The initial profile of a tsunami centre can be calculated by considering the profile as a seabed deformation with fault parameters which can be determined by using the mechanism solution data of the earthquake centre. Deformations of the seabed as outputs of centre modelling were assessed using tsunami L-2008. Furthermore, it was applied in the ocean and run up modelling. The vertical displacement (deformation of the seabed) as an output of source modelling was needed in the ocean modelling using L-2008 application and bathymetric data. The bathymetric coordinate boundaries of this study were between 4°51'23.43"S to 6°43'50.50"S and 102°3'3.30"E to 105° 2'29.63"E. Moreover, the level of the earthquake was 7.8 SR, and the distance of the coast to the epicentral was 241 km. In this study, the input parameter was observed at 739.66 km x 82.41 km of length and width fault, resulting in a fault area and slip of 60955.38 km^2 and 11.036 m respectively.

Table 2. Earthquake input parameters for source modelling

Parameter		Research Model	
Length (km)	Al1	369.83	369.83
	Al2	369.83	369.83
Width (km)	Aw1	82.41	82.41
	Aw2	0	0
Depth (km)		28	28
Strike (Degree)		312	58
Dip (Degree)		223	64
Slip (m)		11.036	11.036
Mw (Richter)		7.8	7.8
Centre Fault Coordinate	Lat.	-7.082527°	-7.082527°
	Long.	103.307943°	103.307943°
Type Fault		Normal Fault	Normal Fault

The result of the model design was vertical displacement in Figure 8, shown as deformation in the seabed. The movement of deformation which occurred in the seabed was followed by a shift in the sea level. The source modelling results illustrated variation in the maximum vertical displacement and event and a minimum vertical displacement value of 3.35 m and -3.35 m, respectively. The maximum vertical value showed an increase in the seabed vertical movement, but the minimum vertical displacement showed a reduction in the seabed vertical movement. The rapid vertical movement of the seabed was detected as a response to the earthquake and had caused the rise and fall of the sea level. The ocean modelling results showed the propagation of tsunami waves in all directions. In this study, the timing and height that were taken for the propagation of a tsunami wave from the epicentre to the beach (target area) were 28.56 min and 2.65 m, respectively shown in the Figure 9. In this study, the target location was Krui on the west coast Lampung. The run-up modelling result from the calculation fault of the length and width is shown in Figure 10. The run-up score was affected by the distance of the epicentre to the affected area, bathymetric, the morphology of the beach, the slope of the beach, and the wave height of the tsunami [32]. Enhancement of the height of the tsunami wave was comparable with the shallowness of the approaching land.

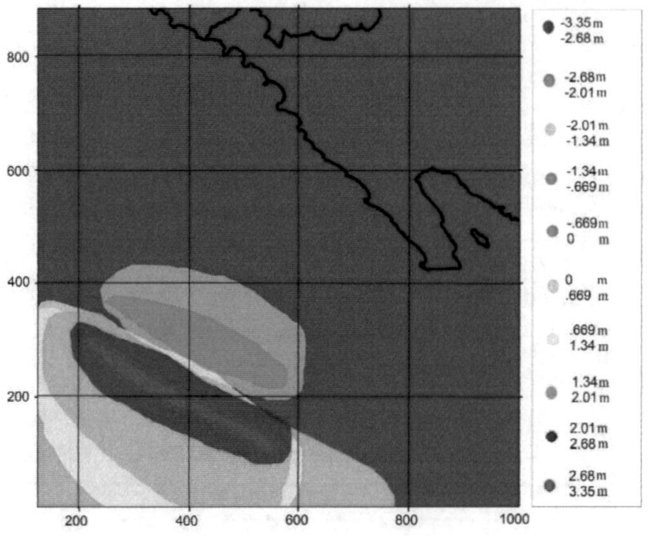

Figure 8. Vertical displacement

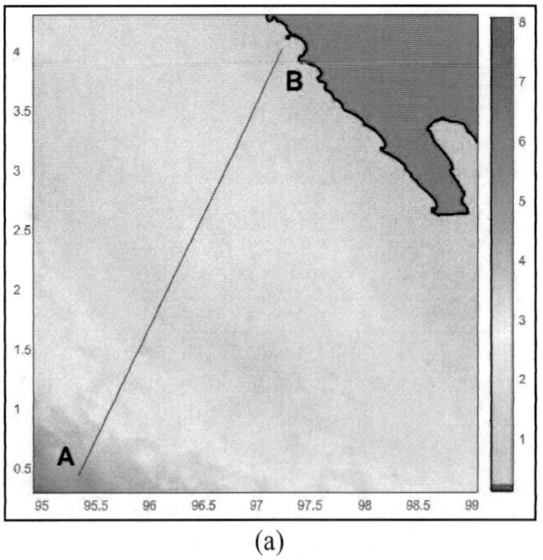

(a)

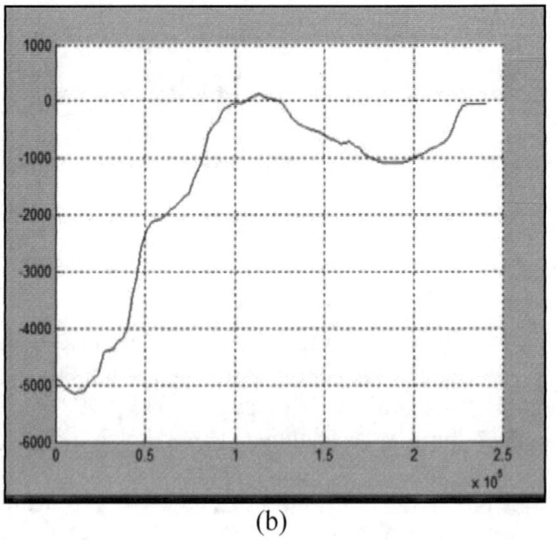

(b)

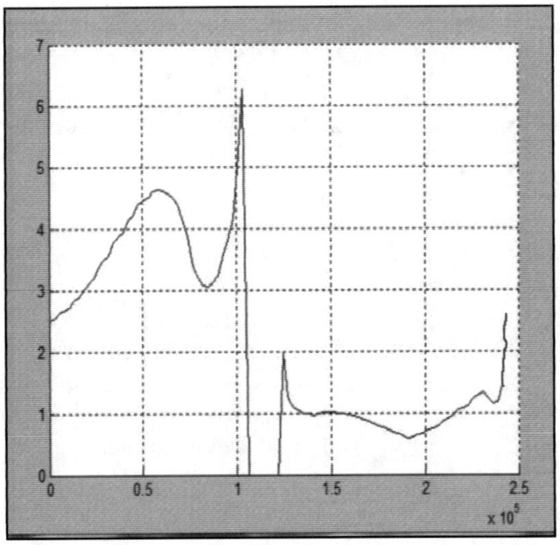

(c)

Figure 9. (a) Cross-section of plot A-B (241 km); (b) Cross-Section of ocean bottom A-B; (c) Cross-section of maximum tsunami height A-B

This phenomenon was followed by refraction, reflection, and diffraction and had brought about a higher tsunami wave. In this study, the height of the run-up and inundation were analysed. The length of the inundation affected the tsunami wave which then affected the population.

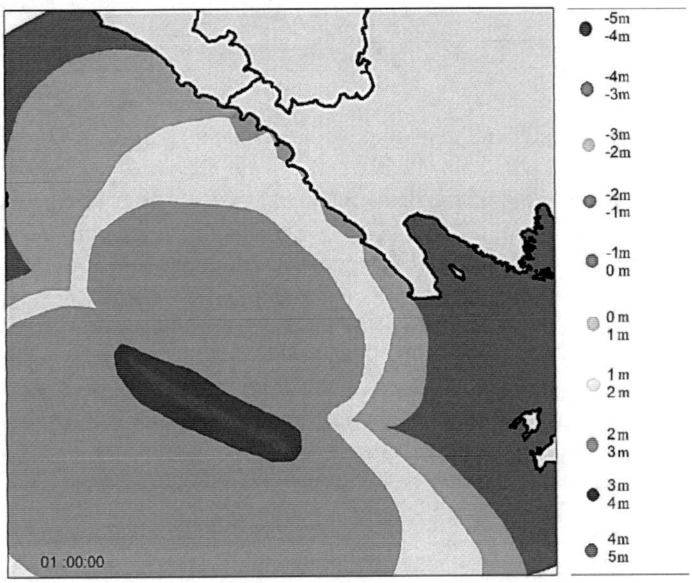

Figure 10. Tsunami height based on calculation

3.2. AHP processing

This study classified all the parameters of tsunami vulnerability into five classes: Number 1, 2, 3, 4, and 5 in which representing low, slightly low, medium, slightly high, and high vulnerability, respectively. All the parameters were assessed using a weighting and scoring system in order to determine the limiting factors of each parameter. The weighting system was also used to determine the different classes of tsunami vulnerability. The AHP was applied to assess the weight of each parameter using pair-wise comparisons. The AHP method created a scaled set of preferences and described the essential parameters in relation to other parameters [33-37]. In addition, assessed the natural hazards of vulnerability. The sequencing number of 1 to 9 as tabulated in Table 3 was used as a comparison scale to create a pair-wise comparison matrix. The number depended on the relative weight of each parameter and the relative weight was produced by pair-wise comparison in Figure 11.

Table 3: Nine-point comparison scale by Saaty [38]

Intensity of importance	Definition
1	Equal importance
3	Weak importance of one over another
5	Essential or strong importance
7	Demonstrated importance
9	Absolute importance
2,4,6,8	Intermediate value between the two adjacent judgements

Prior to the spatial analysis, a consistency level computation was required. Consistency Ratio (CR)) is a procedure that determines the index of consistency. It shows that the probability of the matrix findings was randomly calculated [35]. In other words, AHP allows inconsistency by using the consistency ratio computation. The tolerable consistency ratio is less than or equal to 10 per cent [39]. CR is the ratio between the consistency index (CI) and random consistency index (RI), and can be expressed using equation (7)

$$CR = \frac{CI}{RI'}\ and\ CI = \frac{(\lambda\, max - N)}{(N-1)} \tag{7}$$

where λmax represents the largest eigenvalue, and N is the size of the comparison matrix. In this study $N = 5$. λmax was calculated from the sum of all the parameters, and then multiplied by its eigenvector. RI is based on the random consistency index ($RI = 1.12$ for five parameters). Normalised matrix (Fig. 10) was calculated from the pair-wise comparison in four iterations, and showed that CI was 0.252, while CR was 22.5%. The pair-wise comparison revealed that run-up was the most important factor (50.18%) followed by elevation (19.52%), slope (13.10%), coastline distance (9.53%) and vegetation density (7.67%). AHP results showed that those classes were the most important parameters in deciding vulnerability. Spatial multi-criteria were analysed using the AHP method from which the vulnerability map was created by using weight factors for each parameter. It was found that the run-up parameter showed the highest weight. 0.99% as the data provided in table 4. The area was identified as being slightly high while a high class of tsunami vulnerability was found in the coastal area which has a lower elevation as in Figure 12(b).

Pair-wise comparison matrix

	Elevation	Slope	Coastial Proximity	Rund Up	Vegetation
Elevation	1.00	2.00	3.00	0.33	3.00
Slope	0.50	1.00	2.00	0.25	3.00
Coastial Proximity	0.33	0.50	1.00	0.20	3.00
Rund Up	3.03	4.00	5.00	1.00	5.00
Vegetation	0.33	0.33	0.33	0.20	1.00

Normalized Matrix

	Elevation	Slope	Coastial Proximity	Rund Up	Vegetation
Elevation	0.1924	0.2553	0.2647	0.1667	0.2000
Slope	0.0962	0.1277	0.1765	0.1263	0.2000
Coastial Proximity	0.0641	0.0638	0.0882	0.1010	0.2000
Rund Up	0.5831	0.5106	0.4412	0.5051	0.3333
Vegetation	0.0641	0.0462	0.0294	0.1010	0.0667

4th iteration

0.1952	0.1952	0.1952	0.1952	0.1952	19.52%
0.1310	0.1310	0.1310	0.1310	0.1310	13.10%
0.0953	0.0953	0.0953	0.0953	0.0953	9.53%
0.5018	0.5018	0.5018	0.5018	0.5018	50.18%
0.0767	0.0767	0.0767	0.0767	0.0767	7.67%

Figure 11. Pair-wise comparison and normalised matrix

3.3. Spatial Analysis for Vulnerability Mapping

The weighted overlay was applied using spatial analysis in GIS after creating a parameter map. In addition, the weighted overlay tool was used to solve multi-criteria problems. The weighted overlay method is a protocol that applies a standard scale of values for various parameters. The overall steps of

IOP Conf. Series: Materials Science and Engineering 1062 (2021) 012036 doi:10.1088/1757-899X/1062/1/012036

this process are shown in the figure 12(a). The level of vulnerability starts from 1 (low vulnerability) to 5 (high vulnerability). Furthermore, the statistic of vulnerability map was calculated. The vulnerability classes as in table 4 were created using the subtraction of maximum and minimum values. Moreover, the number of the class was also used to divide the vulnerability classes. Finally, tsunami vulnerability was mapped into five classes as show by figure 12(b). Results showed that the high vulnerability areas were mostly found in the coastal area with the sloping coast type.

Table 4. Vulnerability classes

Vulnerability Class	Area (square kilometre)	Percentage (%)
High (5)	11.26	0.42%
Slightly High (4)	15.28	0.57%
Medium (3)	266.18	9.93%
Slightly Low (2)	112.218	4.19%
Low (1)	2276.06	84.90%

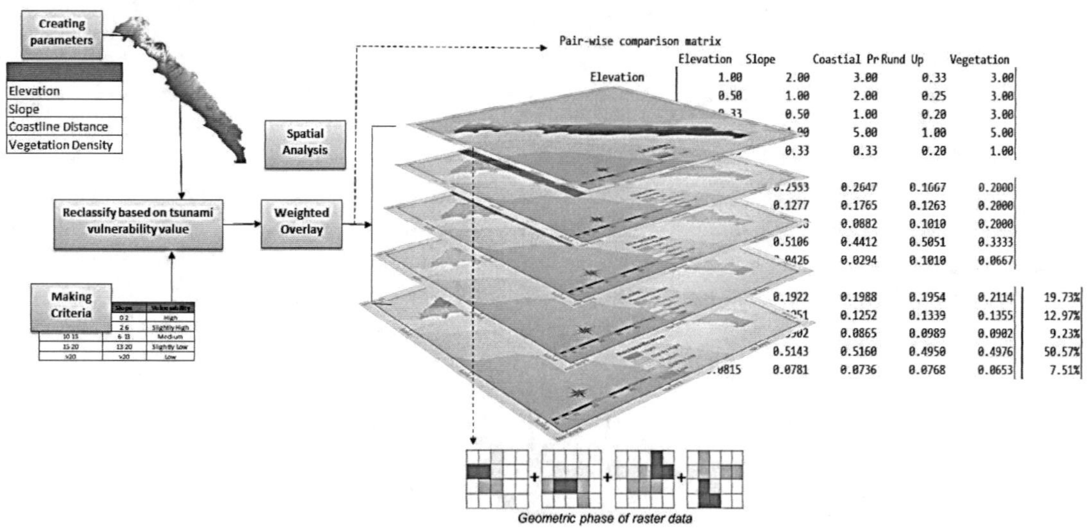

(a)

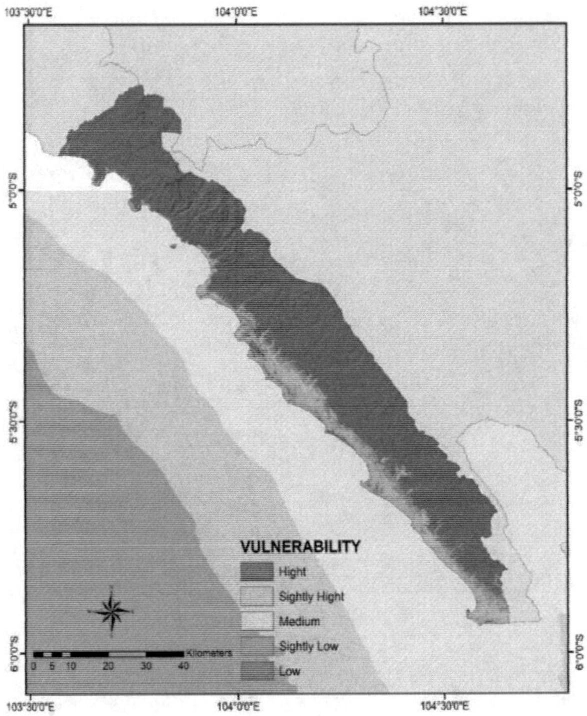

(b)

Figure 12. (a) Spatial analysis step (b) Tsunami vulnerability map of west coast Lampung

3.4. Tsunami inundation and impact assessment

The simulation of tsunami wave results found that the maximum wave height was 2.65 m, inundation area was 26.54 km^2, and vulnerability level was 11.26 km^2. While the building area was estimated at approximately 6.6 km^2 along the West coast of Lampung as in Table 4 and Figure 12, the inundation areas were predicted as high vulnerability and slightly high vulnerability. Meanwhile, a significantly different result of the inundation area (22.09 km^2) was recorded from the previous study [40]. Enhancement of the inundation area in 2019 was 4.45 km^2 compared with the BNPB result in 2016. This might be due to the changing vegetation around the area as vegetation plays an essential role as a tsunami barrier that reduces the impact of tsunami destruction [41-44]. In addition, SAVI values of West coastal Lampung showed the state of vegetation of the inundation area as follows: 6.02 km^2 (22.68%) of low vegetation, 11.85 km^2 (44.65%) of slightly low vegetation and 8.67 km^2 (32.67%) of medium vegetation as in Figure 7 and 12.

Table 5. Vulnerability class of building area

Vulnerability class	Inundation area		3	2	1
	5	4			
Building area (km^2)	0.91	5.69	3.26	2.69	3.88

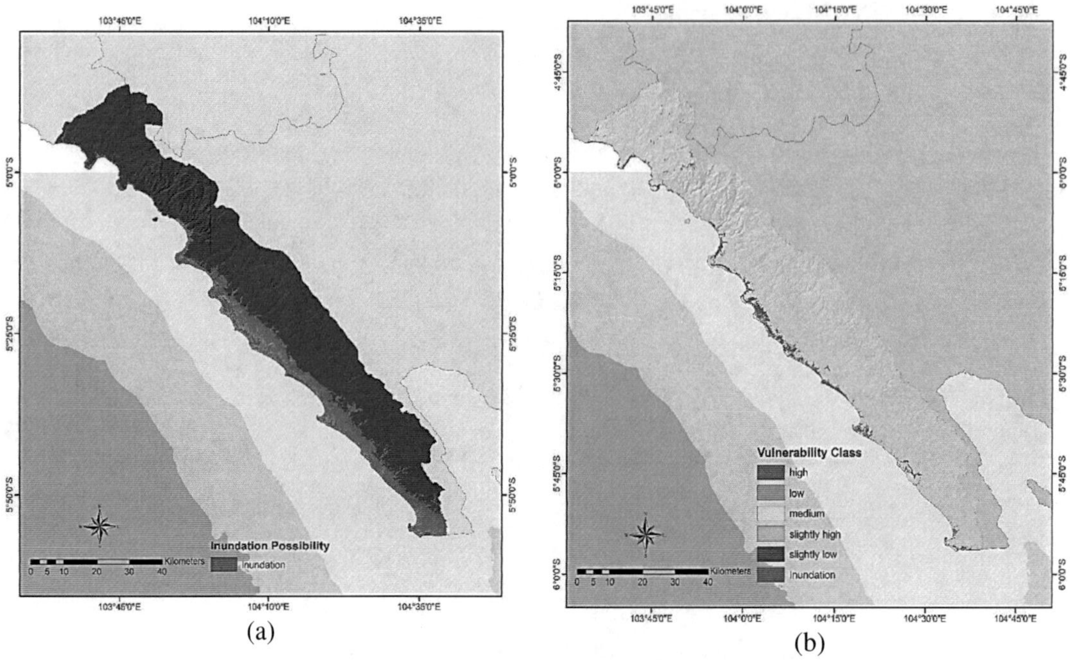

Figure 13. (a) Map of possible inundation area in west coast Lampung (b) Tsunami vulnerability class of building area in west coast Lampung

4. Conclusion

The identification of tsunami vulnerability areas on the west coast of Lampung Province and evaluation of the tsunami impact by mapping possible inundation areas using a geospatial analysis-based tsunami run-up simulation was studied. The simulation was carried out using the L-2008 software. It resulted in a tsunami wave of 2.65 m in height at a timing of 28.56 min when it reached the coast area. There were five parameters that were used to create a tsunami vulnerability map and predict the possibility of an inundation area along the west coast of Lampung, including the run-up, elevation, slope, coastal line, and vegetation. In addition, the analytical hierarchy process (AHP) was used to estimate the weight of the parameters of AHP was calculated using a pair-wise comparison matrix. The results showed CI and CR values of 0.252 and 22.5%, respectively. The pair-wise comparison revealed that run-up was the most important factor (50.18%) which affected vulnerability and inundation area, followed by elevation (19.52%), slope (13.10%), coastline distance (9.53%) and vegetation density (7.67%). The final map of tsunami vulnerability was constructed using the weighted overlay through a spatial analysis in GIS. Furthermore, the classes of vulnerability were identified into five classes with a series of numbers from 1 (low vulnerability) up to 5 (high vulnerability), and the tsunami inundation mapping was analysed. The assessment result indicates that 11.26 square kilometres of the study area falls under the high vulnerability zone due to tsunami and 6.6 square kilometres of the building area comes under the inundation area. Subsequently, our study has shown that the total number of inundation area along the west coast of Lampung was 26.54 km^2. Therefore, the result of this study can be used as a reference by the local government to facilitate improvements in basic data for city planning related to evacuation processes and management strategies during disasters.

References

[1] Birkmann J 2006 *Measuring Vulnerability to Hazards of Nature Origin* (Tokyo: UNU press) p 1
[2] Birkimann J and Wisner B 2005 *Measuring the Un-Measureable* (Germany:UNU Press) p 40
[3] Papathoma M and Dominey-Howes D 2003 *Nat. Hazard Earth Syst. Sci.*3(6) 733-747

[4] Papathoma M, Dominey H, Smith D T M and Zong Y 2003 *Nat. Hazard Earth Syst. Sci.***3(5)** 377
[5] Papathoma M and Dominey-Howes D 2003 *Nat. Hazard Earth Syst. Sci.* **3(6)** 734
[6] United Nations Disaster Relief Coordinator (UNDRO) 1991 *Mitigation Natural Disaster: Phenomena, effects and Options, a Manual for Policy Makers and Planners* (United Nations Publication, New York)
[7] Pelling M 2003 *The Vulnerability of cities: Natural Disaster and Social Resilience* (London: Earthscan)
[8] Bollin C, Cardenas C, Hahn H and Vasta K S 2003 *Natural Disaster Network: Disastere Risk Management by Communities and Local Goverment* (Washington DC: Inter American Development Bank)
[9] Schmidt-Thome P 2006 *Natural and Technological Hazards and Risks Affecting the spatial Development of European Regions* (Finland:Special Paper 42 Geological Survey)
[10] Mahendra R S, Mohanty P C, Bisoyi H, Srinivasa K T and Nayak S 2011 *Ocean Coast. Manag.* **54(4)** 302
[11] Caver S J 1991 *Int. J. Geo. Info. Syst.* **5(3)** 321
[12] Jankowski P 1995 *Int. J. Geo. Info. Syst.***9(3)** 251
[13] Sinaga T P T, Adhi N, Lee Y W and Yongcheol S 2011 *KSCE J. Civ. Eng.* **15(3)** 537
[14] Strunz G, Post J, Zosseder K, Wegsceider S, Muck M, Riedlinger T, Mehl H, Dech S, Birkmann, Gebert N, Harjono H, Anwar H Z, Sumaryono, Khomarudin R M, and Muhari A 2011 *Nat. Hazards Earth Syst. Sci.* **11(1)** 67
[15] Lida K 1963 *Magnitude, Energy and Generation Mechanisms of Tsunami and a Catalogue of Eartquakes Associated with Tsunamis*" (Honolulu:Proceeding of Tsunami Meeting at the 10th Pacific Science Congres) pp 7-18
[16] Van Zuidam R A 1983 *Guide to Geomorphological Aerial Photograpic Interpretation and Mapping* (Enschede:International Institute for Geo Information Science and Eart Observation)
[17] Bretschneider C.L and Wybro P.G 1976 *Coast. Eng. Proc.* **1(5)** 59
[18] Hills J G and Mader C L 2006 *Ann. N. Y. Acad. Sci.* **822(1)** 381
[19] Huete A R 1988 *Rem. Sens. Environ.* **25(3)** 295
[20] Araujo L S, Dos Santos J R and Shimabukuro Y E 2000 *Int. Arch. Photogramm. Remote Sens. Spat. Inf. Sci.* **33** 77
[21] Chander G and Markham B 2003 *IEEE Trasns. Geosci. Remote Sens.***41(11)** 2674
[22] Bouvet M, Chander G, Goryl P and Sater R 2007 *IEEE Int. Geosci. Remote Sens. Symp.* 2673
[23] *Data Users Handbook Data Products, Chapter 11* 2011 (National Aeronautics and Space Administration)
[24] Tucker C J 1979 *Remote Sens. Environt.* **8(2)** 127
[25] Jackson R D, Slater P N and Pinter P J 1983 *Remote Sens. Environ.* **13(3)** 187
[26] Tucker C J, Newcomb W W, Los S O and Prince S D 1991 *Int. J. Remote Sens.***12(6)** 1133
[27] Hansen M C, Defries R S, Townshend J R G and Sohlberg R 2000 *Int. J. Remote Sens.* **21(6-7)** 1331
[28] Gong P, Pu R, Biging G S and Larrieu M R 2003 *IEEE Trans. Geosci. Remote Sens.* **41(6)** 1355
[29] Huete A R, Hua G, Qi J, Checbouni A and Van Leeuwen W J D 1992 *Remote Sens. Environ.* **41(2-3)** 143
[30] Huete A R 1988 *Remote Sens. Environ.* **25(3)** 295
[31] Jensen J R 2000 *Remote Sensing of the Environment An Earth Resource Perspective* (Prentice Hall, Upper Sadle River)
[32] *Peta Sumber Bahaya Gempa Indonesia* 2017
[33] Rahmawati N I, Santosa B J, Setyonegoro W, Bambang S 2017 *Jurnal Sains dan Seni* **6(20)** 2337
[34] Latief, H., Sunendar H, Yuhsananta. P and Riawan E 2006 *Laporan Program Penelitian dan Pengembangan IPTEK* **155**
[35] Saaty T L 1970 *J. Math Psyc.***15(3)** 234
[36] Saaty T L 1980 *The Analytic Hierarchy Process, Planning,Priority Setting, Resource Allocation*

(New York:McGraw Hill)
[37] Saaty T L 2003 *Eur. J. Oper. Res.* **145(1)** 85
[38] Forman E H and Selly M A 2001 *Decision by Objectives* (World Scientific Publishing Company)
[39] Saaty T L 1996 *Decision Making for Leaders: The Analytical Hierarchy Process for Decisions in a Complex World* (Pittsburgh:RWS Publications)
[40] Harada K, Imamura F 2002 *Proc. 12th Int. Offshore Polar Eng. Kitakyushu* pp 652-658
[41] Alongi D M 2008 *Estuar. Coast. Shelf Sci J.* **76(1)** 1
[42] Pranata Y A dan Johnny G P 2014 *Jurnal Teknik Sipil* **13(1)** 25
[43] Husrin S, Jaya K, Aprizon P, Joko P, Yudhi C, Aditya H 2015 *Procedia Earth and Planetary Science* **12(2005)** 20
[44] Susanto D, Faida L R and Sunarto S 2019 *Jurnal Ilmu Kehutanan* **13(1)** 4

Implementing Student Attendance System Using Fingerprint Biometrics for Kolej Universiti Poly-Tech Mara

N Zakiah Lamin[1*], W N Asnida Wan Jusoh[1], Juanita Zainudin[1], Hafiza Samad[1]
[1] Computer Science, Faculty of Computing and Multimedia, Kolej Universiti Poly-Tech Mara (KUPTM), 56100 Cheras, Kuala Lumpur, Malaysia

*Corresponding e-mail: nzakiah@kuptm.edu.my

Abstract. Recording students' attendance has been a major concern at Kolej Universiti Poly-Tech Mara. However, monitoring the attendance manually is a cumbersome issue to lecturers as students tend to manipulate the attendance by signing each other's attendance. It has been found that fingerprint biometrics is capable to monitor the attendance systematically and efficiently. The aim of this study is to verify the students' attendance using fingerprint biometrics. Evolutionary prototyping model were used to develop the students' attendance system. Students are required to thumbprint using the fingerprint device installed in the classroom to record the students' attendance. Their fingerprint images were captured by the fingerprint device and the images were then registered to the server for attendance process. The implementation of fingerprint biometric has helped lecturers to monitor the attendance of the students more systematic, efficient and ethically. By using the system embedded with biometrics, reporting on absenteeism is genuine and easy. Lecturers just need to print out and do necessary action. Therefore, fingerprint biometrics is useful and helpful in keeping track and managing the attendance of the students.

Keywords: Fingerprint; biometrics; student attendance; systematic; trustworthy

1. Introduction

Recording students' attendance manually is a cumbersome issue to many lecturers at Kolej Universiti Poly-Tech Mara as they need to monitor the student attendance as well as relating it to the performance of subjects at the end of the semester. Implementing an automated system will help lecturers to ease the burden of recording and monitoring the student's attendance [1].

Student attendance system is a system that records and monitors the students' presence in each class that they attend every day. Lecturers are able to see the attendance record and analysing the data in order to produce a report at any time. There are many ways to capture data in order to record and keep it in database of a system. Data can be captured in methods such as Optical Character Recognition (OCR), Intelligent Character Recognition (ICR), barcodes and biometrics. Fingerprint biometrics is one of the ways to capture data using an automated method of recognizing a person based on the physiological or behavioural characteristic. Fingerprints are small strips and patterns at the tip of each fingers. They are formed from the pressure on the small fingertips of the growing baby in the womb. No two individuals have been found to have the similar fingerprint. The pattern enhances biometric security by authenticating the fingerprint scanner [2].

Many studies have also shown that using biometric fingerprints can help lecturers to manage and keep the attendance record systematically and efficiently. Fake attendance or the act of signing on behalf of others can be eliminated through the biometric fingerprint device that requires every student to place his or her registered finger to the device in order to record their presence in attending classes [3], [4]. Hence, this study aims to verify the students' attendance using fingerprint biometrics. The evolutionary prototyping model is used as a methodology to develop the students' attendance system. User requirements and feedbacks will be collected and analysed in order to implement a systematic, efficient and ethical system that help lecturers in keeping track and managing the attendance of the students.

2. Research Problem

Taking student attendance manually is an inconvenient issue to lecturers at Kolej University Poly-Tech Mara. Due to audit purposes, every lecturer has to monitor their students' attendance in every subject for the whole semester. There is certain amount of percentage of absentees of the students that will be taken into account for the lecturers to produce warning letter to the students. The categories and percentages are [5]:

1. Warning letter 1 – WAR 1 (7% absenteeism)
2. Warning letter 2 – WAR 2 (14% absenteeism)
3. Warning letter 3 – WAR 3 (21% absenteeism)

For every class subject, lecturers need to print the hardcopy of students' name list from the main system (campus management system) for every slot time based on the time table. In a class, students need to sign in a hardcopy attendance sheet provided by their respective subject lecturer. At the end of the class, the lecturer will collect it and check whether the total signature in the hardcopy sheet matches with the number of the physical students attending the class that day. This practice may not allow for the lecturer to attain a genuine attendance list as students tend to fake the signatures or sign on behalf of their friends without the lecturers' knowledge.

Within a semester, the lecturer needs to monitor the attendance in order to produce warning letter (if any) to the students in the three categories as mentioned above. It is a tedious task as lecturers need to check the hardcopy list one by one weekly and there are possibilities of inaccuracy of data to occur while processing the data. Study shows that there are correlations between attendance and academic performance [6], [7]. That is why monitoring the attendance of students' is necessary and should be improved to assist the lecturers [8].

3. Methodology

The methodology selected in implementing this system is Evolutionary Prototyping. This method allows developers to continue receiving feedback and suggestions phase by phase in order to improve the system prototype. The evolutionary prototyping consists of four phases covering the initial concept, design and implementing initial prototypes, improving prototypes, and delivering complete systems.

Prototyping evolution can help developers to accelerate the development of their systems. In addition, the evolution prototype helps in improving the quality of the final product because it is necessary to undergo some prototypes and eventually to the final version after the functionality and overall requirements are met. The whole phase for evolutionary prototyping were shown in Figure 1 below.

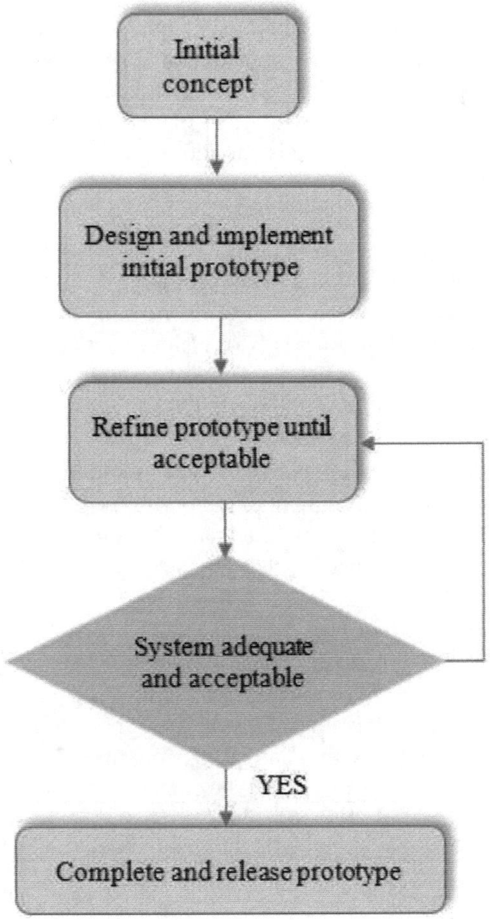

Figure 1. Evolutionary prototyping

3.1. Phase 1 - Initial concept

During this phase, developers will come out with an early concept and identify the required materials that are relevant to the system in this phase. Developers will collect requirements by conducting an interview to those who appear to be involved in the system development such as lecturers, students, and others. Based on results obtained from the interview, the developer will identify the requirement as an initial input from the user in developing this system. In addition, an established project plan will provide some information to developers which covers project schedules, clear explanations of projects, and so on. At the end of this phase, developers need to issue a business plan, project plan, and initial list of user requirements as output for this phase. Therefore, the initial concept of replacing existing student presence systems for computerized and automated student attendance systems has been stated.

3.2. Phase 2 – Design and implement initial prototype

In this phase, developers will carry out design activities and implement initial concepts in developing systems. The development of basic user interface prototypes will be published for the purpose of showing the users to ensure development is on track. The predefined user interface will include fingerprint student attendance system interface to give users a clear image for better understanding of the system. At the same time, developers will perform an initial system design analysis based on the requirements listed in this phase. After developing the interface and identifying the initial system design, the user interface is created and shown to users for purposes of enabling users to assess the prototype of the system interface. Subsequently, the developers will then capture the feedback from the user after the evaluation and modify the requirements according to the information provided by the user in this phase. After all, the basic user interface developed in this phase is archived and will no longer be used. At the end of this phase, developers need to come out with a list of validated requirements, system designs, and ratings and feedback from users about lost needs and more.

3.3 Phase 3 – Refine prototype

In this phase, the developers will change the system design through feedbacks and ratings from users in the previous phase. In the development of the first prototype, developers will ensure that the development of the system meets the requirements that have been verified and the system designs obtained from users. In addition, the developers will also consider the system's quality while developing it as well as running operations such as system programs, performing debugging activities, and testing the system. Therefore, the first prototype of the fingerprint attendance system is created. Subsequently, the first prototype of the system specified in this phase will be shown to the user for evaluation. After the first prototype assessment, the developer will record new or modified needs specified by the user for the purpose of reviewing the subsequent development of the prototype. This phase will be repeated until the specified set of claims specified by the user is fixed. In addition, developers will need to provide with the final condition that has been confirmed.

3.4 Phase 4 – Complete and release prototype

In this phase, the developers will develop a complete student attendance system based on a set of legitimate final requirements obtained from the user. Next, the developers will execute the system and perform the test to the final system. Once all requirements are met, the system will deliver to end users as an approved system with the required functionality and built-in quality. Therefore, the overall objectives of the system will be fulfilled at the end of this phase. The current attendance system will be automatically replaced by computer attendance system and computerized at the end of this phase. Student attendance system using fingerprint will be reached at the end of this phase. The generation of reports on student attendance will be provided at the end of this phase.

4. System Design

This section describes the process of system design by describing the necessary modules, architecture, stakeholders, components, interface and data. All relevant variables are indicated and specified to meet their particular requirements. There are two types of design styles which are 1) Use Case Design - to visualize the user's practical requirement 2) Flow Chart Design - display the steps and decision making during the process.

4.1. Use Case Design

Actors involved in the system are administrators, lecturers and students. Administrators are the administration department's staff while Users are lecturers and students of KUPTM. In the use case diagram below in Figure 2, there are 7 processes involved in the system. For administrators, it will involve in registering lecturers, handling login and password, upload students confirm list and view report in the system. For users as lecturers, it will involve in completing registering process and login and changing the password in the system. The lecturers will involve in recording attendance process, managing the attendance data, update the attendance data with the latest updates from students and be able to view report provided in the system. Students on the other hand will just have to touch the fingerprint biometrics device in order to capture the image to be send in the system for verification process.

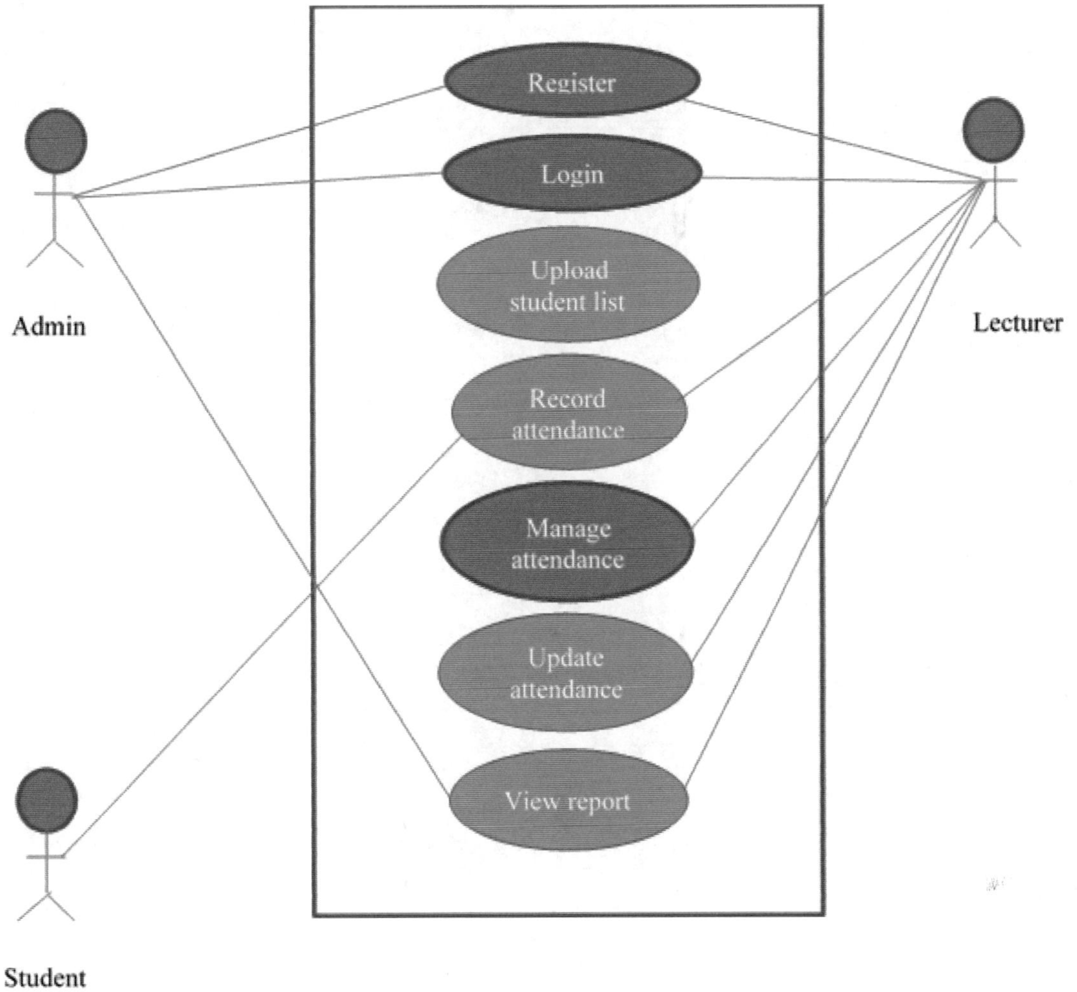

Figure 2. Use case diagram

4.2. Flow Chart Design

In the flow chart design indicated in Figure 3 below, administrators will register the lecturers in the system. Lecturers will receive username and password and need to activate using the register and login process. Once the account is active, the lecturer can start using the system.

As for administrators, he or she will upload the student list accordingly based on lecturers' workload into the system. All lecturers can start recording the attendance using the fingerprint biometrics device that is installed in each class venue. Lecturers will need to set date and duration in the system and all recorded fingerprint will be send to the system. The updated attendance list will be automatically populated in the system after the end of each class. However, the lecturers are still able to update with current status of the students for example MC or excuse letter with permission. Documents such as warning letters will be ready to be printed if needed based on the latest data.

Finally, related reports can be generated from the system at any time depending on the cumulative data recorded in the system. The report may suggest or show the performance of student in attending classes in various subjects in a semester.

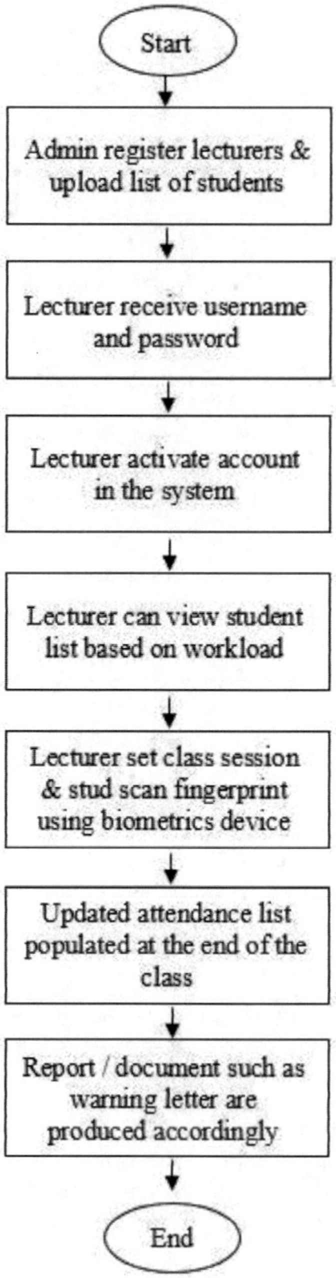

Figure 3. Flow Chart Diagram

5. The Proposed System

A prototype is an early sample of the system built. The user interface prototypes are shown to represent the flow and design of the system before the final system exists. An early user interface prototype allows all the users to visualize and understand the system. With the prototypes of the system, the system was at the final stage of the implementation phase and it will be tested to all users. The system will not be deployed yet as there may be changes that need to be done after the testing phase. The system users are in two categories:

a) Administrators
   - The function for administrators is change or reset password, register lecturers and view in list, update lecturers' data, register and update student list and information related to students.

b) Lecturers
   - The function for lecturers is change password, upload attendance list, view attendance list from device and view student details.

5.1. User interface for all users (Log in System)

Figure 4 shows the login interface for all users to log in to their own accounts. The username and password are needed in order to log in to the system. Before logging into the system, users need to enter the validation code, which is Captcha, to proceed.

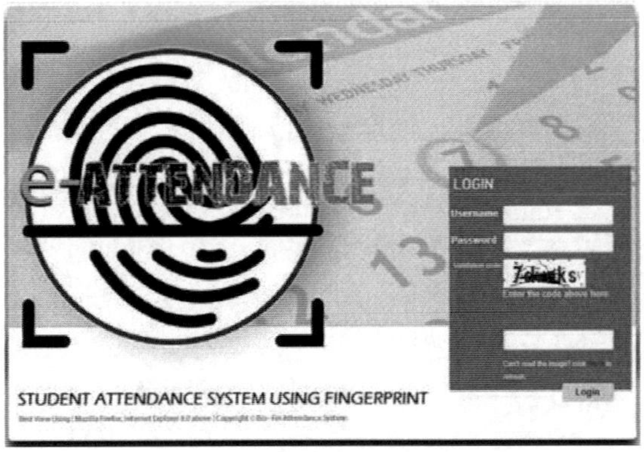

Figure 4. Login interface

5.2. User interface for all lecturer (Change password)

Figure 5 below illustrated the change password interface for lectures where the user can change their current password here.

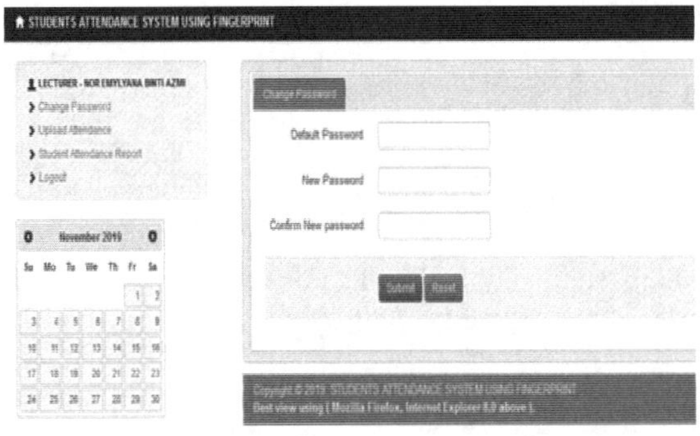

Figure 5. Change Password interface

5.3. Upload Attendance Prototype

The upload attendance file for lectures were depicted in Figure 6. The user can upload the attendance file in excel format only. Before uploading the file, users need to find the class ID then browse the attendance file to upload.

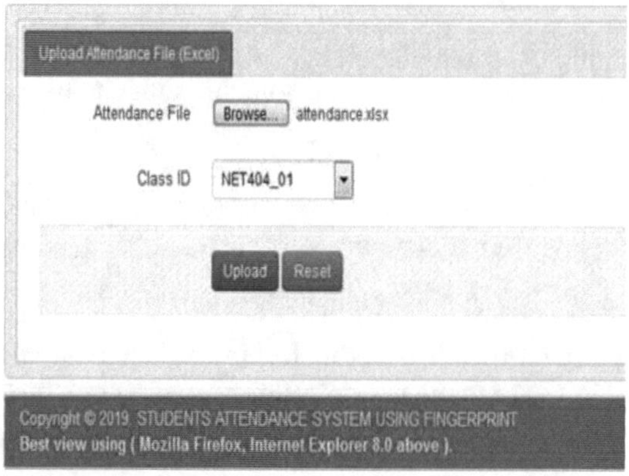

Figure 6. Upload Attendance interface

5.4. *View uploaded attendance prototype*

All the students' attendance report can be viewed by lectures as can be seen in Figure 7. This feature can be used after lecturers have uploaded the excel file, and users need to find course code and search by date. The report will be generated after clicking a button to display the uploaded attendance file in the system

Figure 7. Student attendance report interface

5.5. *Student Details*

As being illustrated in Figure 8, this is the view of the student's details of prototype. This information will be released when the user clicks on the student ID.

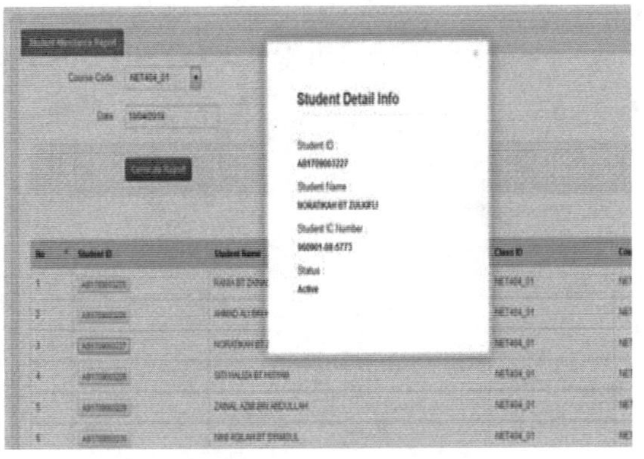

Figure 8. Student details interface

6. Conclusion and Recommendation

The student attendance system using fingerprint biometrics can improve the existing manual practice in KUPTM academics. The system will benefit the academics staff specifically lecturers in order to monitor the attendance of students more systematic, efficient and ethically. Moreover, the advantages of fingerprint biometrics will increase the security as fingerprints are much harder to fake, they also change very little over a lifetime, so the data remains current for much longer than photos and passwords. Other than that, the ease of use is simpler as fingerprints are always with us. It is also non-transferrable but also provides a higher level of accountability as a proof of student's attendance. Thus, the implementation of this system may simplify and ease the burden for the lecturers as reporting and analysing of data will improved. However, several improvements can be done for further development as follow:

1. Upgrade to mobile application:
   Mobile application will be more practical since the emergence of mobile application technology has become a bridge to communication and expand profit in more business industries [9].
2. Integrate with existing system:
   Student attendance system using fingerprint can be integrated with existing Campus Management.
   System (CMS) in providing existing database of staffs (lecturers) and students.
3. Reporting:
   More related reports should be produced to see more information about the students

References

[1] Pireva, K R, Siqeca J and Berisha S. 2013 *IFAC Proceedings* Volumes **46**(8), 137-140

[2] Ivanov V I and Baras J S 2019 *Pattern Recognition,* **94**, 230-249

[3] Banerjee, R, King E M, Orazem P F and Paterno E M 2012 *Economics of Education Review,* **31**(5), 563-574

[4] Dey I 2018 *International Review of Economics Education,* **28**, 29-40

[5] Kolej Universiti Poly-Tech MARA Kuala Lumpur. Academic Regulation Handbook for Foundation, Diploma, Degree and Professional Programmes. *www.kuptm.edu.my*

[6] Landin M and Pérez J 2015 *Currents in Pharmacy Teaching and Learning,* **7**(1), 78-83

[7] Persky A M, Kirwin J L, Marasco C J, May D B, Skomo M L and Kennedy K B 2014 *Currents in Pharmacy Teaching and Learning,* **6**(1), 1-9

[8] Rejnö Å, Nordin P, Forsgren S, Sundell Y and Rudolfsson G 2017 *Nurse Education Today,* **50**, 36-41

[9] Sunaryono D, Siswantoro J and Anggoro R 2019 *Journal of King Saud University - Computer and Information Sciences*

[10] Bhalla A 2020 *Biometric Technology Today, 2020*(8), 5-8
[11] Biometrics for attendance and benefits face challenges in India 2013 *Biometric Technology Today, 2013*(9), 12
[12] Schools learn about the benefits of biometrics 2001 *Biometric Technology Today,* **9**(6), 7-8.

Modelling and Validation of Seismic Performance for Corner and Interior Beam-column Joint Using HYSTERES Program

N H Hamid[1*] and N D Hadi[2]
[1] Institute for Infrastructure Engineering, Sustainable and Management (IIESM), Universiti Teknologi MARA, 40450 Shah Alam, Selangor, Malaysia.
[2] Faculty of Civil Engineering, Universiti Teknologi MARA, 40450 Shah Alam, Selangor, Malaysia.

*Corresponding e-mail: norha454@uitm.edu.my

Abstract. The validation between experimental and modelled hysteresis loops is a preliminary work in determining the dynamic behaviour of the beam-column joints under earthquake excitations using Ruaumoko 2D program. However, this paper only focuses on accurately modelling the hysteresis loops of corner and interior beam-column joint using the HYSTERES Program. The corner and interior beam-column joints were designed using Eurocode 8 and equipped with unbonded fuse-bars which acts as passive energy dissipators. The hysteresis loops were modelled based on the input parameters obtained from the experimental hysteresis loops. From the HYSTERES program, the Pampanin Reinforced Concrete Beam-Column Joint was chosen to represent the hysteretic behaviour (load versus displacement) for overall seismic behaviour of these joints. The lateral strength capacity, stiffness, ductility and equivalent viscous damping from experimental and modelled hysteresis loops were then compared, and the percentage difference from both hysteresis loops was calculated. Based on the outcomes, there are some parameters which have shown good agreement, and this proves the effectiveness of using HYSTERES program in modelling the hysteresis loops of the beam-column joints.

Keywords: corner joint, interior joint, stiffness, ductility, equivalent viscous damping

1. Introduction

Recently, there are a lot of studies have been conducted on non-seismic of beam-column joints under earthquake excitations which contribute to structures damage and collapse. Different damage or failure modes are expected to occur in beam-column joints depending on the types of joint and the adopted structural details (such as the presence of transverse reinforcement in the joint; use of plain round or deformed bars; alternative bar anchorage solutions) [1]. Many laboratories testing had been conducted in the heavy structural laboratory on non-seismic RC beam-column joints under in-plane lateral cyclic loading [2- 5]. Evidently the non-seismic design buildings in Malaysia are vulnerable under seismic load. Most of the buildings in Malaysia were designed and constructed using a non-seismic code of practice known as BS8110, where there is no provision for earthquake load at all [6]. Past studies analyzed the dynamic behaviour of the RC buildings in Malaysia under various real earthquake excitations using Ruaumoko 2D program [7- 9]. It was found that the non-seismic buildings which were design using BS8110 did not sustain Ranau Earthquake (PGA=0.12g), which is the highest recorded earthquake excitation in Malaysia. By assessing the dynamic behaviour of the buildings, researchers can understand and predict the overall seismic behaviour of the building when earthquakes with similar excitations strike. Ruamoko 2D is a notable tool in seismic analysis, and it provides a wide variety of modelling options. One of them is the HYSTERES program which is a program used to determine the best choice of loop parameters to obtain the most suitable hysteresis loop for use in a Ruaumoko analysis [10]. The program provides a list of hysteresis loops model for various types of structures, and it has proven accurate in modelling works. The modelling work done using HYSTERES program is considered as preliminary work in determining the overall dynamic behaviour of the structure. Therefore, it is important to model the hysteresis loops precisely to avoid error in future modelling work.

Generally, most of the beam-column joints were designed in according to Eurocode 8 (EN 1998-1:2004) by considering the seismic loads and applying weak beam and strong column designed philosophy. By applying this design philosophy, the joint will be ductile and therefore, can sustain higher lateral load compared to conventional non-seismic design. However, it was found that the provisions provided in Eurocode 8 for RC beam-column joint lead to "over-strength" design and steel congestion in the beam-column joint region [11- 13]. Therefore, the use of energy dissipator in seismic design beam-column joint is one of the methods to explore to provide additional damping and increase the ductility of the beam-column joints. In this study, the hysteretic behaviour of seismic design corner and interior beam-column joints with unbonded fuse bars was evaluated. The beam-column joints were a part of the RC buildings which were designed per Eurocode 8. The joints were equipped with unbonded fuse bars as Passive Energy Dissipator to enhance the energy dissipation capacity and increase the damping ratio of the structure. The joints were tested under lateral cyclic loading, and their hysteresis loops (load vs displacement curve) were plotted. Following that, the modelled hysteresis loops were established. Both hysteresis loops were then compared, and the difference of lateral strength capacity, stiffness, ductility and equivalent viscous damping were analyzed.

2. A Prototype Two-storey RC Building

A two-storey reinforced concrete school was designed in accordance with Eurocode 8 equipped with fuse bars at the joint were constructed, tested, analyzed, modelling and validation in this study. Figure 1 shows the side view of a prototype two-storey building with a total height of 7000mm, width of 9500mm and 1500mm depth of the stump together with pad footing. Meanwhile, Figure 2 shows the plan view of the two-storey RC school building, which had been designed using Eurocode 8 under Ductility Class Medium (DCM) with PGA=0.3g. The three types of joints are shown in Figure 2 which labelled as EJ-2B, IJ-3B and CJ-2B. EJ-2B means exterior beam-column joints with two beams, IJ-3B denotes as interior beam-column joint with 3 beams perpendicular with each other, and finally, CJ-2B designates as corner beam-column joint with two beams which located at the corner of the first floor of this prototype building. However, only two joints, namely IJ-3B (interior joint) and CJ-2B (corner joint) will be explained, testing, modelling and testing in this paper.

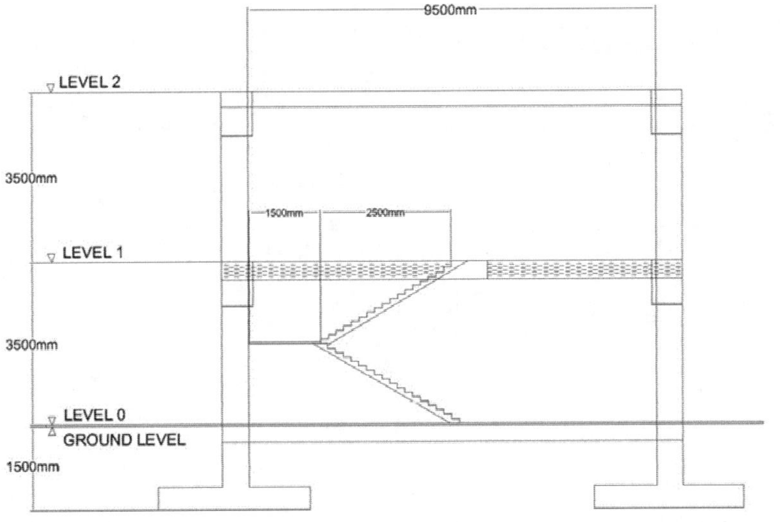

Figure 1. Side elevation of prototype two-storey RC school building

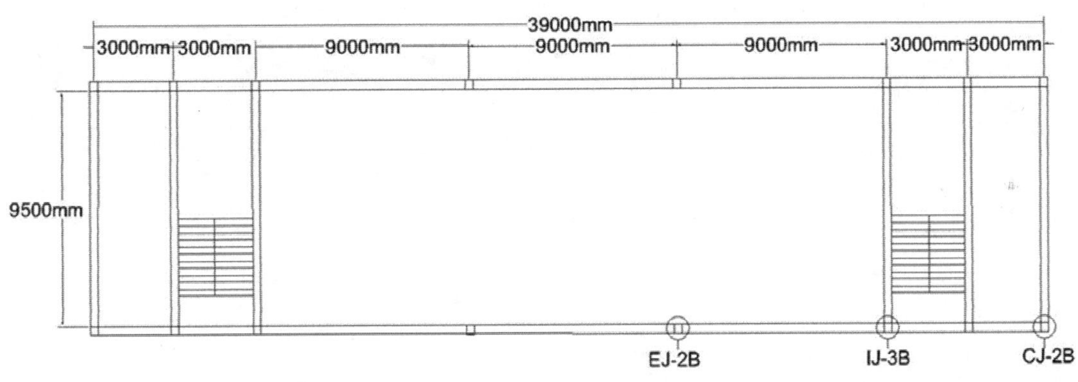

Figure 2. Plan view of the two-storey RC school building

Figure 3 shows the three dimensions and side elevation view of the corner joint, which labelled as CJ-2B joint. The corner joint with two beams and one column was seated on the foundation beam. The dimension of the column is 3500x400x400mm, sizes of the beam are 3500x400x400mm, and a dimension of foundation beam is 1800x900x500mm. Figure 4 shows the three dimensions and front view of the IJ-3B specimen with one column and three beams. The dimensions of column and beams are similar to corner joint labelled as CJ-2B. These corner joint and interior joint with two beams and three beams, respectively seated on foundation beam will be constructed in the heavy structural laboratory starting from ±0.01% until ±2.5% drift as listed in Table 1. These two joints were equipped with fuse-bars as energy dissipators which dissipate to the surrounding when the movement of the column from left to right and back to the original position. The work done by this movement can be determined by calculating the area under the hysteresis loops using area of trapezium between two drifts. Figure 5 shows the fuse-bars with 600mm length and reduction of diameter from 25mm to 20mm at the centre with 200mm length only.

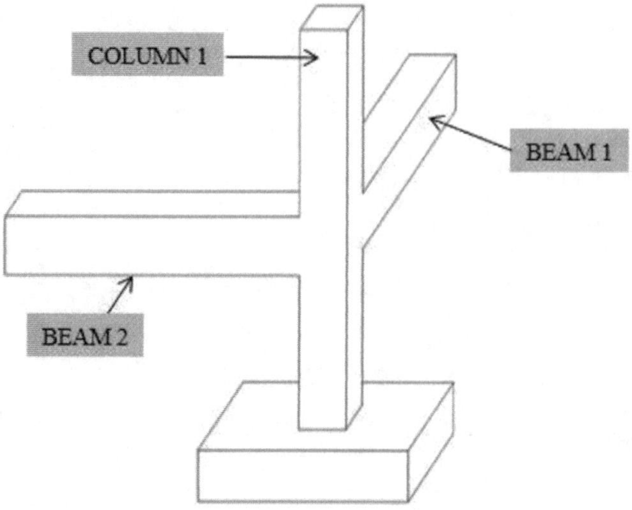

Figure 3. Three dimensions and side elevation view of CJ-2B joint

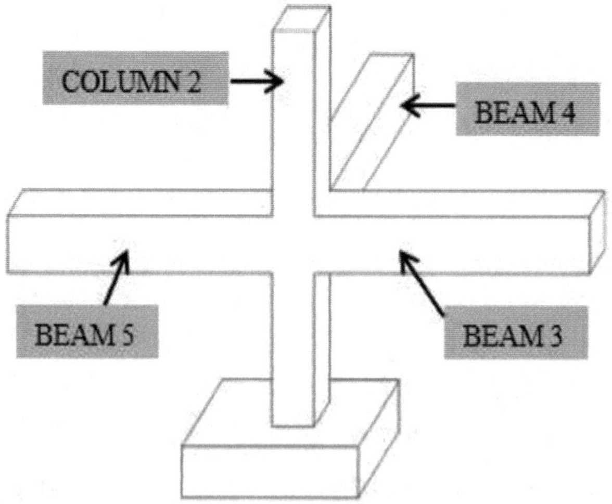

Figure 4. Three dimensions and front view of specimen IJ-3B joint

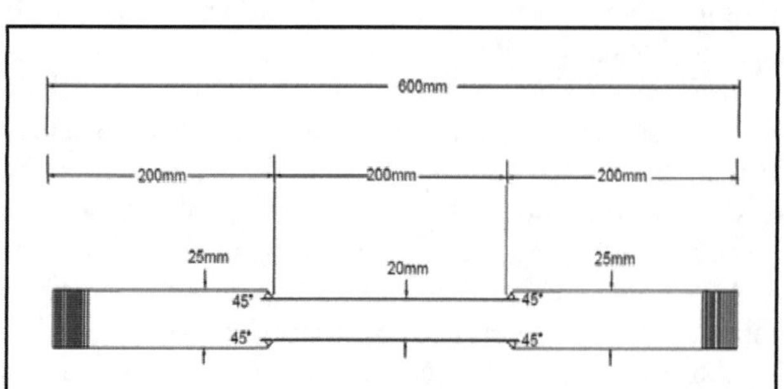

Figure 5. The total length of fuse-bars of 600mm with reduce diameter at centre

Table 1. Specific target drift and displacement in pushing and pulling direction for CJ-2B and IJ-3B

Pushing Direction		Pulling Direction	
Target Drift (%)	Lateral Displacement (mm)	Target Drift (%)	Lateral displacement (mm)
+0.01	+0.38	-0.01	-0.38
+0.05	+1.90	-0.05	-1.90
+0.10	+3.80	-0.10	-3.80
+0.20	+7.60	-0.20	-7.60
+0.50	+19.00	-0.50	-19.00
+0.75	+28.50	-0.75	-28.50
+1.00	+38.00	-1.00	-38.00
+1.15	+43.70	-1.15	-43.70
+1.25	+47.50	-1.25	-47.50
+1.35	+51.30	-1.35	-51.30
+1.50	+57.00	-1.50	-57.00
+1.75	+66.50	-1.75	-66.50
+2.00	+76.00	-2.00	-76.00
+2.25	+85.50	-2.25	-85.50
+2.50	+95.00	-2.50	-95.00

3. Equipment and Instrumentation

Upon the complete construction of specimen CJ-2B and IJ-3B, these two joint samples need to paint with white colour for easy marking of the cracks observed during testing at a different level of drifts. Furthermore, these specimens require installing equipment and instruments such as LVDTs (linear potentiometers), strain gauges, loading cell and data logger. Figure 6 shows a total number of five LVDTs were placed along the column, one LVDT on Beam 1 and two LVDTs on the foundation beam of specimen CJ-2B. LVDT 1 to 5 was located on the column in the x-direction (in-plane) to capture the lateral displacement at different point of the column. LVDT 6 was placed at the end of Beam 1 and LVDT 7 at the end of Beam 2, to observe the movement of the beams in x-direction and y-direction (out-of-plane) respectively.

Meanwhile, Figure 7 shows nine numbers of LVDTs and load cell are installed beside column and at the end of the beams for specimen IJ-3B. During testing, the displacement-controlled method was used and converted in terms of percentage drifts. According to ACI Committee 374 (2005), the drift can be defined as the ratio of lateral displacement over the height of the specimen, multiplied by 100. Whereby the unit is in the percentage of the total height drift is the ratio of lateral displacement over the height of the specimen in terms of percentage, and it is represented in Equation 1.

$$Target\ drifts\ (\%) = \frac{\Delta_x}{H_e} \cdot 100 \qquad (1)$$

Δ_x is the applied in-plane lateral displacement, and H_e is the effective height from top of strong floor to the centre of the double actuator with a value of 3800mm.

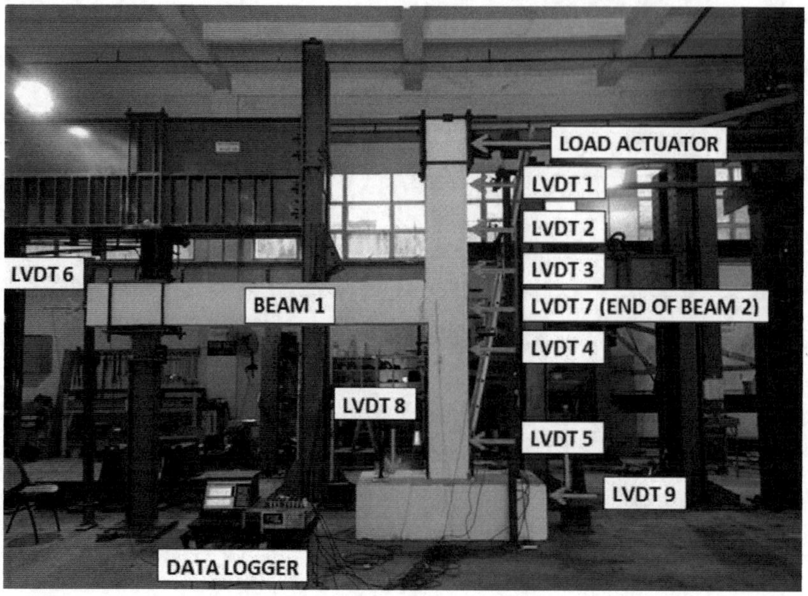

Figure 6. Specimen CJ-2B is equipped with LVDTs, load cell and data logger

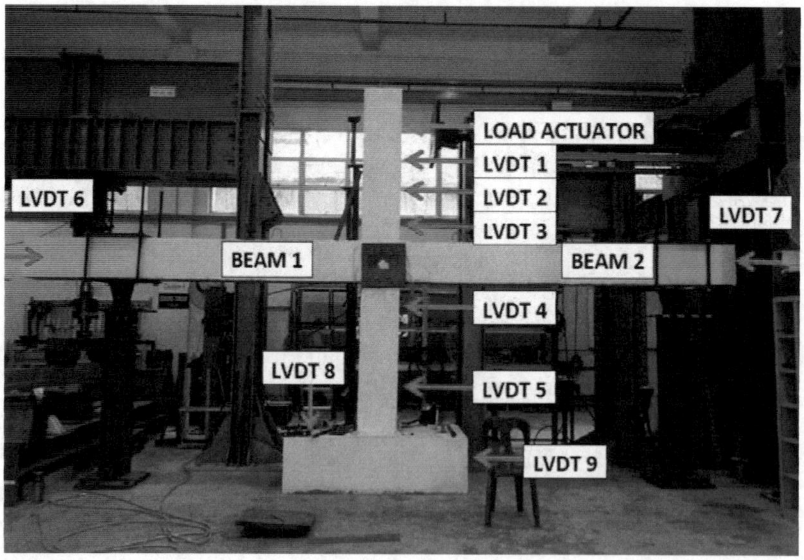

Figure 7. Specimen IJ-3B is equipped with LVDTs, load cell and data logger

4. Experimental Results

From the experimental data, the load versus displacement graph, which is known as hysteresis loops, can be plotted for every drift recorded by LVDTs. However, only data from LVDT 1 was used for analysis and interpretation because it was positioned the closest to the actuator and measured the maximum lateral in-plane displacement. For each drift, two numbers of cycles were recorded and plotted for its hysteresis loop. Figure 8 presents the hysteresis loops using the data obtained from LVDT 1. The yield together maximum in-plane load and displacement, elastic stiffness, secant stiffness, ductility and equivalent viscous damping can be obtained from this hysteresis loops. It was observed that the lateral

IOP Conf. Series: Materials Science and Engineering 1062 (2021) 012038 doi:10.1088/1757-899X/1062/1/012038

displacement and its corresponding lateral load increases as the value of target drift increases. The symmetrical shape in the pushing and pulling direction of the hysteresis loops for both first and second cycle is similar for their respective drifts. Figure 9 shows the hysteresis loops for specimen IJ-3B obtained from data LVDT 1.

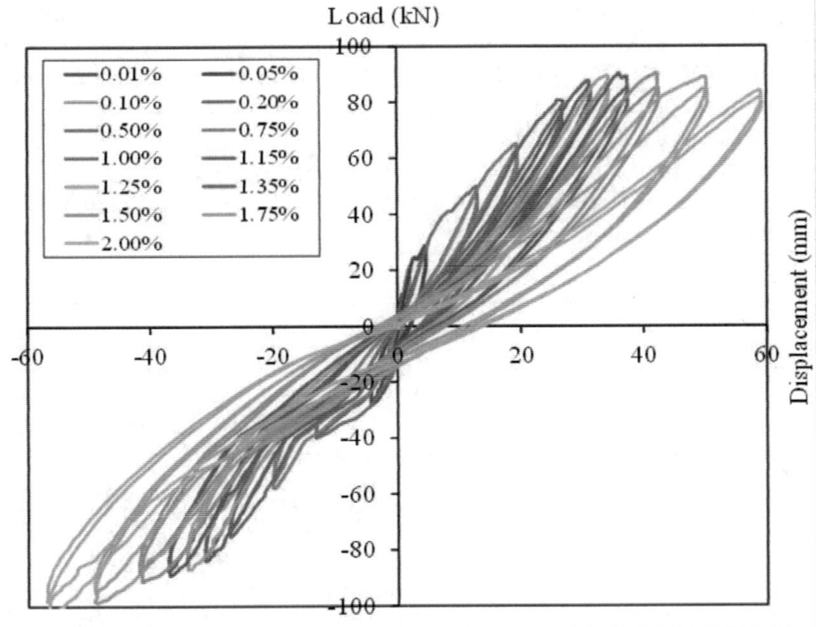

Figure 8. Hysteresis loop specimen CJ-2B obtained from data LVDT 1

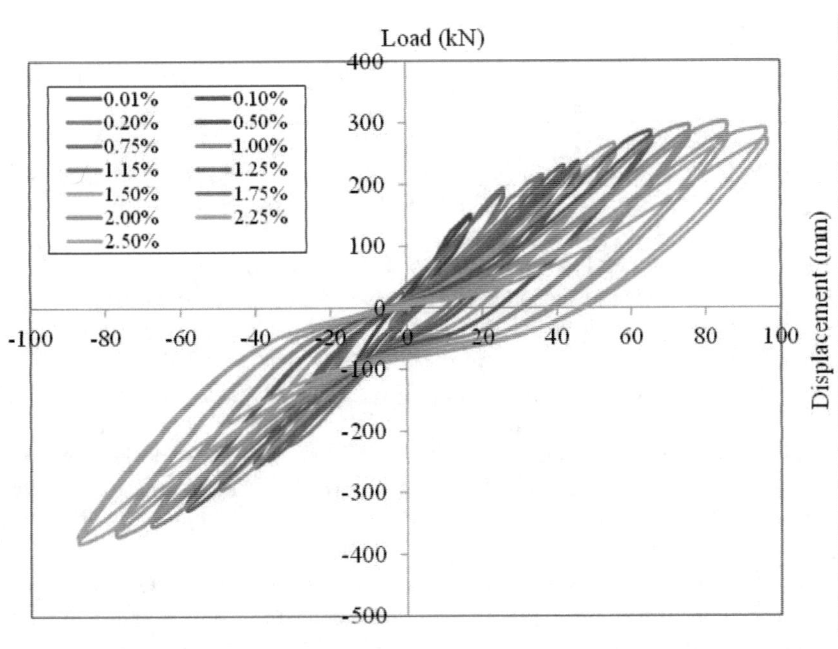

Figure 9. Hysteresis loop specimen IJ-3B which obtained from data LVDT 1

IOP Conf. Series: Materials Science and Engineering 1062 (2021) 012038 doi:10.1088/1757-899X/1062/1/012038

5. Validation of Result and Discussion

The validation process involves analyzing the shape and calculating the parameters of the hysteresis loops for both experimental and modelling results. Selection of the best hysteresis loops representation was made from 57 numbers of Hysteresis Rules which are available in the Manual Appendices of the RUAUMOKO 2D. This involves multiple trial and error with two different Hysteresis Rule, which is Hysteresis Rule 4 (Modified Takeda Hysteresis) and Hysteresis Rule 25 (Takeda with Slip). Ultimately, Hysteresis Rule 44 (Pampanin Reinforced Concrete Beam-Column Hysteresis) was chosen since it is best represented the specimen. The Pampanin RC beam-column hysteresis includes bilinear unloading and slip-on reloading, as shown in Figure 10. Therefore, it is more suitable for defining the behaviour of reinforced concrete sections, as there are parameters that focused on the pinching and strength degradation behaviour. There are six main parameters, namely AlfaS1 (As1), AlfaS2 (Xi), AlfaU1 (Au1), AlfaU2 (Au2), DeltaF (Df) and Beta. Hysteresis Rule 44 was used to modelling both specimen CJ-2B and specimen IJ-3B.

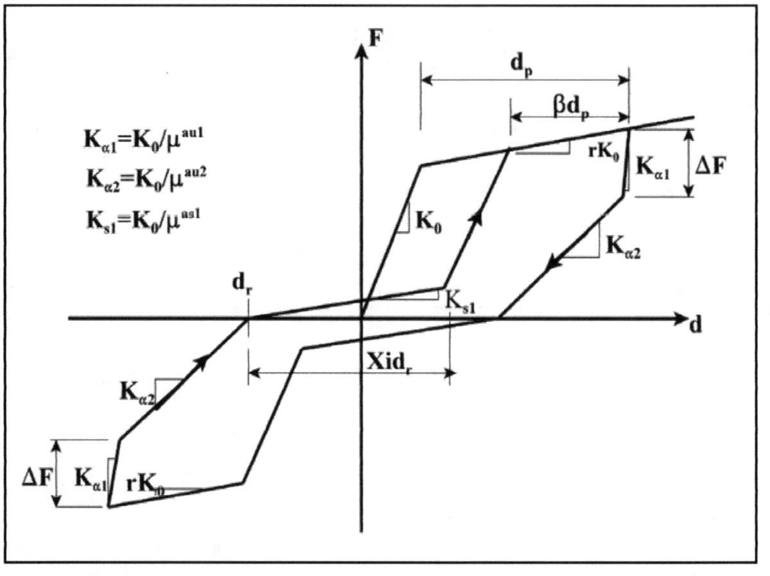

Figure 10. Analytical hysteresis for Hysteresis Rule 44 (Pampanin Reinforced Concrete Beam-Column Joint) [14]

Figure 11 shows the comparison of hysteresis loops between experimental and modelling for specimen CJ-2B which had been tested under in-plane lateral cyclic loading. The solid line represents the modelled hysteresis loops while the dotted line represents the experimental hysteresis loops. The rule incorporated allowed pinching, and therefore a good fit was achieved. Although a similar behaviour was shown between all the hysteresis loops, there are still some discrepancies between them. It was detected that the shape of modelling hysteresis loops is similar to the experimental hysteresis loops. However, the modelling hysteresis loops are a little bit fatter than the experimental hysteresis loops. This phenomenon occurs because during experimental work, a lot of energy was released to the surrounding under friction and kinetic energy. Figure 12 shows the comparison of hysteresis loops between experimental results and modelling results using HYSTERES program for specimen IJ-3B. These two graphs show good agreement and less disparity between experimental hysteresis loops and modelling hysteresis loops. From these graphs will be used to make the comparison between them for seismic performance parameters such as lateral strength capacity, stiffness, ductility and equivalent viscous damping. The comparison in term of percentage for these parameters will be presented in the next section.

IOP Conf. Series: Materials Science and Engineering 1062 (2021) 012038 doi:10.1088/1757-899X/1062/1/012038

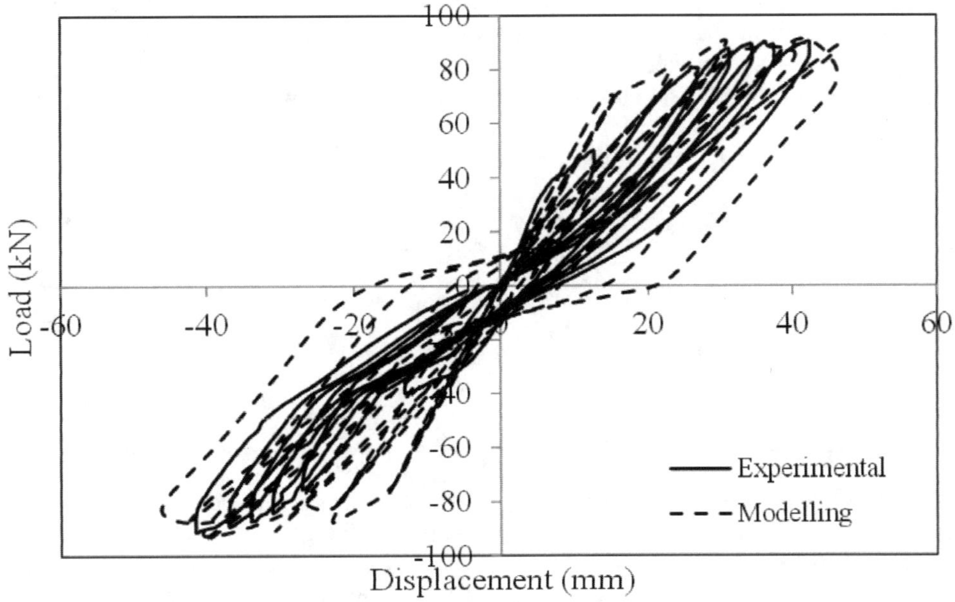

Figure 11. Comparison of hysteresis loops between experimental and modelling results for specimen CJ-2B

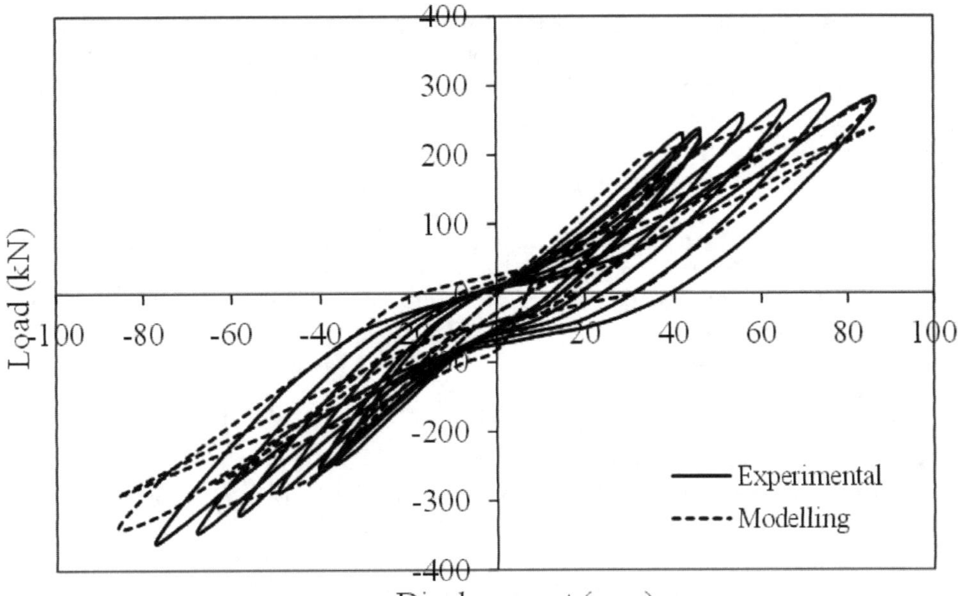

Figure 12. Comparison of hysteresis loops between experimental and modelling results for specimen IJ-3B

6. Comparison between Experimental and Modelling Hysteresis Loops

Table 2 shows the comparison of lateral strength, ductility, stiffness and equivalent viscous damping between experimental hysteresis loops and modelling hysteresis loops in pushing direction for specimen CJ-2B. The lateral strength was taken as the in-plane lateral force, which contributes to the maximum displacement for every drift tested in the laboratory. Based on the results, it can be seen that the percentage difference for lateral strength and stiffness are fairly low. In contrast, the percentage difference for ductility and equivalent viscous damping is higher. There is quite a big disparity between them because the modelling hysteresis loop is bigger than the experimental hysteresis loops which obtained directly from measurement data. However, this limit is within the acceptable range because only one specimen was conducted through this study.

Table 2. Comparison of lateral strength, ductility, stiffness and equivalent viscous damping for specimen CJ-2B

Parameters	Experimental	Modelled	Difference (%)
Lateral Strength (kN)	90.39	93.71	3.54
Ductility	2.22	2.76	19.57
Stiffness (kN/mm)	1.43	1.56	8.33
Equivalent Viscous Damping (%)	8	9.83	18.62

Table 3 shows the comparison of lateral strength, ductility, stiffness and equivalent viscous damping between modelling and experimental for specimen IJ-3B in pushing direction. The results revealed that the percentage difference for lateral strength, stiffness and equivalent viscous damping is fairly low while the percentage difference for stiffness is quite high. This is because the experimental hysteresis loops are steeper than the modelled hysteresis loops. Overall, similar behaviour was shown between all the hysteresis loops but with some discrepancies between them. The HYSTERES Program tends to overestimate the results, however, the results are still reasonable to be used in Ruaumoko 2D for dynamic analysis since the requirement is only from the overall shape of the hysteresis loops. The limitations in the experimental work contributed to the difference between experimental and modeled hysteresis loops. During the experimental work, some of the energy was released to the surrounding under friction and kinetic energy. The untighten screw between a foundation beam and strong floor causes friction, and this contributed to smaller experimental hysteresis loops.

Table 3. Comparison of lateral strength, ductility, stiffness and equivalent viscous damping for specimen IJ-3B

Parameters	Experimental	Modelled	Difference (%)
Lateral Strength (kN)	301.58	275.40	8.68
Ductility	2.74	2.86	4.38
Stiffness (kN/mm)	3.05	3.55	16.39
Equivalent Viscous Damping (%)	7	7.41	5.86

7. Conclusion and Recommendation

Based on experimental work, modelling hysteresis loops using HYSTERES program and validation between them, the following conclusion and recommendation are drawn:

1. It is important to validate the experimental and modelling hysteresis loops so that the seismic parameter can be determined and can be used for the design of RC buildings using Eurocode 8.
2. The parameters obtained from the modelling of hysteresis loops can be used to represent the actual behaviour of RC buildings and simulate them under different level of earthquake excitations using Ruaumoko 2D program.
3. It is recommended to use the parameters for Pampanin Reinforced Concrete Beam-Column Joint Hysteresis Model (Hysteresis Rule 44) in modelling any RC buildings using non-linear time history analysis and Ruaumoko 2D Programme.

Acknowledgement

The authors would like to thank Ministry of Higher Education (MOHE) of Malaysia, Putrajaya, Malaysia for awarding research grant under Fundamental Research Grants Scheme (FRGS) with File Reference No: 600-RMI/FRGS 5/3 (0091/2016) and RMI (Research Management Institute) for managing this research work. Gratitude and appreciation to the laboratory staff members for their invaluable assistance during the experimental research work.

References

[1] Baird A, Palermo A and Pampanin S 2013 *New Zealand Society of Earthquake Engineering* **8**
[2] Jabir H A, Al-Gasham T S and Mhalhal J M 2018 *3rd International Energy and Engineering Congress* **3** 544
[3] Tukiar M A, Kay Dora A G and Hamid N H 2014 *Applied Mechanics and Material* **661** 128–133
[4] Ghani K D and Hamid N H 2013*WIT Transactions on Built Environment* **134** 827–837
[5] Hamid N H, Tarmizi L H and Ghani K D 2015 *AIP Conerence Proceedings* **1660**
[6] Bazzurro P, Mollaioli F, De Sortis A and Bruno S 2006 *8th US National Conference Earthquake Engineering* 6 **1449** 3319–3328,
[7] Karimi N A, Hamid N H and Karimi N 2018 *International Journa Engineering Technology* 7 **4** 6937–6942
[8] Kamaris G S, Hatzigeorgiou G D and Beskos D E 2014 *Bulletin of Earthquake Engineering* 13 **2** 587–612
[9] Abdullah S, Hayati N and Hamid A 2012 **2** 67–76.
[10] Carr A J 2007 4 73
[11] Kristombu Baduge S, Mendis P, Ngo T D and Sofi M 2019 *International Journal of Concrete Structures and Material* **13** 1
[12] Bompa D V and Elghazouli A Y 2017 *Structures* **12** 115–119
[13] Kotsovos G M, Vougioukas E and Kotsovos M D 2013 *ACI Structural Journal* 110 **3** 427–435
[14] Carr A J 2007 **1** 3–4

Parcel Tracking System Using Barcode Scanner with Verified Notification

Juanita Zainudin[1*], Hafiza Samad[1], Faizah Miserom[1] and Shima Sabri[1]
[1]Faculty of Computing and Multimedia Kolej University Poly-Tech MARA, 56100 Cheras, Kuala Lumpur, Malaysia

*Corresponding e-mail: anis_juanita@kuptm.edu.my

Abstract. The emergence of electronic commerce online shopping has significantly increased the use of mail-order within Kolej University Poly-Tech Mara (KUPTM) communities. The tracking system provided by the courier services offers the status of delivery for each parcel until it reaches the designated destination. However, inefficiencies of managing the parcels at KUTPM may causing lost or misplaced. Therefore, the study offers an innovative approach where the study intended to invent a parcel tracking delivery and management systems for KUPTM communities. In particular, this invention relates to computerized tracking of delivered parcels using the barcode or tracking number that will notify recipients via Short Message Service (SMS) notification. The primary purpose of this study is to develop an efficient and secured system called the Parcel Tracking System. The system used a web-based service for managing and tracking the parcel. The development method of this system uses PHP and JavaScript programming languages in Adobe Dreamweaver platform with MySQL programme as the backend of the system. This paper focuses on the development process of the Parcel Tracking System. The findings found that the efficiencies of parcel tracking and security management are significantly improved. The application of Parcel Tracking System is promising and efficient for KUPTM communities. It is evidently that the proposed system was found successful and beneficial for KUPTM communities in tracking and managing parcels efficiently.

Keywords: parcel tracking, barcode scanner, short message services

1. Introduction

A parcel tracking system commonly accompanied by a unique parcel identification which only detected from the central hub to the delivery of the item to the designated address. In Malaysia, Poslaju, GDex or DHL are some of the service providers who offer this kind of services to the customers. However, in some cases, the item may send to the office addresses instead of home address. The present study investigates the efficiencies of parcel tracking system KUPTM. The parcels usually will be received by the office administration department whereby the items will be in stationery for a temporary. However, due to lack of monitoring or managing the parcels, may cause inefficient where it can be potentially lost or unable to track effectively. Therefore, the current study suggesting more systematically approach by designing a system that can benefit the receiver once the item received by the administration office.

Several studies show that the tracking systems require notifications to the receivers to ensure the arrival of the parcel, but this is not the case. In this study, the items managed manually. Thus, these have significant impacts of the parcel's distributions at KUPTM. Also, there are other difficulties experienced by the office administration. These include a lack of monitoring system and the needs to have frequent monitoring of the parcels once it received by the administration office. Thus, notification characteristics need to impose on the tracking system in managing ingoing and outgoing items adequately and proficiently.

Therefore, this research seeks to explain the development of Parcel Tracking System for KUPTM in assisting the staffs and administrators to manage the parcels. In this study, technical features were added such as the tracking procedure and the parcels update collecting status will be provided. Additionally, the system will provide notification through SMS to notify recipients on the arrival of parcel immediately.

The main purpose of this study is to develop an efficient and secured web-based system called the Parcel Tracking System at KUPTM for managing and tracking the parcel. In order to develop the Parcel Tracking System, the Spiral Methodology is used for the software development process while for the programming language, the current study utilised PHP programming. For the server backend, XAMPP, Apache and MyPHP is used and Adobe Dream Weaver serves as IDEs at the development stage. In addition, the present study intended to propose a QR code scanner and verified notification through SMS to the receiver.

The development of the Parcels Tracking System also includes a framework. The suggested framework is to overview the process of an online system which explain steps for the users to access using their portable gadget. Inevitable, the framework as fruitful in improving the efficiency of tracking parcel and security of the items in KUPTM with success. Thus, the study intended to propose a barcode scanner and verified notification through SMS to the receiver. With efficient Parcels Tracking Systems, this study demonstrates the need for better strategies for developing an efficient tracking system at KUPTM

This section has overviewed the significant of tracking system from manually to computerised system where the present study identified key component of developing a Parcels Tracking System.

The next part of this paper will address the main issues of the study following the methodology, developing process of the tracking system, findings and finally the conclusion.

2. Research Problem

The advent of electronic commerce and online business has increased the number of online shoppers yearly [1-5]. Thus, courier services sector such as Poslaju, GDex or DHL has simultaneously expanded their services significantly. Packages are delivered to designated destination via courier services and it has become a major necessity for companies that requires courier services on a regular basis to deliver parcels from different places [6-10].

The KUPTM's communities also requires the usage of courier services to deliver their online purchased goods or items and have them delivered to KUPTM as most of them spend more time at the establishment rather than at home. The deliveries make the transitions or process easier for both parties, i.e; the courier company and the receiver. The motivation of this study was initiated when there were too many parcels delivered at the administration department and some of it was stranded for long time.

The administration department of KUPTM is responsible in accepting letters and parcels addressed to KUPTM's staff and understudies. However, it is not the responsibility of the administration department personnel to oversee individual's parcel, as they are just required to deal with formal letters and parcels. Majority of them feel obliged to receive and deal with the parcels for other staff as well since all works were carried out manually using logbook. Manual work may lead to time-consuming and are prone to human errors [11-13]. Due to this, their remaining task at hand has some way or another expanded definitely as they require additional time in taking care of and dealing with the parcels, in which, now and again the things will take up some of their workspace and disrupts their movements.

The receivers made numerous complaints, as most of them did not get any updates from the administration department in regard to their parcel arrival status. They need to check regularly at the administration office, which some finds it inefficient and cumbersome. Meanwhile, for students, the Student's Council will make an announcement through the social media platforms such as Facebook and Twitter to update on the arrival of their parcel. However, this is not proven efficient as not all students are active on social media platforms and they might not have Facebook or Twitter account.

Considering keeping the items for quite some time may cause distress for the administrative staffs. is due to handling personal belongings which create discomfort among administrative personnel. With limited staff and space contributes to the unorganized parcel management. Ample space also required to store parcels due to limiting the office space and contribute to the inefficiencies of storing the items. On the other hand, information about the parcel's received was recorded manually. Consequently, there is a potential for having misplaced information and increase human errors. By understanding factors that should consider, proposing a Parcel Tracking System is an opportunity to widen the efficiencies of managing the items.

Based on feedback received from administration department (*Bahagian Penyelenggaraan, Pentadbiran & Pengurusan Aset*), regularly, there are an average of 30 parcels delivered and may increase up to 50 packages during the seasonal and promotional month. There were some cases where several items lost due to poorly managed collection records. Based on these feedbacks, the idea of the insecure nature of the traditional method is not reliable. Overall, there are more than five items were not collected by the purchases since October 2018 and is believed the number may increase for the following year. Table 1 shows the analysis of parcels at KUPTM for the year 2019.

Table 1. Analysis of parcel records at KUPTM for year 2019

Category	Number of Parcel
Average delivered parcel per day	30
Uncollected parcel	5
Lost parcel reported	2

3. Methodology

For developing the Parcel Tracking System, the system is using the Spiral Model. The spiral model (Figure 1) has four phases, which is identifying, development, evolution and risk analysis. A software project repeatedly passes through these phases. The baseline spiral, starting in the planning phase, requirements are collected and continuously assessing the risk. Each subsequent spiral build on the baseline spiral. Several advantages are using the spiral model. The spiral model enables to control a high amount of risk analysis thus, avoidance of risk is enhanced. Also, the spiral model suits for large and mission-critical projects. The spiral model is also approval and documentation control. Furthermore, an initial stage of developing the software through a spiral model enables to identifying an additional functionality which can create more efficiencies at an early stage. Therefore, it is suitable to be adopted as their prime development approach for the maintaining, designing, planning and programming of this parcel tracking system.

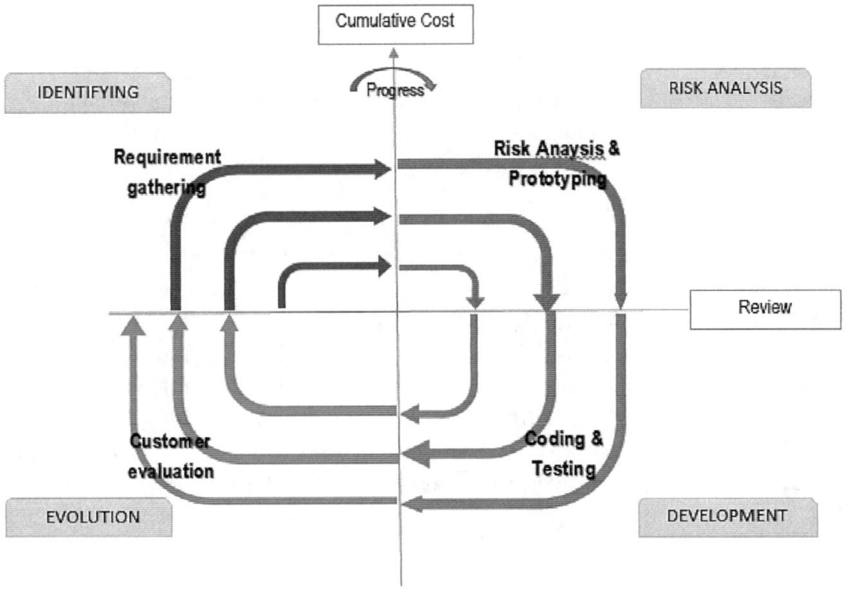

Figure 1. Spiral model

The programming language used for the implementation includes PHP programming was employed for the development of the users' application. XAMPP, Apache, MyPHP were used as server backends and Adobe Dream Weaver served as IDEs at the development stage. Table 2 defines the tasks that were conducted at each phase.

Table 2. List of tasks

Phase	Task
Requirement gathering	Generating a requirement of functional, system, technical and user expectation. Literature analyses form the previous studies was conducted and interview with the KUPTM lecturer and staff of Human Resource Department.
Risk analysis and prototyping	Cultivates creditable outcomes for any issues arise that is confronted during investigation such as previous and current problems, limitation of the system, and KUPTM infrastructure.
Coding and testing	Transform the design of the system to coding by using high-level language application; PHP programming, XAMPP, Apache, and MyPHP. Outcome of the system was tested by providing the program with a set of test inputs and observing the behave of the tracking system.
Customer evaluation	Beta testing by the user and gathering their feedbacks to meet their needs

As illustrated in figure 1, this model is a combination of waterfall and iterative model that supports risk management and is best used when software developers plan to handle the risks at each point. Consequently, failure can be minimized until the user can use it completely.

IOP Conf. Series: Materials Science and Engineering 1062 (2021) 012039 doi:10.1088/1757-899X/1062/1/012039

4. System Design

This section describes the process of system design by describing the necessary modules, architecture, stakeholders, components, interface and data. All relevant variables are indicated and specified to meet their requirements. There are two types of design styles which are use case diagram and flow chart design. Use case diagram is used to visualize the user's practical requirement while flow chart design is to display the steps and decision making during the process.

4.1. Use Case Design

Actors involved in this system are administrator and users. Administrator is the administration department's staff while users are staffs and students of KUPTM. In the use case diagram depicted in Figure 2, users need to register first to enable the email notification with Quick Response Code (QR) code. The parcels can be collected using this verified QR code. There is a tracking activity where Users can enter tracking number and find out the status of parcel. Administrator will add the newly arrived parcels to the system and system will notify the Users through SMS notification. The Update Collector task allows Administrator to update collector information when parcel is collected. All parcel information is manageable by Administrator. Report will be generated providing parcel and collector information for future references. Super Administrator can add and manage all Administrator of the system.

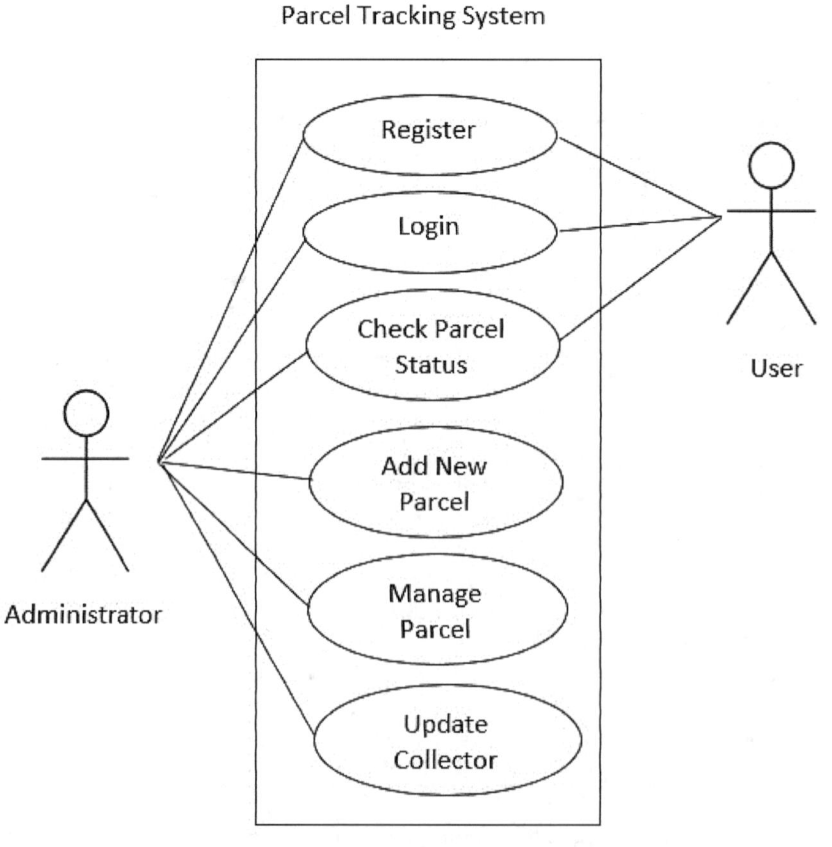

Figure 2. Use case diagram of parcel tracking system

IOP Conf. Series: Materials Science and Engineering 1062 (2021) 012039 doi:10.1088/1757-899X/1062/1/012039

4.2. Flow Chart Design
The proposed program removes all manual works by offering valuable features do not present in current practice. The purpose of this project is to make it easier for users to track and trace parcels that arrived at KUPTM. The flow chart of the proposed system is shown in Figure 3. Administrator will add new parcel information into the system by scanning the barcode number and enter recipient's phone number. Then, the system will automatically send an SMS notification to the parcel's receiver. Users need to register for the first time using the connection given in the SMS notification. The QR code for collecting parcels will be delivered via email. Recipient will be received email providing QR code for collecting purpose. This to ensure the verified recipient would come and collect the respective parcel.

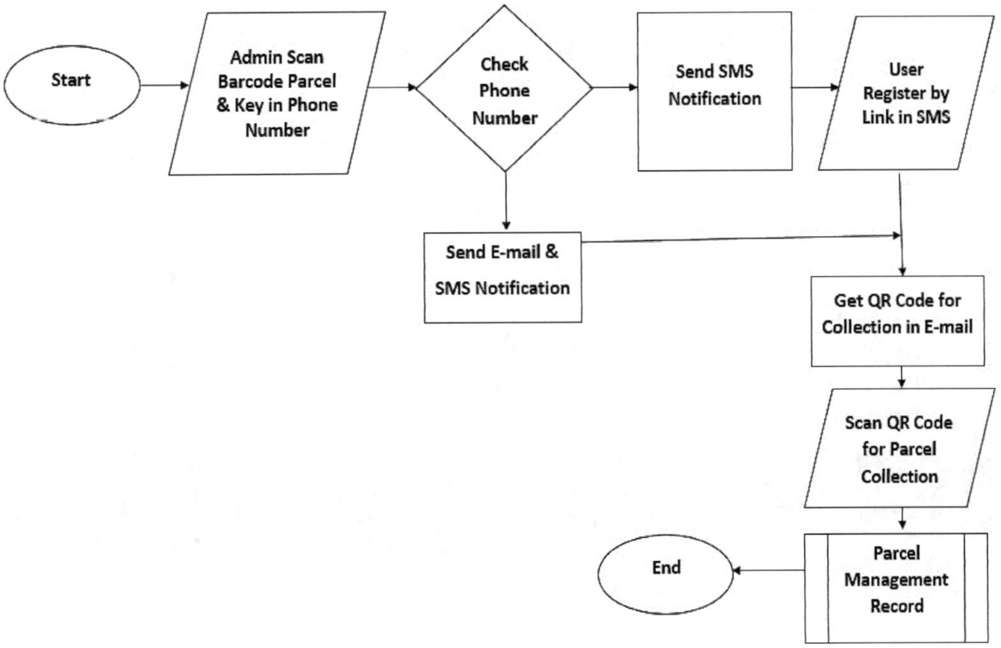

Figure 3. Flow chart diagram of parcel tracking system

5. Results and Discussion
The aim of this study is to develop an efficient and secured Parcel Tracking System using verified notification and barcode technology. In this system, tracking number will be scanned using barcode scanner and receiver information will be recorded into the system. Once recorded, users can trace parcel and the update the parcel collecting status as required. In addition, the system will also provide SMS notifications that will alert users the arrival time without any delay. Verified QR code for collections will also be sent to the email. This to ensure parcels are collected by the owner of the parcel only. According to a research done by University of Nigeria, proper courier service will ensure that customers easily participate effectively and will also build trust in the courier service system [14 - 15].

Figure 4 shows the user's main page. The main features for the main page are the users can track and trace their parcel. The system introduces a simple step by using the barcode scanner to read tracking number or enter manually the tracking number into the text box and clicks the submit button. Users can go to other pages by clicking on Location & Contact or FAQ button.

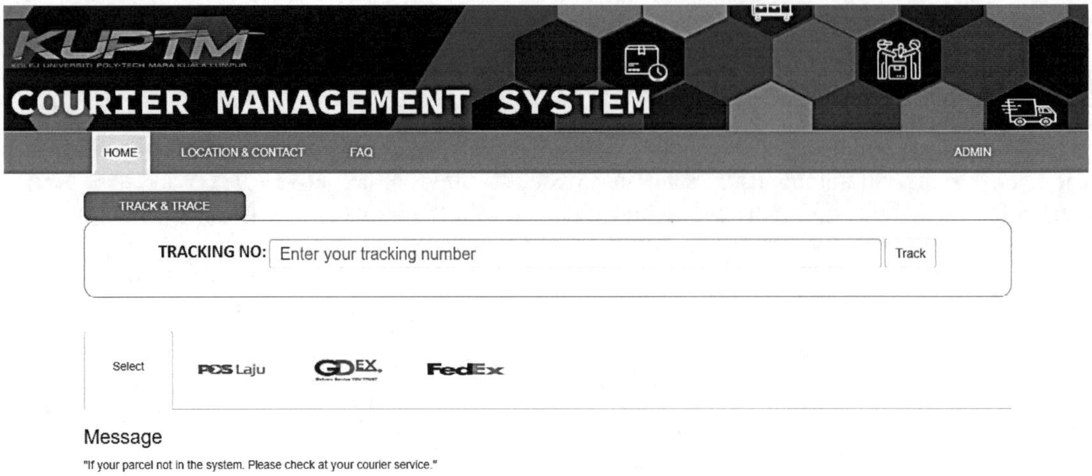

Figure 4. Sample of main page

The security administration page has authentication password and username to login into the system. This to ensure only authorized person can access and control the system. Administration main page provides searching engine for administrators to trace the parcel with a tracking number. Figure 5 shows the main page for system administrator.

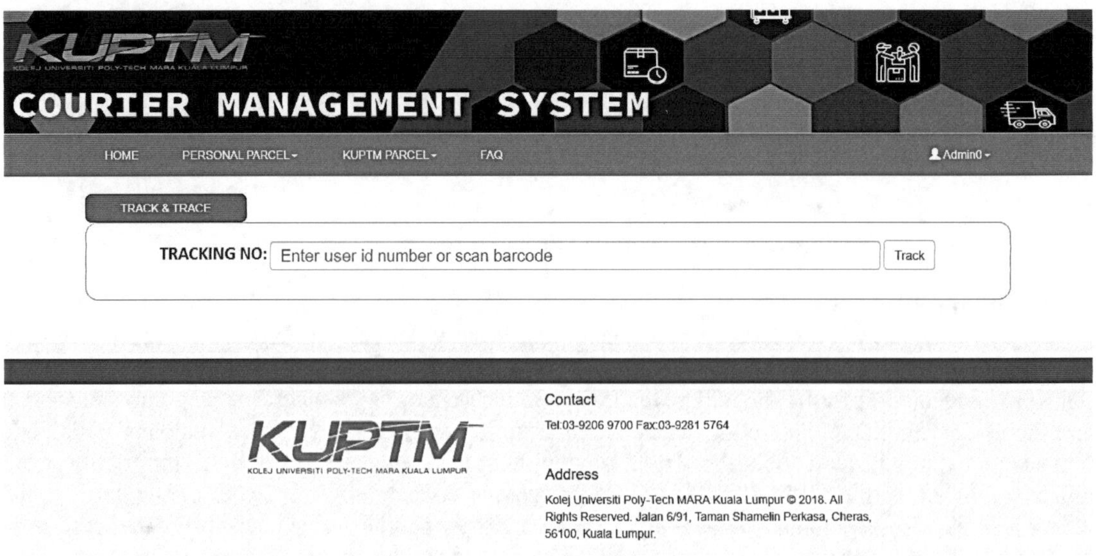

Figure 5. Sample administrator page

The system administrator can insert new received parcel using barcode scanner in Add New Parcel Page as shown in Figure 6. System administrator needs to key in all data provided in the form. System will automatically send notification through SMS to the phone number entered.

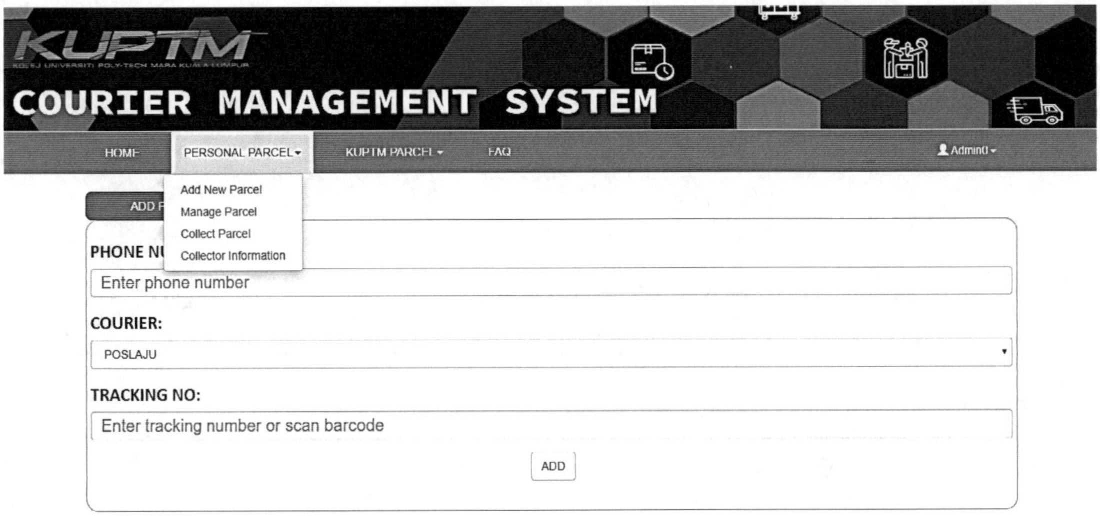

Figure 6. Sample page to add new parcel

On the other hand, the Manage Parcel's Page provides search and filter mechanism that can list data according to the status. System administrator can update or delete the information by clicking the action at the right of the table as shown in Figure 7. Password is required to update or delete information. This to ensure that data is not updated or deleted by unauthorized individuals.

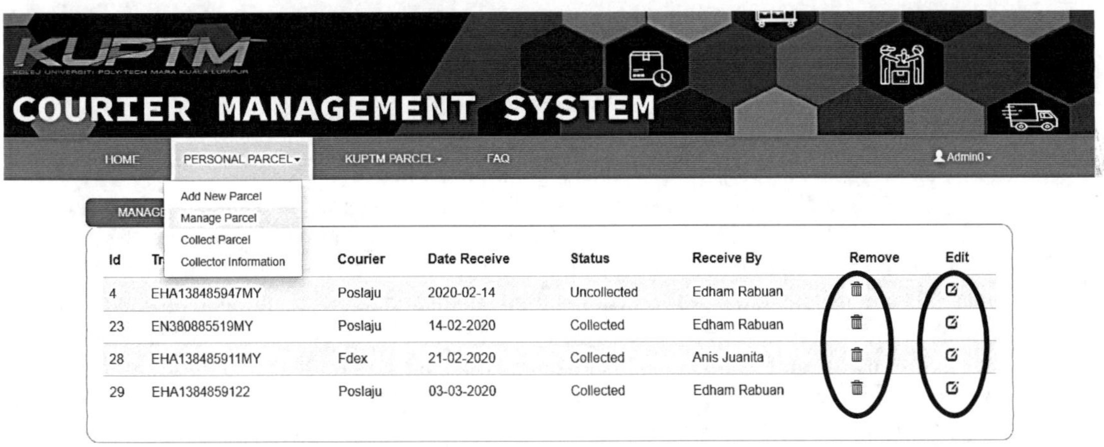

Figure 7. Sample page to manage parcel

Findings from survey has proven that 92.6% of respondents agreed that the proposed system has improved the tracking and management of parcel at KUPTM. The comparison from user feedback in the range of strongly disagree to strongly agree was analysed and measured results is tabulated in Table 3. This comparison of results show that the tracking and management process has been improved and successful. The results also revealed that the proposed system has contributed significantly towards improving management and tracking parcel at KUPTM.

Table 3. Respondent feedback on proposed system can improve parcel data management

Range	Frequency	Percent	Valid Percent	Cumulative Percent	Range Description
1	0	0	0	0	Strongly Disagree
2	0	0	0	0	Disagree
3	9	7.4	7.4	7.4	Neutral
4	18	14.9	14.9	22.3	Agree
5	94	77.7	77.7	100.0	Strongly Agree
Total	121	100.0	100.0		

6. Conclusion

The research undertakes to develop a Parcel Tracking System and implement at KUPTM. It has shown that there are significantly essential in managing items delivered to the administration office. Overall, the Parcel Tracking System enable to increase efficiencies in terms of information storing, provide notification messages, and more importantly, handling and managing the parcels can be achieved successfully. This study has demonstrated the needs of computerized system that enable the users to manage the items. Furthermore, the application of Parcel Tracking System at KUPTM enable eliminate human errors, create a new opportunity for the administration staffs to locate and storing data. In addition, the system demonstrates the capability to transmit the information to the receiver in timely manner. Although the propose system enable to eliminate human errors, several improvements can be suggested to increase the system capabilities. It is recommended that further research be undertaken in the following areas:

1) Additional features for mobile application capabilities: The current project focus on application usage using personal computers. Future works can be done by offering better access to users using mobile phones. The needs of having a mobile app are very much needed for this study. However, adjusting the systems into mobile apps requires added specification such as system interfaces, mobile views and others. It is believed if by offering these additional features to the users, there are more likely the system can be implemented elsewhere.
2) Integrating Parcel Tracking System into KUPTM Main System: The KUPTM Parcel Tracking System is an independent system which designed for notifying parcel status to the receivers. However, it is recommended that the future study enable to integrate with KUPTM system where it is more centralized and accessible by both users, staffs and students. Integrating a system is not new. However, considering a further requirement is very much needed. These include but not limited to the software requirements, designing the compatibility of the system to emerge with the existing system and migrating the database. In addition, the integrative system allows the users to benefiting the functions and notifying the receivers efficiently. Another important practical implication of having the integrated system is to ensure the stability and efficiencies of parcel data to the administration office.

Acknowledgments

The authors gratefully acknowledge the URG/1018/SCTS/01324(16) for providing financial support.

References

[1] Hajli N, Shanmugam M, Papagiannidis S, Zahay D and Richard M O 2017 *J. Bus. Res.* **70** 136-44
[2] Shuen A 2018 Web 2.0: A Strategy Guide: Business thinking and strategies behind successful Web 2.0 implementations: O'Reilly Media.
[3] Mahajan M D and Rane K 2015 *Int. J. Sci. Eng. Res.* **6(10)** 437-439
[4] Ming-Shen W, Chih-Chung C, Su-Chao C and Yung-Her Y 2007 *Int. J. Manag.* **24(2)** 296

[5] Rosqvist L S and Hiselius L W 2016 *J. Clean. Prod.* **131** 163-169
[6] Le T V and Ukkusuri S V 2019 *TRIP* **1** 100008
[7] Rai H B, Verlinde S and Macharis C 2018 *IFAC-PapersOnLine* **51(11)** 588-593
[8] Rajendran S D, Wahab S N, Ling Y W and Yun L S 2018 *Int. J. Sup. Chain. Manag.* **7(5)** 461
[9] Atletiko F J 2017 *Procedia Computer Science* **124** 759-766
[10] Wang Y, Jin M, Chen G, Cui X, Zhang Y and Li M 2019 *J. Adv. Res.* **20** 23-32
[11] Aravind T, Kumar P D and Saketh D 2019 *Int. J. Sci. Res. Comput. Sci. Eng. Inf. Technol.* **5(2)** 420-423
[12] Zhou M, Zhao L, Kong N, Campy K S, Xu G and Zhu G 2020 *J. Retail. Consum. Serv.* **52** 101911
[13] Wu C, Wang X, Chen M and Kim M J 2019 *Adv. Eng. Inform.* **42** 100960
[14] Agu M, Nwoye C and Ogbuokiri B 2015 *IOSR J. Mobile. Comput. Appl.* **2(2)** 56-61
[15] Tyler D P and Subramaniam B 2020 Location tracking using beacons. Google Patents.

Performance of Ocean Thermal Energy Conversion Closed Rankine Cycle Using Different Working Fluids

N Samsuri[1*], N Sazali[2], A S Jamaludin[3] and M N M Razali[3]
[1] Ocean Thermal Energy Centre (OTEC), Ground Floor, Block Q, Universiti Teknologi Malaysia Kuala Lumpur, Jalan Sultan Yahya Petra, 54100 Kuala Lumpur, Malaysia.
[2] Department of Electrical Engineering, College of Engineering, Universiti Malaysia Pahang, 26300, Gambang, Kuantan, Pahang, Malaysia.
[3]Faculty of Manufacturing and Mechatronic Engineering Technology, Universiti Malaysia Pahang, 26600 Pekan, Pahang, Malaysia

*Corresponding e-mail: norazreen1989@gmail.com

Abstract. Ocean Thermal Energy Conversion (OTEC) is a foundation for an appealing renewable energy technology regarding its vast and inexhaustible resources of energy, renewability, stability, and sustainable output. The principle of an OTEC power plant is to exploit the energy accumulated in between the top layer of warm surface seawater (heat source), and the cold layer of deep seawater (heat sink). The plant operates based on a Rankine cycle to produce electricity between the source and the sink at the smallest temperature difference of approximately 20 K. In an OTEC power plant, a commonly utilized working fluid is ammonia since its qualities are suitable for the OTEC cycle. Nevertheless, ammonia poses certain potentially lethal health risks and hazardous fluid. Hence, the effect of the working fluid types, and the subsequent operation conditions may be critical and therefore become the subject of this study. The analysed working fluids, including that of ammonia, are ammonia-water mixture (0.9), propane, and refrigerants (R22, R32, R134a, R143a, and R410a). The results revealed that ammonia-water mixture showed the highest network performance and reliability. Even so, it is essential to continue seeking the suitable working fluids which are safe and economically effective to replace ammonia.

Keywords: ocean, thermal energy, rankine cycle, power plant

1. Introduction

Ocean Thermal Energy Conversion (OTEC) has tremendous prospective in deep ocean water area, in which a sufficiently high temperature difference between the surface water and a specific depth is required to effectively run an OTEC power plant. In 1881, Arsonval's initial concept specified that the optimum temperature difference needed for the installation of an OTEC plant is larger than 20 K [1]. The system will work between the surface seawater at 30°C (known as heat source), and seawater at 1000 m depth with temperature of 4°C (known as heat sink) [2-4]. OTEC power plant technology is developed on a basis of open (OC-OTEC) and closed Rankine cycles (CC-OTEC). Previous research reported that the process has to be founded upon the Uehara cycle for optimal power plant output, implementing ammonia-water mixture as the working fluid with smaller than 20 K of temperature difference, and at 5-6% thermal efficiency [5].

Additionally, the selection of suitable working fluids has a significant impact on the entire system viability and efficiency. Ammonia has been considered as the best working fluid because it has a suitable boiling temperature (28°C - 32°C) for the OTEC purpose [6]. However, it is toxic and therefore can be hazardous to the environment. Recent development of working fluids shows that ammonia can be replaced by other working fluids with zero Ozone Depletion Potential (ODP) and zero Global Warming Potential (GWP). Finding other suitable substitutes is a big challenge to this study. Several studies have been done which have shown better results with the use of other working fluids such as hydrocarbon (HC) and hydrofluorocarbons (HFCs) [7-10]. The other findings reported that R123 displayed the best performance, but the fluid contributes to the ODP and GWP. Therefore, isopentane is suggested as it showed the second-best performance and regarded as environmentally friendly working fluid for the system. However, the applications of their research are for waste heat systems but can still be operated at low temperature.

A paper that reviews about 35 working fluids and analyzes the effect of fluid properties on the cycle efficiency is written by Chen at al. (2010) [11]. They have categorized the working fluids under three characteristics which are dry, isotropic, or wet fluid according to the T-s diagram. Understanding the characteristic of the working fluids eases the process of selecting the appropriate working fluid for the cycle. Calm and Hourahan (2007) [12] have interpreted the data of working fluids into a table with ODP and GWP of selected refrigerant. Figure 1 shows the numbers of OTEC previous research focusing on different working fluids from 1979 to 2016.

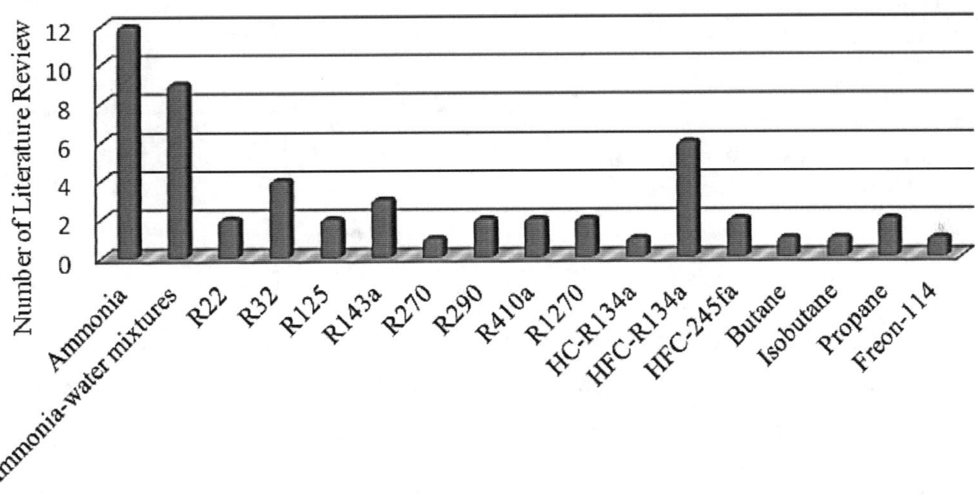

Figure 1. The numbers of OTEC previous research focusing on different working fluids from 1979 to 2016

Therefore, the objective of this paper is to examine the efficiency of the OTEC basic closed Rankine cycle using varying working fluids. At the initial stage of the study, preliminary simulation was conducted to confirm the simulation model with the reference from past OTEC studies. The similar developed model was implemented to analyze the efficiency of the OTEC basic closed Rankine cycle using eight varying working fluids.

2. Methodology

2.1. Introduction

Laboratory Virtual Instrument Engineering Workbench (LabVIEW) is a platform design framework created by National Instruments that is employed as languages of visual programming. Its implementation in numerous fields of engineering (e.g., aeronautical, mechanical, electrical, etc.) has led to the advancement of the world's largest and most complex applications to fulfill future demands. LabVIEW offers the users with flexibility through intuitive graphical programming which helps to reduce the time needed for test development. The thermodynamic model has been created in LabVIEW and linked to the working fluid data base in National Institute of Standards and Technologies (NIST) RefProp 9 and PROPATH. The thermodynamic model of OTEC cycle is created in LabVIEW to run numerical calculation, simulations and compare the working fluids from a thermophysical perspective.

2.2. Analytical Techniques of Thermodynamics

The simulation was based upon the thermodynamic analysis of the OTEC Rankine cycle performance. The Rankine cycle comprises of four major components, which are condenser, coolant pump, turbine, and evaporator. Several assumptions were included to facilitate the simulation analysis and assessment [13,14], which are described as follows:

- Every component is in steady state.
- Any heat loss and pressure drop are disregarded.
- The system is completely insulated.
- All pumps and turbines are given isentropic efficiency.

For the steady state energy balance equation, the total energy entering a system is equal to the total energy exiting the system, as expressed in equation (1)

$$E_{in} = E_{out} \tag{1}$$

or it can be elaborated as in equation (2)

$$W_{in} + Q_{in} + \sum \dot{m}_{in} = W_{out} + Q_{out} + \sum \dot{m}_{out} \tag{2}$$

where $\dot{Q}$ represents heat transfer rate; $\dot{m}_{in}$ and $\dot{m}_{out}$ is inlet and outlet mass flow rate; whereas W_{in} and W_{out} is work inlet and outlet, respectively. By assuming the system is completely insulated and any heat losses are neglected; $Q_{in} = 0$, $Q_{out} = 0$ and $W_{out} = 0$; the energy balance in the pump is expressed as in equation (3)

$$W_{in} + \sum \dot{m}_{in} = \sum \dot{m}_{out} \tag{3}$$

From equation (3), the work supplied is given as in equation (4)

$$W_{in} = \sum \dot{m}_{out} - \sum \dot{m}_{in} \tag{4}$$

Rate of heat supplied to the cycle (evaporator), $\dot{Q}_e$ is expressed as in equation (5)

$$\dot{Q}_e = \dot{m}_{wf} \Delta h_e \tag{5}$$

Rate of heat rejected from the cycle (condenser), $\dot{Q}_c$ is indicated as in equation (6)

$$\dot{Q}_c = \dot{m}_{wf}\,\Delta h_c \tag{6}$$

Rate of heat absorbed from the warm seawater, $\dot{Q}_{e,ws}$ is expressed as in equation (7)

$$\dot{Q}_{e,ws} = \dot{m}_{ws}\,c_p \Delta T_{ws} \tag{7}$$

Rate of heat rejected into the cold seawater, $\dot{Q}_{c,cw}$ is indicated as in equation (8)

$$\dot{Q}_{c,cw} = \dot{m}_{cs}\,c_p \Delta T_{cs} \tag{8}$$

where $\dot{m}_{ws}$ and $\dot{m}_{cs}$ are the mass flow rate of warm and cold seawater, respectively. c_p is the seawater specific heat capacity at constant pressure.

The working fluid pump, $W_{P_{wf}}$ and the turbine work, W_T is written as in equation (9) and equation (10)

$$W_{P_{wf}} = \dot{m}_{wf}\,v(P_2 - P_1) \tag{9}$$

$$W_T = \dot{m}_{wf}\,\Delta h \tag{10}$$

where Δh represents the enthalpy difference in the turbine system.

Referring to Uehara and Ikegami (1990) [14], the working fluid pumping power, P_{Pwf} is given as in equation (11). The pumping power of warm seawater, P_{ws} is indicated as in equation (12); whereas the pumping power of cold seawater, P_{cs} is expressed as in equation (13)

$$P_{Pwf} = \frac{\dot{m}_{wf}\,\Delta H_{wf}\,g}{\eta_{wf,p}} \tag{11}$$

$$P_{ws} = \frac{\dot{m}_{ws}\,\Delta H_{ws}\,g}{\eta_{ws,p}} \tag{12}$$

IOP Conf. Series: Materials Science and Engineering 1062 (2021) 012040 doi:10.1088/1757-899X/1062/1/012040

$$P_{cs} = \frac{\dot{m}_{cs} \Delta H_{cs} g}{\eta_{cs,p}} \quad (13)$$

where ΔH refers to the difference in pressure.

The net power output, P_n is indicated as in equation (14)

$$P_n = P_G - P_{ws} - P_{cs} - P_{p_{wf}} \quad (14)$$

2.3. *Selection of Working Fluid*

An ideal working fluid should have the relevant thermophysical properties corresponding with its application, besides sustaining its chemical stability within the specified range of temperature. Working fluid selection plays a major part on the system in terms of its performance, operating conditions, effects on the environment and economic feasibility. In this section, the parameters for identifying a suitable working fluid for the cycle system are described. Sami (2012) [15] has listed the main factors affecting the properties of thermodynamic and thermophysical of the system, among which are thermal conductivity, chemical stability, specific heat, boiling temperature, latent heat, toxicity, as well as flash point, as outlined in Figure 2.

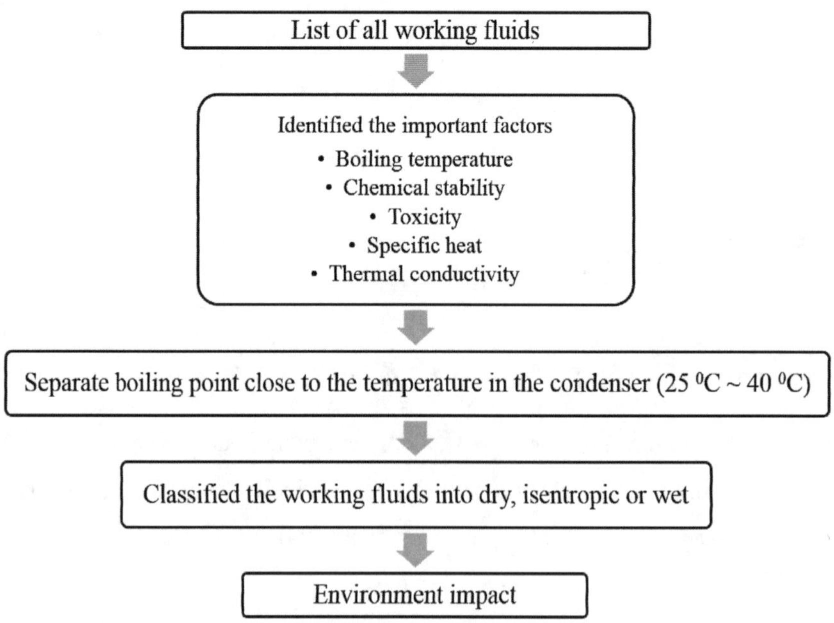

Figure 2 Steps in selecting the working fluids [15]

The OTEC closed Rankine cycle in this study utilized the boiling point of the working fluid near the evaporator operating temperature, that is about 25°C to 40°C [16]. In addition, the fluids were classified as dry, isentropic, or wet relative to the saturation curve (dT/ds). A dry or isentropic fluid is appropriate to be implemented in OTEC closed Rankine cycle [17]. The purpose of separating the type of fluids is to make sure that the fluids are totally superheated after isentropic expansion, intended to avoid the appearance of liquid drops on the blades of the turbine.

2.4. *Types of Working Fluids*

There are dual kinds of working fluid, namely pure fluid (pure compound) and pseudo-pure fluid (a mix of several pure compounds of fluid). Ammonia, propane, R22, R134a, and R143a, are marked as pure fluid, and are not combined with some other compounds. Meanwhile, ammonia-water mixture, R404a, R410a, R470c, and R507a are a pseudo-pure fluid. Figure 3 and Figure 4 show that the highest enthalpy difference can be discovered in ammonia-water mixture, followed by ammonia, propane and R32.

According to Figure 4, in contrast with other working fluids, ammonia-water mixture has the highest quantity of heat applied. This situation is caused by its greater latent heat value, also can be defined as the amount of heat that a liquid absorbs to stay at a constant pressure or temperature throughout the process of vaporization.

2.5. *Preliminary Simulation*

A preliminary design model for simulation of a 1 MWe OTEC closed Rankine cycle was conducted using ammonia as working fluid. This preliminary simulation is to validate the model developed by Yeh et al., (2014) [18]. Apart from that, the preliminary design model allows the estimation for 5 MWe and 10 MWe OTEC closed Rankine cycle.

Table 1. Parameters for three OTEC cycles to be investigated

Parameter	Symbol	Unit	Value
Evaporating temperature	T_E	°C	28
Condensing temperature	T_C	°C	8
Warm seawater inlet temperature	T_{wsw}	°C	30
Cold deep seawater inlet temperature	T_{csw}	°C	5
Working fluid pump efficiency	η_{wf}	%	0.75
Turbine efficiency	η_T	%	0.82
Generator efficiency	η_G	%	0.95
Warm seawater pump efficiency	$\eta_{pump,wsw}$	%	0.80
Cold deep seawater pump efficiency	$\eta_{pump,csw}$	%	0.80

The OTEC closed Rankine cycle simulation based on Uehara and Ikegami (1990) [14] was conducted according to the fixed condition parameters as tabulated in Table 1, in which the ammonia is in a steady state. The graph that represents the simulated model is shown in Figure 5 (b). When comparing the reference case with the preliminary analysis, it was found that ammonia generated the maximum total work output. Such results are reinforced by the point that ammonia possessed the maximum as well as the most appropriate value of latent heat for the OTEC cycle system.

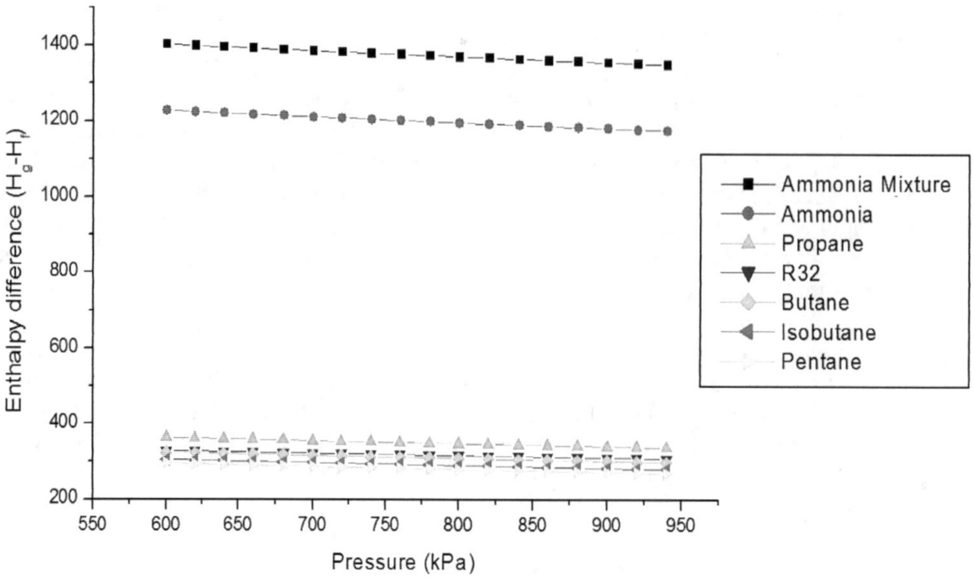

Figure 3 Latent heat-pressure diagram of pure fluid and pseudo-pure fluid

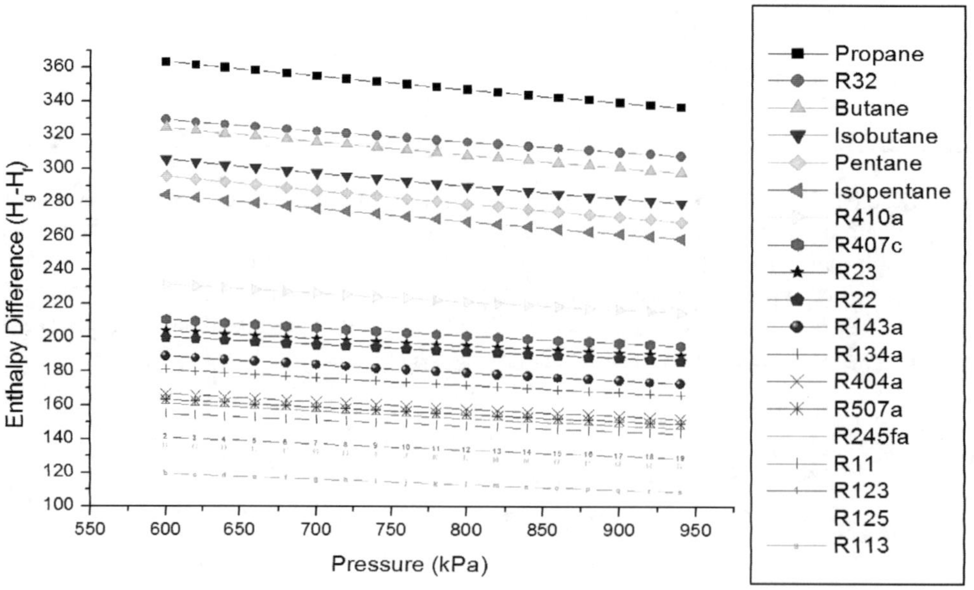

Figure 4 Close up of latent heat-pressure diagram of pure fluid and pseudo-pure fluid

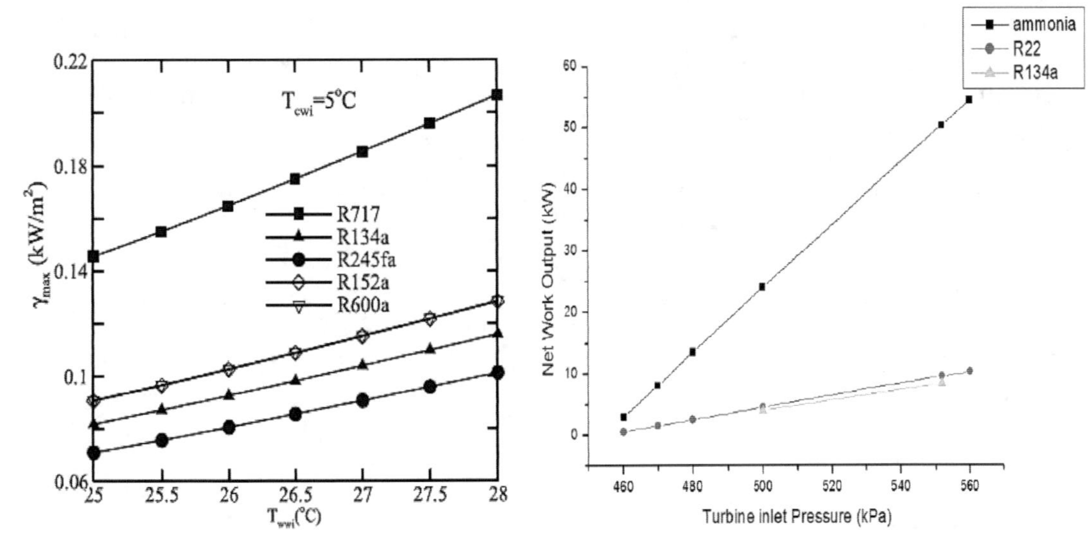

(a) (b)

Figure 5 The network output of closed Rankine cycle using several working fluids; (a) reported by Yeh et al., (2014) [18]; (b) simulation model using LabVIEW and RefProp

As can be seen from Table 2, the net power output increased significantly when the system was scaled up [19]. The preliminary study acts as an initiation to the visualization procedure used in the subsequent assessment in Section 3. On the other hand, the preliminary simulation is shown to explain the sufficiency of the parameters used in this study.

Table 2 Analysis of OTEC Closed Rankine cycle using ammonia as working fluid

	Unit	1MWe	5MWe	10MWe
Q_{in}	kW	19724.40	81375.50	162751.00
Q_{out}	kW	18685.00	76166.90	152334.00
$W_{p(wf)}$	kW	13.22	54.53	109.06
$W_{p(wsw)}$	kW	96.74	399.12	798.24
$W_{p(cws)}$	kW	118.76	484.11	968.21
$\dot{m}_{wf}$	kg/s	15.82	65.25	130.50
$\dot{m}_{wsw}$	kg/s	1793.01	7397.29	14794.60
$\dot{m}_{csw}$	kg/s	1587.67	6471.91	12943.80
W_T	kW	905.00	4525.00	9050.00
W_{net}	kW	676.28	3587.24	7174.48

3. Results and Discussion

The simulated net power output of eight varying working fluids produced by a work pump of deep seawater is shown in Figure 6. It was noticeable that the net power output for ammonia-water mixture was the maximum with 740 kW, and that the power needed for the cooling system to pump deep seawater was also small. A feature which is widely recognized in the OTEC power cycle is the point that ammonia resulted in the second highest net power output value. The third highest net power output was R134a followed by R22 and propane. R134a was the possible candidate to replace the ammonia as it possessed the highest net power output among the other five working fluids; however, it has the biggest value of work pump for deep seawater. Therefore, it required a big pipe to pump from the deep seawater to condense R134a. R22 has the higher net power output but lower pumping power for deep seawater compared to propane. R32 was the fourth possible candidate to replace ammonia. As the graph shown, R32 gaves the lowest pumping power than the other working fluids including pure ammonia. The graph also indicated that R410a and R143a have low pumping power compared to pure ammonia, but it has the lowest net power output. Even so, a substitute working fluid must be introduced to replace ammonia which is detrimental to the ecosystem and needs a special substance to be preserved.

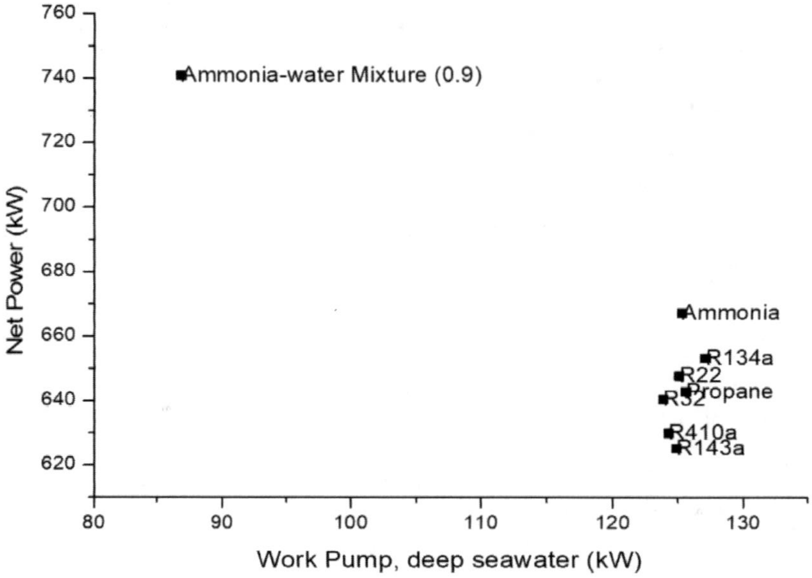

Figure 6. The simulated net power output of eight varying working fluids produced by a work pump of deep seawater

The relationship between the network output and efficiency is shown in Figure 7. Although both ammonia and ammonia-water mixture have greater network output and efficiency in contrast to the other working fluids, they need a separator to make sure that water vapor from the fluid (particularly for ammonia-water mixture) does not affect the blade of the turbine. When propane and R32 were implemented as working fluids, the resulting performance was poorer. However, in comparison with R22, R134a, R143a, and R410a, both propane and R32 have a comparatively broader range of working pressure as well as a more stable working range.

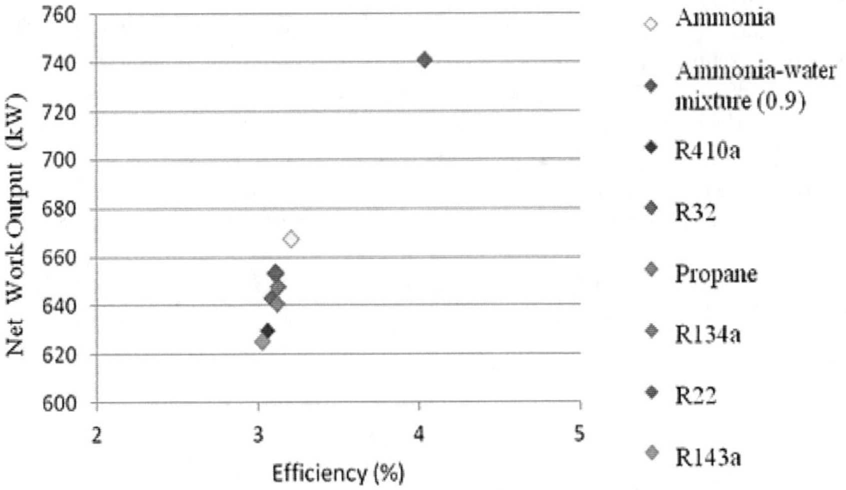

Figure 7 The relationship between the network output and efficiency of eight varying working fluids

4. Conclusion

In conclusion, a model which incorporated LabVIEW and Refprop software's was successfully developed and deployed for a preliminary assessment of the OTEC cycle efficiency. The preliminary analysis of a test run at a net power output of 1 MW showed a close agreement with that of exiting data. The similar developed model was implemented to analyse the efficiency of the OTEC basic closed Rankine cycle using eight varying working fluids. The analysed working fluids, including that of ammonia, are ammonia-water mixture (0.9), propane, and refrigerants (R22, R32, R134a, R143a, and R410a). The results revealed that ammonia-water mixture showed the highest network performance and reliability. Even so, it is essential to continue seeking the suitable working fluids which are safe and economically effective to replace ammonia.

Acknowledgments

The authors express gratitude to the Ocean Thermal Energy Centre (OTEC) lecturers and staffs, Universiti Teknologi Malaysia for their knowledge, laboratory, and equipment supplied. The authors also thanked to Ministry of Higher Education (MOHE) for financially supported this study through The Fundamental Research Grant Scheme for Research Acculturation of Early Career Researchers, FRGS-RACER/1/2019/TK05/UMP//1 - RDU192621.

References

[1] D'Arsonval J 1881 *Reveu Scientifique* 370-372
[2] Nihous G C 2005 *J. Ene. Res. Tech.* **127(4)** 328
[3] Nihous G C 2007 *J. Ene. Res. Tech.* **129(1)** 10
[4] Vega L A 2002 *Ocean Thermal Energy Conversion Primer Marine Technology Society Journal* **36(4)** 25-35
[5] Zhang X, He M and Zhang Y 2012 *Ren. Sust. Ene. Rev.* **16(7)** 5309-5318
[6] Ganic E N and Wu J 1980 *Ene. Conv. Man.* **20(1)** 9-22
[7] Palm B 2008 *Intern. J. Ref.* **31(4)** 552–563
[8] Anderson J H 2009 *ASME 2009 Power Conference*
[9] Perang M R M, Nasution H, Latiff Z A, Aziz A A and Dahlan A A 2013 *Int. J. Tech.* **4(1)** 81-92
[10] Alkhalidi A, Qandil M, and Qandil H 2014 *Int. J. Therm. Env. Eng.* **7(1)** 25-32
[11] Chen H, Goswami D Y, and Stefanakos E K 2010 *Renewable and Sustainable Energy Reviews*

14(9) 3059–3067

[12] Calm J M, and Hourahan G C 2007 Refrigerant data update *Heating/Piping/Air Conditioning Engineering* **79(1)** 50-64

[13] Wu C and Burke T 1998 *App. Therm. Eng.* **18(5)** 295–300

[14] Uehara H and Ikegami Y 1990 *Journal of Solar Energy Engineering* **112-4** 247-256

[15] Sami S M 2012 *International Journal of Ambient Energy* 37-41

[16] Gong J, Gao T and Li G 2012 *Journal of Solar Energy Engineering* **135(2)** 024501

[17] Liu W M, Chen F Y, Wang Y Q, Jiang W J and Zhang J G 2011 *Adv. Mat. Res.* **354-355** 275–278

[18] Yeh R H, Su T Z and Yang M S 2005 *Ocean Eng.* **32(5-6)** 685–700

[19] Upshaw C R 2012 *Thermodynamic and Economic Feasibility Analysis of a 20 MW Ocean Thermal Energy Conversion (OTEC) Power*

IOP Conf. Series: Materials Science and Engineering 1062 (2021) 012041 doi:10.1088/1757-899X/1062/1/012041

The Development of a Web-based Credit Transfer Application (CTA) for Higher Academic Institution: From Feasibility Study to Testing Phase

N Azizan[1*], R Isa[1], F Farzana[1], A Aziz[1] and M Amiruddin[1]
[1] Faculty of Computing and Multimedia, Kolej Universiti Poly Tech MARA Kuala Lumpur, 56100 Kuala Lumpur, Malaysia

*Corresponding e-mail: noraliza@kuptm.edu.my

Abstract. Credit transfer is the process of transferring course credits of the subject to another programme or another institution to support and promote students' flexibility in their study by reducing number of courses or subjects required to complete their studies. Currently, the enrolment of new degree students in Kolej Universiti Poly-Tech MARA Kuala Lumpur (KUPTMKL) is using the manual application forms to apply the subject credit transfer. Understandably, the manual process is tedious and students are pruned to make mistake. The objective of this paper is to develop a web-based Credit Transfer Application system (CTA) for Faculty of Computing and Multimedia (FCOM) of KUPTMKL that can be used for the FCOM degree programme Coordinator and student in managing the application of credit transfer with several security applications and mechanisms such as random password, restriction of unsupported file format and database security. The CTA is developed using Agile methodology and Visual Studio Code is used to build the CTA as the code editor to write the programming languages such as Hypertext Markup Language (HTML), Cascading Style Sheets (CSS), Hypertext Pre-processor (PHP) and XAMMP to manage the MySQL databases. In the development of the CTA system, the feasibility study is performed and three methods are used namely, observation, conducting interviews and delivering questionnaires. Based on 100 respondents of FCOM degree students and 5 Coordinators, the result shows more than 95 percent of evidence indicates that the new system developed is able to store, display and manage all information successfully and performed efficiently compared to manual method. CTA can be easily accessed by students and management via the Web.

Keywords: credit transfer, web-based, feasibility study, system

1. Introduction

In any higher academic institution, new degree students can apply for a credit transfer. An appropriate mechanism designed to perceive previous academic achievement is essential to guarantee a full range of student mobility choices. It is for this reason that credit transfer and student mobility are linked. Credit transfer systems provide the lubricant to ensure seamless academic mobility [1].

While arrangements for credit transfer exist across the higher academic institution, very little is thought about credit-transfer students from other academic institution or why they reconnect with study. The management in education have referred to acknowledge the credit transfer as an instrument for lessening wastage and drop-out in studies [2]. Credit transfer or subject articulation is to prevent unnecessary duplication of study, thus reducing the total number of courses or subjects required by students to graduate. Therefore, it also is seen as a mechanism to motivate students to further their studies.

Kolej Universiti Poly-Tech MARA Kuala Lumpur (KUPTM KL) was formerly known as Kolej Poly-Tech MARA (KPTM) is a private university college managed by Kolej Poly-Tech MARA Sdn. Bhd. KUPTM KL offers high-quality home-grown degree programmes in the area of Accountancy, Communication, Business, English and Computing. Currently, KUPTM KL is using a manual system for the credit transfer application. Therefore, we took this opportunity to carry out a research on this topic and develop a Credit Transfer Application (CTA) which is a web-based credit transfer system for Faculty of Computing and Multimedia (FCOM) students in KUPTM KL.

Based on the observation that has been made, in manual system to apply for credit transfer, the Coordinator needs to distribute the credit transfer form to students by gathering them in one venue. Then, students need to fill in the form and attach their transcripts before submitting to the Coordinator. The manual system puts pressure on the Coordinator as well as to the students. The form is also susceptible to lost as the Coordinator needs to keep all the credit transfer forms with the transcript attachment from each students. It takes more effort and physical space to keep track of the forms. Not only that, the form and the transcript can easily be damaged if natural disasters happen such as flash flood and fire.

Findings are presented from a survey of 100 respondents of FCOM degree students and 5 Coordinators, the result shows more than 95 percent of evidence indicates that the new system developed is able to store, display and manage all information successfully and performed efficiently compared to manual method.

Not only CTA promotes paperless environment, it also can reduce huge number of human management task which will lessen human error. The credit transfer application will be more secure, no loss of credit transfer form & transcript, and only authorised users are allowed to have an access to it. With the positive feedback on this new system, it is believed it can enhance the education management in transferring academic credit.

2. Web-based Credit Transfer Application System

Web-based applications have now gotten one of the ideal advancements being utilized to facilitate the way toward overseeing information and records. A portion of the inspirations that add to the advancement of many online frameworks these days are a direct result of its productivity in taking care of quick access of records and its capacity in supporting multiusers at the same time, in this way sparing a great deal of time and hassle free [3-4].

Transferring academic credits is the process to evaluate the components of qualification in determining the overall or the equivalence with another qualification by establishing credits for individuals' academic achievements [5]. It is a mechanism through which an institution shares students' accomplishments with other institutions to facilitate their horizontal mobility and vertical progression. This system seeks to avoid duplication of studies. Furthermore, it saves time and money for students who decide to embark on an exchange programme. An appropriate system of credit transfer is a key facilitator of student mobility and cooperation among higher education institutions [6]. Therefore, the credit transfer system is an important system for institutions and students.

According to Cornet, there are two types of student mobility which are horizontal mobility and vertical mobility. Horizontal mobility can be defined as a student who is pursuing the next education level without changing their education field [7]. This type of student mobility is beneficial for students as it can reduce many credit hours by transferring subjects to avoid duplication of study. As an example, a student took a Diploma in Computer Science and the student continue to the next education level in BSc (Hons) in Information Security.

For vertical mobility, it can be defined as a student who is pursuing the next education level in a different field. This type of student mobility is quite difficult for the student to apply for credit transfer as the education field is different. For example, a student took a Diploma in Mechanical Engineering and the student pursues the next education level in Bachelor of Education (Honours) in Teaching English as A Second Language (TESL).

3. Methodology

The system's development makes used of Agile Development methodology as shown in Figure 1. This methodology extends the advantages of Iterative Methodology and aims at user satisfaction and product adaptability by the rapid delivering of products. It anticipates change and allows for much more flexibility than traditional methods. Moreover, this methodology offers a lightweight framework, given a constantly evolving functional and technical landscape, maintain a focus on the rapid delivery of business value [8-9].

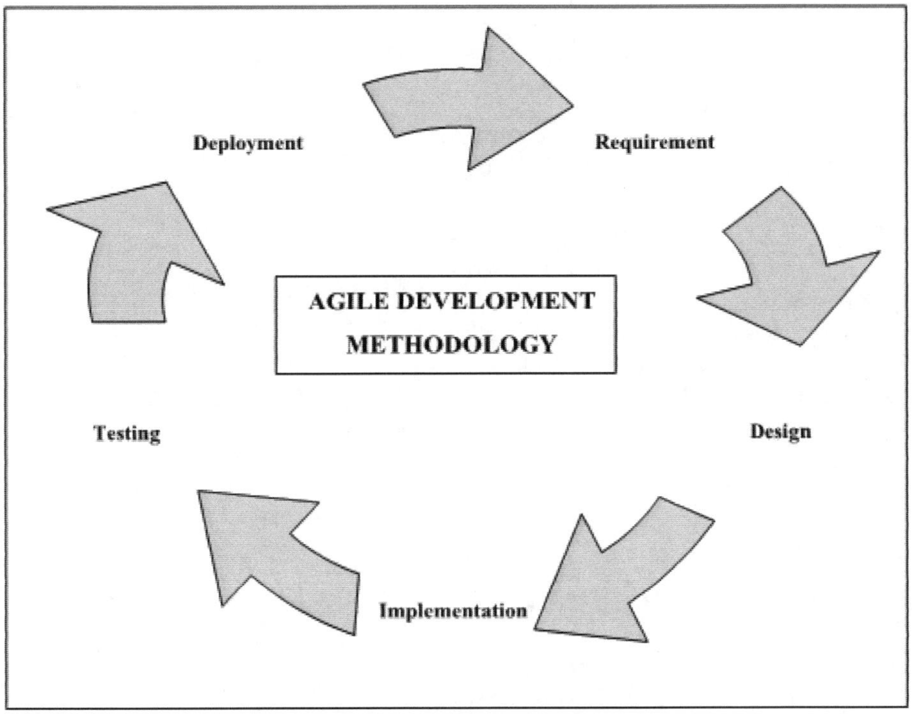

Figure 1. The agile development methodology phases.

The requirement phase mainly focuses on communicating with users to gather and analyse requirements. We conducted a feasibility study to determine whether the new system is a feasible solution. Observation, interview and questionnaires have been used as the methods in this phase to understand the business problems. Therefore, we have selected Kolej Universiti Poly-Tech Mara (KUPTM), Malaysia to be a pilot study where 105 questionnaires have been distributed amongst students and Coordinators. The results of the questionnaires will then be used to analyse the users' requirements and preferences.

In second phase we evaluated and started on the logical design of the system by making use of the information and requirements collected during the requirement phase. We designed the system interfaces using Hypertext Markup Language (HTML) and Cascading Style Sheets (CSS) programming languages. The design of the system must be user-friendly and easy to understand by the users. Putting too many elements on the page of the system may lead to distracting the users from its main purpose thus we keep it simple as simplicity always works in an effective web page system design [10].

After the best or most appropriate design has been selected, implementation started immediately by using Visual Studio Code, XAMPP and PHP. Testing is the last phase we will discuss in this paper as it is a complementary process that needs to be done in order to verify whether all features and scenarios are working as what users expected. We will present the user testing in next section with more details.

4. Result and Discussion

Figure 2 shows the flowchart diagram of the system. The system starts with the sign up of the user. Then, the user can log in through the system. The student needs to fill in the form and submit the form. After that, the Coordinator will approve or reject the subjects applied for transfer credit. The Dean will then approve the application follow by the Admission and Student Record Department who will verify the application. Lastly, the student can check the application status through the system.

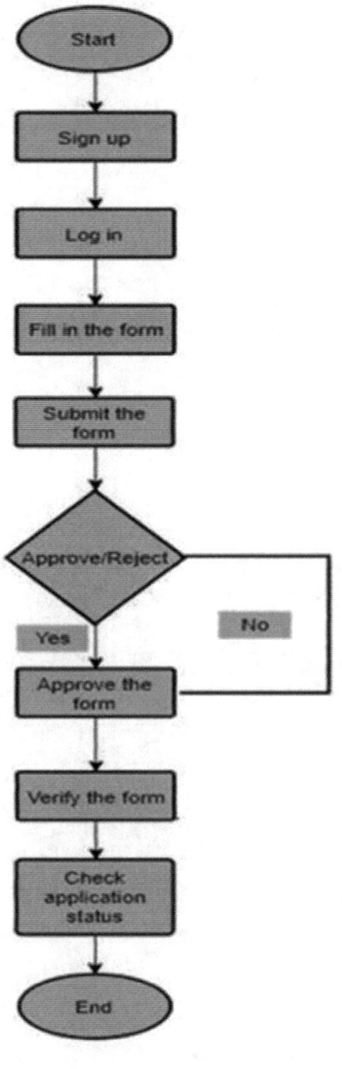

Figure 2. Flowchart diagram.

Figure 3 depicts the flow design of the Credit Transfer Application. The main flow starts with students applying for the credit transfer. Then, the Coordinator will approve or reject the applied subjects. Next, dean will approve the application and the Admission and Student Record Department will verify the application.

The student is able to perform various tasks or activities such as sign up, log in, fill in the form and upload the transcript. After that, they can submit the form. Furthermore, they can view the submitted form and the uploaded transcript. They also can download the submitted forms and check the application status of credit transfer.

The Coordinator is able to sign up and log in to the system. He or she also can add course, institution and subject in the system. Once the student submitted the form, the Coordinator can view the student application's list. Next, the Coordinator also can view and download the form and uploaded transcript by student. Furthermore, the Coordinator can approve or reject the subjects applied for credit transfer.

Once the Coordinator approves or rejects the application, the Dean can view the programme and student application's list. Then, the Dean can view and download the form and the uploaded transcript. Once the Dean has approved the application, the Admission and Student Record department can view the programme and student application's list before verifying the application.

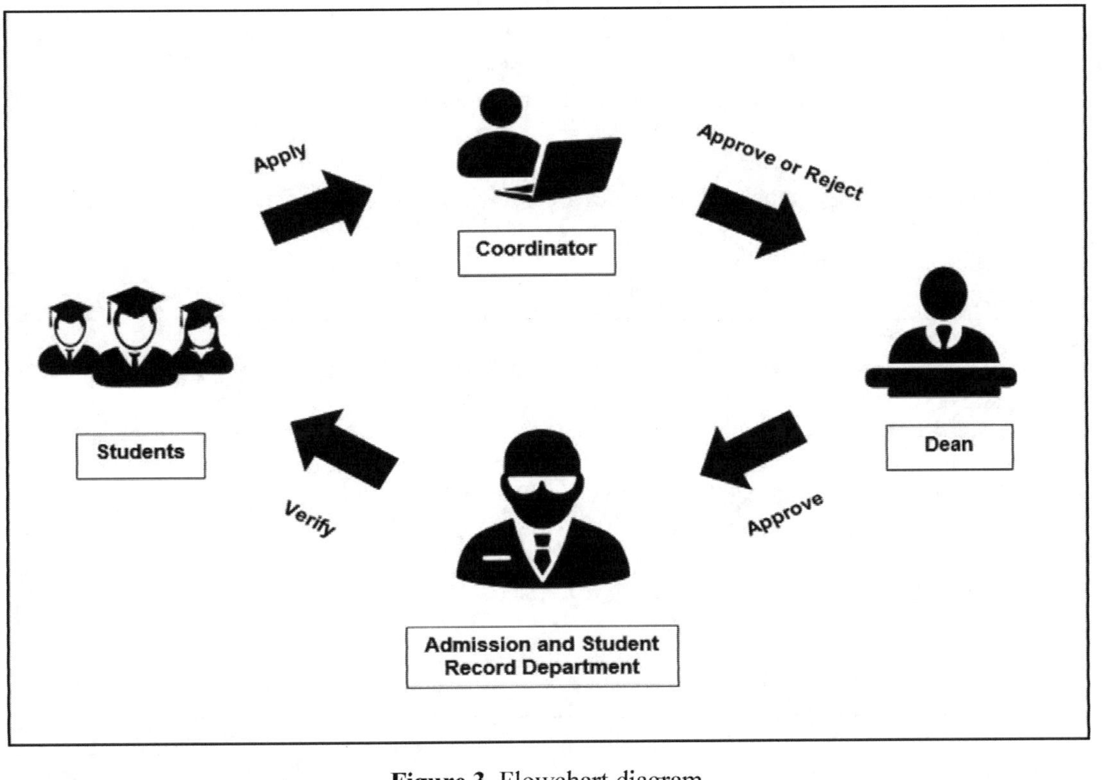

Figure 3. Flowchart diagram.

Table 1 shows the comparisons of percentages among 100 respondents which is students regarding the use of the manual method and the developed system. The highest percentage which is 96% shows that students strongly agreed that the developed system is efficient and effective compared to the manual method. Students also strongly agreed that the developed system makes record keeping is more organized, lower the potential of losing application form and transcript, fastens the application process and the potential of unauthorized access to the private and confidential data is low with the percentage of 95%. Furthermore, 95% of them satisfied with the developed system rather than the manual method for credit transfer application. Meanwhile, 94% of students strongly agreed that the developed system enables them to check the application status easily compared to the manual method. Students also are happier with no more printing and photocopying need to be done with the developed system. Lastly, it shows that in total average, 95% of students strongly agreed that the developed system is better for credit transfer application rather than the manual method.

Table 1. Comparisons between using manual system and online system for students.

No.	Factor	Strongly agree	
		Manual method	Developed method
1	Efficiency & effectiveness	20%	96%
2	Record keeping is more organized	15%	95%
3	The potential of losing application form and transcript is low	15%	95%
4	Student can fill in the application form faster	20%	95%
5	Student can easily check the application status	15%	94%
6	The potential of unauthorized access to the private and confidential data is low	15%	95%
7	Satisfaction with the method	20%	95%
8	Student can submit the application anytime from anywhere (mobility)	No	Yes
9	More printing and photocopy	Yes	No
	Average	17.1	95%

Table 2 depicted the comparison of percentages among 5 Coordinators on the use of the manual method and developed system. The highest percentage which is 100% of Coordinators strongly agreed that developed system makes record keeping is more organized, lower the potential of losing application form and transcript, the list of student application can be easily tracked and view, the application can be approved faster, unauthorized access is low and satisfy with the developed system rather than the manual method. Meanwhile, 80% of Coordinators strongly agreed that the developed system is more efficient and effective compared to the manual method. Lastly, it shows that in total average, 97.1% of Coordinators strongly agreed that the developed system is better for credit transfer application rather than the manual method.

Table 2. Comparisons between using manual system and online system for Coordinator.

No.	Factor	Strongly agree	
		Manual method	Developed method
1	Efficiency & effectiveness	20%	80%
2	Record keeping is more organized	20%	100%
3	The potential of losing application form and transcript is low	20%	100%
4	Coordinators can track and view the list of the student application easily	20%	100%
5	Coordinators can approve the application form faster	20%	100%
6	The potential of unauthorized access to the private and confidential data is low	20%	100%
7	Satisfaction with the method	20%	100%

IOP Conf. Series: Materials Science and Engineering 1062 (2021) 012041 doi:10.1088/1757-899X/1062/1/012041

8	Coordinators can access and view the application anytime from anywhere (mobility)	No	Yes
	Average	20%	97.1%

5. Conclusion

Result and discussion had shown that more than 95 percent of evidence indicates that the new system developed is able to store, display and manage all information successfully and performed efficiently compared to manual method. Many advantages can be seen from this online application such as mobility access and usability.

With the development of CTA, students can fill in the credit transfer form, upload their transcript and submit it through the system securely. Once they submit the form through the system, students can check the status of their application and download the form. The burden of academic management also can be lessening with the used of this system. In case of natural disaster, all records will be kept safely in the database system which can be accessed only by authorized users. CTA also promotes the paperless working environment which can reduce the use of paper thus promotes green computing.

However, there are many more functions that can be added in CTA such as auto-calculation for all credits applied by students to be transferred. This is a real scenario which can minimize Coordinator's task in checking the credits. Apart from that, watermark can be implemented to the uploaded transcript to protect the transcript from unauthorized used as it identifies the rights of the owner. However, the advancement of this system will be considered in the future enhancement.

Acknowledgments

The authors gratefully acknowledge the support provided by KUPTM KL and the supervisor, Associate Professor Ts. Dr Zahrah Yahya, the Deputy Vice Chancelor, Kolej Universiti Poly-Tech MARA (KUPTM). The authors also would like to thanks all the team members involved for their contribution.

References

[1] Di Paolo T and Pegg A 2013 *Journal of Further and Higher Education* **37(5)** 606-622

[2] Junor S and Usher A 2008 Student Mobility & Credit Transfer: A National and Global Survey. *Educational Policy Institute (NJ1)*

[3] Othman M, Othman Z and Aziz M N A 2006 *J. Gading* **10(1)** 87-98

[4] Othman M, Othman Z and Aziz M N A 2008 *Proceedings of the National Seminar on Science, Technology & Social Sciences* 577-586

[5] Pollard E, Hadjivassiliou K, Swift S and Green M 2017 *Credit Transfer in Higher Education: A review of the literature. Department for Education* 10

[6] Santoso D 204 *Riding the Tide: Student Mobility in Cross Border Education* 1-12

[7] Cornet F 2015 *J. Higher Education and Lifelong Learning* 57-66

[8] Shankarmani R, Pawar R and Mantha S 2012 *A International Journal of Computer Applications* 31-37

[9] Petersen K and Wohlin C 2008 *Software Engineering Research and Practice* 1-10

[10] Johnson J 2014 *Designing with the Mind in Mind: Simple Guide to Understanding User Interface (*Elsevier Inc*)* 151-152

IOP Conf. Series: Materials Science and Engineering 1062 (2021) 012042 doi:10.1088/1757-899X/1062/1/012042

Techno-economic efficiencies and environmental criteria of Ocean Thermal Energy Conversion closed Rankine cycle using different working fluids

N Samsuri[1], N Sazali[2*], A S Jamaludin[3], and M N M Razali[3]
[1] Ocean Thermal Energy Centre (OTEC), Ground Floor, Block Q, Universiti Teknologi Malaysia Kuala Lumpur, Jalan Sultan Yahya Petra, 54100 Kuala Lumpur, Malaysia
[2] Department of Electrical Engineering, College of Engineering, Universiti Malaysia Pahang, 26300 Gambang, Pahang, Malaysia
[3] Faculty of Manufacturing and Mechatronic Engineering Technology, Universiti Malaysia Pahang, 26600 Pekan, Pahang, Malaysia

*Corresponding e-mail: norazreen1989@gmail.com

Abstract. Ocean Thermal Energy Conversion (OTEC) is a foundation for an appealing renewable energy technology owing to its vast and inexhaustible resources of energy, stability, and sustainable output. Development of OTEC power plant is to exploit the energy accumulated in between the top layer of warm surface seawater (heat source), and the cold layer of deep seawater (heat sink). It operates based on Rankine cycle to produce electricity between the source and the sink at the smallest temperature difference of approximately 20 K. OTEC power plant commonly utilized ammonia as working fluid. Nevertheless, ammonia poses potential lethal health risks and hazardous fluid. Hence, the effect of the working fluid types and the subsequent operation conditions may be critical and therefore become the subject of this study. In addition to OTEC power plant's thermodynamic efficiencies study, this research also explores the economic efficiencies in term of capital cost per net power output ($/kW) and environmental criteria of different working fluids including that of ammonia, ammonia-water mixture (0.9), propane, and refrigerants (R22, R32, R134a, R143a, and R410a). The results showed that ammonia-water mixture gave the excellent performance with regard to the characteristics of heat transfer with the best thermodynamic efficiency of 4.04% compared to pure ammonia with 3.21%, propane with 3.09%, followed by refrigerants from 3.03% to 3.13%. Capital cost of using propane was economically efficient with 15730 $/kW compared to ammonia-water mixture at 16201$/kW, refrigerants from 16990 $/kW to 21400 $/kW, and pure ammonia being the costliest at 21700 $/kW. Despite being lower in its thermodynamic efficiency, propane gave the lowest capital cost and had the lowest toxicity in contrast to all other working fluids. Therefore, propane has the potential to be used as a clean and safe working fluid that would further enhance the OTEC technology.

Keywords: ocean, thermal energy, renewable, heat sink, power plant

1. Introduction

Ocean Thermal Energy Conversion (OTEC) has tremendous prospective in deep ocean water area, in which a sufficiently high temperature difference between the surface water and a specific depth is required to effectively run an OTEC power plant. In 1881, Arsonval's initial concept specified that the optimum temperature difference needed for the installation of an OTEC plant is larger than 20 K [1]. The system will work between the surface seawater at 30°C (known as heat source), and seawater at

1000 m depth with temperature of 4°C (known as heat sink) [2-4]. OTEC power plant technology is developed on a basis of open (OC-OTEC) and closed Rankine cycles (CC-OTEC). Previous research reported that the process have to be founded upon the Uehara cycle for optimal power plant output, implementing ammonia-water mixture as the working fluid with smaller than 20 K of temperature difference, and at 5-6% thermal efficiency [5].

Additionally, the selection of suitable working fluids has a significant impact on the entire system viability and efficiency. Ammonia has been considered as the best working fluid because it has a suitable boiling temperature (28°C - 32°C) for the OTEC purpose [6]. However, it is toxic and therefore can be hazardous to the environment. Recent development of working fluids shows that ammonia can be replaced by other working fluids with zero Ozone Depletion Potential (ODP) and zero Global Warming Potential (GWP). Finding other suitable substitutes is a big challenge to this study. Several studies have been done which have shown better results with the use of other working fluids such as hydrocarbon (HC) and hydrofluorocarbons (HFCs) [7-10]. The other findings reported that R123 displayed the best performance but the fluid contributes to the ODP and GWP. Therefore, isopentane is suggested as it showed the second best performance and regarded as environmental friendly working fluid for the system. However, the applications of their research are for waste heat systems, but can still be operated at low temperature.

A paper that reviews about 35 working fluids and analyzes the effect of fluid properties on the cycle efficiency is written by Chen at al. [11]. They have categorized the working fluids under three characteristics which are dry, isotropic, or wet fluid according to the T-s diagram. Understanding the characteristic of the working fluids eases the process of selecting the appropriate working fluid for the cycle. Calm and Hourahan [12] has interpreted the data of working fluids into a table with ODP and GWP of selected refrigerant. Figure 1 shows the numbers of OTEC previous research focusing on different working fluids from 1979 to 2016.

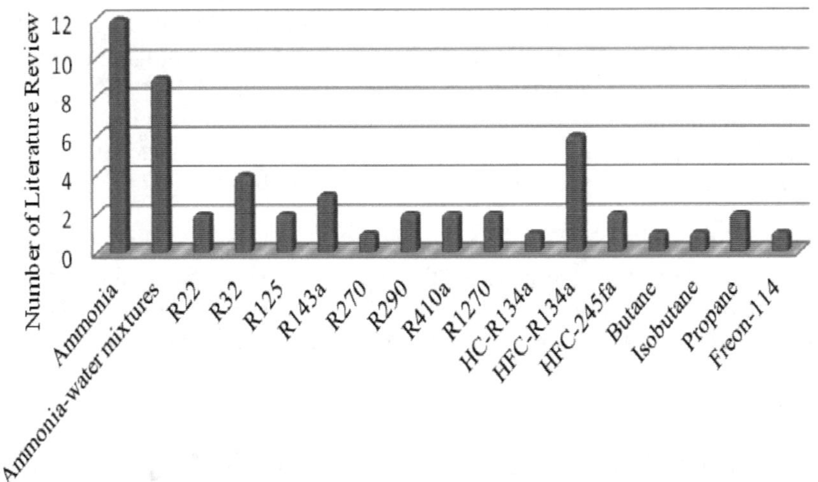

Figure 1. The numbers of OTEC previous research focusing on different working fluids from 1979 to 2016.

Plant operation for OTEC technology does not use any fuel. Therefore, the major cost component could be construction cost of the infrastructure. According to the findings by Lennard [13] a 10 MW capacity of floating OTEC power plant would cost 10,000USD/kW. The closed rankine cycle and ammonia as the working fluid was used. A 10MW floating plant had been designed using plate-fin heat exchanger which estimated to consume the highest percentage of cost component of the plant. This research had been stopped because of the exorbitant estimated cost until further development of the design of the demonstration plant until it would be suitable for production. Similarly, Rogers et. al [14] stated that a land based of 1 MW OTEC plant would cost 18,000 USD/kW. The study conducted to enhance the economic prospect for both open cycle and closed rankine cycle. They highlighted the key importance was the demonstration of technical feasibility of using the OTEC flash evaporation to produce seawater. Thus, from the findings of such 'by-product' of fresh water, notably improves the cost effectiveness in producing electrical energy. Subsequently, from the research of 'Economics of Ocean Thermal Energy Conversion' by Vega [15], it was found that for the 10 MW open cycle plant with second stage water production estimated by 14,700 USD/kW for the capital cost of the plant. The research continues by Vega [16] for 100 MW floating plant for 7900 USD/kW. The capital cost of 53.5 MW is just about 8430 USD/kW for this study. The differential value of the plant cost was calculated by converting to the present day cost using the USA 20-year average for equipment price-index inflation. The estimated cost value also relied on the application of the same technologies, that new generation design will achieve cost savings of as much as 30% because in the past, the OTEC work did not yield a single order because there were no real customers for the technology. According to the previous literature, the capital cost was calculated based on ammonia as working fluid. Evidently, the current research will come out with the comparison for economic efficiency by using different working fluids.

Therefore, the objective of this paper is to examine the efficiency in terms of techno-economic and environmental criteria of the OTEC basic closed Rankine cycle using varying working fluids. At the initial stage of the study, preliminary simulation was conducted to confirm the simulation model with the reference from past OTEC studies. The similar developed model was implemented to analyze the efficiency of the OTEC basic closed Rankine cycle using eight varying working fluids.

2. Methodology

2.1. Introduction

Laboratory Virtual Instrument Engineering Workbench (LabVIEW) is a platform design framework created by National Instruments that is employed as languages of visual programming. Its implementation in numerous fields of engineering (e.g. aeronautical, mechanical, electrical, etc.) has led to the advancement of the world's largest and most complex applications to fulfil future demands. LabVIEW offers the users with flexibility through intuitive graphical programming which helps to reduce the time needed for test development. The thermodynamic model has been created in LabVIEW and linked to the working fluid data base in National Institute of Standards and Technologies (NIST) RefProp 9 and PROPATH. The thermodynamic model of OTEC cycle is created in LabVIEW to run numerical calculation, simulations and compare the working fluids from a thermophysical perspective..

2.2. Analytical Techniques of Thermodynamics

The simulation was based upon the thermodynamic analysis of the OTEC Rankine cycle performance. The Rankine cycle comprises of four major components, which are condenser, coolant pump, turbine, and evaporator. Several assumptions were included to facilitate the simulation analysis and assessment [17,18], which are described as follows:

i. Every component is in steady state.
ii. Any heat loss and pressure drop are disregarded.
iii. The system is completely insulated.
iv. All pumps and turbines are given isentropic efficiency.

For the steady state energy balance equation, the total energy entering a system is equal to the total energy exiting the system, as expressed in Eq. 1

$$E_{in} = E_{out} \tag{1}$$

or it can be elaborated as in Eq. 2

$$W_{in} + Q_{in} + \sum \dot{m}_{in} = W_{out} + Q_{out} + \sum \dot{m}_{out} \tag{2}$$

where $\dot{Q}$ represents heat transfer rate; $\dot{m}_{in}$ and $\dot{m}_{out}$ is inlet and outlet mass flow rate; whereas W_{in} and W_{out} is work inlet and outlet, respectively. By assuming the system is completely insulated and any heat losses are neglected; $Q_{in} = 0$, $Q_{out} = 0$ and $W_{out} = 0$; the energy balance in the pump is expressed as in Eq. 3

$$W_{in} + \sum \dot{m}_{in} = \sum \dot{m}_{out} \tag{3}$$

From Eq. 3, the work supplied is given as in Eq. 4

$$W_{in} = \sum \dot{m}_{out} - \sum \dot{m}_{in} \tag{4}$$

Rate of heat supplied to the cycle (evaporator), $\dot{Q}_e$ is expressed as in Eq. 5

$$\dot{Q}_e = \dot{m}_{wf}\,\Delta h_e \tag{5}$$

Rate of heat rejected from the cycle (condenser), $\dot{Q}_c$ is indicated as in Eq. 6

$$\dot{Q}_c = \dot{m}_{wf}\,\Delta h_c \tag{6}$$

Rate of heat absorbed from the warm seawater, $\dot{Q}_{e,ws}$ is expressed as in Eq. 7

$$\dot{Q}_{e,ws} = \dot{m}_{ws}\,c_p \Delta T_{ws} \tag{7}$$

Rate of heat rejected into the cold seawater, $\dot{Q}_{c,cw}$ is indicated as in Eq. 8

$$\dot{Q}_{c,cw} = \dot{m}_{cs}\,c_p \Delta T_{cs} \tag{8}$$

where $\dot{m}_{ws}$ and $\dot{m}_{cs}$ are the mass flow rate of warm and cold seawater, respectively. c_p is the seawater specific heat capacity at constant pressure.

The working fluid pump, $W_{P_{wf}}$ and the turbine work, W_T is written as in Eq. 9 and Eq. 10

$$W_{P_{wf}} = \dot{m}_{wf}\, v(P_2 - P_1) \tag{9}$$

$$W_T = \dot{m}_{wf}\,\Delta h \tag{10}$$

where Δh represents the enthalpy difference in the turbine system.

Referring to Uehara and Ikegami [18], the working fluid pumping power, P_{Pwf} is given as in Eq. 11. The pumping power of warm seawater, P_{ws} is indicated as in Eq. 12; whereas the pumping power of cold seawater, P_{cs} is expressed as in Eq. 13

$$P_{Pwf} = \frac{\dot{m}_{wf}\,\Delta H_{wf}\, g}{\eta_{wf,p}} \tag{11}$$

$$P_{ws} = \frac{\dot{m}_{ws}\,\Delta H_{ws}\, g}{\eta_{ws,p}} \tag{12}$$

$$P_{cs} = \frac{\dot{m}_{cs}\,\Delta H_{cs}\, g}{\eta_{cs,p}} \tag{13}$$

where ΔH refers to the difference in pressure.
The net power output, P_n is indicated as in Eq. 14

$$P_n = P_G - P_{ws} - P_{cs} - P_{p_{wf}} \tag{14}$$

2.3. Selection of Working Fluid

An ideal working fluid should have the relevant thermophysical properties corresponding with its application, besides sustaining its chemical stability within the specified range of temperature. Working fluid selection plays a major part on the system in terms of its performance, operating conditions, effects on the environment and economic feasibility. In this section, the parameters for identifying a suitable working fluid for the cycle system are described. Sami [19] has listed the main factors affecting the properties of thermodynamic and thermophysical of the system, among which are thermal conductivity, chemical stability, specific heat, boiling temperature, latent heat, toxicity, as well as flash point, as outlined in Figure 2.

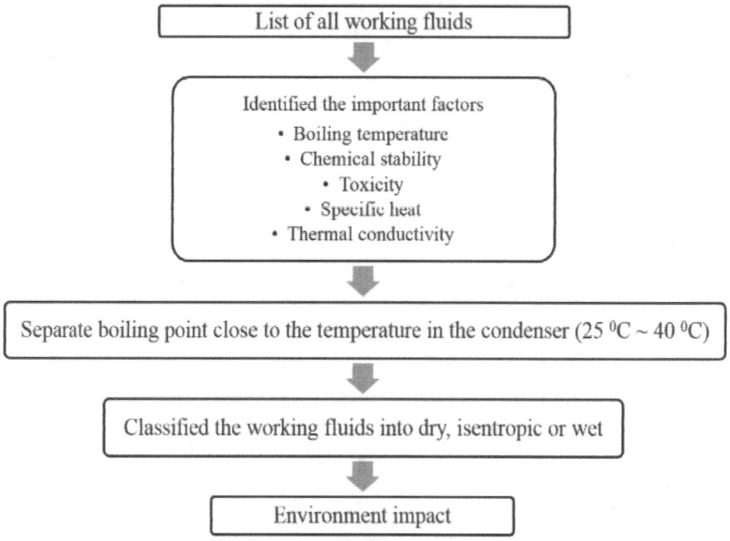

Figure 2. Steps in selecting the working fluids [15].

The OTEC closed Rankine cycle in this study utilized the boiling point of the working fluid near the evaporator operating temperature, that is about 25°C to 40°C [20]. In addition, the fluids were classified as dry, isentropic or wet relative to the saturation curve (dT/ds). A dry or isentropic fluid is appropriate to be implemented in OTEC closed Rankine cycle [21]. The purpose of separating the type of fluids is to make sure that the fluids are totally superheated after isentropic expansion, intended to avoid the appearance of liquid drops on the blades of the turbine.

2.4. Types of Working Fluid

There are dual kinds of working fluid, namely pure fluid (pure compound) and pseudo-pure fluid (a mix of several pure compounds of fluid). Ammonia, propane, R22, R134a, and R143a, are marked as pure fluid, and are not combined with some other compounds. Meanwhile, ammonia-water mixture, R404a, R410a, R470c, and R507a are a pseudo-pure fluid. Figure 3 and Figure 4 show that the highest enthalpy difference can be discovered in ammonia-water mixture, followed by ammonia, propane and R32.

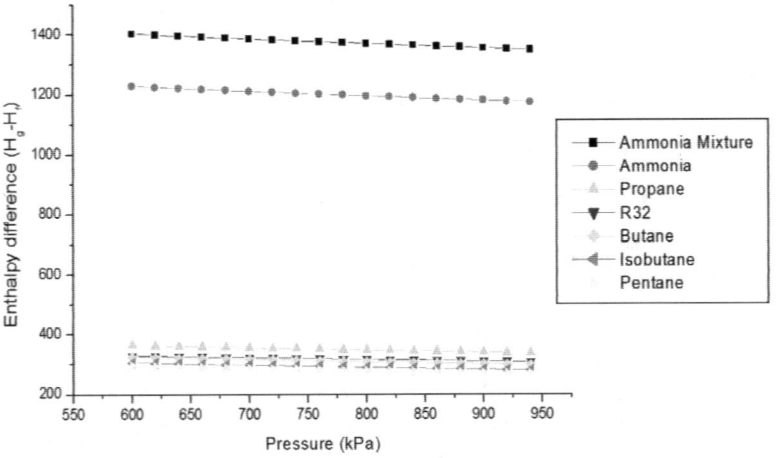

Figure 3. Latent heat-pressure diagram of pure fluid and pseudo-pure fluid.

According to Figure 4, in contrast with other working fluids, ammonia-water mixture has the highest quantity of heat applied. This situation is caused by its greater latent heat value, also can be defined as the amount of heat that a liquid absorbs to stay at a constant pressure or temperature throughout the process of vaporization.

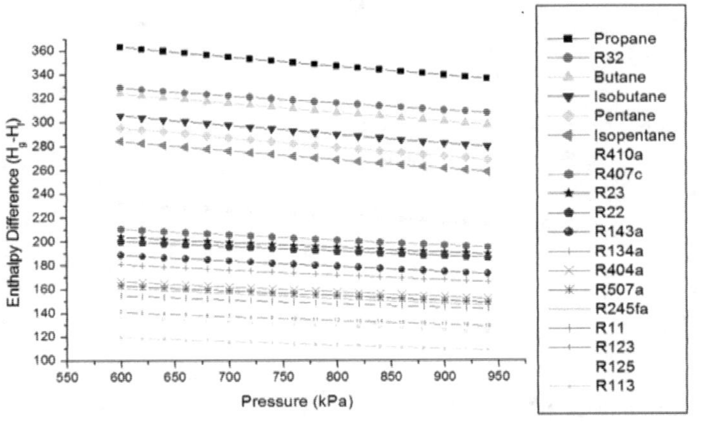

Figure 4. Close up of latent heat-pressure diagram of pure fluid and pseudo-pure fluid.

2.5. Preliminary Simulation

A preliminary design model for simulation of a 1 MWe OTEC closed Rankine cycle was conducted using ammonia as working fluid. This preliminary simulation is to validate the model developed by Yeh et al. [22]. Apart from that, the preliminary design model allows the estimation for 5 MWe and 10 MWe OTEC closed Rankine cycle.

Table 1. Parameters for three OTEC cycles to be investigated.

Parameter	Symbol	Unit	Value
Evaporating temperature	T_E	°C	28
Condensing temperature	T_C	°C	8
Warm seawater inlet temperature	T_{wsw}	°C	30
Cold deep seawater inlet temperature	T_{csw}	°C	5

Working fluid pump efficiency	η_{wf}	%	0.75
Turbine efficiency	η_T	%	0.82
Generator efficiency	η_G	%	0.95
Warm seawater pump efficiency	$\eta_{pump,wsw}$	%	0.80
Cold deep seawater pump efficiency	$\eta_{pump,csw}$	%	0.80

The OTEC closed Rankine cycle simulation based on Uehara and Ikegami [18] was conducted according to the fixed condition parameters as tabulated in Table 1, in which the ammonia is in a steady state. The graph that represents the simulated model is shown in Figure 5 (b). When comparing the reference case with the preliminary analysis, it was found that ammonia generated the maximum total work output. Such results are reinforced by the point that ammonia possessed the maximum as well as the most appropriate value of latent heat for the OTEC cycle system.

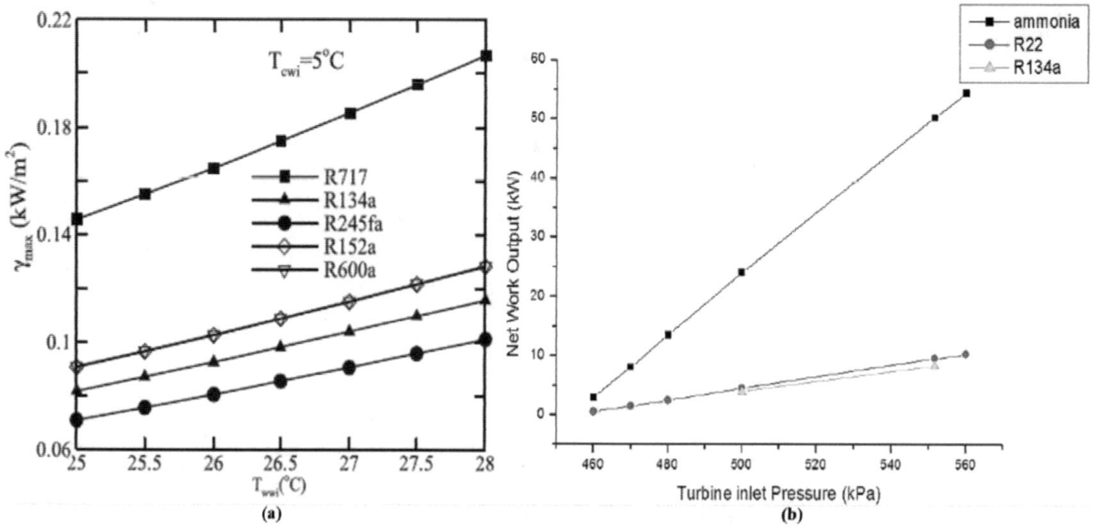

Figure 5. The net work output of closed Rankine cycle using several working fluids; (a) reported by Yeh et al. [22]; (b) simulation model using LabVIEW and RefProp.

As can be seen from Table 2, the net power output increased significantly when the system was scaled up [23]. This will later increase the value of capital cost unless the heat exchangers are able to transfer large amount of energy at minimum pumping power. The preliminary study acts as an initiation to the visualization procedure used in the subsequent assessment in Section 3. On the other hand, the preliminary simulation is shown to explain the sufficiency of the parameters used in this study.

IOP Conf. Series: Materials Science and Engineering 1062 (2021) 012042 doi:10.1088/1757-899X/1062/1/012042

Table 2. Analysis of OTEC Closed Rankine cycle using ammonia as working fluid.

Parameter	Unit	1 MWe	5 MWe	10 MWe
Q_{in}	kW	19724.40	81375.50	162751.00
Q_{out}	kW	18685.00	76166.90	152334.00
$W_{p(wf)}$	kW	13.22	54.53	109.06
$W_{p(wsw)}$	kW	96.74	399.12	798.24
$W_{p(cws)}$	kW	118.76	484.11	968.21
$\dot{m}_{wf}$	kg/s	15.82	65.25	130.50
$\dot{m}_{WSW}$	kg/s	1793.01	7397.29	14794.60
$\dot{m}_{CSW}$	kg/s	1587.67	6471.91	12943.80
W_T	kW	905.00	4525.00	9050.00
W_{net}	kW	676.28	3587.24	7174.48

3. Results and Discussion

3.1. OTEC Closed Rankine Cycle using Different Working Fluids

The simulated net power output of eight varying working fluids produced by a work pump of deep seawater is shown in Figure 6. It was noticeable that the net power output for ammonia-water mixture was the maximum with 740 kW, and that the power needed for the cooling system to pump deep seawater was also small. A feature which is widely recognized in the OTEC power cycle is the point that ammonia resulted in the second highest net power output value. The third highest net power output was R134a followed by R22 and propane. R134a was the possible candidate to replace the ammonia as it possessed the highest net power output among the other five working fluids; however it has the biggest value of work pump for deep seawater. Therefore, it required a big pipe to pump from the deep seawater to condense R134a. R22 has the higher net power output but lower pumping power for deep seawater compared to propane. R32 was the fourth possible candidate to replace ammonia. As the graph shown, R32 gaves the lowest pumping power than the other working fluids including pure ammonia. The graph also indicated that R410a and R143a have low pumping power compared to pure ammonia, but it has the lowest net power output. Even so, a substitute working fluid must be introduced to replace ammonia which is detrimental to the ecosystem and needs a special substance to be preserved.

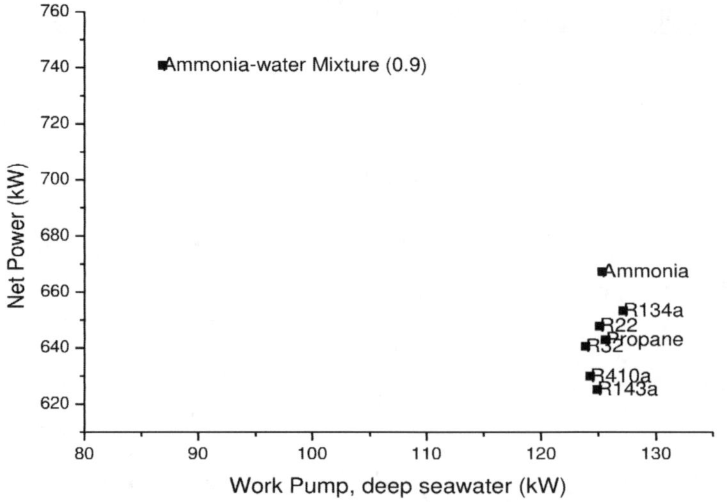

Figure 6. The simulated net power output of eight varying working fluids produced by a work pump of deep seawater.

The relationship between the net work output and efficiency is shown in Figure 7. Although both ammonia and ammonia-water mixture have greater net work output and efficiency in contrast to the other working fluids, they need a separator to make sure that water vapour from the fluid (particularly for ammonia water mixture) does not affect the blade of the turbine. When propane and R32 were implemented as working fluids, the resulting performance was poorer. However, in comparison with R22, R134a, R143a, and R410a, both propane and R32 have a comparatively broader range of working pressure as well as a more stable working range.

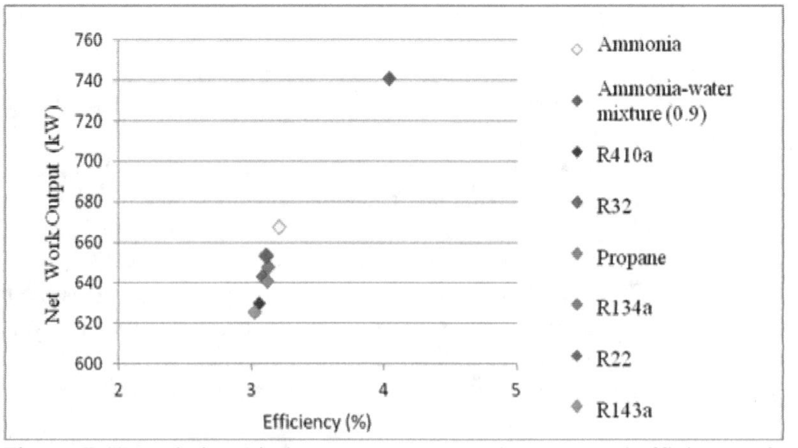

Figure 7. The relationship between the net work output and efficiency of eight varying working fluids.

3.2. Techno-economic Efficiencies and Environmental Criteria of Different Working Fluids

The simulated net power output of eight varying working fluids produced by a work pump of deep Figure 8 shows the simulated result of thermal efficiency and capital cost (USD/kW) for eight different working fluids. Meanwhile, Table 3 is generated from Figure 8 for the exact value of the capital cost per net power output. The data from the simulation results indicated that in contrast to ammonia, ammonia-water mixture has a greater value of net power generated and main product with relatively low cost. Ammonia-water mixture was regarded as the best working fluid owing to its net power efficiency and low capital cost. It is because of these 3 factors which are; ammonia water mixture has the highest efficiency among the other fluids, a lowest pumping power needed and a lowest capital cost to build an OTEC system. However, the ammonia water mixture needs a proper maintenance on the turbine because of water droplet will be occurred in the turbine. Therefore, recommendation for further study on the maintenance cost of using ammonia water mixture is needed for further improvement. Ammonia has become the second highest of thermal efficiency but the highest capital cost because of its higher pumping power needed and the cost of special material to handle with pure ammonia. The net power output of propane is much lower than that of ammonia and ammonia-water mixture. Nevertheless, in contrast to the other working fluids, propane has the least capital cost of main system components, making it the better choice to substitute ammonia.

The simulation result shown in Figure 8 indicated that propane and R32 are possible fluids which can serve as an ammonia substitute, attributed by their lower cost and non-toxic feature. Furthermore, the environmental criteria of different working fluids tabulated in Table 4 showed that propane is regarded as a highly flammable fluid. Nevertheless, such issue can be neglected and does not prevent its implementation in OTEC closed Rankine cycle system as the maximum temperature reaches is only 40 °C. R32 is the second best option to replace ammonia. At normal boiling point, it possesses high specific heat value, and an immediate effect of saturated vapour pressure on temperature. R32 is also distinguished by its high productivity at cold temperature and high energy efficiency, even though these properties are marginally lower than R22 and ammonia. Propane and R32 have the ability to efficiently meet the safety and environmental criteria of the system. They are potentially the ideal choices to be deployed as the working fluid in the OTEC cycle in the future, replacing ammonia or ammonia-water mixture.

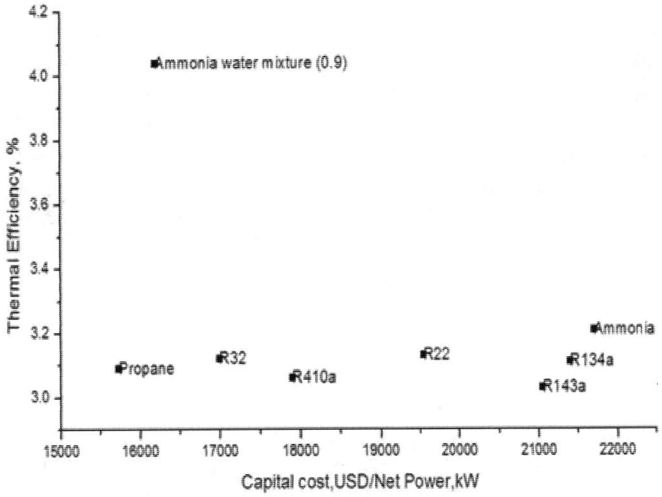

Figure 8. The simulated result of thermal efficiency and capital cost (USD/kW) for eight different working fluids.

Table 3. Calculated result of the different working fluid with the capital cost per net power ($/kW).

Working Fluid	Capital Cost/Net Power ($/kW)
Ammonia	21700
Ammonia-water mixture	16201
Propane	15730
R22	19540
R32	16990
R134a	20025
R143a	21400
R410a	17900

Table 4. Environmental criteria of different working fluids [24].

Working Fluid	Flammability	Toxicity	ODP	GWP
Ammonia	Low	High	0	1.00
Ammonia-water mixture (0.9)	Low	High	0	1.00
Propane	High	Low	0	3
R32	Low	Low	0	675
R22	Non	Low	0.055	1900
R134a	Non	Non	0	1300
R143a	Non	Non	0	4470

4. Conclusions

In conclusion, a model which incorporated LabVIEW and Refprop softwares was successfully developed and deployed for a preliminary assessment of the OTEC cycle efficiency. The preliminary analysis of a test run at a net power output of 1 MW showed a close agreement with that of exiting data. The similar developed model was implemented to analyze the efficiency of the OTEC basic closed Rankine cycle using eight varying working fluids. The analyzed working fluids, including that of ammonia, are ammonia-water mixture (0.9), propane, and refrigerants (R22, R32, R134a, R143a, and R410a). Accordingly, the results showed that ammonia-water mixture gave the excellent performance with regard to the characteristics of heat transfer with the best thermodynamic efficiency of 4.04% compared to pure ammonia with 3.21%, propane with 3.09%, followed by refrigerants from 3.03% to 3.13%. In terms of capital cost, propane was economically efficient with 15730 $/kW compared to cost of ammonia-water mixture at 16201$/kW, refrigerants from 16990 $/kW to 21400 $/kW, and pure ammonia being the costliest at 21700 $/kW. Despite being lower in its thermodynamic efficiency, propane gave the lowest capital cost and had the lowest toxicity in contrast to all other working fluids. Therefore, propane has the potential to be used as a clean and safe working fluid to substitute ammonia, thus further enhance the OTEC technology.

Acknowledgments

The authors express gratitude to the Ocean Thermal Energy Centre (OTEC) lecturers and staffs, Universiti Teknologi Malaysia for their knowledge, laboratory, and equipment supplied. The authors also thanked to Ministry of Higher Education (MOHE) for financially supported this study through The Fundamental Research Grant Scheme for Research Acculturation of Early Career Researchers, FRGS-RACER/1/2019/TK05/UMP//1 - RDU192621.

References

[1] D'Arsonval J A 1881 *Rev. Sci.* **17** 370–2
[2] Nihous G C 2005 *J. Energ. Resour.-ASME* **127** 328
[3] Nihous G C 2007 *J. Energ. Resour.-ASME* **129** 10
[4] Vega L A 2002 *Mar. Technol. Soc. J.* **36** 25–35
[5] Zhang X, He M and Zhang Y 2012 *Renew. Sust. Energ. Rev.* **16** 5309–18
[6] Ganic E N and Wu J 1980 *Energ. Convers. Manage.* **20** 9–22
[7] Palm B 2008 *Int. J. Refrig.* **31** 552–63
[8] Anderson J H 2009 *ASME Power Conf.* p 645–53
[9] Perang M R M, Nasution H, Latiff Z A, Aziz A A and Dahlan A A 2013 *Int. J. Technol.* **4** 81–92
[10] Alkhalidi A, Qandil M and Qandil H 2014 *Int. J. Therm. Environ. Eng.* **7** 25–32
[11] Chen H, Goswami D Y and Stefanakos E K 2010 *Renew. Sust. Energ. Rev.* **14** 3059–67
[12] Calm J M and Hourahan G C 2007 *Heat.-Piping-Air Cond.* **79** 50–64
[13] Lennard D E 1987 *IEE Proc. A* vol 134 p 381–91

[14] Rogers L J, Hays R J, Trenka A R and Vega L A 1990 *Ocean Resources*, eds D A Ardus and M A Champ (Dordrecht: Springer)
[15] Vega L A 1992 *Ocean Energy Recovery: The State of the Art*, ed R J Seymour (New York: American Society of Civil Engineers)
[16] Vega L A 2010 *Proc. of the Annual Offshore Technology Conf.* p 1–18
[17] Wu C and Burke T 1998 *Appl. Therm. Eng.* **18** 295–300
[18] Uehara H and Ikegami Y 1990 *J. Sol. Energ.-T ASME* **122** 247–56
[19] Sami S M 2012 *Int. J. Ambient Energy* **33** 37–41
[20] Gong J, Gao T and Li G 2012 *J. Sol. Energ.-T ASME* **135** 024501
[21] Liu W M, Chen F Y, Wang Y Q, Jiang W J and Zhang J G 2011 *Adv. Mat. Res.* **354-355** 275–8
[22] Yeh R H, Su T X and Yang M S 2005 *Ocean Engineering* **32** 685–700
[23] Upshaw C R 2012 Thermodynamic and Economic Feasibility Analysis of a 20 MW Ocean Thermal Energy Conversion (OTEC) Power *Thesis*
[24] Nouman J 2012 Comparative Studies and Analyses of Working Fluids for Organic Rankine Cycles – ORC *Thesis*

The Implementation of Electronic Voting System for Student Representation Council using reCAPTCHA

Amirul Asyraf Khairul Salleh[1], Harlinawati Abdul Kadir[1*], Aslimariah Ahmad[1], and Mohd Azahari Mohd Yusof[1]

[1] Faculty of Computing and Multimedia, Kolej Universiti Poly-Tech MARA (KUPTM), Jalan 6/91, Taman Shamelin Perkasa, 56100 Kuala Lumpur

*Corresponding e-mail: harlinawati@kuptm.edu.my

Abstract. The electronic voting system (E-voting System) is a web-based application that enables voters to record safe and confidential votes electronically. This research aims to develop an E-Voting System by exploiting the 'reCAPTCHA' security component for a private international college. The Waterfall Model is deemed the most suitable to be used after data was gathered prior to developing the system and it is also considered simple to be employed. The functional testing was conducted with thirty students and results revealed that 63% of the respondents are in agreement that the traditional voting system should be supplanted with electronic voting system, especially to cut down the time in the voting process. Ultimately, the system developed demonstrated that it efficient, reliable and displayed transparency in the E-voting System.

Keywords: E-Voting, Student Representative Council, 'reCAPTCHA', Waterfall Model

1. Introduction

Election is the utmost leading method for associates to have ability to speak in the leadership and direction of their association. People feel a greater sense of value, authority, and obligation when they are allowed to vote in fair and open elections. Other than that, online voting service allow members of associations, peoples, and other independent organizations by making voting easy and useful [1].

According to Fish S [1], online voting decreases voting costs by controlling manufacture costs. In electronic votes, paper, postage and postal rates are all significantly lower than for traditional methods of polling. Also, employees will save time when electronic polls remove the need to collect ballot packets and tabulate votes manually. Moreover, he claims that there are no rejected, mismarked or null votes for online voting. The figures will be measured automatically, removing the need for manual tabulation and reckoning. Computerized tabulation helps administrators of elections to reveal the outcomes easily.

reCAPTCHA has been the most effective and reliable CAPTCHA program in recent years. The main property that distinguishes it from standard text-based CAPTCHAs is the method of creating problems that they are a by-product of digitizing huge quantities of data. Each time the algorithm for optical character recognition (OCR) fails to recognize one word properly, it is considered difficult to digitize and is applied to the reCAPTCHA program after further distortions. It means that the word in question is not audible by robots at once. ReCAPTCHA is used on several of the Internet's most popular sites, including Facebook, Twitter and Craigslist. According to reCAPTCHA, the apps overcome over 200 million tasks every day [2].

Google's reCAPTCHA software is the most commonly deployed captcha program, and many popular websites have implemented it to discourage computer bots from carrying out nefarious activities. Google confirmed that a new reCAPTCHA system is being implemented to make it more human-friendly and safe [3].

1.1. Problem Statement

Time Consumption – The student of (Kolej Teknologi Antarabangsa Cybernatics) KTAC is currently using the traditional system which is vote using paper ballot. The current traditional system is no longer suitable and convenient because there are many students that study in that college. This is not suitable because the organization need to hire others to calculate the vote or paper ballot during the elections. Use paper ballot is time consuming, but it creates a lot of work for the people participating in the counting of ballots. This will increase the time and costing for the institutions during this program.

Cost Increase – The manual system voting also increase the cost to hire more staff during the election process to avoid the fraud and lack of process during the elections. The rate of paper ballot expenditure is much greater than that of electronic voting. For the election commission, the use of electronic voting is quite economical.

Lack of Security – Poor security system because one can get easily the document and change whatever they want, loss of information. The manual election systems may not secure. For example, student can use other student document to proceed with the fake votes during election. Moreover, in few places where the traditional system is corrupt, they can easily insert several fake paper votes in the ballot and then it becomes impossible to track the honest votes. Paper is a flammable material, so, under certain situations, the document in which the votes were registered in the referendum could be lost or damage and then the records of the votes cannot be recovered.

Votes Counting – The manual system that being used is not accurate for the result because there is too many students' votes in one time. Sometimes the staff accidentally skip to count votes and makes the result not tally with the student's votes. The traditional system also not accurate during the calculation when the election finished. Using the computerized system will helps to improve the accuracy and can prevent frauds during the elections.

1.2. Objectives

More Secure – The system will be more secure and prevent from frauds or fake vote from other student. The system must be secure to prevent frauds, malicious attack and hacker. The system login will be completely secure with developer will implement the reCAPTCHA during login the system. The reCAPTCHA security is easy use on human but hard for bots. The reCAPTCHA security implementation is letting the valid user to pass through the system with ease. Bots is an interactive program for executing basic and repetitive tasks that would be time-consuming, uncommon or difficult to do for a human being. Bots can be used for productive tasks, but they are also frequently used for malicious purposes. Moreover, developer implement one time password which is for user enter valid code that have been sent to email. This method also can avoid fraud during election. Lastly, developer implement security for database. This security is called password hashing. Password hashing executes a password with one-way process, transforming the password into another string named the hashed password. Hashing password is for storing password securely in database.

Reduce Cost – The new system will reduce more cost of hiring other staff that will be on duty during the election. Moreover, the system will reduce cost for buying more paper for votes, so students can vote use the new system and paperless. The system helps to reduce cost because anything will process using system and electronically. The data of the students also save in the database and students no need to do a registration before voting using a form paper. The database of E-Voting System connect to the Campus Management System so the staff do not wasting time to do a registration for student.

Result More Accurate – The electronic voting system also can helps staff to view the result and more accurate instead staff need to counts the paper during the election. So the termination will be come out with the result more accurate using electronic voting system compare to the traditional voting system.

This result is faster come out and display in the system compare to traditional that need to be counts first. The result also display live and update so the students and staff can view the result whenever they want.

More Convenient – The system is user friendly and easy to understand and user easy to use. This system is convenient and suitable to implement at the institution. This is because developer made this system used a basic system design that user can understand. The flow of the system also easy and user can understand. The flow is very simple makes everyone use the system can understand very well.

Once Vote Only – The system allow voter to vote one candidate only for each position. After vote have been cast, voters cannot allow to login into the system again. Voter can login unless voter do not cast any vote during election. This method will make sure the register voter tally with the result after election to make sure there is no unregistered student allow to vote.

2. Literature Review

Electronic voting systems can be used to register voters, tally ballots, and record votes [4]. Online voting is a web-based system that promotes the digital election process and surveys. This program has been set up to simplify the organisation of elections and make it convenient for voters to vote remotely from their personal computers while maintaining confidentiality, privacy and auditing capability [5]. Several categories of CAPTCHAs are image CAPTCHA, text CAPTCHA, voice CAPTCHA and video CAPTCHA [6].

Authentication is the mechanism to recognize and verify an entity before granting access to any secured service such as a personal account. This double layer of encryption, named two- factor authentication, provides a route that involves passwords (username / email and password) registration, accompanied by the development and confirmation of the One Time Password (OTP). The OTP is a numeric code that is created spontaneously and uniquely for each authentication case. It adds an additional layer of security, as the produced password is a fresh set of digits each time an authentication is attempted and offers the benefit of being random for the next session provided [7].

One explanation e-voting faces such a protection threat is that any successful attack would be high-profile, an aspect that motivates much of the hacker activity up to date. Even worse, the most extreme assaults will come from someone who is inspired by the ability to change the result without anyone knowing [8].

ReCAPTCHA offers an alternative for the visually impaired. Our standard audio CAPTCHAs are easy to solve for your legitimate users. On the other side, bots can get a much tougher CAPTCHA audio designed to block them [9].

3. Methodology

Methodology that has been used to develop this system is waterfall model as shown in Figure 1. The primary distribution on the cascade model is credited to Walter Royce's article in 1970 [10]. This concept of waterfall is an example of a sequential process. Software development process is divided into different phases in this model, and each step consists of a series of tasks and has different goals. First, we had to concentrate on preparation, time management, target date, expenditures and the execution of a whole program.

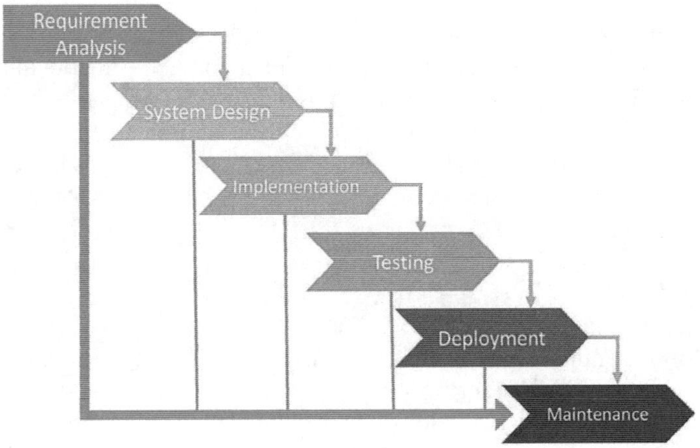

Figure 1. The waterfall model.

Waterfall model can help users to obtain a better understanding of developments in information systems development in the text on the paper. This is because the system specification in text provided is clear and easily understood by users. In this methodology, the involvement of users in the initial phase in the development of systems to ensure the systems achieve the objectives required by the user.

3.1. Requirement Analysis Phase

The first phase of waterfall methodology is system analysis and requirements. The first phase involves understanding what need to be design and what is the function and purpose of the system that will be develop. The developer has supplied a questionnaire to users so the developer can identify the user and client needs. This phase developers need do brainstorming and generate more ideas and to understand the requirement. Moreover, developers need to do research about the previous system similar for what he want to develop to get the ideas. After the requirement collected, developer should identify the overall direction that the project will take through the creation of the project strategy documents. The primary attraction of the analytical process is the selection of criteria. Assessment of specifications, also referred to as requirements development, is the process of determining user needs for a new or modified product.

3.2. System Design

In this step the criteria requirements are learned from the first process and the planned of system design. System Design helps determine the hardware and system specifications and helps define the overall design of the system. The developer need to generate the idea on how the product will look like. Moreover, the developer need to take serious the client needed to make the project successful. The design for this system is very simple and easy to understand.

3.3. Coding Implementation

The basic role of this phase is to convert designing into code using the programming language decided in designing phase. This is one of the longest phases in software development life cycle. This is a few points which we need to take care in this phase. Before begin the actual coding, you should spend some time on selecting development tool, which will be suitable for your debugging, coding, modification and designing needs. There should be a regular review meeting with lecturer need to conduct in this stage. It helps to identify the prospective defects in an early stage and helps to improve product and coding quality.

3.4. System Testing
In this phase, every code that implement need to testing to make sure if it works as expected. Moreover, developer need to perform all the testing activities functional and non- functional to make sure that the system meets the requirements. The developer aim to find defects within the system as well as verifying whether the system behaves as expected and according to what was documented in the requirements analysis phase. It is possible that defects are identified in the testing phase. When defect been identified, the developers will make sure to fix the problem and find a solution to make sure the system running and well function before ready to deployment.

3.5. System Deployment
After the system testing successfully passed, the system can be used and deploy at the client organization. Perform a stability check in the environment after the application is deployed to ensure the application does not break. Make sure system well function and can run smoothly as client needed before deployment. After client satisfied, the system can be deploy at the client organization to be use.

3.6. System Maintenance
The final phase is system maintenance. This step occurs after installation, and involves making modifications to the system or an individual component to alter attributes or improve performance. These modifications arise either due to change requests initiated by the client. Other than that, in case client encounters and defect, make sure to note and fix the issues faced. The application is always enhanced to incorporate more features, update the environment with the latest features.

4. Results
This section discusses the findings and the result of data gathering from dissemination of questionnaires to KTAC students. The questionnaires includes 17 questions and it is about E-Voting system. The total respondents we collected for the data is for 30 students.

Table 1. Shows the response from the student.

Questions	Response	
1. Do you think E-Voting System should be implement in KTAC?	Agree 64%	Disagree 0%
2. Do you think E-Voting System will help student to save their time instead using traditional vote?	Agree 67%	Disagree 0%
3. Do you think E-Voting System need to be a high security system to secure student data and personal details?	Agree 63%	Disagree 0%
4. I think E-Voting System is more productive compare to traditional votes.	Agree 59%	Disagree 0%
5. E-Voting System will reduce the level of fraud compare to traditional system.	Agree 61%	Disagree 3%
6. E-Voting System should prevent student vote more than 1 time for each positions.	Agree 56%	Disagree 3%
7. E-Voting System will makes result more accurate compare to traditional system.	Agree 58%	Disagree 0%
8. E-Voting System will helps student to view instant result during the election.	Agree 58%	Disagree 3%
9. E-Voting System can reduce cost compare to traditional system.	Agree 73%	Disagree 0%
10. I think E-Voting can be more convenient compare to traditional system.	Agree 63%	Disagree 0%

IOP Conf. Series: Materials Science and Engineering 1062 (2021) 012043 doi:10.1088/1757-899X/1062/1/012043

Based from the survey, results revealed that 63% of the respondents agree that the traditional voting system should be supplanted with electronic voting system, especially to cut down the time in the voting process.

5. System Modelling

System modeling is the process of developing intellectual models of a system, with each model offering a different view or perception of that system. Moreover, system modelling helps the analyst to know the functionality of the system and models that are used to communicate with customers. System modeling will also represent the view of the whole system in a visual diagram and clearly shows the functionality of the whole system.

5.1. Flow chart Diagram

A flowchart is a visual representation for the sequence of steps and decisions that need to perform as a process. Each step in the sequence is noted within a diagram shape. Steps are linked by connecting lines and directional arrows. This will allows anyone to view the flowchart and logically follow the process from beginning to end. Figure 2 shows a flowchart for voter. The process flow start when user (voter) need to solve CAPTCHA that show by the system. Then, system will verify CAPTCHA, if it is correct, user need to enter their ID and password but if the answer is incorrect, user need to re-solve CAPTCHA. System will give one time password (OTP) code to verify the user.

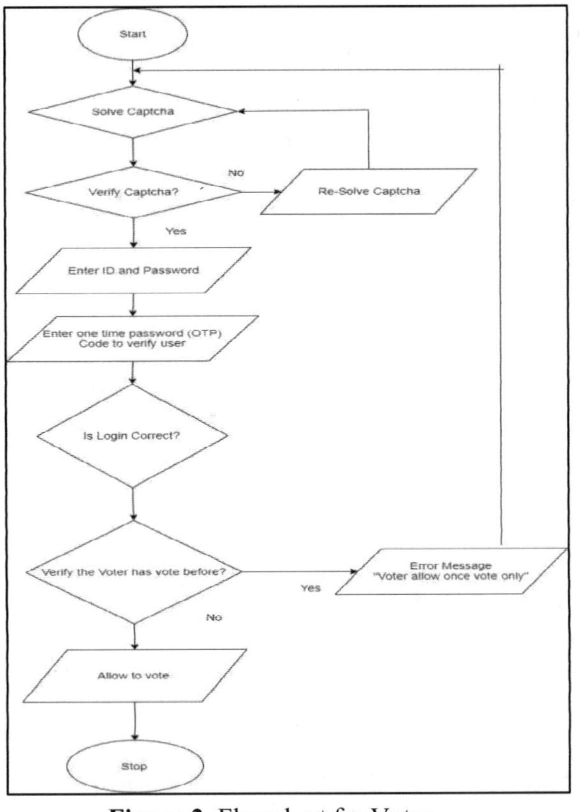

Figure 2. Flowchart for Voter

If the login is correct, system will verify either user has done voting previously. Message box will appear that "voter allow one vote only" if user have done voting previously and user is allow to vote if system cannot find the record about user. Figure 3 illustrates a flowchart for administrator (student welfare). The process flow begin with start symbol to indicate that the process is started after voter enter student ID and password. If the user ID and password is correct the system will view the dashboard. After that, students will be registered as a voters and the system will register the candidate, add position, edit election title, view report and print the result. Admin also can update their account and last step they can log out from the system. Figure 3 shows the flowchart for admin.

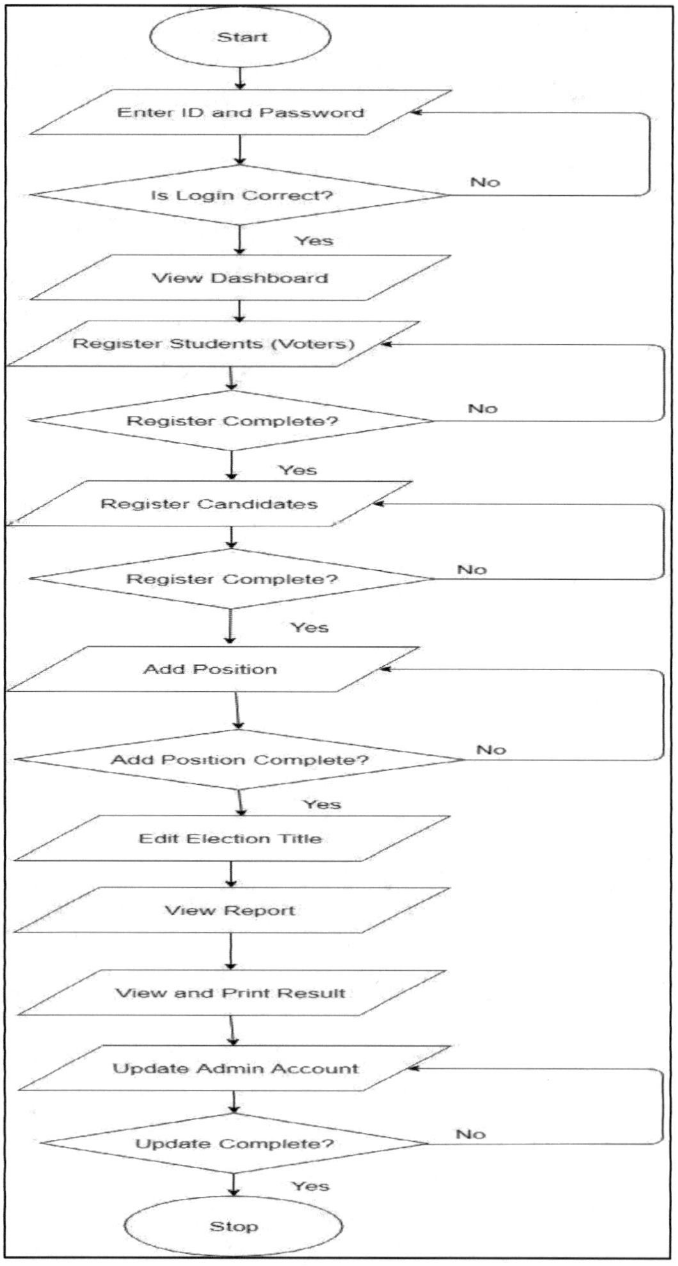

Figure 3. Flowchart for admin

5.2. Use Case Diagram

This will allows anyone to view the flowchart and logically follow the process from beginning to end. Use cases are a set of actions, services, and functions that the system needs to perform. For E-Voting system using reCAPTCHA, there are two actors involved which are voters (students) and administrator (student's welfare).

Based on Figure 4, it shows that voters are able to login to the system by solving CAPTCHA. The system can detect either voters has been done voting previously or not. Figure 5 shows that administrator can manage all the information about candidate.

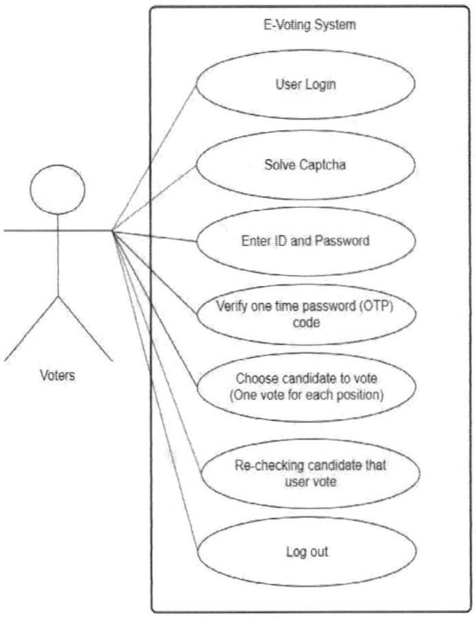

Figure 4. Use case diagram for voter.

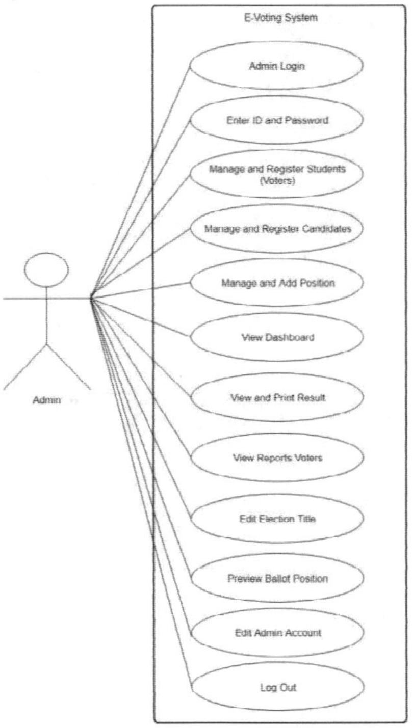

Figure 5. Use case diagram for admin.

6. Conclusions

This research has shown that the developed electronic voting system will detect either the voters have been vote for the students representative candidate twice using reCAPTCHA. Security is very important in developing E-Voting website. This electronic voting system will give the organizations an opportunity to increase the efficiency, reliability and transparency of an elections.

Acknowledgments

The authors gratefully acknowledge the support provided by the supervisor, Puan Harlinawati Abdul Kadir, the Lecturer of Faculty of Computing and Multimedia (FCOM), Kolej Universiti Poly-Tech MARA (KUPTM). The authors also would like to thank all the lecturers from the Faculty of Computing and Multimedia (FCOM) for their contribution, especially Puan Aslimariah Ahmad and Dr. Mohd Azahari Mohd Yusof.

References

[1] Fish, S 2009 Online Voting: Advantages and Challenge for 3

[2] Baecher P 2010 CAPTCHAs: The Good, the Bad, and the Ugly 13

[3] Suphannee Sivakom J P 2016 I'm not a human: Braking the Google reCAPTCHA. *Black Hat*, 12

[4] Sridhar S 2013 Electronic Voting – A Survey Electronic Voting 6

[5] Raja Lakshmi M N 2015 Student Online Voting System. *Voting System*, 3

[6] Rane S S 2016). 3D Socio Graphical CAPTCHA based on *International Conference on Advances in Electronics, Communication and Computer Technology,* 6

[7] Bhatia A 2018 *One Time Password (OTP) Algorithm in Cryptography*. Retrieved from Greek4greek:https://www.geeksforgeeks.org/one-time-password-otp-algorithm-in-cryptography

[8] Abdalla Al-Ameen S A 2013 *International Journal of Network Computing and Advanced Information Management, 10*

[9] Anwar N K 2009 *A Case Study of Estonia* 12

[10] Royce W 1970 *International Proc. IEEE WESCOM. IEEE Computer Society Press, Los Alamitos*

Cutting Forces and Flank Wears Analysis for End Mill Processes Using HSS Tools Cryogenic Treatment When Cutting PFRP Material

F Ridwan[1*] and D Noerhamzah[1]
[1] Department of Mechanical, Engineering Faculty, Andalas University, Padang, Indonesia

*Corresponding e-mail: firmanridwan@eng.unand.ac.id

Abstract. This study aims to characterize the wear and cutting forces of cryogenically treated and non-treated high-speed steel (HSS) tools when milling PFPR work piece. Pineapple fibre mixed with unsaturated polyester resin was prepared for the work piece. While three levels of cryogenic treatment were applied to the HSS tools with time immersion parameters in liquid nitrogen and time duration. The end mill process using the tools was applied to examine wear and cutting forces. Significant improvement of flank wear ensued on the first tool variation with average flank wear of 90 μm where the tool was treated cryogenically with the immersion of 48 hours and raising the heat up to 150o then normalized and repeating twice. Severe flank wears arisen on a non-treated cryogenic tool with 115 μm. The axial force, cutting force, and thrust force the tool variation varied in the range of 0.4 N to 2.5 N, 40 N to 210 N, 5 N to 28 N, respectively. It can be concluded that cryogenic treatment on HSS employing cryogenic treatment on the first variation of the tool is recommended to be applied in workshops.

Keywords: Cutting forces; Flank wear; Treated cryogenic HSS tool, PFRP, End mill

1. Introduction

Products made from natural fibre-reinforced composites are increasingly becoming an industrial concern due to their advantages such as low cost, biodegradability, relatively good mechanical properties, chemical reactions, and corrosion resistance. It is not as easy as metal machining. Machining difficulties due to irregular fibre structure are a challenge to be investigated. The lay-up of fibre would impact cutting forces, tool wear, chatter, and work piece surface deterioration when improper machining parameters are applied.

The literature shows many of the experimental works that have been conducted on cutting forces, tool wear, and selection of appropriate machining parameters. End-milling experiments of GFRP composite to examine their machinability have been carried out by Azmi [1]. Offered ANFIS models have shown the relationship between cutting force and the growth of tool wear. Experiments show that better machinability and higher surface finish in the machining of GFRP has been achieved by selecting the right tool insert [2]. Moreover, the machinability tests were conducted through different cutting conditions. Many studies on improvement in surface roughness, cutting force, cutting temperature, and tool flank wear in machining composite by selecting the proper cutting speed, feed-rate, and depth of cut has been reported [3-7] in literature. On the other hand, research of the HSS tool in the machining of FRP has become attractive since this tool material has been widely used in the manufacturing industry because of its useful properties at a reasonable cost. Indeed, the HSS tool was not reliable enough to cut

FRP material; therefore, many researchers made improvements to the HSS tool's performance by cryogenic treatment. Details reviews of the cryogenic treatment of the HSS tool were conducted, concluding the effectiveness of heat treatment that could widely be applied to various cutting tools and reduced production costs [8-10]. The deep cryogenic treatment has been carried out to improve HSS tool performance [11-13].

Despite many studies on machinability, cutting forces and tool wear for FRP machining are seen in the literature, very few have focused on HSS treated cryogenic tools affecting machinability, cutting forces and wear. This is a problem for the workshop industry to get references in increasing productivity. The purpose of this research is that the characterization of wear and cutting forces has been assessed using four different HSS tools, namely three types of high speed steel (HSS) tools which are cryogenically processed and not processed when milling PFPR work pieces. This study's results are expected to be used as a reference by the workshop industry in reducing production costs and increasing productivity.

2. Preparation of HSS Tool

The purpose of this process is to achieve a better steel microstructure and obtain the most desirable properties. The complete process in sequence consists of austenitizing, quenching, cryo processing, and tempering [8]. In this research, three types of cryogenic tools were prepared by adopting three cryogenic treatment methods [8, 14, 15] and a cryogenically non-treated HSS tool.

2.1. Temperature Variation of the Cryogenic Treatment Process

In the first variation, the cryogenic treatment was applied using shallow and deep cryogenic treatment at a cooling rate of 0.5 °C/min to prevent potential damage in the tools' microstructure. The treatment is the immersion process using cryogenic liquid (liquid nitrogen) at a temperature of -196 °C for 48 hours, then raising the heat up to 150 °C for 10 hours, then normalized and repeating twice. The second variation of cryogenic treatment is a cryogenic process at a temperature of -196 °C for 24 hours and raising the heat up to 204 °C for 1 hour, then normalized and repeating three times. The third variation of cryogenic treatment is a cryogenic process with a temperature of -196 °C for 24 hours and raising the heat up to 200 °C for 2 hours then normalized. The graph of immersion time for each variation can be seen in Figure 1.

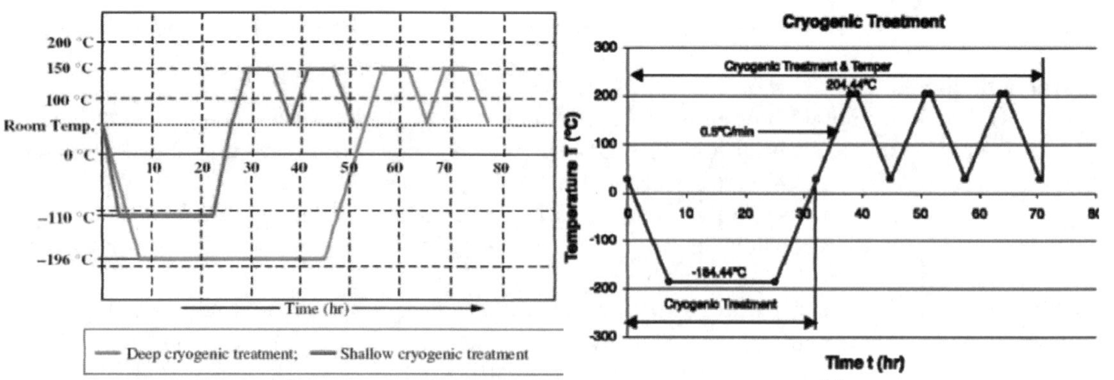

(a) (b)

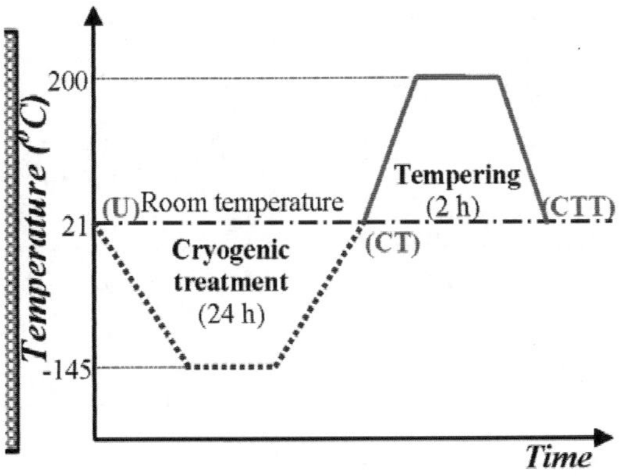

(c)

(d)

Figure 1. (a) The first variation of cryogenic tools [8], (b) The second variation of cryogenic tools [14], (c) The third variation of cryogenic tools [15], and (d) Cryogenically treated HSS tools.

The PFRP work piece material was fabricated using Unsaturated Polyester Resin BQTN-157, Yukalac, and pineapple fibers as reinforcement. The fiber was blended and sizing into 25 µm by screen level. The resin was poured into a box-shaped mold and followed by the fiber to get the work piece's rigid shape. The mixture was stirred repeatedly for 5 minutes so that the resin catalyst was evenly distributed. The size of the work piece can be seen in Figure 2.

Figure 2. PFRP work piece material.

3. Experimental Setup

The PFRP work piece material was mounted on end-milling and orthogonally cut using four different HSS tools. The experiment was conducted at a spindle speed of 283 rpm, feed-rate of 104 mm/min, cutting depth of 1 mm, and cutting length of 40 mm. The cutting force, axial force, and trust force were acquired with strain gauges based dynamometer. For validation purposes, three cutting experiments were applied. The measurement setup can be seen in Figure 3. Summary of used tools are summarized in Table1.

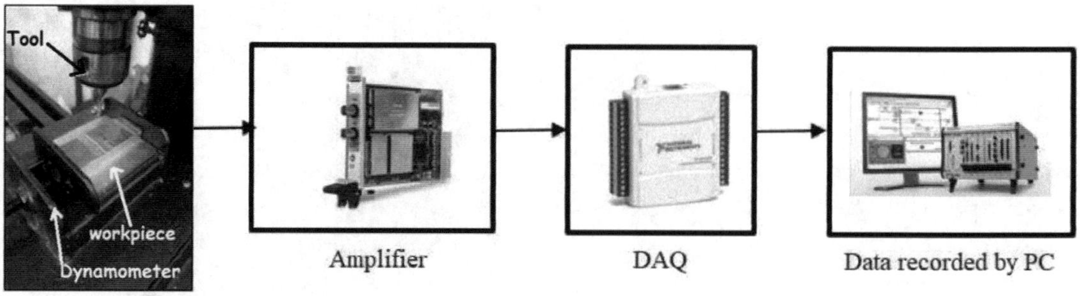

Figure 3. Setup measurement device and PFRP work piece material on the dynamometer.

Table 1. Summary of tools treatment.

HSS Tool	Variation	Cryogenic treatment	Adopted
A	First	A deep cryogenic treatment temperature of -196°C for 48 hours, then raising the heat up to 150°C for 10 hours, then normalized and repeating twice	[8]
B	Second	A cryogenic process at a temperature of -196°C for 24 hours and raising the heat up to 204°C for 1 hour, then normalized and repeating three times	[14]
C	Third	A cryogenic process with a temperature of -196°C for 24 hours and raising the heat up to 200°C for 2 hours then normalized	[15]
D	Forth	Non-treated HSS tool	

4. Results and Discussion

4.1. Axial Force, Cutting Force, and Thrust Force

The axial force, Fz results for A, B, C and D tools can be seen in Figure 4, Figure 5, Figure 6, and Figure 7, respectively. The cutting force, Ft of A, B, C and D tools can be seen in Figure 8, Figure 9, Figure 10, and Figure 11, respectively. A, B, C, and D thrust, Fy force can be seen in Figure 12, Figure 13, Figure 14, and Figure 15, respectively.

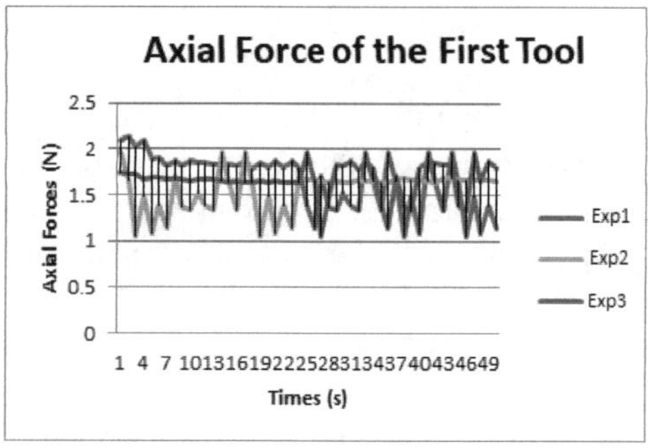

Figure 4. Axial force of the first tool variation.

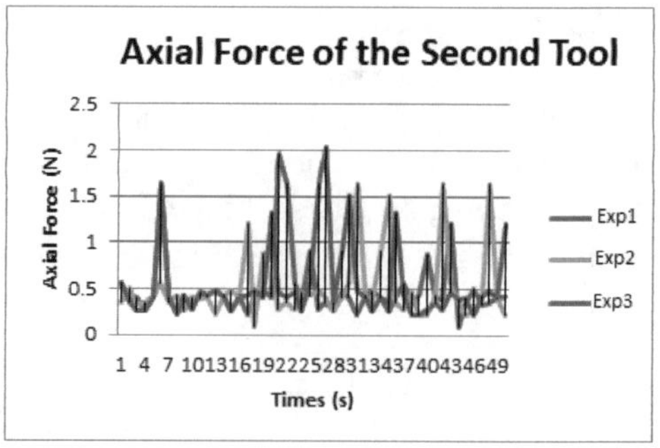

Figure 5. Axial force of the second tool variation.

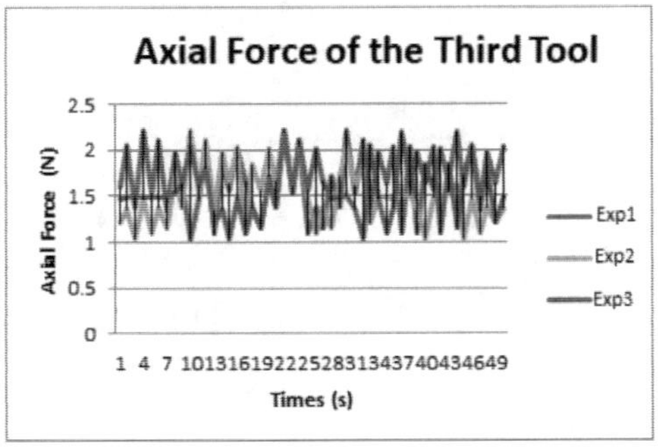

Figure 6. Axial force of the third tool variation.

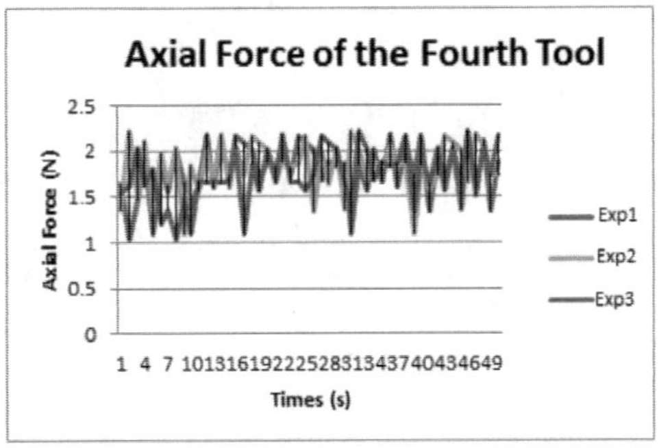

Figure 7. Axial force of the fourth tool variation.

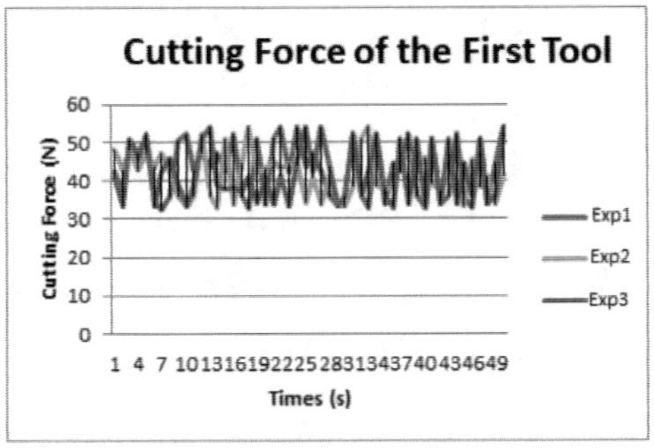

Figure 8. The cutting force of the first tool variation.

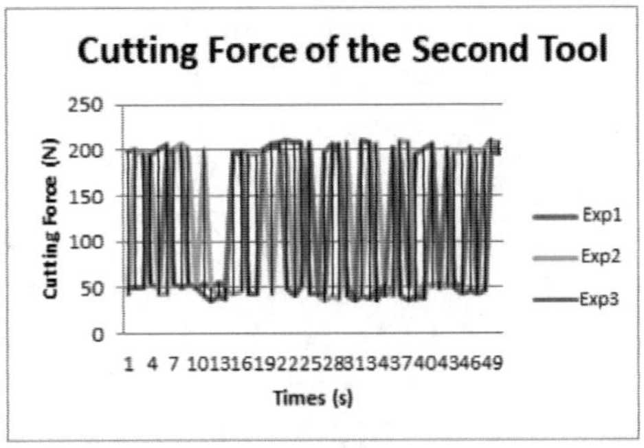

Figure 9. The cutting force of the second tool variation.

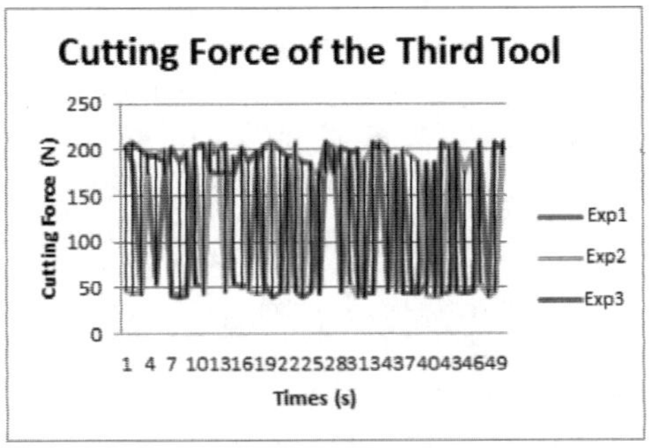

Figure 10. The cutting force of the third tool variation.

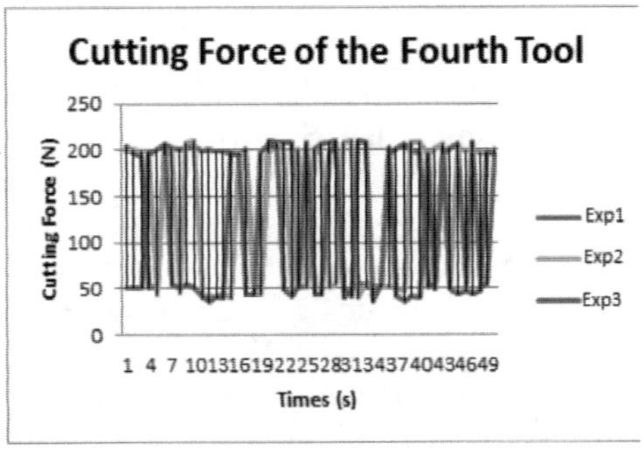

Figure 11. The cutting force of the fourth tool variation.

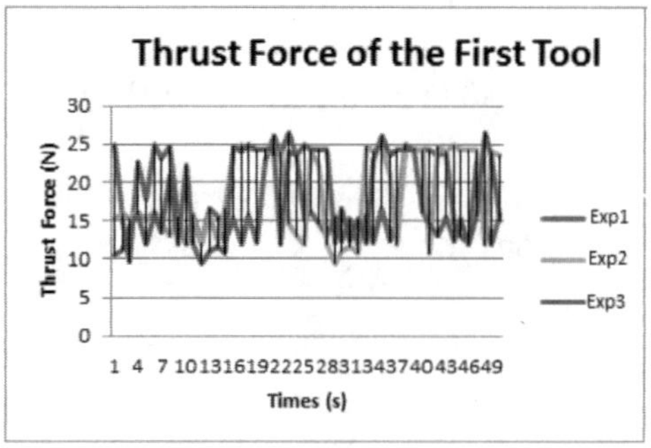

Figure 12. The thrust force of the first tool variation.

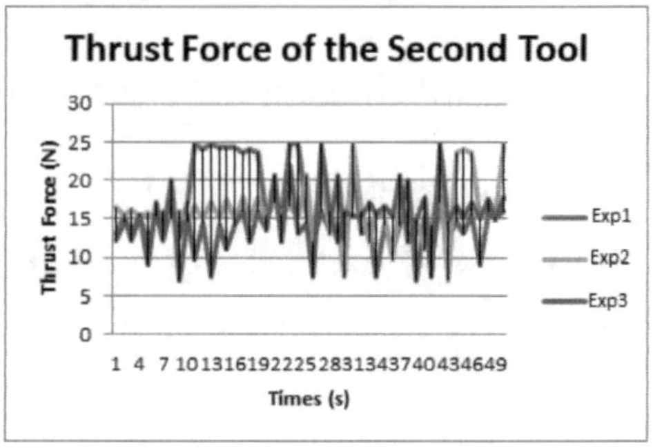

Figure 13. The thrust force of the second tool variation.

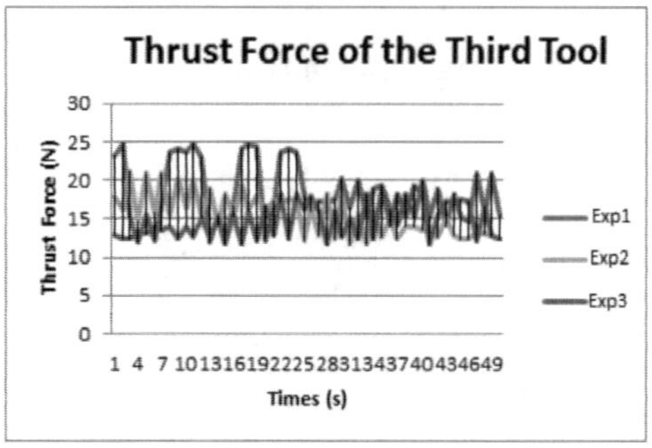

Figure 14. The thrust force of the third tool variation.

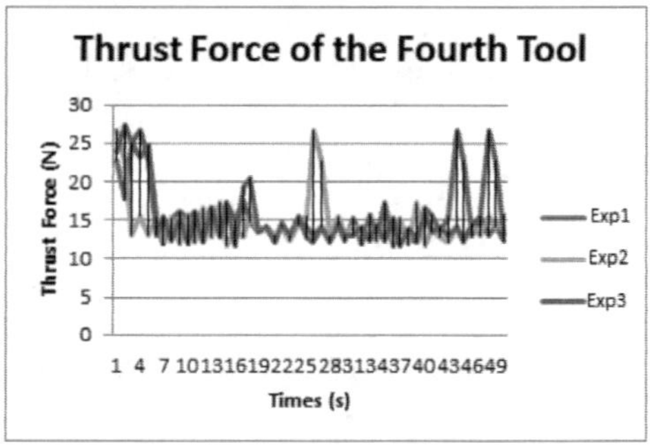

Figure 15. The thrust force of the fourth tool variation.

Figure 8 shows the axial force (Fz) of tool A (the first variation of the HSS tool cryogenically treated) with three times experiments and a depth of cutting of 1 mm each where the axial forces vary in the range of 1 N to 2.5 N. While, Figure 9 shows the axial force (Fz) of tool B (the second variation of cryogenically treated HSS tool) with three times experiments and a depth of cut of 1 mm each where the axial forces ensue in the range of 0.4 N to 2 N. Furthermore, Figure 10 shows the axial force (Fz) of tool C (third variation of cryogenically treated HSS tool) with three experiments and a depth of cut of 1 mm each where the axial forces transpire in the range of 1 N to 2.5 N. Finally, Figure 11 shows the axial force (Fz) of tool D (cryogenically non-treated HSS tool) with three experiments and a depth of cut of 1 mm each where the axial forces occur in the range of 1 N to 2.5 N.

The average axial forces (Fz) in each tool range from 0.4 N to 2.5 N. This force is relatively small. It is natural because, according to the axial force theory (Fz), the end milling does not make any axial motion. Still, the axial force may arise due to vibrations during the cutting process so that the work piece goes up and down.

Average cutting force (Ft) is obtained from the summation of the maximum cutting force divided by the data amount. The average cutting force (Ft) on the first cryogenic tool variation is 42.0127 N, the average cutting force (Ft) on the second cryogenic tool variation is 119.2925 N, the average cutting force (Ft) on the third cryogenic tool variation is 135,157 N. The average cutting force (Ft) on the fourth tool variation cryogenically non-treated is 133,5846 N. It can be seen that the smallest cutting force (Ft) occurs in the first cryogenic tool variation. In contrast, the most extensive cutting force ensues in the third cryogenic tool variations and the fourth tool. The cryogenic process in the first tool allows the tool to be tougher, thus increasing the tool's cutting speed, which results in reduced cutting forces [12].

The average cutting forces (Ft) are fluctuated in the range of 40 N to 210 N. Fluctuations arise due to the spread of fibre in resin, which has different strengths at each point. Consequently, when the tool cuts the soft part, the cutting force ensues low, and when the tool touches the hard part, cutting forces goes high. Fluctuating cutting force (Ft) can also be analysed because of the wear on the tool when cutting the pineapple fibre composite material so that the cutting force occurs quite large. However, the cutting force in the first tool variation is quite low compared to other tool variations. It happens because the wear rate of the first tool variation is lower than other tool variations.

The average thrust forces (Fy) arising in each tool is in the range of 5 N to 28 N. The thrust force (Fy) also fluctuates, but not as much as the fluctuation of the cutting force (Ft). High fluctuations ensue when the cutting edge starts to touch the PFRP work piece's surface, and the smallest fluctuation occurs when the cutting has finished. Fluctuations also happen due to PFRP work pieces' unevenness at each point, so the tool reacts provides resistance when cutting, resulting in vibrations in the work piece

4.2. Tool Wear Measurement Results

Figure 16 shows the wear measurements of four HSS tools in which three tools with three variations of cryogenic treatment and untreated tools. After the milling process, the wear of the cut tool edges was observed using an Olympus SZX10 stereo microscope. The HSS tool is positioned perpendicular to the microscope. The way to see the wear is taking 10 points perpendicular to the line drawn along the cutting edge and then averaging the wear. The measurement uses the Olympus SZX10 stereo microscope. Flank wear measurement calculation method by drawing a 10-point line can be seen in Figure 16.

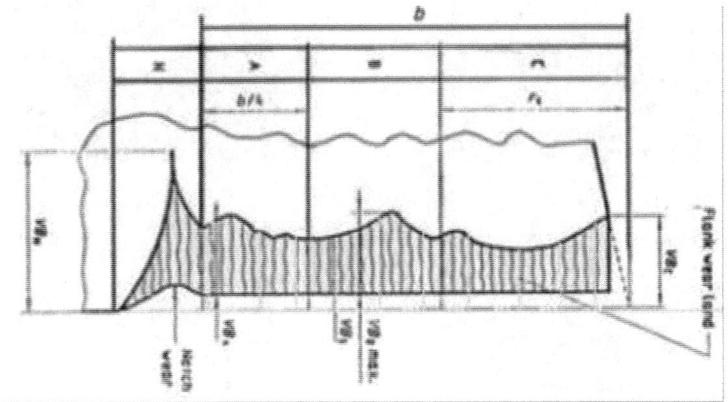

Figure 16. Wear the calculation method by drawing a 10-point line.

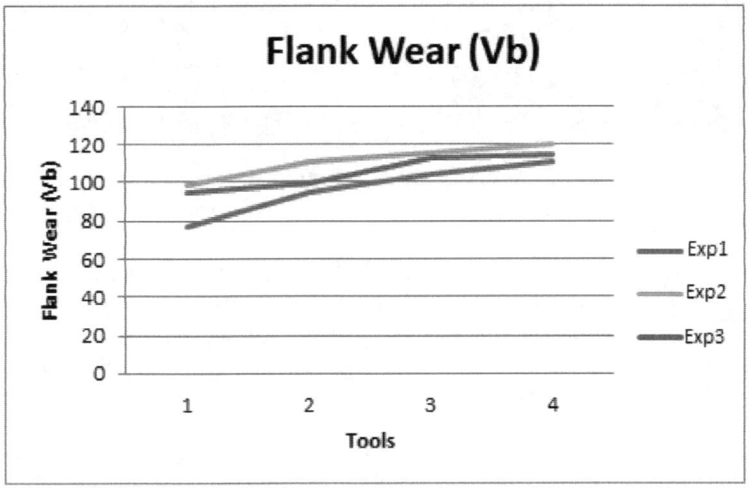

Figure 17. Flank wear graph (Vb) of the four HSS tool variations.

IOP Conf. Series: Materials Science and Engineering 1062 (2021) 012044 doi:10.1088/1757-899X/1062/1/012044

It can be seen in Figure 17 that by three experiments, the graph trend remains consistent for each trial where the flank wear increases with increasing tool variation. The average flank wear (Vb) ensues in the first tool variation is 90 μm, in the second tool is 101 μm, in the third tool variation is 110 μm, and in the fourth variation is 115 μm.

The smallest flank wear (Vb) occurs in the first cryogenic tool variation, an immersion treatment into cryogenic liquid (liquid nitrogen) at -196°C for 48 hours. It then raises the heat up to 150°C for 10 hours, then normalized and repeating two times. In contrast, the most significant wear arises in the fourth tool variation. Thus, it can be concluded that an efficient cryogenic treatment with toughness and reliability for cutting composites occurs in the first variation HSS tool followed by the second variation HSS tool then the third tool variation HSS tool. Whereas the flank wear (Vb) of the fourth variation, HSS tool without treatment, is extensive compared to the cryogenically treated-HSS tool. Figures 18-21 show snapshots of the flank wear (Vb) captured using the SXZ10 stereo Olympus microscope.

Figure 18. Flank wear (Vb) of the first cryogenic tool variation in the second experiment.

Figure 19. Flank wear (Vb) of the second cryogenic tool variation in the second experiment.

Figure 20. Flank wear (Vb) of the third cryogenic tool variation in the second experiment.

Figure 21. flank wear (Vb) of the fourth cryogenic tool variation in the second experiment.

The treatment of various cryogenic methods on the HSS tool provides better machining performance for increased productivity and decreased production costs. Information on the results of this experiment can be used as a workshop industry to improve machine performance. Application in the workshop industry will provide several advantages, including; longer tool life increased productivity and ultimately will reduce production costs.

5. Conclusions

Experimental results have shown a comparison between HSS tools with cryogenic treatment and an untreated HSS tool. The first variation HSS tool provided good performance by giving the smallest flank wear of 90 μm with the axial force, cutting force, thrust force component values of 2.5 N, 42.0127 N, and 17.5 N, respectively. The tool's cryogenic treatment was by immersing into cryogenic liquid (liquid nitrogen) at -196°C for 48 hours, then raising the heat up to 150°C for 10 hours, then normalized and repeating twice. Severe flank wears arisen on a non-treated cryogenic tool with 115 μm. The axial force, the cutting force, and the thrust force fluctuated in the range of 0.4 N to 2.5 N, 40 N to 210 N, 5 N to 28 N, respectively. The cutting force will increase with increasing feed-rate. Increasing the feed-rate will result in high friction and impact loads, thereby increasing the cutting temperature. As a result, there are abrasive and adhesion mechanisms that cause expansion and erosion of the main cutting edges. The greater the feed-rate used, the greater the flank wear and cutting force occurring in the milling process of pineapple fibre-reinforced composite work piece. Information on experimental data on cryogenic treatment on HSS can be used as a recommendation to the workshop industry for better productivity.

Acknowledgments
Publication of this article is supported by the Mechanical Engineering Department, Engineering Faculty, Andalas University, Padang, West Sumatra, Indonesia.

References

[1] Azmi A I 2015 *Advances in Engineering Software* **82** 53–64
[2] Anand K et al 2015 *Applied Mechanics and Materials* **787** 664–668
[3] Ridwan F, Havendri R, Susanti O, 2019 *IOP Conf. Series: Materials Science and Engineering* **602(1)** 012083
[4] El-Kady E Y, Gaafer A M, Ghaith M H G, Khalil T and Mostafa A A 2015 *Advances in Materials* **4(3)** 43–50
[5] Gill S K, Gupta M and Satsangi P S 2013 *Frontiers of Mechanical Engineering* **8(2)** 187–200
[6] Guo D M, Wen Q, Gao H and Bao Y J 2012 *Journal of Engineering Manufacture* **226(1)** 28–42
[7] Wang H et al 2016 *The International Journal of Advanced Manufacturing Technology* **82(9)** 1517–1525
[8] Akincioğlu S, Gökkaya H and Uygur İ 2015 *The International Journal of Advanced Manufacturing Technology* **78(9)** 1609–1627
[9] Gill S S et al 2010 *The International Journal of Advanced Manufacturing Technology* **48(1)** 175–192.
[10] Yildiz Y and Nalbant M 2008 *International Journal of Machine Tools and Manufacture* **48(9)** 947–964
[11] Alava. 2009 *2nd International Conference of the HSS Forum*
[12] Kalin M, Leskovšek V and Vižintin J 2006 *Materials and Manufacturing Processes* **21(8)** 741–746
[13] Li S et al 2010 *Cryogenics* **50(11)** 754–758
[14] Kalsi N S, Sehgal R and Sharma V S 2010 *Materials and Manufacturing Processes* **25(10)** 1077–1100
[15] Kıvak T and Şeker U 2015. Mater. Technol. **49** 949–956

IOP Conf. Series: Materials Science and Engineering 1062 (2021) 012045 doi:10.1088/1757-899X/1062/1/012045

Design of indenter rig for measuring seat cushion dynamic stiffness

AM Hassan[1], KAM Rezali[1*], NA Abdul Jalil[1], A As'arry[1] and MA Azizan[2]
[1] Sound and Vibration Research Group (SVRG), Department of Mechanical and Manufacturing Engineering, Faculty of Engineering, Universiti Putra Malaysia (UPM), 43400 UPM Serdang, Selangor.
[2] Aerospace Section, Universiti Kuala Lumpur-Malaysian Institute of Aviation Technology (UniKL MIAT, Lot 2891 Jalan Jenderam Hulu, 43800 Dengkil, Selangor,

*Corresponding e-mail: khairilanas@upm.edu.my

Abstract. Vibration induced on the car floor is transmitted to human body through seat structure, affecting both comfort and health. Seat structure components are mainly comprised of rigid frame and seat cushion. Previous studies have shown and suggested that the indenter is preferable option for replacing human subject in seat vibration experimental work when assessing the seat cushion dynamic characteristics. Indenter method offers more control, cheaper and less time intensive compare to other methods such as passive dummies, sandbag, and rigid mass. This paper proposes a design of indenter testing rig to measure the dynamic stiffness of seat cushion material. The rig was designed to operate at frequencies between 2 to 100 Hz and applied load of maximum 50 kg. The size of bed is proposed to be able to fit the size of standard sedan car seat. Modal analysis was conducted on indenter rig structure design to ensure the suitability of the design. Structural frame was designed to fit the mounting of the shaker used in this study. Six modes shapes were extracted from the modal analysis and its corresponding frequencies ranging from (121.63Hz – 507.3Hz). The lowest natural frequency for the proposed design of indenter testing rig occurs in the first mode of vibration with peak displacement of 174.04mm. From this result, it can be concluded that the proposed design met the criteria set and produces a fundamental frequency of greater than the operating range of the rig.

Keywords: seat cushion, dynamic stiffness, vibration, frequency

1. Introduction

Seat is one of the components in vehicle's design of the transport system such as cars, trains, commercial vehicles, and aircraft. Under dynamic motion, the direct contact between seat material and human body plays a significant role in attenuating vibration transmission. This exposure can affect body comfort and health risks.

Many methods and experimental techniques have been developed to study the vibration transmitted from the vehicle's structure to human body via seat system and for objective research, the evaluation of seating dynamics can involve several parameters such as dynamic stiffness, biomechanical response of whole body, seat transmissibility and SEAT value [1, 2]. Seat transmissibility has been the common method for determination of seat performance [2]. Measuring seat transmissibility requires subjects to be sitting on the seat. The interaction between human body and seat system acted as couple dynamic system where the vibration transmitted is dependent on the mechanical impedance of the seat and the sitting body [2]. It is preferable to measure vibration transmissibility using human subject although larger vibrator is needed as it must be able to provide vibration with human subject actual weight. However, testing with human subject is time consuming, a slight change in human body impedance and sitting characteristics can affect the overall result [2]. In addition, working with human subjects requires medical and ethical approval which can be difficult to get for certain condition [3, 4]. Seat testing conducted on a small group of subjects can be inconsistent, non-repeatable of test result and objectivity [3, 4].

As an alternative to using human subjects, several researchers have proposed the use of passive or active dummies as replacement for human subjects [3, 4]. The passive dummy method is currently the common method to be used in seat vibration testing. Some passive dummies are constructed using single degree of freedom (SDOF) systems and as an improvement, some passive dummies have been developed with multi-degree of freedom (MDOF). A SDOF based passive dummies for seat vibration testing was previously developed but it was found that at higher resonance frequency, the results were not satisfying [3, 5]. By adding more degree of freedoms, it will improve the results but require extensive construction efforts [4, 6, 7]. It has also been reported that both single and multi DOF of passive dummies will exhibit friction and unwanted nonlinearities; the nonlinearities of human body dynamic response under vibration evinced great differences with nonlinearities of passive dummy subjects [2, 4, 6, 8].

Some researchers propose several other methods such as using sandbag, and rigid mass as replacement for human subjects. L. Wei & Griffin [1] found that using sandbag might not be appropriate especially for larger weight where it might consume greater contact area in comparison with normal human body. The excessive contact area reached beyond or to the edge of the seat will influence the measured seat transmissibility [1]. In the case of subject's replacement using rigid mass, it will produce satisfying outputs and in good agreement with measured transmissibility using human subjects. However, rigid mass on the seat foam can easily be subjected to rotation and movement when exposed to lateral or front-aft vibration.

Another method of classifying seat performance is using indenter test. A rig with indenter is used to apply load before vibration is initiated vertically, and the measured output force and input acceleration are used to calculate seat dynamic performance. Indenter test method requires the measured seat dynamic stiffness to be used with a measured human biomechanical response to vibration, to predict seat transmissibility using mechanical impedance model. Indenter test offers more controllable options, and the variation of loading magnitudes can be applied in, easier and quicker than other seat testing methods. Hence, indenter has become preferred method to replace human subject for quantifying seat performance [1]. The used of indenter for assessing seat vibration testing has produced remarkably useful results since it will eliminate the drawbacks and limitations with human subjects and other types of mass elements [9-14].

The objective of this paper was to propose a design of indenter rig system for the assessment of seat dynamic stiffness. The design was based on the previous indenter test rig. An improvement was made on the structure, a smaller overall design for greater rigidness, to be able to connect on a different vibrator. With smaller sized, the natural frequencies of the rig were expected to be greater than the operating frequencies for the indenter test which is between 2 to 100 Hz.

2. Proposed Indenter Test Rig Design and SIT-PAD Indenter

The design of the indenter test system, components, and testing requirements were chosen and made according to the allowable size and capacity of the equipment available at Sound and Vibration Research (SVRG) lab, Universiti Putra Malaysia. The complete indenter testing rig is comprised of a vibration shaker machine, a structural frame, an intender, an accelerometer, and a force sensor. Force sensor was suggested to be of a donut type (Futek LTH-350) with the maximum capacity of 2.2kN, which is a piezoresistive type, for the measurement of static pre-load as well as dynamic load during vibration exposure.

Vibration magnitude will be generated using ETS Solution L215M shaker machine which has maximum loading capacity of 70kg and supported frequency range of up to 4500Hz. A frame structure was design and to be attached on the vibrator as shown in Figure 1. This setup can be operated using MATLAB and LMS Test Lab software with SCADAS data acquisition system. The type of material used to fabricate structural frame is mild steel with rectangular hollow section with dimension of 50mm x 500mm and a thickness of 3.2mm.

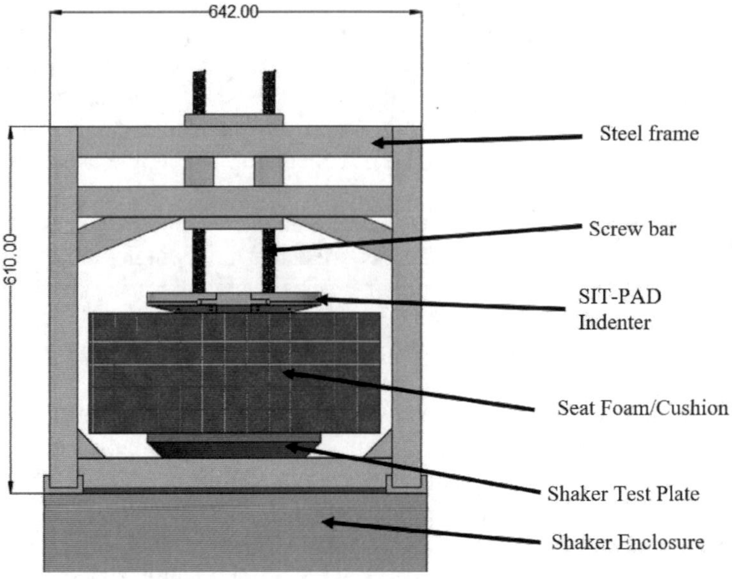

Figure 1. Shaker machine and test frame assembly.

Indenter test rig will be used to run with capacity and frequency range as specified in the British Standard BS2361-Part 1. For the SIT-PAD Indenter (see Figure 1), the material was chosen to be aluminium. Figure 2 shows the assembly of SIT-PAD indenter proposed of this work.

3. Modal Analysis on Structural Frame

Modal analysis on structural frame was conducted to determine the vibration characteristics by extracting the mode shapes and its corresponding natural frequencies. Both mode shapes and natural frequencies are important parameters as reference for designing structural system to be used in dynamic loading conditions [15].

3.1. Finite Element Analysis on Structural Frame

The structural frame was modelled, exported, and imported to ANSYS Workbench for simulation and analysis. The imported geometry was then meshed and applied with boundary conditions. For this simulation, fixed support boundary condition was applied with pre-stressed load on screw rod. Four fixed supports are applied at bolt locations and a pressure of 2.264 N/mm2 (Force = 400N) was applied at both screw bar bottom surface in upward direction as shown in Figure 3.

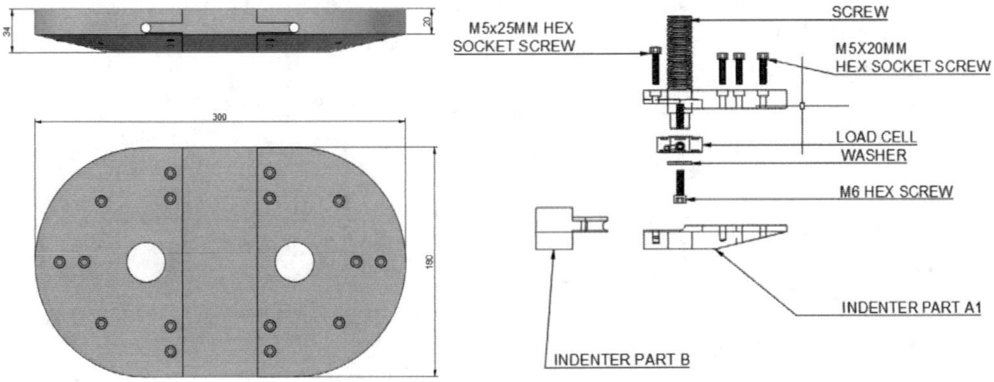

Figure 2. SIT-PAD indenter sizing.

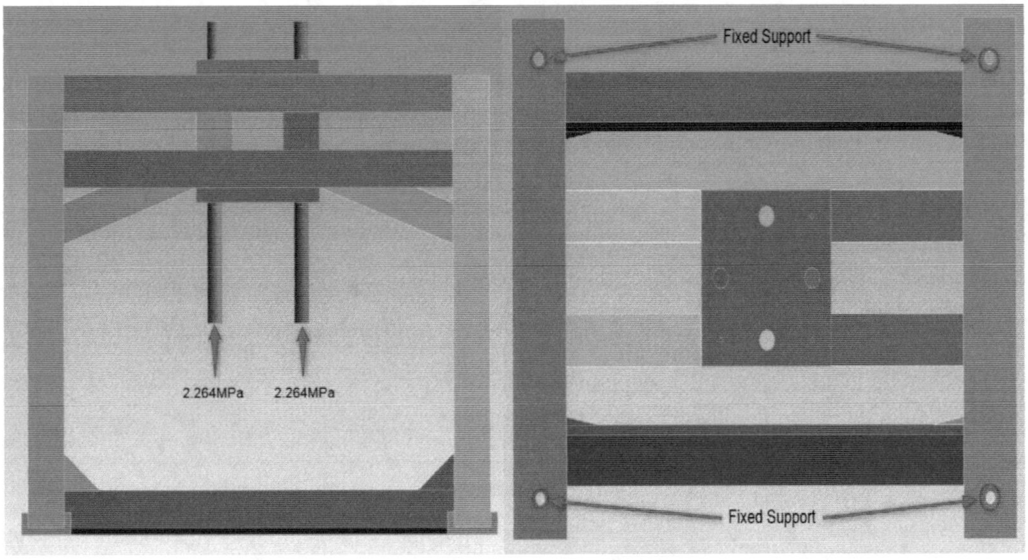

Figure 3. Boundary conditions on structural frame.

The large displacement mode was switch on for the analysis and adaptive meshing was enable with resolution set to 4. This is based on checking by trial and error, the selected meshing resolution produced better stability for meshing and analysis output. The material used to fabricate the frame was mild steel and for the simulation, the material input is as displayed in Table 1.

Table 1. Material property.

Parameter	Detail
Young Modulus, E (MPa)	200e3
Poisson Ratio	0.3
Shear Modulus, MPa	76923
Bulk Modulus, MPa	1.6667e+005

Figure 4 displayed the flowchart of finite element analysis using ANSYS Workbench to obtain the natural frequencies and mode shapes of SIT-PAD indenter frame structure under dynamic condition.

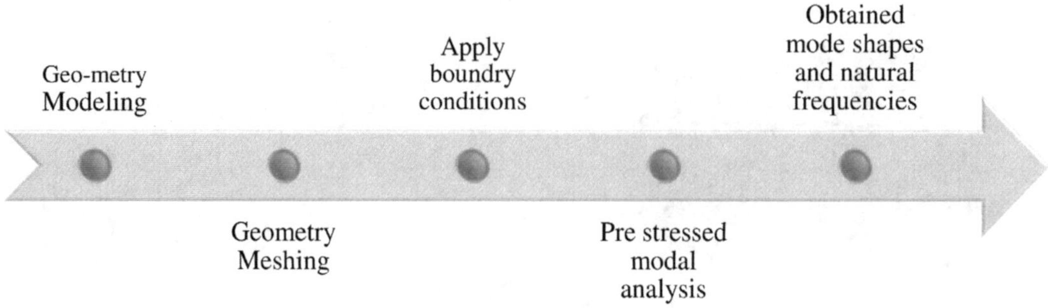

Figure 4. Finite element analysis flowchart.

Table 2 displays the effect of number of elements on first mode natural frequency obtained from the modal analysis. The structural was meshed using Automatic Method which resulted the use of tetrahedron (10 nodes), hexahedrons (20 nodes) and wedge (15 nodes) elements.

Table 2. Effect of meshing sizing on natural frequencies (first mode).

Element Size, mm	Number of elements	Number of Nodes	First mode Natural Frequency, Hz
30	20879	65732	118.61
28	22165	70062	117.44
26	24437	76549	117.41
24	29231	92915	116.88
22	32439	106422	116.89
20	34516	117405	116.96
18	27212	107592	116.83
16	32899	139052	116.38
14	40176	174593	116.08
12	48954	219348	115.95
10	63858	292616	115.5
8	100563	467683	115.56
6	137177	522344	115.17
4	307002	1179723	115.02

The number of elements increased with the decreased of element size and at the same time, the natural frequency exhibited increasing trend as shown in Figure 5.

The use of finer size of element expected to produce better accuracy (not always true), however it can lead to several problems such as longer processing time or processing errors especially in multibody analysis.

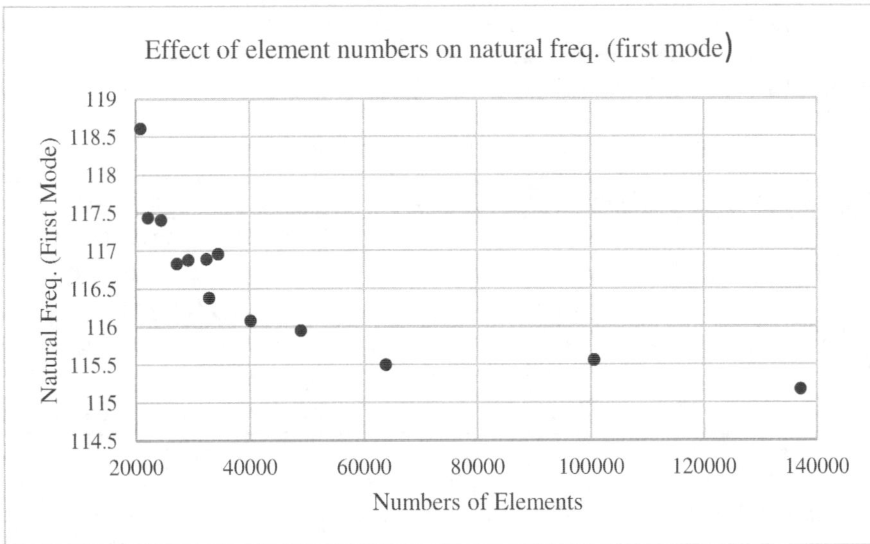

Figure 5. Effect of numbers of elements on natural frequency (first mode).

3.2. Simulation Results

Six mode shapes were extracted from the analysis and the results using element size of 6mm was selected because skewness level of most elements falls in the acceptable range as proposed in ANSYS documentation. Skewness level lower than 0.75 produced acceptable quality and larger size of element will result more elements in poor quality range (0.75-1) [16] as shown in Figure 6 and Figure 7.

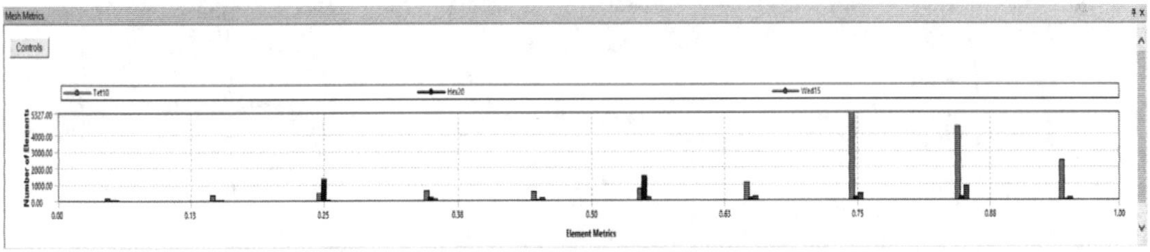

Figure 6. Mesh metric at 30mm element size.

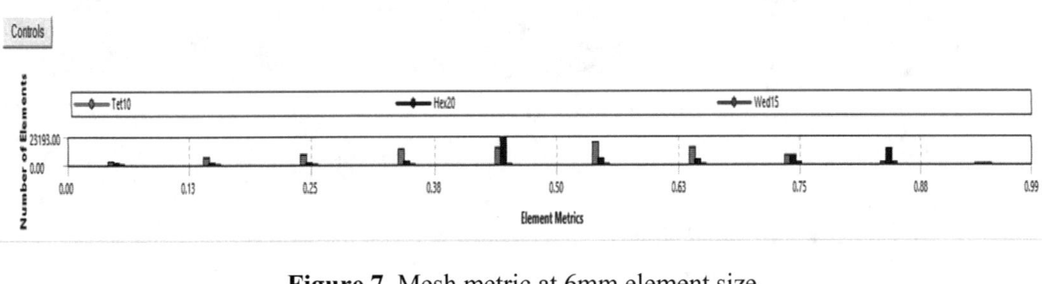

Figure 7. Mesh metric at 6mm element size.

As shown in Figure 7, most elements fall in the skewness region lower than 0.75 which is satisfactory

IOP Conf. Series: Materials Science and Engineering 1062 (2021) 012045 doi:10.1088/1757-899X/1062/1/012045

for this analysis. Further efforts to reduce element size beyond 6mm lead to higher processing time and resulted solver error at 2mm element size. Plus, natural frequency in the first mode for smaller size of 4mm produced small difference 6mm element size. The natural frequency for each mode using 6mm size of element is displayed in Table 3.

Table 3. Mode shapes and its corresponding frequencies.

Mode Shape	Frequencies, Hz	Structural Deformation
1	115.17	Primary (X direction)
2	225.91	Primary (Y direction)
3	398.06	Primary (Z-torsion)
4	418.38	Secondary (Z direction)
5	465.99	Secondary (Y direction -Screw Bending)
6	498.8	Secondary (X direction -Screw Bending)

As displayed in Table 3, there were different structural deformation at each mode shape. At first mode shape (ω = 115.17 Hz), the structural frame translated in X direction and at second mode shape (ω = 225.91 Hz), it resulted translation in Y direction. Third mode shape produced higher natural frequency of 298.06 Hz; the structural frame exhibited torsional deformation around Z axis. At fifth and sixth mode shapes, the deformation on testing rig only imposed on screw rod in bending mode through Y and X axis, respectively. All visual deformations are shown in Figure 8.

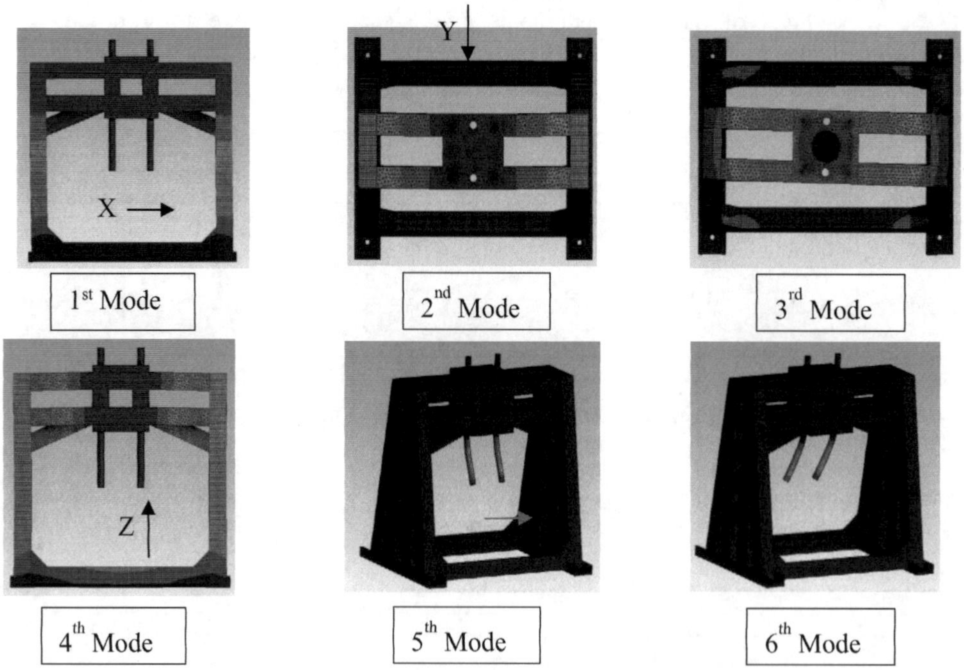

Figure 8. Deformation at each mode shape of indenter testing rig.

3.3. Discussion

Referring to simulation results of the modal analysis as shown in Table 3, the natural frequency should not be coinciding or close to, with the operating frequency, or ranges, of the indenter test rig. The lowest natural frequency out of all modes extracted, set as reference of guideline for structural design [17, 18].

The main criteria for indenter testing rig to operate properly under vibration condition is the natural frequency must not fall in the range of seated human body frequency. For seated human body under vibration condition, it can have significant effect for low frequency up to 50 Hz [19] and some other researchers reported that the fundamental frequency of seated human body under Whole-Body Vibration (WBV) is below 10Hz .In this case, the fundamental frequency for indenter testing rig is 121.63Hz, which way higher than the reported fundamental frequency for seated human body. This result satisfies the criteria of the proposed design for indenter testing rig and can be used in the experimental work.

4. Conclusions

The development of indenter testing rig requires several considerations such as the dimension of the test samples, machine's mounting size or dimension, ease of fabrication, the force cell, and easy access to material quality. The test rig should be able to operate within the required frequency range and capacity as specified in the standardised method. Modal analysis conducted in this work showed that the first natural frequencies found to be greater than the operating range of the testing rig. This condition satisfies the requirement as specified in the standard for indenter testing rig.

References

[1] Griffin M J 1990 *Handbook of human vibration* (Academic Press)
[2] L Wei & Griffin J1998 *Journal of Sound and Vibration* **214**(1) 121–137
[3] Gu Y 1999 *SAE Technical Papers* 724
[4] Suggs C W, Abrams C F & Stikeleather L F1969 **12**(1) 79–90
[5] Huston D R, Johnson C C & Zhao X D N 1998 *Journal of Sound and Vibration* **214**(1) 195–200
[6] Cullmann A & Wölfel H P 2001a *Clinical Biomechanics* **16**(SUPPL. 1) 64–72
[7] Stanczyk T L & Zuska A 2015 *Journal of Theoretical and Applied Mechanics*) **53**(4) 1029–1039 (Poland
[8] Lewis C H & Griffin M J 2002 *Journal of Sound and Vibration* **253**(1) 295–311
[9] Fairley T E & Griffin M J 2010 *SAE Technical Paper Series* 1
[10] Garcia-Mendez Y, Pearlman J L, Cooper R A & Boninger M L 2012 *The Journal of Rehabilitation Research and Development* **49**(1) 7
[11] Toward M G R & Griffin M J 2011 *Journal of Sound and Vibration* **330**(26) 6526–6543
[12] Tufano S & Griffin M J2013*Vehicle System Dynamics* **51**(1) 122–138
[13] Whitham E M & Griffin M J 2010 *SAE Technical Paper Series* 1
[14] Zhang X, Qiu Y & Griffin M J 2015a *Part D: Journal of Automobile Engineering* **230**(6) 736–744
[15] ANSYS 2018b ANSYS Mechanical User's Guide **15317** (August) 724–746
[16] ANSYS 2018a ANSYS Discovery AIM and Workbench Scripting (Issue January)
[17] AUTODESK 2020 **5** 37–45
[18] Hammad B K 2014) *Shock and Vibration*
[19] Bae J J & Kang N 2018 *Shock and Vibration* 1–15

Stress Analysis on Cracking Plate Structure For Reducing Stress Concentration

F Huda[1*], R A Safrudin[1], S Utomo[2] and M Rusli[3]

[1] Department of Mechanical Engineering, University of Riau, Kampus Bina Widya, Pekanbaru 28293, Indonesia
[2] Indonesian Institute of Sciences, Bandung 40135, Indonesia
[3] Department of Mechanical Engineering, Andalas University, Kampus Limau Manis, Padang 25163, Indonesia

*Corresponding e-mail: feblil.huda@eng.unri.ac.id

Abstract. The structures which are subjected to load experience stress. If there is no discontinuity (defect), the stress is distributed in a particular pattern based on the type of stress. The crack is one defect that is frequently found in the plate structure, which causes stress concentration factor. In this paper, the treatments to reduce concentration factors are applied to plate structures under cracked conditions to investigate the treatment effect to reduce the stress concentration. For that purpose, the low-cost strain indicator using strain gauge integrated with Arduino Uno and computer is constructed to measure the stress distribution on cantilever aluminum plate in the normal conditions, crack, and crack with treatment. The treatments to reduce the stress concentration factor are applied in the form of the stop drill hole and additional hole. The result shows that the given treatment could reduce the stress concentration factor by around 11%. Even though the reduction of the factor was not significant, the treatment could be a fast and straightforward solution in decreasing the stress on the cracking plate structure.

Keywords: stress analysis, cracks, plate structure, strain indicator

1. Introduction

Many engineering structures are designed to expose the mechanical loadings. Because of the critical function, the structures are maintained to run its function with full integrity. So the structural health monitoring to detect damage in structures and measurement analysis to solve the problem in structures is carried out to ensure the structure's eligibility to perform the expediency [1- 6].

When the structures are subjected to load, some parts will experience stress. If there is no discontinuity (defect), the stress is distributed in a particular pattern based on the type of stress. The crack is one of the defects found in the plate structure, which causes stress concentration. The stress concentrations will cause excessive stress, and the damage gets even worse.

Some methods have been investigated regarding the under-crack plate structure treatments for holding up or captivating crack growth [7- 12]. Methods of captivating crack propagation such as drilling a hole at the crack tip [7–9], and drilling holes around the crack tip [10– 12] have been proposed.

Murdani et al. [7] investigated the crack-growth arresting technique in the aluminum alloy, showing the effectiveness of additional holes following the drilling of holes at crack tips in retarding crack propagation. It was reported that crack propagation was postponed longer than by a stop-drilled hole alone. The research was then continued by investigating the stress concentrations at stop-drilled holes and additional holes [13]. The models used in the experiment are the plate under pure tension load. However, the plate in mechanical structures may experience various types of load.

In this study, the treatments to reduce stress concentration are applied to under-cracked cantilever aluminum plate structure to investigate the treatment's effectiveness in reducing the stress concentration. The treatments to reduce the stress concentration are applied in the form of the stop-drilled hole (stop hole) and additional hole (add hole). It is hoped that the effectiveness of providing the stop-drilled hole and additional hole on the under-crack cantilever plate structure will be demonstrated well.

2. Methods

The experimental set-up consists of several main components: cantilever aluminum plate as test specimens, fixed supports, strain gauge sensors, strain indicators, Arduino as simple DAQ, and personal computer (PC) are constructed and joined to measure and analyze the stress on crack plate. The set-up is presented in Figure 1. The aluminum plate cantilever dimension is 150 mm in length, 50 mm in width, and 2 mm in thickness.

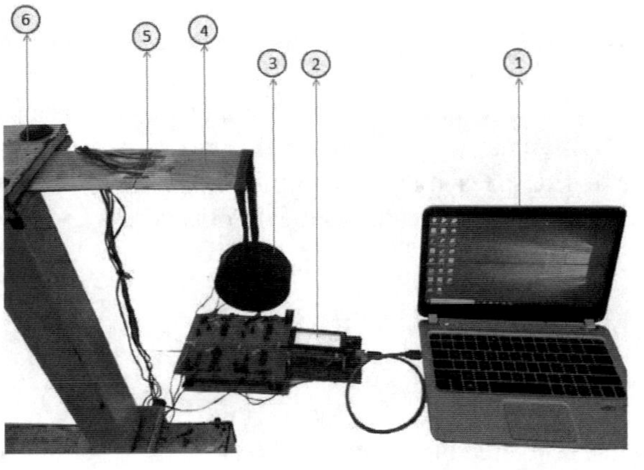

1. Personal Computer
2. Strain Indicator
3. Load
4. Specimen Test
5. Strain-Gauge
6. Fixed Support

Figure 1. Experimental set-up or stress measurement and analysis

The strain gauge sensor used in this research is BF 350-3 AA type, with specifications that can be seen in Table 1. The strain gauge is affixed to the test specimen's surface using an adhesive (glue) made from cyanoacrylate. The strain gauge used has a grid array type, single linear pattern. The strain gauge is connected by two jumper wires soldered to the solder tabs. The jumper cable from the strain gauge is connected to the strain indicator through the available ports.

Table 1. Specification of *Strain Gauge* Tipe BF 350-3 AA [14]

Specification	Value
Nominal resistance (Ω)	350
Tolerance of resistance	<± 0.1 %
Gauge factor	2.00 – 2.2
Strain limit	2.0 %
Fatigue life	>107
Gauge grid (mm) (l x w)	3.2 x 3.1
Gauge backing (matrix) (mm)	7.4 x 4.4

The resistance of the strain gauge changes when it is deformed. Metal and semiconductor materials experience changes in electrical resistance when exposed to strain. The magnitude of the change in resistance depends on strain gauge deformation, material, and strain gauge design. Strain gauge resistance is also a basic form for other transducers, such as load cells, pressure transducers, and torque meters [15].

The stress value's magnitude is determined using the relationship between stress and strain expressed through Hooke's law equation. The stress value from the measurement results is determined by [16]:

$$\sigma = E.\varepsilon \tag{1}$$

E is the material elastic modulus (GPa) value, and ε is the strain value from the strain gauge measurement results. The normal stress caused by bending moment theoretically is calculated using the following equation:

$$\sigma_{\max} = \frac{M.c}{I} \tag{2}$$

Where σ is the normal stress, M is the bending moment, c is the radius from the neutral axis to the surface, and I is the beam's inertia section.

The stress concentration will cause a difference in the value of stress at a point far from discontinuity with the value of the stress at the crack tip. The stress-value ratio is expressed by the stress concentration factor (Kt), as in equation 3. The stress at the crack tip is expressed as the maximum stress (σ_m), while the stress that is far from the defect is expressed as the nominal stress (σ_0) [16].

$$K_t = \frac{\sigma_m}{\sigma_0} \tag{3}$$

The flow chart of the experiments conducted in this research is presented in Figure 2. The experiments will be conducted in three different conditions.

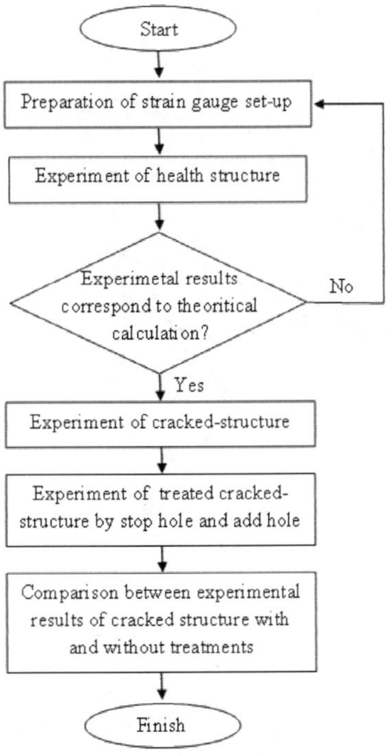

Figure 2. Flowchart of method used in the research

The first condition is the health condition (normal) where there is no damage on the aluminum plate to ensure the measurement system runs appropriately, where the experimental result will be compared with the theoretical calculation. The second condition is the crack's condition, where the single artificial crack is applied to the plate. These conditions are repeated with different sizes of a single crack. The third condition is the condition of crack with treatment. One-stop holes and additional holes are applied to the crack to reduce the stress concentration factor. The distribution of stress on the tip and area after the crack is measured to investigate the effect of treatment in reducing the stress concentration factor. The measurement positions for health and crack conditions as the basic experimental analysis can be seen in Figure 3. The measurement positions for stress distributions are shown in Figure 4.

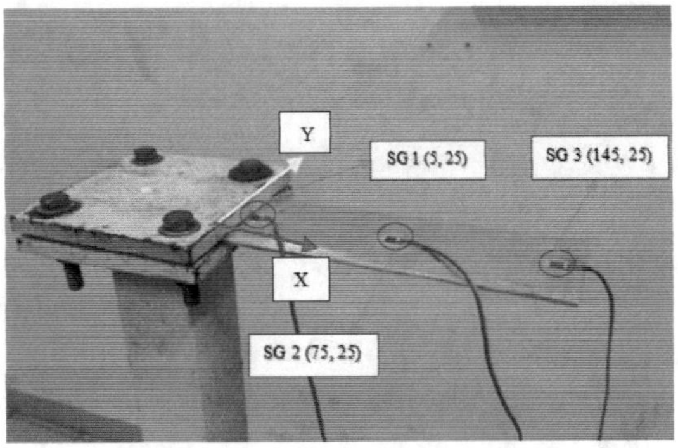

Figure 3 (a). Health condition

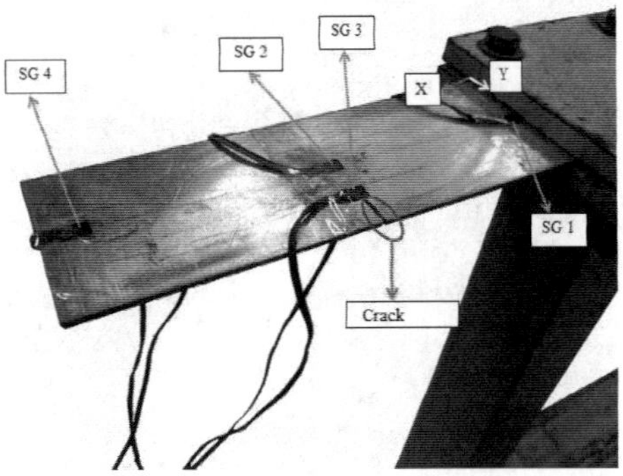

Figure 3 (b). Crack condition

IOP Conf. Series: Materials Science and Engineering 1062 (2021) 012046 doi:10.1088/1757-899X/1062/1/012046

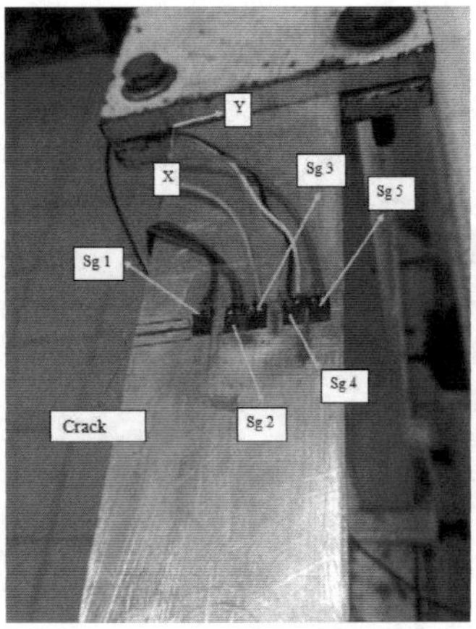

Figure 4 (a). Single crack

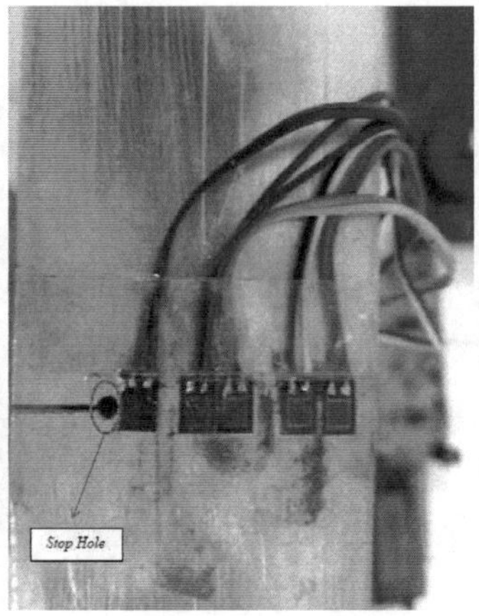

Figure 4 (b). Crack with a stop-drilled hole

Figure 4 (c). Crack with additional holes

3. Results and Discussion

3.1. Measurement in the normal condition

Table 2 shows the normal stress between theoretical calculation and experiment results in the condition without damage. The stress measurement using the strain gauge sensor again compared with the theoretical method. The percentage of errors are quite small, so the experiment results are still within reasonable limits. It means that the experiments by using the set-up explained in chapter 2 are considered to be ideal.

Table 2. Comparison of the normal stress between theoretical calculation and experiment result in the health condition

Measurement Points	Load of 6.7 N		*Error*	Load of 15 N		*Error*
	Stress (MPa)		(%)	Stress (MPa)		(%)
	Theories	Experiment		Theories	Experiment	
SG 1	28.71	29.627	3.09	63.73	66.427	4.06
SG 2	14.85	15.245	2.56	32.96	33.795	2.47
SG 3	0.99	0.942	5.32	2.197	2.116	3.8

3.2. Measurement in the crack condition

The stress measurements carried out in crack conditions are not much different from measurements in normal conditions. The difference only lies in the point of measurement. The measurement point is increased by 1 point for the crack condition, namely the critical area (close to the crack tip). The total point of measurement becomes 4 points. Figure 3 shows that the 4 points are SG 1 (5, 25), SG 2 (75, 25), SG 3 (75, 12.5), SG 4 (145, 25), where SG3 is the measurement point which closes to the crack tip. The sizes of the crack are varied in two forms for the 10 mm crack length. The first crack dimension is 10 mm in length (t) and 0.7 mm in width (a). The second one is t = 10 mm and a = 1 mm. The crack sizes for the length of 15 mm are also varied in two sizes, where the widths are 0.7 mm and 1 mm. The load of 6.7 N (in the form of the mass) is applied at the cantilever's end.

Tables 3 and 4 show the experimental results of Stress measurements in the condition of crack with the crack length of 10 mm and 15 mm, respectively. From the tables, it can be seen that the more sharpness and the more length of crack causes an increase in stress. The high values of stress are measured on SG1, where the maximum bending moment happens at this point. Compared to the crack position measurement, the stress values slightly different for the length of the crack 10 mm, but the stress is raised when the length of the crack extended to 15 mm.

Table 3. Experimental results of stress measurement with the crack length of 10 mm

Measurement Points	$a = 1$ mm	$a = 0.7$ mm
	Stress (MPa)	Stress (MPa)
SG 1	29.07	30.461
SG 2	17.46	17.72
SG 3	27.85	29.35
SG 4	0.95	0.98

Table 4. Experimental results of stress measurement with the crack length of 15 mm

Measurement Points	$a = 1$ mm	$a = 0.7$ mm
	Stress (MPa)	Stress (MPa)
SG 1	30.62	31.12
SG 2	19.89	20.00
SG 3	31.28	31.76
SG 4	1.01	1.02

Figure 5 compares stress measurement with the normal condition and crack condition with different crack sizes. From the graphs can be analyzed that the crack enhances the stress on the plate. The value of stress at crack position rockets and goes beyond the stress point with maximum bending moment.

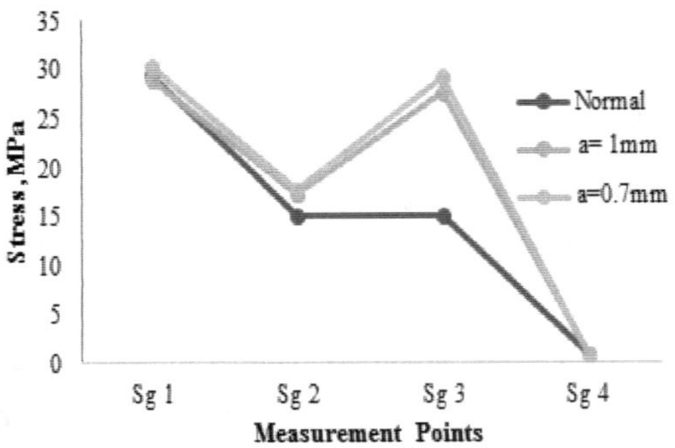

Figure 5 (a). Crack with t = 10mm

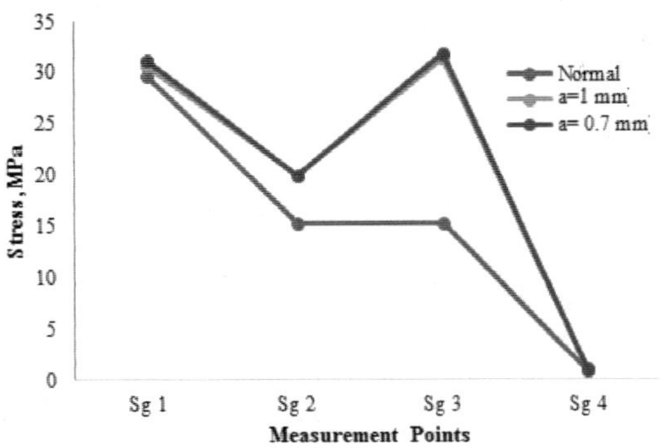

Figure 5 (b). Crack with t = 15mm

3.3. The stress distribution and stress concentration factor

In order to get the stress concentration factor of the cracking plate from the experimental data, the number of measurement points has to be enhanced. The addition of one measurement point is allocated to get the stress distribution under 6.7 N of the load. From the stress distribution, the maximum stress that occurs at the crack's tip, and the nominal stress can be obtained to calculate the stress concentration factor. For that purpose, one more measurement point is added, as can be seen in Figure 4. So the number of measurement points becomes 5 points.The measurement points with coordinate are SG1 (75, 17.5), SG2 (75, 25), SG3 (75, 30), SG4 (75, 40) and SG5 (75, 45). The SG1 position is very close to the crack tip, and the maximum stress is available at this point. The Measurement points of SG2-SG5 are used to get the nominal stress by the average of measured stress from SG2 to SG5.

Table 5 shows The measurement results of stress distribution at the position of the crack. The SG1 is the maximum stress (σ_m), and the average of SG2 to SG5 is the nominal stress (σ_0). In order to calculate the stress concentration factor (K_t), Table 5 can be simplified and completed with the stress data, which are needed to get stress concentration data. Tabel 6 shows the stress concentration factor due to the crack.

Table 5. The measurement results of stress distribution at the position of the crack

Length, *t*	Width, *a*	SG 1 (MPa)	SG 2 (MPa)	SG 3 (MPa)	SG 4 (MPa)	SG 5 (MPa)
10 mm	0.7 mm	29.35	18.81	17.72	17.22	16.93
	1 mm	27.85	19.95	17.46	17.02	16.97
15 mm	0.7 mm	31.76	20	17.73	16.2	12.53
	1 mm	31.28	19.89	18.11	17.13	15.88

Table 6. The stress concentration factor due to the crack

t/a	σ_{max} (MPa)	σ_{nom} (MPa)	*Kt*
10	27.85	18.23	1.5277
14.28	29.35	17.81	1.6479
15	31.28	17.75	1.7622
21.42	31.75	16.62	1.9103

In Table 6 can be seen that values of t/a are proportional to stress concentration factor Kt. Increasing the values of t/a will raise the stress concentration factor, where the rise of crack length and the reduction of crack width increase t/a. So, the sharper notch and the more extended crack size upgrade the stress concentration factor.

3.4. The crack condition with treatment

Table 7 shows the stress measurement result of the crack cantilever plate under the load of 6.7 N with the treatment of a stop-drilled hole and an additional hole. It is shown that both applied methods to reduce stress concentration factors are useful. The stop hole reduces the stress concentration factor (Kt) by 3.76%. By increasing the size of the hole, Kt is reduced by 4.41%. Applying the additional hole drops the value of Kt by 11.23%.

Table 7. Stress measurement results on the treated crack plate

Condition	*Stop hole*	*Add hole*	SG 1 (MPa)	SG 2 (MPa)	SG 3 (MPa)	SG 4 (MPa)	SG 5 (MPa)	Kt	% Kt Reduction
crack	-	-	31.76	20	17.74	16.2	12.53	1.911	-
t=15 mm,	3 mm	-	32.22	20.75	17.84	17.8	13.68	1.839	3.76
a=0.7	4 mm	-	32.36	20.98	17.94	17.98	13.97	1.826	4.41
mm	4 mm	4 mm	30.72	22.24	18.8	18.35	13.04	1.696	11.23

The reduction of the stress concentration factor in cantilever aluminum plate is not significant, but treatment could be a fast and straightforward solution in decreasing the stress on the cracking plate structure.

4. Conclusions

From the analysis that has been conducted for stress measurement and the effort to reduce stress concentration factor, It is shown that the proposed stress measurement method on cracking cantilever plate by using strain gauge and Arduino works well in measuring stress. The crack on the plate tends to propagate by the sign of high-stress value on the crack tip. The longer and sharper the crack, the higher the stress value on the crack tip, caused the rise in the stress concentration factor. For preventing the stress concentration factor magnifying, two kinds of treatment on the crack were conducted; stop-drilled hole and additional hole. The stop-drilled hole reduced the stress concentration factor by 4.4% and 11.23% for the additional hole. Even though the reduction of the factor is not significant, the treatment could be a fast and straightforward solution in decreasing the stress on the cracking plate structure.

References

[1] Huda F, Kajiwara I and Hosoya N 2014 *Journal of Sound and Vibration* **333** 3609–24
[2] Huda F, Kajiwara I and Hosoya N 2014 *Health Monitoring of Structural and Biological Systems*
[3] Huda F, Kajiwara I, Hosoya N and Kawamura S 2013 *Mechanical Systems and Signal Processing* **40** 589–604
[4] Huda F, Kajiwara I, Hosoya N and Kawamura S 2013 *Health Monitoring of Structural and Biological Systems*
[5] Huda F, Anggriawan A and Rusli M 2019 *IOP Conference Series: Materials Science and Engineering* **539** 012034
[6] Bur M, Okuma M, Malta J and Huda F 2010 Poros 8 4
[7] Murdani A, Makabe C, Saimoto A and Kondou R 2008 *Engineering Failure Analysis* **15** 302–10
[8] Domazet Ž 1996 *Engineering Failure Analysis* **3** 137–47
[9] Song P 2004 *International Journal of Fatigue* **26** 1333–9
[10] Thomas S, Mhaiskar M and Sethuraman R 2000 *Theoretical and Applied Fracture Mechanics* **33** 73–81
[11] Goto M, Miyagawa H and Nisitani H 2007 *Fatigue & Fracture of Engineering Materials & Structures* **19** 39–49
[12] Shin C 1996 *International Journal of Fatigue* 18 535–46
[13] Murdani A, Makabe C, Saimoto A, Irei Y and Miyazaki T 2008 *Engineering Failure Analysis* **15** 810–9
[14] AGS-TECH Inc. 2018 Coding System of Strain gauge.
[15] Figliola R S 2012 *Studyguide for theory and design for mechanical measurements by Richard S. Figliola, 5th edition* (Moorpark, CA: Content Technologies, Inc.)
[16] Callister W D and Rethwisch D G 2015 *Fundamentals of materials science and engineering: an integrated approach*, (Hoboken, NJ: Wiley)

Effect of Forging Temperature on Biodegradable Mg-0.7%Ca Alloy Properties for Implant Application

Is Prima Nanda[1], Zahran Ali[2], Nur Farah Bazilah Wakhi Anuar[3], Mohd Hasbullah Idris[2], Andril Arafat[4]
[1]Department of Mechanical Engineering, Faculty of Engineering Universitas Andalas, 25163 Padang, Sumatera Barat, Indonesia
[2]Department of Materials, Manufacturing and Industry Engineering, School of Mechanical Engineering, Faculty of Engineering, Universiti Teknologi Malaysia, 81310 Skudai, Johor, Malaysia
[3]Fakulti Teknologi Kejuruteraan Mekanikal dan Pembuatan, Universiti Teknikal Malaysia, 76100 Durian Tunggal, Melaka, Malaysia
[4]Department of Mechanical Engineering, Faculty of Engineering, Universitas Negeri Padang, 25131 Padang, Sumatera Barat, Indonesia

*Corresponding e-mail: isprimananda@yahoo.com

Abstract. The potential of binary Mg-Ca alloy as biodegradable material is considerable interest in implant application among researchers. This research was conducted to investigate the effect of different forging temperature and forging speed on the hardness, microstructure and corrosion rate of Mg-0.7%Ca. The experiment was established by preparing the alloy sample with 0.7%wt calcium content. The forging process was carried out under four different temperature variations of 140°C, 180°C, 220°C, and 260°C (±10°C) with two different speed;25 and 45 strokes per minute (spm). The samples microstructure was examined by optical microscope and scanning electron microscope (SEM) equipped with energy dispersive X-ray (EDX). The mechanical properties of the forged samples were measured in its hardness and plastic deformation ability along with samples cold-working percentage. The corrosion rate was determined by performing the electrochemical test in simulated body fluid. This research found that increases of forging temperature and forging speed provide a higher rate of recrystallization and Mg2Ca compound precipitation results in greater hardness, increase deformation and reduce the cold-working percentage. However, the investigated factors still led to a high corrosion rate compared to a previous study and consequently, reduce the feasibility of the alloy in implant application for biodegradable material.

Keywords: forging, biodegradable, implant, calcium, temperature, microstructure

1. Introduction

The research involves Mg-Ca binary alloys as biodegradable materials have attracted various investigation due to their biocompatibility feature and mechanical properties that exhibit similarities to natural bone [1]. As binary Mg-Al and Mg-RE (rare earth element) alloys are substantially harmful that could cause nerve toxicity due to Al, and hepatotoxicity for the latter [2][3], Mg-Ca alloy is favourable since it has a similar density to the bone and shown to be degraded within the bone and biocompatible in vivo and vitro [4][5][6]. Exclusive of the stated qualities, Mg-Ca binary alloy having less than 3% of calcium content is fairly an adequate threshold for implant application [4,3]. The forging process has been proposed with increasing the forging temperature to reduce the critical shear stress stored during cold-working [7]. Previous researchers found that increasing forging temperature on Mg-1%Ca alloy result in higher corrosion rate due to high Mg_2Ca precipitation thus lead to faster degradation of the implant before the bone healing process can be achieved [7]. In order to decrease the corrosion rate of Mg-Ca alloy, an investigation is conducted by reducing the Ca addition at 0.7%, decreasing forging temperature and speed than previous study [7] to improve the corrosion quality along with the mechanical properties of Mg-Ca alloy and thus, increase the feasibility of the alloy in implant application by forging process technique.

2. Experimental Procedure

2.1. *Material Fabrication*

The process on preparing the Mg-0.7%Ca alloy involves gravity casting whereby commercial purity magnesium ingot (99.9%) and Mg-30wt%Ca master alloy was melted in a mild steel crucible under the temperature of 670°C in an induction furnace protected with a continuous flow of argon gas to avoid oxidation. The molten alloy was then poured into a 300°C preheated simple steel mould. The samples were obtained from the middle of the cast product by the mechanical cutting process.

2.2. *Forging Process*

The forging process was performed using AIDA NC200 hot forged stamping machine (200 tonnes) whereas the speed was regulated at 25 and 45 strokes per minute (spm) under 1962kN constant force. The alloy was forged at the temperature of 140 °C, 180°C, 220°C and 260 °C (+10 °C). Forged samples were 61 x 22 x 26.5 mm.

2.3. *Microstructure Evaluation*

The microstructure of the alloys was characterise using X-ray diffractometer (CuKα radiation) for identification of the crystal structure of the phases. Optical microscope was used to characterise the surface morphology while the chemical composition was analysed energy diffractometry spectrum (EDS).

2.4. *Hardness Measurement*

The hardness of the samples was tested as-cast and as-forge using Matsuzawa DVK-2 material testing machine, at nine different locations of the samples according to ASTM-E98-82 (Standard test method for Vickers hardness of metallic materials 2003). The samples were cut in 10 x 15 x 15 mm dimension.

2.5. *Corrosion Test*

The corrosion of the forged samples was evaluated using a potentiostat/ galvanostat corrosion test system (Parstat-2263) according to ASTM G102-89. A saturated calomel electrode (SCE) and graphite were used as the reference and counter electrodes, respectively. 500 ml SBF was used as the test solution and the temperature of the test was maintained as 37°C during the experiment.

3. Results and Discussion

3.1. *Microstructure Characterisation*

The production of Mg-0.7%Ca was successfully prepared as described in its chemical composition in Table 1 with EDX analysis in Figure 1 whereas 0.7% of Ca is presented in the binary alloy composition. As shown in Figure 2 and 3, the grain size of the as-cast Mg-0.7%Ca is smaller compared to 99.9%Mg. The microstructure of the forged samples under different temperatures and speed are displayed in Figure 4 and 5. Both figures shows that better grain size reduction is observed at increasing forging temperature except at 220°C and 260°C whereby the grain starts to grow due to the forging temperatures are in the range of recrystallization or warm-working condition and hence, a low strain-hardening was subjected to the alloy creates lesser dislocation for blocking the growth of grains. Low forging speed, on the other hand, is also a factor to be considered whereas the grain, in which already had lesser dislocation, have more time and fewer blockage for the grain to grow along the surface. The microstructure also described twinning planes distribution in the grains. This indicates that twinning system is the dominant deformation factor instead of the slip system. Increasing in forging temperatures provide more driving force of recrystallization but, without sufficient forged speed, the grain size will grow larger due to low strain energy and deformation. Apart from that, precipitation of Mg_2Ca phase in α(Mg)primary grain and grain boundaries can be observed in both Figure 4 and 5 with most on higher forging speed. The SEM images in Figure 6 indicates that the precipitation increase with increasing on forging temperature and the speed as verified in Figure 7 on the identification of Mg_2Ca precipitation by EDS analysis. The precipitation was detected at $2\Theta = 33.797^{\circ}$ corresponding to (1 1 2) crystallographic planes in both as-cast and forged samples, as shown in the XRD pattern in Figure 8. The results show that the presence and formation of α(Mg) matrix and Mg_2Ca phases are in all samples, which mean that Ca is found in the form of Mg_2Ca compound. in the amount of Mg_2Ca phase in the alloy

Table 1. Chemical composition of Mg-0.7%Ca.

Element	Wt (%)	At (%)
MgK	99.29	99.61
CaK	0.71	0.39

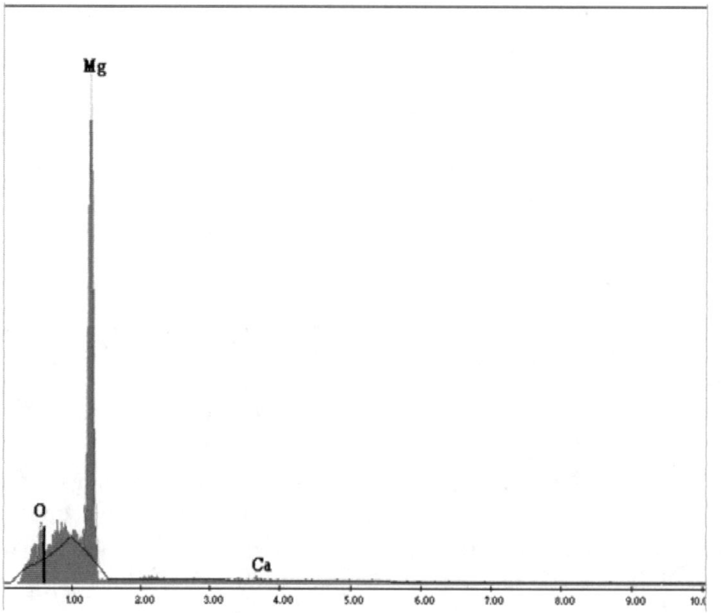

Figure 1. EDX analysis of Mg-0.7%Ca.

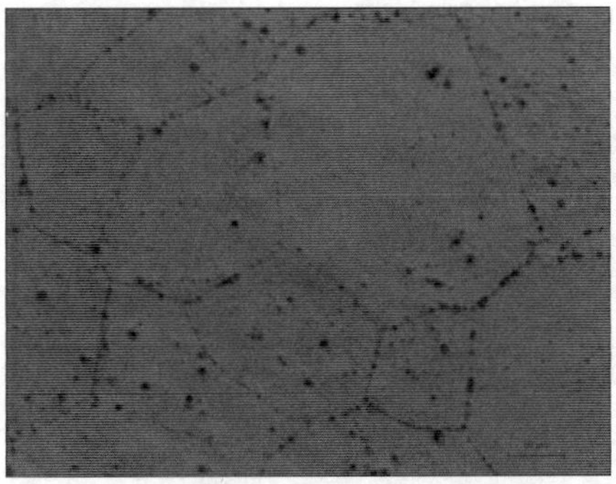

Figure 2. Microstructure of pure 99.9%Mg (50x magnification).

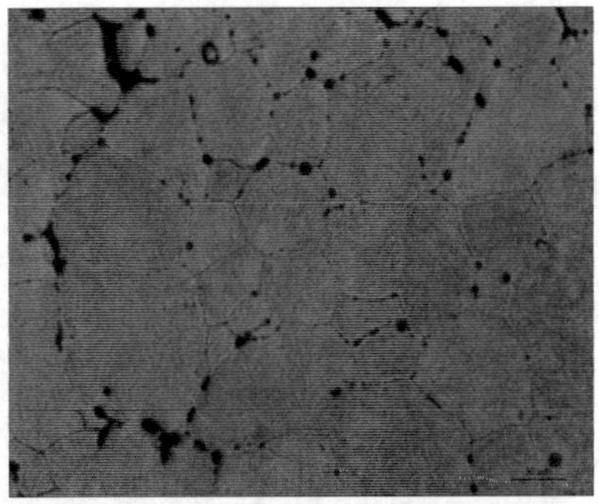

Figure 3. Microstructure of Mg-0.7%Ca (50x magnification).

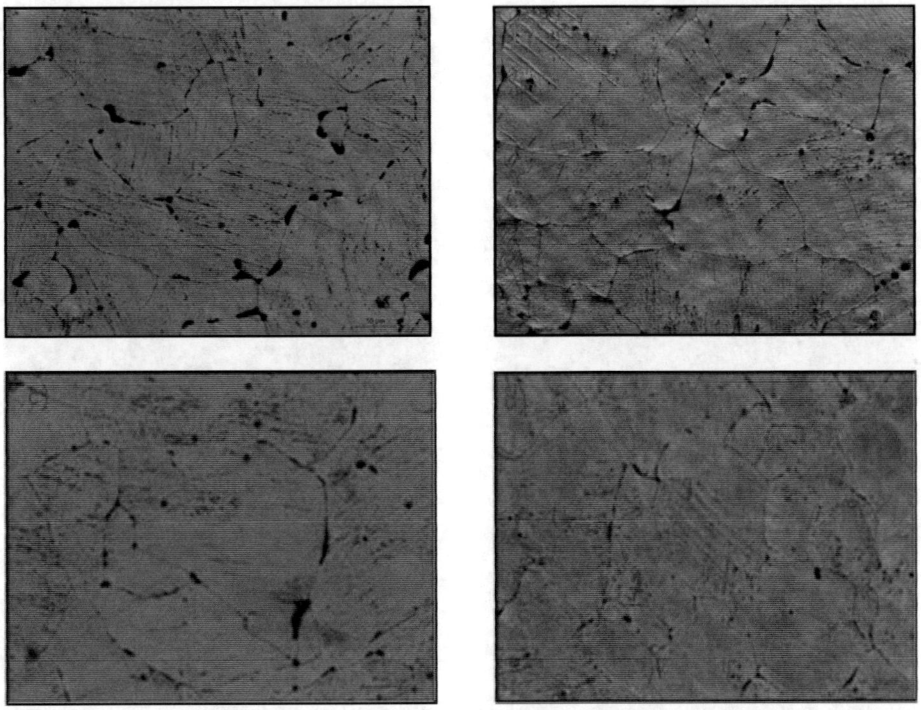

Figure 4. The microstructure of Mg-0.7%Ca alloy forged at (a) 140°C, (b) 180°C, (c) 220°C, and (d) 260°C at the speed of 25spm.

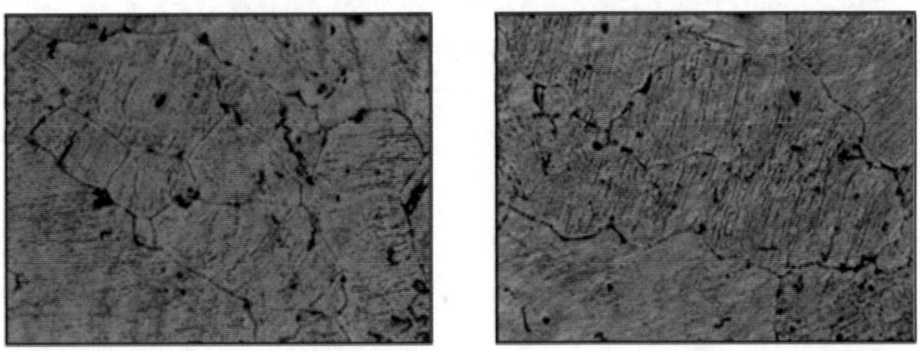

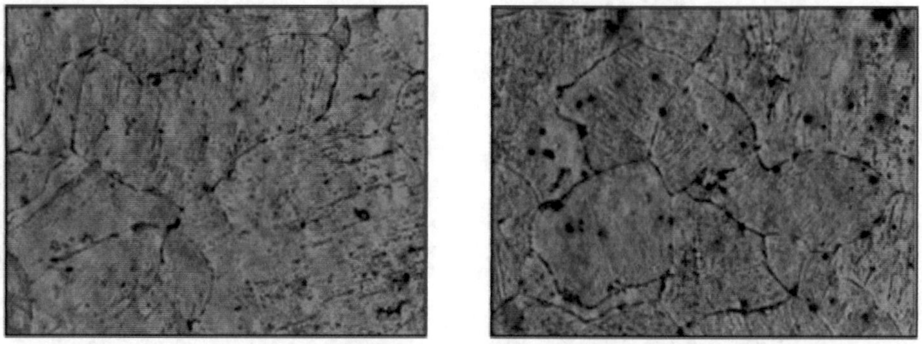

Figure 5. The microstructure of Mg-0.7%Ca alloy forged at (a) 140°C, (b) 180°C, (c) 220°C, and (d) 260°C at the speed of 45spm.

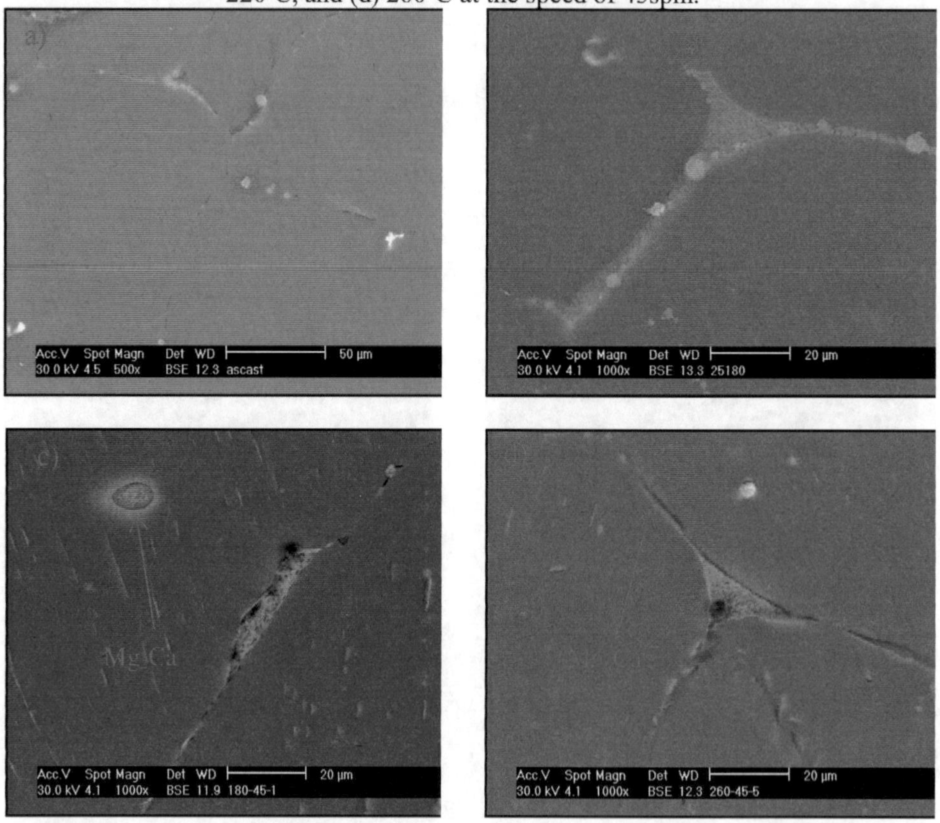

Figure 6. SEM images (a) as-cast, and forged samples at (b) 180°C (25spm), (c) 180°C (45spm) and (d) 260°C (45spm).

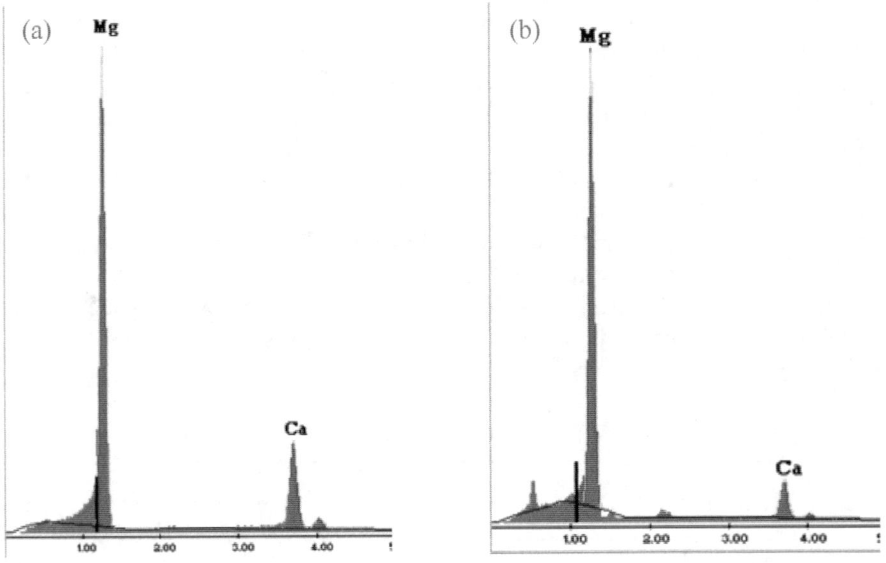

Figure 7. EDX analysis of Mg_2Ca at sample (a) 180°C at 45spm and (b) 260°C at 45spm.

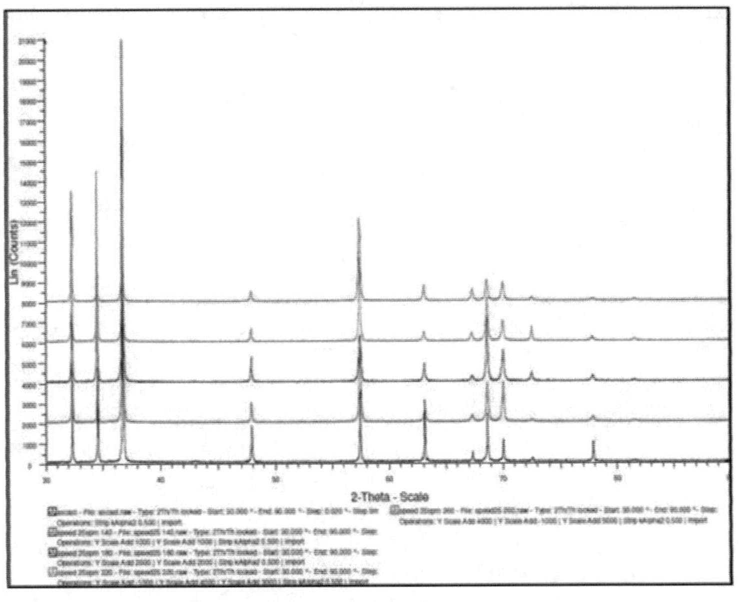

Figure 8. The XRD patterns of the as-cast and as-forged samples at different temperature for speed 25spm.

3.2. Mechanical properties

The hardness of 99.9%Mg sample was measured as 28.9HV while as-cast Mg-0.7%Ca alloy was at 36.2HV. Ca improves the mechanical property of Mg. Table 2 and Figure 9 show the Vickers hardness of forged samples under 25spm and 45spm, respectively for four types of temperatures. The hardness increasing when forged temperature and forged speed increase due to grain refining under recrystallization condition. However, there was a sudden trend drop in value, starting at 220°C until 260°C for both forging speeds. Higher forging temperature in which already in the recrystallization temperature range, decrease the hardness of samples as a result of Mg_2Ca migration into grain boundaries and dissolution in α(Mg) matrix. Therefore, grain size starts to grow faster after recrystallization phase. High forging speed and temperature affect the increase in the deformation of the binary alloy, as shown in Table 3 and graphically visualised in Figure 10. The pronounce of forging speed effects is clear since high force is necessary in cold-working in order to drive the dislocation movement. Cold-working percentage is decreased by the increasing of forging temperature and slower forging speed, as shown in Table 4 and Figure 11. The dislocation density in a metal increases with deformation, due to dislocation multiplication or the formation of new dislocations. As the dislocation density increases, the resistance to dislocation motion by other dislocations becomes more pronounced. Thus, the imposed stress necessary to deform a metal increases with increasing of cold-working. In the recrystallization phase, the mechanical properties increment during cold-working are restored results in weaker material but enhance ductility. Besides that, the rate of recrystallization increases when the cold-working percentage is higher.

Table 2. Vickers hardness of forged Mg-0.7%Ca.

Temperature (°C)	Hardness value (HV)	
	25spm	45spm
140	40.7	43.7
180	42.9	47.7
220	41.0	45.0
260	40.0	44.0

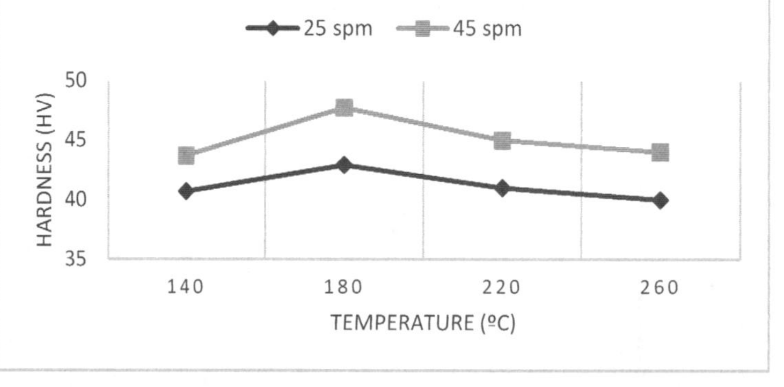

Figure 9. Graph of Vickers hardness of forged Mg-0.7%Ca versus temperature and forging speed.

Table 3. Deformation value of forged Mg-0.7%Ca.

Temperature (°C)	Deformation (mm)	
	25spm	45spm
140	23.30	23.79
180	23.50	24.00
220	23.80	24.30
260	24.20	24.60

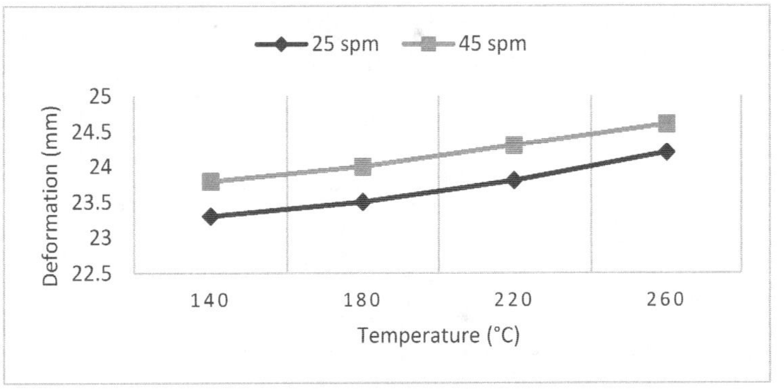

Figure 10. Graph of deformation of forged Mg-0.7%Ca versus temperature and forging speed.

Table 4. Cold-working percentage.

Temperature (°C)	Cold-working percentage (%)	
	25spm	45spm
140	2.80	1.25
180	2.25	0.79
220	1.20	0.81
260	0.50	0.41

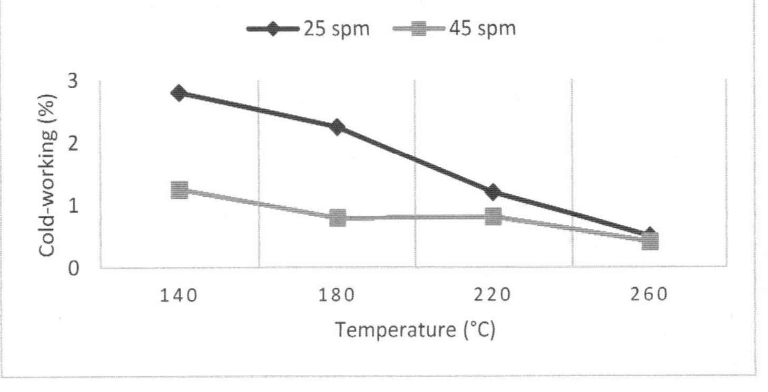

Figure 11. Graph of cold-working percentage versus temperature and forging speed.

3.3. Corrosion electrochemical testing

The potentio-dynamic polarisation curves of the as-cast Mg-0.7%Ca alloy and the forged samples under different temperatures for 25spm and 45spm of forging speed have been obtained and shown in Figure 12 for the latter. The derived current density (I_{corr}), corrosion potential (E_{corr}) and corrosion rate are given in Table 5 and graphically displayed in Figure 13. The corrosion rate is seen to be decreased when subjected to higher forging speed. The increasing in forging temperature increases the corrosion rate but sudden drop starting with 220°C for both forging speeds. However, at 260°C, corrosion rate spikes under 45spm forging speed meanwhile at 25spm, the trend still shows a drop. The precipitation of Mg_2Ca increase the corrosion rate [7] which is true for the samples measured at 180°C and 260°C, but both samples have different grain refinement whereby the samples forged at 180°C is harder due to smaller grain size whereas samples forged at 260°C experienced grain growth thus vulnerable for corrosion to take place.

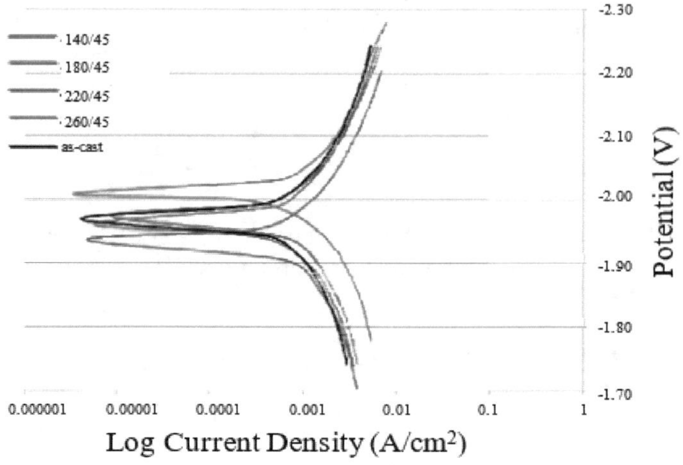

Figure 12. Tafel curves of as-cast and as-forged samples at speed of 45spm

Table 5. Chemical composition of Mg-0.7%Ca.

Sample	E_{corr} (mV)	I_{corr} (μA) x10^3	Corrosion rate (mmpy)
as-Cast	-1969.80	5.63	129
as-forged 140/25	-1978.27	6.53	149
as-forged 180/25	-2009.39	1.05	239
as-forged 220/25	-2010.31	7.70	176
as-forged 260/25	-1962.39	5.67	129
as-forged 140/45	-1974.05	5.27	120
as-forged 180/45	-1961.57	6.83	156
as-forged 220/45	-1937.60	5.92	135
as-forged 260/45	-2008.23	458.00	10500

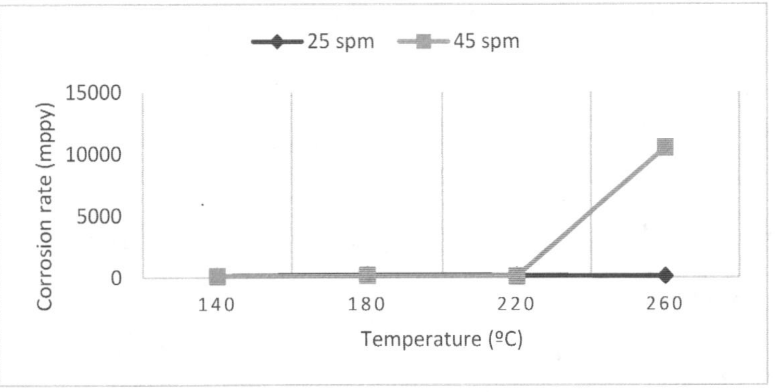

Figure 13. Graph of corrosion rate of forged Mg-0.7%Ca versus temperature and forging speed

4. Conclusion
The following conclusion can be drawn based on these research findings:
- Forging temperature increase the rate of recrystallization, which generates equiaxed and finer grains in Mg-0.7%Ca except when the temperature rose started from 220°C, grain size tends to grow and largely under low forging speed.
- High forging speed and temperature increase the precipitation of Mg_2Ca, which results in higher corrosion rate and deformation. However, the effect of both factors on hardness was different, whereby Mg-0.7%Ca forged at 180°C gives better hardness value than 260°C. This shows that forging speed and temperature are both significantly affect the alloy properties.
- Cold-working percentage is decreased when subjected to higher forging temperature and speed, thus improves material hardness.
- The corrosion rate value of Mg-0.7%Ca obtained in this research is not far from a previous study [7] on Mg-1%Ca although the forging temperature and speed used are lower. This shows that the forging process does not provide the intended corrosion quality for this alloy.

Acknowledgments
The authors wish to place their sincere thanks to Department of Mechanical Engineering, Faculty of Engineering, Universitas Andalas for technical and financial support (Hibah Penelitian UNAND 2020), to Department of Materials, Manufacturing and Industry Engineering, School of Mechanical Engineering, Faculty of Engineering, Universiti Teknologi Malaysia (UTM), and again to Universiti Teknikal Malaysia (UTeM) and to Department of Mechanical Engineering, Faculty of Engineering, Universitas Negeri Padang, and all the staffs involved for the value provided support during this work.

References
[1] Wang H X, Guan S K, Wang X, Ren C X and Wang L G 2010 *Acta Biomater* **6(5)** 1743–1748
[2] Zhang S *et al* 2010 *Acta Biomater* **6(2)** 626–640
[3] Zhang E and Yang L 2008 *Mater. Sci. Eng. A* **497(1–2)** 111–118
[4] Li Z, Gu X, Lou S and Zheng Y 2008 *Biomaterials* **29(10)** 1329–1344
[5] Ilich J Z and Kerstetter J E 2013 *Am. J. Nutr.* **19(6)** 37–41
[6] Serre C M, Papillard M, Chavassieux P, Voegel J C and Boivin G 1998 *J. Biomed. Mater. Res.* **42(4)** 626–633
[7] Harandi S E, Idris M H and Jafari H 2011 *Mater. Des.* **32(5)** 2596–2603
[8] Mirshahi M, Harandi S E, Idris M H, Jafari H and Koleyni S 2012 *Adv. Mater. Res.* **(445)** 301–306

Validation of Modal Analysis of Two-Storey Reinforced Concrete Frame

I F Azmi*[1], N H Hamid[2], W I I Wan Iskandar Mirza[3], M N Abdul Rani[3] and M A Yunus[3]

[1]Faculty of Civil Engineering, Universiti Teknologi MARA, 40450 Shah Alam, Selangor, Malaysia.
[2]Institute of Infrastructure Engineering, Sustainable and Management, Universiti Teknologi MARA, 40450 Shah Alam, Selangor, Malaysia.
[3]Structural Dynamics Analysis & Validation (SDAV), Faculty of Mechanical Engineering, Univesiti Teknologi MARA (UiTM), 40450 Shah Alam, Selangor, Malaysia

*Corresponding e-mail: ilina_azmi@yahoo.com

Abstract. The aim of this paper is to perform experimental modal analysis of two-storey reinforced concrete frame which designed using BS 8110 without seismic load and validate the experimental result with model using finite element analysis software. Experimental Modal Analysis (EMA) is the process of determining the modal parameters such as natural frequencies, damping factors, modal vectors and mode shape of a structure from a set of frequency response functions. The experimental work was conducted on half-scale of two-storey reinforced concrete frame at Heavy Structural Laboratory. Moreover, the Impact Modal Testing and response accelerometer are employed to determine the modes of vibration of a structure during experiments without imposed any in-plane lateral force. The experimental natural frequencies, which are extracted from measurements, exhibit good agreement with analytical model predictions. It shows that experimental modal analysis method can be usefully employed to investigate the dynamic behaviour of reinforced concrete frame due to the minimum percentage error between experimental modal analysis and computer software analysis.

Keywords. Modal Analysis, Natural Frequencies, Mode Shape, Reinforced Concrete Frame

1. Introduction

West Malaysia is considered as low seismic region and East Malaysia is categorized as moderate seismic region. However, Malaysia is located not far away from the active seismic fault zones known as Ring of Fire as shown in Figure 1 [1-2] . Most of the reinforced concrete (RC) buildings in Malaysia were designed in accordance with the BS8110 where not specification for seismic load. Past earthquake records showed that most of RC building which designed using non seismic code of practice proved that they cannot survive under moderate or strong earthquakes [3]. After experiencing several tremors originating from neighbouring countries especially from Sumatra, Indonesia and the 2015 Sabah Earthquake, Malaysians start to ask the question on the integrity of existing structures in Malaysia to withstand the earthquake load. In non-seismic code of practice such as BS8110, there are no specification on beam-column joint detailing, no ductility designed based on target displacement and only 10% nominal load specification for wind load in this standard [3-4]. The 2015 Sabah Earthquake had caused moderate and severe structural damage to a school, mosque and residential house. The total cost of damage after this disaster were around RM100 million which affected 61 buildings, 22 roads, 22 slopes and 200 families in Ranau and Kota Belud, Sabah [5]. However, these structural damages can be prevented or minimized if more attentions were focused on the structural health monitoring, repaired and retrofitting technique. As the consequences, structural health monitoring is an attractive filed to be explored especially in civil engineering especially for bridges, high rise buildings and other infrastructures due to ground motions [6].

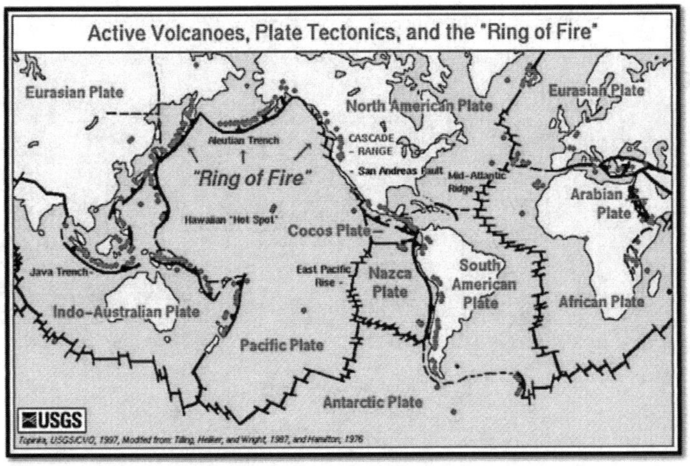

Figure 1. Active volcanoes, plate tectonics and the ring of fire around the world [1-2]

Experimental Modal Analysis (EMA) can be employed as a basis for health monitoring for structures and infrastructures under earthquake excitations. Experimental modal analysis is generally defined as the process of an effective mean for identifying, understanding, and simulating the dynamic characteristics and response of a structure through determining the modal parameters such as natural frequencies, mode shape and damping factors of a linear, time-invariant system [6-9]. Experimental work and analyses usually are very costly and long time consuming to be performed as compared to numerical simulations using the finite element method. However, it is found that the finite element results are frequently not in good agreement with experimental counterparts due to the invalid assumptions made in the finite element modelling. Therefore, the correct assumption and justification needed to minimize the percentage difference between experimental result and finite element analysis result. This is due to the fact that modal parameters of a structure have an important role on its behaviour under dynamic loading. A comparative study of the natural frequencies and mode shape of concrete frame structure designed using BS8110 is performed based on the measured by experimental and predicted results by using finite element analysis using Hyperworks program. Usually the structural damage of the structures depends on the natural frequencies of the buildings [10]. Thus, it is important to determine the natural frequencies of this two-storey building using Hyperworks program before conducting destructive test in Heavy Structural Laboratory, Faculty of Civil Engineering, Universiti Teknologi MARA, Shah Alam, Selangor.

2. Equipment and Instrumentation

In order to achieve the subsequent steps of experimental modal analysis, there are four types of equipment and instrumentation that going to use in the laboratory for conducting experimental work. Prior to testing, a half-scale two-storey reinforced concrete frame was designed using BS 8110 and constructed in the laboratory. The two-storey frame was placed on top of the foundation beam and bolted together to the strong floor. A special based isolation system with superior elastic performance rubber will be used as based isolation system.

2.1. *Impact Hammer*

Impact hammer equipped with a force transducer from PCB Electronics, which allowed the measurement of the applied impulse. The reference sensitivity is 2.73mV/N. A force sensor mounted in the head of the hammer transforms the input force pulse into a waveform that contains the necessary amplitude and phase information to completely describe the forcing function. Impact tip material stiffness helps determine the frequency content of the input forcing function by controlling the impact pulse duration. By defining the frequency and amplitude of the forcing function, impulse hammers present a fast, simple way to excite structures in a well-defined fashion. Figure 2 shows the impact hammer with the red head which can capture the amount of force exerted to the surface of concrete.

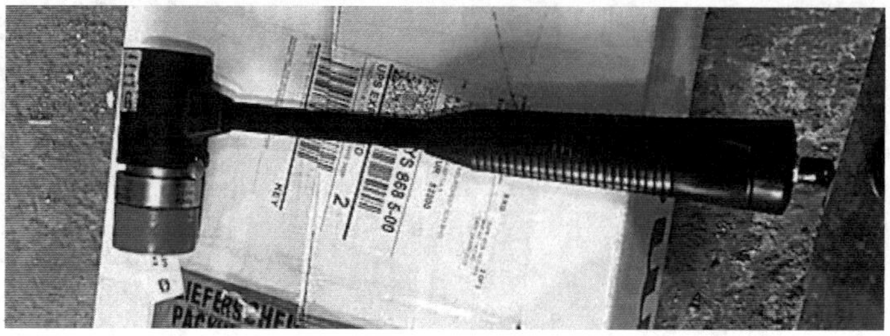

Figure 2. Impact Hammer

2.2. *Accelerometer*

Four accelerometers MMF (KS48C) model were used in the analysis, accelerometers with a sensitivity of about 104.23mV/ms-2 (1022.1mV/g) and frequency range between 1 Hz-100 Hz. One important characteristic of accelerometer is that it includes integrated amplifier that converts the movement of the seismic mass into acceleration signal in the form of voltage. In order to measure the acceleration response of the RC frame, steel anchor was bolted first at side of the concrete frame. Then, the accelerometers were screw to the steel anchor which attached to the two-storey reinforced concrete frame. Figure 3 shows the accelerometer MMF(KS48C) model attached to RC frame and connected to personal computer using special electric cable.

Figure 3. The accelerometer MMF(KS48C) model attached to RC frame

2.3 *Amplifier*

In experiment modal analysis, the amplifier is a necessary device to boost the very small electrical charges which are generated by the accelerometer into a signal. The signals were fed to the data acquisition card after filtering.

IOP Conf. Series: Materials Science and Engineering 1062 (2021) 012048 doi:10.1088/1757-899X/1062/1/012048

2.4 Personal Computer

Personal computer was used to measure the experimental data. Moreover, personal computer with signal processing software was also used to analyses the signals. A special program is used to transform from signal data to amplitude versus frequencies graph.

3. The Measurement Procedure

In this study, experimental modal analysis was performed on a half-scale two-storey reinforced concrete frame structure at the Heavy Structure Laboratory, Faculty of Civil Engineering, UiTM, Shah Alam, Selangor. The height of each floor is 1100mm with the size of the column is 200×200mm and beam size 300x200mm. The upper level of RC frame was constructed using reinforced concrete flat slab. The thickness of the slab is 200mm while the first-floor level is one bay frame without any infill of brick wall or masonry. Figure 4 shows a half-scale one-bay two-storey reinforced concrete frame with steel plate placed on top of foundation beam. The non-destructive test is conducted in order to determine the natural frequencies and mode shape of this frame.

Figure 4. One-half scale of two-storey reinforced concrete frame

Before conducting experimental modal analysis using impact hammer, four numbers of accelerometers were bolted to left hand side of RC frame as shown in Figure 5. The impact hammer with force sensor was applied to the concrete frame as shown in Figure 6. The vibration from the impact hammer was captured by four sensitive accelerometers which attached adjacent to the frame.

Figure 5. Four accelerometers placed at one side of the RC frame.

Figure 6. Impact hammer was exerted on the surface of the concrete

Figure 7 shows the numbers of nodes for 3D two-storey RC frame based on the guidance from the results of modal parameters of the frame obtained from the finite element analysis program. There are a total number of twelve nodes for two-storey RC frame with fixed support at four columns. Prior to perform the experimental work, there are several factors that contribute to the significant results such as the number of accelerometers used, measuring points and amount of force exerted to the surface of concrete.

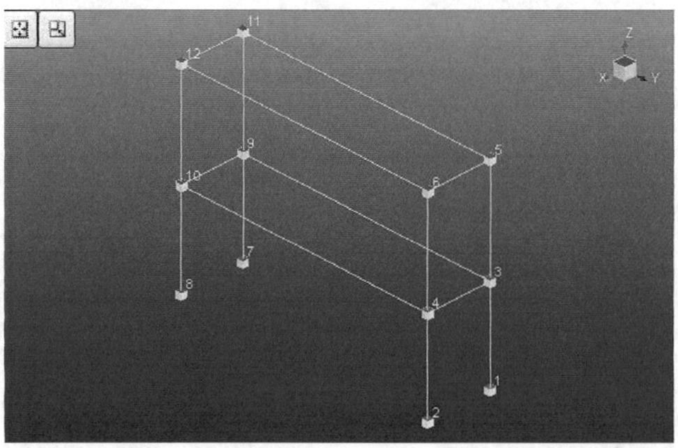

Figure 7. 12 number of nodes for two-storey RC frame

A total number of three significant mode shapes were detect based on the experimental result, which label as Mode Shape 1, Mode Shape 2 and Mode Shape 3 from the computer. The other mode shapes have very small peak and did not include in this analysis. Since only four accelerometers were placed at node 3, 4, 5 and 6, therefore only 4 nodes were involved in the measurement and validation process. The frequency bandwidth of interest was between 0 to 32Hz. In this experimental work, an impact hammer and accelerometers were used to measure the dynamic behavior of the frame structure.

Before the modal analysis experiment conducted in laboratory, the finite element analysis using HYPERWORK program must be perform. The natural frequencies and mode shapes of the finite

IOP Conf. Series: Materials Science and Engineering 1062 (2021) 012048 doi:10.1088/1757-899X/1062/1/012048

element model of the structure were predicted using normal modes analysis in which the model properties of the finite element models were defined as follows: Young's modulus = 30000MPa, Poisson's ratio = 0.2 and density = 2400 kg/m^3. This data input of the finite element defined based on the standard properties of the reinforced concrete. The support of the specimens was assigned as fixed support for total four numbers of columns.

4. Validation of Result and Discussion

Basically, there are three important mode shapes to be analysed, validate and discuss for 3D two-storey reinforced concrete frame. The dynamic properties for this type of buildings will be discussed in term of frequencies, lateral displacement, damping ratio and times with respective of each mode shape. Figure 8,9 and 10 shows the dynamic behaviour of the first, second and third mode shape which obtained from HYPERWORK program.

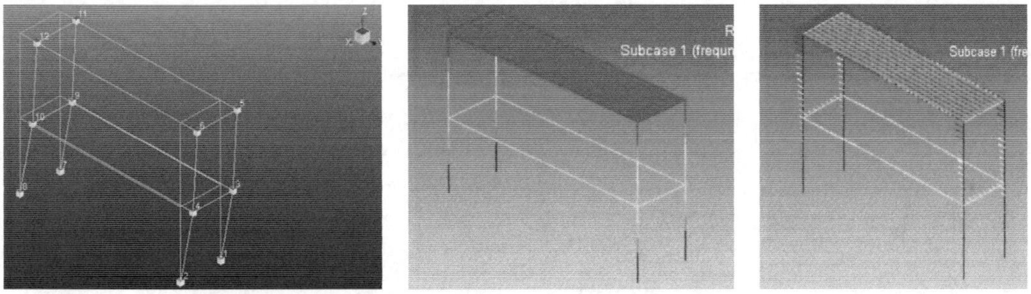

Figure 8. First Mode Shape of two-storey of RC frame

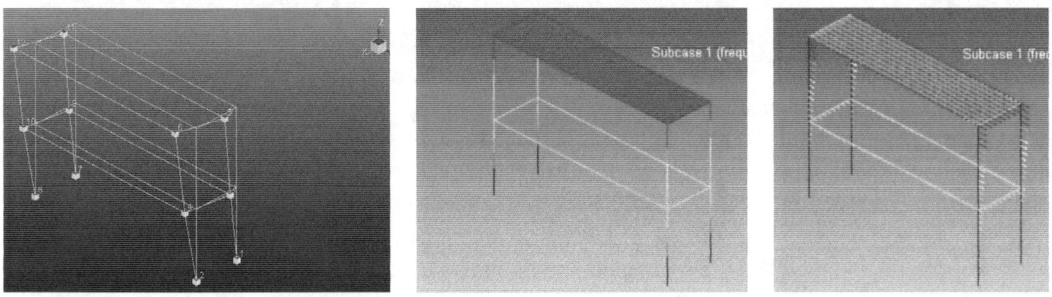

Figure 9. Second Mode Shape of two-storey of RC frame

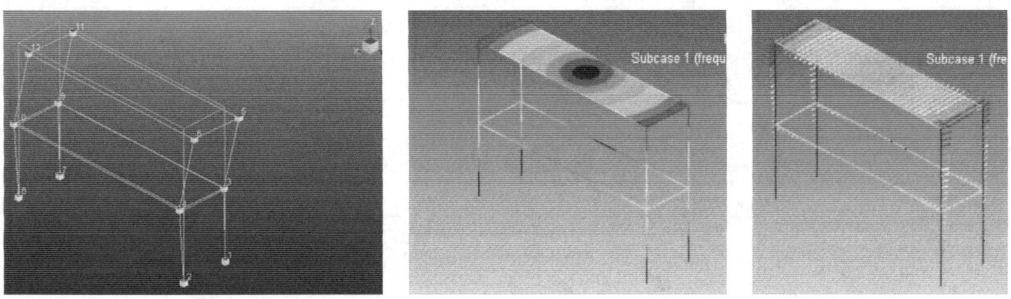

Figure 10. Third Mode Shape of two-storey of RC frame

Three experimental natural frequencies were match with analytical results in order to find the accuracy of the experimental modal analysis as shown in Table 1. The relative error between three

IOP Conf. Series: Materials Science and Engineering 1062 (2021) 012048 doi:10.1088/1757-899X/1062/1/012048

analytical and experiment nature frequencies reported. The percentage error for each mode shape is expressed using equation 1 as follows:

$$Percentage\ Error = \frac{Higher\ Frequency - Lowest\ Frequency}{Total\ Frequency} \times 100 \quad (1)$$

Table 1. Natural Frequency of the RC frame obtained from modal analysis experiment and HYPERWORK software

Mode Shape	Natural Frequency		% Error
	EMA (Hz)	FEA (Hz)	
1	7.94	11.8	19.6
2	14.49	12.8	6.2
3	19.26	19.6	0.9

As mentioned earlier, only three mode shapes are consider in this study. The first mode is what often defines the highest loads in a structure or how that structure will interact with the rest of the system around it when vibrating. For the first mode, the experimental natural frequency of the two-storey reinforced concrete building is 7.94Hz and 11.8Hz obtained from model natural frequency. The percentage difference error for first mode shape is 19.6%. The experimental natural frequency for second mode shape is 14.49Hz and 12.8Hz for model natural frequency with percentage error of 6.2%. Finally, the smallest difference percentage error for third mode shape is 0.9% where experimental and model natural frequencies are 19.26Hz and 19.6Hz, respectively. The comparison of the results clearly reveals that there is a good correlation between the experimental and computer analysis software natural frequencies with percentage of error maximum only 19.6%. Such a difference is quite acceptable due to the complex behavior of RC frame and characteristics of support idealization. More specifically, the finite element computer analysis software model assumes the reinforced concrete frame to be linear, elastic, homogenous and isotropic. Whereas in reality, concrete is heterogeneous and anisotropic material and the specimens inevitably contain imperfections in forms of crack, defects and steel bars misalignment.

Table 2 shows the maximum lateral displacement with respect each of mode shape at the top of the two-storey reinforced concrete frame which obtained from the finite element analysis program. The first mode shape in the modal frequencies is the most important parameters to be considered during earthquake excitations and it is the common failure mode which normally occurred during the disaster. The second and third mode shape not commonly occurred unless the structures did not design in accordance current seismic code of practice. The most common failure for these types of structures is known as soft-storey mechanism. The highest displacement of reinforced concrete two-storey frame is 0.02261mm which obtained from finite element analysis software addressing the first mode shape.

Table 2. Displacement of RC Frame obtained from the HYPERWORK software

Mode Shape	Displacement (mm)
1	0.02261
2	0.0003
3	0.0009

Table 3 shows the modal damping ratio of two-storey reinforced concrete frame with respect to mode shapes which obtained from experimental data. The first mode shape has 1.01% intrinsic damping ratio, second mode shape has 1.11% intrinsic damping ratio and the third mode has 1.64% intrinsic damping ratio. It can be clearly seen that as the percentage intrinsic damping ratio increases as the number of mode shape increases. The equivalent viscous damping comprises of radiation damping, intrinsic damping and hysteresis damping. The equivalent viscous damping for normal reinforced concrete building usually has the value of 5% and this value normally used to model the structures using nonlinear time history analysis. Table 4 shows the time taken to complete one cycle for first, second and third mode shape. From the frequency data, the time taken to complete one cycle or period can be determine for RC frame for mode shape 1, mode shape 2 and mode shape 3 using Equation 2 as follows.

$$Period, T = \frac{1}{frequency, f} \tag{2}$$

Table 3. Damping Ratio of RC Frame obtained from the experimental work

Mode Shape	Damping Ratio ξ %
1	1.01
2	1.11
3	1.64

Table 4. Time of RC Frame by calculation

Mode Shape	Time (s)
1	0.085
2	0.078
3	0.051

5. Conclusion

The main goal of this paper is to validate the natural frequencies between experiment modal analysis and modelling finite element program. From the modelling, the natural frequencies, mode shapes and modal damping of the reinforced concrete frame can be obtained and analysis based on different level of earthquake excitation. To judge the reliability of experimental results of RC frame, a comparison between experimental and finite element analysis natural frequencies had been made. The finite element analysis of natural frequencies was obtained using HYPERWORK program while experimental natural frequencies were determined from laboratory work. The differences in results are relatively small, hence within acceptable limits for the reasons mentioned in the discussion section above. Overall, the experimental work shows successful extraction of experimental modal parameters when the frame is excited by impact hammer. Such promising investigation can be deemed as the right step towards the description of dynamic behaviour of the structures.

Acknowledgement
The authors would like to thank Ministry of Higher Education (MOHE) of Malaysia, Putrajaya, Malaysia for awarding research grant under Fundamental Research Grants Scheme (FRGS) with File Reference No: 600-RMI/FRGS 5/3 (0091/2016) and RMI (Research Management Institute) for managing his research work. Gratitude and appreciation to the laboratory staff members for their invaluable assistance during the course of this experimental research work.

References

[1] Sivanandam H 2016 Another earthquake could strike Sabah *The Star Online*
[2] Mohamad I I, Yunus M Z M and Harith N S H 2018 *MATEC Web Conf.* **250** 01014
[3] Hamid N.H, Azmi I F, Tiong P L Y, Awang H 2019 *IOP Conf. Ser.: Earth Environ. Sci.* **365** 012012
[4] Hamid N H, Adiyanto M I and Mohamad M 2017 *ARPN J. Eng. App. Sci.* **12(22)** 6502

[5] Lee S 2015 Damage from Sabah quake costly *The Star Online* pp. 7–8
[6] Ghani K D A, Hamid N H A and Jaafar M S 2013 *Key Eng. Mater.* **594–595(Jan)** 390
[7] Patwari A U, Faris W F, Nurul Amin A K M and Loh S K 2009 *Adv. Acoust. Vib.* **2009** 508076
[8] Schwarz B and Richardson 2006 *Proc. Int. Modal Analysis Conf* (Missouri: USA) pp 1–9
[9] Izham M H M, Abdullah N A Z, Zahari S N and Sani M S M 2017 *MATEC Web of Conferences* 90 (01043)
[10] Zhao B, Xu Z, Kan X, Zhong J and Guo T 2016 *Shock and Vibration* **2016** 8194549

IOP Conf. Series: Materials Science and Engineering 1062 (2021) 012049 doi:10.1088/1757-899X/1062/1/012049

Buckling Analysis of Curve Stiffened Fuselage Panel of Very Light Jet Aircraft

Devi Chandra[1*], Eka Satria[1] and Nukman Yusoff [2]
[1] Department of Mechanical Engineering, Faculty of Engineering, Universitas Andalas, Padang, West Sumatera, Indonesia 25166
[2] Manufacturing System Integration, Department of Mechanical Engineering, Faculty of Engineering, University of Malaya, Kuala Lumpur 50603, Malaysia

Corresponding e-mail: devichandra@eng.unand.ac.id

Abstract. Buckling characteristic of curve stiffened fuselage panel of very light jet (VLJ) aircraft which use one of rivet or friction stir welding (FSW) joint on its stringer-frame-skin connections has been investigated by employing finite element methods. The linear and non-linear finite element analyses were used to determine the effects of stringer shapes and pitch on the buckling load of the fuselage panel. The buckling modes of all fuselage panels have also been observed. The material, boundary, and geometric non-linearity were applied on the fuselage panel in the analysis. Degradation of material elastic modulus around Heat Affected Zone (HAZ) due to friction stir welding processes has been taken into account. The results show that the buckling load of curve FSW stiffened fuselage panel slightly smaller than that curve rivet stiffened fuselage panel for all stringer shapes and stringer pitch. Besides, the curve stiffened fuselage panel with a smaller stringer pith has a higher critical buckling load and higher loading capacity. Both curve rivet and curve FSW stiffened fuselage panels with close hat stringer shape have a better loading capacity than with the open hat and the Z-stringer shape.

Keywords: Buckling analysis, curve rivet stiffened fuselage panel, curve FSW stiffened fuselage panel, finite element, very light jet aircraft

1. Introduction

To reduce manufacturing cost and weight of aircraft, the aircraft manufacturers have been introduced and developed the integrally stiffened panel (ISP) as an alternative fuselage structure for the conventional stiffened panel (CSP). The emergence of ISP to potentially replace CSP can be attributed to progress made in welding technology such as Friction Stir Welding (FSW) and Laser Beam Welding (LBW). FSW which was patented by The Welding Institute [1] in 1991 is a technique for material joining and processing in which a rotating tool develops sufficient friction heat as it is moved through the joint interface to soften (without melting) and "stir" the two interfaces together. It is a solid-state welding process that occurs below the solidus temperature of the materials being joined to produce a high strength joint with low residual stress and low distortion. FSW can be used to joining the high-strength aerospace aluminum alloys and other metallic alloys that are hard to be welded by conventional fusion welding [2,3]. It also can be applied to various types of joints like butt, lap, T-butt, and fillet joints. The lap joint of skin to the stringer of the fuselage panel can be made by rotating the pin tool moves over the stringer flange to be welded along its length. While the butt joint that connects one fuselage panel with the other is obtained by moving the rotating pin tool at the adjoining edge of the panel.

The strength and buckling characteristic of stiffened fuselage panels is an important aspect in the designing of fuselage structures. A stiffened fuselage panel may buckle in various modes depending on its dimensions, part geometries, part connections, and loading conditions. Usually, however, buckling occurs in stiffened panel structures when they are subjected to compressive stress. Buckling means that the deflection of the structure diverges from the linearly predicted approach suddenly or gradually leading to a lower loading capacity of the structure. Normally, buckling analysis is conducted to calculate the loading value at which the structures begin to buckles. It can be done in many ways such as theoretical, numerical, and experimental analyses. Theoretical analysis can only be used on simple structures under simple loading and boundary conditions. Theoretically, the buckling strength of the stiffened fuselage panel can be calculated by using equation (1) where ν is Poisson's ratio, E is the elastic modulus of the material, t is the thickness of the skin, b is skin width, and k is compressive buckling load coefficient [4]. There are some simplifying assumptions when applying this equation, that is the distance between stringer rivet lines is skin width, the distance between frame rivet lines is skin length and simply-supported on all four edges of the panel.

$$\sigma_{skin} = \frac{\pi^2 . k . E}{12\,(1-\nu^2)} \left(\frac{t}{b}\right)^2 \quad (1)$$

For more complex structures such as curve stiffened fuselage panel, the numerical finite element, and experimental analysis have widely been used as the most effective approach in buckling analysis [5].

Many researchers have been conducted buckling analysis of aircraft components such as center fuselage [6], fuselage skin-frame [7], fuselage composite skin [8], spar panels [9]. Some investigations using the numerical and experimental method have also been done to determine and compare the strength and buckling characteristics of both ISP and CSP fuselage panels. ABAQUS finite element codes have been used to model the 2-stiffener and 3-stiffener types of FSW stiffened panel. It is reported that the presence of the weld on the FSW stiffened panel reduced the maximum buckling load due to the reduction in material strength in the weld zone. It depends on the sectional shape of the stringer [9]. Structural response of curve riveted panel and curve machined ISP panel under uniform axial compression loads have been investigated using PATRAN and MCS_MASC finite element code. The results show that the damage model of the integral panel is the stress of the skin exceeding the yield level, and the damage model of the build-up panel is buckling waves cross the stringers on an oblique angle, which is different from the integral panel [11]. Buckling analysis of FSW and rivet aluminum stiffened panels under axial compression load using NASTRAN finite element codes shows that differences between them can be ignored when considering imperfection-free panel and material heat affected zone level [12]. Analysis of the effects of stiffener shapes on buckling strength using PATRAN and NASTRAN software shows that fuselage stiffened panel with L and Z cross-section of the stiffener is a safe design under bending loads, instead, U shape stiffener [13]. The C-section and Z-section of the frame on curve composite stiffened fuselage panel has effects to buckling modes. The buckling of the C-section frame panel firstly occurs in the middle shape, then local buckling occurs in the middle of the frame, while for the Z-section frame panel only local buckling occurs in the middle of the frame [14]. Besides, an investigation on the effects of the curvature of the stiffened panel to buckling behavior shows that the higher the curvature, the higher the number of half-wave in the circumferential direction [15].

This paper presents numerical simulation results that describe the buckling characteristics of curve riveted and FSW stiffened fuselage panels subjected to uniform compression loads. Two configurations of curve stiffened panels based on its stringer pith with three types of stringer cross-sections (Z, open hat, and closed hat) each are numerically examined by finite element method. The panel geometry represents the fuselage panel structures of a very light jet aircraft (VLJ) that is approved for single-pilot operation, 4 to 8 passengers, and the maximum take-off weight less than 10.000 pounds [16].

2. Stiffened Fuselage Panel

The initial fuselage structure of the VLJ aircraft is shown in Figure 1. For simplification, only one-sixth of the fuselage structure is analyzed, called curve stiffened panel as shown in Figure 2. The schematic curve stiffened fuselage panel being analyzed as depicted in Figure 3 consists of three main parts namely skin, stringer, and frame. The overall dimensions of the panel are arc length of 840 mm, a width of 600 mm, and a radius of 800 mm. Both rivets stiffened fuselage panel as CSP and FSW stiffened fuselage panel as ISP have same dimensions and part configuration but different in joining method of the skin-stringer that is using rivets for riveted panel and using friction stir welds for FSW panel. The skin is stiffened by three longitudinal stringers and two C-section curve frames. The frames are riveted to the stringer by L-shape curve clips and riveted to the skin directly with a rivet pitch of 20 mm and a rivet diameter of 5 mm. The dimensions of all stringers, frames, and clips are shown in Figure 4. The geometrical data of configurations 1 and configuration 2 of both curve rivet and FSW stiffened panels to be analyzed are listed in Table 1 and Table 2 respectively. The stringer, frame, and clip are made of 2 mm thick aluminum alloy (AA) 7075-T6 and the skin panel is made of aluminum alloy (AA) 2024-T3 with a thickness of 1.6 mm. Typical mechanical properties for these material panels are presented in Table 3.

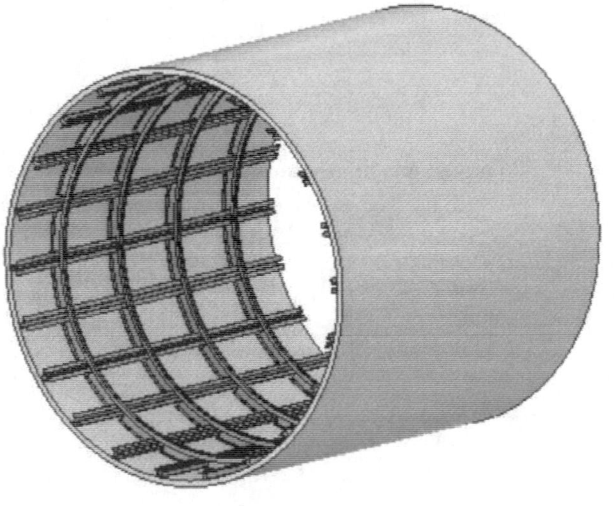

Figure 1. Fuselage structure of a VLJ.

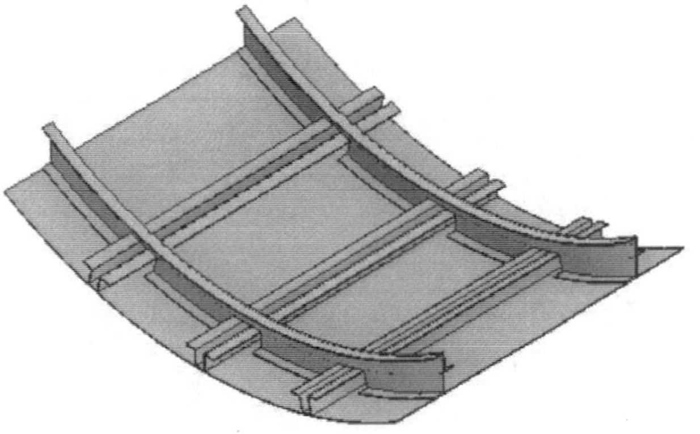

Figure 2. One-sixth fuselage structure of a VLJ.

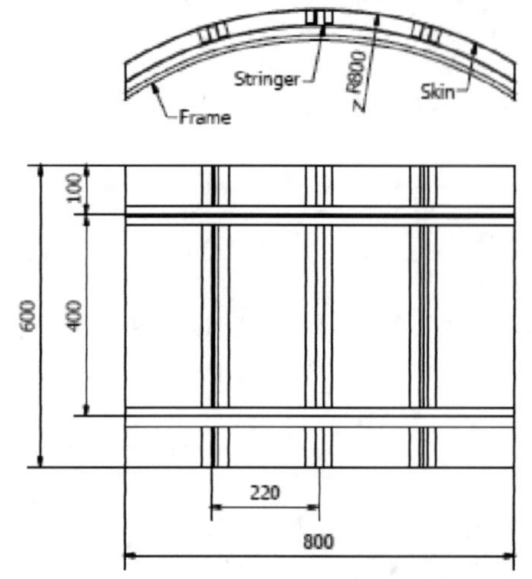

Figure 3. Overall dimensions of the panel.

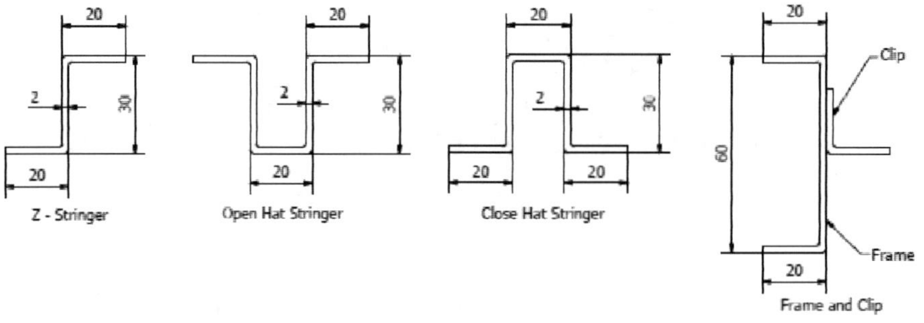

Figure 4. Geometry and dimensions of stringer, frame and clip.

Table 1. Geometry data of configuration 1 stiffened fuselage panel

Panel type	Rivet stiffened panel			FSW stiffened panel		
Stringer type	Z	Open Hat	Closed Hat	Z	Open Hat	Closed Hat
Stringer pitch (mm)	22	22	22	22	22	22
Frame pitch (mm)	40	40	40	40	40	40

Table 2. Geometry data of configuration 2 stiffened fuselage panel

Panel type	Rivet stiffened panel			FSW stiffened panel		
Stringer type	Z	Open Hat	Closed Hat	Z	Open Hat	Closed Hat
Stringer pitch (mm)	32	32	32	32	32	32
Frame pitch (mm)	40	40	40	40	40	40

Table 3. Material properties of fuselage panel

Part name	Material	Elastic Modulus, E (MPa)	Poisson ratio,ν
Skin	AA 2024-T3	73000	0.33
Stringer	AA 7050-T6	72000	0.33
C-Frame	AA 7050-T6	72000	0.33
L-Clip	AA 7050-T6	72000	0.33

3. Finite Element Model

Finite element model and non-linear structural analysis are generated and performed for skin-stringer-frame curve fuselage panels using ABAQUS version 6.10 software code [17]. All parts of the panels were modeled using shell elements with different properties for the un-welded zone and FSW zone elements. The material properties were modeled by an elastic-plastic constitutive model incorporating a Von Mises yield surface and an isotropic strain-hardening flow rule. The plastic strain data was entered as a Table of points for ABAQUS to use in the constitutive model for each material. For the FSW zone, material imperfections due to the welding process are taken into account by the degradation of the HAZ material elastic modulus on the upper flange of the stringer where the degradation is assumed to be 80% and the width of the HAZ is equal to the upper flange width [8].

An appropriate way for representing rivets in an ABAQUS model is the use of the FASTENER feature. FASTENER provides a kinematic connection between two nodes. These nodes are attached to the surfaces which have to be connected. BEAM section connector elements are used to define deformable fasteners. The element set containing the connector elements is referenced using the ELSET parameter on the FASTENER option. To idealize an FSW joint, many researchers have developed a method to model this joint in a finite element model by including the modeling of the degradation of material properties due to the welding process. This study adopts the rigid link multi-point constraints (MPC) approach to model the skin-stringer FSW lap joint [10-12] and the surface to surface CONTACT elements to model the contact between the un-welded skin and stringer flanges.

Element connectivity of the panel model is defined in the mesh module. The stiffened panel is discretized into a sufficient number of elements to allow for the free development of the buckling modes. The mesh density of the panel was chosen based on the size of the stiffener and rivet pith. A first-order curved quadrilateral 4-noded finite strain general-purpose shell element (S4R) that allows for changes in the thickness was selected along with the mesh illustrated in Figure 5. Physical constraints of the panel are defined with boundary conditions in the load module. The panels are fixed in all six degrees of freedom on one end and simply supported along the longitudinal edges. On the other end of the panel, the enforced displacement of 4 mm was exerted in the longitudinal direction U3.

The first stage in the simulation is a linear eigenvalue buckling analysis to determine the buckling mode shape and buckling stress. An incremental loading pattern was defined in the BUCKLE step where a subspace eigensolver is used to extract the linear buckling modes. The simulation includes large displacements, material plasticity properties, and contact relations between multiple deformable bodies such as skin-stringer joint, skin-frame joint, stringer-frame joint, and frame-clip joint. The second stage of the simulation is the non-linear buckling analysis to determine the static buckling load or buckling load-carrying capability of the panel. The modified RIKS method is used to obtain a solution since the problem under consideration is unsTable. The essence of this method is that it finds a single equilibrium path in a space defined by the nodal variables and the loading parameters, and simultaneously solves for displacements and loads. Since the load magnitude is an additional unknown variable, ABAQUS uses the "arc length" along the static equilibrium path which is given on the data line of the STATIC option to measure the progress of the solution [17].

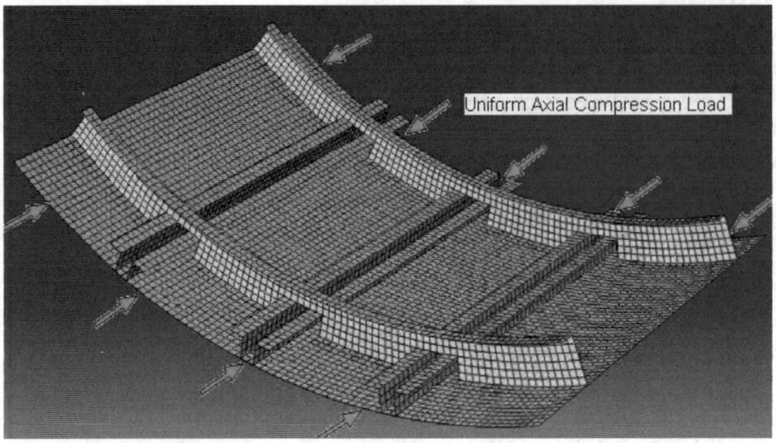

Figure 5. Finite element mesh and load direction for open hat stringer panel Configuration 1.

4. Results and Discussion

4.1. Linear buckling analysis

Eigenvalue buckling analysis was performed to estimate the critical buckling loads (buckling strength) and mode shapes of both FSW and rivet stiffened fuselage panels. About 30 mode shapes were observed for them. Figures 6 and 7 present the first and 19th buckling mode shapes of the rivet stiffened panel for Configuration 1 panel. It can be seen from Figures 6 and 7 that the earliest modes for closed hat stringer panel were local buckling at the skin between one longitudinal edge with adjacent stringer, and local buckling at all skins between stringer and frame for open hat and Z-stringer panel. The local modes were repeated from the first to 18th mode shapes. At the 19th mode shape, the closed hat stringers have different mode shapes compared to both open hat and Z-stringer panel where its mode shapes have three big waves at the skin between stringer along the longitudinal direction while the closed hat and Z-stringer panels have smaller waves along with longitudinal and lateral directions. These differences were possibly affected by the width of panel skin between stringers where the closed hat panel had smaller skin width.

Tables 4 and 5 compare all critical buckling loads of FSW and riveted stiffened panels for both configurations 1 and configuration 2 panel, where *R* in both Tables is the ratio of these two values. It can be observed that the critical buckling loads of these two panels for the same configuration are similar such that the differences of the buckling strength of curve FSW and rivet stiffened panels are negligible. However, the difference is significant when comparing Configuration 1 and Configuration 2 panels where Configuration 1 had the biggest critical buckling load compared to Configuration 2 panel. This is possibly caused by the effective width or stringer pith of the panel. Comparing the buckling eigenvalue analysis results, the closed hat stringer panel has the biggest buckling strength for both configurations while the open hat stringer panel has the smallest buckling strength.

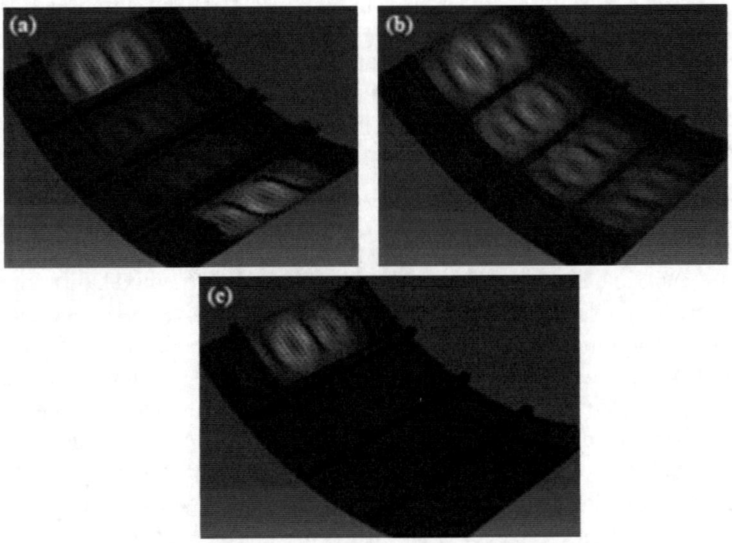

Figure 6. First mode shape of Rivet Stiffened Panel for Configuration 1; (a) open hat stringer, (b) Z-stringer and (c) closed hat stringer.

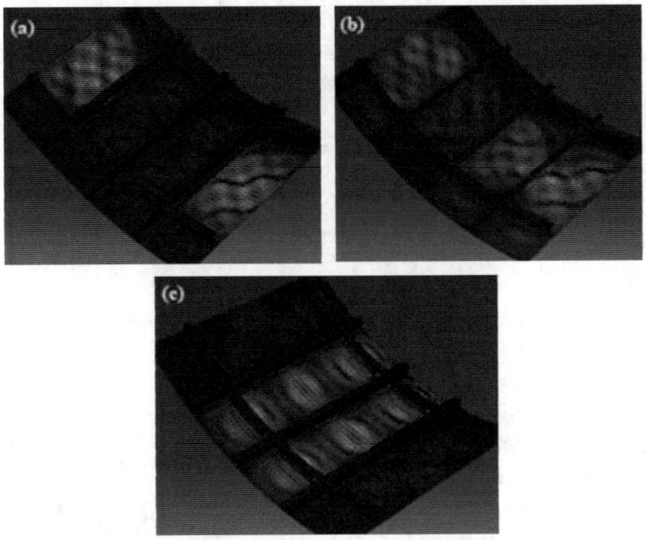

Figure 7. 19th mode shape of Rivet Stiffened Panel for Configuration 1; (a) open hat stringer, (b) Z-stringer and (c) closed hat stringer.

Table 4. Results of linear buckling analysis for Configuration 1.

Stringer type	Buckling strength (MPa)		R
	Rivet panel	FSW panel	
Closed Hat stringer	116	111	1.045
Open Hat stringer	96.7	94	1.029
Z-stringer	108	104.7	1.032

Table 5. Results of linear buckling analysis for Configuration 2.

Stringer type	Buckling strength (MPa)		R
	Rivet panel	FSW panel	
Closed Hat stringer	95.8	92	1.041
Open Hat stringer	87	84.6	1.028
Z-stringer	91.7	89	1.03

4.2. Non-Linear Buckling Analysis

The non-linear static buckling analysis that uses the modified RIKS method has produced results for the deformation and stress-strain response of the panel under compressive load. Normally it is reported in the form of a load versus end shortening graph for various failure modes of the panel as shown in Figure 8 for Configuration 1 panel. An examination of these Figures indicates that at the initial stage of loading or pre-buckling stage the applied stresses are uniform across the skin panel. As the compressive load is increased, the applied stresses become gradually non-uniform but remain symmetric about the stiffener. When the peak load level is approached, however, the portion of the skin near the stiffener carries more load. Since the stiffener was failing to retain its effectiveness, as evidenced by the reduction of load-bearing in the post-buckling range, the post-buckling behavior after panel buckling is much more ductile than in the case of tripping failure. Figure 8 also shows that the FSW panel was able to undergo a bigger strain before failure than the riveted panel especially for the closed hat and Z-stringer panels. Further details of finite element post-buckling solutions for all panels are given in Tables 6 and 7. Comparing the results of the analysis, it can be observed that the damage load of the riveted stiffened panel is higher than the FSW panel for both configurations, such that the riveted panel has better loading capabilities than the FSW panels. Comparing the results for stringer types, the damage load of the closed hat stringer panel is bigger than the open hat and Z-stringer panels.

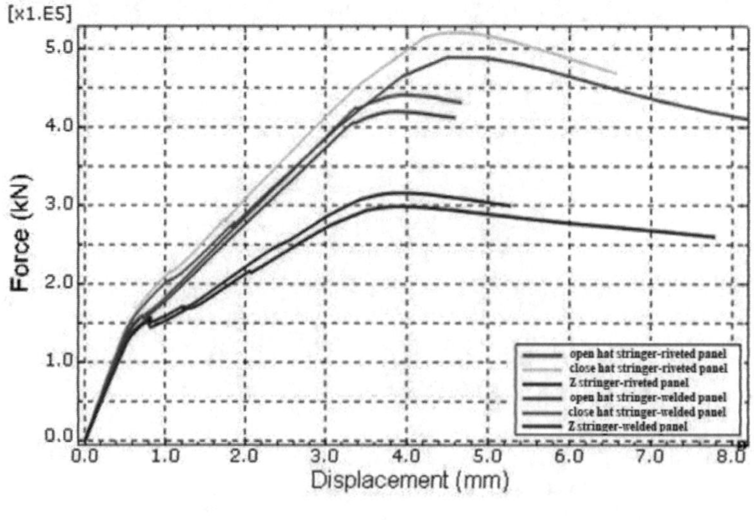

Figure 8. Load versus end shortening of Configuration 1 panel.

Table 6. Results for non-linear buckling analysis of Configuration 1 panel.

	Type of Stringer					
	Z		Open Hat		Closed Hat	
	RSP	FSW	RSP	FSW	RSP	FSW
Damage load (kN)	315.83	298	441	420	520.88	498
End shortening (mm)	3.75	3.9	4.0	3.8	4.6	5.0

Table 7. Results for non-linear buckling analysis of Configuration 2 panel.

	Type of Stringer					
	Z		Open Hat		Closed Hat	
	RSP	FSW	RSP	FSW	RSP	FSW
Damage load (kN)	277	261	429	409	515	471
End shortening (mm)	3.65	3.65	4	5	5	4.6

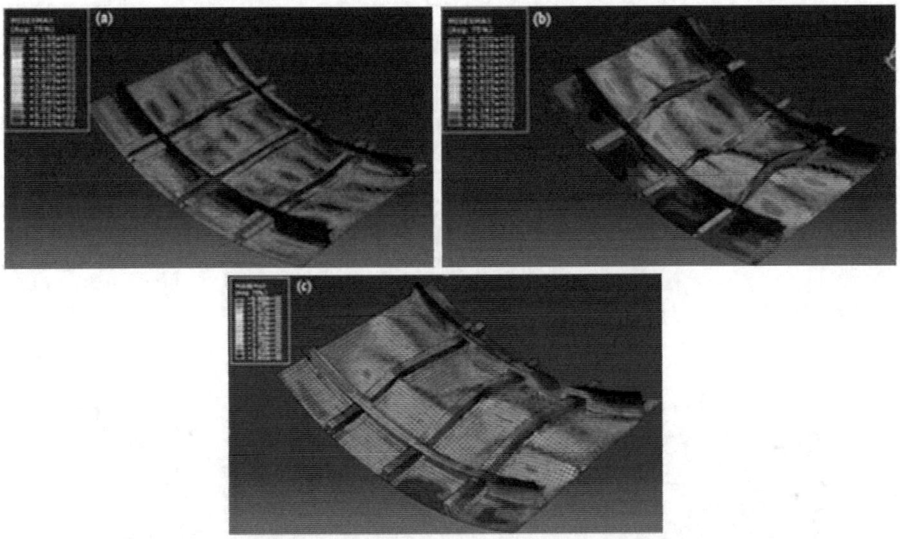

Figure 9. Failure shape of RIKS method for Rivet Panel Configuration 1; (a) open hat stringer, (b) Z-stringer and (c) closed hat stringer.

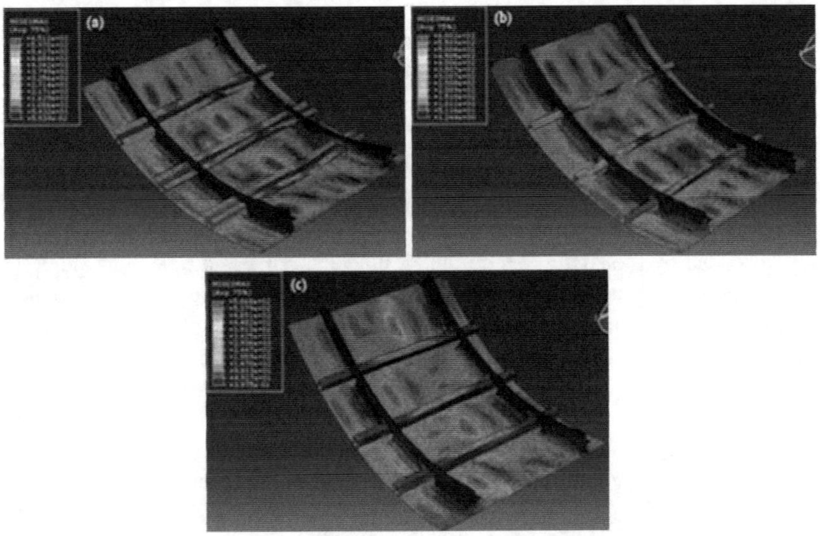

Figure 10. Failure shape of RIKS method for FSW panel Configuration 1; (a) open hat stringer, (b) Z-stringer and (c) closed hat stringer.

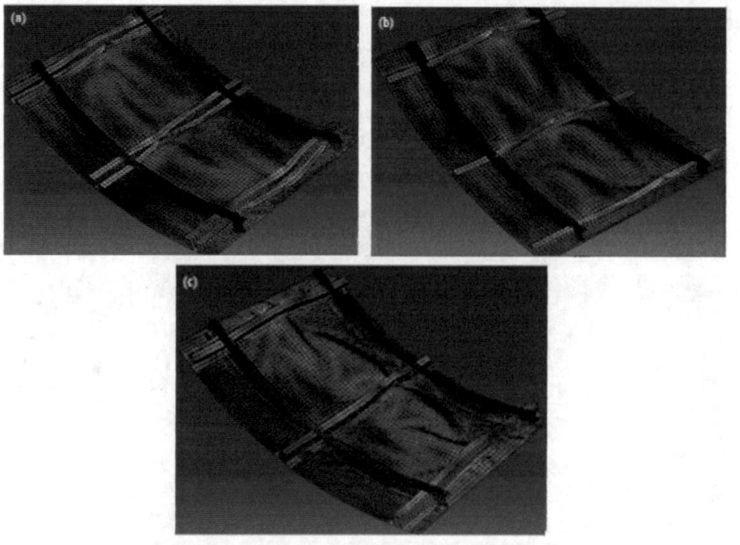

Figure 11. Failure shape of RIKS method for Rivet Panel Configuration 2; (a) open hat stringer, (b) Z-stringer and (c) closed hat stringer.

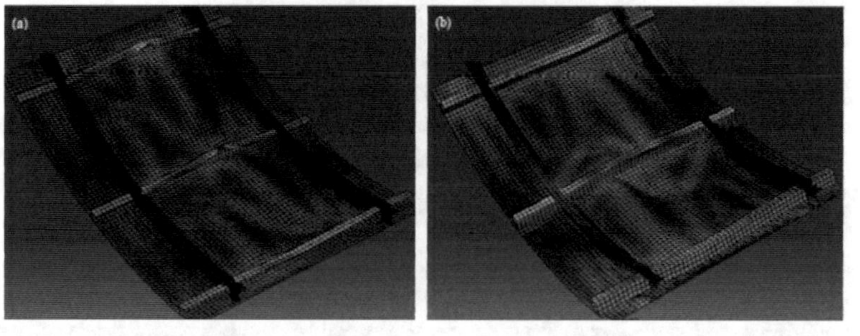

Figure 12. Failure shape of RIKS method for FSW panel Configuration 2; (a) Z-stringer and (b) closed hat stringer.

Figures 9 - 12 above show the failure modes of the curve rivet and FSW stiffened panels for Configuration 1 panels and Configuration 2 panels. For all panels, it was observed that at the initial load stage or small compressive load, the skin was able to absorb this load. However, when the load increased the skin begin to buckle and the load-carrying capacity of the skin became seriously weak. Thus, the load was mostly absorbed by the stringers and the skin near them. As the peak load level was achieved, the stringer stress rose to yield stress, tripping the edge of the stringer, and stringer bending will occur, and thus the panel could not stand the compressive load and the panel underwent total failure. This damage mode is similar to the result reported by [11] where for built-up curve stiffened fuselage panel, the damage modes are buckling waves cross the stringer on the oblique angle

5. Conclusions

The simulation work above aimed to study the buckling behavior of a curve rivet stiffened panel and curve FSW stiffened panel subjected to an axial load. Three types of stringer cross-sections with two configurations of stringer pitch were numerically tested utilizing a simplified finite element model. The results of the analysis can be concluded as listed below:

- Concerning critical buckling load, no significant differences between FSW and rivet stiffened fuselage panels were found when considering material HAZ elastic modulus degradation level of welding.

- The curve rivet stiffened fuselage panel has better loading capability than FSW stiffened fuselage panel.
- Comparing the stringer pith, configuration 1 fuselage panel (small stringer pith) has a higher critical buckling load and higher loading capacity than Configuration 2 fuselage panel.
- Comparing the stringer types, the closed hat stringer panel has better loading capacity than the open hat and Z-stringer panels for both configurations 1 and 2 of the fuselage panels.

Acknowledgments
The authors wish to thank the Faculty of Engineering Universitas Andalas for financial support through publication grant No. 054/UN16.09.D/PL/2020.

References
[1] Friction Stir Welding, FSW, Wayne Thomas and Solid State Welding - TWI (https://www.twi-global.com/what-we-do/research-and-technology/technologies/welding-joining-and-cutting/friction-welding/friction-stir-welding)
[2] Pettit R, Wang J and Toh C 2000 Validated Feasibility Study of Integrally Stiffened Metallic Fuselage Panels for Reducing Manufacturing Costs *NASA CR-2000-209342*
[3] Zahari S N, Sani M S M, Husain N A, Ishak M and Zaman I 2016 *J. Teknol.* **78** 57-65
[4] Timoshenko S P and Gere J M 1961 Theory of elastic stability, 2nd. ed. McGraw-Hill Book Company Inc.
[5] Li J, Hung K and Cen Z 2002 Shell element of Relative Degree of Freedom and its Application on Buckling Analysis of Thin Walled Structures *Thin Walled Struct* **40** 865–876
[6] Hussain S and Chandan R 2016 Static and Buckling Analysis of Center Fuselage Structure of a Transport Aircraft Through FEA Approach *International Research Journal of Engineering and Technology* **03(08)** 1354-1357
[7] Sreenivasa R, Kallesh S, Naveen K and Sandeep S 2016 Study the Effect of Buckling on Aircraft Fuselage Skin Panel with or without Airframe *International Journal of Scientific Development and Research* **1(6)** 346-353
[8] Clint J, Kumar S and Shaik N 2014 Buckling Analysis on Aircraft Fuselage Structure Skin. *International Journal of Engineering Development and Research* **2(4)** 3461-3474
[9] Ganeshprasad B, Mathew A, Janakare A and Prathima N 2019 Linear Static and Shear Buckling Analysis of Spar Panels *International Journal of Scientific and Research Publication* **9(3)** 252-257
[10] Yoon J, Bray G, Valente R and Childs T 2009 Buckling Analysis for an Integrally Stiffened Panel Structure with a Friction Stir Weld *Thin Walled Struct* **47** 1608–1622
[11] Weimin S, Mingbo T, Liang G and Dengke D 2008 Post-buckling Simulation of an Integral Aluminum Fuselage Panel Subjected to Axial Compression Load *Asia Simulation Conference 2008-7th International Conference on System Simulation and Scientific Computing* Chengdu
[12] Neto L, Monteiro F and Ruela H 2010 A Comparison of Buckling Performance of Rivet and Friction Stir Welding Stiffened Panels *Mecanica Computacional* **29** 4967-4975
[13] Rajan N, Raja B, Ramanan G and Sherin M 2018 Buckling Analysis of Stiffened Panel under Fuselage Bending *International Journal of Engineering & Technology* **7** 147-149
[14] Hao C, Nan Y, Peng Z and Lei L 2019 Research on Buckling and Post Buckling Characteristics of Composite Curved Stiffened Fuselage Panel under Hoop Bending Load. *IOP Conf. series: Materials Science and Engineering* 531
[15] Tran K, Douthe C, Sab K, Dallot J and Davaine L 2014 Buckling of Stiffened Fuselage Panels under Uniform Axial Compressions *Journal of Constructional Steel Research* **103** 140-147
[16] Very Light Jet – VLJ (https://www.globalsecurity.org/military/world/very-light-jet.htm)
[17] ABAQUS/Standard Version 6.9. Hibbitt, Karlsson and Sorenson, 2008

Filter Component Impact on EGS002-based Inverter Circuit Performances

Hardisal[1], R A Candra[1], D N Ilham[1], D P Saragi[2], R Harahap[2], S Suherman[2] and Z. Yahya[3*]
[1] Politeknik Aceh Selatan, Jl. Merdeka Komplek Reklamasi Pantai, Tapaktuan, Aceh, Indonesia
[2] Universitas Sumatera Utara, Department of Electrical Engineering, 2 Almamater Road, 20155 Medan, Indonesia
[3] Faculty of Computing and Multimedia, Kolej Universiti Poly-Tech MARA, Kuala Lumpur, Malaysia

*Corresponding e-mail: zahrah@kuptm.edu.my

Abstract. This paper examines filter component impact on inverter output. The inverter prototype was designed by using pulse width modulation (PWM) signal generated by EGS002 module, combined with DC voltage source, voltage regulator, MOSFET bridge and low pass filter. As component assembled, the output voltage quality is determined by the choice of filter components. Experiment shows that the calculated values may not suitable as soldering and connection imperfection may change component values. As adjustments were performed, filter components of 12.63 H and 6.5uf generated the finest sinusoidal output of 30.8 Vpp with frequency of 50,05 Hz.

Keywords: Inverter, EGS002 module, low pass filter, sinusoidal voltage

1. Introduction

Inverter compliments alternative energy sources such as refilled battery and solar cell. Inverter is often used in remote area where electric supply from the electric company does not exist. Inverter is also applied for stand-alone equipment such as water-pump, street-lighting and other equipment [1].

There are three types of inverter: square-wave, modified sinusoid and sinusoid inverters. The square-wave inverter has the lowest quality but is the lowest in price. Meanwhile, sinusoid inverter is expensive and complicated [2]. Current research has arrived in multilevel inverter [3] and application specific inverter [4].

This paper designed and implemented a sinusoid inverter to convert 12 V DC to 220 V AC by using EGS002 SPWM Driver Board. The EGS002 SPWM Driver Board is a pulse wave modulated (PWM) inverter driver which requires filter to obtain pure sine wave voltage. After designing and implementing the circuit, component adjustments were performed to obtain the expected voltage shape. The following sections discussed the inverter design and implementation, followed by the output signal analysis.

2. Inverter Design

In order to realize the expected inverter, the EGS002 SPWM Driver Board is used as the main component. The module employs PWM switching which compare the reference wave to a triangle carrier wave [5]. EGS002 uses an application specific integrated circuit (ASIC) CMOS EG8010 to

IOP Conf. Series: Materials Science and Engineering 1062 (2021) 012050 doi:10.1088/1757-899X/1062/1/012050

manage the circuit. Voltage and current protection, temperature sensor, fan control and indicators are performed by IR2110S. Frequency choices are selected via jumper. EGS002 has been improved from predecessor EGS001 by adding some protection schemes, and liquid display interfacing [6].

A full H bridge converter is used to change stable DC voltage to AC ripple. The bridge is made from some MOSFET that is controlled EGS002 to switch on and off. A low pass filter is added to reduce high frequency components of the bridge output. Inductor and capacitor combination [7] were employed to perform this task. Figure 1 shows the simplified diagram.

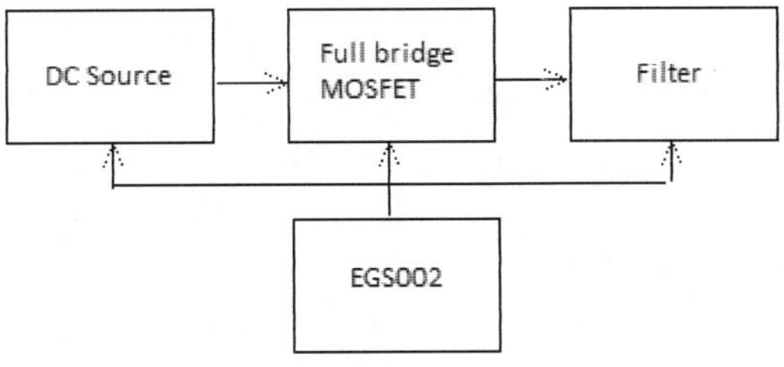

Figure 1. Inverter scheme.

In order to realize the inverter circuit, IC 7805 and 7812 were employed for voltage regulator to activate the IC. MOSFET 40N60C3 were employed for the full bridge. The full circuit is shown in Figure 3.

Measurement was performed by setting the circuit as in Figure 2, where output wave form was displayed in an oscilloscope. A 40 AH 12 V Battery was used as the DC voltage source, converted to 220 V AC by the inverter.

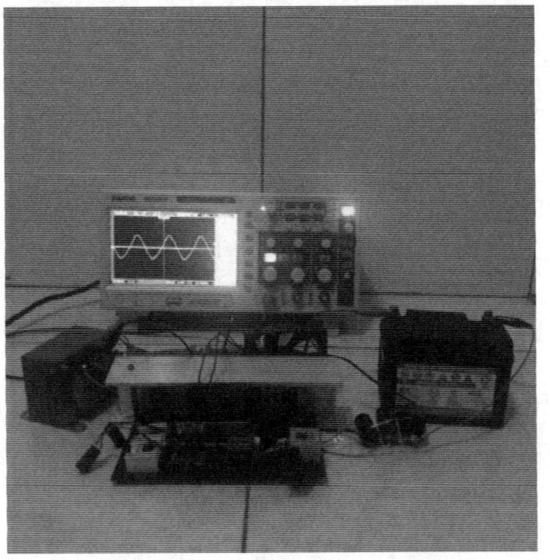

Figure 2. Measurement set up.

IOP Conf. Series: Materials Science and Engineering 1062 (2021) 012050 doi:10.1088/1757-899X/1062/1/012050

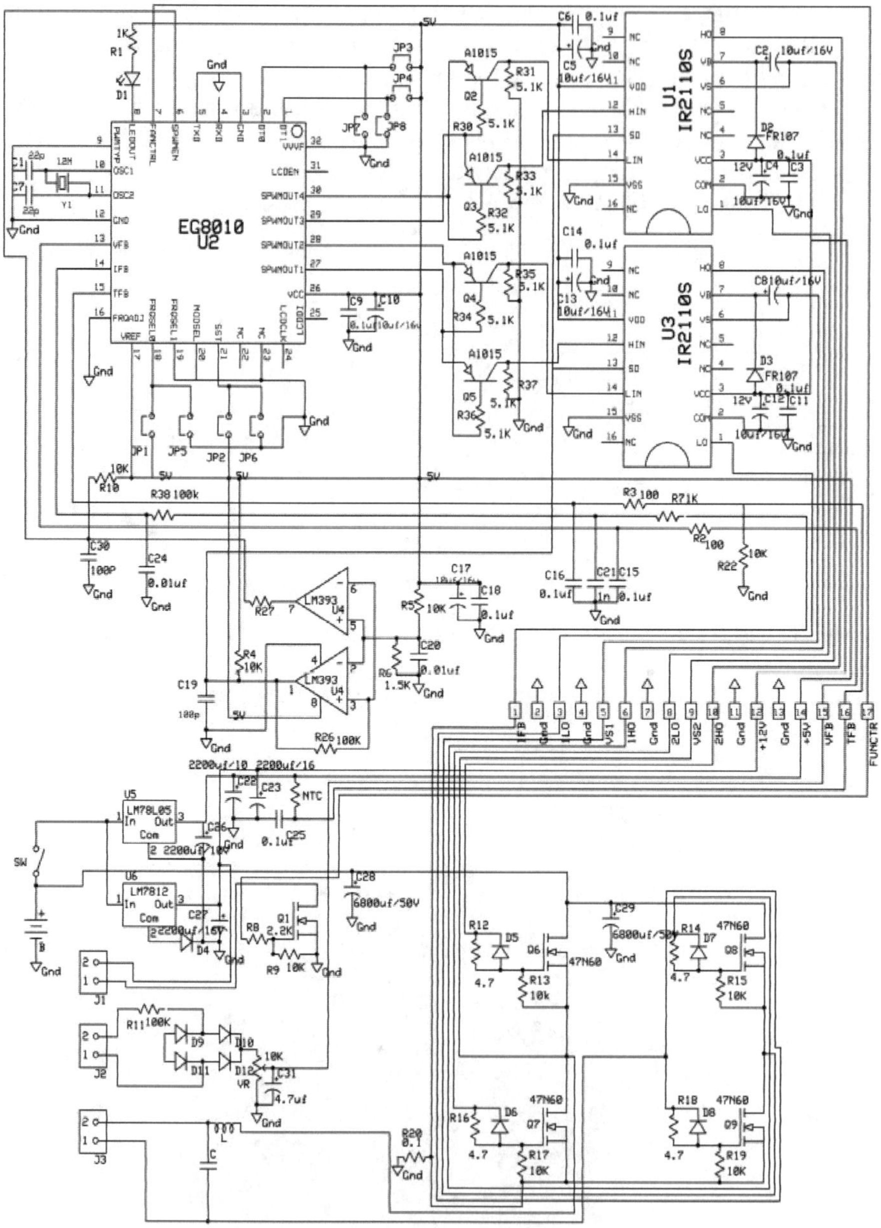

Figure 3. Inverter circuit.

3. Capacitor Impact to Inverter Performances

Since voltages to activate IC EGS002 and IR2110S influence the inverter performances, the regulated voltage should be ensured sufficient. Measurement showed that regulated voltages are 5 V and 12.1 V. The battery source was 12.8 V. The PWM voltage generated by MOSFET had frequency of 25 KHz, 23.6 Vpp, minimum voltage was -12.8 V and maximum voltage was 10.2 V.

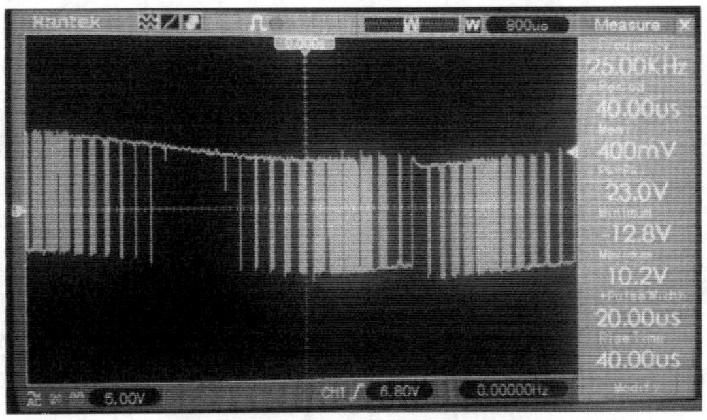

Figure 4. PWM output.

Filter circuit within Figure 2 determined the inverter performance. Figure 5 shows second order LC filter. Output input voltage comparison is given by Equation (1) and the resonance frequency in Equation (2):

$$\frac{V_{out}}{V_{in}} = \frac{1}{1 - \omega^2 LC} \tag{1}$$

$$f_c = \frac{1}{2\pi\sqrt{LC}} \tag{2}$$

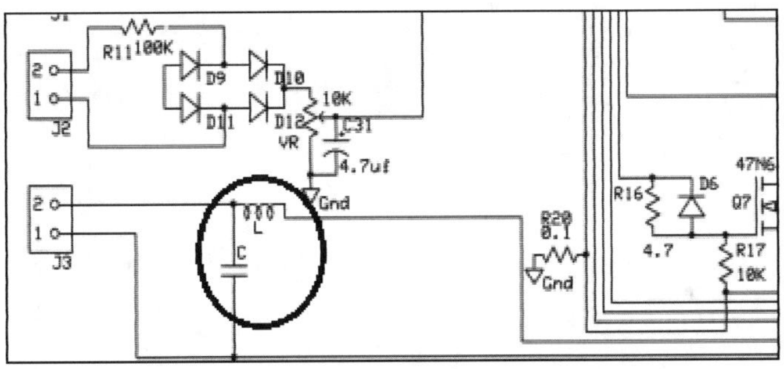

Figure 5. Second order LC filter.

The measurement produced various waveform as LC filter adjusted. Figure 6 shows the variations. The best performance was obtained when LC filter component: L=12.63 H and C=6.5 μf generating 30.8 Vpp (Figure 6d). At this frequency gives 17.57 Hz according to Equation (2). This frequency is far from 50 Hz. This may cause the imprecise component values. Suppose these component values were right, the bode plot was given by (OKAWA Electric Design, 2019) [8] plotted in Figure 7. The gain is about 35 dB at 17.57 Hz and 16.3 dB at 50 Hz. This means, if the L and C are selected so that f_c=50Hz, the output may reach 66.13 Vpp.

IOP Conf. Series: Materials Science and Engineering 1062 (2021) 012050 doi:10.1088/1757-899X/1062/1/012050

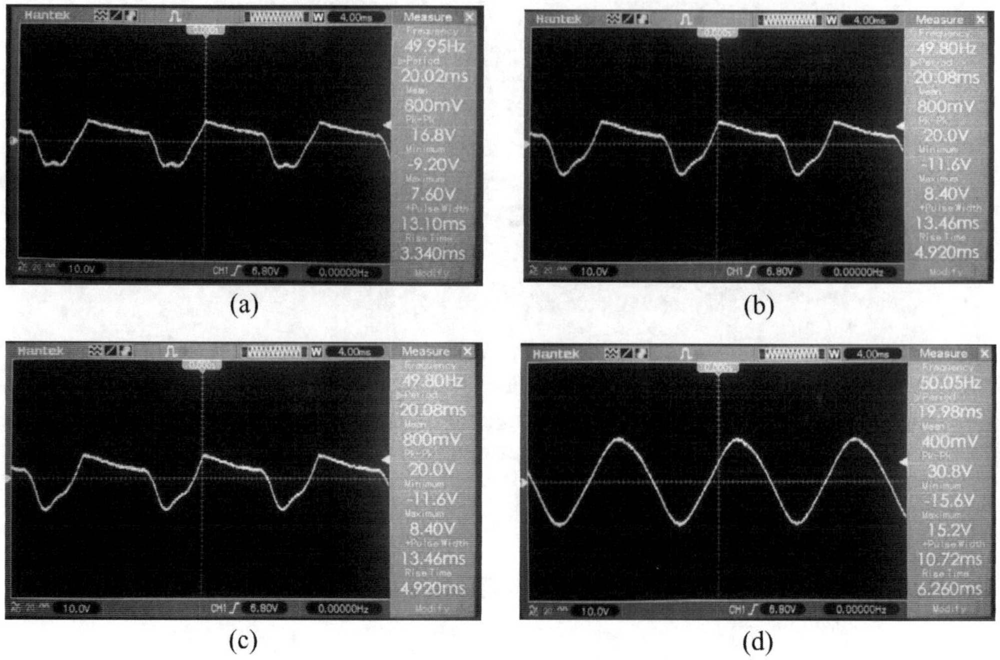

(a) (b)

(c) (d)

Figure 6. Inverter output to LC filter variations.

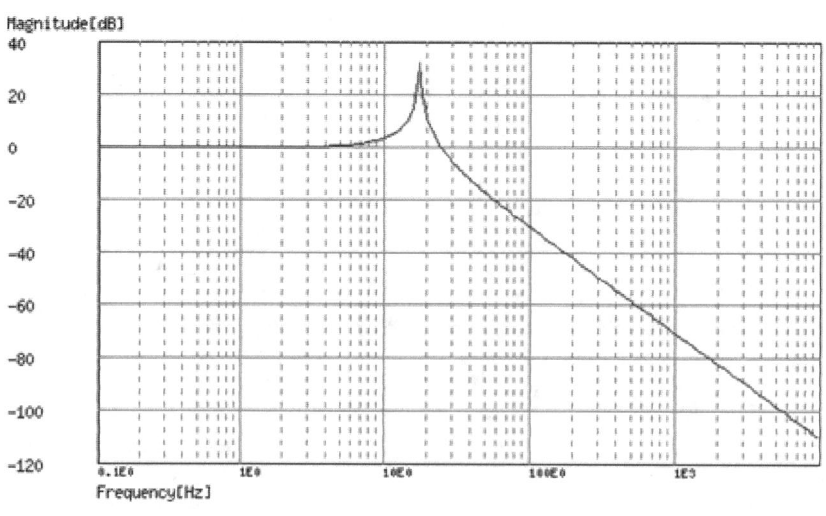

Figure 7. LC Filter bode plot.

4. Conclusions

This paper designed an inverter circuit based on EGS002 where DC voltage was recycled using 40N60C3 H-Bridge MOSFET and the output was filtered by using LC circuit. LC tank was adjusted to get perfect sinusoid shape. The best form was when L=12.63 H and C=6.5μf generating 30.8 Vpp. Although the plot shows that at these values, circuit was not at 50 Hz, instead it was working at 17.57 Hz, so that output voltage was only 30.8 Vpp. Despite the imperfection, circuits produced sinusoid waveform and filter components held the important role.

IOP Conf. Series: Materials Science and Engineering 1062 (2021) 012050 doi:10.1088/1757-899X/1062/1/012050

Acknowledgments
This research has been sponsored by Universitas Sumatera Utara through TALENTA scheme 2019.

References

[1] Sandeep P 2014 *International Journal of Science and Research,* **3(8)** 1793–1798
[2] Mutua J B 2014 *Project Report F*, **17**
[3] Gupta K K, Ranjan A, Bhatnagar P, Sahu L K and Jain S 2015 *IEEE Transactions on Power Electronics* **31(1)** 135–151
[4] Chakravarthy B K 2019 *CVR Journal of Science and Technology,* **16(1)** 66–71
[5] Doucet J, Eggleston D and Shaw J 2007 *PFC Worcester Polytecnic Institute*
[6] EGS002 data sheet, EGS002 Sinusoid Inverter Driver Board User Manual.
[7] Arman M F 2011 *An Active Passive Filter Topology for Low Power DC/AC Inverters* (Doctoral dissertation, Brunel University School of Engineering and Design PhD Theses)
[8] Okawa-denshi, http://sim.okawa-denshi.jp/en/RLCtool.php.

Nanovoids in fracture surface of unsaturated polyester/vinyl ester blends resulting from disruption of the cross-linking of the polymer chain networks

Hairul Abral[1], Rahmat Fajrul[1], Melbi Mahardika[1], Dian Handayani[2], Eni Sugiarti[2], Ahmad Novi Muslimin[2], Santi Dewi Rosanti[3]

[1]Laboratory of Biocomposites Department of Mechanical Engineering, Andalas University, Padang, Indonesia
[2]Laboratory of Sumatran Biota, Faculty of Pharmacy Andalas University, Padang, Sumatera Barat, Indonesia
[3]Laboratory of High-Temperature Coating, Research Center for Physics Indonesian Institute of Sciences (LIPI) Serpong, Indonesia

*Corresponding e-mail: abral@eng.unand.ac.id

Abstract. A weakness of unsaturated polyester is brittle due to the cross-linking of the UP chain network. Improving the toughness may be carried out by blending with appropriate additives. This work aims to characterize the fracture surface of unsaturated polyester (UP)/vinyl ester (VE) blends with a ratio of 70% and 30% respectively. Addition of VE in UP increases the toughness of the blends resulting from disruption of the cross-linking of the UP chain network. The disrupted network was characterized by the FESEM morphology from fracture surface of the blend which exhibits three different regions i.e., the smooth zone corresponding to the slow crack growth rate (1); a transition zone, where the surface roughness steadily increases (2); a final failure zone with much rougher marks (3). The presence of more nanovoids in region 2 and 3 confirmed the bigger disruption of the crosslinking in UP networks increasing UP chain mobility and toughness of the blends. The nanovoids result in a dimpled appearance of the fracture surface and become points of high-stress concentration from which cracks can initiate. When subjected to a sudden load, many cracks can grow simultaneously forming beach marks where the cleavage from several cracks meets.

Keywords: nanovoids, fracture, polyester, crack, morphology

1. Introduction

Thermosetting like unsaturated polyester resin is widely used in automotive construction as coatings [1]. This polymer has advantages in term of high chemical and water resistance, high tensile strength, and low cost [2]. But, UP resin was brittle due to its rigid chain structure [3]. Therefore, this weakness limits the commercial applications of UP resin. A simple attempt has been conducted to enhance the toughness of this material by mixing it with VE resin [4][5]. The blending aims to destroy the crosslinking chain structure [6]. The disruption of this polyester chain network increased tensile strength and fracture strain [7]. However, there was no previous work presenting the disrupted chain structure which can be observed from the fracture surface morphology of the UP/VE blends. Thus, this work aims to characterize tensile properties of the UP resin mixed with VE resin and to relate the measured differences in the properties to the changes in polymer structures.

IOP Conf. Series: Materials Science and Engineering 1062 (2021) 012051 doi:10.1088/1757-899X/1062/1/012051

2. Materials and Method

2.1 Materials

Polyester resin (YUKALAC BQTN 157-EX type), vinyl ester resin (RIPOXY R-802 EX-1, methyl ethyl ketone peroxide (MEKEP) as a catalyst and methyl methacrylate (MM) were obtained from Justus Sakti Raya Inc, Indonesia.

2.2 Method

Preparation of UP and VE blends was similar as reported in our previous work [4]. The ratio of additives in UP resin was presented in Table 1. Homogenization of the mixture was conducted using a magnetic stirrer (Daihan MSH-20D) at 400 rpm and 25 °C for 10 min. The catalyst of MEKEP of 4 mL was mixed to the solution and stirred for a 5 min. This blend was then poured onto a die made from silicon material at 25 °C for 72 h. The die dimension for tensile testing was based on ASTM D638-14 standard (Type I) [8].

Table 1. Sample with composition and abbreviations used.

Samples	VE (mL)	UP (mL)	MM (mL)
1-UP	-	100	
1-VE	100	-	
0.9-UP/0.1-VE	10	90	10
0.8-UP/0.2-VE	20	80	10
0.7-UP/0.3-VE	30	70	10
0.6-UP/0.4-VE	40	60	10

2.3 Morphology investigation using field emission scanning electron microscopy (FESEM)

JIB 4610 FE-SEM model from JEOL was used for observation of the fracture surfaces from the tensile samples. An accelerating current of 5 kV and probe current of 8 mA was set up for this testing. All samples were coated with carbon and then followed by gold for two minutes using an argon plasma metallizer (sputter coater K575X) (Edwards Limited, Crawley, United Kingdom).

2.4 Tensile test

A Universal Testing Machine Gotech (GT-7001-LC-30) was used for tensile testing with a tensile test speed of 4 mm/min. All specimens before testing were stored in a closed chamber for 48 h under 50 ± 5% relative humidity (RH) at 25°C. Tensile tests were carried out five times for each sample.

3. Results and discussion

3.1 Fracture surface

The fracture surfaces of selected tensile samples were presented in Figure 1. Figure 1a and Figure 1b are smooth surfaces of UP and VE respectively. This smooth appearance corresponds to low toughness probably resulting from rigid cross-linked chain network. The fracture surface was perpendicular to the tensile stress direction resulting from a growing crack which cleaves the polymer structure along with the weakest atomic bonds [4]. This phenomenon is in agreement with the tensile testing curve (Figure 2) showing a lack of plastic deformation for both pure resin. After mixing with a blending ratio as shown in Table 1 the fracture surface of the UP/VE blend had a rougher appearance (Figure 1c, number 1 and 2). Some defect sections appeared on the surface (marked by the red arrow) probably as a result of the presence of VE polymer as a filler in UP chain network. This results in a disruption of the UP polymer chain structure causing a decrease in the number of linked chain networks which increase in the mobility of the polymer chains, consequently, improving the plastic deformation as presented in the stress-strain curve marked with a white arrow (Figure 2). For further loading, the mobility of free cross-linking chain became slower resulting from restriction of the available rigid cross-linked chain network. This phenomenon creates a nano-sized void in UP polymer chain structure. The void promotes an initial crack for a catastrophic fracture of this polymer and results in a rough fracture surface as shown in Figure 1d.

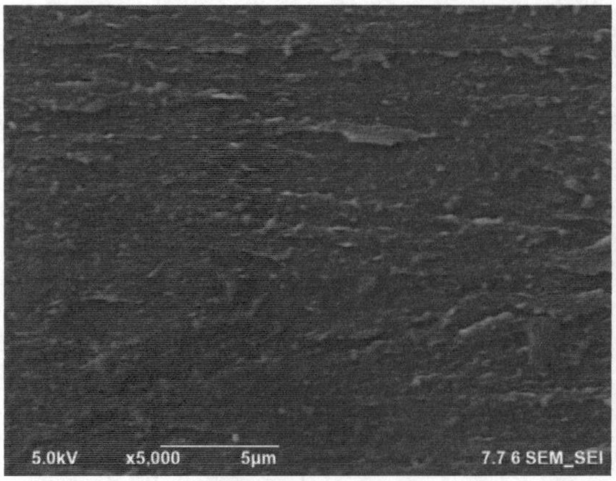

a. 1-UP

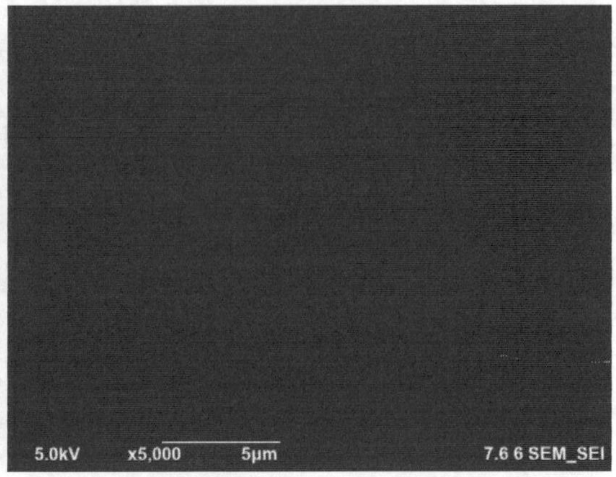

b. 1-VE

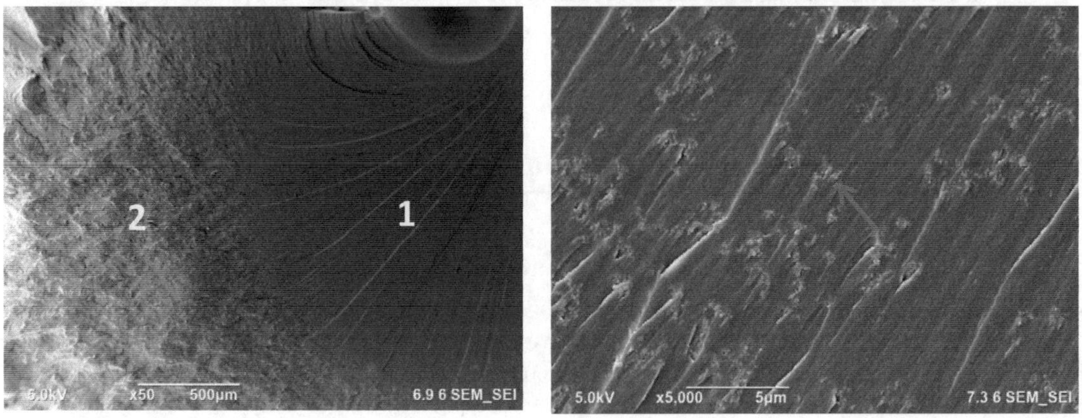

c. 0.7-UP/0.3-VE

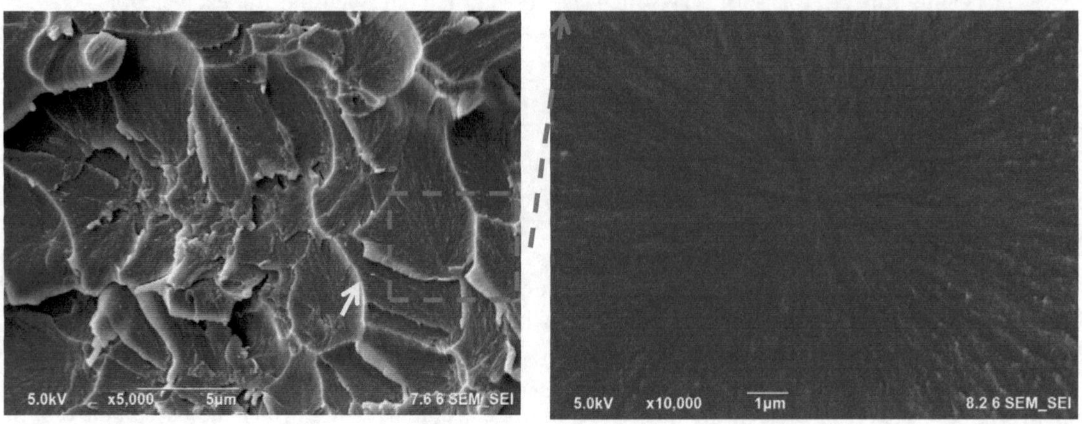

d. 0.6-UP/0.4-VE

Figure 1. FESEM in a section of the fractured surface from the tensile sample.

3.2 Tensile properties

Figure 2 displays the tensile strength for each sample. Pure UP and VE stress-strain curves did not have plastic deformation. This agrees with a smooth fracture surface (Figure 1a and 1b). After blending with VE resin, the 0.7-UP/0.3-VE stress-strain curve shows an elastic and plastic deformation. This sample had tensile strength (64 MPa) and elongation at break (13.3%) are higher than those for pure VE or UP resin. This result indicates the toughness and ductility became higher. A similar phenomenon was also presented with increasing the TS of the polyester resin after blending with vinyl ester resin [7].

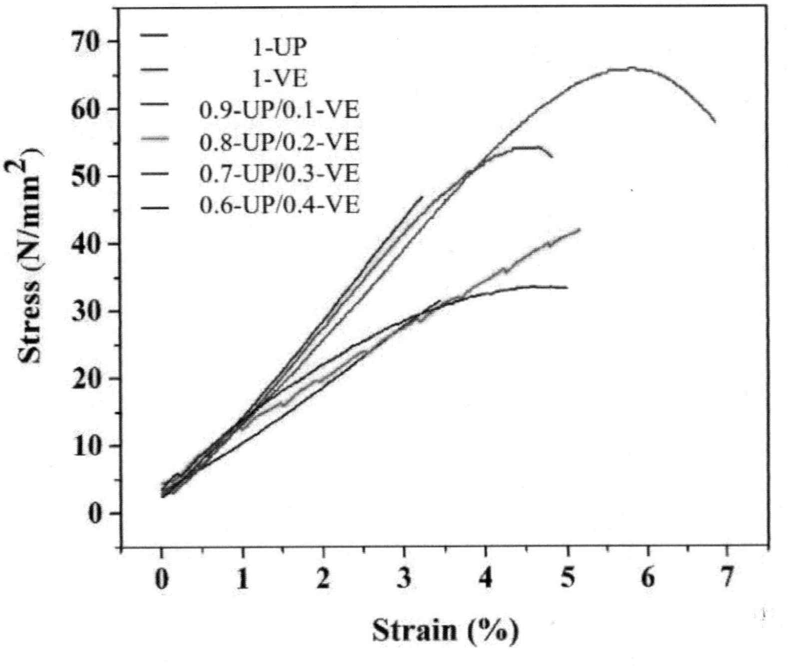

Figure 2. Tensile properties of UP, VE resin and UP/VE blend

4. Conclusion

This work reported the morphology of fracture surface from pure UP or VE, and the blend of both these resin materials. The blending UP and VE resin resulted in rougher fracture surface in comparison to the pure resin. The presence of VE disrupted the polyester chain structures causing an increase in mobility of UP chains. Consequently, the structural rigidity was reduced and increased in the toughness of UP resin. The area in the plastic deformation section presented the smooth surface. This is a result of slow tortuous crack growth through the zones with the lowest rigidity. The rapid rupture occurred resulting from the presence of nano-sized voids.

Acknowledgements

This project was financially supported by the Mechanical Engineering Department, Faculty of Engineering, Andalas University in the year of 2020, contract number 058/UN16.09.D/PL/2020.

References

[1] Yadav S K, Schmalbach K M, Kinaci E, Iii J F S and Palmese G R 2018 *Eur Polym J* **98** 199–215
[2] Ali U, Karim K J B A and Buang N A A 2015 *Polym Rev* **55** 678–705
[3] Abral H, Kadriadi D, Rodianus A, Mastariyanto P, Ilhamdi, Arief S, et al. 2014 Mater Des **58**

125 - 129.
[4] Abral H, Fajrul R, Mahardika M, Handayani D, Sugiarti E, Muslimin A N, et al. 2019 *Polym Test* 106193
[5] Hari S P, Mohana R Y V and Hemachandra R K 2015 *Fibers Polym* **16** 443 - 448
[6] Dinakaran K and Alagar M 2002 *J Appl Polym Sci* **86** 2502 -2508
[7] Ardhyananta H, Puspadewa F D, S T Wicaksono W, Wibisono A T, Kurniawan B A, Ismail H, et al. 2017 *IOP Conf. Ser. Mater. Sci. Eng.* p. 012088
[8] ASTM Standard: D638-14 Standard Test Method for Tensile Properties of Plastics

IOP Conf. Series: Materials Science and Engineering 1062 (2021) 012052 doi:10.1088/1757-899X/1062/1/012052

Microstructure and Corrosion Rate of the Hot Rolled Mg-1.6Gd Alloys as Prospective Degradable Implant Materials

O Susanti* and D Dwiputra
Metallurgy Laboratory, Mechanical Engineering Department, Universitas Andalas Padang, Indonesia

*Corresponding e-mail: oknovia.s@eng.unand.ac.id

Abstract. At present, degradable materials implants with the capability of high loads for the temporary replacement of bones and biocompatible are not been available much more. Mg-Gd alloys have been intensively observed as appropriate degradable materials, since they combine a good strength and corrosion rate. Thus, the aim of this study is development of Mg-Gd alloys as materials implant with biomechanical tailored properties. Mg-1.6Gd alloys of hot rolled at severe reduction of 80% at 40 % per stage with a speed of 10 mm/min. The hot rolled was employed by cross rolling with 0° rolling direction first and followed by changing the rolling direction to 90° each pass (for two time). The hot rolled of the alloys have been investigated in the temperature range recrystallization (400- 550 °C). Meanwhile, the corrosion experiments also were research after 3, 7 and 14 days of immersion in Ringers solution. The phenomena of corrosion was observed using optical and weight of the samples before and after immersion. The Results show that the hot rolling techniques reveals refine microstructure with the grain size range of 50-60 μm. The number of precipitates have been founded on the cross rolling, particularly at the higher temperature and the highest corrosion rate occurs on the short immersion time for 3 days.

Keywords: microstructure, corrosion, bones, biocompatible, implant

1. Introduction

The small addition of Gd in Mg have been significantly improved microstructure and mechanical properties by thermo-mechanical process. Susanti *et al* showed that the Mg-1.6Gd alloys have been exhibited values of its mechanical properties through the hot extrusion. One of ways is the micro-alloying addition of Gd (1.6 wt%) in magnesium. The strength can increase due to precipitates hardening. The hot extrusion of the Mg-1.6Gd alloying also showed the recrystallization grain size and the other excellent mechanical properties. This discussion already explained in the paper before [1-4]. Before, it well known that the elastic modulus of Mg is close to that of human bone properties which it is suitable and is used as the requirement of implant materials. Therefore, magnesium alloys have been great attention as a degradable implant. However, many efforts must be employed to improve mechanical properties and degradation characteristic of the alloy. In the term of mechanical properties that can be related to the microstructure control to purpose for applications such as strength, hardening, ductility and elastic modulus etc [5].

Another paper, Susanti et al clearly reported that the effect of the micro alloying on microstructures by the hot rolling with a total reduction of 95 % [6]. This study explained that the microstructures of the Mg-1.6Gd alloys reveals nearly similar grain morphology with tendency fine equated grain on each temperature for the both of the methods. Especially for cross rolling (CR) which has larger grain size compared to the unidirectional rolling (UR). It has visually more pretending homogenized with the size of about 60-100 μm. The number of precipitates found at the higher temperature of the rolling. The grain size and shape of the alloy that were caused by the mechanical processing to that related the mechanical process. This study is to prove the previously research which Luo *et al.* showed that the addition of Gd in to Mg can improved the microstructure and the mechanical properties with the ASTM standard, Mg-xGd (x= 0.8, 1.5, and 2.5) [7].

The effect of rolling on microstructure can change the grain size of the MgGd which is correlated on the corrosion rate. The grain size increases with the increasing the temperature of rolling. The grain size can be also changed by the addition of rare earth elements (REE), such as Ce, La, Nd Mo or Gd as showed on the papers [8, 9]. Moreover, it can be summarized that the addition of REE can from corrosion resistance by reducing the grain size and the formation of surface layer as protection. The other way is to give the use of REE in order to form surface layer by oxide as reported by Susanti *et al* [10]. This study explained that the hot rolled of Mg-1.6Gd alloys can be improved the grain size and can be also founded $Mg(OH)_2$ surface layer as corrosion resistance for the MgGd alloys after the longer immersion time.

2. Experimental

Mg-1.6Gd alloys is prepared the number of specimen for hot rolling which were machined by the initial size of 50 mm in length, 25 mm in width and 10 mm in thicknesss. Specification of specimens are the same with the paper before [6]. The Mg-1.6Gd alloys was employed cross rolling with reduction of 80 % which is 0° rolling direction first and followed by changing the rolling direction to 90° each pass (for two times). This cross rolling was carried at a temperature range recrystallization (400- 550°). All the samples were observed by standard metallography procedure and then followed the corrosion process which all the procedures nearly the same explanation with the before paper [10]. The samples were weighed and the corrosion rate was also calculated with using equation (1):

$$CR = \frac{W\,K}{\rho\,A\,t} \tag{1}$$

where, *CR* refers to corrosion rate, W is the weight loss from sample, ρ is density of the sample, A is area of sample and t is time in Ringers solution.

3. Result and discussion

The microstructures of cross rolling of Mg-1.6Gd alloys result nearly the same the grain size for all the samples with tendency fine equiaxed as shown in Figure1. A lot of twin grow spread on each the microstructure and also found the formation of precipitates. The highest temperature exhibits the number of precipitates much more compared the others. As well known that these precipitates in grain are Mg_5Gd as explained by Susanti et al [10] and then the formation of precipitates grow with the increasing temperature of rolling.

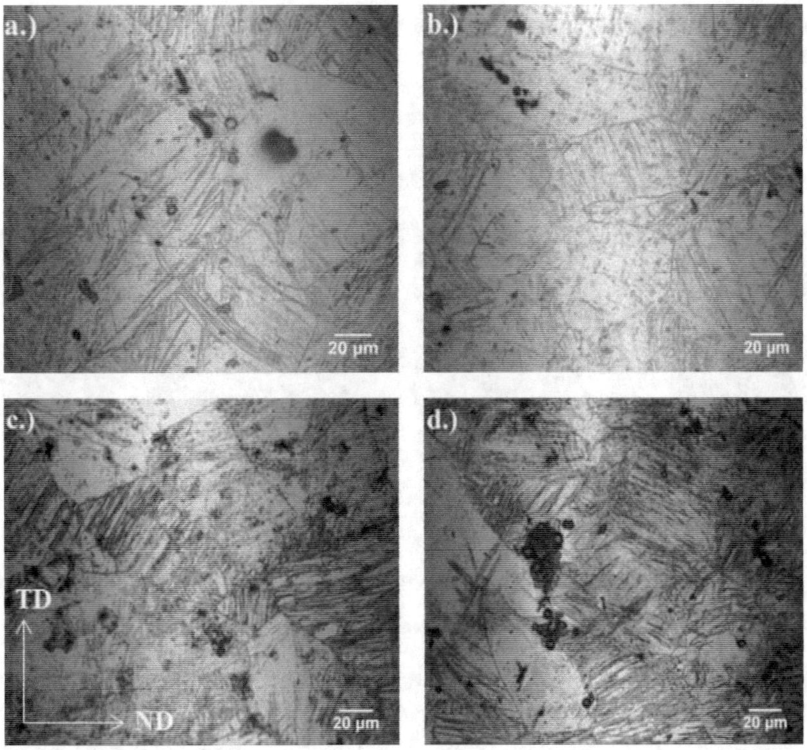

Figure 1. Microstructure of the hot rolled Mg-1.6Gd Alloys at various temperature with reduction of 80%; a) 400°C, b). 450°C, c). 500°C, d). 550°C of cross rolling.

The strength of the Mg-1.6Gd alloys is very influenced by the grain size and shape of the as-rolled. At the early temperature of the rolled Mg-1.6Gd alloys (400°C) has about the grain size of 65 µm. The grain size decrease to be about 48 µm at a temperature of 450 °C and then drop to be about 46 µm at a temperature of 400°C and the increasing to be 52 µm at a temperature of 550 °C as shown in Figure 2.

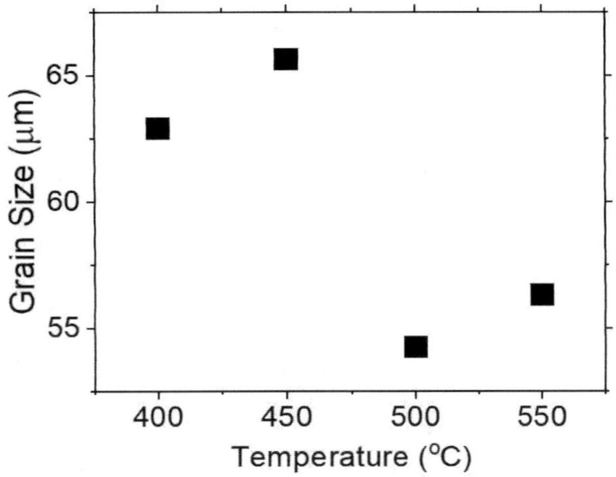

Figure 2. The grain size of the cross rolling Mg-1.6Gd at various temperature.

The degradation of the Mg-1.6Gd alloys are tested with use of immersion test with Ringers solution. The immersion test with using Ringer solution for 3 days, 7 days and 14 days to get corrosion rate on the each sample as shown in Figure 3. It can be seen that the early immersion (3 days) occurs pitting corrosion on around the surface of samples and result the white colour as the corrosion product. And then the number of degradation spread and reaches a depth of samples for 7 days and the degradations of sample reach the peak on the immersion of 14 days.

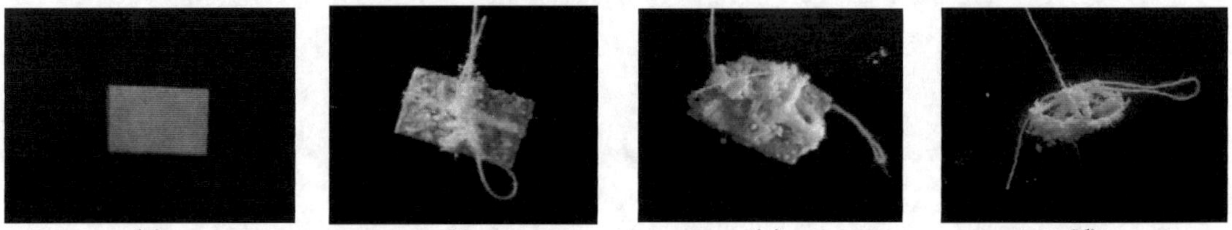

(a) (b) (c) (d)

Figure 3. Typical corrosion morphologies of the Mg-1.6Gd alloys sample: (a) sample before immersion; (b). immersion for 3 days; (c) immersion for 7 days; (d) immersion for 14 days.

Generally, the highest corrosion rate occurs at the short immersion duration. In this case, the initial corrosion stage no occur the protection on samples yet. The formation of surface layer as a protection is obtained by MgO in water to form $Mg(OH)_2$ layer after the longer immersion time as explained by Song et al [11]. Moreover, the protection can reduce due to attack Cl^- from the water and then gradually the thick protection layer can be loss and the samples can be finish. Meanwhile, the corrosion rate occurs higher at the low temperature compared the others which is very correlation to grain size and precipitate. At the low temperature of rolling result the corrosion rate of about 28 mmpy and then drop to about 22 mmpy at a temperature of 550 °C. On the each temperature always find the highest corrosion rate after the immersion time for 3 days as shown on Figure 4.

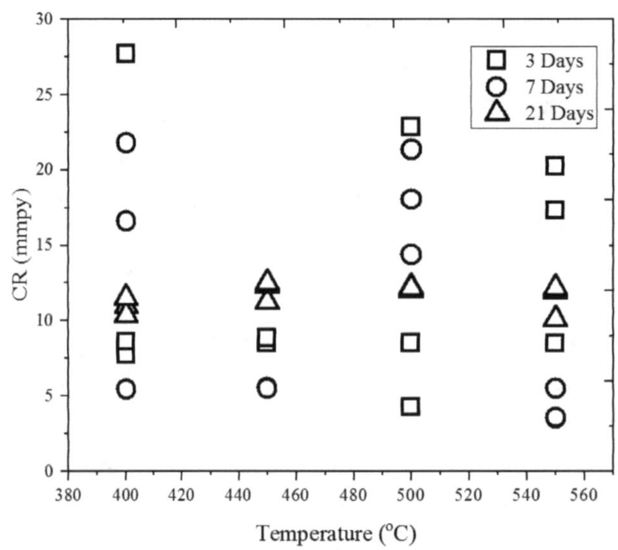

Figure 4. Corrosion rate of cross rolled Mg-1.6Gd alloys at different temperature.

4. Conclusions

In this study, the microstructure and corrosion rate observation of the hot rolled Mg-1.6Gd with reduction of 80 % at temperature range of 400-550 °C. It is resulted that the The number of precipitation are found on microstructure and with the increasing temperature of rolling has more precipitation which spread in the grain. A lot of twin also found in the grain on every temperature of rolling which it can improve toughness of the alloys. Following that the corrosion rate occur higher on the immersion time and then the corrosion rate is slowly at the longer immersion time. This is caused by the formation of $Mg(OH)_2$ as protection surface layer. The corrosion rate is influenced by the number of grain boundaries.

Acknowledgments
The work presented in this paper was funded by DIPA-2020 of Faculty of Engineering, Universitas Andalas, West Sumatera, Indonesia. To also Thanks to my students, Afif Rahmadino and Dezy on the assistance to get and calculate of data.

References

[1] Bohlen J, Yi S, Letzig D and Kainer K U 2010 Effect of rare earth elements on the microstructure and texture development in magnesium–manganese alloys during extrusion *Mater. Sci. Eng. C: A* **527** 7092-7098
[2] Peng Q, Wu Y, Fang D, Meng J and Wang L 2006 Microstructures and Properties of Melt-Spun and As-Cast Mg-20Gd Binary Alloy *J. Rare Earths* **24** 466-470
[3] Hort N, Huang Y, Fechner D, Störmer M, Blawert C, Witte F, Vogt C, Drücker H, Willumeit R, Kainer K U and Fayerabend F2010 Magnesium alloys as implant materials – Principles of property design for Mg–RE alloys *Acta Biomater.* **6** 1714-1725
[4] Susanti O, Harjanto S and Mochtar M 2015 Microstructure and Mechanical Properties of Extruded Mg-1.6Gd as Prospective Degradable Implant Materials *Adv. Mater. Res.* **1112** 592
[5] Witte F, Hort N, Vogt C, Cohen S, Kainer K U, Willumeit R and Fayerabend F 2008 Degradable biomaterials based on magnesium corrosion *Curr. Opin. Solid State Mater. Sci.* **12** 63-72
[6] Susanti O, Mochtar M and Harjanto S 2017 Microstructure Transformation of Mg-1.6 Gd during Hot Rolled at High Deformation Ratio *IOP Conference Series: Mater. Sci.Eng.* **662** 012071
[7] Luo S, Yang G, Liu S, Wang J, Li J and Jie W 2016 Microstructure evolution and mechanical properties of directionally solidified Mg-xGd (x=0.8, 1.5, and 2.5) alloys *Mater. Sci. Eng. C: A* **662** 241-250
[8] Yang L, Feyerabend F, Kainer K U, Willumeit W and Hort N 2011 Corrosion Behavior of As-Cast Binary Mg-Dy Alloys *Mater. Sci. Forum* 417-421
[9] Ding Y, Wen C, Hodgson P and Li Y 2014 Effects of alloying elements on the corrosion behavior and biocompatibility of biodegradable magnesium alloys: a review *J. Mater. Chem. B* **2** 1912-1933
[10] Susanti O and Harjanto S 2019 Corrosion behavior of hot rolled Mg-1.6 Gd alloys as degradable materials implant *AIP Conf. Proc.* 020052
[11] Song G and Song S Z 2007 A possible biodegradable magnesium implant material *Adv. Eng. Mater.* **9** 298-302

Thermogravimetric Analysis of Densified Coco Peat Briquettes

A Pratoto[1*] and A J Abhen[1]

[1] Mechanical Engineering Department, Andalas University Padang 25163, West Sumatra, Indonesia

*Corresponding e-mail: adjar.pratoto@eng.unand.ac.id

Abstract. The present work deals with the investigation of thermal decomposition of densified coco peat briquettes using the thermogravimetry method to elucidate the effect of finite size of the biomass briquette on the decomposition profile. The cylindrical samples are varied in term of its density and its shape (sphericity). The sample is heated in a nitrogen atmosphere at different heating rates, ranging from 5°C/min to 20°C/min. The TG curves show three distinct phases of decomposition, i.e. preheating and drying, devolatilization, and char reduction. For all samples, thermal decomposition shifts to higher temperatures with the increase in the heating rate. From the TG curves, it is shown that the initial degradation temperature varies from 181°C to 302°C and the final degradation temperature varies from 328°C to 608°C. The shift to higher temperatures is also observed for the less dense briquettes. The shape of the briquettes affects also the decomposition profile. The less spherical (less of contact area) sample will decompose at higher temperatures. From the DTG curves, it can be seen that the maximum thermal degradation rate takes place at varying temperature ranging from 243°C to 419°C. The peak temperatures depend on the heating rate. However, the density and the shape of the biomass have no appreciable effect on the peak temperature.

Keywords: thermogravimetric, coco peat briquettes, biomass, decomposition

1. Introduction

Indonesia represents the world leader of coconut (*Cocos nucifera L.*) producer. It is estimated that the production reached around 2.9 million tons in 2018 [1]. Various products may be derived from the coconut fruit. The main product of coconut is copra (endosperm) which is widely utilised in both food and non-food industries. Apart from copra, coconut fruits also engender shells and husks. Shells can be used as raw material for activated carbon production and husks can be utilized for the production of floor covering, mats and matting, etc. Each coconut fruit yields about 35% by weight of husk [2]. The coconut husk consists of about 30% fibre and 70% coco peat or coco pith [3]. Coco peat constitutes commonly a by-product of coir fibre extraction. In Indonesia, the fibre extraction is mostly carried out by smallholders. It is often that coco peat is dumped on site. This poses serious disposal and environmental problems.

Coco peat has a high potential for renewable energy production. The utilization of coco peat for energy use will alleviate the problem of waste disposal and at the same time uphold the reduction of greenhouse gases by substituting to some extent fossil fuels. The potential as renewable energy feedstock is promising by the fact that coconut is a non-seasonal crop and, hence, its availability would be year-round.

Like majority of agricultural wastes, coco peat is commonly available in a loose packed form that engenders a problem of handling and transportation. Biomass pelletizing or briquetting could be adopted to cope with such kind of problem. Pelletizing or briquetting also increases the energy density of the biomass. Physical and burning properties of pellets or briquettes made from coco peat are given in [4], [5]. Another method of preparing coco peat as energy feedstock is by converting it into biochar [4], [6].

A fundamental understanding of the thermal properties of the cocopeat is of importance in considering its use as fuel feedstock. Various thermal analysis techniques have been developed to characterize the thermal behaviour of biomass feedstock, such as direct scanning calorimetry, thermogravimetry analysis, mechanical thermal analysis. The detail description of thermal analysis techniques can be found elsewhere [7]. Thermogravimetric analysis (TGA) is the simplest one. This technique measures the mass of a sample in a controlled atmosphere as a function of temperature. The TGA can be used to analyse the thermal decomposition of substances, corrosion of metals, determination of moisture, volatiles and ash content, evaporation rates and sublimation, distillation and evaporation of liquids, reaction kinetics studies, compound identification, and heat vaporization and vapour pressure [8]. Olatunji et al. [9] also reported that the TGA has been applied to different studies, such as devolatilization, reaction region and kinetics, combustion, pyrolysis, gasification, and thermal decomposition. The study of biomass thermal decomposition by TG analysis has been also covered a wide range of different biomass feedstocks such as wooden biomass [10], [11], biomass-derived oil [12], [13], and marine biomass [14] to name a few. The information from thermogravimetric analysis can contribute to the understanding of pyrolysis reaction mechanism which is constructive in the process optimization or reactor design at industrial scale [15].

TGA analysis uses in general a sample of small size to prevent the bulk effect and thermal gradient developed within the sample. A knowledge of thermal characteristics of biomass feedstocks of finite size is equal importance in view of its utilization as a renewable energy resource. To the knowledge of the author, the thermogravimetric analysis of the biomass of finite size is still limited. The application of TGA to samples of finite size has been performed by Jiang et al. [16], among others. They investigate the effect of the polystyrene particle size on the reaction rate. The particle size ranges from 5 to 50 μm. They reported that size presented influence to the first and second steps reactions. The biomass fuel feedstock is generally available in the form of pellets or briquettes at larger sizes. The knowledge of the decomposition profile of such fuel feedstock is useful in the design of the reactor at industrial scale. The current study aims at characterizing thermal decomposition of cocopeat briquettes of finite size at this stage with different densities and shapes by using thermogravimetric analysis.

2. Experimental

2.1. Samples

The cocopeat sample was taken from a smallholder in West Sumatra, Indonesia. The structural composition of the cocopeat is given in Table 1. It can be seen from the Table 1 that the main constituent of the cocopeat is lignin; it counts more than 50%. Different briquette samples were prepared. The first were samples having the same dimension but differing in density (sample A and B). The second were samples having the same density but differing in shape or size (sample A and C). The dimension and density of samples used in the experiment are summarized in Table 2.

Table 1. Structural composition of cocopeat

Composition	Mass Fraction (%)
Hemicellulose	9.40
Cellulose	32.90
Lignin	54.00
Extractive	2.28

Table 2. Dimension and density of the briquette samples

Sample	Diameter (cm)	Height (cm)	Density (g/cm^3)
A	4.0	2.0	1.3

B	4.0	2.0	2.3
C	6.0	1.5	1.3

2.2. Experimental set-up and procedure

The experimental set-up is schematically depicted in Figure 1. A programmable muffle furnace is used as reactor. A rigid lever bar is inserted through the furnace exhaust port that is located at the rear wall. The biomass sample dish is mounted on one end of the lever with the help of stainless wires. To record the change of the sample mass in the course of experiment, a tension/compression load cell was employed. The load cell was attached to the other extremity of the lever. An amplifier was connected to load cell with the objective of enhancing the signal output. Calibration of the load cell output was made by using pre-weighed masses. Four pre-weighed masses were selected such to correspond the biomass sample mass range in the experiment. A mathematical expression relating the load cell voltage outputs and the pre-weighed masses was established by employing a regression method.

Three levels of heating rate were selected during the run of experiment, namely 5, 10, and 20°C/min. The nitrogen flow was kept small in purging air from the reactor chamber and maintaining the chamber atmosphere inert. The runs were concluded whenever the change in sample mass is indiscernible. Mass data recorded were smoothed using a moving average technique as described by Chen et al. [17]. For every data point, smoothing was made by averaging a span of 11 data points. A smaller number of data points was used for endpoints. The DTG curves were deduced from TG data by employing five-points numerical differentiation.

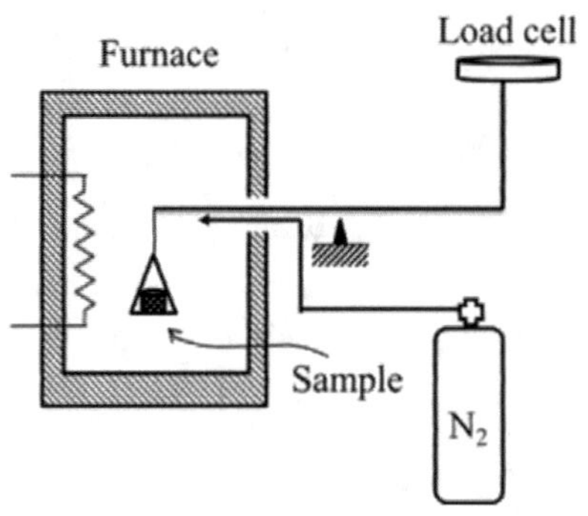

Figure 1. Simplified schematic of experimental set-up

3. Results and Discussion

3.1. Effects of heating rate

Figure 2 shows the TG and DTG curves for different heating rates. It can be seen from the figure that all samples show three discernible thermal degradation phases. However, at higher heating rates, the points of inflection are less obvious. For all samples, the higher heating rate shifts the decomposition to higher temperatures. It is to also note that the shift of temperature is even more notable. This is due the size of the sample. At higher heating rate, the surface of the sample is subject to excessive heating over the inner parts of the biomass and no sufficient time is available for every part of the biomass reaching thermal equilibrium. The outer layer will reach first the temperature required by volatiles for decomposition. At the same time, char with low thermal properties is formed at the outer surface which in turn impedes the heat diffusion towards inside of the biomass [18].

For the samples used in the experiment, the onset temperature varies between 181°C – 302°C, the completion temperatures from 328°C to 608°C and the peak temperatures from 243°C to 419°C. Table 3 summarizes the temperature degradation. Yang et al. [19] showed that hemicellulose decomposes mainly at 220°C – 315°C, cellulose at 315 °C – 400 °C, and lignin at wider range, i.e. 160 °C – 900 °C.

IOP Conf. Series: Materials Science and Engineering 1062 (2021) 012053 doi:10.1088/1757-899X/1062/1/012053

However, Panamgama & Peramune [20] indicated that thermal behaviour of lignin varies, depending on the chemical extraction protocols and energy supply modes.

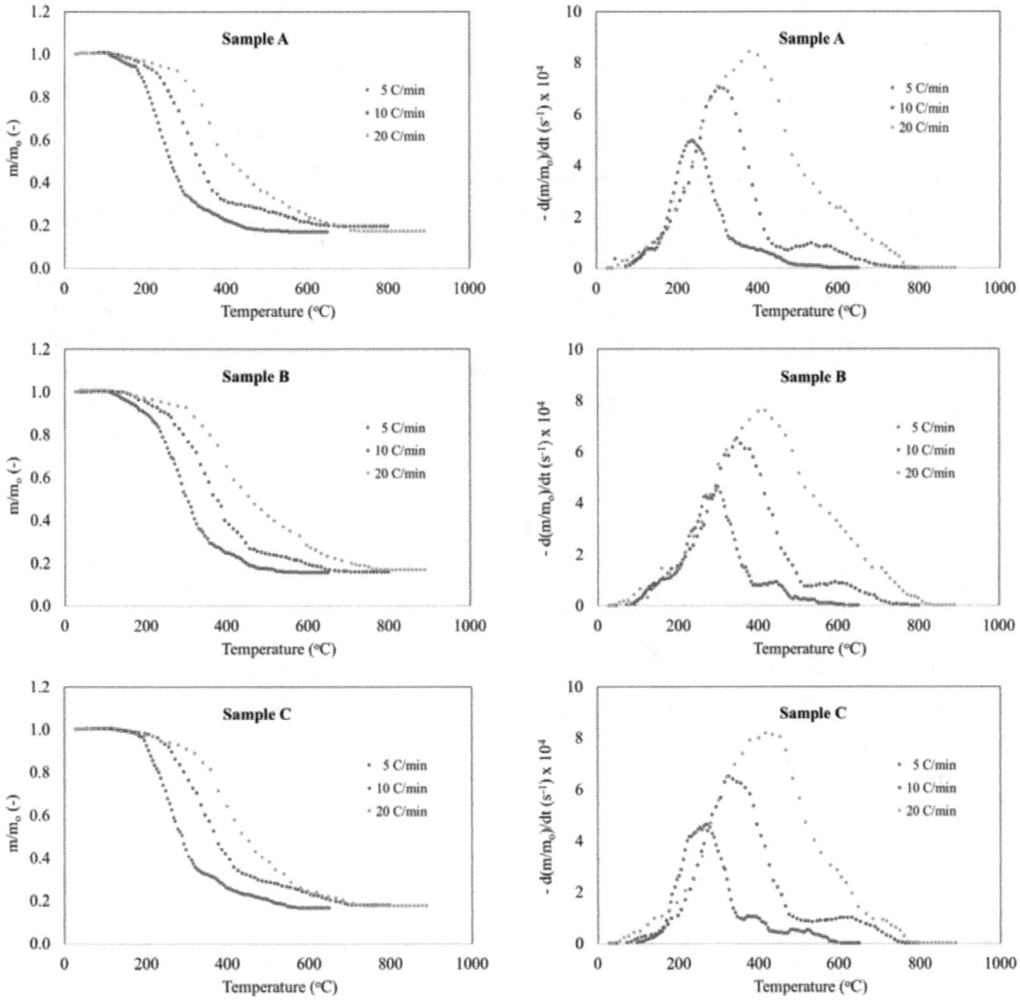

(a) TG curves (b) DTG curves

Figure 2. TG and DTG curves at different heating rates

DTG curves show shifts of peak temperatures to higher values with the increase in the heating rate. The thermal degradation rates also increase with the heating rate. A relatively linear increase in peak temperatures with respect to the heating rate is observed in the range of 5 – 20°C/min. However, such tendency should be verified for a greater range of the heating rate.

Table 3. Degradation temperatures

Sample	Heating Rate (°C/min)	Onset Temperature (°C)	Completion Temperature (°C)	Peak Temperature (°C)
	5	181	328	243
A	10	248	407	307
	20	282	503	380

	5	192	378	301
B	10	248	477	347
	20	302	608	419
	5	192	359	269
C	10	258	458	327
	20	282	581	419

3.2. *Effects of density*

Figure 3 shows TG and DTG curves of samples of different density. Figure 3 is extracted from Figure 2 with the objective to elucidate the effects of density to the decomposition profile. The figure compares the sample A and B. Both samples have the same dimension, but differ in density and hence differ in mass. Both TG and DTG curves are normalized to the initial mass, m_o. For all heating rates, the temperature of degradation shifts to higher temperature for the dense sample. Since the biomass has low thermal properties, the dense biomass will impart higher thermal resistance. From the DTG curves, it is seen that the absolute rate of degradation of the dense sample is superior over the less dense one, while the peak temperatures just differ to a lesser degree for all heating rates. This suggests that the density of the samples has less pronounced effect to the peak temperature than the heating rate does.

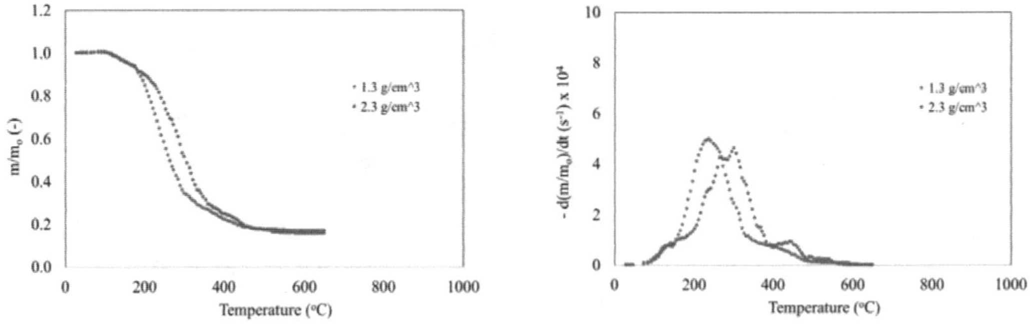

(a) Heating rate: 5°C/min

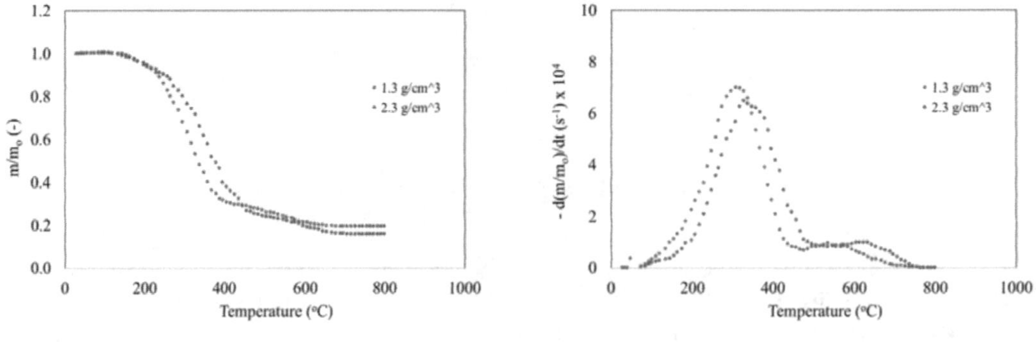

(b) Heating rate: 10°C/min

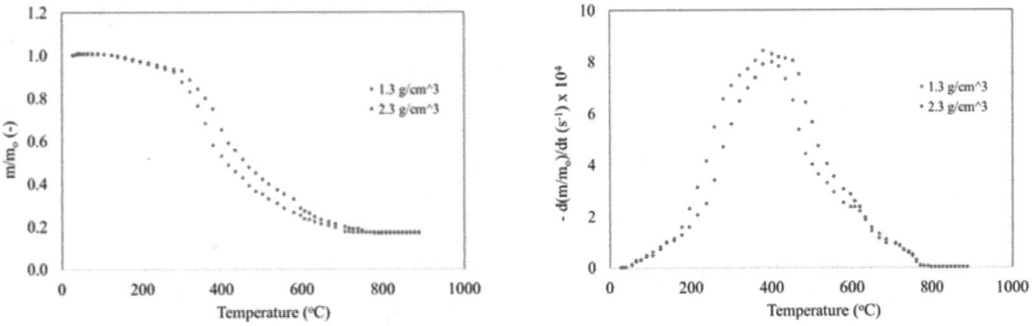

(c) Heating rate: 20°C/min

Figure 3. TG and DTG curves of samples with different density

3.3. *Effects of shape*

Figure 4 compares the TG and DTG curves between the sample A and sample C. Both samples have the same density, but differ in dimension, and hence differ in both volume and mass. The shape of the sample is described by its sphericity; the sample A and C have sphericity of 0.83 and 0.69, respectively. A greater sphericity indicates a greater surface area per unit of volume. Figure 4 reveals that greater sphericity biomass sample tends to decompose at lower temperatures. The sample with greater sphericity has greater contact area that promotes higher heat exchange. This result is in an agreement with findings of various literature [21- 22]. Xiao, et al. [23] also showed the same results in the pyrolysis of rice straw and pine sawdust, but an opposite result was found for the case of Phoenix tree leaves. As with the peak temperature, there is no appreciable difference with respect to the shape of the biomass.

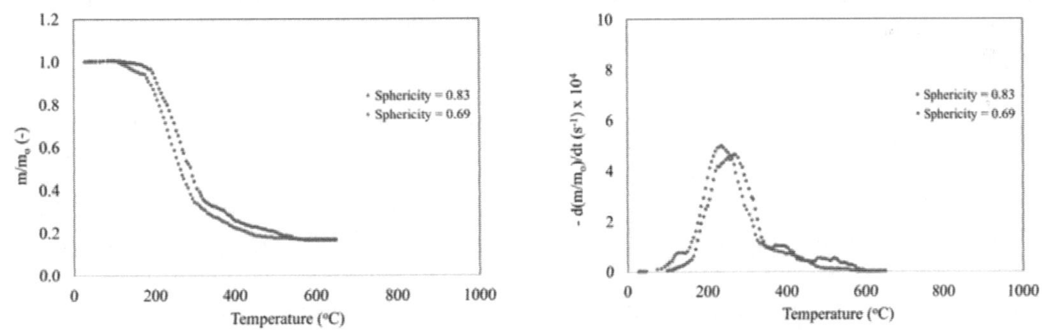

(a) Heating rate: 5°C/min

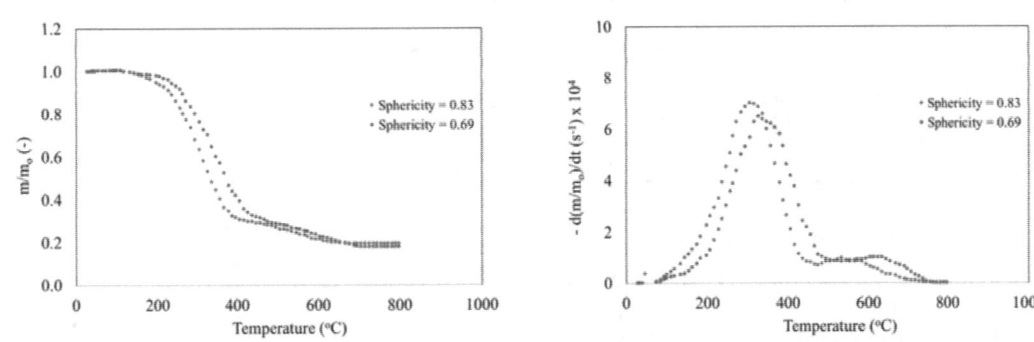

(b) Heating rate: 10°C/min

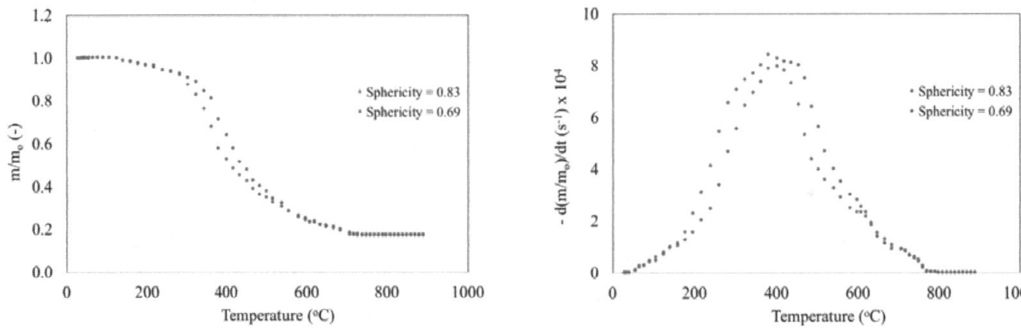

(c) Heating rate: 20°C/min

Figure 4. TG and DTG curves of samples with different shape

4. Conclusions

An experimental study has been carried out to elucidate the thermal behaviour of coco peat briquette using thermogravimetric analysis. From the TG and DTG curves, the temperatures of degradation were identified. The effects of heating rate, density of the biomass, and shape of the biomass briquettes were also exposed. The heating rate has effect on the mass loss profile and the temperature of degradation. The decomposition is remarkably shifted to higher temperatures with the increase in the heating rate due to the sample size that hinder the heat diffusion towards inside of the sample. The density of the sample also affects the thermal decomposition. The denser sample shifts the decomposition to the higher temperatures but its effect is less pronounced to the peak temperature than that of the heating rate. The sample with greater sphericity tends to decompose at lower temperatures since it has greater contact area per unit volume. The greater contact area will facilitate the heat to diffuse towards inside the material.

Acknowledgements

Authors would like to acknowledge the financial support from the Mechanical Engineering Department, Andalas University.

References

[1] Hirschmann R 2020 https://www.statista.com/statistics/706968/coconut-production-in-indonesia/
[2] Festin T F and Jose W I 1980 *Conservation & Recycling* **3** 383
[3] van Dam J E G 2002 *Coir Processing Technologies – Improvement of Drying, Softening, Bleaching and Dyeing Coir Fibre/Yarn and Printing Coir Floor Coverings* (Rome, Amsterdam: FAO & CFC)
[4] Alamsyah R, Siregar N C and Hasanah F 2017 *IOP Conf. Series: Earth and Environmental Science* **65**
[5] Islam Md H, Hossain Md M, and Momi M A 2014 *Int. Journal of Renewable Energy Development* **3** 119-123
[6] Supriya A, Samantray R and Mishra S C 2017 *IOP Conf. Series: Materials Science and Engineering* **178**
[7] Brown M E (Ed.) 2001 *Introduction to Thermal Analysis: Techniques and Applications* (New York, USA: Kluwer Academic Publishers)
[8] Jentoft R 2006 Teaching winter semester 2006_2007
[9] Olatunji O O, Akinlabi SA, Mashinini M P, Fatoba S O, and Ajayi O O 2018 *IOP Conf. Series: Materials Science and Engineering* **423** (2018) 012175
[10] Sobek S and Werle S 2020 *Fuel* **261** 116459

[11] Jain S Sebio-Puñal T, Naya S, López-Beceiro J, Tarrío-Saavedra J, and Artiaga R 2012 *J. Therm. Anal. Calorim.* **109** 1163
[12] Jain S and Sharma M P 2012 *Fuel* **93** 252
[13] Tutunea D, Ilie D, Racila L, Geonea I, Otat O, and Matei L 2019 *IOP Conf. Series: Materials Science and Engineering* **568** (2019) 012075
[14] Hong Y, Xie C, Chen W, Luo X, Shi K, and Wu T 2020 *Renewable Energy* **145** 2159
[15] Wang S, Dai G, Yang H, and Luo Z 2017 *Progress in Energy and Combustion Science* **62** 33
[16] Jiang L, Yang X-R, Gao X, Xu Q, Das O, Sun J-H and Kuzman MK 2020 *Polymers* **12**, 421
[17] Chen H X, Liu N A, Shu L F, and Zong R W (2004) *Journal of Thermal Analysis and Calorimetry* **78** 1029
[18] Cheng Q, Jiang M, Chen Z, Wang X, and Xiao B 2016 *Energy Sources, Part A: Recovery Utilization, and Environmental Effects* **38** 3383
[19] Yang H, Yan R, Chen H, Lee D H, Zheng C 2007 Characteristics of hemicellulose, cellulose and lignin pyrolysis *Fuel* **86** 1781
[20] Panamgama L A and Peramune P R U S K 2018 *Intl. J. Biological Macromolecules* **113** 1149
[21] Niu H and Liu N 2014 *The Eleventh International Symposium on Fire Safety Science* Christchurch, New Zealand 9-14 February 1393
[22] Sharma R and Sheth P N 2018 *Energy* **151** 1007
[23] Xiao R, Yang W, Cong X, Dong K, Xu J, Wang D, and Yang X 2020 *Energy* **201** 117537

Mechanical Properties and Machinability of Magnesium Alloy AZ31 and AZ91 – A Comparative Review

Louis Denis Kevin Catherine[1] and Darulihsan Abdul Hamid[2]*
[1]Manufacturing Engineering Technology, Universiti Kuala Lumpur (MIDI), Malaysia
[2]Kolej Universiti Poly-Tech MARA, Kuala Lumpur

*Corresponding e-mail: darulihsan@kuptm.edu.my

Abstract. Throughout the last decade magnesium alloys have slowly overtaken aluminium alloys, especially in the automotive and aerospace industry. Due to its attractive weight-resistance and weight-hardness characteristics, the global production of Magnesium alloys has grown from 360,000 tons in 1999 to 850,000 tons in 2013. Magnesium alloys AZ91 and AZ31 are the most widely used grades.AZ 91 exhibits a Knoop's hardness (HK) of 76.2 that is higher to the AZ31 with a 51.1 HK. The manufacturing of AZ 91 is through a casting process whereas AZ31 is manufactured by using an extrusion process. The microstructure of AZ31 and AZ91 are highly influenced by the heat generated during the machining process that also affects the surface hardness due to the high thermal conductivity of the material that range from 62 W/m.k to 96W/m.k. Both materials have a very good machinability where milling, drilling, and turning are commonly used. However, extreme care should be taken in terms of machining conditions as Magnesium alloy is highly flammable. However, a safe machining process can be performed by avoiding very fine chips, magnesium dust and also the usage of water-based coolants that react with the raw material. This paper has widely reviewed the studies that have been conducted since the evolution of magnesium alloys in terms of machining parameters, machining consumables and their effect on the mechanical properties, surface conditions and limitations of the various machining processes.

Keywords: Magnesium alloy AZ31, magnesium alloy AZ91, microstructure hardness

1. Introduction

It Magnesium is in the group 2 in the periodic table and its symbol is Mg. The density of pure magnesium is 1.745 g/cm^3 which are currently the lightest metal on the market [1]. At room temperature, pure magnesium is weak and lack strength that is needed to be considered as a structural base material. This condition is enhanced with the addition of alloys like Aluminium (Al), Zinc (Zn) and other elements. Shi et al. stated that the addition of aluminium brings more strength and hardness to the final alloy [2]. It also contributes to the smooth casting process as it stretches the overall freezing range of the alloy. Gziut et al. observed in their study that an excess of 6 wt% of Al has shown a positive effect on the quality of heat treatment of the alloy even though a lower value also offers a good combination of strength and ductility [3]. On the other hand, the same authors noticed that the addition of Zinc improves both the strength at room temperature and corrosion resistance against impurities such as iron and nickel. In addition, Manganese element have a minor role in the composition of Mg alloy as it will only create an intermetallic with some potential corrosive impurity elements such as iron and other heavy earth material to reduce their corroding ability in the alloy [3]. In a further study, Atwell et al. concluded that the addition of Manganese in the alloy chemical composition is merely when its application is met for a very corrosive environment such as being in contact with sea water [4]. Next, Yang et al. highlighted in their experimental works that the hardenability of the mechanical properties of Mg alloy is enhance with the inclusion of silver. Silicon element is widely use in most alloy composition and its main purpose in Mg alloy is to increase the overall fluidity of the molten metal as well as the creep resistance of the alloy with the formation of Mg2Si particles that will stick to the grain boundaries. The presence of the rare earth element in Mg alloy is to decrease the appearance of porosity and crack in castings by reducing the freezing range of the alloy. Magnesium alloy is use in many industries such as in aviation, medical, electronics and automobile. The main reason that it has been use in the transportation industry is to produce a lighter body structure that is a very critical factor to increase fuel economy and at the same time reduce pollution. The density of magnesium is 33% lighter than aluminium and four times lighter than steel. Apart from being light, Magnesium alloy have an excellent specific strength and stiffness, exceptional dimensional stability, high damping capacity, and high recycle ability [5]. Akyuz found out that Mg alloys can be classified as AM (Al, Mn), AZ (Al, Zn), and AS (Al, Si) series Mg alloys. Nevertheless, in terms of machining of Mg alloys, the author observed that by using the most convenient type of cutting tools and combination of process parameters, the risk of ignition can be reduced. On the other hand, applying a higher cutting speeds in the turning process have shown a high possibility of chip ignition but lowering the cutting forces may contribute to reducing this risk [6]. The same author also mentioned that the production process for Mg is mostly by casting and extrusion process. Next, the author also stated that machining of parts that are made of Mg alloy is quite common especially for those dedicated to the aviation and automotive industry. A very high cutting speed can be applied but serious concern has arisen that there is a high risk of flank build up (FBU) due to adhesion between the cutting tool and the workpiece together with the high risk of ignition.

2. Mechanical Properties

The mechanical properties of Mg alloys AZ31 and AZ91 are quite different to each other due to their chemical composition. Table 1 shows some of the main mechanical properties between Mg alloys AZ31 and AZ91. In a study by Sunil et al, that was investigating on the characteristics of both Mg alloys, they found out that AZ91 shows higher brittleness and hardness due to the high level of aluminium that created a compound known as Mg12Al12. The Vickers hardness for the AZ91was in the range of 80-120 Hz and that of AZ31 was only in the range of 70-95 Hz. According to this experiment, it was suggested that the formation of Mg12Al12 was the main cause of the high hardness. It was also observed that the length of the indent was 33μm when applied to a region dominated by the ß-phase compared to other regions containing α-phase and α+ß phase that were only in the range of 36-48 μm. This data confirm that ß-phase is harder than other phases [7].

Table 1. Elements for AZ31 and AZ91 alloys

Material	Elements (wt%)							
	Al	Zn	Mn	Si	Cu	Ni	Fe	Mg
AZ31	3.05	0.82	0.40	0.020	0.003	0.0012	0.0023	Bal.
AZ91	8.80	0.71	0.19	0.029	0.002	<0.001	0.0010	Bal.

Table 1 shows the chemical composition for Mg alloy AZ31 and AZ91.The major difference between these two grades is the highest wt. % of aluminium content in the AZ91 which enhance the hardness of the alloy when it is combined with the other elements.

Aluminium (Al) is one of the most important elements in magnesium alloy. Al is among the few material that can easily dissolve in magnesium. In addition, the solubility limit of aluminium at eutectic temperature is 11.5 at wt.% (12 mass %) and falls to about 1% at room temperature [9]. The aluminium content in both AZ31 and AZ91 contribute largely to providing a better mechanical property, corrosion resistance, and dies-castability. However, it was observed that at a temperature of above 125°C, the creep resistance tends to be lower that limit the application of this alloy for some components [9].

Currently, Mg12Al12 is not suitable for some automotive parts such as automatic transmission cases due to its high operating temperature that can go up to 175°C whereas an engine block operating temperature can go up to 200°C and an engine piston, even higher than [10-12]. According to Yang et al., the creep resistance is another essential mechanical property to be taken into consideration specially when dealing with selected automotive parts that generate high temperature during their operation. Further development is underway to create a more effective heat resistant magnesium alloy. They can be divided into three categories which are the alloying of Mg-Al based alloys with Calcium (Ca), the combination of Mg based alloys with Zirconium (Zr) and Rhenium (Re) and the addition of zinc (Zn) to magnesium. Moreover, it was reported in a US patent whose authors were Bronfin et al., that the addition of Ca to Mg-Al was met to improve the creep resistance and it was tested in another works conducted by Hollrigl-rosta et al. which concluded that an improvement was observed when 1% of Ca was added to AZ81 [13,14]. In a further study, it was reported that a calcium level of about 1% in Mg alloy can create casting defects such as cold-shuts, hot cracking, and die-sticking. On the other hand, when the Ca levels increases to 2 % the problems were less significantly or completely reduced [15]. The authors also observed that the addition of strontium (Sr) was very efficient in enhancing the corrosion resistance when combine with Mg alloy.

The increase of the aluminium content to 5% and the addition of 1.2 to 1.8% of strontium were demonstrated by Pekguleryuz and Barai that it increases the tensile properties of magnesium alloy as well as the creep resistance of the overall magnesium alloy compare to conventional alloy [16]. Further research on the mechanical properties also highlight the potential of the addition of tin (Sn) in Mg alloy. It was reported by Alikawa and Taketani, that the inclusion of tin in the alloy contributes to an increase in the strength especially for the tensile yield strength [17]. This claim was also supported by Bakke and Westengen that concluded that Mg alloys with high Sn content effectively contribute to a gain in tensile strength that is above AZ91D couple with an increase in creep resistance [18]. Antimony (Sb) is another element that was investigated by Yuan et al. to evaluate its behaviour when combine into Mg alloy [19]. The authors observed that an addition of up to 0.35 wt. % of Sb results in a significant increase in both yield strength and creep resistance. On the other hand, a slight decrease of the alloy ductility was also noted. Furthermore, cerium (Ce) was another element that was subject to a study to observe the final microstructure of Mg alloy especially on AZ91. This study claim that a substantial grain refinement was noted when less than 1.0 wt. % of Ce was added to the alloy which also indicate an increased in tensile strength [20] The effect of Bismuth (Bi) on the aging process on AZ91 alloy was evaluated by Yuan et al. and they concluded that the inclusion of Bi in Mg alloy tends to slowdown the aging process [21]. In addition, Drits et al. found out that Bi has a positive impact on the overall behaviour of Mg alloy in term of heat resistance at elevated temperature [22]. But a decrease in the strength properties was recorded when Bi was included to Mg-Nd alloys in both room temperature and at 250° C. The authors assume that this decrease in strength of Mg alloy can be connected to the decrease of the Nd solubility in Mg solid solution. In another study it was also noted that Ca, Li, Ba and Co have no effect on the tensile strength of Mg alloy [23].

2.1. Machinability
Magnesium alloy is among the types of material that has a good machining ability mainly due to its low hardness and modulus of elasticity. The cutting power recorded during machining operations such as sawing, punching, drilling, milling and turning for magnesium is low and the surface quality is exceptional. In addition, given the softness of magnesium alloys such as AZ 31 and AZ 91, the machining parameters such as the spindle speed can be set at an extremely high level. In addition, with a high thermal conductivity, the tool life is extended and stay sharp for a longer machining time [8]. In terms of power consumption, Figure 1 demonstrate that machining magnesium consume less power compare to other materials listed such as aluminium alloys or nickel alloys [24].

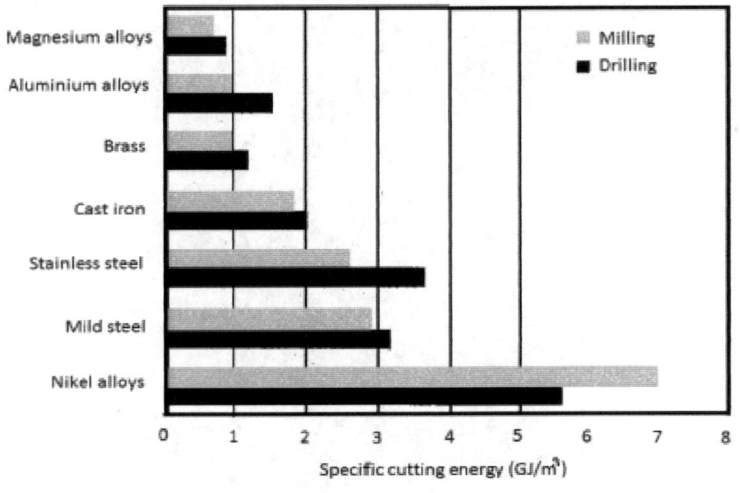

Figure 1. Values of specific power consumption for a variety of metals [24]

Moreover, with the less power requirements in the machining of magnesium alloys, a deeper cuts and higher feed rate are achievable thus creating a more efficient machining operation compare to other materials. Most of the machining operation relation to magnesium are done without any coolant but if necessary, a very light mineral oil is use and its highly recommended not to make use of any water-based coolant due to the risk of ignition. The most common cutting tools used for the machining of magnesium are high speed steel (HSS) or carbide tipped tools. It is also advisable to maintain a sharp cutting tool tip to prevent any unwanted rise of cutting temperature [24]. Moreover, other type of cutting tool such as diamond cutter was also experimented in machining of magnesium alloy and the results were quite promising but unfortunately the cost is relatively high. Next, milling operation using magnesium alloy as base material is very common but the cutting tool should have less teeth compare to standard tools so that the larger chips between the teeth will prevent a high degree of friction resulting in a lower cutting temperature. In addition, this specific type of cutting tool will also allow a higher cutting speed, lower power consumption and finally a better surface finish. To perform a smooth drilling operation on magnesium alloy, a twist drills made of High-speed steel is recommended. In addition, for Die threading of magnesium, a cutting-edge relief of 0.10 mm is generally suggested [24].

2.2. Microstructure

The microstructure of magnesium alloy was found by Czerwinski to be quite complex to understand due to its reaction to the environment that as a result form oxidation [25]. This author claim that the rate of oxidation is increase due to the additions of the alloying elements such as Zn, Sn, Ni, Cu, In, Ag, Cd, and TI. However, a study suggests that the addition of some reactive element such as rare earth metals can decrease oxidation in magnesium alloy [26]

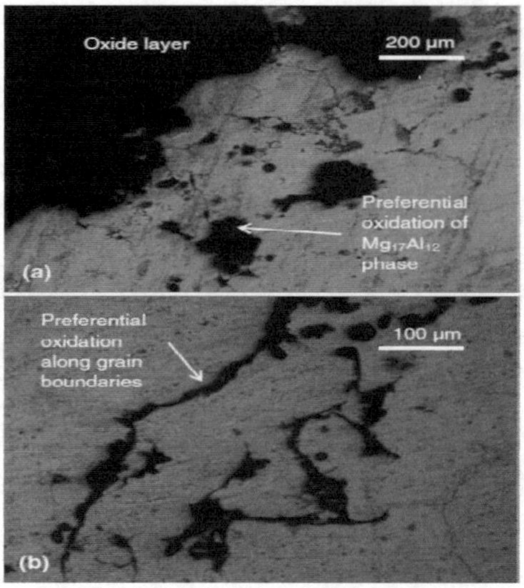

Figure 2. Effect of alloy phase composition on preferential oxidation at high temperatures: (a) subsurface oxidation; (b) selective oxidation of grain boundaries in AZ 91 [27]

Furthermore, in an experimental study on the evolution of oxidation of AZ 91 as shown in figure 2, the authors observe that the reaction of the alloy to oxidation is not dependent on the thickness of the scale but assume that oxygen can easily penetrate the metal surface and create oxidation. In addition, it was also reported that the oxidation is also directly related to the temperature as at a temperature of 430ºC, a catastrophic oxidation was form. At this specific stage, the nodular growth is triggered by a different phase composition and inhomogeneity of the alloy which in the end create a liquid island of metal [27]. The nodular growth can also be prevented by the addition of rare earth material as stated earlier. As stated in in another findings, the rare earth ions tend to diffuse to native oxide grain boundaries and at the same time blocking the outward diffusion of metals cations. Therefore, an internal diffusion of oxygen ions is the only option for the oxidation to build-up. The thin-oxide layer form due to the reaction of Magnesium oxide (MgO) is not a permanent solution to prevent oxidation because they are very sensitive to environmental factors and may be subjected to severe damage. On the other hand, if the MgO is combine to a reactive element such as rare earth, it was reported that the magnesium alloy was given a better protection against the environment or even during in service in high temperature conditions [27].

3. Conclusion

The review on magnesium alloy briefly expose some of the key elements involve in the current manufacturing industries such as mechanical properties, microstructure, and machinability. All these elements are interrelated to each other starting from the chemical composition of the material where the control of the grain structure is very essential to achieve a required level of strength, ductility, and toughness. The control of the grain structure has been very successful in other material such as aluminium alloy, but more study is needed for magnesium alloy to be fully accepted by the manufacturing industries. In addition, the high probability of oxidation has also been discussed and much more study is needed to expand the application of magnesium alloy in other highly corrosive environment. Concerning the machinability of magnesium alloy, this review highlight that most machining operation are conducted in a smooth way by using the proper cutting tools and cutting parameters to prevent the risk of ignition that is very common when dealing with magnesium alloy.

REFERENCES

[1] Huang Y, Hajo D, Kainer K and Norbert H. 2014 *J. Mg. Al.* **2** 124-132
[2] Shi K, Zhang D, Ren J, Yao C and Huang X 2016 *Adv. Mech. Eng.* **8(1)** 1-9
[3] Gziut O, Józef K and Ireneusz Z 2015 *Manag. Prod. Eng. Rev.* **6(1)** 4-9
[4] Atwell D L, Barnett M R and Hutchinson W B 2012 *Mat. Sci. Eng.* **A (549)** 1-6
[5] Yang Z, Li J P, Zhang J X, Lorimer G W and Robson J 2008 *Acta Metallurgica Sinica (English Letters)* **21(5)** 313-328
[6] Akyuz B 2013 *Transactions of Nonferrous Metals Society of China* **23(8)** 2243-2249
[7] Sunil B R, Ganesh K V, Pavan P, Vadapalli P, Swarnalatha C, Swapna P, Bindukumar P and Reddy G P K 2016 *J. Mg. Al.* **4(1)** 15-21
[8] Buldum B B, Aydın S and Iskender O 2012 *Int. J. Electron. Mech. Mechatron.* **2** 261-268
[9] Suman C 1991 *SAE Technical paper No.910416* (Warrendale, PA, Society of Automotive Engineers)
[10] Luo A A 2004 *Magnesium Technology* (TMS, Warrendale)
[11] Lyon P, King J F and Nuttall K 1996 *Proc. 3rd Int. Magnesium Conf.* (Manchester: UK) pp 99-108
[12] Bronfin B and Aghion E 2001 *Magnesium Technology* (J N Hryn Pub, London)
[13] Bronfin B, Aghion E, Von Buch F, Schumann S and Katzir M 2006 US patent No 7041179
[14] Hollrigl-rosta F, Just E, Kohler J and Melzler H 1980 *Light Metals Age* **1980** 22-29
[15] Powell B R, Rezhets V and Luo A A 2001 US Patent No. 6,264,763
[16] Pekguleryuz M O and Barai E 2001 *Magnesium Technology* (TMS, Warrendale) p 119
[17] Alikawa K and Taketani K 1994 *US patent No.5304260*
[18] Bakke P and Håkon W 2016 *Essential Readings in Magnesium Technology* (Springer International Publishing) p 313-318
[19] Yuan G Y, Li L Z, Dong W Q and Ding W J 2002 *Mater. Lett.* **56(1)** 53-58
[20] Yang Z, Li. J P, Li G H and Yang J M, *Mater. Sci. Forum*, pp. 488-489 (2005) 219.
[21] Yuan G Y, Sun Y S and Zeng X Q 2001 *J. Shanghai Jiaotong Univ. Chinese Edition* **35(3)** 451-456
[22] Drits M E, Sviderskaya Z A and Rokhlin L L 1962 *Zhurnal Neorganicheskoi Khimii* **7** 2771-2777
[23] Rokhlin L L 2003 *Magnesium Alloys Containing Rare Earth Metals: Structure and Properties*. (Crc Pres)
[24] Smith W F 1993 *Society of Manufacturing Engineers, Forces at The Cutting Tool, Tooling, and Manufacturing Engineers Handbook* (McGraw-Hill)
[25] Czerwinski F and Szpunar J 1998 *Acta Materialia* **46(4)** 1403-1417
[26] Czerwinski F 2002 *Acta Materialia* **50(10)** 2639-2654
[27] Czerwinski F 2012 *JOM* **64(12)** 1477-1483.

Pattern Recognition of Facial Electromyography (FEMG) Signal for Aceh Language Speech using Naïve Bayes and Learning Vector Quantization (LVQ)

Darma Setiawan Putra[1], Yuril Umbu Woza Weru[1], Fardiansyah[1], Fitriady[2], Fakhruddin[2] and Z.Yahya[3*]
[1] Politeknik Aceh Selatan, Infromatics Engineering Department, Jalan Merdeka, Kompleks Reklamasi Pantai, Tapaktuan, Aceh Selatan, Indonesia
[2] Politeknik Aceh, Industrial Electronics Engineering Department, Jalan Politeknik Aceh Gp. Pango Raya, Banda Aceh, Indonesia
[3]Faculty of Computing and Multimedia, Kolej Universiti Poly-Tech MARA, Kuala Lumpur

*Corresponding e-mail: zahrah@kuptm.edu.my

Abstract. Facial electromyography is myoelectric signals that formed by human facial muscles. This signal can be acquired by attaching an electrode to the facial muscle that has been connected with an electromyography sensor. When human say certain words, the articulation muscles contract and facial electromyography signals appear in the muscles. This study aims to recognize patterns in facial electromyography signals by classifying signals using naïve bayes and learning vector quantization classifier. Feature extraction used one-dimensional discrete wavelet transforms. Wavelet transform type used wavelet daubechies2 level 5. The transformation produces a level 5 approximation coefficient called a5 and five detail coefficients called d1, d2, d3, d4, and d5. The result of this study show that the average classification accuracy for *ho neuk jak* sentence using naïve bayes and LVQ classifier was 62.5% and 92.5% respectively. The average classification accuracy for *ja'* word using naïve bayes and LVQ classifier was 70% and 92.5% respectively. The average classification accuracy for *ja' wo* sentence using naïve bayes and LVQ classifier was 52.5% and 90% respectively. The average classification accuracy for *pane* word using naïve bayes and LVQ was 70% and 90% respectively. The average classification accuracy for *soe* word using naïve bayes and LVQ classifier is 85% and 95% respectively. Thus, this study shows that when humans say the words, facial electromyography signals that appear on facial muscles difference for each subject.

Keyword: Pattern Recognition, Facial Electromyography, Aceh Language Speech, Naïve Bayes, Learning Vector Quantization

1. Introduction

The development of science and technology has influenced human's life, especially in the field of pattern recognition. Human have electromyography (FEMG) signal that occurs due to muscle contraction. In addition to the signals that appear in the muscles of the arms and lower body, the human also have electromyography signals in the face. Signals that appear on the face are called facial electromyography (FEMG) signal.

Facial electromyography is usually used to measure the expression of emotions on human faces. This FEMG signal is recorded by attaching an electrode to facial skin. This expression of emotion is usually an expression of pleasure, sadness, and anger. The facial muscles commonly used to measure this expression are corrugators supercili, zygomaticus major and orbicularis oculi [1,2]. This emotion expression can be interpreted with the magnitude of facial electromyography signal

FEMG signals can also be used to recognize the human facial gestures for assistive and rehabilitation technology [3-4]. In addition, to measure emotional expressions and facial gestures, FEMG signals can also be used to distinguish speech in communicating through the signal patterns. When humans do speech-language, the articulation muscles around the mouth contraction that causes FEMG signals. FEMG signals that appear in certain muscles analysed specifically for being used in human speech patterns recognition. The difference in language speech was indicated by differences in FEMG signal patterns. Thus FEMG signal pattern can be used as a unique identity. In addition, this FEMG signal can also be used as a medium to convey information in understanding the language conveyed to deaf people.

2. Related Studies

Speech recognition based on facial electromyography (FEMG) signals has been carried out by several researchers. Research conducted by [5] about silent speech interfaces on the introduction of Spanish syllables based on EMG signals that are on facial muscle. The syllables used are vowels, labials, dentals, palatals, velars, and alveolar. The validation method used 10 fold cross-validations. The results showed that average 70% of the 30 syllables could be recognized. The next research conducted by [6] about the introduction of Thai language for classifying five tones based on electromyography signals recorded from six electrode positions placed on the face and neck muscles when the participant is speaking 21 words of Thai with five tones for each word. Artificial neural networks were used to classify the EMG signals. The feature of EMG signal was a signal that has the 5 highest values from the RES index. The results show that the accuracy was 56.2% for classifying five Thai tones. The next research conducted by [7] about the introduction of vowels in the spelling of the 11 letters of Bangli. The EMG signal feature selection uses the minimum Redundancy Maximum Relevance (mRMR) method and the signal classification uses an artificial neural network (ANN). The results show that the accuracy value was 82.3%. The next research conducted by [8] about how speech synthesis techniques were directly derived from surface electromyography signals in facial articulation muscles. Four methods of capturing the features used are gaussian mixture model (GMM), deep neural network (DNN), long short term memory (LSTM) and unit selection. Among the four methods discussed, the DNN method has shown the best performance.

In this study, researcher conducted a classification of facial electromyography signals in Acehnese speech using two classifiers, namely naïve bayes and learning vector quantization. The feature extraction method used wavelet transform type daubechies2 level 5 by calculating the mean for the approximation coefficient (a5) and the detail coefficient (d1-d5).

3. Methodology

The method used in this study is carried out in several stages. There are FEMG signal acquisition, FEMG signal extraction using wavelet transform and classification using naïve bayes and learning vector quantization (LVQ).

3.1. Research Flowchart

Research flowchart of this study is shown in Figure 1.

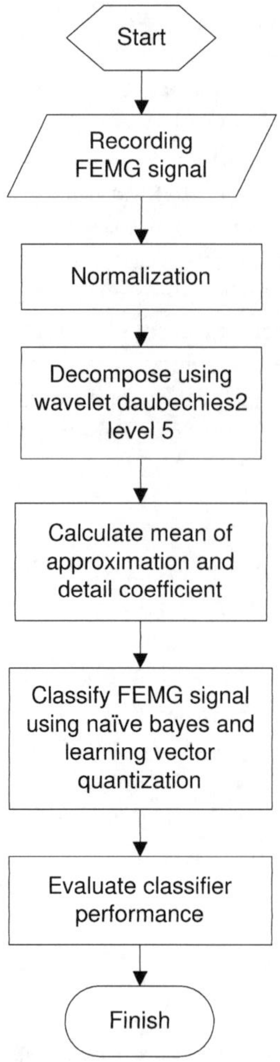

Figure 1. Research flowchart.

3.2. Data Acquisition

The data acquisition by using several devices namely surface electrode, FEMG sensor, and arduino uno. Surface electrode and FEMG sensor are shown in Figure 2. The FEMG signal can be obtained by attaching a surface electrode to masseter, risorius and depressor muscles. The reference electrode is attached to the masseter muscle. The electrode placement point of FEMG is shown in Figure 3.

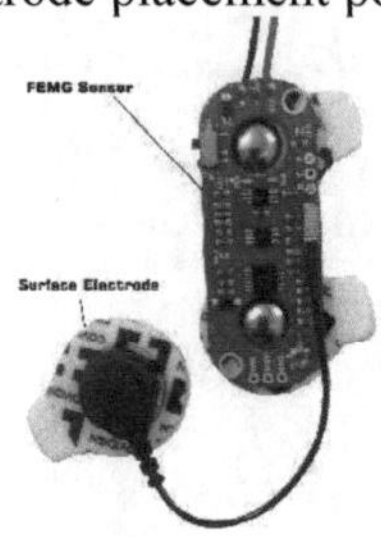

Figure 2. The FEMG sensor and surface electrode.

IOP Conf. Series: Materials Science and Engineering 1062 (2021) 012055 doi:10.1088/1757-899X/1062/1/012055

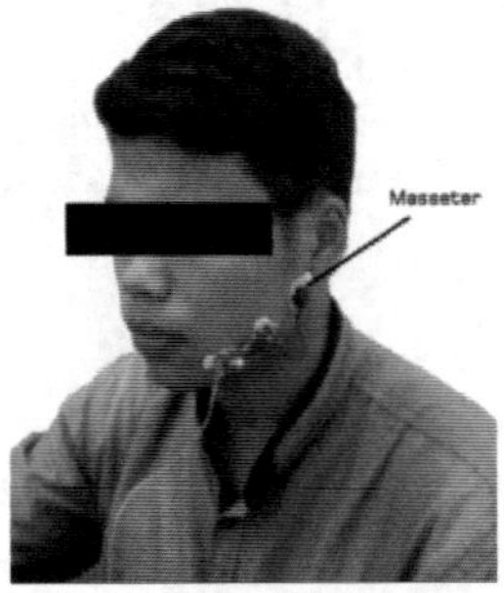

Figure 3. The electrode placement point.

The FEMG signal recording uses CoolTerm software. Every word spoken will be standardized by giving initialization time at the beginning and termination time in the end. This initialization and termination time is needed to ensure the normal conditions. The initialization and termination time is 5 seconds.

The number of subjects involved in this study was 4 native Acehnese speakers. They said as many as 5 words as shown in Table 1. Each subject was taken for 10 times of the pronunciation of the word.

Table 1. English and Aceh Language Vocabulary.

No.	English	Aceh
1.	Where	Ho neuk jak
2.	Go	Ja'
3.	Go home	Ja' wo
4.	Come from	Pane
5.	Who	Soe

3.3. Data Extraction

The FEMG signal was normalized using zero mean and extracted using wavelet transforms. An overview of the wavelet transform decomposition process is shown in Figure 4.

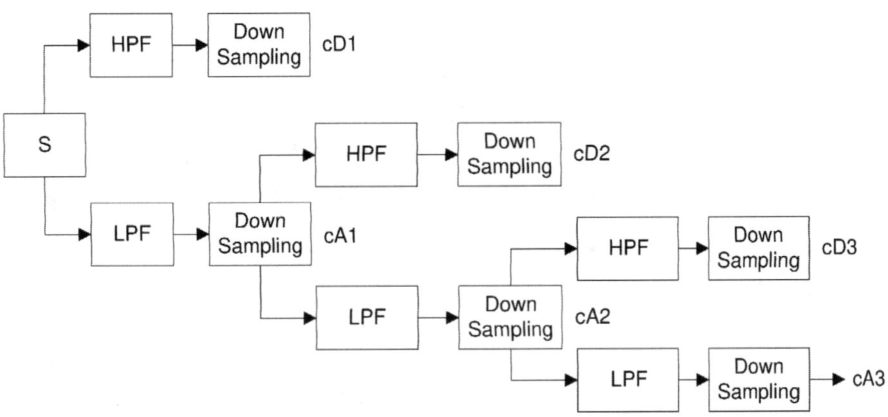

Figure 4. An overview of wavelet transform decomposition for FEMG signal.

Figure 4 above shows that the wavelet transformation decomposition uses two filters, low pass filter and high pass filter. The low pass filter decomposes the signal to produce approximation components and the high pass filter decomposes the signal to produce detailed components. In this study, the wavelet transform used daubechies2 level 5. It means that it produces an approximation component (a5) and detail components (d1 – d5).

Mean value was calculated for each approximation and detail component so that the signal feature for classifying is six units. The LVQ network simulation uses weka opensource software. The method for training and testing data in LVQ dan naïve bayes classifier uses 5 – cross validation. Learning rate is 0.1 and iteration is 10.

4. Result and Discussion

The features of EMG signal were trained and tested using naïve bayes and learning vector quantization. The result of training and testing data can be obtained as shown in Table 2. The accuracy of the value can be calculated from true positive, true negative, false positive and false negative obtained [9].

Table 2. The accuracy of FEMG signal classification.

No.	Vocabulary	Classifier	Subject	Accuracy (%)
1.	Ho neuk jak	Naïve bayes	#1	90
			#2	70
			#3	30
			#4	60
		LVQ	#1	100
			#2	80
			#3	90
			#4	100
2.	Ja’	Naïve bayes	#1	80
			#2	30
			#3	90
			#4	80
		LVQ	#1	100
			#2	100
			#3	70
			#4	100
3.	Ja’ wo	Naïve bayes	#1	30
			#2	100
			#3	60
			#4	20
		LVQ	#1	100
			#2	100
			#3	70
			#4	90
4.	Pane	Naïve bayes	#1	80
			#2	90
			#3	90
			#4	20
		LVQ	#1	100
			#2	100
			#3	70
			#4	90
5.	Soe	Naïve bayes	#1	80
			#2	100
			#3	80

	#4	80
LVQ	#1	100
	#2	100
	#3	80
	#4	100

Table 2 shows that every word pronounced by each subject has a different classification accuracy. For the pronunciation of the word *ho neuk jak*, LVQ classifier has better accuracy than naïve bayes classifier with the greatest classification accuracy in the amount of 100% for subject#1 and subject#4. For the pronunciation of the word *ja '*, LVQ classifier has better accuracy than naïve bayes classifier with the greatest classification accuracy in the amount of 100% for subject#1, subject#2 and subject#4. For the pronunciation of the word *ja' wo*, LVQ classifier has better accuracy than naïve bayes classifier with the greatest classification accuracy in the amount of 100% for subject#1 and subject#2. For the pronunciation of the word *pane*, the LVQ classifier has better accuracy than the naïve bayes classifier with the greatest classification accuracy in the amount of 100% for subject#1 and subject#2. And for the pronunciation of the word *soe*, the LVQ classifier has better accuracy than the naïve bayes classifier with the greatest classification accuracy in the amount of 100% for subject#1, subject#2 and subject#4.

In general, the description of the classification accuracy for each word pronunciation for naïve bayes classifier is shown in Figure 5 while the classification accuracy for each word pronunciation for learning vector quantization classifier is shown in Figure 6

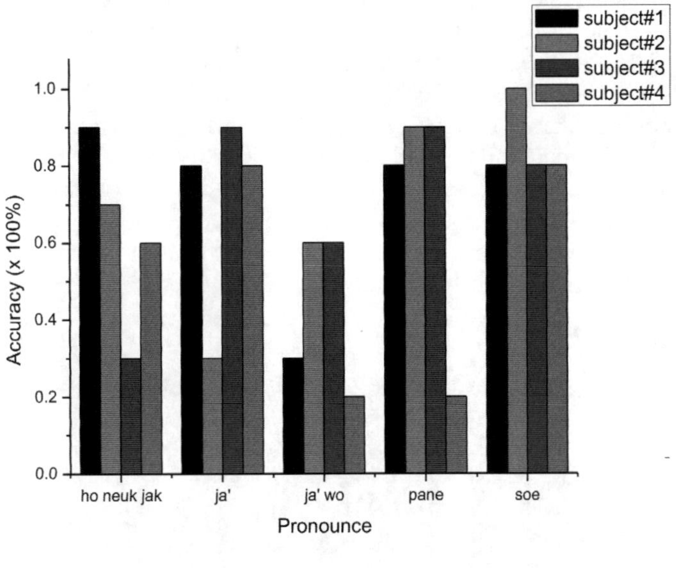

Figure 5. Accuracy of FEMG signal classification for naïve bayes classifier.

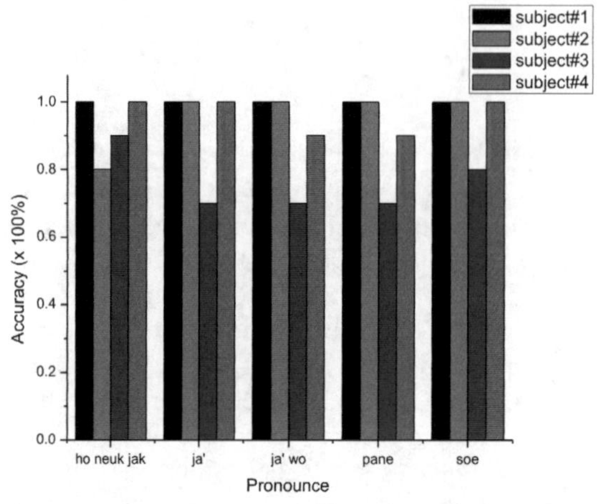

Figure 6. Accuracy of FEMG signal classification for LVQ classifier.

Figure 5 explains that subject#2 has only the greatest classification accuracy using naïve bayes classifier for pronouncing the word *soe* by 100%. While Figure 6 explains that subject#1 and subject#4 have classification accuracy using LVQ classifier for pronouncing the word *ho neuk jak* by 100%. Subject#1, subject#2 and subject#4 have classification accuracy for pronouncing the word *ja'* by 100%. Subject#1 and subject#2 have classification accuracy for pronouncing the word *ja' wo* by 100%. Subject#1 and subject#2 have classification accuracy for pronouncing the word *pane* by 100%. Subject#1, subject#2 and subject#4 have classification accuracy for pronouncing the word *soe* by 100%. The classification accuracy of 100% explains that all tested FEMG signals are very well recognized. This indicates that the FEMG signal has a different pattern between one subject and another so that this FEMG signal can be used as a unique identity for every human.

The ROC area value for each subject is shown in Figure 7. The maximum ROC value is 1. There is only 1 subject that has an ROC value below 0.7, naive bayes classifier for subject#3 (*ho neuk jak* and *ja' wo*). In this study, the performance of the classifier was found to be very good for classifying the FEMG signal.

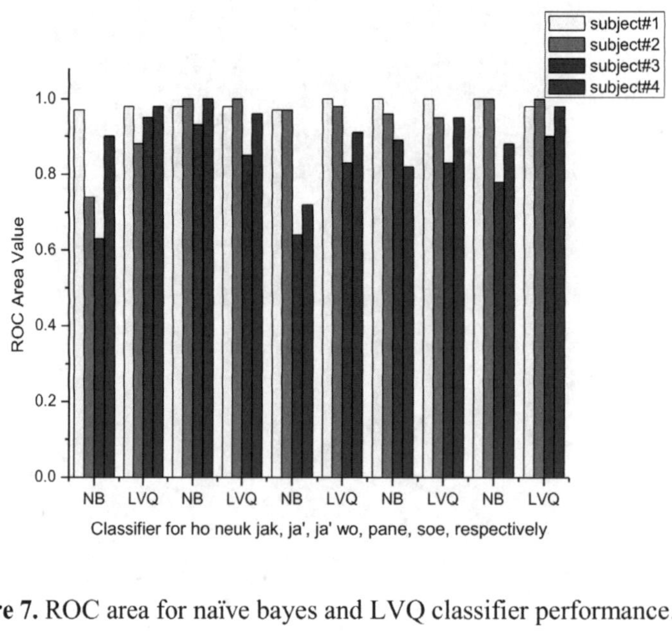

Figure 7. ROC area for naïve bayes and LVQ classifier performance.

5. Conclusions

The results of the study also suggest that the facial electromyography signal in facial muscle can be used to identify speech recognition. The best facial electromyography signal classification accuracy is LVQ classifier. The classification accuracy for *ho neuk jak*, *ja'*, *ja' wo*, *pane* and *soe* are 92.5%, 92.5%, 90%, 90% and 95% respectively. In this study has also shown that the feature selection and classifier method was sophisticated for studying the FEMG signal pattern recognition. Further research might investigate the FEMG signal classification involving more participants so that the classification can achieve better accuracy.

Acknowledgments

Thanks to the Indonesia Ministry of Research, Technology and Higher Education for financial support in Beginner Lecturer Research scheme No. D/79.D/POLTAS/IV/2019.

References

[1] Read G L 2017 Facial Electromyography (EMG) *The International Encyclopedia of Communication Research Methods* 1-10
[2] Mavratzakis A, Herbert C and Walla P 2016 Emotional facial expressions evoke faster orienting responses, but weaker emotional responses at neural and behavioral levels compared to scenes: A simultaneos EEG and facial EMG study *NeroImage* **124** 931-946
[3] Cai Y, Guo Y, Jiang H, Huang M C 2017 Machine-learning approaches for recognizing muscle activities involved in facial expressions captured by multi-channels surface Electromyogram *Smart Health* **5(6)** 15-25
[4] Orguc S, Khurana H S, Stankovic K M, Leel H S and Chandrakasan A P 2018 EMG-based Real Time Facial Gesture Recognition for Stress Monitoring *2018 40th Annual International Conference of the IEEE Engineering in Medicine and Biology Society (EMBC)* 2651-2654
[5] Lopez-Larraz E, Mozos O M, Antelis J M and Minguez J 2010 Syllable-Based Speech Recognition Using EMG *International Conference of the IEEE Engineering in Medicine and Biology* **4699-702**
[6] Srisuwan N, Phukpattaranont P and Limsakul C 2013 Three Steps of Neuron Network Classification for EMG-based Thai Tones Speech Recognition *International Conference on Electrical Engineering/Electronics, Computer, Telecommucation and Information Technology* 1-6
[7] Mostafa S S, Awal M A, Ahmad M and Rashid M A 2016 Voiceless Bangla vowel recognition using sEMG signal *Springerplus* **5(1)** 1522
[8] Janke M and Diener L 2017 EMG-to-Speech : Direct Generation of Speech From Facial Electromyographic Signals *ACM Transaction on Audio, Speech, and Language Processing* **25(12)** 2375-2385
[9] Putra D S, Weru Y U W and Fitriady 2019 Pattern recognition of electromyography (EMG) signal for wrist movement using learning vector quantization (LVQ) *IOP Conference Series: Materials Science and Engineering* **506** 12020

Candlenut Shell Charcoal Yield in Every Chamber Produced by Candlenut Shell Carbonization Tool Using a Vertical Multi Chamber

Asbahrul Amri[1], Oktalia Triananda Lovita[1], Devi Satria Saputra[1], Ihsan[1], Fitriani[2], Mahdi Arifan[2], Miki Saputra[2] and Z.Yahya[3*]

[1] South Aceh Polytechnic, Department of Industrial Engineering Jalan Merdeka – Komplek Reklamasi Pantai, Tapaktuan, Indonesia
[2] Industrial Engineering Department, South Aceh Polytechnic
[3]Faculty of Computing and Multimedia, Kolej Universiti Poly-Tech MARA, Kuala Lumpur

*Corresponding e-mail: zahrah@kuptm.edu.my

Abstract. Candlenut shell contained organic compound, such as hemicelluloses, cellulose, and lignin which can be converted to solid, liquid, and gaseous products by thermochemical methods. One of the products is charcoal. One of the tool that can carbonize the candlenut shell is candlenut shell carbonization tool using a vertical multi chamber. The tool can carbonize candlenut shell without having any additional energy. Some variables from the tool has been identify from the previous research, such as its maximum capacity and process time, while its charcoal yield remains unknown. Therefore, this research is aimed to determine the charcoal yield that can be produced by candlenut shell carbonization tool using a vertical multi chamber. Three carbonization experiments were performed with 8 kg candlenut shell filled in every carbonization chamber (CC). The carbonization process temperature was measured for each chamber by using a thermocouple in order to know the effect of temperature toward charcoal yield. As the result, the average candlenut charcoal yield that can be produced by candlenut shell carbonization tool using a vertical multi chamber is 24.99 %. Every chamber was experiencing deferent capability in producing charcoal. CC3 is the highest charcoal yield chamber with 31.38 % and then followed by CC2 with 25.04 % and CC1 with 18.79 %. Based on temperature that experiencing by every chamber, it indicates that the temperature has a significant impact on charcoal yield. When the temperature high, the charcoal yield is low and also goes the same in contrary. Therefore, candlenut shell carbonization tool using a vertical multi chamber has capability in producing candlenut shell charcoal at 24.99 % yield.

Keyword: Candlenut shell, carbonization, charcoal yield, candlenut shell carbonization tool

1. Introduction

Candlenut/Aleurites moluccana (L.) Willd which is one type of industrial plant from the family of Euphorbiaceae [1] has been planted for decades in Indonesia with both commercially and non-commercially purposes. This kind of plant has some advantages: its seed can be used as a media for a light, food and medicine, while its tree is used for timber [2]. The seeds are contained within a hard, black, rough shell ellipticalin shape and about 2.5-3.5cm (1-1.4 inches) long [3]. Hence, Candlenut consists of two main parts which are candlenut seed and candlenut shell. Candlenut and its parts are shown in Figure 1.

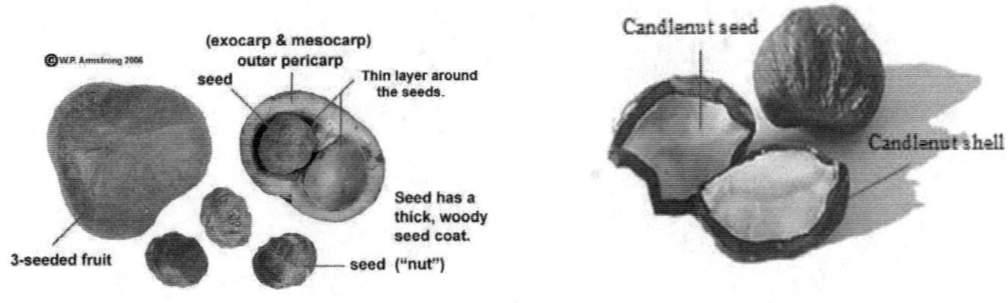

(a) (b)

Figure 1. Candlenut and its parts;(a) Candlenut fruit [4];(b) Candlenut seed and its shell [5].

Candlenut seeds were taken by breaking the shells as the solid waste from the candlenut fruit. Candlenut shell is 3-4 mm thick [6] containing organic compound, such as hemicelluloses, cellulose, and lignin [7] and one of biomass waste which is a renewable energy source. Its potential has to be utilized as a part of minimizing the waste and also a new option for energy by converting its form. The biomass can be converted using biochemical, chemical, and thermochemical processes. These days, thermal conversion is found widely used in the community and has received special attention since it leads to useful products and simultaneously contributes to solving pollution problems arising from biomass accumulation [8]. Biomass materials can be converted to solid, liquid, and gaseous products by thermochemical methods. One of the products is charcoal. Charcoal is produced by slow heating wood (carbonization) in airtight ovens or retorts, in chambers with various gases, or in kilns supplied with limited and controlled amounts of air [9]. Therefore, some tools were developed to produce charcoal using carbonization method.

One of biomass conversion tool that used carbonization method in converting biomass into charcoal is Candlenut Shell Carbonization Tool using a Vertical Multi Chambers. The tool can carbonize the candlenut shell without any additional energy and process 24 kg candlenut shell in 7.75 hours [10], while the amount of charcoal that produced by the tool remain unknown. Hence, this research is aimed to determine the yield of the charcoal produced by Candlenut Shell Carbonization Tool Using a Vertical Multi Chambers with temperature consideration.

2. Methodology

This section will discuss about the method in answering research objectives. They are material, research design and how to calculate charcoal yield produced by Candlenut shell carbonization tool using a vertical multi chambers.

2.1. Material

Candlenut shell is taken from candlenut collectors in South Aceh District. They get candlenut from surrounding candlenut farmers in the area. Candlenut shell is obtained from a manual process of taking candlenut seed done by collectors. This process resulted candlenut shell in irregular shape and varies in size. Crushed candlenut shell is shown in Figure 2.

Figure 2. Crushed candlenut shell.

2.2. Candlenut Shell Carbonization Tool Using a Vertical Multi Chambers

Candlenut shell carbonization tool using a vertical multi chambers is invented in 2019 as a respond of candlenut farmers need. The tool is designed using 3 vertical carbonization chambers, carbonization chamber I, II, and III, which aims to provide space for adequate air supply. It will attain a perfect burning process that generates enough heat to facilitate other chambers, carbonization chamber II and III, having carbonized. Each chamber has Ø60cm x 30cm in size. The main components are a metal drum, a reservoir, two chamber barriers, and an exhaust pipe [10]. Figure 3 will give a better understanding about the tool

The tool carbonizes the shell in every chamber after initial burning done in chamber I. It is starting from chamber I, chamber II and finally chamber III. The carbonization process was carried out through predetermined stages which are burning and cooling stage [10].

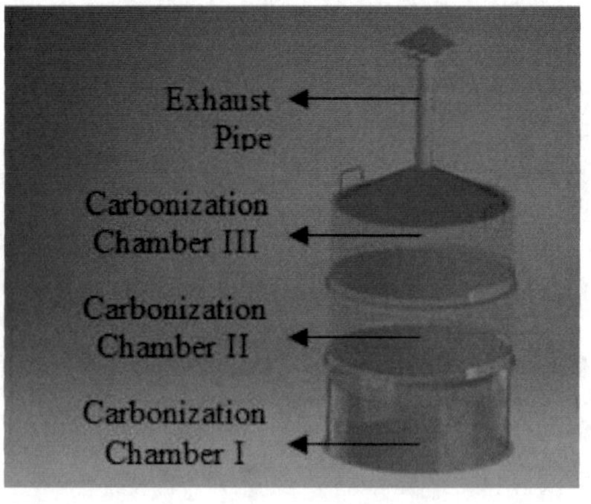

Figure 3. Candlenut shell carbonization tool using a vertical multi chamber.

2.3. Research Design

The research design is made to answer the objectives which are carbonization process temperature and charcoal yield produced by candlenut shell carbonization tool using a vertical multi chambers. The research will be performed based on the maximum capacity of the tool which is 24 kg candlenut shell in a process [10]. It is mean that every chamber will carbonized 8 kg candlenut shell. The research will be running 3 time experiments in order to have enough data to be analyzed. Table 1 lists how the experiment will be done.

IOP Conf. Series: Materials Science and Engineering 1062 (2021) 012056 doi:10.1088/1757-899X/1062/1/012056

Table 1. Characteristic of every experiment.

Experiment	The amount of candlenut Shell (Kg)		
	Chamber I	Chamber II	Chamber III
I	8	8	8
II	8	8	8
III	8	8	8

2.4. Charcoal Yield

After having charcoal from the carbonization process, it is weighed to identify the quantity of the charcoal gained in every carbonization chamber. Both candlenut shell and charcoal quantity are used to determine charcoal yield. Hence, the charcoal yield is determined according to the equation (1) below [11]:

$$y_{char} = \frac{m_{char}}{m_{bio}} \times 100 \quad (1)$$

where y_{char} is charcoal yield (%), m_{char} is mass of charcoal (g) and m_{bio} is the mass of biomass (g).

2.5. Research Flow

The following is a flow diagram in determining candlenut shell charcoal yield produced by candlenut shell carbonization tool using a vertical multi chambers as shown in Figure 4.

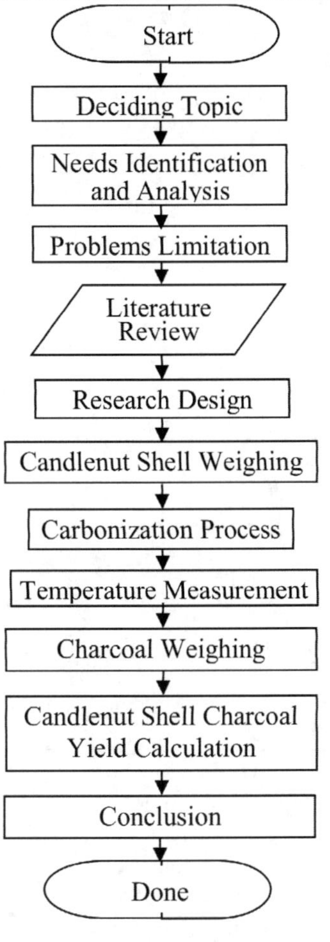

Figure 4. Candlenut shell carbonization tool using a vertical multi chamber.

3. Result and Discussion

3.1. Candlenut Shell Carbonization Temperature

As the carbonization process consisted of two predetermining stages, burning and cooling, so the temperature measured separately. In burning stage, temperature was measured gradually in every 15 minutes and the first measurement was done 15 minutes after initial burning in the carbonization chamber (CC) 1. When burning stage finished, cooling stage was started by closing all air inlet on the tool. After 30 minutes since the cooling stage started, the first temperature measurement was completed and it was frequently done for every next 30 minutes until the temperature in every chambers equal with the ambient temperature which is 32 °C. The following Table 2 shows all temperature measured for all experiments in every chambers.

Table 2. Candlenut shell carbonization temperature in Experiment I.

Time	Experiment II (8 kg/CC)									Rata-rata
	1st Run			2nd Run			3rd Run			
	CC 1	CC 2	CC 3	CC 1	CC 2	CC 3	CC 1	CC 2	CC 3	
Burning Process										
0 m	32	32	32	32	32	32	32	32	32	32.0
15 m	210	92	77	212	95	78	211	94	78	127.4
30 m	311	133	94	315	135	96	310	134	95	180.3
45 m	330	265	183	333	268	185	330	265	185	260.4
60 m	297	344	258	330	345	258	298	345	260	303.9
75 m	230	269	209	232	270	210	230	270	211	236.8
Cooling Process										
0.5 h	82	77	67	83	77	68	82	78	69	75.9
1 h	58	55	50	58	56	51	59	56	52	55.0
1.5 h	42	43	39	43	44	40	43	44	41	42.1
2 h	37	40	36	37	42	37	38	42	38	38.6
2.5 h	35	38	35	36	39	36	36	39	37	36.8
3 h	33	35	35	35	38	36	33	37	36	35.3
3.5 h	33	35	34	33	35	35	33	35	33	34.0
4 h	32	34	34	33	34	34	32	35	34	33.6
4.5 h	32	34	34	32	34	34	32	34	34	33.3
5 h	31	32	33	31	33	33	31	33	34	33.0
5.5 h		32	33		32	33		33	33	32.7
6 h		32	32		32	33		32	33	32.3
6.5 h		31	32		31	32		31	32	31.5
7 h			31			31			31	31.0

Table 2 revealed that the burning stage is able to generate its maximum temperature at 345 °C which is gain in Experiment II and III, while Experiment I was only at 344 °C. All of those highest temperature was found in CC2. It may occur because CC2 is placed in between CC1 and CC3 vertically where the heat from the CC1 and CC3 were positively contribute to the increased temperature. In addition, the tool's maximum average temperature is occurred after 60 minutes of burning process in CC2 at 303.9 °C. Figure 5 below is showing how the temperature increased in each chambers at the burning stage.

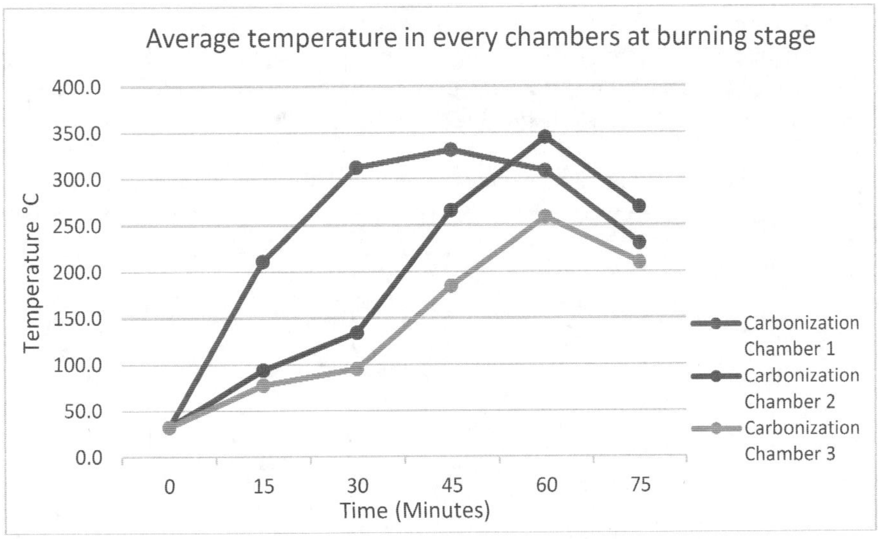

Figure 5. Temperature in every chambers at burning stage.

In the cooling stage heat gradually decrease by time. Although each experiment was experiencing identic cooling process, each chamber reached ambient temperature differently. CC1 completed the cooling process in 4.5 hours, CC2 in 6 hours and CC3 in 6.5 hours. Hence, CC1 is the first chamber that reach ambient temperature compared to other chambers. The following Figure 6 will give a better understanding how the temperature decreased.

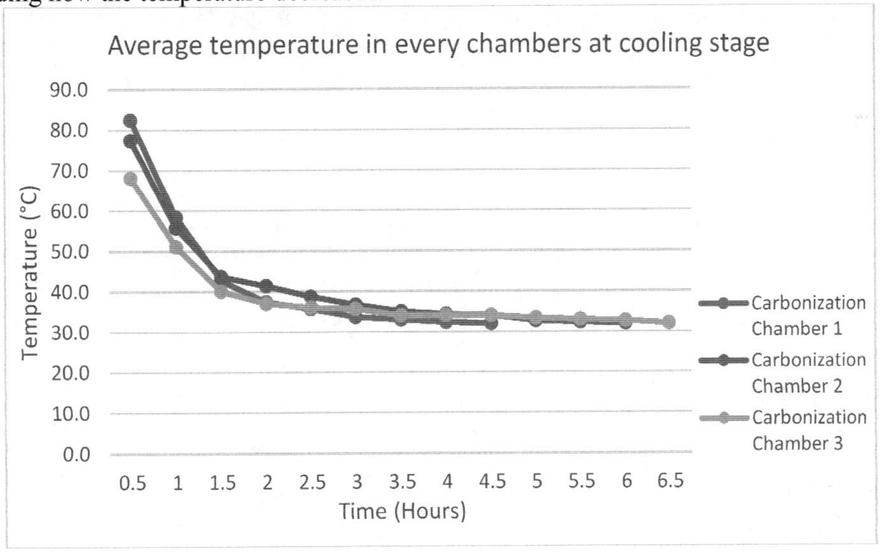

Figure 6. Temperature in every chambers at cooling stage.

The result demonstrated that the carbonization process takes 7.75 hour to carbonized 24 kg candlenut shell that distributed evenly to 3 carbonization chambers with maximum temperature 345 °C and all maximum temperature is found in CC2 with average 303.9 °C.

3.2. Candlenut Shell and Candlenut Shell Charcoal Quantity

After carbonization process, all candlenut shell charcoal in every chambers were place separately. Soon, those charcoal weighed to identify the quantity of the charcoal. Those data will be used to determine candlenut shell charcoal yield. Table 3 is listing the quantity of charcoal in every carbonization chamber.

Table 3. Candlenut shell and candlenut shell charcoal quantity.

Experiment	CC	Quantity	
		Candlenut Shell (kg)	Candelnut Shell Charcoal (kg)
I	CC1	8	1,50
	CC2	8	2,00
	CC3	8	2,50
II	CC1	8	1,49
	CC2	8	2,00
	CC3	8	2,52
III	CC1	8	1,50
	CC2	8	1.90
	CC3	8	2.51

3.3. Candlenut Shell Charcoal Yield

Candlenut shell charcoal yield produced by candlenut charcoal carbonization tool using a vertical multi chambers is calculated based on data in Table 2. The yield calculation is done by utilizing equation 1 and the results are given in Table 4.

Table 4. Candlenut shell charcoal yield.

Experiment	CC	Candlenut Shell Charcoal Yield (%)	Average Candlenut Shell Charcoal Yield (%)	Total Average Candlenut Shell Charcoal Yield (%)	Average Candlenut Shell Charcoal Yield in CC1 (%)	Average Candlenut Shell Charcoal Yield in CC2 (%)	Average Candlenut Shell Charcoal Yield in CC3 (%)
I	CC1	18.75	25.00	24.99	18.79	24.79	31.38
	CC2	25.00					
	CC3	31.25					
II	CC1	18.63	25.04				
	CC2	25.00					
	CC3	31.50					
III	CC1	19.00	24.92				
	CC2	24.38					
	CC3	31.38					

From the result listed in Table 4 it is clear that every experiment resulted in identical number of candlenut shell charcoal yield. They are 25 % in Experiment I, 25.04% in Experiment II and 24.92% in Experiment III. These results can be concluded that there is no significant yield difference from all experiments that has been done and the average candlenut shell charcoal yield produced by candlenut shell carbonization tool using a vertical multi chambers is 24.99 %.

Even though all experiments' result is identical, Table 4 revealed that the yield in every chambers are significantly different. CC1, CC2 and CC3 produced 18.79 %, 25.04 % and 31.38 % of candlenut shell charcoal yield. The range is significantly wide which is from 18.79 % to 31.38 %. As indicates in Table 2 and Figure 5, the differences among those three chambers are in temperature. It is possibly cause the yield produced differently. Figure 7 below displays a comparison between the yield and average temperature generating in every chambers.

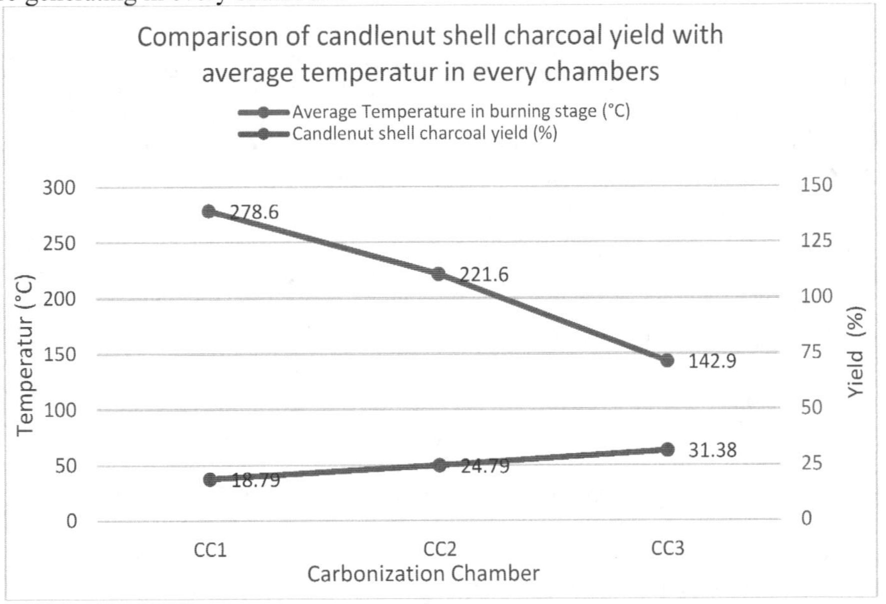

Figure 7. Comparison of candlenut charcoal yield with average temperature in every chambers.

As illustrated in Figure 6, candlenut shell charcoal yield in every carbonization chambers tends to increase when the temperature decrease. The highest average temperature was experienced by CC1 with 278.6 °C which is producing 18.79 % yield of charcoal. In the other side, CC3 was experiencing the lowest average temperature among the chambers with only 142.9 °C, but capable of producing 31.38 % yield of charcoal. It is mean that temperature affects the amount of candlenut shell charcoal yield significantly. Therefore, carbonization chamber that experiencing the lowest temperature will generate higher candlenut shell charcoal yield. In this case, the highest candlenut shell charcoal was produced by CC3 and followed by CC2 and CC1.

Since the cambers are placed vertically, where CC1 is at the bottom of the tool, followed by CC2 and CC3 at the top of the tool, chamber at the bottom of the tool will be the lowest chamber that can produce candlenut shell charcoal and chamber at the top will be in the opposite, while CC2 remains intermediately. As the result, candlenut shell charcoal yield will increase as the position of chambers increase vertically.

4. Conclusions

Candlenut shell carbonization tool using a vertical multi chamber is able to produce candlenut shell charcoal with the average yield of 24.99 %. It is 10% lower than what a process of carbonization can be generated which is 35 %. Every chambers are varying in its capability in producing candlenut shell charcoal. The main variable that has been identify is temperature. From the results, it is found that the highest candlenut shell charcoal yield is produced by the lowest temperature experienced by a chamber. The research revealed that CC1 was experiencing the lowest temperature among the chambers. Hence, CC1 is the highest chamber that can generate candlenut shell charcoal yield and followed by CC2 and CC1. In addition, the candlenut shell charcoal yield will go higher when the position of the chamber is placed higher at the tool.

Acknowledgments
This research was supported by Ministry of Research, Technology and Higher Education of the Republic of Indonesia and candlenut farmers and collector in Kluet Timur Sub-district, South Aceh District. We thank our colleagues from South Aceh Polytechnic who provided insight and expertise that greatly assisted the research.

References

[1] Rosman R and Djauhariya E 2005 Status dan Teknologi Budidaya Kemiri (Bogor: Balai Peneletian Obat dan aromatika)
[2] Krisnawati H, Kalio M and Kanninen M 2011 Aleurites Mollucanna (L) Wild (Bogor: Center for International Forestry Research)
[3] Elevitch C R and Manner H I 2006 Aleurites Molucanna (kukui) Specific profiles for Pacific Island agroforestry (Holualoa, Hawai: Permanent Agricultural Resources (PAR))
[4] Armstrong 2006 Tung oil and kukui nut candlenuts (www.Economic-plant-photographs.com)
[5] Darma R and Amri A 2019 Simple Candlenut Shell Carbonization Tool *IOP Conf. Ser. Mater. Sci. Eng.* **506** 012052
[6] Sinaga R, Desrial and Wulandani D 2016 Physical and Mechanical Characteristics of Candle Nut (Aleurites moluccana Wild) *Jurnal Keteknikan Pertanian* **4(1)** 97-106
[7] Sulhatun, Hasibuan R, Harahap H, Iriani and Fithra H 2018 Improving Production of Liquid Smoke from Candlenut Shell by Pyrolisis Process *Proc. MICoMS 2017 (Emerald Reach Proc. Ser.)* **1** 143-149
[8] Rocha P A D, Cerrella E G, Bonelli B R and Cukierman A L 1999 Pyrolysis of hardwood residues: On kinetics and chars characterization *Biomass Bioenergy* **16** 79-88
[9] Demirbas A, Ahmad W, Alamoudi R and Sheikh M 2016 Sustainable charcoal production from biomass *Energ. Source Part A* **38(13)**
[10] Amri A, Maulina D, Muhibban 2019 Development of Candlenut Shell Carbonization Tool Using a Vertical Multi Chambers *IOP Conf. Ser.: Mater. Sci. Eng.* **506** 012019
[11] Antal M J and Grønli M 2003 The art, science, and technology of charcoal production *Industrial Engineering Chemistry Research* **42**

Distillation Equipment Design of Nutmeg Oil using Hybrid System

Nuzuli Fitriadi[1], Balkhaya[1], Edi Saputra[2] and S. Othman[3*]

[1] Mechanical Engineering of Aceh Selatan Polytechnic, Merdeka Street, Tapaktuan, South Aceh 23716
[2] Mechanical Engineering of Politeknik Negeri Lhokseumawe, Lhokseumawe
[3] Faculty of Computing and Multimedia, Kolej Universiti Poly-Tech MARA, Kuala Lumpur

*Corresponding author: nuzuli@poltas.ac.id

Abstract. The quality of nutmeg oil is determined by the quality of the nutmeg and the technology that are used in the purification process of the nutmeg. For instance, the distillation process will affect the yield and the essential quality of the production of the nutmeg oil are involving the boiler material, cooling system, and boiler design. The existing nutmeg oil distillation system is time consuming because the operators are highly depending on the firewood and the operators who must monitor closely throughout the time during the nutmeg oil refining process. The new distillation process and technique is based on the new type of boiler that is made from Stainless steel 304 with a diameter of 70 cm, length 120 cm, and with the thickness of 3 mm. The Boiler also comes with a new control system to precisely control the pressure and temperature. Meanwhile, the excess pressure of the boiler is controlled by pressure safety valve because the boiler is designed to withstand up to 3 bar maximum pressure capacity. The kettle is made of 304 Stainless steel equipped with a multi-storey basket to polish and to enhance the circulation and vapour pressure inside the boiler. The condenser pipes are made from the aluminium pipe to maximise the condensation process of the raw nutmeg oil. For the new distillation process, the hybrid system is utilised, where the sources of firewood and electricity are used as the source of energy to operate the boiler. It can be seen that hybrid kettle design reaches 2.8 tons with a 30% operating-saving cost. On top of that distillation time can be shortened to 48 hours from the previous 72 hours. It is stated that this hybrid boiler system is much more effective when compared to conventional systems. Laboratory test results stated that the distilled nutmeg oil has fulfilled the Indonesian national standard.

Keywords: distillation, boiler, kettle, nutmeg oil

1. Introduction

The Southern Aceh District is known as one of the producing areas of nutmeg in Indonesia. The production of essential oils from nutmeg is done through refining. There are four methods of distillation commonly used, namely the method of steam distillation, water distillation, steam distillation, and extraction using solvents. Judging by the speed and capacity of the oil production, the method of distillation commonly used to obtain nutmeg oil is the method of steam distillation [1]. The distillation process is done by using steam from ripe beans and dried seeds and nutmeg flowers. Steam distillation to extract nutmeg oil is conditioned under atmospheric pressure because Nutmeg has a component of fatty oil. Nutmeg oil is obtained after the fat contained therein are excreted first. High pressure distillation can lead to the possibility of following fatty oils thereby reducing the quality of essential oils [2].

The quality of nutmeg oil is determined by the quality of the basic ingredients and distillation processes used [3]. Drying techniques for raw materials and distillation processes that include boiler material, cooling system and boiler design for distillation affect the yield and quality of essential oils produced [4], [5].

Southern Aceh Regency is one of the regions producing nutmeg in Indonesia. According to the editor of Binoculars Aceh, in 2017, there are three locations of breeding the largest nutmeg plants in South Aceh district, namely in the village of Air Pinang and Air Berudang, sub Tapaktuan and Blang Kuala Village, Meukek regency, with 15,821 hectares of total Breeding nutmeg in Aceh. [6].

Based on data input by the central Statistical Agency of Southern Aceh Regency, Nutmeg production in Southern Aceh Regency in 2017 is 6,510 tons. Table 1 shows the area and self-production of nutmeg in Southern Aceh Regency [7].

Table 1. Area and self-production of nutmeg in Southern Aceh

	Area and Nutmeg Production Pattern of 2017				
Sub District	TBM (Ha)	TM (Ha)	TR (Ha)	Total	Production
Trumon	8	4	2	14	3
Trumon timur	32	12	3	47	5
Trumon tengah	50	33	10	93	20
Bakongan	4	5	1	10	3
Bakongan timur	55	46	10	111	30
Kota bahagia	30	13	5	48	7
Kluet selatan	38	14	5	57	4
Kluet timur	47	39	6	92	21
Kluet utara	198	98	6	302	75
Pasie raja	565	400	7	970	310
Kluet tengah	50	10	10	70	5
Tapaktuan	718	970	90	1.778	1.010
Sama dua	625	960	75	1.660	850
Sawang	754	2.529	374	3.657	830
Meukek	754	2.529	374	3.657	2.070
Labuhan haji	800	510	125	1.435	410
Labuhan haji timur	912	650	125	1.687	525
Labuhan haji barat	780	440	15	1.235	330

Some of the research that will be used as reference in this research namely, the energy efficiency and performance of the prototype wood-fired nutmeg oil refining equipment. From the distillation with this prototype obtained oil with characteristic that is in accordance with SNI 06-2388-2006 about nutmeg oil. The combined response of the weighted value of type 0.904, the bias index 1.478, the solubility of ethanol 90% at a temperature of 20 °C by 1:1, the optical rotation (+) 16.8 °, and the remaining evaporation of 0.7%. Standard nutmeg oil on SNI for 0.880-0.910 type, bias index 1.470-1.497, ethanol solubility 90% at 20 ° C 1:3 and so on clear, optical rotation (+) 8 °-(+) 25 ° and residual evaporation 0.7% [8]. To optimize the use of energy is done by designing an electronic-based control system to the dominant parameters of temperature, steam pressure and the weight of nutmeg that will be distilled. Nutmeg oil Refining System

The results of this research have been able to control temperature and pressure automatically. Control done by controlling the fuel valve (used gas) so that the energy use optimization is 14.87% [9].

Due to the increasing demand of world nutmeg oil, it is necessary to effectively enhance the performance of the purification techniques in order to increase the quantity and quality of nutmeg oil. The problems in the distillation process of nutmeg oil can be briefly explained as follows:

1. The existing nutmeg oil distillation system is still using firewood and takes a long time during the distillation process.
2. The distillation process has not used an electronic-based regulatory system, so it still takes operators who must monitor carefully throughout the time of the nutmeg oil refining process.

The purpose of this research in general is to produce good quality nutmeg oil so that this commodity will have high selling value in local and world market. Furthermore, to keep the continuity of the supply of the nutmeg oil continuously and to increase the quantity of nutmeg oil production. In order to achieve this general goal, there are a few researches that related to this nutmeg oil production process. One of them is research that are related to distillation process with hybrid system. The specific objectives of the study namely:

1. Designing a distillation system of nutmeg oil hybrid systems that can use firewood and electricity.
2. It can be control electronically and the settings can be reset if needed parameter changes such as temperature and duration of distillation time.
3. Reduce the cost for energy needs in the distillation process (distilling) nutmeg oil, the length of distillation time optimised according to the weight/amount of nutmeg to be extracted and the vapour pressure given.
4. Increase the productivity and quality of the resulting nutmeg oil.

The main goal of the refining process (distillation) nutmeg oil, is the designing of the distillation system with boiler water tube, boiler with multilevel baskets and utilising the energy sources from the firewood and electricity. Therefore, for energy optimisation consumption the electrical energy is mainly required in order to reduce production costs, other than that, is the system of arrangement on the enhancement of the process is done electronically by a microcontroller that is easy to use and suitable applied to the regulatory (control) systems.

2. Methodology

This research was conducted based on the results of the study of nutmeg distillation process. The results of the study were shown in the Fishbone diagram in Figure 1.

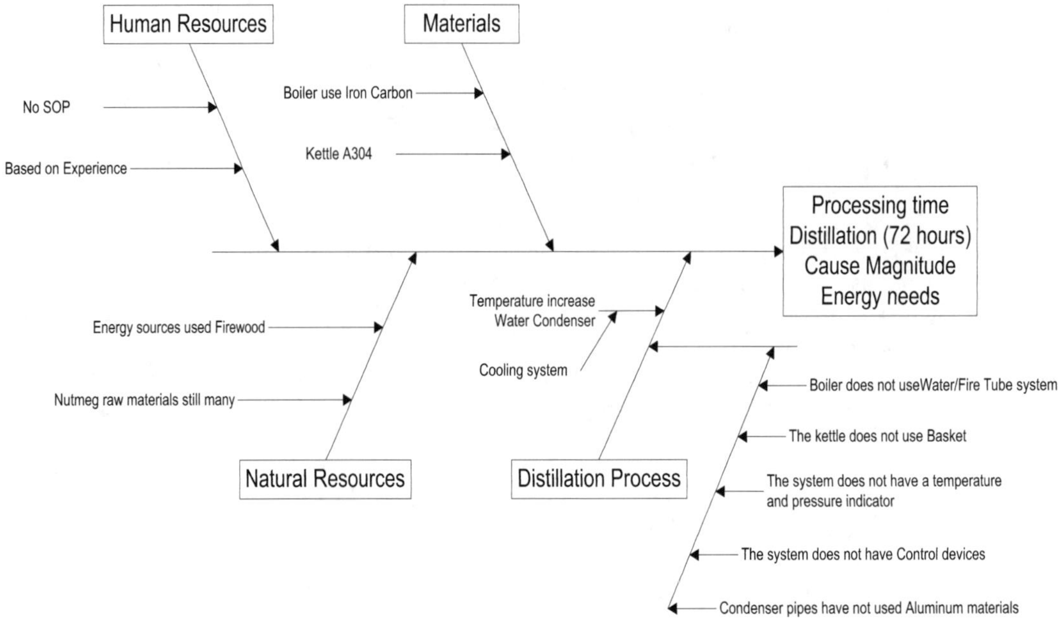

Figure 1. Fishbone diagram

Based on the Fishbone diagram in Figure 1, the study was conducted for 2 years period, in which the first year it started by designing and creating a gas-fuelled distillation system, tool components such as stainless-steel boilers, boiler with stainless steel material baskets, pipes and stainless-steel condenser tub. Meanwhile, in the boiler equip with a temperature and pressure indicator and an exhaust valve for the excess gas pressure. While the boiler is only equipped with temperature indicator only. In the second year, the research was conducted by designing and creating wood and electrical distillation equipment as well as modifying boilers with water tube system and adding water cooling system to cool water in the condenser tub. At the stage of this research has been equipped also with a control system against temperature and pressure during the distillation process. Further from the results obtained then the design of nutmeg oil refinery scales industrial scale with hybrid system.

The main components of the nutmeg oil refiners made in this study are the boiler, kettle, and condenser. From the first and second stage research has been built and tested the performance of this tool to produce the quality of the National Standard of Nutmeg oil.

2.1. Boiler

The boiler is made of stainless steel A304 with a diameter of 40 cm, height 71 cm and with a thickness of 3 mm. The boiler also features a manometer/pressure gauge to control the pressure in the boiler and the thermometer to control the temperature. In addition, there is also a pressure safety valve that will function if there is excessive pressure. The kettle is designed with a pressure capacity of 3 bars. Process to produce steam is by using a water tube system. The water tube is made of 1/2-inch stainless steel pipe with a thickness of 3 mm. The process of making the boiler with the water tube system as shown in Figure 2.

Figure 2. Boiler Water Tube Manufacture

2.2. Kettle

The distilled kettle used in the study is also made of Stainless steel A304 with a diameter of 72 cm and a height of 100 cm and a thickness of 3 mm plate. In addition, Baldem with a diameter of 50 cm is used as a cover for a distilled kettle. The installation process is shown in Figure 3.

Figure 3. Kettle with Beldam's

2.3. Condenser

The condenser is made of stainless-steel pipe with a size of 1/2 inches A304 in Figure 3. The condenser pipes are made in spiral form with a height of 80 cm. The condenser making process is shown in Figure 4. After all the process of design and manufacture, nutmeg distillation equipment is done by the water tube method in Figure 4. Overall, the nutmeg oil refining system in this study was shown in Figure 5 and Figure 6.

Figure 4. Manufacture of condenser

Figure 5. The process of assembling and installing distillation tools.

1. Steam Boiler
2. Distillation Kettle
3. Steam Conductor Pipe
4. Condenser Tank
5. Cooling Tower

Figure 6. Prototype nutmeg oil distillation

3. Results and Discussion

3.1. Number of low-yield

Nutmeg oil refining capacity of 40 kg is distilled for 12 hours which spends 12.5 kg of wood and obtained a different amount of yield between phase 1 without water tube and stage 2 is with water tube as shown in Figure 7. The best results are shown by the system that use boiler water tube so that the steam flowing to the boiler is always constant.

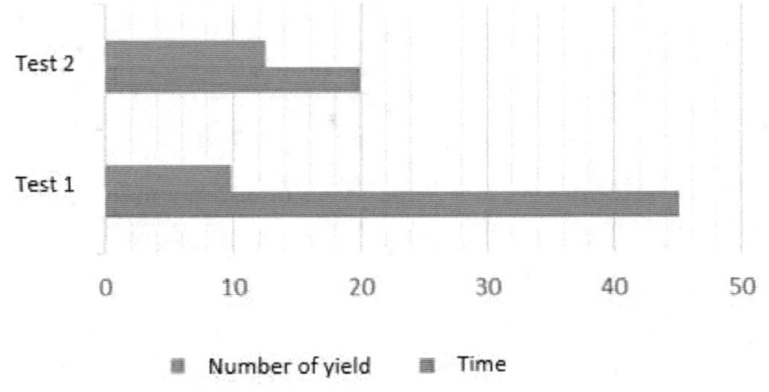

Figure 7. First stage nutmeg oil distillation

3.2. *The results of the nutmeg oil characteristic*

Testing characteristics of nutmeg oil is done in the laboratory of Palm Research Centre with the results as in Table 2.

Table 2. Test results

Parameter	Unit	Test results (1)	Test results (2)	Method
Remaining Evaporation	%	3,48	1,52	SNI 06.2388.2006
Specific gravity	g/cm^3	0,8847	0,8295	ASTM D 4052
Colour	-	Pale yellow	Clear	SNI 06.2388.2006

From the results of the testing showed that the quality of nutmeg oil is increased in the use of the second stage distillation system where the boiler uses water tube system, kettle with a basket of materials to improve the vapor circulation evenly, the condenser are using the aluminium pipes and the water cooling systems in a condenser container.

3.3. *Industrial capacity designing*

Based on the research results for 2 (two) years, then the design of industrial capacity distillation equipment that will be applied to the Nutmeg Society Handicraft Industry Centre in the Meukek Village in Southern Aceh District. The production capacity for this distillery is 2.8 tons for distillation.

3.3.1. *Boiler*

Boilers that serve as steam producers using wood and electricity fuels as heat sources. This boiler has 40 pipe water tube that serves to accelerate the process of steam formation especially if the source of heat from wood, while if using electricity has been available 4 (four) rod heater that each heater is capacity 400 Watt. In addition, this boiler has a control system for temperature and pressure settings that the data acquisition can be directly record by the computer. All parts of this boiler are made of A304 stainless steel with a 3 mm thick plate and isolated by fire brick for a more maximum warming process as shown in Figure 8.

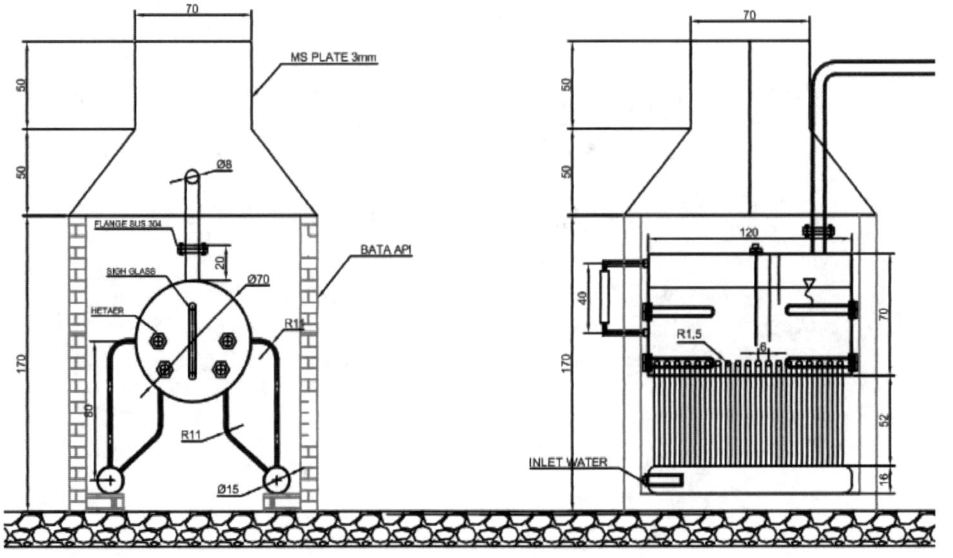

Figure 8. Design of boiler

3.3.2. Kettle

The kettle is designed to have a multi-level basket that serves to maximize steam circulation in the boiler as shown in Figure 9. In addition, the use of baskets can also facilitate the filling and demolition of nutmeg raw materials. The kettle cover is Bold Dum shaped tied by clamp. In this cover, paired pressure and temperature indicators can be read by the computer. The schematic diagram is shown in Figure 10.

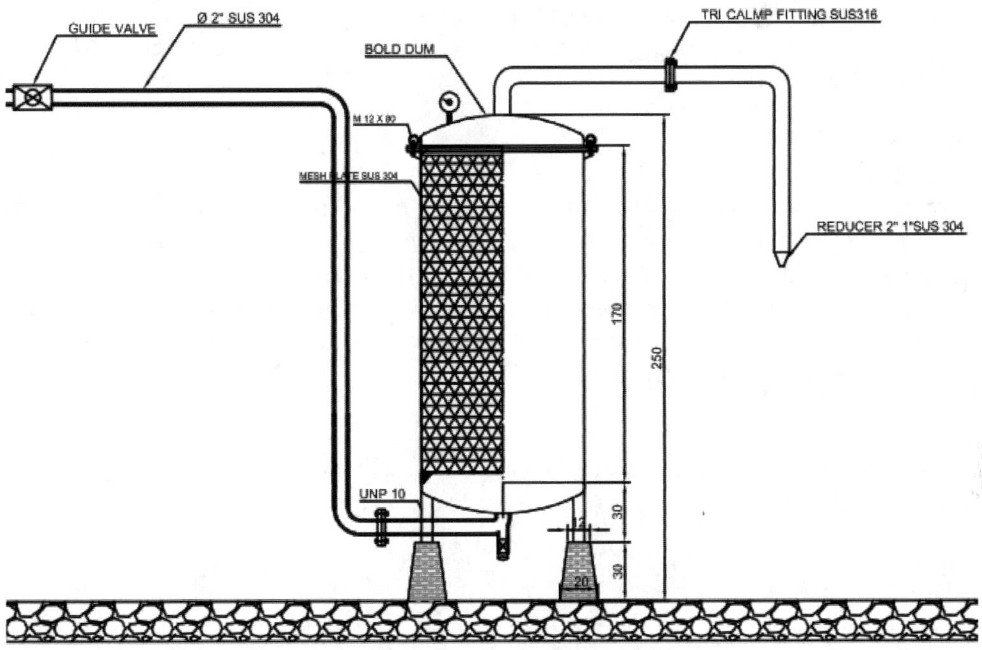

Figure 9. Design of kettle

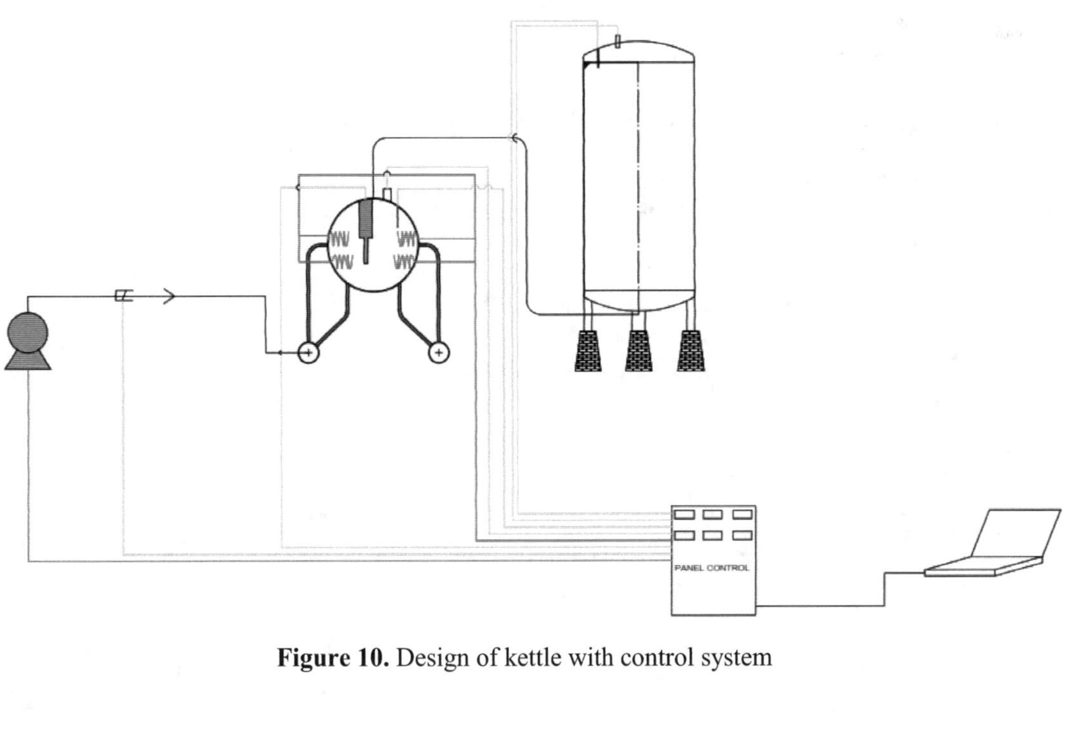

Figure 10. Design of kettle with control system

3.3.3. Condenser
Condenser pipes that are found above the water coolant surface using stainless steel A304. While the whole condenser pipe is immersed in cooling water using aluminium material because this material is relatively faster cooler if it is in water, so the process of phase change of steam to liquid is relatively more constant

4. Conclusion
The conclusions gained from this study are as follows:
1. Have done research to design build nutmeg kettle prototype with hybrid system. The result of this prototype will be used as reference in designing nutmeg boiler industry capacity.
2. The time required for the heating process to produce steam and achieve a pressure of 3 bar between 15-20 minutes, much more effective when compared to conventional systems.
3. Laboratory test results stated that nutmeg oil, which was distilled using a water-tube boiler, meets Indonesian national standard.

Acknowledgements
Author's thanks Aceh Selatan Government for finding this research entitled "Distillation Equipment Design of Nutmeg Oil using Hybrid System".

References
[1] Indonesia.go.id 2007 Sumber Daya Alam Provinsi Nanggroe Aceh Darussalam. http://www.indonesia.go.id/en/regional-government.
[2] Amos dan Purwanto W 2002 *Jurnal sains dan teknologi Indonesia* Vol **4**(5):1-6.
[3] Hidayati N, Ilmawati H dan Sara E 2015 *Fakultas Teknik, Universitas Muhammadiyah Surakarta.*
[4] Suprihatin, Ketaren S, Ngudiwaluyo S dan Friyadi A *Jurnal Teknologi Industri Pertanian* Volume **17**(1), 23-28.
[5] Sipahelut S G dan Telussa I 2011 *Jurnal Teknologi Hasil Pertanian*, Vol. **IV**, No. 2
[6] Redaksi Teropong Aceh 2016 Dirjen Perkebunan Sertifikasi Bibit Pala Aceh Selatan, Teropongaceh.com
[7] BPS Kabupaten Aceh selatan. Luas area produksi pala. http://acehselatankab.bps.go.id.
[8] Syamsul, Dkk, 2016 *National Conference of Applied Sciences, Engineering, Business and Information Technology.* Politeknik Negeri Padang
[9] Prasetyo D A 2009 *Fakultas Teknologi Pertanian Institut Pertanian Bogor*, Bogor

Movement Parameters for Evaluating the Two-legged Humanoid Robot

Dirja Nur Ilham[1], Rudi Arif Candra[1], Hardisal[1] Erwinsyah Sipahutar[2] Abdullah Abdullah[3], Suherman Suherman[3] and S Othman[4*]

[1] Politeknik Aceh Selatan, Jl, Merdeka Komplek Reklamasi Pantai, Tapaktuan, Aceh, Indonesia
[2] Politeknik ATI Padang Jl. Bungo Pasang Tabing Padang, Indonesia
[3] Universitas Sumatera Utara, Department of Electrical Engineering 2 Almamater Road, Medan, Indonesia 20155
[4]Faculty of Computing and Multimedia Kolej Universiti Poly-Tech MARA, Kuala Lumpur

*Corresponding e-mail: suherman@usu.ac.id

Abstract. Many robot prototypes have been made and control algorithms have been proposed. The performances are measured through experiments. This paper presents a two-legged robot prototype with and without Kalman filter in processing sensor data and proposes some movement evaluation parameters to measure the robot performances. Experiments show that the evaluated performances are able to differentiate method achievements. Movement parameters: movement angle, distance travelled, and failed frequency have been successfully identified robot performances.

Keywords: humanoid robot, two-legged robot, Kalman filter, movement parameters

1. Introduction

Robot has been used in industrial sector for years. Currently, not only for industrial purposes, robot is also use for service and specialized tasks [1]. Robot development now involves the integration of intelligent system [2] where the mechanical engineering meets information and communication technologies. The first issue initially was about electro-mechanical devices such as motor and its controller. Now it moves to human-like characteristics such as image, language and other learning processes.

Robot is defined as a programmable mechanism in multi axes, moving around its surrounding to perform the intended tasks [3] Robot with intelligence shoul be able to perform environment perception and make decision. Intelligent robot can simulate human behaviors like emotion and thinking. Robot assessment is not only for covering the mechanical function but also how far it follows human commands and finishes the predefined tasks, learns environment and improves behaviors.

In order to measure robot performances, some measurements were set up. Some researchers used goal accomplishment as the main parameter [4] others use sub task success rates such as task localization, object detection, specific movement, task detachment, cycle time, and failure rates [4] Meanwhile, additional performance parameters for industrial or manipulator robots consider metrics for safety, task quality and task quantity [5].

This paper focuses on performance measurement of the walking performance of humanoid robot. A prototype was developed, three performance parameters were proposed. Since walking robot requires self-balancing, the parameters proposed in this paper include movement angle, travelled distance, and failed frequency. In order to see the effectiveness of the parameter to see performance increment, Kalman filter was inserted to enhance robot performances.

2. Robot Prototype

The humanoid robot prototype that will be evaluated was designed using 5 degree of freedom (DoF) in each leg: 2 DoFs in ankle, 1 DoF in knee and 2 DoFs in hip as shown in Figure 1.

Electrical design includes microcontroller circuit and programming, sensors and drivers, servo control and drivers, and power supplies. Kalman filter is inserted in microcontroller program to evaluate the effectiveness of the proposed measured parameters. Kalman filter is complex mathematical expression but able to improve robot performance [6, 7]. Kalman filter reduces noise generated by sensor circuits when performing maneuver.

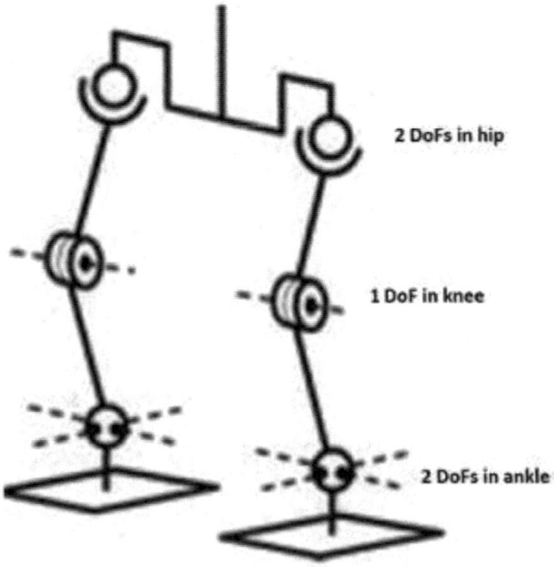

(a)

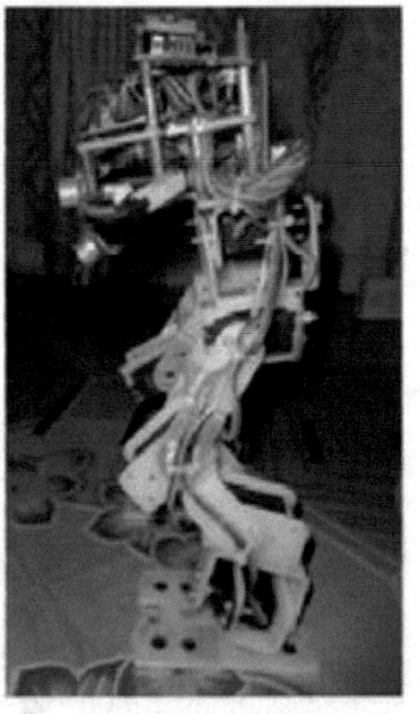

(b)

Figure 1. 5 DoFs prototype.

Robot was controlled by ATmega32 and ATmega8 microcontrollers. ATmega32 processes navigation sensors while ATmega8 controls servo motors. Figure 2 shows the circuit. Two microcontrollers were employed to enhance the processing power. ATmega8 and Atmega32 exchanges information through the parallel data line. Microcontrollers were supplied by 5V power supply while the motors were driven by 12V power supply.

Initially, ATmega32 microcontroller checks the standing position from sensors to inform ATmega8 how motor should work. If it is accomplished, navigation sensors are read to ensure smooth movement. Output of those sensors are processed by using Kalman filter before adjusting the movement angles as results of ATmega32 and ATmega8 coordination. In order to measure movement angles, trip time, and the failure rates, robot was tested to travel about 50 times.

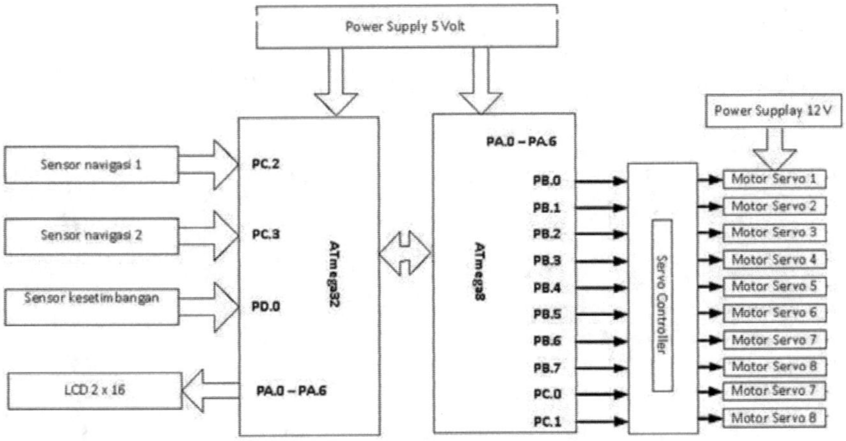

Figure 2. Controller circuit.

3. Evaluation Results

Figure 3 shows the movement angle of robot. The graph shows distinguishable performance parameters where without Kalman filter, robot experiences unstable movement angle that changes very frequently with large amplitude. Kalman filter can make robot hold in its position more steadily. Average angle is 25.46o for robot without Kalman, reduced to average 23.13o with Kalman filter. The performance difference achieves 9.2%. Both experiments and average value are distinguishable, which means this evaluation parameter is working as expected.

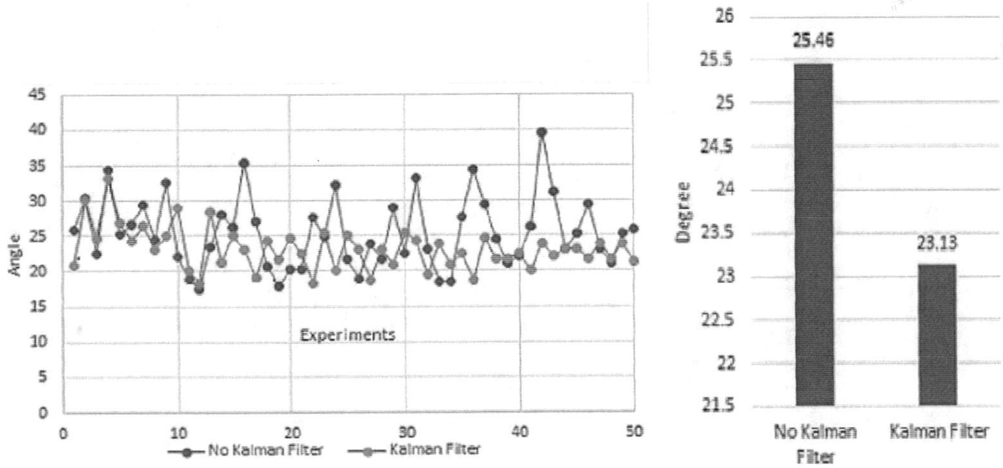

Figure 3. Movement angle.

Meanwhile, the travelled distance parameter is also able to differentiate robot performance as the robot with Kalman filter experiences zero failure. The plain robot was failed to complete the target which reached only 39.2% of the overall distance (Figure 4). This performance parameter produces in average 60.8% performance achievement differences.

IOP Conf. Series: Materials Science and Engineering 1062 (2021) 012058 doi:10.1088/1757-899X/1062/1/012058

The last parameter is the failed frequency which produced in average 4.2 against 0. This parameter can distinguish the robot performance as can be seen in Figure 5.

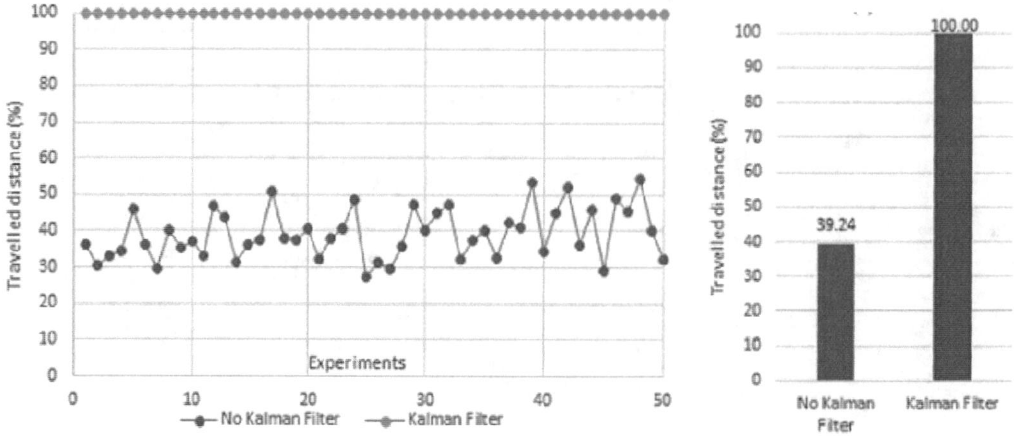

(a) (b)

Figure 4. Travelled distance.

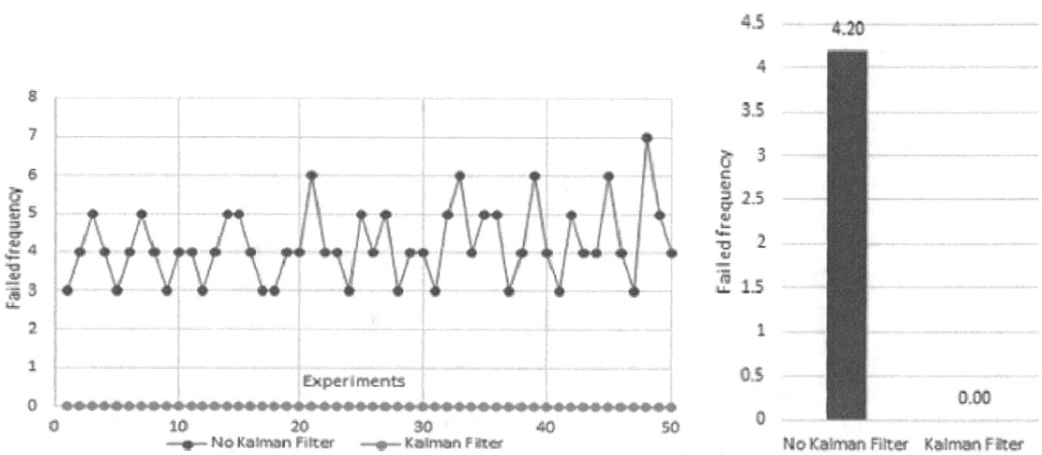

(a) (b)

Figure 5. Failed frequency.

4. Conclusions

This paper also proposed three evaluation parameters which are movement angle, travelled distance and failed frequency. The experiment shows that three parameters that can distinguish robot performances. Movement angle distinguished robot performance 9.2%, while travelled distance gives 60.8% differences. Finally, failed frequency produces 4.2 against 0 for system without and with Kalman filter.

References

[1] Berg J, Lottermoser A, Richter C, & Reinhart G 2019 *Procedia CIRP* **79** 614-619
[2] Tian-Miao Wang, Yong Tao, Hui Liu 2018) *International Journal of Automation and Computing* **15**(5) 525-546
[3] International Federation of Robotics. *Industrial Robotics Standardization.* (Online)
[4] Bac CW, Hemming J, Van Tuijl B A J, Barth R, Wais E & van Henten E J 2017 J*ournal of Field Robotics* **34**(6) 1123-1139
[5] Bostelman R, Hong T, & Marvel J 2016 *Journal of Research of the National Institute of Standards and Technology* **121** 342-366
[6] Moore T & Stouch D 2016 *In Intelligent autonomous systems* **13** (335-348 (Springer Cham.)
[7] Congwei Hu, Wu Chen, Yongqi Chen and Dajie Liu 2003 *Journal of Global Positioning Systems* **2**(1)

Finite Element Modelling and Updating of a Thin Plate Structure using Normal Mode Analysis

A R Bahari[1], M A Yunus[1]*, M N Abdul Rani[1], W I I Wan Iskandar Mirza[1], M A S Aziz Shah[1] and Z Yahya[2]

[1]Structural Dynamics Analysis & Validation, Faculty of Mechanical Engineering, Universiti Teknologi MARA (UiTM), 40450, Shah Alam, Selangor, Malaysia

[2]Faculty of Computing and Multimedia, Kolej Universiti Poly-Tech MARA, 56100, Kuala Lumpur

Corresponding e-mail: mayunus@uitm.edu.my

Abstract. Structural design and analysis require accurate and precise finite element model in order to have a structure with high safety conditions such as aerospace, offshore and civil structures. However, the predicted results of the finite element model are often considerably different from the experimental results due to the modelling assumptions and simplifications on the properties and parameters of the actual structure. This paper presents the finite element modelling and updating methods on the thin plate structure. The modal parameters obtained from the plate are measured by experimental modal analysis by using an impact hammer and roving accelerometers method under free-free boundary conditions. The initial modal parameters are predicted by numerical analysis using normal mode analysis via NASTRAN SOL 103. The accuracy of the natural frequencies and the mode shapes obtained from the initial finite element model are compared with the measured counterparts. The modal based updating methods using NASTRAN SOL 200 is applied to update the properties of the plate. In conclusion, updating the material properties of the plate is successfully achieved. The total error of the initial finite element model of the plate is reduce from 45.12 percent to 6.28 percent respectively.

Keywords: structural design, model updating, experimental modal analysis, normal mode, natural frequency

1. Introduction

In engineering field practices, structural design and analysis is an essential task that deals with the use of structures to support and resist loads during the operations. From the large-scale engineering structures such as bridges, platforms and aircrafts to the medium scale engineering structures such as vehicles and industrial equipment, the design and analysis processes require accurate and effective finite element (FE) model [1]. Structural analysis encounter modelling and load analysing which are depending on the quality of the model. In addition, an implementation of the advance structural design and analysis of the components and structures are crucial in order to verify the functional and operations with the structural safety condition and serviceability.

Finite element method is a tool for the numerical prediction using computational processes on the discrete model to obtain the objective solutions. This finite element model is normally utilizing uncertain system parameters such as geometry, material properties, boundary conditions and kinematic interactions [2,3]. Several types of sources which contribute to the inaccuracy of the model are model structure errors, model parameters errors and model order errors [4]. There is a considerable discrepancy between the numerical predictions and the corresponding test results [5]. Modifications on the modelling assumptions and parameters are crucial until the correlation satisfy the requirements [6,7]. A process of adjusting and refining certain parameters and incorrect assumptions of the finite element model based on the experimental results is known as finite element model updating [8]. This reliable method is widely used to solve and minimize numerous sources of modelling error which can be divided into three categories: idealization errors, discretization errors and erroneous assumptions for the modal parameters [9].

In this paper, the modal parameters (natural frequencies and mode shapes) of the thin bending plate structure are measured using the experimental modal analysis (EMA) by impact testing. Meanwhile the finite element method by normal mode analysis is used to predict the dynamic behaviour of the initial model. The objective function and the sensitivity analysis are discussed and the finite element model updating are performed to produce accurate updated finite element model.

2. Experimental Set Up and Modal Testing

Modal testing was conducted to determine the modal parameters of the thin plate that is fabricated from a cold rolled mild steel. The modal parameters involve natural frequencies, mode shapes and damping ratios [10,11]. The EMA set up is shown in Figure 1. The plate was suspended on the test rig by using soft bands to simulate free-free boundary conditions. Figure 2 shows the equipment used for the modal testing. An impulse excitation force was applied to the plate by using an impact hammer for the broadband vibration excitation. The applied force was measured using the force transducer installed at the impact hammer. The vibration response of the plate was measured using three accelerometers. The method of roving accelerometers was used to avoid mass loading issue during the experimental modal analysis [12,13]. Two accelerometers were used for roving and one accelerometer was used as a reference at the driving point to measure the acceleration response.

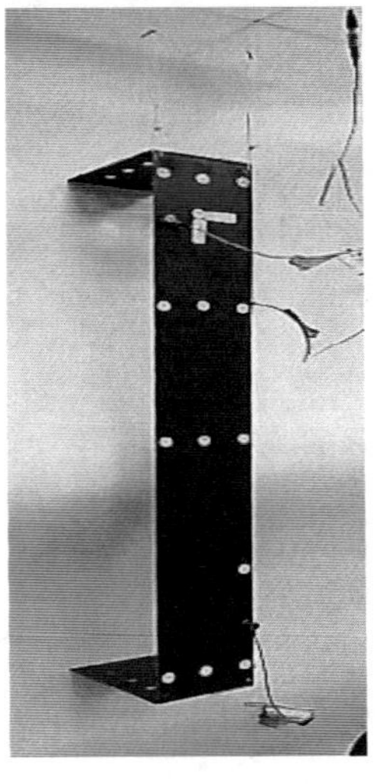

Figure 1. Modal testing setup for the plate

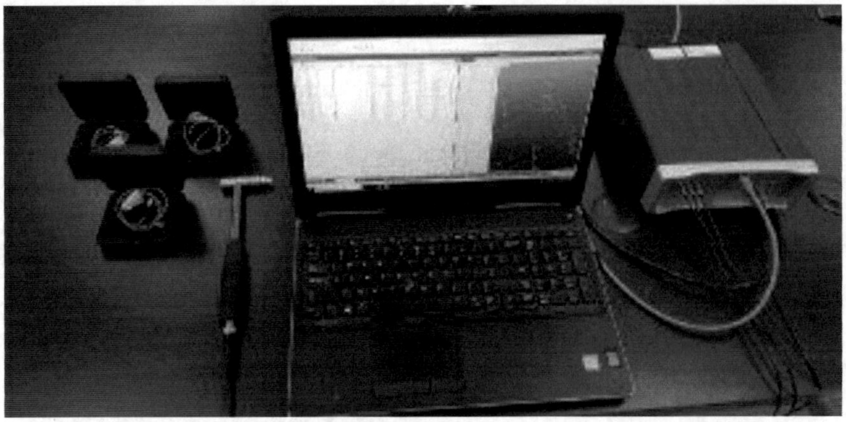

Figure 2. Testing equipment for experimental modal analysis

The schematic diagram of the modal testing is shown in Figure 3. The data acquisition equipment used to calculate the modal parameters is LMS SCADAS data acquisition system. The measured force and acceleration responses were processed using LMS Test.Lab analyser. The bandwidth for the frequency of interest is below 200 Hz.

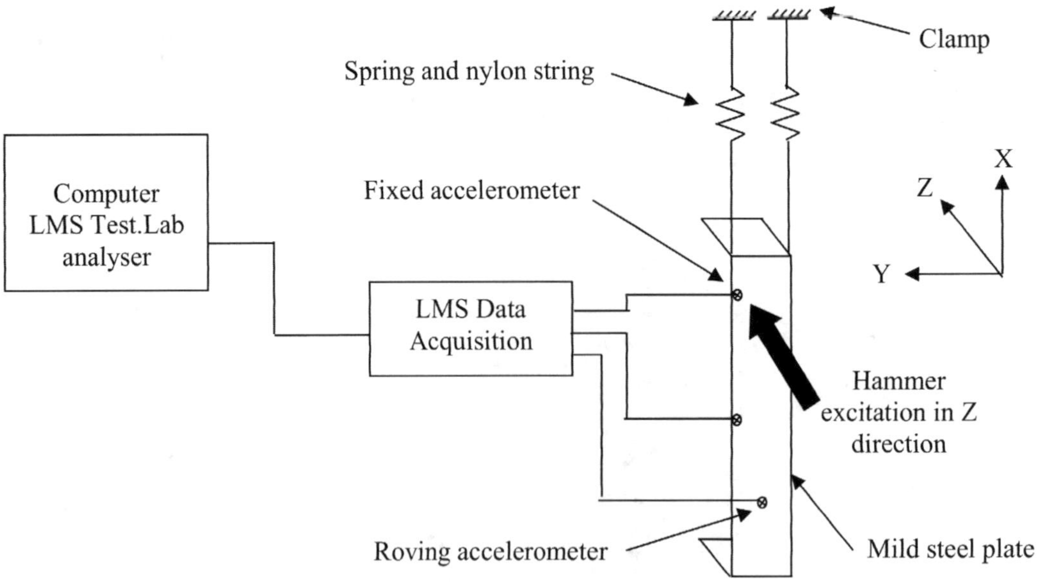

Figure 3. Schematic diagram of the modal testing of the plate

3. Finite Element Analysis

The three main steps in the numerical simulation works which are pre-processing, processing and post-processing were performed using OptiStruct from HyperWorks in order to obtain natural frequencies and mode shapes. The FE model of the plate is shown in Figure 4. The model was developed which consists of 10,172 elements of 2D mixed mesh (combination of triangle and quadrilateral) and 10,555 nodes. The property of the mesh is thin shell elements of size 5 mm. The nominal value of material properties [14] that was assigned to the FE model are tabulated in Table 1. Normal modes analysis with free–free boundary conditions was utilized to determine the modal parameters which are natural frequencies and mode shapes. The bandwidth for the frequency of interest is from 1 Hz to 200 Hz.

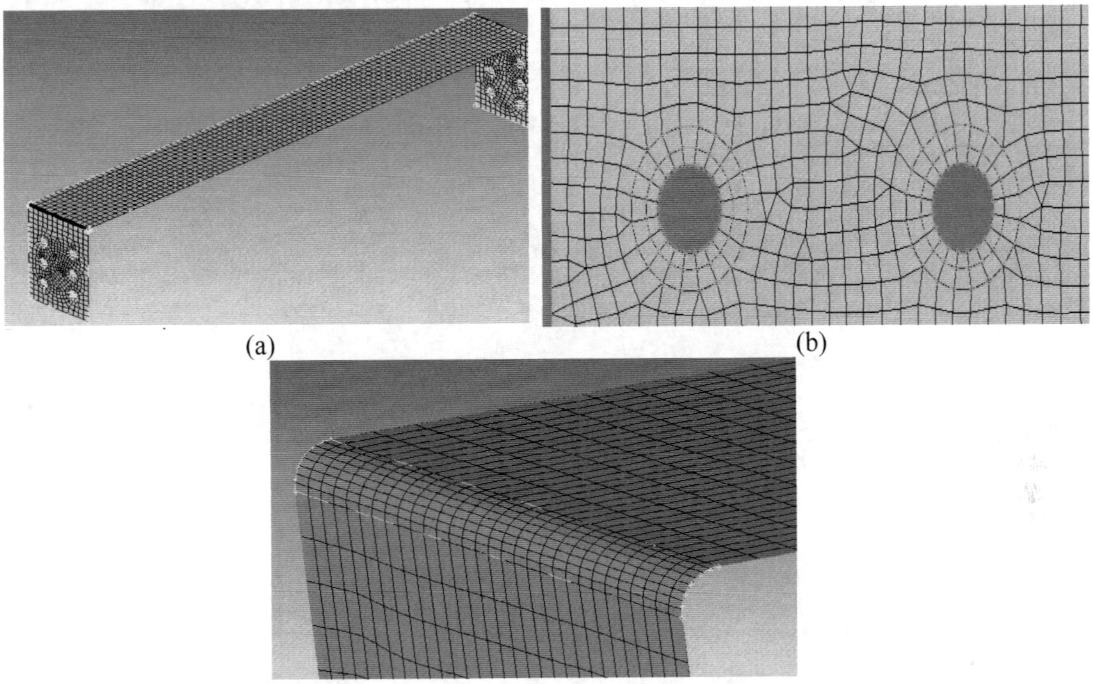

(a) (b) (c)

Figure 4. Finite element model (a) plate model (b) meshing of the bolt hole (c) meshing at the bend location

Table 1. Material properties of the plate for the FE normal mode analysis

Properties	Values
Young's Modulus, E (MPa)	200,000
Poisson's Ratio, υ	0.3
Density, ρ (kg/m^3)	7800
Thickness, t (mm)	1.5

The equation of motion for the free vibration of an undamped finite element model [15] can be calculated using Eq. (1)

$$M\ddot{x}(t) + Kx(t) = 0 \tag{1}$$

where K and M are the stiffness and mass matrices. Meanwhile, $\ddot{x}$ and x are the vector of acceleration and displacement respectively. These system matrices are calculated based on the geometry and properties of the FE model. The natural frequencies and mode shapes of the structure are obtained by solving Eq. (2) where ω is a natural frequency and ϕ is a mode shape [16,17].

$$(K - \omega^2 M)\varphi = 0 \tag{2}$$

In this study, Lanczos method was used for algorithm computing free vibration modes. The free vibration characteristics depend on the magnitude and distribution of the structural stiffness and mass [18,19]. All structures display several different mode shapes at different frequencies, both at a local and a global level.

4. Finite Element Model Updating

Normal mode analysis using Lanczos method was used to determine the modal parameters. The natural frequencies and mode shapes obtained from the FE model are compared with the measured counterparts in order to validate the accuracy of the initial FE model. Table 2 shows the comparison of natural frequencies between EMA and initial FE analysis. It can be seen that the predicted natural frequencies from the FE analysis are higher compare to the measured natural frequencies from the EMA. The range value of the error is between 6.0 per cent and 10.0 per cent which contributing to the high total error of 45.12 per cent.

Table 2. Comparison of EMA and initial FE results for the plate

Mode	EMA (Hz)	Initial FE Analysis (Hz)	Error (%)
1	22.49	24.17	7.42
2	57.56	62.09	7.87
3	64.02	68.13	6.42
4	85.34	92.55	6.80
5	136.18	145.62	6.92
6	185.89	203.79	9.69
	Total Error		45.12

Table 3 shows the mode shapes from EMA and FEA corresponding to each natural frequency. It is observed that four modes which are mode 1, mode 2, mode 4 and mode 5 are elastic bending modes in *x* and *z* direction. Two modes are the elastic torsional modes which are mode 3 and mode 6 that is in *y* and *z* direction. The accuracy of the mode shapes between the experimental and finite element model has been evaluated by pairing each mode correctly. The modal assurance criterion (MAC) can be calculated using Eq. (3) where experimental (ϕ_m) and finite element (ϕ_a) mode shapes in matrix form. MAC is used to quantify level of correlation of mode shapes between the FE models and the experimental models. Table 3 also shows the MAC value for each mode which has a good correlation above 0.8.

$$MAC = (\varphi_m \varphi_a) = \frac{|\varphi_m{}^T \varphi_a|^2}{(\varphi_a{}^T \varphi_a)(\varphi_m{}^T \varphi_m)} \tag{3}$$

Table 3. Comparison of the mode shapes of the plate between the measured and predicted natural frequencies

Measured EMA Mode Shape	Predicted FEA Mode Shape	MAC
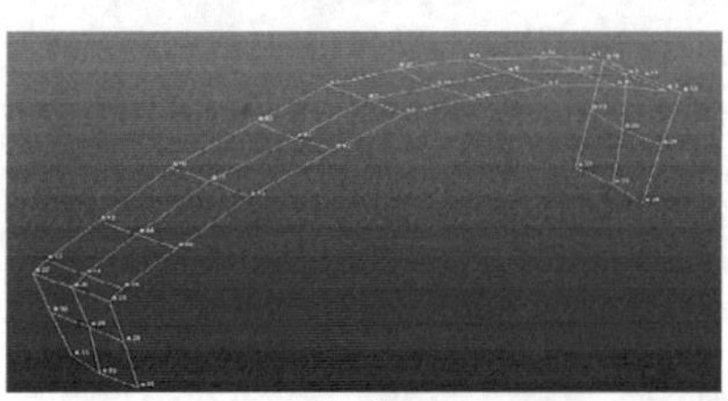 Mode 1	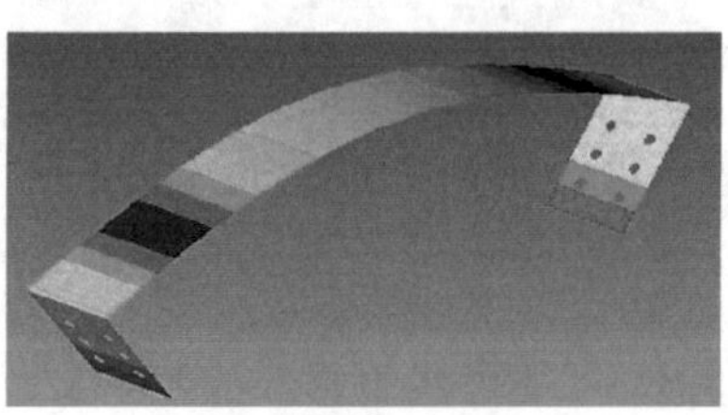 Mode 1	0.98

IOP Conf. Series: Materials Science and Engineering 1062 (2021) 012059 doi:10.1088/1757-899X/1062/1/012059

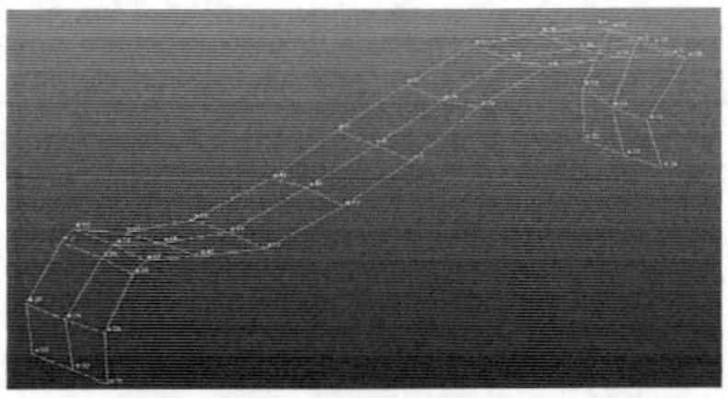 Mode 2	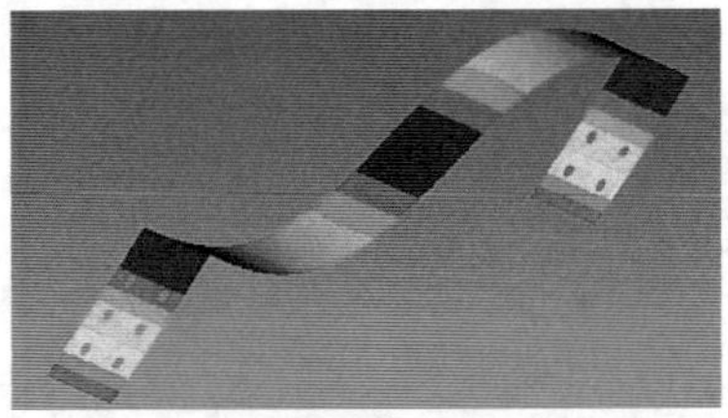 Mode 2	0.93
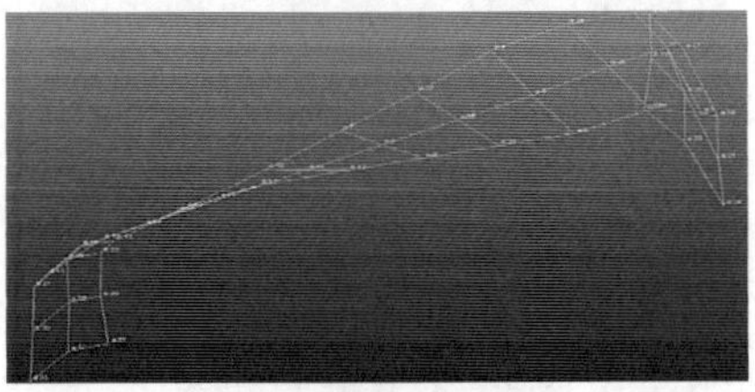 Mode 3	 Mode 3	0.95
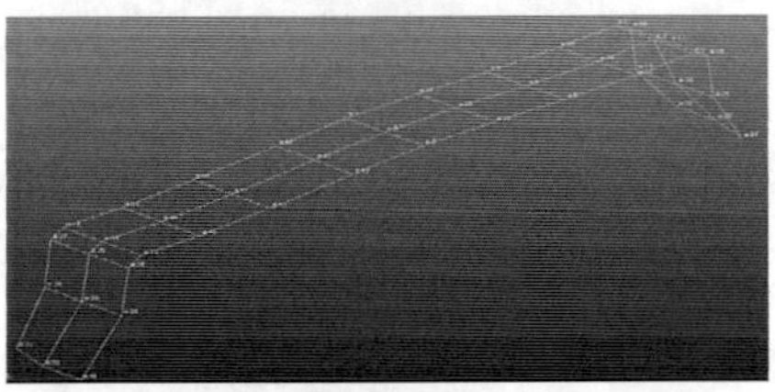 Mode 4	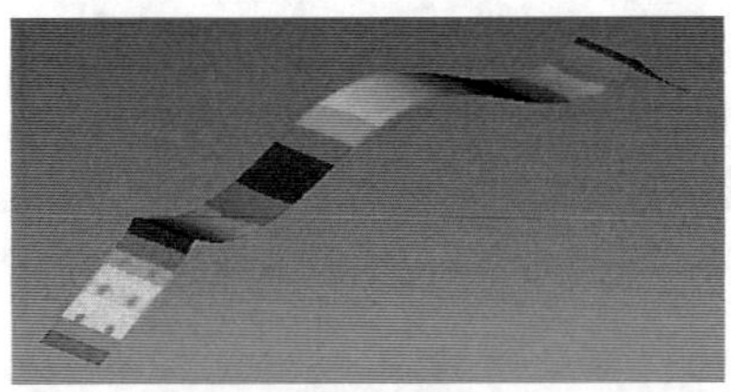 Mode 4	0.91
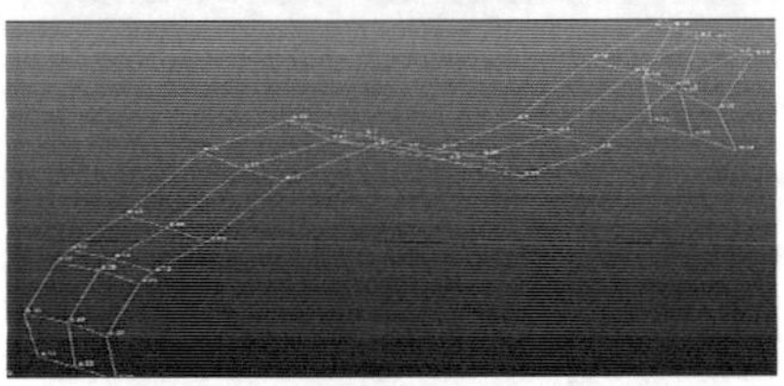 Mode 5	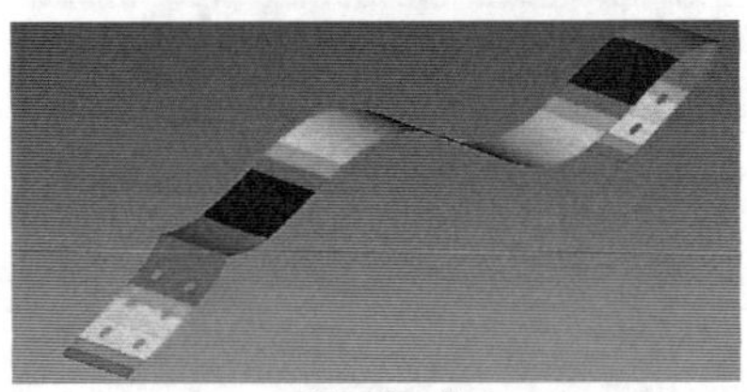 Mode 5	0.92
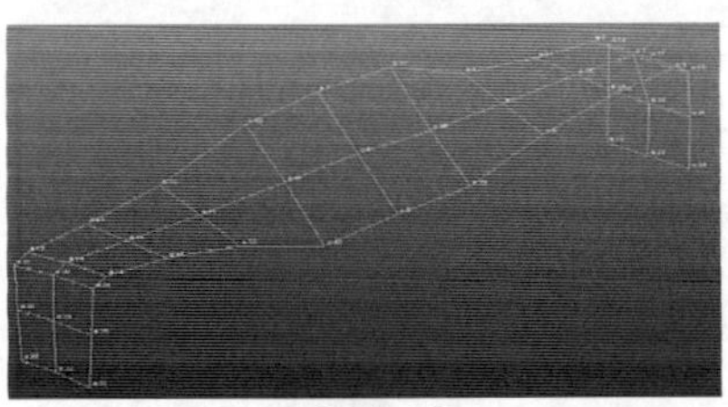 Mode 6	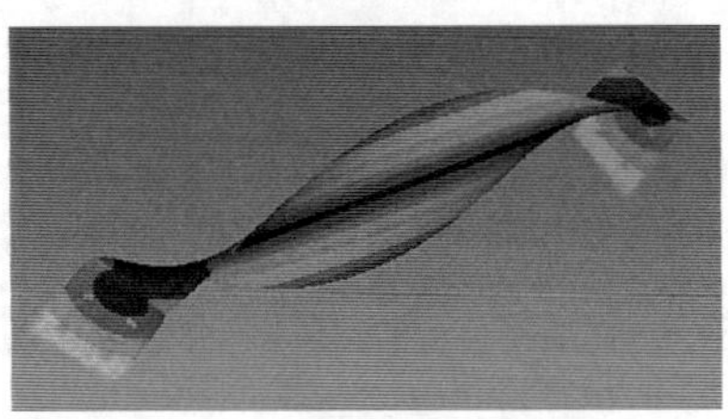 Mode 6	0.87

From the high value of total error shown in Table 2, the model updating processes has been performed on the initial FE model to minimise the error. This is important for the finite element model to be as close enough to the test structure prior to be used for further structural engineering analysis. Model updating is widely used to improve the accuracy of finite element models by correcting the invalid assumptions about the finite element model. This is carried out by changing some uncertain finite modelling parameters, which have the potential to influence modal properties and subsequently to improve the accuracy of the models. Finite element model updating is performed using NASTRAN optimisation code (SOL 200). The optimisation algorithm, which is a sensitivity-based iterative procedure, allows the objective function (J) to be minimised by adjusting the eigenvalues of the initial finite element model until the objective function is converged. The objective function based on eigenvalues is calculated using Eq. (4) where λ_i^{exp} is the ith experimental eigenvalue and λ_i^{fe} is the ith predicted eigenvalue from the finite element model and n is the number of eigenvalues involved in the updating procedure [20].

$$J = \sum_{i=1}^{n\Sigma} W_i \left(\frac{\lambda_i^{fe}}{\lambda_i^{exp}} ()^2 \right) \tag{4}$$

Computation of sensitivity analysis (SOL 200) is performed to identify the sensitive parameters of the initial FE model. Four potential parameters are listed in this sensitivity analysis which are Young's modulus, Poisson's ratio, density and thickness of the plate. Comparison of experimental and updated FE analysis was performed to study the accuracy of the finite element model updating. Table 4 shows the discrepancies between the updated FE analysis and the measured natural frequencies of the plate. The high value of natural frequencies from the initial FE analysis were decrease to the values which close to the measured value from EMA. For the total error of the six modes, the large reduction in the discrepancies is seen from 45.12 per cent to 6.28 per cent. The updated values of the updating sensitive parameters are shown in Table 5. An increment on the updated values have been obtained on Young's modulus (from 200,000 MPa to 203,800 MPa) and density (from 7800 kg/m^3 to 7998 kg/m^3). Meanwhile, the initial thickness of the plate has been reduced from 1.5 mm to 1.43 mm, and also the reduction of the Poisson's ratio value from 0.30 to 0.29.

Table 4. Comparison of experiment and updated FE analysis of the plate

Mode	EMA (Hz)	Initial FE Analysis (Hz)	Error (%)	Updated FE Analysis (Hz)	Error (%)
1	22.49	24.17	7.42	22.43	0.31
2	57.56	62.09	7.87	57.75	0.33
3	64.02	68.13	6.42	63.27	1.17
4	85.34	92.55	6.80	86.71	0.06
5	136.18	145.62	6.92	137.06	0.64
6	185.89	203.79	9.69	192.79	3.77
		Total Error	45.12	Total Error	6.28

Table 5. Updated values of the parameters of the plate

Parameters	Initial Value	Updated Value	Unit
Young's modulus	200,000	203,800	MPa
Poisson's ratio	0.30	0.29	
Density	7800	7998	kg/m^3
Thickness	1.5	1.43	mm

5. Conclusions

The finite element model updating of a thin bend plate has been performed successfully in this study. A 2D shell element FE model has been constructed using HyperMesh OptiStruct software. A numerical simulation work via NASTRAN SOL 103 normal mode analysis has been carried out on the model which was assigned with nominal values of material properties. The measured modal parameters (natural frequencies and mode shapes) have been obtained from the conducted experimental modal analysis on the real plate under free-free boundary conditions with lightly damped conditions. The six natural frequencies from the FE analysis showed large discrepancies in comparison with the measured results showing 45.12 per cent relative error. Sensitivity analysis via NASTRAN SOL 200 has been utilised and listed the sensitive parameters. The discrepancies have been reduced by updating the value of the material properties of the initial FE model of the plate which are Young's modulus, Poisson's ratio, thickness and density. The relative error has been reduced to 6.28 per cent, thus produce better validated FE model for further engineering analysis.

Acknowledgements

The research has been funded by the Malaysian Ministry of Higher Education (MOHE) and Research Management Centre (RMC) of Universiti Teknologi MARA (UiTM) for the Fundamental Research Grant Scheme with the file number 600-IRMI/FRGS 5/3/ (465/2019). The authors would like to express their gratitude to the members of the Centre of Excellence (CoE) Structural Dynamics Analysis & Validation (SDAV), UiTM Shah Alam for their technical assistance and laboratory facilities.

References

[1] Zhou L, Wang L, Chen L and Ou J 2016 *Adv. Struct. Eng.* **19(9)** 1446-1462
[2] Petersen Ø W and Øiseth O 2017 *Eng. Struct.* **150** 573-584
[3] Zárate B A and Caicedo J M 2008 *Eng. Struct.* **30(12)** 3724-3730
[4] Friswell M I and Mottershead J E 1995 *Finite Element Model Updating in Structural Dynamics, Illustrate* (Netherlands: Springer Science & Business Media)
[5] Reynders E, Teughels A and De Roeck G 2010 *Mech. Syst. Signal Process.* **24(5)** 1306-1323
[6] Hernandez E M and Bernal D 2013 *Mech. Syst. Signal Process.* **34(1-2)** 39-46
[7] Lin R M and Zhu J 2006 *Mech. Syst. Signal Process.* **20(8)** 2200-2218
[8] Adel F, Shokrollahi S, Jamal-Omidi M and Ahmadian H 2017 *Journal of Sound and Vibration* **396** 172-185
[9] Mottershead J E, Link M and Friswell M I 2011 *Mech. Syst. Signal Process.* **25(7)** 2275-2296
[10] Cara J 2016 *J. Sound Vib.* **370** 94-110
[11] Ewins D J 2000 *Modal testing: Theory, Practice and Application. 2nd ed* (Hertfordshire: Research Studies Press Ltd.)
[12] Abdul Rani M N, Kasolang S, Othman M H, Yunus M A, Wan Iskandar Mirza W I I and Ouyang H 2016 *20th Int. Congress on Sound and Vibration* (Bangkok: Thailand) pp 1–8
[13] Bahari A R, Yunus M A, Abdul Rani M N and Wan Iskandar Mirza W I I 2019 *AIP Conference Proceedings* **2019** 020054
[14] Hibbeler R C 2011 *Mechanics of Materials 8th ed* (Prentice Hall)
[15] Yunus M A, Ouyang H, Abdul Rani M N and Mat Isa A A 2013 *20th Int. Congress on Sound and Vibration* **2013** (Bangkok: Thailand) pp 1161-1167
[16] Inman D J and Singh R C 2014 *Engineering Vibration* (Pearson)
[17] Rao S S and Yap F F 2011 *Mechanical Vibrations* (Prentice Hall)
[18] Kamgar R and Saadatpour M M 2012 *App. Math. Model.* **36(10)** 4918-4930
[19] Ringsberg J W and Ernholm P 2011 *ASME 2011 30th Int. Conf. on Ocean, Offshore and Arctic Engineering* 67-74
[20] Bahari A R, Yunus M A, Abdul Rani M N, Nalisa A and Aziz Shah M A S 2019 *IOP Conference Series: Materials Science and Engineering* **2019** 012043

AUTHOR INDEX

Abdullah, Abdullah 517
Abdullah, Mohd Faiz Bin 1
Abdullah, N.A.Z. 65
Abdullah, Nik Rosli 100, 111, 181
Abhen, A J 476
Abral, Hairul 466
Adenam, Norfatihah Mohd 93
Adli, H K 165, 174
Adli, Hasyiya Karimah 93
Affi, Jon 268
Afriani, Lusmelia 315
Ahmad, Aslimariah 396
Ahmad, Mohd Azrul 181
Ahmad, Taufik Ejaz 83
Ahmad, Z A B 8
Ahmad, Z. A. B 18
Aida, Fatin Farina 259
Ali, Sinan Abdulghafar 53
Ali, Zahran 430
Alias, E A 41
Amin, Norliyati Mohd 286
Amiruddin, M 377
Amri, Asbahrul 498
Anuar, Nur Farah Bazilah Wakhi 430
Arafat, Andril 430
Arifan, Mahdi 498
Aslfattahi, Navid 194
As'arry, A 414
As'arryi, Azizan 1
Azhari, Dhea Rahma 83
Aziz, A 377
Aziz, Hafiz Aji 157
Azizan, MA 414
Azizan, N 231, 377
Azmi, I F 440
Bahari, A R 522
Balkhaya 507
Baratian, I 203
Bo, Z 165, 174
Bur, Mulyadi 277
Candra, R A 460
Candra, Rudi Arif 517
Catherine, Louis Denis Kevin 484
Chandra, Devi 449
Doh, S I 76
Dwiputra, D 471
Erdiwansyah 132
Faizal, W M 165, 174
Fajri, Hidayatul 268
Fajrul, Rahmat 466
Fakhruddin 490
Fardiansyah 490
Farzana, F 377
Febrina, Rina 315
Fikri, M A 165, 174
Fitriadi, Nuzuli 507
Fitriady 490
Fitriani 498
Gani, Asri 132
Ghani, Amir Radzi Ab 305
Ghobadian, B. 214
Ghobadian, B 203
Gunawarman 268
Hadi, N D 345
Hagos, Ftwi Y 139
Hamid, Darulihsan Abdul 484
Hamid, N H 345, 440
Hamid, Norhayati Abdul 286
Handayani, Dian 466
Harahap, R 460
Hardisal 460, 517
Haron, Hafiza 223
Hartanto, Muchamad 83
Harun, W S W 119
Hasan, M Danial A 18
Hashim, Siti Zaiton Mohd 93
Hassan, AM 414
Hassan, Hazman Abu 139
Hazzim, Amir 252
Hee, L M 8, 18
Hermawati, Elvira 148
Hidayat, D 238, 245
Huda, F 422
Husin, Ku Azmie Ku 93
Hussin, Supyan 223
Idris, Mohd Hasbullah 430
Ihsan 498
Ilham, D N 460
Ilham, Dirja Nur 517
Ilhamdi 238, 245
Isa, R 377
Ishak, Izzul Syazwan 286
Ismail, N I K 41
Ismail, Noraini 252, 259
Jalaludin, Helmisyah Ahmad 100, 111
Jalil, NA Abdul 414

Jamali, F. 214
Jamaludin, A S 294, 366, 384
Jiang, X X 139, 165, 174, 203
Jiang, X.X 214
Juliadmi, Dian 268
Jusoh, W N Asnida Wan 335
Jusoh, Z. 231
Kadarohman, Asep 132, 139
Kadhim, Saleem Khalefa 53
Kadir, Harlinawati Abdul 396
Kadir, Hasannuddin Abdul 181
Kadirgama, K. 119, 194
Khoerunnisa, Fitri 139
Kiat, K L J 41
Lamin, N Zakiah 335
Leman, M N 28
Leong, M Salman 18
Leong, M S 8
Lestari, Della Ayu 83
Lovita, Oktalia Triananda 498
Mahardika, Melbi 466
Mahidin 132
Manjas, Menkher 268
Martoprawiro, Muhamad Abdulkadir 157
Mazlan, Mohamad Hafsham 111
Mirza, W I I Wan Iskandar 440, 522
Miserom, Faizah 356
Miserom, Siti Faizah 259
Mohamaddan, S 28
Mohamed, Z. 34
Mohtar, A M A Abang Mohamad 28
Muhibbuddin 132
Muslimin, Ahmad Novi 466
Mustaqimah 214
Najafi, G. 214
Najafi, G 203
Najiha, A S Intan 76
Nanda, Is Prima 430
Noerhamzah, D 405
Nugroho, A 245
Nur, R A 245
Nurhayati, Mita 157
Nuswantoro, Nuzul Ficky 268
Othman, Mohd Zaid 305
Othman, S. 507
Othman, S 517
Prastio, Tri 315
Pratiwi, Amelinda 148
Pratoto, A. 476
Priyandoko, G 34
Putra, Darma Setiawan 490
Putri, Dian Mustika 268
Rahim, Asiah Ab. 111
Rahim, Asiah Ab 181
Rahman, Arif 277
Rais, Amien 83
Ramadhan, A I 165, 174
Ramasamy, D 119, 194
Ramly, Ramzyzan 305
Rani, M N Abdul 440, 522
Razak, N F D 53
Razali, F 231
Razali, M N M 294, 366, 384
Rezali, KAM 414
Ridwan, F 405
Rosanti, Santi Dewi 466
Roslee, Muhammad Syafiq 100
Rosli, N 41
Rosli, R 34
Rusli, M. 422
Sabri, Shima 252, 259, 356
Safrudin, R A 422
Saidur, R 194
Salleh, Amirul Asyraf Khairul 396
Salleh@Omar, A 231
Salwani, M.S. 65
Samad, Hafiza 223, 335, 356
Samsuri, N. 294, 366, 384
Samykano, M 194
Samylingam, L 194
Sandhya, M 119
Sani, M S M 53
Sani, M.S.M. 65, 132
Saputra, Devi Satria 498
Saputra, Edi 507
Saputra, Miki 498
Saragi, D P 460
Sardjono, R E 139
Sardjono, R.E 132
Satar, Reem Musab Abdul 1
Satria, Eka 449
Saufi, S R 8
Sazali, N 294, 366, 384
Sepriyani, D E 238
Shah, M A S Aziz 522
Sharudin, Hazim 111
Sipahutar, Erwinsyah 517
Son, Lovely 277
Sudhakar, K. 119
Suffian, M S Z Mohamad 28
Sugiarti, Eni 466
Suherman, Suherman 517
Suherman, S 460
Susanti, O 471

Susiloningtyas, Dewi 83
Susilorini, Rr M I Retno 315
Sutanto, Agus 268
Syah, Yana Maolana 148
Tjong, Djong Hon 268
Utomo, S 422
Weru, Yuril Umbu Woza 490
Wong, Khairul Nizar Syazwan Wan Salihin 93
Yahya, Z. 460, 490, 498
Yahya, Z 522
Yasar, Muhammad 203
Yunin, Muhamad Yuzaini Azrai Mat 93
Yunus, M A 440, 522
Yusof, Ahmad Rizal Mohd 223
Yusof, Hafidzan 223
Yusof, Mohd Azahari Mohd 396
Yusoff, Nukman 449
Zain, Mohd Zarhamdy Bin MD 1
Zainudin, Juanita 335, 356
Zaki, M. 132
Zawawi, Zaky Zuhairi 181
Zhang, Bo 139, 203, 214
Zulfikry, M Amirul Syafiq 305